Handbook of Lipid Membranes

Handbook of Lipid Membranes
Molecular, Functional, and Materials Aspects

Edited by
Cyrus R. Safinya
Joachim O. Rädler

Boca Raton London New York

CRC Press is an imprint of the
Taylor & Francis Group, an **informa** business

First Edition published 2021
by CRC Press
6000 Broken Sound Parkway NW, Suite 300, Boca Raton, FL 33487-2742

and by CRC Press
2 Park Square, Milton Park, Abingdon, Oxon, OX14 4RN

© 2021 Taylor & Francis Group, LLC

CRC Press is an imprint of Taylor & Francis Group, LLC

Reasonable efforts have been made to publish reliable data and information, but the author and publisher cannot assume responsibility for the validity of all materials or the consequences of their use. The authors and publishers have attempted to trace the copyright holders of all material reproduced in this publication and apologize to copyright holders if permission to publish in this form has not been obtained. If any copyright material has not been acknowledged, please write and let us know so we may rectify in any future reprint.

Except as permitted under U.S. Copyright Law, no part of this book may be reprinted, reproduced, transmitted, or utilized in any form by any electronic, mechanical, or other means, now known or hereafter invented, including photocopying, microfilming, and recording, or in any information storage or retrieval system, without written permission from the publishers.

For permission to photocopy or use material electronically from this work, access www.copyright.com or contact the Copyright Clearance Center, Inc. (CCC), 222 Rosewood Drive, Danvers, MA 01923, 978-750-8400. For works that are not available on CCC, please contact mpkbookspermissions@tandf.co.uk

Trademark notice: Product or corporate names may be trademarks or registered trademarks and are used only for identification and explanation without intent to infringe.

ISBN: 978-1-466-55572-3 (hbk)
ISBN: 978-1-032-01441-8 (pbk)
ISBN: 978-0-429-19407-8 (ebk)

Typeset in Times
by codeMantra

Contents

Preface vii
Editors xi
Contributors xiii

Chapter 1 A Short History of Membrane Physics 1

Erich Sackmann and Avinoam Ben-Shaul

Chapter 2 Structures and Interactions in Freely Suspended Multilayer Membranes and Dilute Lamellar Fluid Membranes from Synchrotron X-Ray Scattering 33

Gregory S. Smith and Cyrus R. Safinya

Chapter 3 Structures of Lipid Membranes: Cubic and Inverse Hexagonal Phases 49

Charlotte E. Conn and John M. Seddon

Chapter 4 Structure of Lipid Membranes by Advanced X-Ray Scattering and Imaging 65

Tim Salditt

Chapter 5 Adhesion Protein Architecture and Intermembrane Potentials: Force Measurements and Biological Significance 83

Deborah E. Leckband

Chapter 6 Charged Membranes: Poisson–Boltzmann Theory, the DLVO Paradigm, and Beyond 99

Tomer Markovich, David Andelman, and Rudolf Podgornik

Chapter 7 Membrane Shape Evolution *In Vitro* 129

Alexandra Zidovska

Chapter 8 Mechanisms of Membrane Curvature Generation by Peptides and Proteins: A Unified Perspective on Antimicrobial Peptides 141

Michelle W. Lee, Nathan W. Schmidt, and Gerard C. L. Wong

Chapter 9 Lipid Membrane Shape Evolution and the Actin Cytoskeleton 161

David R. Slochower, Yu-Hsiu Wang, Ravi Radhakrishnan, and Paul A. Janmey

Chapter 10 Effects of Osmotic Stress on Topologically Closed Membrane Compartments 177

James C. S. Ho, Bo Liedberg, and Atul N. Parikh

Chapter 11 Cationic Liposomes as Spatial Organizers of Nucleic Acids in One, Two, and Three Dimensions: Liquid Crystal Phases with Applications in Delivery and Bionanotechnology 195

Cyrus R. Safinya, Kai K. Ewert, Youli Li, and Joachim O. Rädler

Chapter 12 Lipids in DNA, RNA, and Peptide Delivery for *In Vivo* Therapeutic Applications 211

Tyler Goodwin and Leaf Huang

Chapter 13 Electrostatics of Lipid Membranes Interacting with Oppositely Charged Macromolecules 223
Guilherme Volpe Bossa, Klemen Bohinc, and Sylvio May

Chapter 14 Lipid-Based Bioanalytical Sensors 241
Marta Bally, Hudson Pace, and Fredrik Höök

Chapter 15 Lipids in Dermal Applications: Cosmetics and Pharmaceutics 271
Jérôme Bibette and Abdou Rachid Thiam

Chapter 16 Supported Lipid Bilayers 293
Theo Lohmüller, Bert Nickel, and Joachim O. Rädler

Chapter 17 Artificial Membranes Composed of Synthetic Copolypeptides 305
Timothy J. Deming

Chapter 18 Synthetic Membranes from Block Copolymers, Recombinant Proteins, and Dendrimers 323
Daniel A. Hammer, Zhichun Wang, Ellen Reed, Chen Gao, and Kevin B. Vargo

Chapter 19 Amphiphilic Self-Assembly and the Origin of Life in Hydrothermal Conditions 337
Christos D. Georgiou and David W. Deamer

Index 349

Preface

Lipid membrane science has evolved from its roots in biochemistry and biophysics to become an essential component in the emerging interdisciplinary fields of nanobioscience and nanobiotechnology, with new everyday applications appearing at an ever-increasing rate. This shift calls for a new type of introductory-level reading material spanning multiple disciplines and appropriate for both beginners and advanced researchers. The *Handbook of Lipid Membranes: Molecular, Functional, and Materials Aspects* offers a multifaceted perspective, leading the reader through membrane-related processes in reconstituted and living matter systems. It teaches how to learn from nature to build artificial and functional membrane-based systems, including those not based on lipids. The emphasis is on the science and technology of lipids and artificial membranes and includes chapters on advanced X-ray and force measurement techniques. The 19 chapters are simultaneously self-contained yet connected with common themes running between the chapters. Included are chapters covering the role of lipids as structural components in determining distinct membrane shapes, topologies and inter-membrane interactions. By making a direct connection to cellular systems, membrane curvature generation, mediated by peptide and protein binding, and biological signaling lipids in concert with cytoskeletal proteins are presented. Beyond the science in lipid biology and biophysics, the chapters reveal how mastering lipid interactions enables novel biomedical and nanotechnological applications with fine-tuned functionality. Subject-based chapters discuss current applications in biomedicine, cosmetics and nanotechnology. This includes lipid vectors in nucleic acid and drug delivery, in dermal applications, and in artificial biointerfaces and lipid-based sensors. A final engaging chapter takes the reader to the beginning of where it all started and explores the role of lipids in the origin of life.

The discovery of liposomes by A. D. Bangham and R. W. Horne in the 1960s with efficient encapsulation properties and striking similarities in electron micrographs to cell plasma membranes led to their hypothesis that lipids are the primary structural components responsible for the permeability barriers of biological membranes. This major discovery led the way for a rapid worldwide increase in research on the biophysical properties of lipid vesicles and on the biological function of membrane-associated proteins, reconstituted in liposomes. The realization that liposomes contain sites for incorporation of hydrophobic, hydrophilic and amphiphilic molecules, led researchers early on to investigate, on the one hand, their promise as components of personal care products and, on the other hand, as carriers of drugs and genes for therapeutic applications. The latter includes over 100 lipid-based currently-ongoing clinical trials worldwide targeting a single gene, such as cystic fibrosis, and multigene cancer diseases. Most remarkably, we now see a flourishing of lipid-based science opening new approaches to vaccines with the Moderna and Pfizer/BioNTech lipid-mRNA nanoparticles saving millions of lives in the worldwide pandemic in 2020 and 2021 due to COVID-19.

This handbook brings together contributions from world experts in lipid and artificial membrane science and applications. In the first chapter, Erich Sackmann and Avinoam Ben-Shaul provide a historical survey of membrane physics from the perspective of two-dimensional self-assembling systems consisting of interacting lipids and membrane-associated proteins. The development of certain key novel experimental techniques and theoretical approaches over the last few decades, which have contributed to our current understanding of membrane physics, is discussed. The significance of lipid mobility, membrane defects and membrane curvature elasticity, to membrane thermomechanical properties, and the impact on biological membrane structure and function are described.

The following two chapters start out by introducing lipid membrane structure and phase behavior. Gregory Smith and Cyrus Safinya introduce the phase diagram of gel-like ordered membranes, consisting of lipids with phosphocholine headgroups, in Chapter 2. X-ray scattering and reflectometry techniques are described that determine the intra- and intermolecular lipid arrangement and probe interlayer interactions by analysis of thermal diffuse scattering. The chapter outlines how structural studies of highly oriented freely suspended multilayer preparations have revealed that the gel phase is comprised of three phases distinguished by the direction of chain tilt within the two-dimensional ordered membrane. Stacks of fluid membranes are analyzed within the context of a Landau–De Gennes type elastic energy displaying characteristics of a low-dimensional Landau–Peierls system. Synchrotron X-ray scattering shows that such very dilute fluid lamellar phases are stabilized by entropically mediated undulation forces elucidated by Wolfgang Helfrich.

Chapter 3 by Charlotte Conn and John Seddon continues with a comprehensive description of the structure and properties of cubic and inverse hexagonal phases. The authors describe how the stability of these non-lamellar phases is derived from forces underlying curvature and topological transformations in lipid membranes. Further, the effect of hydrostatic pressure on phase behavior is discussed and shown to be relevant in marine biology. Numerous illuminating examples are presented that show the occurrence of non-lamellar phases in living matter and in a wide range of basic science studies and bionanotechnological applications. This includes cubic and hexagonal phase structures enabling specific functions, *in vivo*, and, as structured drug delivery vectors, cubic phases to enable membrane protein structure determination.

In Chapter 4, Tim Salditt gives a summary of state-of-the-art X-ray analysis of highly oriented lipid membranes employing X-ray optics, nanoscale focusing, time-resolved X-ray diffraction, and lensless coherent imaging. Illustrations of these powerful new synchrotron X-ray techniques are shown as applied to a wide range of systems of high scientific interest, including nonequilibrium dynamics of membranes, short-lived intermediate structures of membrane fusion, and the asymmetric structures of synaptic vesicles and myelin multilayers. The chapter concludes by presenting coherent X-ray imaging as a powerful new method for membrane structure analysis.

Deborah Leckband introduces the reader to the surface forces apparatus (SFA) in Chapter 5, a unique technique that enables direct measurement of forces between surface-bound membranes containing membrane-associated proteins, including intrinsically disordered proteins. The SFA technique, with biological material contained between surfaces, naturally mimics the confined environment *in vivo*. Thus, measured protein-mediated forces, both attractive and repulsive, are expected to elucidate membrane–protein function at biological interfaces. Forces mediated by adhesion proteins, including lectins that bind carbohydrates, neural cell adhesion molecules, and cadherin adhesion proteins, ubiquitous in most tissues, are highlighted. The chapter further describes cell-binding kinetics employing micropipette measurement techniques.

Chapter 6 by Tomer Markovich, David Andelman and Rudolf Podgornik is an entirely self-contained theoretical chapter covering our current understanding of electrostatic forces in charged membrane systems. Poisson–Boltzmann (PB) theory of charged membranes, in the presence of no added salt and added salt, is described. Ion profiles near single and between two membranes are considered. As becomes evident, the electrostatic forces described in the chapter are relevant to many of the biological systems described in the handbook. The chapter further reviews van der Waals forces and briefly discusses the limitations of PB theory, including the limit of high surface charge and multivalent counterions, where the PB theory breaks down due to its mean-field nature. The material is presented in a manner suitable for instructors teaching at the first-year graduate or upper-division undergraduate levels.

The key physical concepts underlying distinct vesicle shapes are lipid molecule shape, membrane composition, and intermolecular interactions. In Chapter 7, Alexandra Zidovska takes the reader on a broad review of membrane shape evolution in vesicles (i.e., liposomes) as discovered in laboratories worldwide over the last few decades. The chapter includes sections connecting modern elasticity models to shape evolution observed experimentally. This includes the hugely successful Helfrich curvature elastic model and beyond Helfrich theory with models that incorporate the intrinsic area differences between the outer and inner membrane leaflets due to the finite curvature of liposomes. The chapter concludes by connecting *in vitro* to *in vivo* observations where many similar shapes have been observed.

Michelle Lee, Nathan Schmidt and Gerard Wong take the concept of membrane curvature further in Chapter 8 and discuss how peptides modulate membrane morphology. Membrane binding peptides play a vital role as antimicrobial agents, enabling budding and release of viruses and enabling membrane tubulation and fission. The chapter provides a survey of membrane curvature generation mechanisms and illustrates the case of antimicrobial peptides. The authors explain how certain cationic and hydrophobic structural motifs of antimicrobial peptides lead to selective pore-forming activity against bacterial membranes. Thus, as the authors explain, intelligently designed alterations in amino-acid sequence, intended to enhance peptide membrane-permeating ability, may be implemented in the development of novel antibiotics to fight persistent bacterial strains.

In Chapter 9, David Slochower, Yu-Hsiu Wang, Ravi Radhakrishnan and Paul Janmey describe how the shape of eukaryotic cell membranes is controlled by both the plasma membrane lipid bilayer and the underlying cytoskeletal network. At a molecular level, the role of phosphoinositides in regulating the membrane–cytoskeletal interface, by binding cytoskeletal proteins and proteins that cause or sense membrane curvature, is described. In parallel, mechanisms of producing membrane curvature driven by the cytoskeleton are introduced. Combining this information within the context of recent experimental results and multiscale simulations, the authors present an integrated view of how these fascinating lipids perform their many cellular functions and help orchestrate the dynamic changes in membrane curvature.

The effect of osmotic stress on cell membrane topology is the focus of Chapter 10 by James Ho, Bo Liedberg and Atul Parikh. Starting with giant unilamellar vesicles, as model membrane compartments, the permeation of water and the generation of osmotic gradients by hindered permeation of solutes is introduced. The chapter gives experimental and theoretical insights into shape changes of flaccid vesicles in hypertonic media and transient pore formation in response to hypotonic media. Shape deformations due to osmotic stress coupling with phase separation of multicomponent vesicles are also discussed. Finally, the potential relevance of stress relaxation mechanisms is discussed within the context of ubiquitous osmotic challenges that primitive protocells most likely experienced in early life.

Lipid-nucleic acid complexes are self-assembled systems that incorporate key concepts of lipid molecule shape and membrane curvature discussed in previous chapters. Their scientific perspective and applications *in vitro* and *in vivo* are described in three consecutive chapters. Chapter 11, by Cyrus Safinya, Kai Ewert, Youli Li and Joachim Rädler, introduces the self-assembled structures of cationic liposome (CL)-nucleic acid complexes used in gene delivery. The authors explain how concepts of membrane curvature and electrostatic interactions explain the formation of liquid crystalline inverse hexagonal, lamellar and hexagonal phases with DNA residing in one, two, or three dimensions, respectively. The chapter describes how transfection efficiency, measuring the expression of DNA transferred by CL complexes into cells, depends on the underlying structures of the complexes. The concepts presented in the chapter are expected to apply to ongoing efforts to understand the physics of self-assembly in cationic ionizable lipid-mRNA formulations used in cancer therapeutics and vaccine applications (e.g., Moderna and Pfizer/BioNTech nanoparticle vaccines).

In Chapter 12, Tyler Goodwin and Leaf Huang cover lipids as delivery vehicles of DNA, RNA and peptides in *in vivo* therapeutic applications. The authors describe how lipid vectors may be designed to be highly efficient in the delivery of biomacromolecules. Based on rational design over many years, lipid vectors were developed to overcome extracellular and intracellular barriers. The chapter reviews the improvements and the remaining challenges of current lipid vectors in *in vivo* applications. The survey highlights promising ionizable lipids and composite nanocore-lipid vectors and discusses the progress made in clinical trials.

Chapter 13 turns to theory, where electrostatic models underlying self-assembly of charged membranes and oppositely charged biomacromolecules, including DNA, proteins and peptides, are reviewed by Guilherme Volpe Bossa, Klemen Bohinc and Sylvio May. The influence of lipid mobility, protein-mediated lipid phase separation and charge regulation on the interactions stabilizing proteins adsorbed on membranes is discussed. Theoretical models of the phase behavior of lipid-DNA complexes are reviewed for cationic and zwitterionic lipids. Discussions on important local interactions between neutral zwitterionic lipid headgroups and divalent cations and DNA and membranes containing zwitterionic lipids are presented. The chapter concludes with a general discussion of the stability of membrane pores resulting from the interactions between amphiphilic peptides and membranes.

Lipid membranes are powerful in functionalizing or protecting surfaces. Marta Bally, Hudson Pace and Fredrik Höök give an account of the use of planar lipid membranes as bioanalytical sensors in Chapter 14. The chapter focuses on a broad range of surface-based biosensing techniques. Immobilization of lipid assemblies at a sensor surface is achieved by a supported lipid bilayer, tethered bilayers, or a free-spanning bilayer of surface-tethered vesicles. Moreover, patterned planar membranes are used for high-throughput screening and lab-on-a-chip devices. The authors show how membrane-associated proteins may refine biosensing to detect interactions between membrane proteins and ligands. Discussions are also presented about how measurements of membrane–membrane interactions in biosensors have been enhanced by the invention of single-vesicle assays that determine equilibrium-binding constants from statistics of residence times. An outlook on the use of lipid-based nanoreactors and force-driven manipulation of membrane components in supported lipid bilayers for chip-based bioanalytics is presented.

Chapter 15, by Jérôme Bibette and Abdou Rachid Thiam, consists of a comprehensive overview of the structure and function of the skin and novel dermal applications in cosmetics and pharmaceutics. The role of lipids in dermal health and the downside consequences of lipid disorders on skin function are reviewed. The chapter contains a wealth of information on the molecular components of cosmetic and pharmaceutical products. The authors describe the state of the art in cosmetic and therapeutic dermal applications, through rational design of lipid and surfactant-based vectors, optimized for delivery through the skin. Current lipid nanovectors used worldwide are described. This includes liposome formulations, nano- and micro-emulsions, and solid lipid nanoparticles.

Chapter 16 by Theo Lohmüller, Bert Nickel and Joachim Rädler is dedicated to the properties and applications of so-called supported lipid bilayer systems, which represent planar lipid bilayers that form at the solid–fluid interface through vesicle fusion and in-plane spreading. In these systems, lipids and incorporated proteins retain high lateral mobility. Hence, charged macromolecules that bind to oppositely charged planar-supported bilayer systems exhibit ideal two-dimensional properties and can be manipulated in-plane through external electric fields. In polymer-supported lipid bilayer systems, the membrane is elevated from the solid, providing room for larger transmembrane proteins. The chapter concludes with examples of the use of supported lipid bilayer systems as novel platforms for presenting molecules to living cells.

In the last two decades, increasing research efforts have been directed toward developing artificial membranes that are not based on lipids.

Chapter 17 by Timothy Deming summarizes advances in the synthesis of well-defined block and statistical copolypeptides that can be assembled into stable membrane structures. Polymerization chemistries are described that allow precision copolypeptide synthesis with control over chain length, chain length distribution, and chain-end functionality. The author explains how well-defined copolypeptides of controlled dimensions, including molecular weight, sequence, and composition, can be prepared, which spontaneously assemble into vesicles with polypeptide membranes. Examples of polypeptide membrane vesicles possessing unique properties (due to the amino acid building blocks and ordered conformations of the polypeptide segments) are presented in the context of biomedical application.

In Chapter 18, Daniel Hammer, Zhichun Wang, Ellen Reed, Chen Gao and Kevin Vargo give a comprehensive overview of membranes constructed with synthetic amphiphilic polymers and biopolymers with novel architectures. The focus of the chapter is on the science and technological applications of artificial vesicles made from spontaneous assembly of block copolymers (polymersomes) and copolypeptides, recombinant proteins and amphiphilic dendrimers. The methods for producing self-assembling amphipathic proteins employing modern recombinant technology are described. The chapter concludes with the latest developments in the construction of artificial vesicles from glycan-based Janus dendrimers. The novel artificial vesicles described in this chapter promise to open new opportunities for the intelligent design of functional nanoparticles with a broad range of applications, including, in molecular delivery and sensing.

The final Chapter 19, by Christos Georgiou and David Deamer, gives a fascinating presentation of some central aspects of the origin of life and, in particular, emphasizes the role played by single-tailed amphiphilic fatty acids and double-tailed lipids. The authors consider alkaline hydrothermal vents and hydrothermal fields as possible geological sites for the prebiotic environment for the spontaneous formation of life around four billion years ago. Discussions are presented about the physical and chemical properties of lipids, which would have been essential in the spontaneous assembly and formation of very simple early membranes with their required barrier properties in comparison with the plasma membrane of current living organisms containing a range of associated proteins designed for a diverse set of functions. The authors conclude with a thought-provoking discussion of the positive and negative aspects of hydrothermal fields versus hydrothermal vents as locations for the onset of the first cellular life on earth.

Cyrus R. Safinya
Joachim O. Rädler

Editors

Cyrus R. Safinya is a Distinguished Professor at the University of California, Santa Barbara (UCSB). His primary appointment is in the Materials Department in the College of Engineering and he has joint appointments in the Molecular, Cellular, and Developmental Biology Department, and, by courtesy, in the Physics Department and the Biomolecular Science and Engineering Program. He received a B.S. in Physics and Mathematics from Bates College in 1975 and a PhD in Physics from the Massachusetts Institute of Technology in 1981 for his studies on liquid crystal phase transitions under the guidance of Robert J. Birgeneau (currently Chancellor Emeritus and Distinguished Professor of Physics and Materials Science at the University of California, Berkeley). Safinya joined the Exxon Research & Engineering Company immediately after obtaining his PhD and started his research studies on the structure of complex fluids and biological membranes before moving to UCSB in 1992. He was a Rothschild Fellow and a Visiting Directeur de Recherche at the Curie Institute in 1994 and, between 2009 and 2013, a Distinguished Visiting Professor at the Korean Advanced Institute of Science & Technology. Safinya is the author or co-author of more than 220 publications, many of which have appeared in journals on research topics in biophysics, chemical physics, physical chemistry, bioengineering, and biomedical sciences. His current research aims to elucidate structures and interactions in soft and biological matter systems, including, in lipid-nucleic acid mixtures and protein assemblies derived from neurons. In parallel, his group works on the development of novel lipid vectors as carriers of nucleic acid (DNA and RNA) and hydrophobic drugs in gene and cancer therapeutics.

Joachim O. Rädler is a Professor of Experimental Physics at Ludwig-Maximilians-University (LMU), and holds the Chair for Soft Condensed Matter. He studied Physics at the Friedrich Wilhelms University in Bonn, at Cambridge University (UK) and at the Technical University of Munich. In 1993, he received his doctoral degree in Biophysics for his work on vesicle adhesion under the guidance of Erich Sackmann. As a post-doctoral fellow during 1993–1996, he studied cationic lipid-DNA complexes with Cyrus Safinya at UC Santa Barbara. Rädler received his habilitation in Experimental Physics at Technical University of Munich, where he worked on supported membranes. In 2000, he was appointed as a senior group leader at the Max Planck Institute for Polymer Research. He was named full professor at LMU in 2001 and became a member of the Center for NanoScience in the same year. From 2008 to 2011, he held a temporary consulting and teaching position for experimental NanoBio physics at the University College Dublin. He served as spokesperson of the Center for NanoScience at LMU, of the collaborative research center "Nanoagents" and of various graduate programs. He is the author or co-author of over 180 publications on research topics in soft matter and biophysics, physical chemistry, chemical physics, bioengineering, and bionanotechnology. His current research focuses on the self-assembly of siRNA and mRNA lipid nanoparticles, the interaction of nanomaterials with living cells, time-resolved studies of single-cell gene expression, and the physics of cell migration.

Contributors

David Andelman
School of Physics and Astronomy
Tel Aviv University
Tel Aviv, Israel

Marta Bally
Department of Physics Chalmers University of Technology
Gothenburg, Sweden
Department of Clinical Microbiology & Wallenberg
 Centre for Molecular Medicine
Umeå University
Umeå, Sweden

Avinoam Ben-Shaul
Institute of Chemistry and Fritz Haber Research Center
Hebrew University of Jerusalem
Jerusalem, Israel

Jérôme Bibette
Laboratoire de Colloïdes et Matériaux Divisés
École Supérieure de Physique et de Chimie Industrielles
 de la Ville de Paris
Université Paris Sciences et Lettres (PSL)
Paris, France

Guilherme Volpe Bossa
Department of Physics
North Dakota State University
Fargo, North Dakota, USA

Klemen Bohinc
Faculty of Health Sciences
University of Ljubljana
Ljubljana, Slovenia

Charlotte E. Conn
School of Science, College of Science Engineering
 and Health
RMIT University
Melbourne, Australia

David W. Deamer
Department of Biomolecular Engineering
Baskin School of Engineering
University of California Santa Cruz
Santa Cruz, California, USA

Timothy J. Deming
Department of Bioengineering and Department of
 Chemistry and Biochemistry
University of California Los Angeles
Los Angeles, California, USA

Kai K. Ewert
Materials Department and Materials Research Laboratory
University of California Santa Barbara
Santa Barbara, California, USA

Chen Gao
Department of Chemical and Biomolecular Engineering
University of Pennsylvania
Philadelphia, Pennsylvania, USA

Christos D. Georgiou
Department of Biology
University of Patras
Patras, Greece

Tyler Goodwin
Division of Molecular Pharmaceutics, Eshelman School of
 Pharmacy
Department of Biomedical Engineering
University of North Carolina
Chapel Hill, North Carolina, USA

Daniel A. Hammer
Department of Bioengineering
Department of Chemical and Biomolecular Engineering
University of Pennsylvania
Philadelphia, Pennsylvania, USA

James C. S. Ho
Centre for Biomimetic Sensor Science
School of Materials Science & Engineering
Nanyang Technological University
Singapore

Fredrik Höök
Department of Physics
Chalmers University of Technology
Göteborg, Sweden

Leaf Huang
Division of Molecular Pharmaceutics and
  Center for Nanotechnology
Eshelman School of Pharmacy
Department of Biomedical Engineering
University of North Carolina
Chapel Hill, North Carolina, USA

Paul A. Janmey
Departments of Physiology and Physics & Astronomy
University of Pennsylvania
Philadelphia, Pennsylvania, USA

Michelle W. Lee
Department of Bioengineering
University of California Los Angeles
Los Angeles, California, USA

Deborah E. Leckband
Department of Chemistry
Department of Chemical and Biomolecular Engineering
University of Illinois at Urbana-Champaign
Urbana, Illinois, USA

Youli Li
Materials Research Laboratory
University of California Santa Barbara
Santa Barbara, California, USA

Bo Liedberg
Centre for Biomimetic Sensor Science, School of
  Materials Science & Engineering
Nanyang Technological University
Singapore

Theo Lohmüller
Photonics and Optoelectronics Group, Nano-Institute
  Munich
Faculty of Physics, Ludwig-Maximilians-Universität
Munich, Germany

Tomer Markovich
School of Physics and Astronomy
Tel Aviv University
Tel Aviv, Israel

Sylvio May
Department of Physics
North Dakota State University
Fargo, North Dakota, USA

Bert Nickel
Faculty of Physics, Soft Condensed Matter Group
Ludwig-Maximilians-Universität
Munich, Germany

Hudson Pace
Department of Integrative Medical Biology
Umeå University
Umeå, Sweden

Atul N. Parikh
Department of Biomedical Engineering
Department of Chemical Engineering and
  Materials Science
University of California
Davis, California, USA
and
Centre for Biomimetic Sensor Science, School of
  Materials Science & Engineering
Nanyang Technological University
Singapore

Rudolf Podgornik
School of Physical Sciences and Kavli Institute for
  Theoretical Sciences
University of Chinese Academy of Sciences
Beijing, China
and
Department of Theoretical Physics, J. Stefan Institute and
  Department of Physics
Faculty of Mathematics and Physics
University of Ljubljana
Ljubljana, Slovenia

Ravi Radhakrishnan
Department of Bioengineering
University of Pennsylvania
Philadelphia, Pennsylvania, USA

Ellen Reed
Department of Chemical and Biomolecular Engineering
University of Pennsylvania
Philadelphia, Pennsylvania, USA

Erich Sackmann
Physik-Department
Technical University Munich
Garching, Germany

Tim Salditt
Institut für Röntgenphysik
Universität Göttingen
Göttingen, Germany

Nathan W. Schmidt
Department of Bioengineering
University of California Los Angeles
Los Angeles, California, USA
and
Ginkgo Bioworks
Boston, Massachusetts, USA

Contributors

John M. Seddon
Department of Chemistry
Imperial College London
London, UK

David R. Slochower
Skaggs School of Pharmacy and Pharmaceutical Sciences
University of California San Diego
La Jolla, California, USA

Gregory S. Smith (Retired)
Neutron Scattering Division Neutron Sciences Directorate
Oak Ridge National Laboratory
Oak Ridge, Tennessee, USA

Abdou Rachid Thiam
Laboratoire de Physique de l'École Normale
 Supérieure, ENS
Université PSL, CNRS
Sorbonne Université, Université de Paris
Paris, France

Kevin B. Vargo
Department of Chemical and Biomolecular Engineering
University of Pennsylvania
Philadelphia, Pennsylvania, USA

Yu-Hsiu Wang
Department of Biochemistry and Molecular Biology
The University of Texas Medical Branch
Galveston, Texas, USA

Zhichun Wang
Department of Bioengineering
University of Pennsylvania
Philadelphia, Pennsylvania, USA

Gerard C. L. Wong
Department of Bioengineering
Department of Chemistry & Biochemistry
California NanoSystems Institute
University of California Los Angeles
Los Angeles, California, USA

Alexandra Zidovska
Center for Soft Matter Research, Department of Physics
New York University
New York, New York, USA

1 A Short History of Membrane Physics

Erich Sackmann
Technical University Munich, Garching, Germany

Avinoam Ben-Shaul
Hebrew University of Jerusalem, Jerusalem, Israel

CONTENTS

1.1 Introduction 2
1.2 Polymorphism of Lipid–Water Systems 3
    1.2.1 Lyotropic Polymorphism 4
    1.2.2 Thermotropic Polymorphism of Lipid Bilayers 4
    1.2.3 The $L_\beta \to L_\alpha$ and Other Bilayer Transitions 4
    1.2.4 Lipid Monolayers 5
    1.2.5 Lipid Chain Order in the L_α Phase 6
1.3 The Lipidome: Membranes as Multicomponent Lipid Mixtures 6
1.4 Lipid–Protein Sorting, Interactions, and Domain Formation 9
    1.4.1 Lipid–Protein Interaction 9
    1.4.2 Lipid and Protein Sorting and the Formation of Functional Microdomains (Rafts)....11
1.5 Membranes as Electrified Interfaces: Electrostatic Switching of Functional Proteins 11
    1.5.1 Role of Lipid Charge on the Chain-Melting Transition..11
    1.5.2 Underlying Thermodynamics and Kinetics..14
    1.5.3 Macroion Aggregation on the Membrane Surface 14
    1.5.4 Counterion Release...15
1.6 Membrane Dynamics: Lipid Lateral Diffusion and Its Impact...15
1.7 Membrane Defects: Physics and Biological Function 17
1.8 Biological Membranes as Elastic Shells 18
    1.8.1 Morphology and Shape Transitions of Soft Elastic Shells 18
    1.8.2 Shape Transitions of Stratified Soft Shells: The Red Blood Cell 19
    1.8.3 Membrane Bending Excitations and Their Physiological Significance...............20
    1.8.4 Red Blood Cell Membranes as Actively Driven Semiflexible Statistical Surfaces....21
    1.8.5 The Physiological Role of Membrane Flickering 21
    1.8.6 Molecular Aspects of Membrane Elasticity 22
1.9 Cell Adhesion: A Membrane-Based Process..22
    1.9.1 Molecular Aspects 23
    1.9.2 Insights Gained by Biomimetic Systems..23
    1.9.3 Synopsis..24
1.10 A Short Outlook: Where We Stand and Where to Go..25
List of Abbreviations..25
Lipid Abbreviations and Nomenclature 25
Types of Lipids...25
Examples..25
Acknowledgements..25
Bibliography 26

1.1 INTRODUCTION

A critical step during biological evolution has been the enclosure of biochemical reactions inside vesicles made of semipermeable membranes composed of lipid bilayers, or in the case of archaea bacteria of monolayers of bipolar bola-lipids. Integrating into the membranes, light-driven proton pumps with pH-dependent activity, such as Bacteriorhodopsin, enabled the reactions to proceed under controlled acidic conditions. The incorporation of these pumps into lipid membranes by hydrophobic peptide sequences is a striking example of nature's ability to transform genetic information into biological nanomachines by making use of basic physicochemical principles of self-assembly, such as the hydrophobic effect. Much of our present understanding of the principles underlying membrane self-assembly and the correlation between membrane structure and functions (such as proton pumping) is based on insights gained by designing sophisticated model systems such as giant vesicles, black lipid membranes (BLMs), lipid monolayers, and supported membranes. Correlations between the molecular organization of lipid–protein bilayers and their basic biological function were established by comparative studies of artificial and natural membranes. Our main goal in this brief historical survey of membrane physics is to point out how our present understanding of membrane biophysics has evolved through the development of novel experimental tools and theoretical concepts of soft matter physics.

In 1925, comparing the surface area of erythrocytes with the monolayer area formed by spreading their constituents, Gortel and Grendel arrived at the conclusion that the thickness of cell envelopes corresponds to that of a lipid bilayer [1]. Ten years later, largely based on the electrical properties of erythrocyte suspensions discovered by K. Fricke and other electrophysiologists, Danielli and Davson suggested that cell envelopes are trilaminar films composed of a lipid bilayer sandwiched between two layers of protein, as illustrated in Figure 1.1a, [2]. It took nearly 40 years before this picture of the biomembrane has been independently challenged by more realistic models. In the United States, based on thermodynamic considerations pertaining to the strong preferences of ionic groups to be in contact with water, and hydrophobic groups to reside in oil-like environment, Singer and Nicolson proposed their Fluid Mosaic Model (Figure 1.1b; [3]). This model, generally favoured by biologists and thus often referred to as "the standard picture of a biomembrane", depicts the lipid bilayer as a passive two-dimensional (2D) fluid, enabling lateral mobility of integral and peripheral proteins, as needed to fulfil their biological functions. A second model, the Heterogeneous Shell Model [4], favoured by many physicists, attributed a more active role to the lipid matrix (Figure 1.1c). This model emphasized the tendency of proteins to assemble in functional microdomains of specific composition by lateral phase separation that are driven by specific lipid–protein interactions in the multicomponent membrane mixture. The model was suggested by studies of auxotrophs of *Escherichia coli* bacteria which can only grow when phospholipids are added to the nutrition medium. One can thus replace the natural lipids by synthetic lipids with well-defined transition temperatures. The experiments showed for the first time that biological membranes can undergo temperature-driven phase separation below the lipid chain-melting transitions and, second, that membrane fluidity enabling lateral diffusion of constituents is indispensable for the division and survival of the cells [4].

Two groundbreaking developments that have greatly motivated the interest of physicists in membrane biophysics were the Huxley–Hodgkin model of nerve conduction by voltage-dependent ion channels and the discoveries by Luzzati and coworkers in the 1960s and 1970s of the rich and fascinating lyotropic and thermotropic polymorphism of lipid/water systems [5,6]. The complex phase behaviour of these systems, and the notion that membranes are two-dimensional smectic phases, has naturally attracted the attention of many physicists from the liquid crystal community. Soon afterwards, stimulated by Frank's theory of liquid crystals, Helfrich presented his curvature elasticity theory of cell shape changes, relating the splay deformation of smectic liquid crystals to the bending deformation of lipid bilayers [7]. Around the same time, many physicists were inspired by Marcel Bessis' wonderful book *The Red Blood Cells* [8]. In parallel, electron microscopy and biochemical studies revealed that the red blood cell envelope is

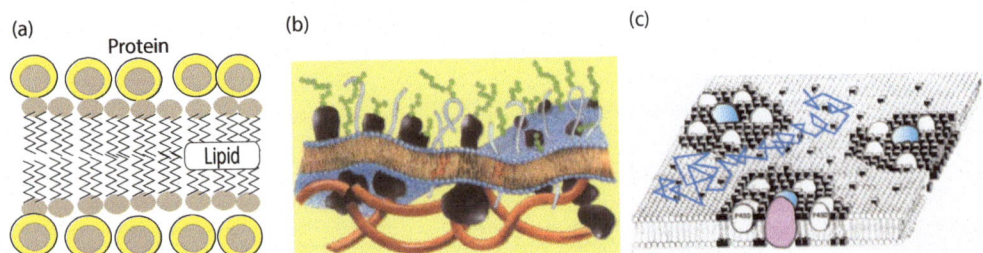

FIGURE 1.1 Early models of biological membranes. (a) Danielli and Davson model [2]. The membrane-associated proteins adsorbed on both surfaces were assumed to control ionic permeability from either side of the lipid bilayer. (b) The fluid mosaic model of Singer and Nicholson [3]. Lipids are assumed to play a passive role, forming a two-dimensional fluid layer embedding the proteins and enabling their lateral transport. (c) The heterogeneous shell model. Lipids play an active role and serve the formation of local functional lipid–protein domains (image reproduced from [4]).

A Short History of Membrane Physics

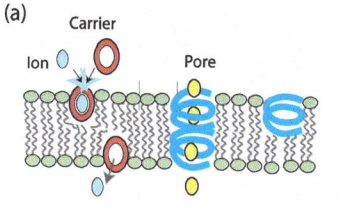
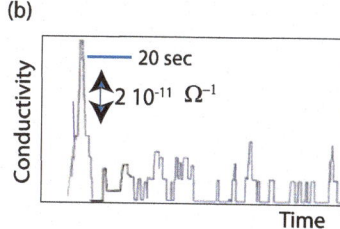

FIGURE 1.2 (a) Schematic view of ion translocation by ion carriers (left) and pore-forming amphiphilic peptides, such as gramicidin (right). (b) Current fluctuations through BLM in the presence of small concentrations of gramicidin. The jumps in conductivity increase by constant increments, indicating that one or several channels open simultaneously. Image reproduced from the work of D. Haydon and S. Hladky [11].

a stratified elastic shell made up of the lipid–protein bilayer and the associated spectrin–actin network [9] – stimulating the formulation of the composite shell model of cell envelopes [4]. Today, we know that the spectrin–actin network coupled to the plasma membrane (PM) is a universal design principle that stabilizes cell envelopes, e.g., of hair cells in the auditory systems [10].

The development of the BLM technique by P. Mueller, M. Montal, and coworkers in 1962 (reviewed in Ref. [11]) allowed for the first time systematic studies of the physics of ion translocation across lipid membranes, mediated by ion carriers (e.g., valinomycin) or through membrane pores formed by the oligomerization of proteins such as gramicidin or alamethicin (Figure 1.2). The subsequent development of voltage jump current relaxation methods and ion current fluctuation spectroscopy provided detailed insights into the kinetics of diffusion-limited formation of ion channels by complex formation of amphipathic polypeptides [12,13].

Notwithstanding the fading interest in BLM studies, the contribution of this technique to our understanding of the physics of ion translocation through membranes has been enormous. It provided the first experimental proof of the Huxley–Hodgkin postulate of voltage-dependent ion channels, paved the way for the interpretation of patch clamp experiments (introduced by Neher and Bert Sakmann), and showed that simple ion carriers, such as valinomycin, exhibit stunning ion selectivity. For instance, the channel-forming alamethicin oligomers were shown to exhibit cation or anion selectivity, depending on the charge of the amino acid side groups facing the channel [12,13], thus providing the first experimental proof that the charged amino acids lining the surface of a membrane channel pore determine their ion selectivity. Based on Fourier spectroscopy of current fluctuations and voltage jump relaxation experiments, sophisticated methods for the analysis of ion translocation kinetics through integral membrane channels or those formed by the assembly of amphipathic proteins (e.g., alamethicin) were developed between 1970 and 1980. The simple design of BLM-based systems played a key role in the development of these techniques and the demonstration of their reliability. It thus paved the way for the kinetic analysis of single channel conductivity in cell membranes by the patch clamp technique (reviewed in Ref. [14]). BLMs have also been successfully used to study various other fundamental membrane processes, among which are vesicle fusion [15] and the control of ion conductivity by lateral phase separation [16]. In the last 20 years, the BLM techniques have been widely replaced by solid-supported membranes [17], fabricated by deposition of lipid–protein bilayers on solid-state devices such as SiO_2- or GaAs-based semiconductors. Similar to BLMs, these model systems offer many (still unexplored) advantages for studying fundamental membrane properties, such as the fusion of endosomes with model membranes [16].

In the following sections, we shall mention various aspects of membrane physics, focusing mainly on the early contributions to each of the topics considered. More recent and modern developments will undoubtedly be described in more detail in other chapters of this volume. Accordingly, we shall briefly discuss some of the early theoretical studies of membrane phase transitions, lipid chain order, lipid–protein interactions, and membrane elasticity, yet we shall barely mention the many recent large-scale and multiscale molecular dynamics (MD) simulations of such systems and phenomena. Obviously, we shall not be able to discuss all the many relevant aspects of membrane physics, nor to properly cite all the outstanding studies that contributed to our current understanding of membrane structure and function. We apologize in advance for any unjustified omission of a relevant reference, as well as for our possible personal biases.

1.2 POLYMORPHISM OF LIPID–WATER SYSTEMS

In 1968, based on sophisticated X-ray scattering studies, Luzzati and coworkers were the first to reveal the rich lyotropic and thermotropic polymorphism of phospholipid–water systems [5,6,18]. They observed that natural lipids, such as egg lecithin, exhibit similar lyotropic phases to those formed by synthetic surfactants. They also found that lipid bilayers undergo a first-order chain-melting transition. Since then, numerous experimental and theoretical studies have been reported, extending and enhancing our understanding of lipid–water assemblies and their phase behaviour. A few of the early findings are briefly outlined below.

1.2.1 Lyotropic Polymorphism

The lyotropic lipid assemblies of the greatest biological relevance are the multilamellar (L) phase and the cubic (Q) phase including its disordered bicontinuous version known as the sponge phase (L_3). Tightly packed multilayers are present, for instance, in the nervous system, playing a key role as low capacitance layers in myelin sheath of axons. A unique feature of the Q-phase is that its constituent lipid molecules form an ordered 3D network of interconnected, water-filled, bilayer tubes [18]. Upon swelling, the cubic symmetry may disappear, but the network topology remains intact, resulting in the formation of the L_3 phase. Interest in this phase increased recently following the discovery by Rapoport and coworkers that the endoplasmic reticulum (ER) can form tubular networks extending throughout the entire cell [19,20].

The early findings about the polymorphism of lipid–water assemblies indicated the possible role of lipid-phase transition in controlling biological membrane processes. For instance, the L_β to L_α transition discussed below allows cells to adjust the elasticity and fluidity of their membranes in response to changing environmental conditions [21]. These notions stimulated the development and application of various new experimental techniques, as well as a burst of theoretical studies. Important insights into the molecular order and dynamics of fluid phases were gained by measurements of lipid orientational order parameters using Deuterium nuclear magnetic resonance (NMR) [22,23], infrared [24], and spin label electron spin resonance (ESR) spectroscopy [25]. Chain segment distributions and dynamics were derived by neutron scattering methods [26–28], and correlations between vesicle shape changes and phase transitions were gained by freeze-fracture electron microscopy (EM) (for review, see reference [29]). In parallel, thermodynamic properties could be measured with high precision using dilatometry [30–32], light scattering, Fourier transform infrared spectroscopy (FTIR) [21,30], and sensitive calorimetric methods [33].

1.2.2 Thermotropic Polymorphism of Lipid Bilayers

In addition to the rich polymorphism of lyotropic 3D lipid phases, lipid bilayers exhibit a number of thermotropic phase transitions taking place within the 2D membrane plane. Their number and character are even richer and more intricate in multicomponent lipid bilayer and lipid–protein membranes [18,29,34–37]. At low temperatures, the lipid molecules are immobile, forming a "gel" phase, in which the lipid head groups are generally densely packed, organizing typically in orthorhombic crystalline order. The hydrocarbon lipid tails in the gel phase are stretched, with their main axes parallel to each other and directed either parallel or tilted with respect to the membrane normal, with the respective phases denoted as L_β and $L_{\beta'}$. The $L_{\beta'}$ structure is generally observed in bilayers composed of lipids with large polar head groups, such as phosphatidylcholines (PCs), whose cross-sectional area is larger than that of their (fully stretched) two-chain hydrocarbon tails. The tilting enables tighter packing of the tails, enhancing chain cohesion within the hydrophobic membrane core.

Below, we briefly outline a few of the numerous pioneering experimental and theoretical studies of bilayer phase transitions and lipid order in membranes. Related topics, such as lipid–protein interactions, as well as membrane defects, dynamics, and curvature elasticity, are discussed in subsequent sections.

1.2.3 The $L_\beta \rightarrow L_\alpha$ and Other Bilayer Transitions

Common to all lipid bilayers is a first-order chain-melting transition ($L_\beta \rightarrow L_\alpha$) from the gel phase to a 2D fluid phase, taking place at a well-defined temperature T_m. Above this temperature, in the L_α phase, the membrane is a 2D fluid, enabling lateral translation of the lipids in the membrane plane, as well as limited conformational freedom of their flexible hydrocarbon tails (see Figure 1.3). The extent of chain conformational freedom depends sensitively on the cross-sectional area per lipid head group a, whose value depends on the balance of repulsive and attractive interlipid forces: The effective attraction between lipid tails, which tends to minimize the unfavourable surface energy associated with exposing the hydrocarbon region to contact with water, versus the repulsive forces due to electrostatic and excluded volume interactions between head groups, as well as the entropic repulsion between the hydrocarbon lipid chains in order to increase their conformational freedom. This balance of forces is schematically described later in Figure 1.4.

Widely varied in their mathematical approaches, most theories of the $L_\beta \rightarrow L_\alpha$ transition agree that its primary thermodynamic driving force is the gain in conformational (trans/gauche) entropy of the hydrocarbon lipid chains above T_m, compensating for the partial loss of cohesive energy of the hydrocarbon chains in the passage to the (~4%) less dense L_α phase. This interplay between interchain attraction energy and conformational entropy is reflected in the chain length dependence of T_m. Thus, for example, in the case of bilayers composed of PC lipids with two saturated hydrocarbon tails, T_m increases from 23°C to 41°C to 55°C as the lipid chain length increases from $n=14$ to 16 to 18 methylene groups, respectively [29,36]. While both the enthalpy and entropy changes in the transition increase with n, the monotonic (roughly linear) increase of $T_m = \Delta H_m/\Delta S_m$ with n indicates a stronger chain length dependence of the transition enthalpy. The presence of a double bond along the hydrocarbon chains has a dramatic effect on the chain-melting transition. Depending on the position of the C=C bond, T_m may be 1°C–60°C lower upon shifting the position of this bond from the vicinity of the head group to the centre of the chain and then up again by about 50°C when the double bond is close to the chain end, as observed for dioctadecanoyl-phosphatidylethanolamine (PE) (C18:1-PC) bilayers [38]. A possible explanation to this behaviour is that when the double bond is located near the beginning or end of the chain, most of the chain can still behave as a saturated one enabling relatively efficient packing in the gel phase, while its presence in the middle of the chain introduces chain tilting resulting in a perturbation to lipid packing and thus a lower transition temperature.

Among the early theories of the chain-melting transition are exact mathematical solutions to approximate bilayer models, e.g., Nagle's dimer model of hydrocarbon chains on a 2D lattice [39], Scott's [40] and Pink's [41] "few-states" chain models, as well as Landau theories using the chain segment density (ρ), the average orientational order parameter (η), or the area per head group (a) as the relevant thermodynamic order parameter [42–44]. Realistic, albeit approximate, molecular-level theories have also been proposed. First among these is Marcelja's pioneering theory [45,46], where all rotational isomeric states of the hydrocarbon chains are taken into account, and their interactions and bond orientational order parameters are treated in analogy to the Maier–Saupe theory of the isotropic–nematic transition. In addition to the chain-melting transition, his theory and subsequent molecular-level models, followed by a growing number of increasingly sophisticated and extensive MD simulations, have provided detailed information about chain conformational properties in the fluid phase, as discussed in little more detail in Section 1.2.5. Many of these approaches were later extended and applied to analyse the phase behaviour of multicomponent lipid bilayers and lipid–protein membranes (Section 1.4) and to explain and predict the molecular aspects of membrane elasticity (Section 1.8).

Another topic of physicists' interest has been the phase behaviour of the lipid bilayer below T_m [47]. While crystalline, the gel phases prevailing at these temperatures are remarkably flexible, resembling the glassy state of polyethylenes. The hydrocarbon chain dynamics in the bilayers involves kink formation and rapid diffusion along the hydrocarbon chains, as discussed in Section 1.6 (see Figure 1.11). Among the challenging issues pertaining to the crystalline phases is their rich defect structure, which is largely due to the frustration of lipid packing owing to two reasons: first, because the fatty acid chains cannot rotate freely about the carboxyl groups linking them to the head group phosphates (in contrast to ether lipids) and, second, because the hydrocarbon chains prefer forming a triangular lattice while the head groups prefer orthorhombic packing (see Figure 1.3c). The defect structures associated with these frustrations [29] became popular paradigms for studying the physics of flexible two-dimensional solids such as polymerized membranes. These solids exhibit a novel type of a phase transition with diverging specific heat, termed the *crumpling transition*, which is a consequence of the dissociation of dislocations [48]. The spatial long-range order of the crystalline lattice of the low-temperature phase is lost in this transition, while the long-range orientational order (i.e., the

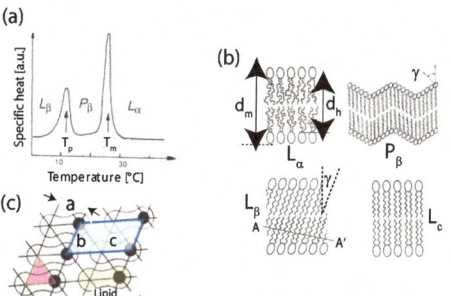

FIGURE 1.3 (a) The heat capacity of a suspension of vesicles composed of dipalmitoylphosphatidyl-glycerol (DPPG, C:16,0PG), showing the chain-melting transition ($L_\alpha \to L_\beta$) and the pretransition ($L_\beta \to P_{\beta'}$). (b) Schematic illustration of the molecular architecture of the fluid (L_α) and the three solid phases ($P_{\beta'}$, L_β and L_c). The nontilted crystalline L_c-phase forms by annealing the L_β-phase at temperatures below the pretransition temperature T_P. (c) The solid phases are frustrated because the head groups favour the orthorhombic symmetry, while the chains prefer the triangular lattice (of lattice constant $a = 0.42$ nm).

orientation of the crystal axes) is maintained. This new type of a solid state is known as the *hexatic phase*. Among others, it was suggested to characterize the structure of the spectrin–actin network of erythrocytes (see, e.g., [29]).

In many cases, predominantly in membranes composed of lipids with large head groups, the chain-melting transition is preceded by a weak pretransition, taking place at a temperature $T_p < T_m$, whereby the bilayer transforms from the crystalline L_β phase to a corrugated phase commonly known as the *ripple phase* and denoted as $P_{\beta'}$ [36,47]. Unravelling the origin of the ripple phase has challenged many experimentalists and theorists. One rather consistent conclusion from the various studies is that the lipids in the long edge of the ripples are densely packed – resembling the L_β order, while the shorter edge is thinner and lipid packing there resembles the L_α phase. The theories suggested range from continuum models, through molecular-level calculations, to MD simulations, yet a full picture is still missing and the nature of the pretransition is still a matter of controversy. Early models of the ripple phase can be found in Ref. [29] (for a recent study and survey of the ripple phase, see, e.g., [49]).

1.2.4 Lipid Monolayers

In the early 1970s, lipid monolayers were intensively studied as potential model systems of biomembranes, offering unique possibilities to establish phase diagrams of lipid mixtures as functions of temperature and lateral pressure. Monolayers are also useful to study the nucleation and growth (Ostwald ripening) of microdomains of lipid mixtures [50–52]. Furthermore, monolayer experiments provide direct determinations of the thermal expansivity and compressibility of lipid layers, enabling parallel characterization of their molecular architecture by high-resolution X-ray and neutron surface scattering [53].

Lipid monolayers have been extensively used to study the adsorption and enzymatic activity of proteins as a function of the lipid packing density. Striking examples are the cleavage of phospholipids by phospholipases and the exchange of proteins between vesicles of different composition mediated by lipid exchange proteins, showing that enzymatic activity in the fluid–solid coexistence region is far more efficient than in either the fluid or the gel phase [53]. Underlying this behaviour is the diverging membrane compressibility in the phase transition, enabling the enzyme to efficiently grab lipids in the domain boundaries and membrane defects accompanying the transition (see Section 1.7). Monolayer studies are also expected to provide important insights into the mechanisms associated with the recruitment of oppositely charged lipids to the vicinity of peripherally bound macromolecules such as proteins or DNA.

1.2.5 Lipid Chain Order in the L_α Phase

Information concerning chain ordering and conformational freedom in the (biologically relevant) L_α phase has been the goal of numerous experimental, theoretical, and computer simulation studies. Of particular interest are the orientational bond order parameter profiles of the lipid chains, which are usually provided by deuterium NMR measurements [22,54] and the spatial distributions of lipid chain segments obtained by X-ray and neutron scattering [26,55,56]. The period from the late 1970s to the 1990s witnessed the appearance of numerous theories and simulation studies of lipid chain order, including elegant lattice models [57,58], pioneering MD simulations [59], as well as a number of mean field theories [45,60–62] showing very good agreement with experiment (for a review of the early theories, see, e.g., [63]).

The orientational bond order profile, $\{S_k\}$, represents the average relative orientations of the C_k–C_{k-1} bonds along the hydrocarbon lipid tails relative to the membrane normal. They are directly related to the C–H bond orientations, which can be determined by selective deutration of the C_k–H bond, yielding $S_k = (1/2)\langle 3\cos^2\theta_k - 1\rangle$, where θ_k is the angle between bond C_k–D bond along the hydrocarbon chain and the membrane normal. The angular brackets denote averaging over all possible chain conformations. In Marcelja's pioneering theory [45], $\{S_k\}$ appears explicitly in the self-consistency equations for the nematic-like molecular field, $\Phi = V_0 \langle (n_{tr}/n) \sum_k S_k \rangle$, with n_{tr}/n denoting the fraction of C–H bonds in *trans* conformation. $V_0 \approx 700$ cal/mole was used as the average van der Waals energy per chain in the corresponding polyethylene liquid. His solution, for any given chain length, n, yields both $\{S_k\}$ and the average area per head group, a, as a function of temperature.

Other molecular field theories [61–63] evaluated the probability distribution of chain conformations based on the "hydrocarbon droplet" assumption, according to which the hydrocarbon segment density within the hydrophobic core is spatially uniform and equal to that in liquid alkanes [64,65]. An explicit expression for $P(\alpha)$ – the probability of finding the lipid chain in conformations α – can be derived (for any given area per lipid head group a) by minimizing the chain free energy subject to the constraint of constant segment density [62,63]. The result is

$$P(\alpha) = \frac{1}{q}\exp\left\{-[\varepsilon(\alpha) + \int \pi(z)\varphi(\alpha;z)\mathrm{d}z]/k_B T\right\} \quad (1.1)$$

where $\varepsilon(\alpha)$ is the chain conformational (e.g., trans/gauche) energy in conformation α; $\varphi(\alpha;z)\mathrm{d}z$ is the number of chain segments whose centres in this conformation fall within the layer $z, z + \mathrm{d}z$ parallel and at distance z from the hydrocarbon–water interface; k_B is Boltzmann's constant; and T is the temperature. The $\pi(z)$ are Lagrangian parameters representing the lateral pressure profile (see Figure 1.4), evaluated by solving the self-consistency equations: $\langle\varphi(z)\rangle = \sum_\alpha P(\alpha)\varphi(\alpha;z) = \rho(z)$, with $\rho(z) = \rho$ denoting the local chain density in the bilayer core. Given $P(\alpha)$, one can calculate any desired single chain property, e.g., the order parameter profile, as illustrated in Figure 1.4. Furthermore, thermodynamic quantities, in the mean field approximation, can be derived using statistical thermodynamic relationships involving the partition function, q, and the lateral pressure profile, $\pi(z)$. Such properties include lipid–protein interaction free energies and membrane elastic moduli, as briefly mentioned in subsequent sections.

1.3 THE LIPIDOME: MEMBRANES AS MULTICOMPONENT LIPID MIXTURES

In the 1970s, biochemists began to analyse the complex lipid composition of biomembranes and to explore the mechanism of lipid synthesis (see Refs. [29,69] for reviews). Their studies showed that each cell organelle exhibits a characteristic lipid composition, which is maintained during the cell's lifetime, despite the rapid exchange of material via endocytosis, exocytosis, and intracellular vesicle trafficking. In another series of elegant biochemical experiments, the van Deenen group showed that the lipids are asymmetrically distributed between the two leaflets of the erythrocyte PM: the sphingomyelins (SPHMs) and oligosugar-carrying gangliosides residing only in the outer leaflet, along with PCs. The inner leaflet was found to contain most of the PEs and all the anionic lipids, specifically phosphatidylserines (PSs) that comprise about 7% of the acidic membrane lipids and phosphatidylinositol PI-4,5-P2 comprising about 2%.

Significant differences were also found with respect to the lipid composition of different organelles. Sphingomyelins and oligosugar-carrying gangliosides reside mainly in the PM and the late endosomes, amounting to about 20% of the total phospholipid content of these organelles, as compared with only 3%–10% of these lipids in other organelles. Deveaux and coworkers provided evidence

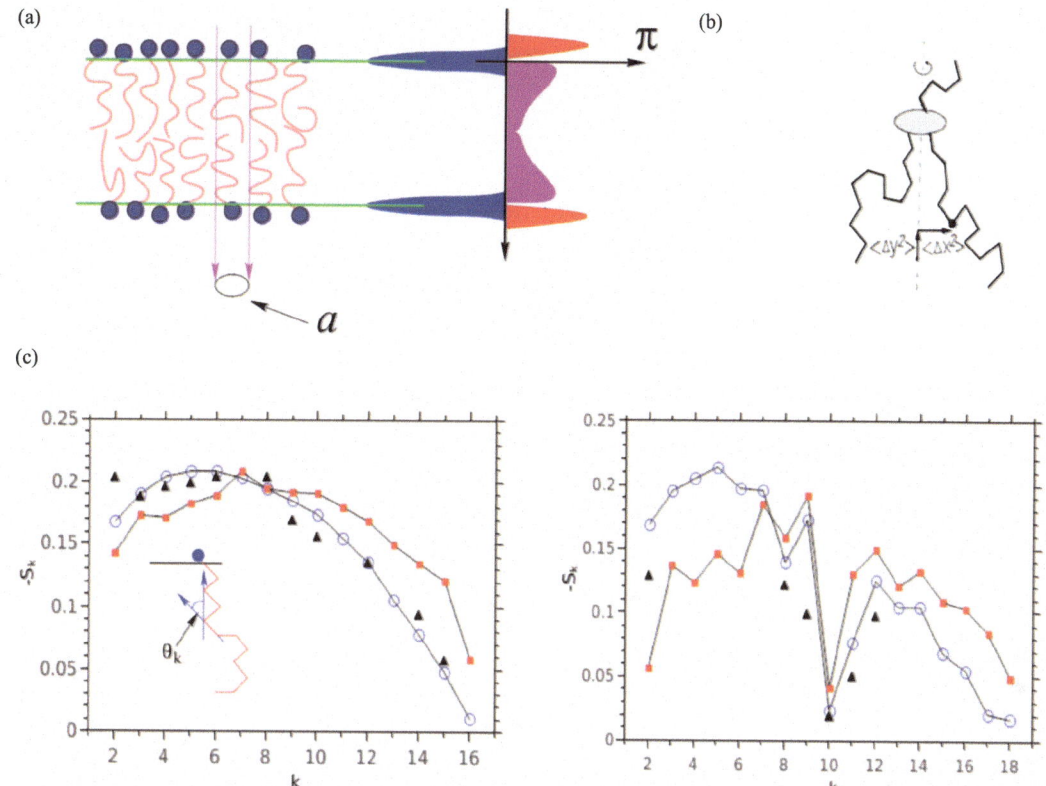

FIGURE 1.4 (a) The lateral pressure profile in a lipid bilayer. Head group repulsion resulting from electrostatic and excluded volume interactions and interchain repulsion due to conformational restrictions are balanced by the tendency of the bilayer to minimize water–hydrocarbon contact. (b) Characterization of dynamics of chain conformations in terms of mean square displacement of chain segments in horizontal and normal directions. The local diffusion coefficients in the two directions increase linearly with the distance from the head group. This gradient in segment motion is equivalent to the gradient of static order parameter [27]. (c) Orientational bond order parameters of the C_k–H bonds along the palmitoyl chain (left), and the oleoyl chain (right) of POPC. The palmitoyl chain is fully saturated, while the oleoyl chain contains one double bond (between carbons 10 and 11), which perturbs chain order, resulting in the sharp dip in the order parameter profile. The inset illustrates the angles θ_k appearing in the definition of the order parameter. (For an all-trans chain along the membrane normal, $\theta_k = \pi/2$, implying $S_k = -1/2$.) The triangles describe the experimental results of Seelig and Waespe-Šarčević [66], the squares are from the molecular dynamics simulations by Heller et al. [67], and the circles are from mean field calculations based on Eq. (1.1) [68].

that the asymmetric lipid distribution is maintained by specific ATP-dependent translocases called flippases [70]. Differences in cholesterol concentrations are also striking; the molar concentration of cholesterol in the nuclear membrane, the ER, and Golgi is less than 10%, as compared with 15% in lysosomes and 42% in the PM which is close to the solubility limit (see Figure 1.5b). Steck and coworkers showed that this gradient of cholesterol distribution is actively maintained [71]. The fine-tuning of lipid homeostasis is mediated by specific exchange proteins, transferring lipids between the outer leaflets of intracellular organelles (see Figure 1.12).

Correlations between the thermodynamic phase state and the biological function of the membrane became evident following the observation that cells can adapt to the ambient temperature by proper choice of lipid chain length, the fraction of unsaturated lipids, and the number of double bonds of these lipids. In particular, specific enzymes were found to help plant cells adjust to low environmental temperatures by increasing the fraction of unsaturated lipids [72].

Taken together, these observations showed that the lipid composition is an intrinsic property of each cell type, that their synthesis is genetically controlled, and that lipid and protein biosyntheses are closely regulated. In analogy to the terms genome and proteome, we may therefore regard the sets of lipids in the various organelles as the lipidome.

Unravelling the homeostatic control of lipid composition of cell organelles triggered numerous physicochemical studies of the phase behaviour of multicomponent lipid and lipid–cholesterol layers. A wealth of experimental techniques were developed for this purpose, including ESR spectroscopy [21,34], NMR techniques [34,73], FTIR and Raman spectroscopy [24], differential calorimetry [74], densitometry [75], and freeze-fracture EM (see [29] for references). All of these techniques determine phase boundaries by scanning the temperature, which means that the lipid composition of the segregated phases changes continuously

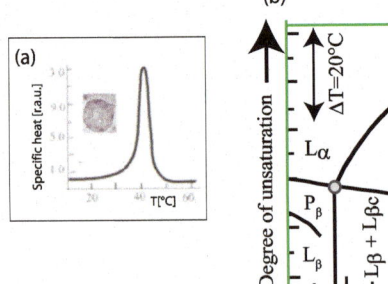
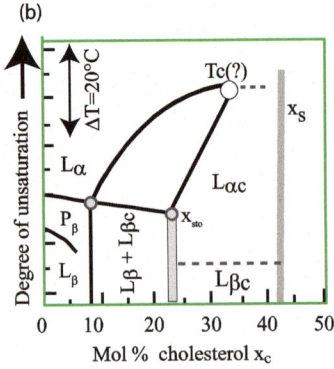

FIGURE 1.5 (a) Demonstration of a phase transition of vesicles of brain sphingomyelin (SPHM), reproduced from [85]. (b) Phase diagram of a binary mixture of cholesterol and PC lipid, in this example DMPC with $T_m = 23°C$. Below T_m, a solid–solid miscibility gap forms, which separates two homogeneous low-temperature phases described in the text. The vertical bar marks a stoichiometric mixture containing 75% PC. Above T_m, two fluid phases coexist. They are separated by a miscibility gap that is expected to end in a critical point at T_c. It separates the cholesterol-poor L_α phase and the cholesterol-rich condensed $L_{\alpha c}$ phase. Image adapted from Ref. [86].

while scanning through the coexistence region. Therefore, these methods do not allow to determine equilibrium-phase diagrams or to directly observe the segregation of phases of equal symmetry, such as solid–solid or fluid–fluid immiscibility. At present, small-angle neutron scattering (SANS) in combination with contrast matching is the only method enabling the establishment of phase diagrams of mixed bilayers of mixed vesicles in thermodynamic equilibrium [76].

An ongoing issue since the early days of membrane physics has been whether nature makes use of lipid-phase separation in order to maintain the characteristic lipid and protein composition of cellular organelles. Two major questions that arise in this context are (1) why is cholesterol mainly accumulated in the PM? and (2) how are proteins in the PM recycled during membrane turnover?

Numerous studies on binary lipid–lipid mixtures and ternary lipid–lipid–cholesterol mixtures, in parallel to the search for specific lipid–protein interaction mechanism, provided partial answers to these questions. Research along these lines was guided by the expectation that phase separation processes in phospholipid mixtures, liquid crystals, and metal alloys are determined by the same thermodynamic concepts, as outlined in the regular solution theory of Hildebrand or the Bragg–Williams lattice models (see Ref. [29]). These concepts predicted that at constant temperature, phase separation is expected to occur if the components exhibit different symmetries (called symmetry rule in liquid crystal physics) or if the molecular shapes of two components (forming the same type of phase) differ remarkably. An example of the first case is the decomposition of DMPC and DPPC mixtures into fluid (L_α) and solid (L_β) phases (fluid–solid demixing). Examples of the second situation are fluid phospholipid–cholesterol mixtures (as shown in Figure 1.5b) and fluid mixtures composed of nonsaturated PC and PE lipids, as studied by Wu and McConnell [73]. Examples of the third case are mixtures of saturated lipids differing in chain length by more than two CH_2 groups, such as DMPC (C14:0-PC) and DSPC (P18:0-PC). This mixture exhibits a peritectic phase diagram with the solid–solid miscibility gap penetrating into the fluid–solid region, resulting in precritical phenomena above the liquidus temperature [77,78].

The phase behaviour of binary and ternary mixtures of phospholipids with cholesterol has been the focus of numerous studies. Mixtures of zwitterionic saturated lipids (such as PCs or SPHMs) with cholesterol exhibit phase diagram similar to the one shown in Figure 1.5 for the special case of DMPC/cholesterol. SANS experiments showed that, below the chain-melting transition ($T<T_m$), this mixture exhibits a miscibility gap between $x_{chol} \approx 0.08$ and $x_{chol} \approx 0.25$. At $x_{chol} < 0.08$, a solid solution is formed, while at $x_{chol} > 0.25$, a homogenous liquid solution of cholesterol in DMPC forms, which saturates at $x_{chol} \approx 0.42$. To fulfil the phase rule, the phase line at $x_{chol} \geq 0.25$ is assumed to mark a stoichiometric mixture. Several experiments providing evidence for this interpretation are summarized in Refs. [77,78].

The packing density in the saturated solution ($x_{chol} > 0.2$) is 20% higher than that of the solid solution. It is thus often referred to as a "condensed ordered phase", despite being a 2D fluid. Above T_m, there is a fluid–fluid miscibility gap, separating a cholesterol poor expanded phase, L_α, from a cholesterol-rich and condensed, yet liquid, phase $L_{\alpha c}$ (where the index c stands for "condensed"). Such fluid–fluid coexistence of phospholipid–cholesterol mixtures was also suggested by theoretical studies [79,80]. The molecular structure and phase changes in lipid–cholesterol systems have been studied in great detail, both experimentally [81], and by MD simulations [82].

The biological relevance of the phase diagram in Figure 1.5b becomes evident when we consider the fatty acid chain distribution of membrane lipids (see Table 1.1 in Ref. [29]). It shows that SPHMs expose mainly long-chain lipids with few double bonds (in particular C16:0, C18:0, C24:0, and C24:1), Therefor the natural SPHM fraction exhibits a chain-melting transition near physiological temperature, as shown in Figure 1.5a [75,76]. In contrast, the glycerol-based lipids, such as PCs, contain mainly C16:0, C18:0, C18:1, C18:2, and C20:4 and melt well below 10°C. Other notable facts are that cholesterol associates preferentially with saturated lipids [83,84] and acts as fluidizer of crystalline membranes [29,76].

Taken together, these findings suggest that the lipid moiety of mammalian cells can be regarded as quasi-binary mixture composed of glycerol lipid (which melts at $T \leq 0°C$) and SPHM (with $T_m \approx 40°C$). Since the glycerol lipids exhibit a high content of unsaturated fatty acids, the cholesterol content ($x_c \leq 10$ mole %) and the lipid packing density are small [83]. In contrast, the high-melting temperature SPHM fraction is saturated with cholesterol and thus forms a condensed (glass-like) state called $L_{\alpha c}$.

A possible role of the pseudo-binary lipid mixtures for the generation of functional domains in cell membranes and lipid–protein sorting became only evident after the establishment of elastic and electrostatic mechanisms of selective lipid–protein interaction, which will be discussed in the following.

1.4 LIPID–PROTEIN SORTING, INTERACTIONS, AND DOMAIN FORMATION

Biological functions depend critically on the interaction of the lipid membrane with peripheral and integral proteins and most often with proteins composed of both hydrophobic and polar domains. The interaction between peripheral proteins and lipid membranes is primarily electrostatic, involving generally attractive forces between polybasic domains and anionic lipids. These interactions will be addressed in the next section, while here we shall comment on the interactions between hydrophobic integral proteins and their surrounding environment of lipid chains.

1.4.1 Lipid–Protein Interaction

The free energy changes associated with the partitioning of polypeptide chains onto the hydrocarbon–water interface of lipid membranes depend sensitively on their length and amino acid composition. Detailed measurements of interfacial hydrophobicity scales provide important information pertaining to the folding free energies of the peptides and their propensity to insert into the lipid hydrophobic core [87]. While preferring the hydrocarbon lipid chains over the aqueous environment, the presence of a hydrophobic protein inclusion in the membrane core incurs a nonnegligible free energy penalty associated with the perturbation of the local lipid order of their surrounding lipid environment. The perturbation energy is proportional to the number of lipid molecules affected by the presence of the integral inclusion and hence to the circumference of the protein. Indeed, experiments with spin-labelled lipids reveal that the fraction of lipids associated with intramembranous proteins is proportional to their perimeter [88]. It also depends on the *hydrophobic mismatch*, $\delta h = h_P - h_L$, which measures the difference between the hydrophobic height of the protein and the thickness of the unperturbed lipid bilayer [89], as illustrated in Figure 1.6.

In multicomponent lipid membranes, the presence of integral inclusions may lead to lipid sorting, attracting lipids of matching height and chemical affinity to their immediate vicinity. Furthermore, in analogy to the hydrophobic interaction – whereby two nonpolar solutes in water are attracted to each other because in contact they cause a lesser perturbation of the water structure around them – the interaction between two hydrophobic inclusions in a lipid membrane is attractive because, together, they perturb a smaller number of neighbouring lipids as compared with their being far apart of each other. If the perturbation of lipid order is large enough, as in the case

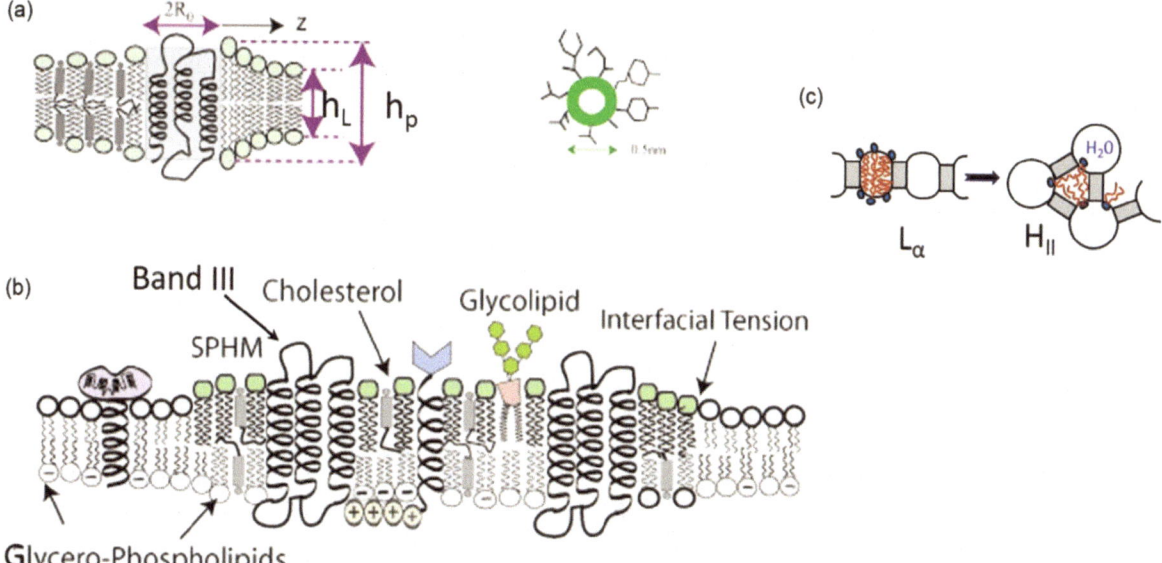

FIGURE 1.6 (a) Elastic distortion of a lipid bilayer induced by the mismatch of the hydrophobic thicknesses of the bilayer (h_L) and the membrane spanning domain of the protein (h_P). (b) Model of the plasma membrane suggested by the pseudo-binary phase diagram of Figure 1.5 and the Bretcher–Munroe analysis of hydrophobic length distribution of proteins [107]. The membrane spanning parts of band III comprise 25 amino acid residues ($h_P \approx 3.8$ nm) and is expected to be surrounded in the outer monolayer primarily by SPHM with C24:0 chains and PE or PC with C18:0 chains. The cholesterol is expected to accumulate in the outer leaflet. (c) Above a threshold concentration of short peptides ($h_P < h_L$) in a bilayer containing long-chain lipids and/or lipids with negative spontaneous curvature, a transition takes place to the inverted hexagonal phase, allowing more conformational freedom to the long lipid chains.

of large $\delta h = h_P - h_L$, the lipid-mediated attraction between proteins can overcome their translational (mixing) entropy, resulting in their 2D condensation to form protein-rich 2D domains.

Treating the lipid–protein membrane as a 2D liquid mixture of N lipids of cross-sectional area a, and N_P proteins of area ga, its free energy in the mean-field approximation is given by

$$\Delta G_{LP} = Nk_BT\{\varphi_L \ln \varphi_L + (1/g)\varphi_P \ln \varphi_P + \chi\varphi_L\varphi_P\}$$

(1.2)

with φ_L and φ_P denoting the area fractions of lipid and proteins, respectively. Proteins whose hydrophobic domain consists of a single α-helix occupy about the same hydrocarbon–water surface area as that of one lipid molecule (typically around 70Å^2) implying $g \approx 1$. Larger proteins, such as the band III ion channel, occupy larger membrane areas amounting to as many as $g \approx 20$ lipids. As usual, the nonideality interaction parameter χ is proportional to the difference between the lipid–protein interaction potential and the average of the lipid–lipid and protein–protein potentials. It should be noted, however, that unlike the case of flexible polymers in solutions, integral proteins in membrane are rigid, so that χ depends on their circumference length rather than on their length. The critical point implied by the mixing free energy Eq. (1.2) is specified by a lower critical protein concentration than that of ordinary $(g = 1)$ mixtures, namely $\varphi_P^c = 1/(1+\sqrt{g}) \to 1/\sqrt{g}$, and a smaller critical interaction strength, $\chi = (1+\sqrt{g})^2/2g \to 1/2$. Resembling polymer solutions, the binodal curve for large g is strongly skewed, corresponding to very low protein solubility. For example, the solubility limit of Band III at room temperature is $\tilde{\varphi}_P = 10^{-4}$.

The strength of lipid–protein interactions can be determined by measuring the shift in the lipid chain-melting transition temperature caused by small amounts of membrane proteins, or by the shift of the solidus and liquidus lines [29]. Using light scattering and EM imaging, Riegler and Möhwald [90] studied the gel–liquid crystal transition of vesicles composed of PC lipids of different chain lengths containing integral (photosynthesis reaction centre) proteins. Consistent with the "mattress model" of Mouritsen and Bloom [89], they found that the lipid-mediated attraction between the proteins increases with the magnitude of the hydrophobic mismatch, $\delta h = h_P - h_L$. In parallel, several molecular-level models and quite a few continuum elastic theories have been proposed to account for the interactions between membrane inclusions. In 1975, using liquid crystal elasticity theory, Gruler pointed out the changes imposed on membrane curvature due to conical inclusions [91]. A year later, Marcelja applied his molecular field theory to the case $\delta h = 0$, where the only perturbation to lipid order is the presence of the rigid protein wall [46]. He found that the width of the perturbation annulus surrounding an isolated inclusion involves about three lipid diameters. The range of perturbation and its magnitude are larger for nonzero hydrophobic mismatches because the lipid chains adjacent to the inclusion must stretch ($\delta h = h_P - h_L > 0$) or compress ($\delta h < 0$) in order to avoid exposing hydrophobic protein regions to water [92]. The lipid-mediated interaction between hydrophobic inclusions has also been studied by quite a few continuum elastic theories. Thus, for example, Schröder [44] predicted that the attraction between inclusions decreases exponentially as a function of their distance R, while Goulian et al. [93] found a $1/R^4$ dependence. Safran and coworkers emphasized the importance of the lipid spontaneous curvature, revealing the possible existence of barriers to protein attraction resulting in nonmonotonic interaction potentials [94].

In addition to in-plane protein aggregation, the elastic deformation of the lipid bilayer by hydrophobic inclusions can sometimes lead to rather dramatic structural transformations of the lipid–protein matrix. For example, it was found that beyond a certain critical concentration of short hydrophobic peptides ($\delta h < 0$) in a lipid bilayer, a morphological transition of the lipid–peptide mixture can take place from the lamellar (L_α) to the inverse-hexagonal (H_{II}) phase, with the peptides serving as bridges between neighbouring water tubes (see, e.g., Refs [95,96]). The critical peptide concentration depends on the spontaneous curvature of the lipids (preferred by lipids with negative spontaneous curvature) and the bending rigidity of the embedding lipid matrix [97]. Another well-known peptide-induced morphological transition that has been extensively studied, both experimentally (see, e.g., Refs [98,99]) and theoretically (see, e.g., Refs [100–103]), is the formation of membrane pores by amphipathic, antimicrobial, peptides. Similarly, many studies have been concerned with the kinetics and thermodynamics of gramicidin channel formation [100,101,104,105].

Like many other membrane properties, lipid–protein interactions have also been studied using computer simulations, including both Monte Carlo (MC) simulations and especially and in rapidly growing numbers, MD simulations. In recent years, with the advent of fast computers and the development of sophisticated computational algorithms, most of the theoretical–computational studies of lipid–protein membranes involve MD simulations. The qualitative insights that have been provided by earlier (both molecular-level and continuum) theory approaches are nevertheless invaluable, especially when large-scale cooperative phenomena such as protein-mediated phase transitions or biomembrane elastic properties are concerned.

The adsorption of proteins exposing hydrophobic membrane anchors can mediate the formation of local buds by lateral expansion of one monolayer. It has been postulated that such mechanisms of induced budding can trigger the formation of coated pits [86] and caveoli [106].

1.4.2 Lipid and Protein Sorting and the Formation of Functional Microdomains (Rafts)

A detailed survey of the length distribution of the hydrophobic thickness of integral proteins residing in the PM and the Golgi apparatus due to Bretcher and Munro provided new insights into the physical mechanisms of lipid and protein sorting during membrane turnover by endocytosis and exocytosis [107]. They found that the hydrophobic domains of integral proteins (consisting of 15 amino acid residues; $h_P \approx 2.5$ nm) in the Golgi apparatus are about 0.5 nm shorter than those present in the PM ($h_L \approx 3$ nm). Together with the selective lipid–protein interaction by hydrophobic matching shown in Figure 1.6a, the Munroe Bretcher analysis suggests that protein sorting within the PM as well as between the PM and intracellular organelles could be controlled by lateral phase separation. As illustrated in Figure 1.6b, SPHM and the gangliosides reside in the outer leaflet of the PM. Comprising only 12%–30% of the lipid population in the PM, SPHM can only cover part of the cell envelope. This suggests that the cholesterol-rich SPHM fraction forms lateral domains within the PM which can accommodate proteins with long hydrophobic domains, as illustrated in Figure 1.6b. One biologically important example is the transferrin receptor [29,108].

This type of domain should not be confused with "rafts" [109]. Rafts are generally observed as Triton-insoluble membrane fragments, and it cannot be excluded yet that they are artefacts generated following the modification of the lipid structure by Triton (108B).

Another case where lipid–protein sorting plays a key role is the recycling of receptors after endocytosis by coated pits and caveoli. A well-studied example is the coated vesicle–mediated transfer of iron ions into cells via the Fe carrier transferrin bound to transferrin receptors (reviewed in Ref. [86]).

Evidence has been accumulated indicating that some cell surface receptors are directly recycled from the early endosomes to the PM by the Rab GTPases Rab 4 and Rab 5, thus avoiding the detour about the multibody particles [110]. For these special (but vital) cases, the lipid–protein sorting could be controlled by the phase separation of the sphingomyelines and the glycerol lipid. The transferrin receptor exhibits a particularly long hydrophobic domain ($h_P \approx 4.5$ nm [107]) and could be recycled through the above mechanism. A major step of the recycling is the formation of microdomains of different curvatures by the two segregated phases, which grow via Ostwald ripening until the endosomal vesicle becomes unstable and eventually decays into two vesicles [85,111]. Model membrane studies suggest that fission of vesicles from large endosomes may also be triggered by cholesterol at molar fractions close to the solubility limit of $x_c \sim 0.42$ [85].

The phase separation–induced shape changes, and vesicle fission is an ongoing theme of membrane physics. It was extensively studied experimentally, first by freeze-fracture EM (reviewed in Ref. [29] and later by microfluorescence techniques [111]). Theoretical studies were based on the combination of the bending elasticity model and the Cahn–Hilliard theory of spinodal decomposition [112].

Unfortunately, the interest in the homeostasis of membrane lipid composition and the physics of structure formation by specific lipid–protein interactions has faded considerably in recent years, perhaps because these are very complex fields of research. We must remember, however, that proper physiological function of cells and hence our health depend critically on the control of lipid and membrane protein patterns. Diseases are often associated with dysfunctions of the lipidome. Thus, for example, erythrocytes of patients suffering from diabetes exhibit a 1–2 mole% higher cholesterol content than those of healthy people, indicating a correlated metabolic disorder. The higher cholesterol content reduces the erythrocytes deformability and could possibly result in blood flow dysfunction.

1.5 Membranes as Electrified Interfaces: Electrostatic Switching of Functional Proteins

The electrical surface potential of biomembranes is dictated by two contributions: first, the presence of about 10% of charged lipids exposing their acidic head groups to the cytoplasmic surface and, second, the electric dipole layer formed by the carboxyl groups of the glycerol backbone, which is located about 0.5 nm below the lipid–water interface (reviewed in Ref. [29]). The 2D density and spatial distribution of surface charges affect the structure and thermodynamics of isolated membranes and play a crucial role in the recruitment, binding, and activation of functional proteins with polybasic domains, as discussed below.

1.5.1 Role of Lipid Charge on the Chain-Melting Transition

One of the first questions that fascinated physicists has been concerned with the influence of the lateral interaction between charged lipid head groups on the gel-to-liquid crystal transition temperature, T_m. As first shown by Träuble and Eibl [113] and further elaborated by Jähnig [114]), the surface lateral pressure associated with the electrical double layer, π_{el}, can be regulated by varying the pH or the ionic strength of the embedding solution. The corresponding shift in transition temperature is given by

$$\delta T_m = \frac{\Delta F_{el}}{\Delta S} \approx -\frac{\Delta A}{\Delta S} \pi_{el} \qquad (1.3)$$

where $\pi_{el} = -\partial F_{el}/\partial A$ with ΔF_{el} denoting the excess electrostatic free energy of the charged membrane relative to the neutral one, with ΔS and ΔA denoting the changes in molar entropy and molar area in the chain-melting transition, respectively, both of which are positive. The transition temperature of the charged membrane is lower because the electrostatic pressure, which tends to increase the area per molecule, favours the more expanded fluid state, thus enhancing the transition.

Application of the classical Gouy–Chapman–Overbeek theory of electrical double layers led to important quantitative relationships between the change in transition temperature and the degree of dissociation of the lipid head groups. Theories developed by Jähnig [114] and others (as discussed in detail by Cevc [115] and Andelman [116,117]) enabled the calculation of the surface pH as a function of the electrostatic surface potential and the ionic strength of the aqueous phase. An important consequence of the negative membrane surface charge revealed by these early studies is the lowering of the surface pH, which plays a key role in controlling the degree of dissociation of lipids and adsorbed proteins [118]. At low ionic strengths ($I \leq 100$ mM), the effect can be dramatic, inducing significant pH shifts; $\Delta pH \geq 2$ [29]. The effects are, however, small at physiological ionic strengths of $I \approx 400$ mM, explaining the fading interest in electrically induced phase transitions.

The electrostatic interactions between lipid membranes containing charged lipids and oppositely charged macromolecules (such as proteins exposing polybasic sequences or DNA) are of great biological relevance. Many of the early studies of these interactions have focused on the calcium-induced membrane binding of annexin to PS-containing membranes [119]. Other early studies demonstrated the charge-induced switching of protein function. One intriguing example is the electrostatic switching of the orientation of amphiphatic helices exposing charged head groups in pH-dependent manner from a horizontally adsorbed state into a membrane spanning state, whereby peptide oligomers organize into ion channels, enabling their function as antimicrobial peptides [120].

The interest in electrostatic membrane–protein interaction increased dramatically in recent years following the discovery that acidic lipids, in particular phosphoinositides (PI), such as PI-4,5-P2 and PI-3,4,5-P3, play a key role in the activation of many of the proteins involved in signal transmission across membranes and the coupling of the actin cortex to the PM. These electrostatically switchable proteins contain polybasic sequences comprising up to about 10 basic residues (generally lysines) and posttranslationally coupled fatty acid chains (see Figures 1.7 and 1.8).

In the resting state of cells, these proteins reside in a sleeping conformation within the cytoplasm, hiding their charged residues and fatty acids. They are activated by exposing their membrane binding sequences (for instance, through phosphorylation) resulting in their electrostatic–hydrophobic binding to the membrane surface, as illustrated in Figure 1.7. Important examples are the molecular switches of the GTPase family and their associated helper proteins, such as GTP exchange proteins (GEF), which activate the GTPases by replacing GDP by GTP. The GTPases can be switched on either directly by the electrostatic hydrophobic binding (as in the case of the Ras proteins in Figure 1.7a) or indirectly by coupling to the membrane-anchored GEF. The active GTPase attracts and triggers the function of one or several enzymes, e.g., kinases.

The delicate balance of forces underlying the physics of the electrostatic–hydrophobic membrane recruitment of natural proteins has been first systematically explored by McLaughlin and coworkers [122–124] who studied the binding of the myristoylated alanine-rich C-kinase substrate (MARCKS) protein to partially charged membranes. MARCKS – serving as a paradigm of proteins recruited to membranes by electrostatic–hydrophobic forces – consists of a polybasic "effector" domain (ED) containing 13 lysine residues and 4 intercalated serine groups. The ED is flanked by two flexible chains, both comprising ~150 amino acid residues. The N-terminus flexible domain ends with myristoyl anchor (see Figure 1.8b). Similar to the case of K-Ras binding described in Figure 1.7, the membrane-binding free energy of MARCKS involves a

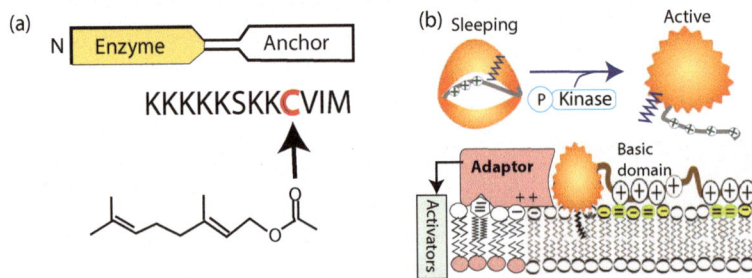

FIGURE 1.7 (a) Domain of structure of the Ras-proteins p21Ras (K-RAS) as a paradigm of protein activation by recruitment to membranes through electrostatic hydrophobic forces. The Ras-protein family consists of a variable domain at the C-terminus containing the polybasic domain and the functional N-terminus. Note that the sequence of lysine groups (K) is interrupted by a polar serine group (S) which can be negatively charged by phosphorylation (see the example of MARCKS). The cysteine group (C) is attached to a fatty acid serving as a hydrophobic anchor. (b) Enzyme activation by electrostatic–hydrophobic switching. The protein (shown at the top) is activated by phosphorylation, resulting in the exposure of fatty acid anchors and polybasic peptide sequences followed by binding to the membrane by electrostatic and hydrophobic forces. It acts as switch that recruits and activates specific adaptor proteins which in turn attract and activate proteins involved in signal transduction (see also Ref. [121]).

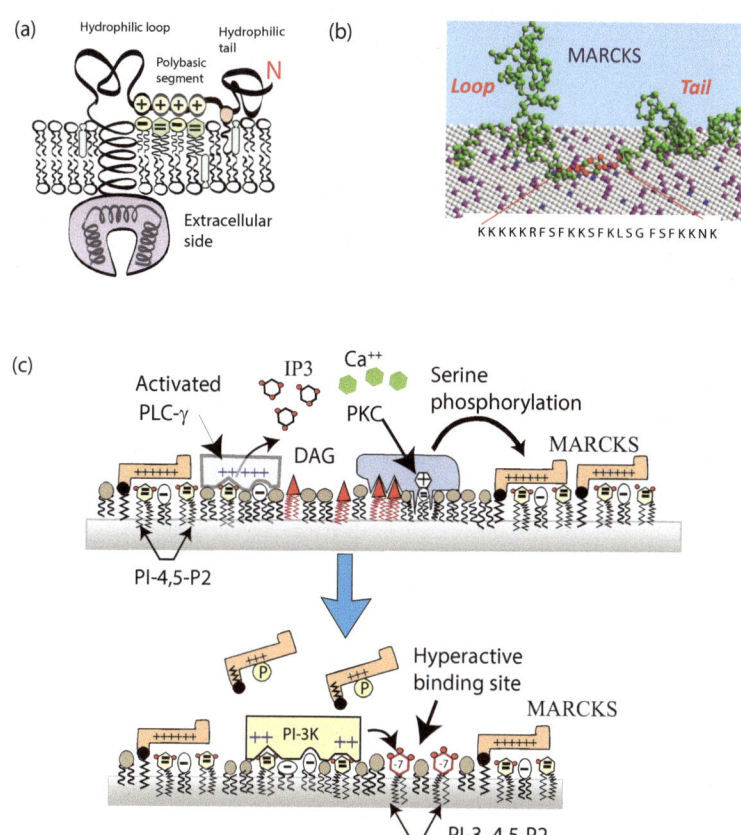

FIGURE 1.8 (a) Schematic illustration of protein binding to membranes by hydrophobic–electrostatic forces. (b) Snapshot from a Monte Carlo simulation of MARCKS adsorption, showing the enrichment of the binding region by polyvalent acidic lipids (purple), and the hydrophobic binding of the myristoyl anchor at the N terminus [127]. (c) Possible control of the access of the MARCKS-covered phosphoinositides by functional proteins through protein kinase C (PKC)–mediated unbinding of the protective layer of MARCKS (shown on the right side). PKC is activated by local bursts of Ca++ and DAG anchors, which are generated by phospholipase Cγ (PLC-Cγ). The activated PKC phosphorylates the serine groups of MARCKS resulting in the explosive displacement of the protective MARCKS layer and the exposure of PI-4,5 anchors. The right side shows the membrane recruitment and activation of the PI-3,4,5-P3 generating protein PI-3K. It generates a burst of the highly selective lipid anchor PI-3,4,5-P2 which enhances the membrane recruitment of many functional proteins, including PLC-Cγ and PI-3K. Note that in the absence of MARCKS, the lipid anchor generators PLC-Cγ and PI-3K would be constantly active, which could result in overexcitation of cells.

subtle interplay between (1) the electrostatic attraction of the basic domains to acidic lipids, ΔF_{el}, which provides the major contribution to the binding free energy, (2) the hydrophobic force due to the fatty acid anchor, ΔF_{tail}, and (3) the entropic free energy loss associated with the flexible – intrinsically unfolded – amino acid chains extending into the aqueous phase, ΔF_{loops}. ΔF_{tail} can be estimated by the empirical law $\Delta F_{tail} = (11 - 3n)\,kJ\,M^{-1}$ [29] or from measurements of unbinding forces by single molecule force spectroscopy [125,126]. The electrostatic binding free energy of the effector domain and the entropy loss of the flexible chain extending into the cytoplasm were estimated by MC studies [127].

Consistent with experiment and theory, efficient binding of the basic domain requires the colocalization of a matching number (namely, 13) of acidic lipid charges to the small ED–membrane contact zone. Importantly, most of the protein charge neutralization is achieved through the recruitment of the polyvalent PI-4,5-P2 lipid (generally carrying a charge of −3 or −5) which comprises only ~2% of the membrane lipids (as compared with 10%–30% of the monovalent PS) [123,127–129]. MARCKS–membrane interaction is just one of many examples of the very strong and selective binding of functional proteins to phosphoinositides (PI-4,5-P2 and PI-3,4,5-P3). Such binding is generally mediated by specific protein homology domains. Important examples, besides pleckstrin homology domains, are the C1 and C2 domains. The former harbours specific binding pockets for phosphoinositides, while the C-domains penetrate the lipid layer with two hydrophobic loops [130]. The binding of the C2 domain is calcium dependent.

A possible role of MARCKS is to prevent the unwanted membrane recruitment and activation of enzymes (such as phospholipases) by inducing the segregation of the phosphoinositides and their shielding from enzymatic attack. Indeed, about 10 μM of MARCKS suffices to inhibit the phospholipase Cγ-mediated generation of the second messengers diacylglycerol (DAG) and inositol triphosphate

(IP3). The electrostatically induced segregation of charged lipids by basic polypeptide sequences such a polylysine has indeed been demonstrated earlier [131].

The screening of the PI-4,5-P2 lipids is abolished by phosphorylation of MARCKS through protein kinase C (PKC), resulting in their unbinding from the membrane. The exposure of PI-4,5-P2 then enables the activation of molecular switches (such as GTPases) following the schemes illustrated in Figures 1.7 and 1.8. One important example is the generation of the specific anchor PI-3,4,5-P3 by phosphoinositol triphosphate-kinase (PI3K) shown in Figure 1.8c. PI-3,4,5-P3 acts as a second messenger which strongly enforces the specific binding of molecular switches of the GTPase family or of the phospholipase-Cγ, which generates the lipid anchor DAG. Since DAG mediates the membrane binding and activation of protein kinase C, the protective shield of MARCKS can be rapidly displaced to expose PI-4,5-P2.

A vital role of MARCKS has been established recently. Together with protein kinase C, MARCKS plays a key role for the plasticity of synapses, by enabling the rapid remodelling of dendritic spines at the tip of axons. Knock-out of its gene is lethal [132].

In summary, numerous vital membrane-based processes, associated with the restructuring of the actin cortex, such as cell locomotion and cell adhesion, are controlled by the recruitment of actin-binding proteins to the cytoplasmic leaflet through electrostatic–hydrophobic forces. These proteins swim in the cytoplasmic space in a sleeping conformation and are recruited to specific sites at the cell envelope in a logistic way by external cues, such as cytokines or hormones. Most important is the very strong selective binding to the PI-3,4,5-P3 lipids by the specific binding pockets other than pleckstrin homology domains [130,133], such as the C2 domain mentioned above. In this way, the functional protein may selectively bind to PI-3,4,5-P3, although this lipid anchor is by a factor 100 less abundant than PI-4,5-P2.

1.5.2 Underlying Thermodynamics and Kinetics

Two kinds of entropy play principal roles in the electrostatic binding of peripheral "macroions" such as proteins, DNA, and other biopolymers to lipid membranes. The first, common to the attraction between any two oppositely charged macroions, is the gain in 3D translational entropy of the mobile counterions originally surrounding the isolated macroions upon their release to the bulk solution (see Section 1.5.4). The second is loss of 2D translational ("demixing") entropy of the oppositely charged mobile "counterlipids" upon their migration and segregation near the macroion-binding site in order to balance its charge, thus enhancing its membrane binding. In the abovementioned case of MARCKS adsorption, the neutralization of its basic ED is primarily due to preferential sequestration of the polyvalent PI-4,5-P2, whose clustering involves a significantly smaller entropy loss in comparison with that of monovalent acidic lipids [127,128].

In parallel to the thermodynamic analyses of the electrostatic–hydrophobic protein binding, the dynamics of this process has also been studied, both experimentally and theoretically. Fluorescence correlation spectroscopy (FCS) measurements revealed that peptide diffusion on the surface of membranes containing PI-4,5-P2 is significantly slower than on those containing only monovalent lipids. It was also found that PI-4.5-P2 diffusion is substantially hampered [129], indicating correlated diffusion of the peptide and the multivalent counterlipids. Theoretical modelling combining the Cahn–Hilliard theory to describe the diffusion of lipid components in the inhomogeneous peptide-dressed membrane, and MC simulations to model peptide diffusion provide further insights into the coupled peptide–lipid diffusion dynamics. Specifically, strongly attracted by the electrochemical potential gradient due to the peripheral peptide, the PI-4.5-P2 lipids rapidly diffuse toward the interaction zone, while monovalent lipids are too slow to follow the motion of the peptide. Furthermore, the relatively strong, albeit reversible, binding of the multivalent lipids to the adsorbing peptides introduces an effective drag that slows down the diffusion of this "complex" [134].

1.5.3 Macroion Aggregation on the Membrane Surface

In analogy to the lipid-mediated aggregation of integral proteins, the perturbation of lipid organization by peripherally bound macroions can lead to their 2D aggregation on the membrane surface [131,135]. The underlying driving forces are quite different, however. One interesting case is that of proteins adsorbing onto a membrane composed of a nonideal, yet subcritical, binary lipid mixture of charged and neutral lipids. That is, the nonideality interaction parameter of the lipid mixture (corresponding to χ in Eq. (1.2), for the case $g = 1$) is not large enough to induce phase separation of the mixture. However, when macroions adsorb onto the membrane surface, they attract oppositely charged lipids to their binding zone, resulting in the formation of small membrane patches rich in charged lipids, as illustrated in Figure 1.9. Each of these patches involves a line energy proportional to $n\chi$, where n is the number of interlipid contacts at the perimeter of the segregated lipid patch [136]. To avoid the unfavourable line energy of the macroions–lipid patches, they attract each other with an effective attraction energy of order $n\chi$, which can lead to 2D phase separation of the dressed membrane into macroion-rich and macroion-poor phases, as illustrated in the bottom panel of Figure 1.9.

Interestingly, under certain conditions, the adsorption of macroions onto oppositely charged membranes can result in charge inversion of the membrane. Namely the net surface charge of the dressed membrane is of opposite sign to that of the bare membrane. This charge inversion phenomenon

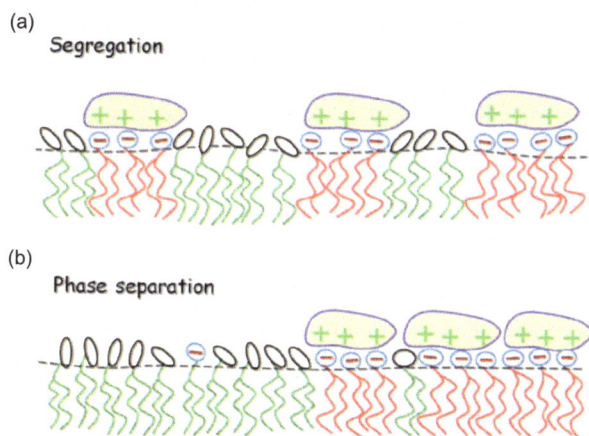

FIGURE 1.9 Adsorbed macroions attract counterlipids to the interaction zone (a). On nonideal lipid mixtures, this can induce lateral phase separation of the macroion dressed membrane (b).

is commonplace in polyelectrolyte adsorption onto surfaces, as well as in several cases involving biological molecules, e.g., the charge inversion of DNA saturated by bound polyvalent cations [137]. Analogously, charge inversion is expected when globular basic proteins bind to acidic lipid layers [138] or when DNA adsorbs onto a lipid membrane containing cationic lipids. Charge inversion may also take place upon the adsorption of intrinsically disordered proteins onto lipid membranes.

1.5.4 Counterion Release

Charged proteins, charged membranes, DNA, and other macroions are surrounded by clouds of mobile, generally monovalent, counterions whose confinement to these clouds involves an unfavourable loss of translational entropy. When a charged protein approaches a membrane containing oppositely charge lipids, the mobile counterions originally surrounding the macroions are no longer needed because charge neutralization can be provided by counterlipids; especially so in the case of fluid membranes where the counterlipids can diffuse and segregate in the protein adsorption zone to provide efficient charge matching, as in Figure 1.9. The mobile counterions can now be released into the bulk solution, providing a major, in some cases the principal, contribution to the electrostatic free energy of macroion–membrane binding. The magnitude of the entropy associated with the release of the mobile counterions diminishes upon increasing the salt concentration in solution, consistent with the well-known fact that protein–membrane binding weakens at elevated levels of salt in the aqueous solution. The counterion release mechanism has been demonstrated in protein–DNA binding [139] and various other systems [140]. A rather dramatic example of this mechanism is the formation of "lipoplexes" – those spontaneously formed complexes of cationic lipid membranes and DNA [141–143] – as illustrated schematically in Figure 1.10a. Conductivity measurements and Poisson–Boltzmann theory calculations suggest that in cases where the cationic membrane charge density matches the charge density of DNA, the entropic contribution to the binding free energy may be as high as ~95%, corresponding to the release of practically all the initially immobile counterions (Figure 1.10b [144]).

1.6 MEMBRANE DYNAMICS: LIPID LATERAL DIFFUSION AND ITS IMPACT

In a pioneering study on membrane dynamics published in 1970, Hermann Träuble [146] showed that water permeability of membrane is driven by the diffusion of gauche-trans-gauche (gtg) kink chain defects along the hydrocarbon fatty acid chains (see Figure 1.11a). In the same year, Frye and Edidin reported the first experimental evidence of lateral mobility of proteins in cell envelopes [147]. Soon afterwards, by analysing the ESR line narrowing due to spin–spin exchange of spin-labelled lipids, Träuble and Sackmann provided the first direct data about the lateral

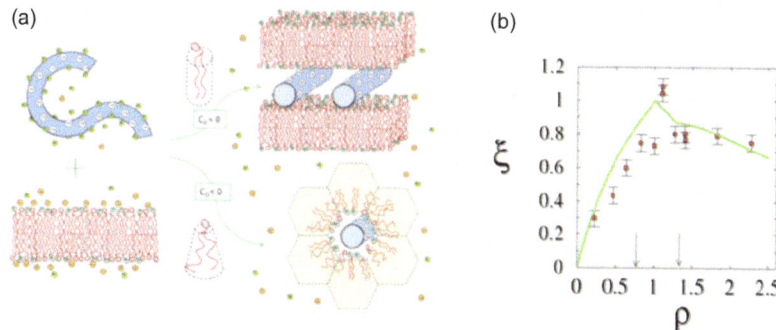

FIGURE 1.10 (a) Schematic illustration of the counterion release mechanism in the formation of DNA–cationic–lipid complexes. The entropy gained by the released mobile counterions provides a major contribution to the complexation free energy. The topology of the lipoplexes formed depends on the spontaneous curvature of the lipids. They are lamellar when the spontaneous curvature is small, and inverse-hexagonal when the lipid bilayers used contain negative-curvature loving ("helper") lipids like PE [145]. (b) The fraction of counterions released to solution in the formation of lamellar lipoplexes, reaching a maximum when the charge density on the DNA and the lipid membranes are equal in magnitude [144].

diffusion of lipids in membranes, finding $D_{lat} \approx 1\text{--}10^{-12}\,\text{m}^2\text{s}^{-1}$ for the lateral diffusion coefficient [148]. Around the same time, evidence for the high lateral lipid mobility has also been demonstrated by NMR spectroscopy [149]. Later on, short-range diffusion coefficients (over ~100 nm) were derived by analysing the kinetics of excimer formation and subsequent fluorescence following collisions between pyrene-labelled lipids [150]. The application of this method to lipid cholesterol mixtures led to the development of the free area model of molecular diffusion in membranes. This model accounts for the fact that (unlike in solids) molecular diffusion in liquids is not an activated process but rather driven by density fluctuations in the liquid. Nevertheless, the diffusivity in the membrane obeys an Arrhenius-like law, where the temperature dependence of D_{lat} is dictated by the thermal expansivity of the fluid [150]. The quantitative analysis of measurements of diffusion coefficients in phospholipid cholesterol mixtures by this model provided one of the first experimental proofs of the condensing effect of cholesterol. The model was extended by Almeida and Vaz [151] and Saxton [152] and applied to explore the membrane heterogeneity by Monte Carlo studies.

The next technical advancement arrived with the development of single-molecule tracking techniques, which were developed following the invention of high-sensitivity CCD cameras, and enabled measuring the time dependence of the mean square displacement of single lipids or proteins, $\langle |x(t) - x(t+\tau)|^2 \rangle = A\tau^\beta$. Deviations from the random walk model ($\beta = 1$) yielded new insights into the heterogeneous organization of cell membranes [153]. Additional new and attractive perspectives opened up following the development of the FCS technique, pioneered by the group of Webb (reviewed in Ref. [154]). By analysing the temporal correlation function of fluorescence signals, FCS enables to distinguish between lateral motional processes and chemical reactions in nanometre-size areas. Among the early applications of this method is the abovementioned measurement of the on-off kinetics of the binding of the MARCKS protein to charged vesicles by electrohydrophobic forces [155].

Lateral diffusion became most prominent among physicists following the seminal work of Saffman and Delbrück who showed that D_{lat} depends only logarithmically on the radius of integral proteins [156]. This law implies that large objects diffuse nearly as fast as lipids, rendering diffusion limited enzyme reaction in membranes very efficient. This is another beautiful example demonstrating how smart nature is in using physical principles to advance biological processes.

The Saffmann–Delbrück model was later extended by Evans and Sackmann to more realistic situations where integral proteins interact weakly with their environment, e.g., through the frictional coupling of membrane proteins to the actin cytoskeleton or the diffusion of proteins in supported membrane [157]. Using Einstein's relationship between the diffusion (D) and the frictional (λ) coefficients, $D_{lat} = k_B T / \lambda$, the model proved useful in measuring the friction coefficient between juxtaposed monolayers of supported membranes or between integral proteins in supported membrane and the solid surface. In the case of large frictional coupling of an integral protein (of radius R) with the surface, the diffusion coefficient becomes $D_{lat} = k_B T / \pi b_c R^2$, where $b_c = \eta_w / H$; η_w denotes the frictional coefficient of the fluid layer, and H is the distance between the protein and the surface [157,158]. Recent progress was made by Komura et al. who studied the diffusion of proteins in membranes coupled to viscoelastic media [159]. Their model opens new possibilities to explore the impediment of membrane protein diffusion by frictional coupling of the extrinsic domains to the actin cytoskeleton or the glycocalyx.

The most versatile technique for measuring molecular motions in model membranes is quasi-elastic neutron scattering (QENS). Owing to spin–spin exchange between spins of equal quantum number ($S = \pm 1/2$), the neutron scattering function of protonated material is incoherent (the scattering length density becomes negative). This unique feature allows us to analyse the motion of single lipid molecules (or portions thereof) by embedding fully (or partially) protonated lipid molecules in a bilayer of fully deuterated lipids. The incoherent scattering provides information on the single proton self-correlation function:

$$G^{\text{inc}}(|\vec{r}-\vec{r}'|,t) = \left\langle \sum_i \delta(\vec{r}' - \vec{R}_i(0))\,\delta(\vec{r} - \vec{R}_i(t)) \right\rangle - \langle n^2 \rangle \quad (1.4)$$

By studying oriented lipid multilayers, one can determine the direction of lipid motion with respect to the membrane normal. If the motion is unlimited, $G^{\text{inc}}(|\vec{r}-\vec{r}'|,t)$ decays to zero for $t \to \infty$. In contrast, the self-correlation function reaches a finite value G_∞ if the motion is restricted. G_∞ is called the elastic incoherent scattering factor (EISF). EISF measurements yield quantitative information on the lateral extent of the motion of molecules or of molecular segments. Application of different QENS instruments enables studying molecular motions in the time range of $10^{-7}\text{--}10^{-9}$ sec. By combining QENS with NMR, the dynamic range can be extended to 1 Hz.

Several important new insights about membrane dynamics that were provided by QENS are depicted in Figure 1.11. These are as follows:

(i) The lipid lateral diffusion involves fast local motions ($D_{loc} \approx 10^{-10}\,\text{m}^2\text{s}^{-1}$) and long-range lateral motion ($D_{lat} \approx 10^{-12}\,\text{m}^2\text{s}^{-1}$), in agreement with the free area model of lateral diffusion. (ii) In the course of the $L_\beta \to L_\alpha$ transition, the number and mobility of chain defects increase continuously, whereas only long-wavelength motions with $Q < 1.9$ Å^{-1} (or $\Delta \geq 3$ Å) show a discontinuity at T_m. (iii) The dynamic range of QENS ($10^{-13} < \nu < 10^{-8}$) can be extended to the millisecond time range by the combination of QENS and NMR relaxation experiments, providing new insights into the local motions of the water molecules bound to lipid head groups [161]. (iv) Measurements of out-of-plane motions

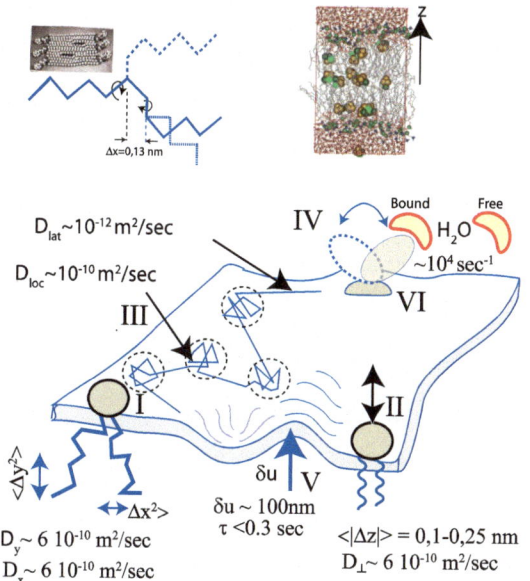

FIGURE 1.11 Top left: model of gtg-kink diffusion. Simultaneous hindered rotations about two C–C bonds (separated by one C–C bond) enable the kink performing diffusional jumps of length $\Delta L \approx 0.13$ nm. The jump frequency of $\nu_j \approx 10^{-11} s^{-1}$ corresponds to kink diffusivity of $D_{kink} \approx 10^{-5} cm^2 s^{-1}$. Top right: snapshot of membrane as observed by MD calculations. The clusters of yellow and green beads indicate molecules of the anaesthetics halothane, demonstrating the role of the hydrocarbon chain region as an apolar organic solvent of small hydrophobic solutes (image reproduced from Ref. [160] with permission of the authors). Bottom: Summary of membrane dynamics from the nanosecond to the ten second time regime as revealed by QENS and NMR or ESR relaxation experiments.

of single lipids support the mechanism of short-range protrusion fluctuations suggested by Israelachvili and Wennerström [65,162].

1.7 MEMBRANE DEFECTS: PHYSICS AND BIOLOGICAL FUNCTION

The role of chain defects in fluid liquid crystalline bilayers has been an ongoing theme since the early 1970s. In natural membranes, defects can come into play in various ways. The transport of water across membranes and the diffusion of small hydrophobic molecules (such as local anaesthetics) within membrane by kink diffusion are among the examples. Chain packing defects are likely to form in highly curved bilayers and at the rim of SPHM-rich microdomains (see Figure 1.6). The detailed analysis of the topology of defect structures of vesicles by freeze-fracture EM showed for the first time that tilted phases, such as the $P_{\beta'}$ and the $L_{\beta'}$ phases, or the nontilted L_c-phase, form in single bilayer vesicles. These studies provided a direct link between liquid crystal physics and membrane physics and played a key role for the recognition of membrane research by physicists. These early studies of defect formation in membranes have been reviewed in Ref. [163]. Below, we briefly mention some pertinent examples.

One example is that of the line defects separating fluid and gel-like domains. These defects provide pathways for rapid lateral diffusion of membrane solutes (analogous to "short-circuit" diffusion in solid-state physics [163–165]). Another example is that of the chain packing defects around conically shaped integral proteins such as the voltage-dependent Na$^+$ channels [166]. These defects provide "cavities" – attracting small solutes such as local anaesthetics which can fill the voids and impede the conductivity (see Figure 1.12). This function of defects as attractive centres is reminiscent of the Cottrell clouds of solutes formed around dislocations in solids which play a key role in the conditioning of material properties of metal alloys.

Density fluctuations in the lipid matrix ($\delta \rho$) may play a key role in enzyme activity and the conductivity of ion channels [166] which are particularly pronounced in liquid–solid domain boundaries, or around SPHM-rich domains (Figure 1.6). The lateral density fluctuations may induce correlated changes in the cross-sectional area (ΔA) of the embedded proteins. In the case of ion channels, for instance, the changes in Na$^+$ conductivity were shown to obey an Arrhenius-type behaviour, $k = k_0 \exp\{-K\Delta A/k_B T\}$, where K is the membrane compression modulus and ΔA is the area difference of the hydrophobic domain of the ion channel between the nonconducting and the conducting state. The area compression modulus is inversely proportional to the density fluctuations, $\langle \delta \rho^2 \rangle / \rho^2 = K^{-1} \rho k_B T$, which increases (hence K decreases) drastically around the SPHM-rich domains and diverges at the boundaries of coexisting liquid–solid domains. In the case of sodium channels of the heart muscle, $\Delta A \approx 0.04$ nm^2. For densely packed lipid bilayers, $K \approx 0.8$ Nm^{-1} and hence $K\Delta A/k_B T \approx 8$. Defect-related local density fluctuations can thus drastically lower K, resulting in significantly higher ion conductivities.

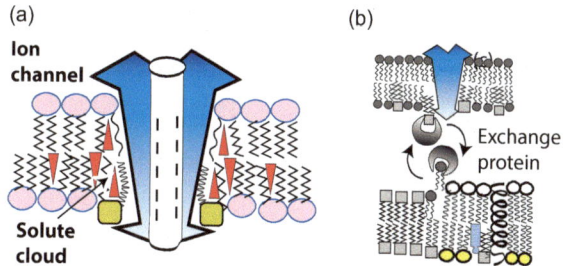

FIGURE 1.12 (a) Defect formation by local splay deformation of the lipid bilayer adjacent to conically shaped integral protein, such as Na$^+$ channels. The stress caused by the deformation can be relaxed when small lipophilic molecules such as local anaesthetics migrate towards the defect, filling up the void, thereby possibly impeding the ion conductivity. (b) Defects formed at lipid halos around integral proteins (top) or the rims of microdomains (bottom) enhance the exchange of lipids between outer leaflets of membranes.

Another physiologically important role of membrane defects is illustrated in Figure 1.12. Studies on model membranes indicate that lateral membrane inhomogeneities, such as those associated with asymmetrical protein inclusions or the boundaries between different lipid domains, can largely accelerate the exchange of specific lipids between outer leaflets of intracellular organelles [163]. Pronounced defect-enhanced activities of this kind were reported for phospholipases, which appear to be most active at fluid–solid phase boundaries. Fluorescence microscopy studies showed that the enzyme is most active at the interface between solid and fluid domains (see [29,167]).

Yet another physiologically important effect of chain packing defects is the increase in the passive transmembrane permeation of molecules and ions, which appears to increase with the square of the density fluctuations, $\langle |\delta\rho^2| \rangle$. Recently, Heimburg et al. suggested that the propagation of defects along the axon could control action potential propagation [168].

Lipid packing defects associated with membrane regions of high curvature can strongly affect enzyme activity in such regions as compared with planar membranes. The role of curvature in controlling enzymatic activity was first demonstrated by Paul Janmey and his coworkers who studied the generation of PI-3,4,5-P3 lipids by phosphoinositol-3-kinase (PI-3K), finding that vesicles with an average diameter of 100 nm are phosphorylated 100 times faster than vesicles with an average diameter greater than 300 nm [163]. New interest in curvature-related membrane defects was boosted recently by studies concerning the binding of BAR proteins to the necks connecting coated pits to the cell envelope. The neck formation is known to be controlled by BAR proteins (such as endophilins [170]). Recent experimental studies provided evidence that the curvature-induced packing defects resulting from the high curvature of the necks may account for the curvature sensing capacity of endophilin. The low lipid packing density in the defects enhances the binding of BAR proteins to the necks by facilitating the penetration of their amphipathic alpha helices [171]. Finally, we note that the role of defects as regulators of the dynamics of membrane processes is reminiscent of the role of defects for the plastic deformation of metals, suggesting analogies between the strategy of metallurgists to control the microstructure of metals and of nature to control the microanatomy of membranes.

1.8 BIOLOGICAL MEMBRANES AS ELASTIC SHELLS

1.8.1 Morphology and Shape Transitions of Soft Elastic Shells

The bending elasticity model of cell shape changes proposed by Helfrich [7] stimulated the interest of physicists and engineers working at the interface between physics and biology, resulting in the development of new concepts of soft shell elasticity. One example is the nonclassical theory of scale-dependent elastic properties of solid-like (tethered) membranes due to thermal excitations of long-wavelength in-plane phonons and bending deformations [172]. It stimulated the development of numerous new techniques, enabling the measurement of the bending, shear, and compression modulus elastic moduli and the Young modulus (E) of cell envelopes. Great enthusiasm was raised by the discovery of undulation forces [173] and their role as regulators of cell adhesion, as discussed in Section 1.9.

The Helfrich theory of membrane elasticity is a special case of the general Föppl-von Karman theory of shells. It was stimulated by Frank's theory of liquid crystal elasticity and led to relating membrane curvature deformations to the splay deformation of smectic liquid crystals. The membrane bending energy ΔG_{el} was expressed in terms of the principal curvatures, $1/R_1$ and $1/R_2$, of the bilayer's neutral plane, with the inside–out asymmetry of the bilayer accounted for by its mean spontaneous curvature C_0,

$$\Delta G_{el} = \frac{1}{2}\kappa \oiint dO(2H - C_0)^2 + \bar{\kappa} \oiint dO H_G \qquad (1.5)$$

Here, $H = (1/R_1 + 1/R_2)/2$ and $H_G = 1/R_1 R_2$ are the mean and Gaussian curvatures, respectively. κ and $\bar{\kappa}$ correspond to the splay and saddle-splay elastic moduli introduced in liquid crystal physics [174]. The integration extends over the membrane surface. Based on this equation, vesicle shapes and phase diagrams have been derived using two generic parameters: the spontaneous curvature, C_0, and the ratio A/V between the vesicle area to its volume, as discussed in detail by Seifert and Lipowsky [175].

One major success of the Helfrich homogeneous shell model was the prediction of some shape changes of erythrocytes, such as the discocyte to stomatocyte transition (see Figure 1.13a). Direct experimental evidence for the application of the model to red blood cells was provided by the observation that this transition can be mimicked by single lipid component vesicles through temperature variations of

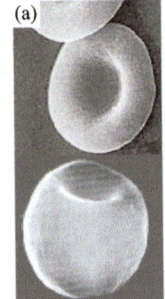
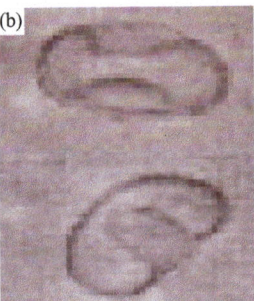
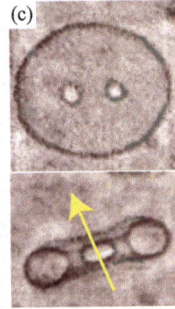

FIGURE 1.13 Experimental evidence for the simple Helfrich model of erythrocyte shape changes. (a) Discocyte-to-stomatocyte transition of red blood cells. (b) Discocyte-to-stomatocyte shape transition of vesicles of DMPC triggered by increasing the area-to-volume ratio A/V, achieved by increasing the temperature from 41°C to 42°C. (c) Demonstration of the generation of toroidal vesicles (image reproduced from Ref. [29]).

the area-to-volume ratio (see Ref. [176] and Figure 1.13b). Another success of the Helfrich model was the interpretation of conformational transitions of toroidal vesicles (originally discovered by Bensimon and coworkers [177], see also Ref. [178]) resulting from the interplay between the mean (H) and Gaussian curvatures (H_G) [179] (see Figure 1.13c). It was shown that shapes of toroidal topology may involve one or several holes and that for a given number of holes (called the genus of toroidal shapes), transitions between different shapes can occur without changing the bending energy [175,180]. A third success was the interpretation of the tension-induced pearling instability of membrane tubes in terms of the Rayleigh instability [180].

The homogeneous shell model could not explain shape changes of vesicles formed from long-chain lipids, which showed mainly spontaneous budding and formation of branched tubules but no discocyte to stomatocyte transition. Svetina and Zeks showed that this shortcoming of the simple Helfrich model is remedied upon accounting for the coupling between the two monolayers of the shells [181]. In fact, the important role of the coupling between the monolayers had already been recognized by Evans in 1972 [182] but was ignored by the membrane community for several years. The theory of Evans accounted for the asymmetry of the membrane by assuming that the shape transitions of vesicles are determined by the initial difference ΔA_0 of the areas occupied by the inner (A_i) and outer (A_o) monolayers.

The coupling between the opposing monolayers has important consequences. If vesicles are prepared by swelling of bilayers in water, the area per lipid is roughly the same in the inner and outer monolayers. Therefore, the inner monolayer has to be compressed to form a closed shell. Since the exchange of lipids between the two monolayers is very slow (of the order of 12 hours for long-chain lipids such as DPPC or DOPC), freshly formed vesicles are initially not in mechanical equilibrium. Consequently, vesicles of long-chain lipids exhibit only inside budding upon increasing the area-to-volume ratio [183].

In 1997, a generalized elastic model accounting for the prestress generated by the tight coupling of the monolayers and the hindered flip flop was developed. The concept of spontaneous curvature was extended by introducing an effective spontaneous curvature consisting of two contributions, namely C_o (called local spontaneous curvature) and a term accounting for the intrinsic area difference $\Delta A_0/A_0$. The spontaneous curvature model and the area difference model were unified into the area difference elasticity (ADE) model, which interpolates between the two models. The ADE model has been extensively reviewed elsewhere [175]. Below we point out some major predictions of the models.

- The spontaneous curvature and the ADE models predict about the same shape transitions but differ concerning the prediction of the order of the transitions and the existence of metastable phase lines.
- The prolate and oblate shapes are stable over a large range of excess areas, which could explain the robustness of the discocyte shape of erythrocytes against changes in osmolarity.
- The phase boundaries between two regimes of vesicle shapes may be separated by a spinodal line (SP). Shape transitions, such as budding, may therefore be drastically delayed, a fact that should be considered in experimental studies of vesicle shape changes.

1.8.2 Shape Transitions of Stratified Soft Shells: The Red Blood Cell

The compelling beauty of the Canham–Helfrich–Evans–Svetina–Zeks model delayed the interest of physicists in biologically important aspects of cell membranes structuring, such as the control of cell shapes by the intimate coupling of the lipid–protein bilayer to the membrane-associated cytoskeleton or the calculation of asymmetric cell shapes [184]. To overcome these shortcomings, the homogeneous shell model was extended in two directions:

Asymmetric cell shapes could be mimicked by considering the coupling between membrane curvature elasticity and in-plane phase separation. This issue has been studied by Andelman and Leibler [112], who combined the elastic energy functional ΔG_{ela} of the bending elasticity model with the free energy expression of the classical Cahn–Hilliard–Langer theory (see also Ref. [29] for a more extensive discussion). A few years later, Godzs and Gompper showed by MC simulations that the disk-like cisternae of the ER could be formed by mixtures of lipid components, one preferring positive curvature and the other preferring negative curvature [185].

Shortcomings of the Helfrich model for the calculation of nonsymmetric cell shapes were clearly pointed out in the PhD work of Stokke [186]. He argued that nonsymmetric red cell shapes, such as echinocytes and elliptocytes (in Figure 1.14), can be formed by the coupling of bending and shear deformation. With the shear elastic modulus, a new length scale $\Lambda = \sqrt{\kappa/\mu}$ is introduced. For typical values of the bending modulus $(\kappa = 5 \times 10^{-19} \, J)$ and the shear modulus $(\mu = 7 \times 10^{-6} \, J \, m^{-2})$, one finds $\Lambda \approx 300$ nm.

A rigorous elasticity theory of cell envelopes of general shapes, based on principles of differential geometry, has been first developed by Mark Peterson [187,188]. The theory allowed for the first time to calculate changes of the contour of cells induced by external forces, such as those applied by high-frequency electric fields [189] or by shear deformation fields generating tank treading motions [190].

A comprehensive elastic theory of erythrocyte shapes was developed 30 years after the pioneering Canham–Helfrich by Wortis and collaborators [191]. These authors calculated cell shapes by minimizing the total energy functional comprising (i) the bending elasticity expression of the ADE model ΔG_{ADE} and (ii) the sum of the extensional

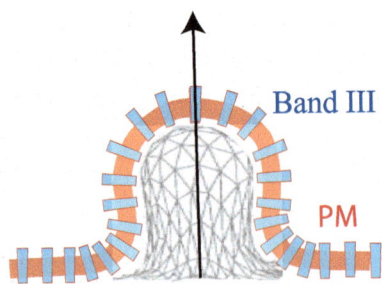
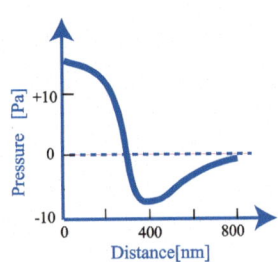

FIGURE 1.14 Variation of surface pressure of the lipid–protein bilayer of red blood cell in the region of the cell surface forming a spicule. The network is stretched at the tip of the spicule and pulls at the membrane towards the inside. It is compressed at the base and exerts a pressure directed towards the outside. Note that the deformation of the cytoskeleton is associated with the redistribution of the membrane proteins, such as band III.

deformation and shear deformation ΔG_{ext} of the cytoskeleton, so that $\Delta G_{tot} = \Delta G_{ADE} + \Delta G_{cyt}$.

The elastic energy of the cytoskeleton is expressed in terms of the rubber elasticity model of Mooney (first introduced by Skalak and Evans [192]) in the field of membrane physics, using λ_1 and λ_2 to denote the principal stretching ratios.

$$\Delta G_{cyt} = \frac{K}{2} \oiint do (1 - \lambda_1 \lambda_2)^2 + \frac{\mu}{2} \oiint do \left(\frac{\lambda_1}{\lambda_2} + \frac{\lambda_2}{\lambda_1} - 2 \right) \quad (1.6)$$

In conclusion, the Wortis model predicts that the shape changes of erythrocytes during their passage through blood vessels require the ongoing reorganization of the spectrin–actin network. Concomitantly, the membrane proteins Band III and glycophorin are expected to be constantly redistributed during the cell transport through the blood vessels, which is an ATP consuming process.

1.8.3 Membrane Bending Excitations and Their Physiological Significance

The interest of many soft matter physicists in biomaterials was stimulated by the studies of cell membrane fluctuations by Brochard and Lennon [193] and Helfrich [173]. In the former study, it was shown that the long-known flickering of erythrocytes was activated by bending excitations of the soft cell envelopes. Helfrich's work led to the beautiful picture of a membrane as a dynamically rough surface, which can be regarded as being composed of cushions of dimension $\zeta_p \times \zeta_p$ performing independent Brownian motions in the normal direction to the membrane [194]. Each collision of random local protrusions with adjacent surfaces was assumed to involve an energy of the order of $k_B T$, in analogy to molecules of an ideal gas hitting a solid wall. For tension-free membranes at distance h from the wall, the repulsive membrane pressure (which is equal to the transmitted energy per unit area) was shown to be given by [173,194]

$$p_{disj} \approx k_B T / \zeta_p^2 h \propto (k_B T)^2 / \kappa h^3 \quad (1.7)$$

Many functions of membrane undulation were discovered after Helfrich's publication. It became evident that the disjoining pressure plays a key role for the swelling of lipid multilayers and that it controls the adhesion of vesicles and cells, compensating the Van der Waals attraction of lipid vesicles at distances $< h > \geq 20 - 30$ nm. The tension-induced freezing-in of the undulation was shown to trigger cell adhesion [195,196]. This effect was applied to measure the membrane bending modulus κ using the micropipette technique [197].

Further theoretical basis of the entropy-driven interfacial forces was laid about 8 years later by Lipowsky and Leibler [198]. To account for the confinement of membrane fluctuations, they extended Helfrich's model by adding a harmonic interaction potential $V(h) = V_0 + V''(h - h_0)^2 / 2$ that restricts these fluctuations. The total free energy functional can then be expressed as

$$\Delta G_{adh} = \oiint dO \left\{ \frac{\kappa}{2} (\Delta u)^2 + \frac{\sigma}{2} \nabla u^2 + \frac{1}{2} \frac{\partial^2 V}{\partial x^2} \{h - h_o\}^2 \right\} \quad (1.8)$$

This yields two observable length scales: the capillary length λ_c and the persistence length ζ_p given by

$$\lambda_c = \sqrt{\kappa / \sigma} \quad \text{and} \quad \zeta_p = \sqrt[4]{\kappa / V''} \quad (1.9)$$

Here, λ_c is a measure of the maximum wavelength over which the membrane excitations are dominated by the bending excitations, and ζ_p accounts for the lateral extension of the membrane deformation induced by a local point force.

Experimental tests of the theories of membrane bending excitations became possible following the development of the Fourier spectroscopy of bending excitations of adhering giant vesicles and erythrocytes by reflection interference contrast microscopy (RICM) combined with fast image processing [199]. The spatial correlation function of the excitation amplitude $u(x,t)$, i.e.,

$$\langle u(x)u(x+\xi)\rangle = \zeta_\perp \exp\{-(x-x)^2/\zeta_p^2\} \quad (1.10)$$

and hence, ζ_p could then be directly measured. By simultaneous measurement of the bending modulus ζ_p, the force constant V'' of the interfacial potential could also be determined.

In 1987, Millner and Safran published a rigorous theory of shape fluctuations of vesicles by accounting for the boundary condition of constant volume in terms of the membrane tension γ [200]. Their model paved the way for high-precision measurement of membrane bending moduli by the Fourier analysis of the contour of giant vesicles [183] with 0.5 μm wavelength resolution. The high-precision method allowed to determine the number of shells of giant vesicles or the modification of κ by small concentrations of solutes and cholesterol.

1.8.4 Red Blood Cell Membranes as Actively Driven Semiflexible Statistical Surfaces

A surprising finding of the high-precision RICM measurements of bending elasticity of erythrocyte membranes was their astonishingly small bending modulus: $\kappa \approx 7k_BT$ as compared with the value of $\kappa \approx 50k_BT$ expected for the PM containing 50 mole% cholesterol [199,201]. Indeed, a parallel measurement of κ by the Fourier analysis of the cell contour derived by phase contrast microscopy yielded a value of $50k_BT$ [202], suggesting that the thermal excitation of the bending undulations is enforced by random chemical forces. These are expected to be caused by local fluctuations of lateral tension by the constituent coupling and uncoupling of the spectrin links from glycophorin, which is associated with ATP turnover (Figure 1.15).

A sophisticated and rigorous two-shell model of the bending elasticity of stratified cell envelopes was developed by Safran and coworkers [203] allowing to evaluate the wavelength dependence of the mean square fluctuation amplitude in the membrane wavelength range $\Lambda \approx 0.5-2.5\,\mu m$ measured by Zilker's Fourier spectroscopy [199]. Korenstein and coworkers later provided direct evidence that the enhanced dynamic surface roughness is associated with ATP turnover [204,205], which provides some evidence for the model [201] of the enhancement of bending excitations by tension fluctuations proposed in Figure 1.16.

It has been suggested that the overexcitation of the erythrocyte shell by fluctuating chemical forces can be described in terms of an effective temperature T_{eff}. $T_{\text{eff}} - T$ is a measure of the excess undulation amplitude excited by stochastic chemical forces. It can be generally determined by simultaneously measuring the bending modulus by flicker analysis and other micromechanical techniques. Another method is based on the measurement of excess membrane tension by the micropipette technique [197]. This approach has been adopted by Bassereau and coworkers to demonstrate the enhancement of bending membrane excitation by active chemical noise generated by ion translocation through membranes. Studying giant vesicles doped with reconstituted bacteriorhodopsin, it was shown that light-driven proton pumping activity amplifies the bending undulation drastically [206]. The chemical noise generated by proton influx can give rise to higher excess temperatures.

The following discussion of the physiological role of membrane flickering shows that more detailed studies of active excitation of flickering are expected to provide valuable insights into dynamic control of membrane processes.

1.8.5 The Physiological Role of Membrane Flickering

In addition to erythrocytes, membrane bending undulations have been observed in various other cells, including endothelial cells and macrophages. These cells show mainly short-wavelength excitations of $\Lambda \leq 1\,\mu m$ with ~10 nm root mean square amplitudes [207]. This is remarkable, considering the fact that the bending and shear moduli of these composite shells are much larger ($\kappa \approx 1000\,k_BT$ and $\mu \approx 4 \times 10^{-4}\,\text{Jm}^{-2}$) than those of erythrocytes. A challenging question is therefore whether membrane bending excitations may serve some purposes. Several possible answers are outlined below.

First and most importantly, the bending excitations prevent the nonspecific adhesion of erythrocytes and macrophages to solid surfaces (see Ref. [207]) or the walls of the blood vessels. Second, erythrocytes do not possess intracellular transport mechanisms, and the hydrodynamic flow induced by flickering in the cytoplasm (see Ref. [193]) could accelerate the exchange of haemoglobin between the rim of the cell and its centre. Third, the disjoining pressure between the lipid–protein bilayer and the actin cytoskeleton controls the distance between the two subshells of cell envelopes [203] and may inhibit the sticking of the actin cortex

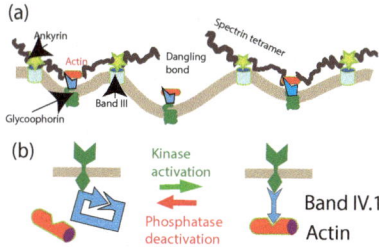

FIGURE 1.15 (a) Schematic view of composite cell envelope of erythrocytes with the lipid–protein bilayer exhibiting thermally and actively driven bending excitations. The undulatory excitations are limited by about 10% excess area of the bilayer with respect to the average area of the spectrin–actin network. They determine the distance between the two subshells and generate an excess membrane tension [194,203]. (b) The ongoing ATP-driven binding and unbinding of band IV.1 from glycophorin associated with the formation of dangling bonds enhances the bending amplitudes and could account for the nonthermal contributions to the undulations.

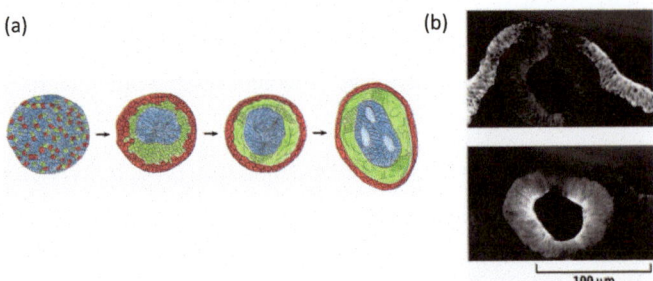

FIGURE 1.16 Experimental evidence for the homophilic nature of cadherin-mediated intercell adhesion. (a) *In vitro* self-organization of a cell mixture into an embryo-like structure. Following the dissociation (by alkali) of a piece of embryonic (amphibian neurola) tissue, the cells were allowed to reassemble. Initially, the cells reassemble to form a uniformly mixed globule. After a while, due to the homophilic nature of the adhesion, segregation takes place, with neural cells (blue) constituting the inner core, mesodermal (green) in the middle layer and epithelial cells (red) in the outer shell. (A coloured version of the figure adapted from the original landmark paper of Towns and Holtfreter [218]). (b) Cadherin segregation during the early stages of embryo morphogenesis. N-cadherins are expressed in the neural tube developing from the epithelial tissue, where cells are held together by *trans* binding of E-cadherins [214].

to the membrane. Fourth, the thermal excitations generate attractive isotropic forces between neighbouring membrane–cytoskeleton linkers, which is balanced by the tension of the entropic springs of the spectrin tetramers. This may serve the isotropic distribution of the spectrin–membrane linkers glycophorin and Band III and the dynamic softening of the cell envelope, as suggested by the stiffening of elliptocytes.

1.8.6 Molecular Aspects of Membrane Elasticity

In parallel to the continuum theories of membrane elasticity, cell shape changes, undulations forces, and the various elegant experimental methods developed to measure elastic membrane moduli, many molecular-level models and theories have been proposed to calculate those moduli. These models range from simple pictures to rather elaborate theories employing realistic chain models and/or sophisticated electrostatic analyses. A simple yet very useful criterion for assessing the spontaneous curvature corresponding to a given lipid molecule is provided by the "packing parameter" $p = al/v$, where a, v, and l are the optimal area per lipid head group, the volume of the tail, and its length, respectively [65]; large p favours positive curvature, whereas small p, as in the case with PE for instance, favours negative curvature, as observed in the inverse hexagonal (H_{II}) phase. Phenomenological "compression models" have yielded insightful scaling relationships relating the bending rigidity to the lipid chain length and area per head group (see, e.g., [63,208,209] and references therein). More sophisticated theories have also been proposed, taking into account the conformational (e.g., the rotational isomeric) statistics of the hydrocarbon lipid chains and/or the electrostatic and excluded volume interactions between the head groups. Any theory or computer simulation that yields information regarding the lateral pressure profile described in Figure 1.4 can be used to calculate the elastic moduli using general relationships, such as

$$\kappa C_0 = \int z\pi(z)\,dz, \quad \bar{\kappa} = -\int z^2 \pi(z)\,dz \quad (1.11)$$

Using the lateral pressure profile derived from Eq. (1.1), these elastic moduli, as well as an explicit expression for κ, were derived, enabling to calculate the bending rigidities of single- and multi-component membranes [209,210], showing very good qualitative and quantitative agreement with experiment. Over the years, additional molecular-level theories and lattice models of membrane elasticity have been proposed, yet in recent years – in the spirit of time – most of the effort is invested on developing and applying large-scale MD computer simulations.

1.9 CELL ADHESION: A MEMBRANE-BASED PROCESS

Among the first questions that motivated the interest of physicists in cell adhesion was whether intermembrane adhesion can be understood in terms of the classical DLVO theory of interfacial forces between charged surfaces. Studies of this question by Israelachvili and coworkers using the surface force apparatus (SFA) [65] and by Parsegian and coworkers using their osmotic stress technique [211] resulted in the first measurements of the Hamaker constants that measure the strength of the interaction between lipid bilayers and between membranes and solid surfaces [65]. A second approach, pioneered by Bell and coworkers [212,213], attempted to explain cell-to-cell adhesion in terms of the competition between specific attractive forces involving ("lock and key") adhesion receptors, and generic repulsion forces mediated by the glycoproteins forming the glycocalyx. Analysing the adhesion-induced deformations of adhering cell envelopes, they concluded that cells can be regarded as liquid droplets whose adhesion strength is determined by the Young–Laplace equation. This pioneering and fundamental study pointed out analogies between the adhesion of cells and partial wetting of solids by fluid droplets.

Below, we briefly comment on two complementary points of view of membrane adhesion. (i) In Section 9.2, we outline the physicists' approach, emphasizing the mechanical aspects of cell adhesion and the interplay between membrane curvature elasticity, surface tension, and the role of "linker" and "repeller" molecules. (ii) In Section 9.1, we briefly comment on the cell biologists' point of view where much emphasis is given to the molecular details and mutual interactions of the adhesive receptors.

1.9.1 Molecular Aspects

Early biological studies were largely concerned with embryogenesis and tissue development, with particular emphasis on cadherin-mediated cell–cell adhesion. Takeichi [214], Steinberg [215,216], and others (see, e.g., [217] and references therein) have demonstrated that cadherin-mediated adhesion is strongly homophilic. Two early examples of this important aspect of cell adhesion are demonstrated in Figure 1.17. The classical experiment of Towns and Holtfreter was actually carried out before the role of adhesive receptors such as cadherins, for instance, has been realized. The homophilic nature of receptor-mediated adhesion was cast in simple physicochemical terms by Steinberg and coworkers, an approach known as the differential adhesion hypothesis (DAH), [215,216]. The DAH asserts that cell mixtures behave analogously to liquid mixtures of barely miscible molecules. Thus, for example, cells expressing epithelial (E) cadherins prefer adhering to cells expressing E-cadherins, rather than to, say, cells expressing neural (N) cadherins. Underlying this behaviour is the fact that although the interaction between E-cadherin and N-cadherin receptors is attractive, its strength (w_{NE}) is smaller than the average strength of the homophilic attractions (i.e., $w_{NE} < (w_{NN} + w_{EE})/2$). The differences between the attractive energies of the different kinds of receptors are small (typically just a few $k_B T$), yet since each cell is covered typically by tens to hundreds of thousands of molecules, the adhesive affinities between like cells are far stronger than those between unlike cells. Consequently, a mixture of cells expressing E- and N-cadherins strongly prefers to "phase segregate", resembling the behaviour of a liquid mixture of immiscible molecules. This has been verified in various *in vivo* systems and *in vitro* assays, [214,216,217].

Additional physical insights into the liquid droplet concept and more quantitative information on the cohesive forces between cells were gained by Brochard and coworkers [219] by studying spherical cell assemblies by the micropipette aspiration technique. They showed that the cohesive force between spherical cell clusters increases monotonically with the cadherin expression. Studies of the dynamics of spreading of cells on surfaces showed some interesting analogies with the spreading of viscoplastic fluid droplets.

While confirming the homophilicity of cell adhesion via selective (lock and key) forces among cell surface receptors, the adhesive energies associated with these forces do not explain the magnitude of cell–cell or cell–surface energies, as they do not account for the very important role played by the coupling of the adhesive receptors to the cytoskeleton of the composite membrane, nor for the important effects due to membrane curvature fluctuations. In other words, these assays provide little insight into the role of the composite cell membrane or into the formation of focal adhesion complexes generally formed at the cell–cell and cell–tissue interface.

1.9.2 Insights Gained by Biomimetic Systems

Important theoretical steps towards understanding the physical basis underlying the control of cell adhesion by soft elastic cell envelopes was made by Lipowsky and Seifert [220], Bruinsma [221], and Evans [222]. These authors showed that the adhesion strength of fluid membranes can be quantified in terms of the spreading pressure W (the free energy of adhesion per unit area) given by the following equations:

$$W = \sigma(1 - \cos\theta_c), \quad W = \frac{\kappa}{2}\left(\frac{1}{R_c}\right)^2 \quad (1.12)$$

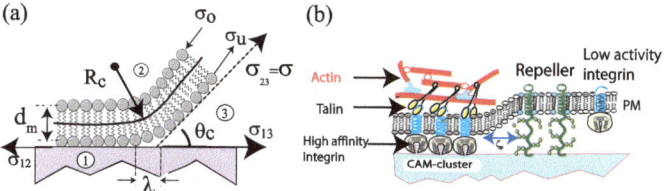

FIGURE 1.17 (a) Mechanical equilibrium of the interfacial tension and the bending moment at the contact line of the adhesion zone. The surface tensions σ_{ij} are measures of the interfacial energies. R_c is the radius of membrane curvature at the contact line (see also [220]). (b) Decay of the cell substrate interface into domains of tight adhesion (formed by integrin clustering) and weakly adhering zones, which are separated from the substrate by long-range forces mediated by glycoproteins of the glycocalyx and undulation forces. The binding strength of the initially formed clusters of integrin is enforced first, by the drastic increase of the affinity of the CAMs for ligands of the tissue (such as fibronectin) and secondly, by the increase of the membrane bending stiffness induced by coupling of actin gel patches to the cytoplasmic domains of integrin (following Eq. (1.12)). The width of the transition zone between the two states of adhesion is equal to the persistence length ζ_p appearing in Eq. (1.7). (Images adopted from Ref. [178].)

In this equation, σ is the membrane tension, κ is the bending modulus introduced in Eq. (1.5), θ_c is the contact angle, and R_c is the radius of curvature of the shell at the contact line L_c (defined in Figure 1.17). The first equation on the left side is the classical Young equation of capillary forces. The second equation accounts for the elastic energy associated with the bending of the membrane at the transition from the adhering to the free membrane area. The contact angle θ_c and the contact curvature R_c can be measured with high precision by using interferometric techniques, such as RICM [196,221].

Experimental physics of cell adhesion research was boosted by the development of planar solid-supported membranes or patterned polymer films as models of cell and soft tissue surfaces (reviewed until 2005 in Ref. [17]). This design opened up the possibility of reconstructing the contour of adhering soft shells, such as vesicles with nanometre spatial resolution, by the RICM technique, enabling high-precision measurements of the geometric parameters R_c and θ_c and hence of the free energies of adhesion (W in Eq. (1.5)).

Simple model systems were designed to gain quantitative insight into the control of the state of adhesion by interplay of short-range attraction forces mediated by specific linker pairs, long-range repulsion forces mediated by glycoproteins and undulation forces, and the adhesion-induced membrane deformation. Giant vesicles doped with ligands of tissue (such as fibronectin or the antigenic oligosaccharide Sialyl Lewis X) served as test cells. Supported membranes exposing a conjugate cell adhesion molecule (CAM, such as integrin and selectins) mimicked the tissue or cell surface. To account for the repulsion forces mediated by the glycoproteins, lipids exposing hydrophilic macromolecular head groups were added to one of the membranes. Unbinding forces between the linker pairs were measured by magnetic tweezer force microscopy. The state of the art of model membrane studies was reviewed recently [178]. We therefore summarize below only few important lessons learned from these model membrane studies:

- An inevitable consequence of the balance of interfacial forces and elastic stresses summarized above is the decay of the adhesion zone into microdomains of tight adhesion (formed by clusters of bound pairs of linkers), which are separated by nonadhering zones.
- Cell adhesion is a two-step process. The first step consists in the formation of clusters of linker pairs by the mechanism described above. In a secondary step, these clusters are stabilized by coupling of actin gel patches to the cytoplasmic tails of the CAMs. In the case of integrin, this coupling is mediated by the integrin–actin linkers talin and ezrin (illustrated in Figure 1.17), which results in a drastic enhancement of the affinity of integrins for ligands (such as ICAM-I). This step plays a key role for the onset of lymphocyte transgression through the endothelial cell layers (see Refs. [223,224]). In the case of cadherin-mediated cell–cell adhesion, clusters of the linkers are stabilized by binding of their cytoplasmic tails to actin by β- and α-catenin. An important consequence of this linkage is a drastic increase of the membrane bending modulus and thus – according to Eq. (1.12) – of the binding strength W.
- A most surprising but important result is the following. The values of the adhesion energies measured by contour analysis, namely $W = 10^{-5} - 10^{-6}\,\mathrm{J\,m^{-2}}$, are several orders of magnitude smaller than the energies estimated from the known density and binding energy of the linker pairs, yielding $W \approx 10^{-3}\,\mathrm{J\,m^{-2}}$ [224]. It was shown that this reduction of W is a consequence of the two-dimensional osmotic pressure exerted by the CAMs and repellers expelled from the adhesion domain. A rigorous theory of the initial process of adhesion domain formation by interplay of elastic and entropic forces was developed by Seifert and Smith [225].

Taken together, the biomimetic studies show that the repulsive force mediated by the glycoproteins of the cell surface plays a key role for the softening of the adhesion strength and the rapid formation and dismantling of adhesion domains. It should be noted, however, that many of the glycoproteins (such as Syndcans and CD44) can act as CAMs, which mediate the adhesion of stromal cells to macromolecules of the extracellular matrix, such as collagen networks. Most CAMs expose cytoplasmic domains which mediate their coupling to the actin cortex through specific linkers, such as talin and ezrin. The global distribution of the linkers can be controlled by the actin–microtubule crosstalk, which plays a key role for the global polarization of cells. Several physiologically relevant examples of the control of cell adhesion by microtubule–actin crosstalk have been recently reviewed in Ref. [178].

1.9.3 Synopsis

The design of sophisticated model systems of cell adhesion, together with the development of new experimental techniques and thermomechanical theories have provided new insights into the control of cell adhesion by intermolecular forces and elastic stresses [178]. It is certainly a long way to design model systems, which can mimic the active control of cell adhesion mediated by coupling of CAM–clusters to the actin cortex.

At present, a more promising strategy is to apply the methods and concepts developed over the last 40 years to study the adhesion of living cells on biofunctionalized surfaces. The bending moduli, the membrane tensions, and the adhesion strengths can be measured by applying lift forces and hydrodynamic shear forces [213,226]. Insights into the control of the adhesion strength mediated by coupling of the actin cortex to the intracellular domains of the CAM–clusters can be gained by mutations of the proteins regulating actin membrane coupling, such as talin (see Ref. [226]). An intriguing example of the benefit of combined studies of biomimetic and natural systems is mentioned in the next section.

The combined study of *in vitro* systems and natural cells has shed new light on the long-standing question: how does myelin sheath form tightly packed multilayers of collapsed cell lobes around axons? The first insights were provided by Israelachvili and coworkers who measured the forces between the inner leaflets of the PMs mediated by the electrostatic interaction of the myelin basic protein [227]. Recent experiments by Janshoff and coworkers on natural systems solved another essential part of the enigma. They showed that membrane lobes protruding from oligodentrocytes (cells forming myelin sheets in the brain) adhere strongly to axon surface after downregulation of the genetic expression of both the repellent glycoproteins and the glycolipids [226]. The intermembrane distance of 2 nm is determined by the van der Waals attraction, which is balanced by the repulsion mediated by uncharged integral proteins.

1.10 A SHORT OUTLOOK: WHERE WE STAND AND WHERE TO GO

In the last 40 years, our understanding of the physical basis of the self-organization of biological membrane has advanced dramatically, owing to the comparative study of model systems and biological membranes. This breakthrough was made possible by the interactive development of new physical techniques and novel theoretical concepts and computational tools. There are still many unsolved problems. Many of these are due to our difficulties to deal with the thermomechanics of multicomponent systems under nonequilibrium conditions.

It appears that there are two directions to go. One strategy is to mimic the structure and function of real composite membranes by reconstitution of active actin and microtubule networks into giant vesicles with reconstituted ion channels and CAMs. Such artificial cell organelles are also of growing interest as smart drug delivery systems.

An easier strategy may be to study natural membranes and their structural and functional modifications by specific mutations or signalling molecules. To this end physicists should realize that biological membranes are composite shells and that their molecular architecture is constantly remodelled by biochemical processes to adapt the cell structure and function to the cell's physiological needs.

Studies of natural membrane processes teach us how to regulate the structure and physical properties of complex materials by local changes of the chemical composition in a logistic way. A future benefit of such painstaking studies may be that we learn how to design smart materials with the capacity to adapt their physical properties to environmental conditions.

LIST OF ABBREVIATIONS

ADE model: area difference elasticity model
BLMs: black lipid membranes
DLVO theory: Derjaguin–Landau–Verwey–Overbeek theory of the stability of colloidal suspensions
EISF: elastic incoherent scattering factor
FCS: fluorescence correlation spectroscopy
FTIR: Fourier transform infrared spectroscopy
MARCKS: myristoylated alanine-rich C-kinase substrate
PLC-Cγ: phospholipase-Cγ (an enzyme generating inositol 1,4,5-trisphosphate and diacylglycerol (DAG) from PI-4,5-P2)
PI-3K: phosphoinositid-3-kinase (generator of PI-3,4,5-P3)
RICM: reflection interference contrast microscopy (also called interference reflection microscopy [IRM]).
SFA: surface force apparatus

LIPID ABBREVIATIONS AND NOMENCLATURE

The structure of each type of lipid is characterized by the structure of the head groups and the number of C-atoms (n) and double bonds (m) of the hydrocarbon chains.

TYPES OF LIPIDS

PC: phosphatidylcholine
PE: phosphatidylethanolamine
PS: phosphatidylserine
PI: phosphoinositides (including PI-4,5-P2 and PI-3,4,5-P3, also abbreviated as PtdIns(4,5)P2 and PtdIns (3,4,5)P3)
DAG: diacylglycerol

EXAMPLES

PC: C16:0; C16:0 and PC: C16:0, C18;1 stand for dipalmitoylphosphatidylcholin and palmitol-strearoyl phosphatidylcholine, respectively.

ACKNOWLEDGEMENTS

E.S. gratefully acknowledges the financial support from the Excellence Initiative of the Technical University Munich and the Center for NanoScience (CeNS) at the Ludwig-Maximilian-University (LMU) Munich. A.B.S. thanks the support of the Fritz Haber Research Center supported by the Minerva Foundation.

BIBLIOGRAPHY

1. Gorter, E., and F. Grendel. 1925. On bimolecular layers of lipoids on the chromocytes of the blood. *J. Exp. Med.* 41: 439–43.
2. Danielli, J.F., and H. Davson. 1935. A contribution to the theory of permeability of thin films. *J. Cell. Comp. Physiol.* 5: 495–508.
3. Singer, S.J., and G.L. Nicolson. 1972. The fluid mosaic model of the structure of cell membranes. *Science* 175(80): 720–731.
4. Sackmann, E. 1983. Physical foundation of the molecular organization and dynamics of membranes in biophysics. In: Hoppe, W., Lohmann, W., Markl, H., Ziegler, H., editors. *Biophysics.* Springer Verlag. Berlin, pp. 425–457.
5. Luzzati, V. 1968. X-ray diffraction studies of lipid-water systems. In: Chapman D., editor. *Biological Membranes*, Vol. 1. Academic Press, New York, pp. 71–123.
6. Luzzati, V., and a Tardieu. 1974. Lipid phases: Structure and structural transitions. *Annu. Rev. Phys. Chem.* 25: 79–94.
7. Helfrich, W. 1973. Elastic properties of lipid bilayers: Theory and possible experiments. *Z. Naturforsch.* 28: 693–703.
8. Bessis, M. 1973. *Living Blood Cells and Their UltrastructueNo Title.* Berlin-Heidelberg-New York: Apringer Verlag.
9. Shen, B.W., R. Josephs, and T.L. Steck. 1986. Ultrastructure of the intact skeleton of the human erythrocyte membrane. *J. Cell Biol.* 102: 997–1006.
10. Legendre, K., S. Safieddine, P. Küssel-Andermann, C. Petit, and A. El-Amraoui. 2008. alphaII-betaV spectrin bridges the plasma membrane and cortical lattice in the lateral wall of the auditory outer hair cells. *J. Cell Sci.* 121: 3347–3356.
11. Hladky, S.B., and D. a Haydon. 1972. Ion transfer across lipid membranes in the presence of gramicidin A. I. Studies of the unit conductance channel. *Biochim. Biophys. Acta.* 274: 294–312.
12. Bamberg, E., and P. Läuger. 1973. Channel formation kinetics of gramicidin A in lipid bilayer membranes. *J. Membr Biol.* 11: 177–194.
13. Kolb, H.-A., and G. Boheim. 1978. Analysis of the multipore system of alamethicin in a lipid membrane. *J. Membr. Biol.* 38: 151–191.
14. Conti, F., and Wanke, E. 1975. Channel noise in nerve membranes and lipid bilayers. *Q. Rev. Biophys.* 8: 451–506.
15. Chernomordik, L., A. Chanturiya, J. Green, and J. Zimmerberg. 1995. The hemifusion intermediate and its conversion to complete fusion: Regualtion by membrane composition. *Biophys. J.* 69: 922–929.
16. Li, F., F. Pincet, E. Perez, C.G. Giraudo, D. Tareste, and J.E. Rothman. 2011. Complexin activates and clamps SNAREpins by a common mechanism involving an intermediate energetic state. *Nat. Struct. Mol. Biol.* 18: 941–946.
17. Tanaka, M., and E. Sackmann. 2005. Polymer-supported membranes as models of the cell surface. *Nature.* 437: 656–663.
18. Seddon, J.M., and R.H. Templer. 1995. Polymorphism of lipid-water systems. In: Lipowsky R, E Sackmann, editors. *Structure and Dynamics of Membranes.* Amsterdam: Elsevier. pp. 98–160.
19. Voeltz, G.K., W.A. Prinz, Y. Shibata, J.M. Rist, and T.A. Rapoport. 2006. A class of membrane proteins shaping the tubular endoplasmic reticulum. *Cell.* 124: 573–586.
20. Sackmann, E. 2014. Endoplasmatic reticulum shaping by generic mechanisms and protein-induced spontaneous curvature. *Adv. Colloid Interface Sci.* 208: 153–160.
21. Sackmann, E., H. Träuble, H.J. Galla, and P. Overath. 1973. Lateral diffusion, protein mobility, and phase transitions in *Escherichia coli* membranes. A spin label study. *Biochemistry.* 12: 5360–5369.
22. Seelig, J. 1977. Deuterium magnetic-resonance theory and applications to lipid membranes. *Q. Rev. Biophys.* 10: 353–418.
23. Mely, B., J. Charvolin, and P. Keller. 1975. Disorder of lipid chains as a function of their lateral packing in lyotropic liquid crystals. *Chem. Phys. Lipids.* 15: 161–173.
24. Mantsch, H.H., and R.N. McElhaney. 1991. Phospholipid phase transitions in model and biological membranes as studied by infrared spectroscopy. *Chem. Phys. Lipids.* 57: 213–226.
25. Esmann, M., and D. Marsh. 1985. Spin-label studies on the origin of the specificity of lipid-protein interactions in Na+,K+-ATPase membranes from Squalus acanthias. *Biochemistry.* 24: 3572–3578.
26. König, B.W., S. Krueger, W.J. Orts, C.F. Majkrzak, N.F. Berk, J. V Silverton, and K. Gawrisch. 1996. Neutron reflectivity and atomic force microscopy studies of a lipid bilayer in water adsorbed to the surface of a silicon single crystal. *Langmuir.* 12: 1343–1350.
27. König, S., W. Pfeiffer, T. Bayerl, D. Richter, and E. Sackmann. 1992. Molecular dynamics of lipid bilayers studied by incoherent quasi-elastic neutron scattering. *J. Phys. II.* 2: 1589–1615.
28. Buldt, G., H.U. Gally, A. Seelig, and J. Seelig. 1978. Neutron diffraction studies on selectively deuterated phospholipid bilayers. *Nature.* 271: 182–184.
29. Sackmann, E. 1995. Physical basis of self-organization and function of membranes: Physics of vesicles. In: R. Lipowsky, E. Sackmann, editors. *Structure and Dynamics of Membranes.* Amsterdam: Elsevier. pp. 213–304.
30. Overath, P., and H. Trauble. 1970. Phase transitions in cells, membranes, and lipids of *Escherichia coli* by fluorescent probes, light scattering and dilatometry. *Biochemistry.* 12: 2625–2634.
31. McIntosh, T.J., and S.A. Simon. 1986. Area per molecule and distribution of water in fully hydrated dilauroylphosphatidylethanolamine bilayers. *Biochemistry.* 25: 4948–4952.
32. Wilkinson, D.a, and J.F. Nagle. 1981. Dilatometry and calorimetry of saturated phosphatidylethanolamine dispersions. *Biochemistry.* 20: 187–92.
33. Privalov, G., V. Kavina, E. Freire, and P.L. Privalov. 1995. Precise scanning calorimeter for studying thermal properties of biological macromolecules in dilute solution. *Anal. Biochem.* 232: 79–85.
34. Shimshick, E.J., and H.M. McConnell. 1973. Lateral phase separation in phospholipid membranes. *Biochemistry.* 12: 2351–2360.
35. Mouritsen, O.G. 1991. Theoretical models of phospholipid phase transitions. *Chem. Phys. Lipids.* 57: 179–194.
36. Mouritsen, O.G. 2005. *Life–as a Matter of Fat: The Emerging Science of Lipidomics.* Heidelberg: Springer.
37. Silvius, D.R. *Thermotropic Phase Transitions of Pure Lipids in Model Membranes and Their Modifications by Membrane Proteins.* New York: John Wiley & Sons, Inc. 1982.
38. Coolbear, K.P., C.B. Berde, and K.M. Keough. 1983. Gel to liquid-crystalline phase transitions of aqueous dispersions

of polyunsaturated mixed-acid phosphatidylcholines. *Biochemistry.* 22: 1466–1473.
39. Nagle, J.F. 1973. Theory of biomembrane phase transitions. *J. Chem. Phys.* 58: 252.
40. Scott Jr., H.L. 1975. Some models for lipid bilayer and biomembrane phase transitions. *J. Chem. Phys.* 62: 1347–1353.
41. A. Caillé, D. Pink, F. De Verteuil, and M.J.Z. 1980. Theoretical models for quasi-two-dimensional mesomorphic monolayers and membrane bilayers. *Can. J. Physics,* 58(5): 581–611. Doi: 10.1139/p80-083.
42. Jähnig, F. 1981. Critical effects from lipid-protein interaction in membranes. I. Theoretical description. *Biophys. J.* 36: 329–345.
43. de Gennes, P.G. 1974. General features of lipid organization. *Phys. Lett. A.* 47: 123–124.
44. Schröder, H. 1977. Aggregation of proteins in membranes. An example of fluctuation-induced interactions in liquid crystals. *J. Chem. Phys.* 67: 1617–1619.
45. Marčelja, S. 1974. Chain ordering in liquid crystals II. Structure of bilayer membranes. *Biochim. Biophys. Acta.* 367: 165–176.
46. Marčelja, S. 1976. Lipid-mediated protein interaction in membranes. *Biochim. Biophys. Acta.* 455: 1–7.
47. Cevc, G. 1991. Polymorphism of the bilayer membranes in the ordered phase and the molecular origin of the lipid pretransition and rippled lamellae. *Biochim. Biophys. Acta - Biomembr.* 1062: 59–69.
48. Nelson, D. 1987. Theory of the crumpling transition. In: Nelsonj, D, Piran, T and Weinberg S, editor. *Statistical Mechanics of Surfaces and Membranes.* Singapore: World Scientific, pp. 137–155.
49. Akabori, K., and J.F. Nagle. 2015. Structure of the DMPC lipid bilayer ripple phase. *Soft Matter.* 11: 918–926.
50. Albrecht, O, Gruler, H, and Sackmann, E. 1978. Polymorphism of phospholipid monolayers. *J. Phys.* 39: 301–313.
51. McConnell, H.M., and V.T. Moy. 1988. Shapes of finitetwo-dimensional lipid domains. *J. Phys. Chem.* 92: 4520–4525.
52. McConnell, H.M. 1989. Theory of hexagonal and stripe phases of monolayers. *Proc. Natl. Acad. Sci.* 86: 3452–3455
53. Möhwald, H. 1995. Lipid Monolayers. In: Lipowsky, R and Sackmann E, editor. *Handbook of Biological Physics* Vol 1A. Amsterdam: North Holland. pp. 161–211.
54. Seelig, J., and W. Niederberger. 1974. Two pictures of a lipid bilayer. A comparison between deuterium label and spin-label experiments. *Biochemistry.* 13: 1585–1588.
55. König, S., T.M. Bayerl, G. Coddens, D. Richter, and E. Sackmann. 1995. Hydration dependence of chain dynamics and local diffusion in L-alpha-dipalmitoylphosphtidylcholine multilayers studied by incoherent quasi-elastic neutron scattering. *Biophys. J.* 68: 1871–1880.
56. Wiener, M.C., and S.H. White. 1992. Structure of a fluid dioleoylphosphatidylcholine bilayer determined by joint refinement of x-ray and neutron diffraction data. III. Complete structure. *Biophys. J.* 61: 434–447.
57. Dill, K.A., and P.J. Flory. 1980. Interphases of chain molecules: Monolayers and lipid bilayer membranes. *Proc. Natl. Acad. Sci.* 77: 3115–3119.
58. Leermakers, F.A.M., and J.M.H.M. Scheutjens. 1988. Statistical thermodynamics of association colloids. III. The gel to liquid phase transition of lipid bilayer membranes. *J. Chem. Phys.* 89: 6912.
59. van der Ploeg, P. Berendsen, H.J.C., and P. van der Ploeg. 1982. Molecular dynamics simulation of a bilayer membrane. *J. Chem. Phys.* 76: 3271.
60. Jähnig, F. 1979. Molecular theory of lipid membrane order. *J. Chem. Phys.* 70: 3279.
61. Gruen, D.W.R. 1985. A model for the chains in amphiphilic aggregates. 1. Comparison with a molecular dynamics simulation of a bilayer. *J. Phys. Chem.* 89: 146–153.
62. Ben-Shaul, A., I. Szleifer, and W.M. Gelbart. 1985. Chain organization and thermodynamics in micelles and bilayers. I. Theory. *J. Chem. Phys.* 83: 3597–3611.
63. Ben-Shaul, A. 1995. Molecular theory of chain packing, elasticity and lipid protein interaction in lipid bilayers. In: Lipowsky R, E Sackmann, editors. *Structure and Dynamics of Membranes.* Amsterdam: Elsevier. pp. 359–402.
64. Israelachvili, J.N., J. Mitchell, and B.W. Ninham. 1976. Theory of self-assembly of hydrocarbon amphiphiles into micelles and bilayers. *J. Chem. Soc. Farad. 2.* 72: 1525–1568.
65. Israelachvili, J.N. 2011. Intermolecular and surface forces. 3rd Edition. London: Academic Press.
66. Seelig, J., and N. Waespe-Sarcevic. 1978. Molecular order in cis and trans unsaturated phospholipid bilayers. *Biochemistry.* 17: 3310–3315.
67. Heller, H., M. Schaefer, and K. Schulten. 1993. Molecular dynamics simulation of a bilayer of 200 lipids in the gel and in the liquid crystal phase. *J. Phys. Chem.* 97: 8343–8360.
68. Fattal, D.R., and A. Ben-shaul. 1994. Mean-field calculations of chain packing and conformational statistics in lipid bilayers: Comparison with experiments and molecular dynamics studies. *Biophys. J.* 67: 983–995.
69. Op den Kamp, J. a, M.T. Kauerz, and L.L. van Deenen. 1975. Action of pancreatic phospholipase A2 on phosphatidylcholine bilayers in different physical states. *Biochim. Biophys. Acta.* 406: 169–177.
70. Seigneuret, M., and P.F. Devaux. 1984. ATP-dependent asymmetric distribution of spin-labeled phospholipids in the erythrocyte membrane: Relation to shape changes. *Proc. Natl. Acad. Sci. U. S. A.* 81: 3751–3755.
71. Steck, T.L., and Y. Lange. 2010. Cell cholesterol homeostasis: Mediation by active cholesterol. *Trends Cell Biol.* 20: 680–687.
72. Zheng, G., B. Tian, F. Zhang, F. Tao, and W. Li. 2011. Plant adaptation to frequent alterations between high and low temperatures: Remodelling of membrane lipids and maintenance of unsaturation levels. *Plant. Cell Environ.* 34: 1431–42.
73. Wu, S., and H. McConnell. 1975. Phase separations in phospholipid membranes. *Biochemistry.* 14: 847–854.
74. Vist, M.R., and J.H. Davis. 1990. Phase equilibria of cholesterol/dipalmitoylphosphatidylcholine mixtures: 2H nuclear magnetic resonance and differential scanning calorimetry. *Biochemistry.* 29: 451–464.
75. Mabrey, S., P.L. Mateo, and J.M. Sturtevant. 1978. High-sensitivity scanning calorimetric study of mixtures of cholesterol with dimyristoyl- and dipalmitoylphosphatidyl-cholines. *Biochemistry.* 17: 2464–2468.
76. Evans, E., and D. Needham. 1986. Giant vesicle bilayers composed of mixtures of lipids, cholesterol and polypeptides. Thermomechanical and (mutual) adherence properties. *Faraday Discuss. Chem. Soc.* 81: 267–280.
77. Knoll, W., G. Schmidt, and E. Sackmann. 1983. Critical demixing in fluid bilayers of phospholipid mixtures. A neutron diffraction study. *J. Chem. Phys.* 79: 3439–3442.

78. Knoll, W., G. Schmidt, K. Ibel, and E. Sackmann. 1985. Small-angle neutron scattering study of lateral phase separation in dimyristoylphosphatidylcholine-cholesterol mixed membranes. *Biochemistry.* 24: 5240–5246.
79. Komura, S., H. Shirotori, P.D. Olmsted, and D. Andelman. 2004. Lateral phase separation in mixtures of lipids and cholesterol systems. *EPL (Europhysics Letters)* 321: 7.
80. Elliott, R., I. Szleifer, and M. Schick. 2006. Phase diagram of a ternary mixture of cholesterol and saturated and unsaturated lipids calculated from a microscopic model. *Phys. Rev. Lett.* 96: 098101.
81. Reinl, H., T. Brumm, and T.M. Bayerl. 1992. Changes of the physical properties of the liquid-ordered phase with temperature in binary mixtures of DPPC with cholesterol. *Biophys. J.* 61: 1025–1035.
82. Tu, K., M.L. Klein, and D.J. Tobias. 1998. Constant-pressure molecular dynamics investigation of cholesterol effects in dipalmitoylphosphatidylcholine bilayer. *Biophys. J.* 75: 2147–2156.
83. Brzustowicz, M.R., V. Cherezov, M. Caffrey, W. Stillwell, and S.R. Wassall. 2002. Molecular organization of cholesterol in polyunsaturated membranes: Microdomain formation. *Biophys. J.* 82: 285–298.
84. Maulik, P.R., and G.G. Shipley. 1996. N-palmitoyl sphingomyelin bilayers: Structure and interactions with cholesterol and dipalmitoylphosphatidylcholine. *Biochemistry.* 35: 8025–8034.
85. Döbereiner, H.G., J. Käs, D. Noppl, I. Sprenger, and E. Sackmann. 1993. Budding and fission of vesicles. *Biophys. J.* 65: 1396–403.
86. Sackmann, E. 2006. Thermo-elasticity and adhesion as regulators of cell membrane architecture and function. *J. Phys. Condens. Matter.* 18: R785–R825.
87. Wimley, W.C., and S.H. White. 1996. Experimentally determined hydrophobicity scale for proteins at membrane interfaces. *Nat. Struct. Biol.* 3: 842–848.
88. Marsh, D. 1990. Lipid-protein interactions in membranes. *FEBS Lett.* 268: 371–375.
89. Mouritsen, O.G., and M. Bloom. 1984. Mattress model of lipid-protein interactions in membranes. *Biophys. J.* 46: 141–153.
90. Riegler, J., and H. Möhwald. 1986. Elastic interactions of photosynthetic reaction center proteins affecting phase transitions and protein distributions. *Biophys. J.* 49: 1111–1118.
91. Gruler, H. 2015. Chemoelastic effect of membranes. *Z. Naturforsch. C.* 30: 608–614.
92. Fattal, D.R., and A. Ben-Shaul. 1993. A molecular model for lipid protein interaction in membranes: The role of hydrophobic mismatch. *Biophys. J.* 65: 1795–1809.
93. Goulian, M., R. Bruinsma, and P. Pincus. 1993. Long-range forces in heterogeneous fluid membranes. *Eur. Lett.* 22: 145–150.
94. Aranda-Espinoza, H., A. Berman, N. Dan, P. Pincus, and S.A. Safran. 1996. Interaction between inclusions embedded in membranes. *Biophys. J.J.* 71: 648–656.
95. Killian, J.A. 1998. Hydrophobic mismatch between proteins and lipids in membranes. *Biophys. Biochim.* 1376: 401–416.
96. Siegel, D.P., and R.M. Epand. 1997. The mechanism of lamellar-to-inverted hexagonal phase transitions in phosphatidylethanolamine: Implications for membrane fusion mechanisms. *Biophys. J.* 73: 3089–3111.
97. May, S., and A. Ben-Shaul. 1999. Molecular theory of lipid-protein interaction and the Lalpha-HII transition. *Biophys. J.* 76: 751–767.
98. Ludtke, S.J., K. He, W.T. Heller, T.A. Harroun, L. Yang, and H.W. Huang. 1996. Membrane pores induced by magainin. *Biochemistry.* 35: 13723–13728.
99. Epand, R.M., Y. Shai, J.P. Segrest, and G.M. Anantharamaiah. 1995. Mechanisms for the modulation of membrane bilayer properties by amphiphatic helical peptides. *Biopolym. (Peptide Sci.* 37: 319–338.
100. Huang, H.W. 1986. Deformation free energy of bilayer membrane and its effect on gramicidin channel lifetime. *Biophys. J.* 50: 1061–1070.
101. Zemel, A., D.R. Fattal, and A. Ben-Shaul. 2003. Energetics and self-assembly of amphipathic peptide pores in lipid membranes. *Biophys. J.* 84: 2242–2255.
102. Zuckermann, M.J., and T. Heimburg. 2001. Insertion and pore formation driven by adsorption of proteins onto lipid bilayer membrane-water interfaces. *Biophys. J.* 81: 2458–2472.
103. Lin, J.-H., and A. Baumgaertner. 2000. Stability of a melittin pore in a lipid bilayer: A molecular dynamics study. *Biophys. J.* 78: 1714–1724.
104. Helfrich, P., and E. Jakobsson. 1990. Calculation of deformation energies and conformations in lipid membranes containing gramicidin channels. *Biophys. J.* 57: 1075–1084.
105. Goulian, M., O.N. Mesquita, D.K. Fygenson, C. Nielsen, O.S. Andersen, and a Libchaber. 1998. Gramicidin channel kinetics under tension. *Biophys. J.* 74: 328–337.
106. Sens, P., and M.S. Turner. 2004. Theoretical model for the formation of caveolae and similar membrane invaginations. *Biophys. J.* 86: 2049–2057.
107. Bretscher, M.S., and S. Munro. 1993. Cholesterol and the golgi apparatus phe. *Science* 261(80): 1280–1281.
108. Kurrle, A., P. Rieber, and E. Sackmann. 1990. Reconstitution of transferrin receptor in mixed lipid vesicles. An example of the role of elastic and electrostatic forces for protein/lipid assembly. *Biochemistry.* 29: 8274–8282.
109. Simons, K., and E. Ikonon. 1997. Functional rafts in cell membranes. *Nature.* 387: 569–572.
110. Schwartz, S.L., C. Cao, O. Pylypenko, A. Rak, and A. Wandinger-Ness. 2008. Rab GTPases at a glance. *J. Cell Sci.* 121: 246–246.
111. Baumgart, T., T.S. Hess, and W.W. Webb. 2003. Imaging coexisting fluid domains in biomembrane models coupling curvature and line tension. *Nature.* 425: 821–824.
112. Leibler, S., and D. Andelman. 2013. Ordered and curved meso-structures in membranes and amphiphilic films. *Journal de physique.* 48: 2013–2018.
113. Träuble, H., and H. Eibl. 1974. Electrostatic effects on lipid phase transitions: Membrane structure and ionic environment. *PNAS.* 71: 214–219.
114. Jähnig, F. 1976. Electrostatic free energy and shift of the phase transition for charged lipid membranes. *Biophys. Chem.* 4: 309–318.
115. Cevc, G. 1990. Membrane electrostatics. *Biochim. Biophys. Acta.* 1031: 311–382.
116. Andelman, D. 1995. Electrostatic properties of membranes: The Poisson-Boltzmann theory. In: Lipowsky R., E. Sackmann, editors. *Structure and Dynamics of Membranes.* Amsterdam: Elsevier. pp. 603–642.
117. Markovich, T., D. Andelman, and R. Podgornik. 2016. Charged membranes: Poisson-Boltzmann theory, DLVO paradigm and beyond. In: Safinya C, J Rädler, editors. *Handbook of Lipid Membranes.*
118. Parsegian, V.A., and D. Gingell. 1972. On the electrostatic interaction across a salt solution between two bodies bearing unequal charges. *Biophys. J.* 12: 1192–1204.

119. Gerke, V., C.E. Creutz, and S.E. Moss. 2005. Annexins: Linking Ca2+ signalling to membrane dynamics. *Nat. Rev. Mol. Cell Biol.* 6: 449–461.
120. Vogt, B., P. Ducarme, S. Schinzel, R. Brasseur, and B. Bechinger. 2000. The topology of lysine-containing amphipathic peptides in bilayers by circular dichroism, solid-state NMR, and molecular modeling. *Biophys. J.* 79: 2644–56.
121. Sackmann, E. Lecture note: Physics of Functional Microdomains. www.biophy.de.
122. McLaughlin, S., and A. Aderem. 1995. The myristoyl-electrostatic switch: A modulator of reversible protein-membrane interactions. *TIBS.* 20: 272–276.
123. Murray, D., B. Honig, and S. McLaughlin. 1998. Sequestration of PIP2 into lateral domains formed by basic peptides. *FASEB J.l.* 12: A1395–A1395.
124. Wang, J.Y., A. Gambhir, G. Hangyas-Mihalyne, D. Murray, U. Golebiewska, S. McLaughlin, G. Hangyás-Mihályné, D. Murray, U. Golebiewska, and S. McLaughlin. 2002. Lateral sequestration of phosphatidylinositol 4,5-bisphosphate by the basic effector domain of myristoylated alanine-rich C kinase substrate is due to nonspecific electrostatic interactions. *J. Biol. Chem.* 277: 34401–34412.
125. Evans, E., and F. Ludwig. 2000. Dynamic strengths of molecular anchoring and material cohesion in fluid biomembranes. *J. Phys. Condens. Matter.* 12: A315–A320.
126. Stetter, F.W.S., L. Cwiklik, P. Jungwirth, and T. Hugel. 2014. Single lipid extraction: The anchoring strength of cholesterol in liquid-ordered and liquid-disordered phases. *Biophys. J.* 107: 1167–1175.
127. Tzlil, S., D. Murray, and A. Ben-Shaul. 2008. The "electrostatic-switch" mechanism: Monte Carlo study of MARCKS-membrane interaction. *Biophys. J.* 95: 1745–1757.
128. Haleva, E., N. Ben-Tal, and H. Diamant. 2004. Increased concentration of polyvalent phospholipids in the adsorption domain of a charged protein. *Biophys. J.* 86: 2165–2178.
129. Golebiewska, U., A. Gambhir, G. Hangyás-Mihályné, I. Zaitseva, J. Rädler, and S. McLaughlin. 2006. Membrane-bound basic peptides sequester multivalent (PIP2), but not monovalent (PS), acidic lipids. *Biophys. J.* 91: 588–599.
130. Stahelin, R. V, D. Karathanassis, K.S. Bruzik, M.D. Waterfield, R.L. Williams, W. Cho, J. Bravo, R.L. Williams, W. Cho, and Stahelin, R. 2006. Structural and membrane binding analysis of the phox homology domain of phosphoinositide 3-kinase-C2α. *J. Biol. Chem.* 281: 39396–39406.
131. Hartmann, W., H.-J. Galla, and E. Sackmann. 1977. Direct evidence of charge-induced lipid domain structure in model membranes. *FEBS Lett.* 78: 169–172.
132. Calabrese, B., M.S. Wilson, S. Halpain, B. Calabrese, S. Margaret, and S. Halpain. 2010. Development and regulation of dendritic dendritic spines: What are they? *Physiology* 21: 38–47.
133. Sackmann, E. 2015. How actin/myosin crosstalks guide the adhesion, locomotion and polarization of cells. *Biochimica et Biophysica Acta (BBA)-Molecular Cell Res.* 1853: 3132–3142.
134. Khelashvili, G., H. Weinstein, and D. Harries. 2008. Protein diffusion on charged membranes: A dynamic mean-field model describes time evolution and lipid reorganization. *Biophys. J.* 94: 2580–2597.
135. Denisov, G., S. Wanaski, P. Luan, M. Glaser, and S. McLaughlin. 1998. Binding of basic peptides to membranes produces lateral domains enriched in the acidic lipids phosphatidylserine and phosphatidylinositol 4,5-biphosphate: An electrostatic model and experimental results. *Biophys. J.* 74: 731–744.
136. May, S., D. Harries, and A. Ben-Shaul. 2002. Macroion-induced compositional instability of binary fluid membranes. *Phys. Rev. Lett.* 89: 268102–268105.
137. Grosberg, A.Y., T.T. Nguyen, and B.I. Shklovskii. 2002. Colloquium: The physics of charge inversion in chemical and biological systems. *Mod. Rev. Phys.* 74: 329–345.
138. May, S., D. Harries, and A. Ben-Shaul. 2000. Lipid demixing and protein-protein interactions in the adsorption of charged proteins on mixed membranes. *Biophys. J.* 79: 1747–1760.
139. deHaseth, P.L., T.M. Lohman, and M.T. Record. 1977. Nonspecific interaction of lac repressor with DNA: An association reaction driven by counterion release. *Biochemistry.* 16: 4783–4790.
140. Record, J.M.T., C.F. Anderson, and T.M. Lohman. 1978. Thermodynamic analysis of ion association or release, screening, and ion effects on water activity. *Q. Rev. Biophys.* 11: 103–178.
141. Rädler, J.O., I. Koltover, A. Jamieson, T. Salditt, and C.R. Safinya. 1997. Structure of DNA-cationic liposome complexes: DNA intercalation in multilamellar membranes in distinct interhelical packing regimes. *Science* 275(80): 810–814.
142. Koltover, I., T. Salditt, J.O. Rädler, and C.R. Safinya. 1998. An inverted hexagonal phase of cationic liposome-DNA complexes related to DNA release and delivery. *Science* 281(80): 78–81.
143. Safinya, C.R. 2001. Structures of lipid-DNA complexes: Supramolecular assembly and gene delivery. *Curr. Opin. Struct. Biol.* 11: 440–448.
144. Wagner, K., D. Harries, S. May, V. Kahl, J.O. Rädler, and A. Ben-Shaul. 2000. Counterion release upon cationic lipid-DNA complexation. *Langmuir.* 16: 303–306.
145. Harries, D., S. May, and A. Ben-Shaul. 2013. Counterion release in membrane–biopolymer interactions. *Soft Matter.* 9: 9268.
146. Träuble, H. 1971. The movement of molecules across lipid membranes: A molecular theory. *J. Membr. Biol.* 4: 193–208.
147. Frye, L.D., and M. Edidin. The rapid intermixing of cell surface antigens after formation of mousehuman heterokaryons. *J. Cell Sci.* 7: 319–336.
148. Träuble, H., and E. Sackmann. 1971. Studies of the crystalline-liquid crystalline phase transition of lipid model membranes: III. Structure of a steroid-lecithin system below and above the lipid phase transiotion. *J. Am. Chem. Soc.* 115: 4499–4510.
149. Devaux, P., and H. McConnell. 1972. Lateral diffusion in spin-labeled phosphatidylcholine multilayers. *J. Am. Chem. Soc.* 94: 4475–4481.
150. Galla, H.J., W. Hartmann, U. Theilen, and E. Sackmann. 1979. On two-dimensional passive random walk in lipid bilayers and fluid pathways in biomembranes. *J. Membr. Biol.* 48: 215–236.
151. Almeida, P.F.F., and W.L.C. Vaz. 1995. Lateral diffusion in membranes. In: Lipowsky R, E Sackmann, editors. *Handbook of Biological Physics: Structure and Dynamics of Membranes.* Amsterdam: Elsevier. pp. 305–358.
152. Saxton, M.J. 2008. A biological interpretation of transient anomalous subdiffusion. II. reaction kinetics. *Biophys. J.* 94: 760–771.
153. Schütz, G.J., G. Kada, V.P. Pastushenko, and H. Schindler. 2000. Properties of lipid microdomains in a muscle cell

membrane visualized by single molecule microscopy. *EMBO J.* 19: 892–901.
154. Schwille, P., and E. Haustein. 2007. Fluorescence correlation spectroscopy: novel variations of an established technique. *Annu Rev Biophys. Biomol. Struct.* 36: 151–169.
155. Rusu, L., A. Gambhir, S. McLaughlin, and J. Rädler. 2004. Fluorescence correlation spectroscopy studies of peptide and protein binding to phospholipid vesicles. *Biophys. J.* 87: 1044–1053.
156. Saffman, P.G., and M. Delbruck. 1975. Brownian motion in biological membranes. *Proc. Natl. Acad. Sci. USA.* 72: 3111–3113.
157. Evans, E., and E. Sackmann. 1988. Translational and rotational drag coefficients for a disk moving in a liquid membrane associated with a rigid substrate. *J. Fluid Mech.* 194: 553–561.
158. Merkel, R., E. Sackmann, and E. Evans. 1989. Molecular friction and epitactic coupling between monolayers in supported bilayers. *J. Phys.* 50: 1535–1555.
159. Komura, S., S. Ramachandran, and K. Seki. 2012. Anomalous lateral diffusion in a viscous membrane surrounded by viscoelastic media. *EPL (Europhysics Lett.* 97: 68007.
160. Tu, K., D.J. Tobias, and M.L. Klein. 1995. Constant pressure and temperature molecular dynamics simulation of a fully hydrated liquid crystal phase dipalmitoylphosphatidylcholine bilayer. *Biophys. J.* 69: 2558–2562.
161. König, S., and E. Sackmann. 1996. Molecular and collective dynamics of lipid bilayers. *Curr. Opin. Colloid Interface Sci.* 1: 78–82.
162. Israelachvili, J.N., and H. Wennerström. 1990. Hydration or steric forces between between amphiphilic surfaces? *Langmuir.* 6: 873–876.
163. Sackmann, E. 1983. Physical foundations of the molecular organization and dynamics of membranes. In: Hoppe W, W Lohmann, H Markl, H Ziegler, editors. *Biophysics.* Berlin: Springer. pp. 425–457.
164. Kapitza, H.G., D. a Rüppel, H.J. Galla, and E. Sackmann. 1984. Lateral diffusion of lipids and glycophorin in solid phosphatidylcholine bilayers. The role of structural defects. *Biophys. J.* 45: 577–587.
165. Schneider, M.B., W.K. Chan, and W.W. Webb. 1983. Fast diffusion along defects and corrugations in phospholipid P beta, liquid crystals. *Biophys. J.* 43: 157–65.
166. Heinemann, S.H., F. Conti, W. Stuhmer, and E. Neher. 1987. Effects of hydrostatic presure on membrane processes: Sodium channels, clcium channels, and exocytosis. *J. Gen. Physiol.* 90: 765–778.
167. Grainger, D.W., a. Reichert, H. Ringsdorf, and C. Salesse. 1989. An enzyme caught in action: Direct imaging of hydrolytic function and domain formation of phospholipase A2 in phosphatidylcholine monolayers. *FEBS Lett.* 252: 73–82.
168. Gallaher, J., K. Wodzińska, T. Heimburg, and M. Bier. 2010. Ion-channel-like behavior in lipid bilayer membranes at the melting transition. *Phys. Rev. E - Stat. Nonlinear, Soft Matter. Phys.* 81: 1–5.
169. Hübner, S., a D. Couvillon, J. a Käs, V. a Bankaitis, R. Vegners, C.L. Carpenter, and P. a Janmey. 1998. Enhancement of phosphoinositide 3-kinase (PI 3-kinase) activity by membrane curvature and inositol-phospholipid-binding peptides. *Eur. J. Biochem.* 258: 846–53.
170. Renard, H.-F., M. Simunovic, J. Lemière, E. Boucrot, M.D. Garcia-Castillo, S. Arumugam, V. Chambon, C. Lamaze, C. Wunder, A.K. Kenworthy, A. a Schmidt, H.T. McMahon, C. Sykes, P. Bassereau, and L. Johannes. 2015. Endophilin-A2 functions in membrane scission in clathrin-independent endocytosis. *Nature.* 517: 493–6.
171. Bhatia, V.K., K.L. Madsen, P.-Y. Bolinger, A. Kunding, P. Hedegård, U. Gether, and D. Stamou. 2009. Amphipathic motifs in BAR domains are essential for membrane curvature sensing. *EMBO J.* 28: 3303–3314.
172. Nelson, D.R., and L. Peliti. 1987. Fluctuations in membranes with crystalline and hexatic order. *J. Phys.* 48: 1085–1092.
173. Helfrich, W. 1978. Steric interactions of fluid membranes in multilayer systems. *Z. Naturforsch. A.* 33: 305–315.
174. de Gennes, P.G., and J. Prost. 1995. *The Physics of Liquid Crystals.* Oxford: Oxford University Press.
175. Seifert, U., and R. Lipowsky. 1995. Morphology of vesicles. In: Lipowsky R, E Sackmann, editors. *Structure and Dynamics of Membranes.* Amsterdam: Elsevier. pp. 403–463.
176. Käs, J., and E. Sackmann. 1991. Shape transitions and shape stability of giant phospholipid vesicles in pure water induced by area-to-volume changes. *Biophys. J.* 60: 825–44.
177. Mutz, M., and D. Bensimon. 1991. Observation of toroidal vesicles. *Phys. Rev. A.* 43: 4525–4527.
178. Sackmann, E., and A.-S. Smith. 2014. Physics of cell adhesion: Some lessons from cell-mimetic systems. *Soft Matter.* 10: 1644.
179. Jülicher, F., U. Seifert, and R. Lipowsky. 1993. Phase diagrams and shape transformations of toroidal vesicles. *J. Phys. II.* 3: 1681–1705.
180. Nelson, P., T. Powers, and U. Seifert. 1995. Dynamical theory of the pearling instability in cylindrical vesicles. *Phys. Rev. Lett.* 74: 3384–3387.
181. Svetina, S., and B. Zeks. 1989. Membrane bending energy and shape determination of phospholipid vesicles and red blood cells. *Eur. Biophys. J.* 17: 101–111.
182. Evans, E.A. 1974. Bending resistance and chemically induced moments in membrane bilayers. *Biophys. J.* 14: 923–931.
183. Häckl, W., U. Seifert, and E. Sackmann. 1997. Effects of fully and partially solubilized amphiphiles on bilayer bending stiffness and temperature dependence of the effective tension of giant vesicles. *J. Phys. II.* 7: 1141–1157.
184. Deuling, H.J., and W. Helfrich. 1976. Red blood cell shapes as explained on the basis of curvature elasticity. *Biophys. J.* 16: 861–868.
185. Góźdź, W., and G. Gompper. 1999. Shapes and shape transformations of two-component membranes of complex topology. *Phys. Rev. E.* 59: 4305–4316.
186. Stokke, B.T., a Mikkelsen, and a Elgsaeter. 1986. Spectrin, human erythrocyte shapes, and mechanochemical properties. *Biophys. J.* 49: 319–327.
187. Peterson, M.A. 1985. Shape dynamics of nearly spherical membrane bounded fluid cells. *Mol. Cryst. Liq. Cryst.* 127: 257–272.
188. Peterson, M. 1992. Linear response of the human erythrocyte to mechanical stress. *Phys. Rev. A.* 45: 4116–4131.
189. Engelhardt, H., and E. Sackmann. 1988. On the measurement of shear elastic moduli and viscosities of erythrocyte plasma membranes by transient deformation in high frequency electric fields. *Biophys. J.* 54: 495–508.
190. Fischer, T.M., M. Stöhr-Lissen, and H. Schmid-Schönbein. 1978. The red cell as a fluid droplet: Tank tread-like motion of the human erythrocyte membrane in shear flow. *Science.* 202: 894–896.

191. Mukhopadhyay, R., G. Lim H W, and M. Wortis. 2002. Echinocyte shapes: Bending, stretching, and shear determine spicule shape and spacing. *Biophys. J.* 82: 1756–1772.
192. Evans, E., and R. Skalak. 1980. *Mechanics and Thermodynamics of Biomembranes.* Boca Raton, FL: CRC Press.
193. Brochard, F., and Lennon J. F. 1975. Frequency spectrum of the flicker phenomenon in erythrocytes. *J. Phys. Fr.* 36: 1035–1047.
194. Helfrich, W., and R.M. Servuss. 1984. Undulations, steric interaction and cohesion of fluid membranes. *Nuovo Cim. D.* 3: 137–151.
195. Albersdorfer, A., T. Feder, and E. Sackmann. 1997. Adhesion-induced domain formation by interplay of long-range repulsion and short-range attraction force: A model membrane study. *Biophys. J.* 73: 245–257.
196. Rädler, J.O., T.J. Feder, H.H. Strey, and E. Sackmann. 1995. Fluctuation analysis of tension-controlled undulation forces between giant vesicles and solid substrates. *Phys. Rev. E.* 51: 4526–4536.
197. Evans, E., and W. Rawicz. 1990. Entropy-driven tension and bending elasticity in condensed-fluid membranes. *Phys. Rev. Lett.* 64: 2094–2097.
198. Lipowsky, R., and Leibler S. 1986. Unbinding transition of ineracting membranes. *Phys. Rev. Lett.* 56: 2541–2544.
199. Zilker, A., H. Engelhardt, and E. Sackmann. 1987. Dynamic reflection interference contrast (RICM) microscopy : A new method to study surface excitations of cells and to measure membrane bending elastic -moduli. *Journal De Physique* 48: 2139–2151.
200. Milner, S.T., and S.A. Safran. 1987. Dynamical fluctuations of droplet microemulsions and vesicles. *Phys. Rev. A.* 36: 4371–4379.
201. Ben-Isaac, E., Y. Park, G. Popescu, F.L.H. Brown, N.S. Gov, and Y. Shokef. 2011. Effective temperature of red-blood-cell membrane fluctuations. *Phys. Rev. Lett.* 106: 238103.
202. Strey, H., M. Peterson, and E. Sackmann. 1995. Measurement of erythrocyte membrane elasticity by flicker eigenmode decomposition. *Biophys. J.* 69: 478–488.
203. Auth, T., S. Safran, and N. Gov. 2007. Fluctuations of coupled fluid and solid membranes with application to red blood cells. *Phys. Rev. E.* 76: 051910.
204. Bitler, A., and R. Korenstein. 2004. Nano-scale fluctuations of red blood cell membranes reveal nonlinear dynamics. *Biophys. J.* 86: 582A.
205. Monzel, C., D. Schmidt, C. Kleusch, D. Kirchenbüchler, U. Seifert, A.-S. Smith, K. Sengupta, and R. Merkel. 2015. Measuring fast stochastic displacements of bio-membranes with dynamic optical displacement spectroscopy. *Nat. Commun.* 6: 8162.
206. Manneville, J.B., P. Bassereau, S. Ramaswamy, and J. Prost. 2001. Active membrane fluctuations studied by micropipet aspiration. *Phys. Rev. E. Stat. Nonlin. Soft Matter Phys.* 64: 021908.
207. Zidovska, A., and E. Sackmann. 2006. Brownian motion of nucleated cell envelopes impedes adhesion. *Phys. Rev. Lett.* 96: 048103.
208. May, S., and A. Ben-Shaul. 1995. Spontaneous curvature and thermodynamic stability of mixed amphiphilic layers. *J. Chem. Phys.* 103: 3839–3848.
209. I. Szleifer, D. Kramer, A. Ben-Shaul, D. Roux, W.M.W.M. Gelbart, I. Szleifer, D. Kramer, A. Ben-Shaul, D. Roux, and W.M.W. M. Gelbart. 1988. Curvature elasticity of pure and mixed surfactant films. *Phys. Rev. Lett.* 60: 1966–1969.
210. Szleifer, I., D. Kramer, A. Ben-Shaul, W.M. Gelbart, and S.A. Safran. 1990. Molecular theory of curvature elasticity in surfactant films. *J. Chem. Phys.* 92: 6800–6817.
211. Parsegian, V. a, N. Fuller, and R.P. Rand. 1979. Measured work of deformation and repulsion of lecithin bilayers. *Proc. Natl. Acad. Sci. U. S. A.* 76: 2750–2754.
212. Bell, G.I. 1978. Models for the specific adhesion of cells to cells. *Science.* 200: 618–627.
213. Bell, G.I., M. Dembo, and P. Bongrand. 1984. Cell adhesion competition between nonspecific repulsion and specific bonding. *Biophys. J.* 45: 1051.
214. Hatta, K., and M. Takeichi. 1986. Expression of N-cadherin adhesion molecules associated with early morphogenetic events in chick development. *Nature.* 320: 447–449.
215. Steinberg, M.S. 1975. Adhesion-guided multicellular assembly: A commentary upon the postulates, real and imagined, of the differential adhesion hypothesis, with special attention to computer simulations of cell sorting. *J. Theor. Biol.* 55: 431–443.
216. Foty, R.A., C.M. Pfleger, G. Forgacs, and M.S. Steinberg. 1996. Surface tensions of embryonic tissues predict their mutual envelopment behavior. *Development.* 1620: 1611–1620.
217. Katsamba, P., K. Carroll, G. Ahlsen, F. Bahna, J. Vendome, S. Posy, M. Rajebhosale, S. Price, T.M. Jessell, a Ben-Shaul, L. Shapiro, and B.H. Honig. 2009. Linking molecular affinity and cellular specificity in cadherin-mediated adhesion. *Proc. Natl. Acad. Sci. U. S. A.* 106: 11594–9.
218. Townes P. S. and Holtfreter J. 1955. Directed movements and selective adhesion of embryonic amphibian cells. *J. Exp. Zool.* 128: 53–120.
219. Douezan, S., K. Guevorkian, R. Naouar, S. Dufour, D. Cuvelier, and F. Brochard-Wyart. 2011. Spreading dynamics and wetting transition of cellular aggregates. *Proc. Natl. Acad. Sci. U. S. A.* 108: 7315–20.
220. Seifert, U., and R. Lipowsky. 1990. Adhesion of vesicles. *Phys. Rev. A.* 42: 4768–4771.
221. Bruinsma, R., A. Behrisch, and E. Sackmann. 2000. Adhesive switching of membranes: Experiment and theory. *Phys. Rev. E.* 61: 4253–4267.
222. Evans, E. 1995. Chapter 15 Physical actions in biological adhesion. *Handb. Biol. Phys.* 1: 723–754.
223. Cinamon, G., V. Shinder, and R. Alon. 2001. Shear forces promote lymphocyte migration across vascular endothelium bearing apical chemokines. *Nat. Immunol.* 2: 515–522.
224. Smith, A.-S., E. Sackmann, and U. Seifert. 2004. Pulling tethers from adhered vesicles. *Phys. Rev. Lett.* 92: 1–4.
225. Nardi, J., R. Bruinsma, and E. Sackmann. 1998. Adhesion-induced reorganization of charged fluid membranes. *Phys. Rev. E.* 58: 6340–6354.
226. Simson, R., E. Wallraff, J. Faix, J. Niewöhner, G. Gerisch, and E. Sackmann. 1998. Membrane bending modulus and adhesion energy of wild-type and mutant cells of Dictyostelium lacking talin or cortexillins. *Biophys. J.* 74: 514–22.
227. Bakhti, M., N. Snaidero, D. Schneider, S. Aggarwal, W. Möbius, A. Janshoff, M. Eckhardt, K.-A. Nave, and M. Simons. 2013. Loss of electrostatic cell-surface repulsion mediates myelin membrane adhesion and compaction in the central nervous system. *Proc. Natl. Acad. Sci. U. S. A.* 110: 3143–8.

2 Structures and Interactions in Freely Suspended Multilayer Membranes and Dilute Lamellar Fluid Membranes from Synchrotron X-Ray Scattering

Gregory S. Smith (Retired)
Oak Ridge National Laboratory, Oak Ridge, Tennessee

Cyrus R. Safinya
University of California, Santa Barbara, California

CONTENTS

2.1 Introduction 33
2.2 X-Ray Scattering Techniques 36
    2.2.1 X-Ray Reflectivity 37
2.3 Phospholipid Membranes 38
    2.3.1 The Lamellar L_α Phase of Chain-Melted Lipid Bilayers 38
    2.3.2 X-Ray Scattering Studies of Freely Suspended Multilayer Films of DMPC Reveal That the L_β, Gel Phase Comprises Three Phases (Lamellar $L_{\beta F}$, $L_{\beta L}$, and $L_{\beta I}$) Distinguished by the Direction of Tilt of the Lipid Tails in the Bilayer Membrane..39
2.4 Dilute Multilayer Fluid Membranes Stabilized by the Helfrich Undulation Forces 42
    2.4.1 The Origin of the Undulation Forces in Multilayer L_α Phase Membranes..........................42
    2.4.2 Synchrotron X-Ray Scattering Studies of a Lamellar L_α Phase in a Quaternary Mixture: X-Ray Structure Factor of Stacked Fluid Membranes and the Landau–Peierls Effect 42
2.5 Concluding Remarks 45
Acknowledgements 45
References 46

2.1 INTRODUCTION

Phospholipid bilayer membranes are major components of the semi-impermeable plasma membrane of the living cell (Singer and Nicolson 1972). They play important roles in basic cell functions from selective permeation for nutrients and waste, fixing of proteins for selective transport, containing carbohydrates for cellular recognition, and active participation in shape-changing cellular functions such as endocytosis and mitosis. All of these functions relate to the basic structure of the bilayer membrane. In general, lipids are surfactant molecules with a charged or polar uncharged hydrophilic head group bound through a backbone linker to two nonpolar hydrophobic tails. Figure 2.1 shows the chemical and space filling structure for 1,2-dimyristoyl-sn-glycero-3-phosphocholine (DMPC), a common phospholipid in biological membranes. The zwitterionic phosphocholine (PC) head group consists of an anionic phosphate group with a cationic choline end group. The hydrophobic interaction drives self-assembly of lipids, which orient to prevent contact between the tails and the nearby aqueous environment (Israelachvili et al. 1977, Israelachvili 2011). In many cases, the resulting structures have spherical topology with the vesicles consisting of a single lipid bilayer or a few concentric multibilayers with the head groups facing water. Determining the detailed interactions of the lipid molecules in the bilayers through their structure is fundamental to understanding their properties, behaviour, and functions.

Among the earliest examples of the use of X-ray diffraction, to study the structure of bilayer lipid membranes, is a seminal paper published in June 1941, by Richard Bear, Ken Palmer, and Francis Schmitt, on lipids derived from nerve tissue (Bear et al. 1941). This was followed by others, in particular, Vitorrio Luzzati in his studies of structures formed by lipids extracted from a variety of tissues (Luzzati 1968).

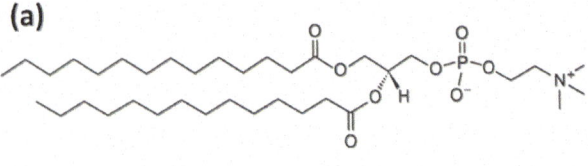

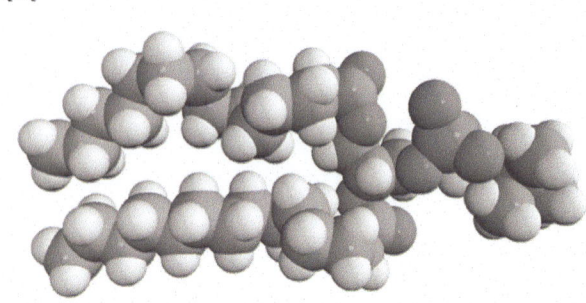

FIGURE 2.1 (a) The chemical structure of 1,2-dimyristoyl-sn-glycero-3-phosphocholine (DMPC) and (b) the space filling model showing the degrees of freedom in the DMPC chains. (Adapted from (Avanti Polar Lipids).)

Indeed, since the late 1960s, X-ray scattering has been a fundamental tool used to study the phase diagram of lipids from the early use of natural phospholipid mixtures to more detailed studies of pure lipid structures (Janiak et al. 1976, Tardieu et al. 1973). In more recent times, lipids with PC head groups have been the subject of many studies designed to understand general lipid behaviour. This is because PC-based lipids are one of the major constituents in the cell membrane. Among the most studied model systems are PCs with two hydrocarbon chains of equal length containing only saturated carbon–carbon bonds. The naming convention for the PCs indicates the number of carbon atoms in the hydrocarbon chains with dilauroyl (DL), dimyristoyl (DM, in Figure 2.1), dipalmitoyl (DP), and distearoyl (DS), referring to lipids with two fully saturated chains each of 12, 14, 16, and 18 carbon atoms, respectively.

In this chapter, we describe experiments to determine the structure and phase behaviour of DMPC in highly oriented freely suspended films maintained under controlled temperature and relative humidity, which determines the amount of water between lipid bilayers (Smith et al. 1988, 1990). The phase diagram of natural lipid mixtures extracted from egg cell membranes (referred to as lecithin or egg PC) was determined using X-ray scattering measurements from oriented (Torbet and Wilkins 1976) and unoriented (Tardieu et al. 1973) mixtures of lipids. The binary lipid–water phase diagram for *purified PC* as a function of hydrocarbon chain length was first determined in unoriented mixtures; that is, with multilayer domains randomly oriented with respect to each other (Janiak et al. 1979, Tardieu et al. 1973). The temperature versus concentration phase diagram of this early study is shown in Figure 2.2 for DMPC–water mixtures. In the lamellar L_α phase, which is the most biologically relevant state of the membrane, the lipid molecules are in the liquid state allowed to freely diffuse within each two-dimensional bilayer sheet (referred to as the chain-melted state). From a structural view point,

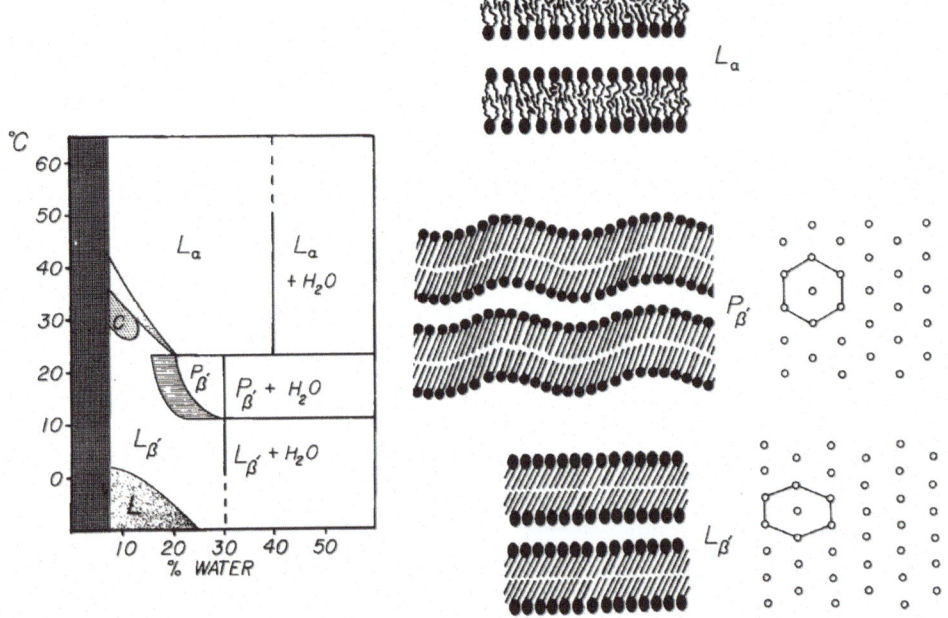

FIGURE 2.2 Phase diagram of 1,2-dimyristoyl-sn-glycero-3-phosphocholine (DMPC)/water determined by X-ray diffraction of unoriented powder samples. Three primary one-phase regions are the fluid multilamellar L_α (with chain-melted lipids), the ordered multilamellar $L_{\beta'}$ (with chain frozen lipids exhibiting in-plane positional ordering, commonly referred to as the gel phase), and the so-called rippled $P_{\beta'}$ phase. (Reprinted with permission from Janiak et al. 1979. Copyright 1979 by the American Society for Biochemistry and Molecular Biology.)

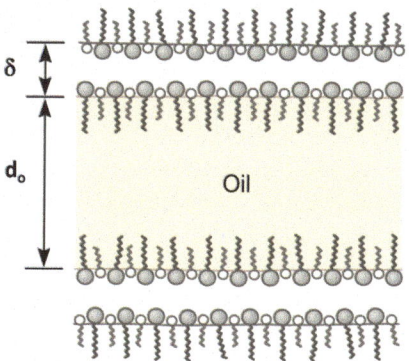

FIGURE 2.3 Multilamellar L_α phase. The "inverted" membranes (a layer of water coated by surfactant/cosurfactant molecules) are separated by oil (hydrophobic molecules). The membrane is effectively "inverted" compared with a biological membrane, which consists of lipid bilayers (see Figures 2.2 and 2.4). Cosurfactant (short-tail) molecules are shown at the oil/water interface. Cosurfactants typically partition between water and oil. The membrane thickness is denoted by δ. The oil region thickness is denoted by d_o. The interlayer distance between the membranes is given by $d = d_o + \delta$.

the L_α phase consisting of a stack of fluid membranes has the same symmetry as the smectic-A phase of thermotropic liquid crystals (Chaikin and Lubensky 1995, de Gennes and Prost 1993). The $L_{\beta'}$ phase, which occupies a large fraction of the phase diagram at lower temperatures, consists of hydrocarbon chains primarily in the all-trans conformation (referred to as the chain-ordered state). The $P_{\beta'}$, known as the "rippled phase", has a mix of frozen and melted chains and greater in-plane order than the L_α, but its most distinguishing feature is a long-wavelength (100–200 Å) in-plane modulation of the layers (Sirota et al. 1988, Wack and Webb 1989, Zasadzinski et al. 1988). Most biologically active membranes consist of multicomponent lipids and tightly coiled biologically active polymers, e.g., proteins, which function as receptors, molecular pumps, enzymes, and so on.

In addition to presenting structural studies of oriented freely suspended films of hydrated DMPC, in this chapter, we further present a study of the structure and interactions in the fluid lamellar L_α phase in a quaternary mixture of surfactant sodium dodecyl sulphate (SDS), cosurfactant (pentanol), water, and dodecane. The lamellar phase in this system has an "inverted" membrane structure; that is, one which consists of a water layer coated with surfactant and cosurfactant as shown in Figure 2.3 with oil separating the membranes (Roux and Bellocq 1985). We note that the water/surfactant/cosurfactant inverted membrane is overall charge neutral where all Na^{+1} counterions to the anionic sulphate group of SDS are confined to the water domain. The study we discuss will focus on the L_α phase along the dodecane dilution pathway in the phase diagram where increasing amounts of added oil push the membranes apart at nearly constant membrane thickness (Safinya et al. 1986, Safinya 1989).

The surfactant/cosurfactant L_α phase described in Section 2.4 consists of stacks of flexible fluid membranes with membrane bending rigidity $\kappa \approx k_B T$. Because fluid membranes have very low surface tension (de Gennes and Taupin 1982, Helfrich 1973), their free energy is governed by membrane curvature fluctuations; thus, thermal fluctuations may result in highly wrinkled membranes. Fluctuating membranes stacked in multilayers may lead to very dilute L_α phases with intermembrane distances of order hundreds (di Meglio et al. 1985a, b, Roux and Safinya 1988, Safinya et al. 1986, 1989, Safinya 1989), or even a few thousand ångstroms (Bassereau et al. 1987, Larche et al. 1986, Porte et al. 1987).

The thermodynamic stability of dilute lamellar L_α phases was elucidated by Wolfgang Helfrich in seminal papers (Helfrich 1978, Helfrich and Servuss 1984). The so-called Helfrich undulation force between membranes is an effective long-range repulsive interaction, which arises from the difference in entropy between a fluctuating "free" membrane and a "bound" membrane in a multilayer system (Figure 2.4). The strength of the interaction increases with decreasing κ and overcomes the attractive van der Waals (vdW) interaction when $\kappa \approx k_B T$ stabilizing multilayer membranes at large separations.

The behaviour found in membranes stabilized by undulation forces is in striking contrast to the L_α phase of the DMPC/water system (Figure 2.2) where the fluid biological lipid DMPC membranes are relatively rigid with κ of order $20\, k_B T$ (Bivas et al. 1987, Schneider et al. 1984). In these systems, the intermembrane forces are dominated

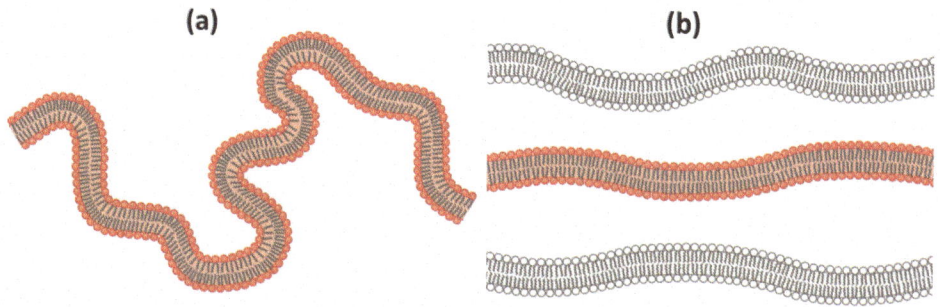

FIGURE 2.4 Schematic drawing of a free membrane (a) compared with a membrane (red) confined between neighbours in a multilamellar L_α phase (b). The free membrane has more space to explore and is in a higher entropic state.

by hydration repulsion balanced by the van der Waals attraction, which prevents the layers from separating more than 20–30 Å (Mahanty and Ninham 1976, Parsegian et al. 1979, Rand 1981, Smith et al. 1987). Before we begin our discussion of the experimental results, we present a brief discussion of X-ray scattering techniques used in membrane research.

2.2 X-RAY SCATTERING TECHNIQUES

X-ray scattering is a powerful tool for the study of the structure of surfactant molecules including lipids. X-rays produced for materials research studies either by laboratory-based X-ray machines or by multi-user synchrotron radiation facilities have wavelengths on the order of the spacing between atoms in molecular materials. Being electromagnetic radiation, when X-rays impinge on a sample, they interact with the atoms by scattering from the electronic charge density. Since the X-ray wavelengths are comparable to atomic spacings, the scattered X-rays interfere with each other constructively or destructively to produce diffraction patterns. These patterns can be analysed to reconstruct the atomic structure.

An X-ray travelling through space can be described as a plane wave with a wave vector $\vec{k}_i$, where the direction of $\vec{k}_i$ is along the direction of travel of the wave packet and its magnitude is $|\vec{k}_i| = 2\pi/\lambda$ where λ is the wavelength of the X-ray photon. The momentum of the X-ray is given by $\hbar\vec{k}$. When the X-ray encounters the charge of the atom, it is scattered. If we consider elastically scattered X-rays, the momentum changes, but the energy remains the same such that the magnitude of $\vec{k}_i$ remains constant, but the direction changes (Figure 2.5a). The wave vector transfer is defined as $\vec{q} = (\vec{k}_i - \vec{k}_f)$ where $\vec{k}_i$ and $\vec{k}_f$ are the incident and scattered wave vectors, respectively, and the magnitude of $|\vec{q}| = 2|\vec{k}_i|\sin(\theta) = 4\pi\sin(\theta)/\lambda$.

For an incident beam of X-rays with intensity I_0, the ratio of the number of scattered photons per unit area and time into solid angle $d\Omega$ to the incident number of photons per second per unit area (Figure 2.5a) defines the differential scattering cross section

$$\frac{I}{I_0} \propto \frac{d\sigma}{d\Omega} \tag{2.1}$$

and $d\sigma/d\Omega$ is the probability that an X-ray incident on a sample will scatter into volume element $d\Omega$ and I is the number of particles scattered into solid angle $d\Omega$. In the Born Approximation for scattering (Merzbacher 1970), the incident and scattered X-ray wave functions are approximated by plane waves. The scattering intensity can then be calculated by summing all of the contributions from the waves scattered by the individual atoms at positions $\vec{r}_i$ in the sample. In this case, the probability for scattering can be written as a sum over the scattered plane waves of the i atoms in the sample so:

$$I(\vec{q}) \propto \left| \sum_i b_i \, e^{i\vec{q}\cdot\vec{r}_i} \right|^2 \tag{2.2}$$

Where b_i is the scattering power of the atom. In the case of X-rays scattered from the electrons on the i_{th} atom, b_i is the atomic number, Z, of the i_{th} atom times the classical radius of the electron $r_0 = 2.818 \times 10^{-5}$ Å. This is the fundamental equation for calculating the scattering from a set of atoms. If the atoms form repeating sub lattices or sub-cells, we can rewrite Eq. (2.2) as

$$I(\vec{q}) \propto \left| \sum_j \left(\sum_k b_{jk} e^{i\vec{q}\cdot\vec{r}_k} \right) e^{i\vec{q}\cdot\vec{r}_j} \right|^2 \tag{2.3}$$

where the sum k runs over the atoms in each kth subcell and j runs over all of the cells. The term within parentheses represents the scattering from the repeating geometrical shape of each cell (this could be a repeating molecular shape for example)

$$f_j(\vec{q}) = \left(\sum_k b_{jk} e^{i\vec{q}\cdot\vec{r}_k} \right) \tag{2.4a}$$

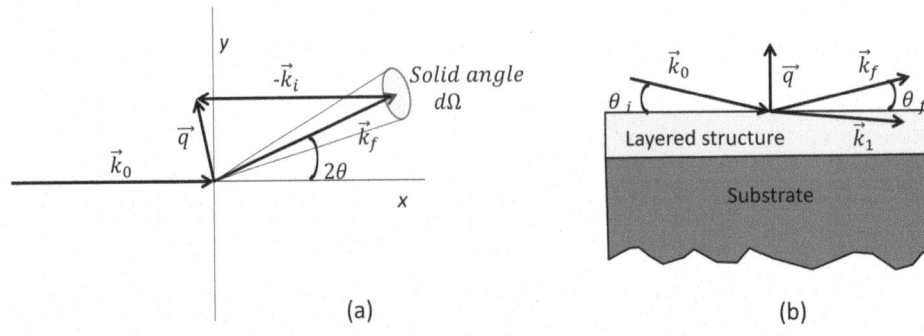

FIGURE 2.5 (a) Scattering geometry. Scattering at angle θ is measured over an area of the detector $d\Omega$. The wave vector transfer in the process of scattering is $\vec{k}_f - \vec{k}_i = \vec{q}$. (b) Reflection geometry. The wave vector transfer for specular scattering is in the z direction normal to the interface.

or in integral form

$$f_j(\vec{q}) = \int_{subcell} d\vec{r}^3\, \rho_j(\vec{r}) e^{i\vec{q}\cdot\vec{r}} \qquad (2.4b)$$

where

$$\rho(\vec{r}) = \sum_k b_{jk} \delta(\vec{r} - \vec{r}_k) \qquad (2.4c)$$

and $\rho(\vec{r})$ is the scattering length density (which is the electron density for X-ray scattering) (Chen 1986) and $f_j(\vec{q})$ is called the form factor. From Eq. (2.4a), we see that the form factor of the unit cell is the Fourier transform of the scattering length density (or the electron density in X-ray scattering) for that unit cell. If we have identical subunits forming crystalline organization of the structure, then $f_j(\vec{q}) = f(\vec{q})$ for all subunits. Upon inserting this term and completing the square of the complex conjugates, Eq. (2.4a) becomes

$$I(\vec{q}) \propto |f(\vec{q})|^2 \sum_{jk} e^{i\vec{q}\cdot(\vec{r}_j - \vec{r}_k)} = P(\vec{q}) \times S(\vec{q}) \qquad (2.5)$$

or

$$I(\vec{q}) \propto \text{geometrical form factor} \times \text{the structure factor} \qquad (2.6)$$

Because Eq. (2.3) represents the square of a Fourier transform of the total scattering length density, the information about the phase change of the plane wave upon scattering is lost which precludes simple inversion of a set of scattering data to obtain the relative atomic or molecular positions of the sample. In this case, typically, a model is constructed consistent with the knowledge of the system (e.g., from the known chemical structure, from other experimental techniques, or from simulations) and the scattering data is fit to that model. If the sample consists of a single crystal, then we can use Eq. (2.3) or (2.5) to calculate the scattered intensity pattern. However, often the sample itself is not a large single crystal but instead is composed of a solution, powder of crystallites, or partially oriented samples. In that case, the result is that the X-rays probe many orientations of the sample at a time, so we need to perform an orientational average of the calculated intensity to compare with the measured intensity. In this case, we rewrite the equations as

$$\frac{d\sigma}{d\Omega} \propto \left\langle \left| \sum_j \left(\sum_k b_{jk} e^{i\vec{q}\cdot\vec{r}_k} \right) e^{i\vec{q}\cdot\vec{r}_j} \right|^2 \right\rangle \qquad (2.7)$$

and

$$\frac{d\sigma}{d\Omega} \propto \left\langle |f(\vec{q})|^2 \sum_{jk} e^{i\vec{q}\cdot(\vec{r}_j - \vec{r}_k)} \right\rangle \qquad (2.8)$$

where the $\langle\,\rangle$ represents the appropriate rotational average of the calculated intensities. For example, for particles in solution with spherical symmetry, the average would consist of integration over all angles. On the other hand, for a two-dimensionally aligned sample, the integration only consists of a rotation about the normal to the plane of the 2-dimensional crystal.

2.2.1 X-Ray Reflectivity

If we consider an X-ray photon polarized parallel to the surface of a sample, in free space, it can be described by a plane wave with momentum $\vec{k}_0$ as described above. Upon impinging on a sample at grazing incidence, the component of the wavelength of the photon along the surface-normal, z, direction is much smaller (Figure 2.5b) with magnitude, $k_{z0} = 2\pi \sin(\theta)/\lambda$, where θ is the angle of incidence and λ is the wavelength of the photon. Like visible light waves, X-rays obey the Fresnel laws of reflection and refraction with the index of refraction

$$n = 1 - \delta - i\gamma \qquad (2.9)$$

For X-rays, $\delta = 2\pi\beta/k_0^2$, $\gamma = -\mu/2k_0$, $\beta = \rho r_0$, r_0 is the classical radius of the electron, ρ is the electron density, and μ is the linear absorption coefficient for the X-rays. In the sample, this z-component of the wave vector is modified by the material such that (Born and Wolf 1965, Russell 1991, Smith and Majkrzak 2006)

$$k_{z1} = \left(k_{z0}^2 - 4\pi\beta\right)^{1/2} \qquad (2.10)$$

By matching wave functions at the interface, we find that the ratio of the scattered wave amplitude to that of the incident wave, defined as the reflectance r, is

$$r = \left(\frac{k_{z0} - k_{z1}}{k_{z0} + k_{z1}}\right) \qquad (2.11)$$

Applying the equation of continuity, we find that ratio of the probability density for the reflected photons to the incident photons is

$$I/I_0 = |r|^2 = \left|\frac{k_{z0} - k_{z1}}{k_{z0} + k_{z1}}\right|^2 \equiv R(k_{z0}) = R\left(\frac{q_z}{2}\right) \qquad (2.12)$$

where $R(k_{z0})$ is defined as the reflectivity. Equation (2.12) is derived for a single surface and shows that at the critical edge defined at $k_{z0}^2 = k_{crit}^2 = 4\pi\beta$ and below, the reflectivity is unity and the incident waves are totally reflected. Also, for a single surface, at k_{z0} greater than k_{crit}, $R(k_{z0}) \sim k_{z0}^{-4}$, the familiar Fresnel law for reflectivity. Equation (2.12) is an exact solution for the reflection from a single smooth surface. For multiple layers, the reflectance from the interface between the i_{th} and $i_{th} + 1$ layers can be computed using the relationship:

$$r_{i,\,i+1} = \left(\frac{k_{zi}^2 - k_{zi+1}^2}{k_{zi}^2 + k_{zi+1}^2}\right) \quad (2.13)$$

where

$$k_{zi+1} = \left(k_{z0}^2 - 4\pi\beta_i\right)^{1/2} \quad (2.14)$$

Starting from the substrate towards the surface, the reflectance can be calculated recursively until the reflectance between the top surface and vacuum is obtained. Then using Eq. (2.12), the reflectivity may be calculated. This method lends itself to defining an electron density profile, $\rho(z)$, and then approximating that profile by a series of uniform layers to calculate the reflectivity as a function of the z-component of the incident wave vector.

2.3 PHOSPHOLIPID MEMBRANES

2.3.1 THE LAMELLAR L_α PHASE OF CHAIN-MELTED LIPID BILAYERS

We now turn to X-ray scattering from the L_α phase in DMPC–water mixtures at high temperatures with the lipid tails in the chain-melted state (see the phase diagram in Figure 2.2 and schematic of L_α phase in Figure 2.4). Small-angle X-ray scattering (SAXS) investigations of powder mixtures (Janiak et al. 1979, Tardieu et al. 1973) show several sharp diffraction peaks at low q. The low-q diffraction peaks may be indexed on a set of regularly spaced (in q-space) diffraction peaks. From Eq. (2.8), if we consider the structure factor for a one-dimensional stack of layers with a repeat spacing, d the structure factor is

$$S(\vec{q}) \propto \sum_{j,k} e^{i\vec{q}\cdot(\vec{r}_j - \vec{r}_k)} = \sum_{j,k} e^{i(n_j - n_k)q_z d} = S(q_z) \quad (2.15)$$

which peaks at $q = m(2\pi/d)$ = reciprocal lattice vectors (i.e., Bragg's law of diffraction for this one-dimensional lattice along the direction normal to the membrane layers). In Eq. (2.15), "m" is an integer and is referred to as the harmonic number of the diffraction peak. In this way, the interlayer repeat spacing in real space, d, can be deduced from the peaks in reciprocal space. For hydrated PCs, $2\pi/q$ for the first-order diffraction peak yields a d-spacing ≈50 Å. The electron density profile extracted from the data by fitting the form factor plus the structure factor for DLPC (Torbet and Wilkins 1976) is consistent with a water–head group–tail–tail–head group–water configuration consistent with the interlayer repeat spacing. Looking at the behaviour of DMPC at 37°C below 40 wt.% concentration, as the water content is decreased in the L_α phase, the lamellar repeat spacing also decreases (Figure 2.6a) (Janiak et al. 1979, Smith et al. 1987). In the one-phase L_α region, the lipid thickness can be derived from the relation $\delta = \Phi_L d$, where Φ_L is the volume fraction of lipid in the

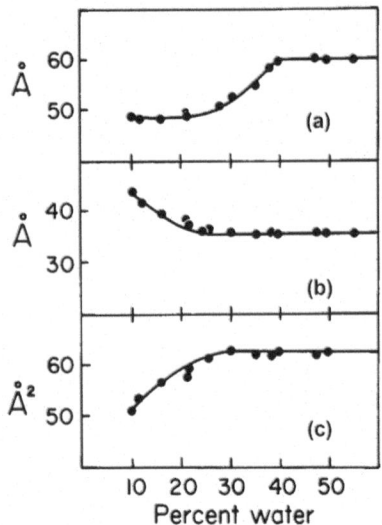

FIGURE 2.6 The water concentration dependence in the L_α phase at 37°C of the (a) interbilayer repeat spacing, (b) the lipid bilayer thickness, and (c) the in-plane area per lipid molecule. (Reprinted with permission from Janiak et al. 1979. Copyright 1979 by the American Society for Biochemistry and Molecular Biology.)

lipid–water mixture. Using this expression, we see that in the L_α phase, the bilayer thickness increases as the water concentration is decreased (Figure 2.6b). This is consistent with the in-plane area per lipid decreasing with decreasing water content (Figure 2.6c) as the lipid chains are becoming more ordered approaching the phases with larger in-plane positional ordering (see Figure 2.2).

In addition to the low-angle bilayer diffraction peaks, there is a distinct, high-angle diffraction peak observed for PC in the L_α phase. The q-centre for this peak corresponds to a spacing of ≈4.5 Å in real space (Janiak et al. 1979, Smith et al. 1987). Using the Sherrer formula (Guinier 1994), $L = 4\pi/(\Delta q_z)_{\text{FWHM}}$, one may determine L, the correlation length or "effective size" of the crystalline domain from $(\Delta q_z)_{\text{FWHM}}$ the full width at half maximum of the diffraction peak. The high-angle peak has a broad width, which implies that the in-plane structure has correlation lengths of order 2–3 neighbours. Thus, the lipid chains are liquid-like in the bilayer plane with short-range positional order in the L_α phase.

We will expand on our discussion of the structure factor for the lamellar L_α phase in Section 2.4.2 where we develop the appropriate expression to be used in line-shape analysis, which incorporates the complete effect of thermal fluctuations (see the discussion leading to Eq. (2.21)). Thermal fluctuations are enormously enhanced in low-dimensional systems such as the L_α phase, compared with fluctuations in 3D crystals, where positional ordering occurs only in one direction normal to the layers with 2D liquid like in-plane ordering.

2.3.2 X-Ray Scattering Studies of Freely Suspended Multilayer Films of DMPC Reveal That the $L_{\beta'}$ Gel Phase Comprises Three Phases (Lamellar $L_{\beta F}$, $L_{\beta L}$, and $L_{\beta I}$) Distinguished by the Direction of Tilt of the Lipid Tails in the Bilayer Membrane

At intermediate temperatures (Figure 2.2 in the $L_{\beta'}$ phase), the X-ray diffraction patterns from unaligned lipid–water mixtures again show low-angle diffraction peaks consistent with a lamellar repeat (Janiak et al. 1979, Smith et al. 1987, Tardieu et al. 1973). In addition, the wide-angle peaks change dramatically exhibiting sharpening indicative of enhanced in-plane ordering compared with that in the L_α phase. Since the peak is sharp, it implies a much longer in-plane positional correlation or crystalline structure. The electron density normal to the bilayer planes was determined from detailed fits to the intensities from the lamellar peaks. This density map plus the two in-plane diffraction peaks yield the picture for the $L_{\beta'}$ phases where the lipid chains pack in a two-dimensional crystalline, distorted, hexagonal lattice.

As noted in Eq. (2.7), the scattering from liquid crystalline phases in powder mixtures is averaged over all possible angles of rotation of the sample. In that case, distinct peaks from an aligned sample may overlap with one another making them hard or impossible to distinguish, thus losing some of the details contained in the scattering information. To capture some of the more detailed information on the lipid structure, the sample may be aligned. For lipids, this can be achieved either by spreading the sample on a solid substrate (Katsaras 1995) or by creating freely suspended films of lipid–water mixtures (Smith et al. 1987, 1988, 1990). Examples of freely suspended multilayer lipid films can be seen in the photographs depicted in Figure 2.7. In either case, to control the amount of water in the films, the chemical potential of the water in the gas phase surrounding the sample must be controlled and known. This is done by controlling the relative humidity of the air around the lipid sample (Gruner 1981, Smith et al. 1987, 1988, 1990). Calculating the scattering from such an aligned stack of multilayers only requires rotationally averaging the calculated intensities about the normal direction to the surface of the layers.

If we consider our representative, freely suspended DMPC system, to model the scattering from a tilted hexagonal phase of oriented chains, we need to calculate the structure factor for the lattice and the form factor for each chain, and then calculate the product of the two terms (Eq. 2.8). For the aligned DMPC tails, we can simplify the model by considering each tail as a cylinder of charge density containing the same number of electrons as in a single chain. If the charge density is uniform in the radial direction, then we can separate the radial (r) and surface normal (z) contributions to the charge density such that (Smith et al. 1990)

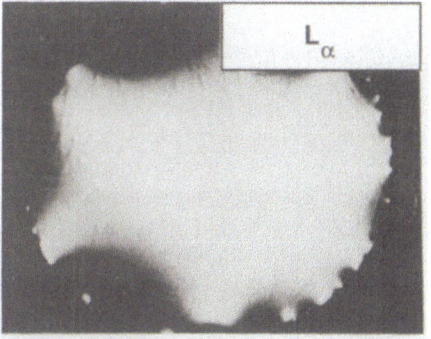
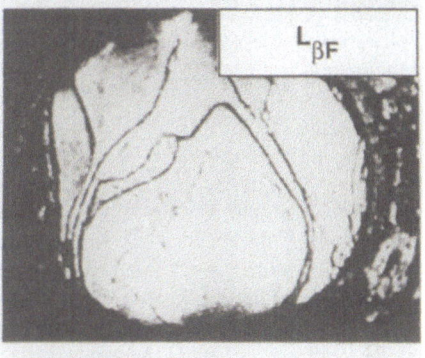
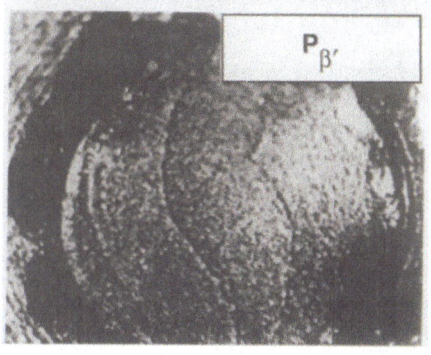

FIGURE 2.7 Pictures of freely suspended DMPC films under controlled temperature and hydration. The photographs were obtained in reflected white light. Films are shown for three distinct phases: L_α, $L_{\beta F}$, and $P_{\beta'}$. DMPC, 1,2-dimyristoyl-sn-glycero-3-phosphocholine. (Reprinted with permission from Smith et al. 1990, Copyright 1990 by the American Institute of Physics, and reprinted from Sirota et al. 1988 with permission from AAAS.)

$$\rho(r) = \rho(z), \quad r \leq a \quad (2.16a)$$

$$= 0, \quad r > a \quad (2.16b)$$

If we consider a close-packed, planar stack of these cylinders forming a single bilayer (Figure 2.8a), the centres of each cylinder would form a two-dimensional hexagonal lattice of points. The structure factor from Eq. (2.8) of the two-dimensional hexagonal lattice (i.e., the Fourier transform of that lattice) is a two-dimensional hexagonal reciprocal space lattice in the x–y plane. If the spacing between the lattice points in real space (corresponding to the cylinder centres) is d, then in reciprocal space, the

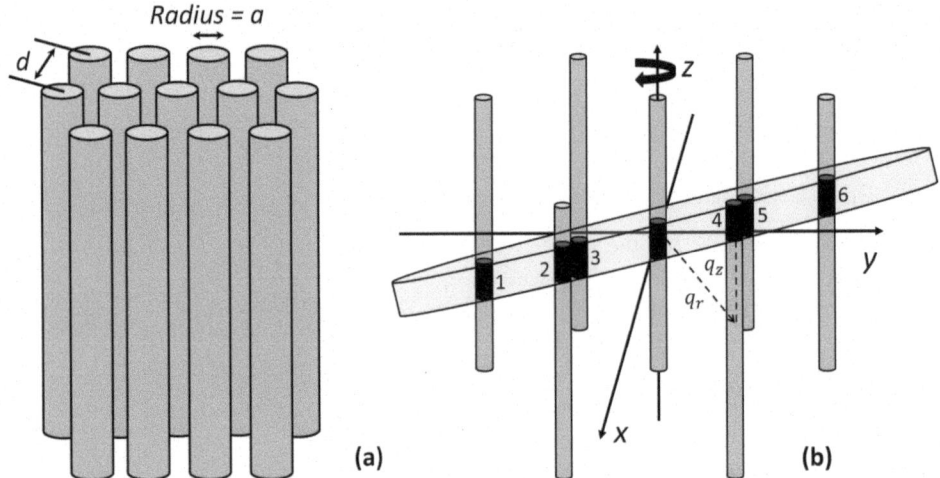

FIGURE 2.8 (a) Schematic model of a stack of oriented charge density for lipid tails forming an in-plane hexagonal lattice. The distance between the tails is d, and the radius of each tail is a. (b) Graphical representation of the disc-shaped form factor intersecting with the infinitely tall Bragg rods (i.e., the structure factor) in the $L_{\beta F}$ phase. In real space, the rods are tilted between the next nearest neighbours. However, in reciprocal space, they tilt towards their nearest neighbours. Excluding the intersection at $(x, y, z) = (0,0,0)$, this yields a set of six distinct spots. In this case, after rotating about the z-axis (i.e., performing an in-plane powder average), spots 2 and 6 merge and spots 3 and 5 merge. This yields four distinct spots in the $L_{\beta F}$ phase. Weaker higher-order terms result for the additional structure normal to the disc forming the truncation of the cylindrical scattering length density.

distance between points is $4\pi/d\sqrt{3}$. Since the lattice forms a set of two-dimensional points in real space, then in reciprocal space the z-direction is infinite. Thus, the reciprocal space lattice is a set of "Bragg rods" with radii inversely proportional to the radius, a, of the real space cylinder of charge density defined in Eq. (2.16). If the rods are tilted, then the hexagonal lattice is distorted by contracting the lattice spacing in the tilt direction.

Now let us consider the form factor of the chains. In our model of a cylinder, the form factor from a cylinder can be separated into the radial (along q_r) and longitudinal (along q_z) parts; using Eq. (2.16), one finds (Smith et al. 1990)

$$f(\vec{q}) = \left[\int \rho(z) e^{iq_z z} \, dz\right] \frac{2\pi q_r a}{q_r^2} J_1(q_r a) \quad (2.17)$$

where $\rho(z)$ is the electron density normal to the bilayer and represents the water–head–tail–tail–head–water repeat, a is the radius of the cylinder, and J_1 is the first-order Bessel function. Since the electron density near the centre of the bilayer corresponds to the location of the terminal CH_3 chain ends, there is a dip in electron density. In the simplest form, this can be modelled by gradual (e.g., Gaussian) function with a linear charge density falloff near the water layer (Smith et al. 1990). If we consider that this is the Fourier transform of a tall cylinder in real space, in reciprocal space to first order, the function looks like a disc although there are also some higher-order oscillations along q_z due to the sharp cutoff in the length of the cylinder. The scattered intensity is calculated at the intersection of the disc-shaped form factor with the structure factor. To represent a tilted hexagonal lattice, we can perform a rotation of the cylinder about a chosen axis in the x–y plane yielding a tilted disc in reciprocal space. The intensity is then calculated as the product of the tilted disc form factor and the hexagonal structure factor graphically illustrated in Figure 2.8b. The measured scattering intensity is independent of the angle of rotation in the plane of the sample (Smith et al. 1990). That suggests that the sample is polycrystalline in the plane of the film or that, from one plane to another, there is no registry. Also, since the model cylindrical form factor that fits the data has a length consistent with the thickness of a single bilayer, this suggests that the lipid tails are correlated from one side of each bilayer to the other, but not between bilayers. For such a sample which is aligned along the layer stacking direction but acting like an in-plane powder, the scattered intensity is finally determined by performing an angular integration of the calculated intensity. Again, graphically, this can be understood as a rotation about the z-axis of the intersecting points of the disc-shaped form factor and the Bragg rods (Figure 2.8b).

The scattered intensity from aligned freely suspended thin films of DMPC at three different relative humidities is shown in Figure 2.9 (Smith et al. 1988, 1990). One sees that there are three distinct patterns in the scattering. If we define the direction between next nearest neighbours in the x–y plane as $0°$ and the direction towards nearest neighbours at $30°$, then from low humidity to high, the scattering patterns correspond to the three different $L_{\beta'}$ phases where the chains are tilted between their next nearest neighbours at $0°$ $(L_{\beta F})$, at an angle between $0°$ and $30°$ $(L_{\beta L})$ and towards next nearest neighbours at $30°$ $(L_{\beta I})$ (Figure 2.9). In the $(L_{\beta L})$ phase, the pattern constantly changes as a function of humidity corresponding to a continuous change in the tilt direction as a function of increasing humidity from $0°$ to $30°$.

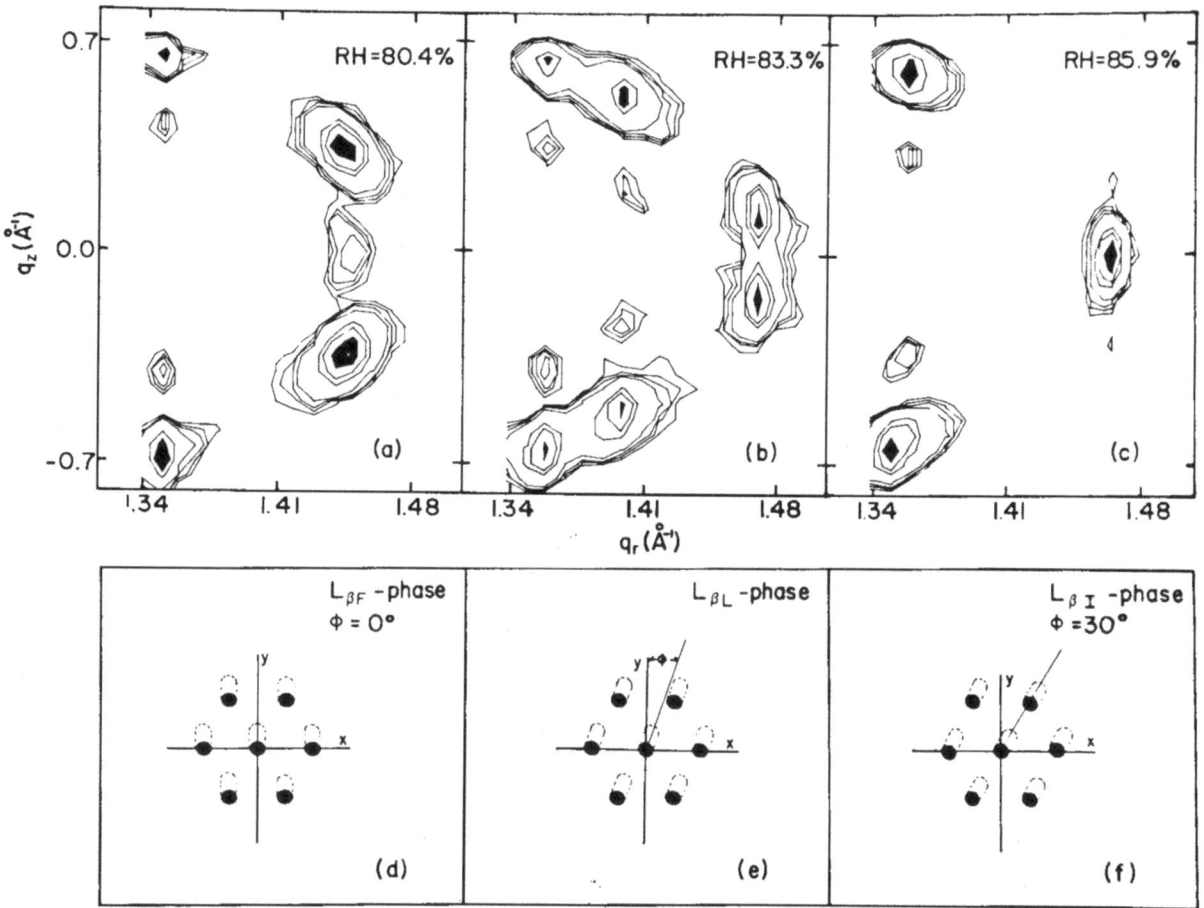

FIGURE 2.9 The measured high angle scattering from an aligned, freely suspended film of DMPC at 23.5°C. The patterns from three distinct phases are immediately apparent. These patterns correspond to the different tilt directions of the DMPC tails. The scattering is represented as a contour plot of constant intensity versus the radial component, q_r, and the z-component, q_z, of the wave vector. The three phases are found to be the (a) $L_{\beta F}$ phase at 80.4% relative humidity (RH), (b) the $L_{\beta L}$ phase at 83.3% RH, and (c) the $L_{\beta I}$ phase at 85.9% RH. The tilt directions in real space are also shown for the (d) $L_{\beta F}$, (e) $L_{\beta L}$, and (f) $L_{\beta I}$ phases (Smith et al. 1988, 1990). DMPC, 1,2-dimyristoyl-sn-glycero-3-phosphocholine. (Reprinted with permission from Smith et al. 1988. Copyright 1988 by the American Physical Society.)

The full temperature-relative humidity phase diagram for the DMPC–water system is shown in Figure 2.10 where comparing with Figure 2.2 (phase diagram obtained from X-ray studies of unoriented samples), we see that the region that has historically been labelled as the $L_{\beta'}$ gel phase actually consists of three distinct $((L_{\beta F}), (L_{\beta L}),$ and $(L_{\beta I}))$ phases. We emphasize that the observation of these distinct phases was only made possible because of the freely suspended preparations (or equivalent preparations on solid surfaces), which orient the layer normal.

Since the three $L_{\beta'}$ phases were identified in a freely suspended film, several studies have been performed to examine the scattering from aligned stacks of lipid multilayer films on solid substrates. Once again, the water content was controlled in the films by controlling the humidity around the lipid multilayers and thermal gradients reduced to a minimum (Katsaras 1998). These studies widened the range of detailed phase space explored and discerned new phases in the DMPC phase diagram (Tristram-Nagle et al. 2002) and verified that DPPC also exhibits the same $L_{\beta'}$ phases seen in DMPC (Katsaras 1995). Most recently (Miller et al. 2008, Watkins et al. 2014) using high-energy X-rays that can penetrate into thick solid substrates, the in-plane structure of single lipid bilayers at the solid–water interface has been studied. For the single bilayers on a solid substrate deposited using the Langmuir–Blodgett technique, it was determined that the substrate has a significant effect on one of the bilayer leaflets of a single dipalmitoyl phosphoethanolamine (DPPE) bilayer and that the electron density deduced from the scattering is more consistent with the lipid tails scattering independently instead of being correlated from one side of the bilayer to the other (Miller et al. 2008). On the other hand, Langmuir–Blodgett deposited DPPC showed a correlation between leaflets and exhibited in-plane structure with tilt angle rotated to an angle of 5° with respect to the nearest neighbour directions similar to the $(L_{\beta L})$ phase (Watkins et al. 2014).

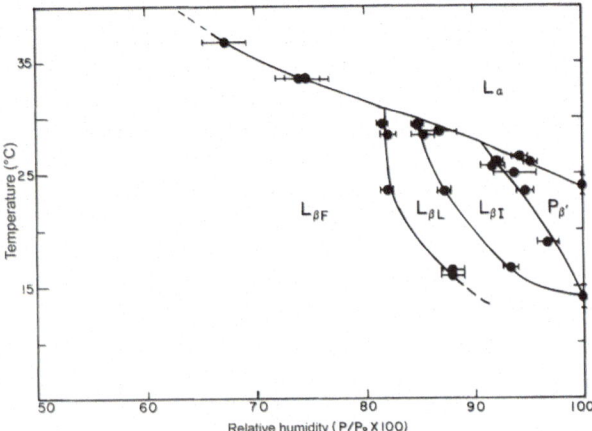

FIGURE 2.10 Phase diagram of DMPC/water system plotted as a function of temperature (T) and relative humidity (RH). Five distinct phases are identified (lamellar L_α, $L_{\beta F}$, $L_{\beta L}$, and $L_{\beta I}$ phases and the $P_{\beta'}$ rippled phase). Note that because the x (T) and y (RH) axes are both fields, there are no two-phase regions (as occurs when the phase diagram is plotted in the temperature and concentration plane, see Figure 2.2). DMPC, 1,2-dimyristoyl-sn-glycero-3-phosphocholine. (Reprinted with permission from Smith et al. 1988. Copyright 1988 by the American Physical Society.)

2.4 DILUTE MULTILAYER FLUID MEMBRANES STABILIZED BY THE HELFRICH UNDULATION FORCES

After publication of his landmark paper on a model of curvature elasticity for fluid membranes (Helfrich 1973), Wolfgang Helfrich realized that membranes that have a low bending rigidity, $\kappa \approx k_B T$, will naturally be wrinkled with height fluctuations at long and short wavelengths. He went on to show that flexible membranes, when bound between neighbouring membranes, as happens in multilayers (Figure 2.4), should exhibit a long-range repulsive interaction, labelled undulation forces, that may overwhelm the van der Waals attractive forces and stabilize multilayers at interlayer spacing (d) much larger than the membrane thickness (δ) (Helfrich 1978, Helfrich and Servuss 1984). Because biological lipids usually form bilayer membranes with $\kappa \approx 20\,k_B T$ (including lipids with phosphocholine head groups with 14, 16, or 18 hydrocarbon chains (Bivas et al. 1987, Schneider et al. 1984)), the realization of the undulation force remained elusive for many years.

2.4.1 THE ORIGIN OF THE UNDULATION FORCES IN MULTILAYER L_α PHASE MEMBRANES

To understand the physics behind the undulation forces, we follow a paper by Helfrich and Servuss (Helfrich and Servuss 1984). At finite temperature, height fluctuations of a membrane of area A, with negligible surface tension, may be calculated from the equipartition theorem $\langle h^2 \rangle = 1/4\pi^3 (k_B T/\kappa) A$ ($h(r)$ is the height or layer displacement in the z-direction normal to the layers) (Safinya 1989, 2006). The strong divergence with the size of the membrane is due to the inherent statistical mechanical properties of tensionless 2D objects embedded in 3D. The bending modulus κ provides the restoring force to height deviations. In a multilayer with interlayer spacing d, thermally driven height fluctuations of a membrane will be cut off by neighbouring membranes through collisions when $\langle h^2 \rangle^{1/2} = d$. Thus, a large enough patch of the membrane with area $A_p = 4\pi^3(\kappa/k_B T)d^2 < A$ will, on average, experience a collision. Because of the steric constraint that membranes cannot cross each other, every membrane–membrane collision will lead to a loss of configuration (compared with a free membrane), thus decreasing the entropy by $\approx k_B T$, or, equivalently, increasing the free energy by $\approx k_B T$. This can be seen schematically in Figure 2.4, where a free membrane has more phase space to explore compared with a bound membrane between its neighbours. Thus, the increase in free energy per unit area of a multilayer (compared with an equivalent number of free membranes) is proportional to $k_B T(A/A_p)(1/A)$, which, in turn, is proportional to $(k_B T)^2/\kappa d^2$. This is the physical origin of the undulation force, namely, the difference in entropy between a fluctuating "free" membrane and a "bound" membrane in a multilayer system. Helfrich derived the undulation interaction for a multilayer system using the elastic free energy per unit volume of the L_α phase (Helfrich 1978), which also describes the elastic energy of a smectic-A phase (de Gennes and Prost 1993):

$$\frac{F}{V} = \frac{1}{2} \times \left\{ B\left(\frac{\partial h}{\partial z}\right)^2 + K\left[\left(\frac{\partial^2 h}{\partial x^2}\right) + \left(\frac{\partial^2 h}{\partial y^2}\right)\right]^2 \right\} \quad (2.18)$$

Here, B and K are the bulk moduli for layer compression (energy/volume) and layer curvature (energy/length). The bulk modulus K is related to the layer (or membrane) bending rigidity: $K = \kappa/d$. The undulation interaction per unit area for membranes with thickness δ is given by

$$\frac{F_{\text{undulation}}}{A} = 0.23 \times \frac{(k_B T)^2}{\kappa(d-\delta)^2} \quad (2.19)$$

We see that the undulation interaction is long range. For membranes with $\kappa \approx k_B T$, the repulsive force overwhelms the attractive long-range van der Waals force between membranes and leads to very dilute membranes with $d \gg \delta$ (Roux and Safinya 1988, Safinya et al. 1986, 1989).

2.4.2 SYNCHROTRON X-RAY SCATTERING STUDIES OF A LAMELLAR L_α PHASE IN A QUATERNARY MIXTURE: X-RAY STRUCTURE FACTOR OF STACKED FLUID MEMBRANES AND THE LANDAU–PEIERLS EFFECT

Evidence of undulation forces was first established in studies exploring the stability of dilute lamellar L_α phases of a quaternary mixture of SDS (the surfactant), pentanol (cosurfactant), water, and dodecane as a function of dodecane dilution (Safinya et al. 1986). Figure 2.11 shows the ternary phase diagram plotted at constant water/SDS weight ratio equal to 1.55. This phase diagram was discovered and mapped out by Didier Roux and Ann-Marie Bellocq (1985). The five one-phase regions observed are the isotropic microemulsion (I_2) (Andelman et al. 1987) and sponge (L_3) (Cates et al. 1988) phases, and the lyotropic liquid crystal hexagonal (E), rectangular (R), and lamellar (L_α) phases. The columnar E and R phases consist of cylindrical and elongated micelles with elliptical cross sections, respectively, and may also be labelled as H_I and R_I. In the L_α phase, the membrane consists of a water layer coated by two monolayers of SDS and pentanol (see Figure 2.3, with membrane thickness δ and membrane wall-to-wall spacing d_o). The pentanol, shown at the water/oil interface in Figure 2.3, partitions between water and oil ≈30/70 mol/mol. The focus of our discussion is on the L_α phase, which may be diluted with dodecane (oil) up to large weight fractions of dodecane ≈ x = 0.80 corresponding to intermembrane separations ($d = \delta + d_o$) ranging between ≈4.0 nm and larger than 50.0 nm. This cut (i.e., with water/SDS (wt/wt) constant and pentanol weight fraction nearly constant) was cleverly chosen by Didier Roux because it maintains a nearly constant membrane thickness δ, while d_o is increased by addition of dodecane.

Synchrotron X-ray scattering profiles for longitudinal scans through the first harmonic (q_{001}) of the structure factor for x = 0, 0.07, 0.13, 0.18, 0.23, 0.35, and 0.54, in the L_α phase, are plotted in Figure 2.12. For each x, the scattering profile is normalized to the peak intensity. The intermembrane spacing $d = 2\pi/q_{001}$ increases from 3.82 (x = 0) nm to 11.5 nm (x = 0.54). In the mixtures studied, the dilution corresponds to a path where the membranes (surfactant/cosurfactant-coated water layers, see Figure 2.3) with water layer thickness $\delta \approx 18$ Å are pushed apart with d_o/δ varying between ≈1 and ≈7 (for x = 0, the tails of SDS and pentanol are counted as part of the oil region and make up the value of d_o). A striking feature of the scattering profiles with increasing amounts of dodecane (i.e., increasing x) is the tail scattering, which becomes dramatically more pronounced as d increases.

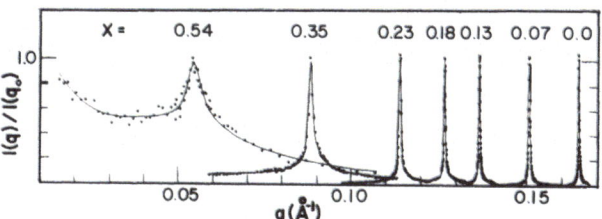

FIGURE 2.12 Synchrotron X-ray profiles of the first harmonic for seven different mixtures along the dodecane dilution path. The weight fraction of dodecane in the mixtures (x) is indicated above each profile. All peak intensities are normalized to unity to enable a qualitative inspection of tail to peak intensities, as the interlayer spacing increases with increasing weight fraction of dodecane. The solid lines are fits to the powder-averaged Caillé structure factor given by Eq. (2.21). (Reprinted with permission from Safinya et al. 1986. Copyright 1986 by the American Physical Society.)

To analyse the scattering data for a lamellar phase (i.e., a stack of fluid membranes), we turn to papers by Rudolf Peierls (1934, 1935) and independently Lev Landau (1937a, b, 1965). These authors argued that 3D liquids whose densities are periodic in only one direction (i.e., smectic-A and L_α phases with elastic energy described by Eq. (2.18)) are only marginally stable at finite temperature because thermally induced layer height fluctuations diverge logarithmically with sample size and thus destroy long-range positional order. Starting with the elastic curvature energy (Eq. (2.18)), Caillé showed that the logarithmic divergence replaces the normal delta-function structure factor peaks of the smectic-A or lamellar L_α phase (i.e., what would be expected for an infinite system of stacked layers with long-range order) with weaker power-law singularities at every reciprocal lattice vector $(0,0,q_{00m})$ ($q_{00m} = mq_{001} = m2\pi/d$, d = interlayer spacing and m = 1, 2, … is the harmonic number) (Caillé 1972). Technically, weaker algebraic singularities signify quasi-long-range positional order. In real materials, the divergence in the power-law singularities predicted by Caillé is cut off by finite size effects (i.e., the coherent lamellar domain sizes). The observation of power-law behaviour, indicative of algebraic decay of layer–layer correlations in a smectic-A liquid crystal, was first described in a landmark paper resulting from a collaboration between three laboratories headed by Jens Als-Nielsen, Robert J. Birgeneau, and J. David Litster, respectively, in the early 1980s (Als-Nielsen et al. 1980), nearly half a century after this remarkable state of matter was first imagined by Landau and Peierls.

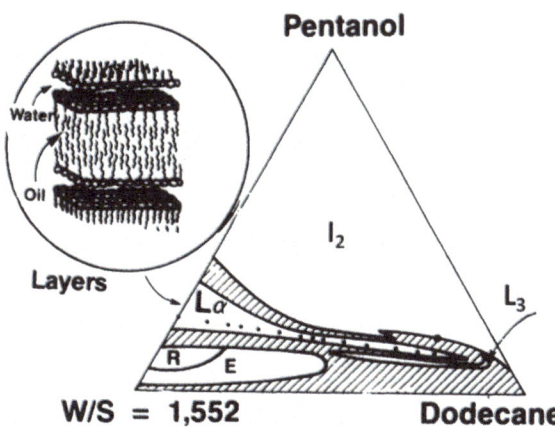

FIGURE 2.11 A cut of the phase diagram of the quaternary mixture of surfactant SDS, cosurfactant pentanol, water, and dodecane shown in the plane with a constant water/SDS weight ratio = 1.55 (Roux and Bellocq 1985). Filled circles in the lamellar L_α phase show location of mixtures studied along the dodecane dilution line. SDS, sodium dodecyl sulphate. (Reprinted with permission from Safinya et al. 1986. Copyright 1986 by the American Physical Society.)

The asymptotic form of the structure factor derived by Caillé is described by power laws:

$S(0,0,q_z) \propto (q_z - q_{00m})^{-2+\eta_m}$ and $S(q_\rho, q_{00m}) \propto q_\rho^{-4+\eta_m}$

where q_ρ and q_z are components of the wave vector parallel and normal to the layers and the power-law exponent η_m is given by

$$\eta_m(d) = m^2 \frac{(q_{001})^2 k_B T}{8\pi \sqrt{BK}} = m^2 \eta_1(d) \quad (2.20)$$

See Eq. (2.18) for definitions of B and K. While the asymptotic forms for $S(\mathbf{q})$ are simple, in order to analyse the scattering profiles *quantitatively*, we fit to the exact expression for the structure factor (Als-Nielsen et al. 1980, Caillé 1972), powder averaged over all solid angles in reciprocal space due to the presence of randomly oriented domains in the sample, and we also incorporate the effect of coherent lamellar domain size (L) (Safinya et al. 1986):

$$\langle S(\mathbf{q}) \rangle = \text{constant}$$

$$\times \int_{-\infty}^{\infty} dz \int_{0}^{\infty} \rho \, d\rho \, S(z,\rho) e^{-\frac{R^2 \pi}{L^2}} [(\sin qR)/qr] e^{-iq_{00m}z} \quad (2.21)$$

$S(\mathbf{R}) = S(z,\rho) = \text{constant} \times \left(\frac{1}{\rho}\right)^{2\eta_m} e^{-2\gamma\eta_m} e^{-E_1(\rho^2/4\lambda z)\eta_m}$ is

the layer–layer correlation function derived by Caillé (1972), $R^2 = z^2 + \rho^2$, γ is Euler's constant, $E_1(x)$ is the exponential integral function, and $\lambda = \sqrt{K/B}$. Thus, we see from the expression for $S(\mathbf{R})$ that the exponent η_m describes the algebraic decay of layer–layer correlations for a stack of fluid membranes. The exponential term in Eq. (2.21) incorporates the finite size effect typically between $L \approx 2000 \, \text{Å}$ and $\approx 10,000 \, \text{Å}$. The analysis of the data consists of fits of Eq. (2.21) to the scattering profile consisting of one, two, or three harmonics where the scaling of $\eta_m = m^2 \eta_1$ with harmonic number (m) is incorporated (harmonics are more visible at smaller values of x). The powder averaging results in asymptotic power-law behaviour for $S(\mathbf{q}) \propto (q - q_{00m})^{-P_m}$ where the exponent P_m is approximately equal to $1 - \eta_m$ (Safinya et al. 1986).

Returning back to the synchrotron X-ray scattering profiles, Figure 2.13a depicts a plot of the profile at $x = 0.23$ on a log–log intensity versus $q - q_{001}$ scale, where the solid line is a result of the fit to Eq. (2.21) yielding $\eta_1 = 0.25$ and $L = 8640 \, \text{Å}$. Two features in the scattering profile and the theoretical cross section are evident. First, at large $q - q_{001} > 1/L$, the scattering exhibits power-law behaviour where the slope on the log–log scale is ≈ 0.75 consistent with $P_1 \approx 1 - \eta_1$. Second, the profile, as q approaches q_{001}, shows a rounding off for $q - q_{001} < 1/L$, due to the finite lamellar domain size. Scattering profiles and fits for the first and second harmonics at $x = 0.07$ on a normalized logarithmic intensity scale are shown in Figure 2.13b. For this sample, $\eta_1 = 0.14 \pm 0.02$, $\lambda = 8.59 \pm 1$ for the first harmonic, and $\eta_2 = 0.57 \pm 0.02$, $\lambda = 8.13 \pm 2$ for the

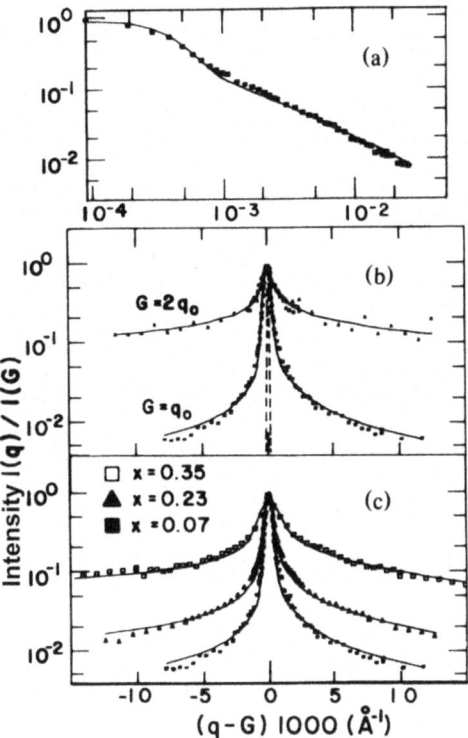

FIGURE 2.13 (a) Profile for the first harmonic, $G = q_{001}$, for a sample at $x = 0.23$ on a log–log scale, which shows finite domain size rounding at small $q - q_{001}$ and power-law behaviour at larger $q - q_{001}$. (b) Profile of the first ($G = q_{001}$) and second ($G = q_{002}$) harmonics for a sample at $x = 0.07$ on a log intensity scale. (c) Profile of the first harmonic, $G = q_{001}$, for three samples at $x = 0.07, 0.23,$ and 0.35 on a log intensity scale. All scattering intensities are normalized to the peak scattering intensities. The solid lines are fits to Eq. (2.21). (Reprinted with permission from Safinya et al. 1986. Copyright 1986 by the American Physical Society.)

second harmonic. Thus, this confirms the scaling, $\eta_m = m^2 \eta_1$, predicted by the Caillé structure factor.

Comparison of scattering profiles for three samples along the dodecane dilution line at $x = 0.07, 0.23,$ and 0.35 is shown in Figure 2.13c. The significant difference in the profiles over the entire dilution range is now immediately clear (including comparing the profiles at $x = 0.35$ and $x = 0.54$ plotted on linear scale in Figure 2.12). It is qualitatively clear that η_1, which characterizes the asymptotic scattering profile and is a measure of the ratio of the tail to peak intensity scattering, is increasing as d increases. This significant increase in η_1, the exponent describing the algebraic decay of layer–layer correlations, signals strong enhancement of layer undulations as the interlayer spacing increases with increasing dilution. Another feature of the scattering profile, seen in Figure 2.13c, is the slight asymmetry evident around $q = q_{001}$ where the high q intensity is larger than the low q intensity. The asymmetry results from the powder averaging and is dependent on the parameter $\lambda = \sqrt{K/B}$, which enters the real-space correlation function $S(\mathbf{R})$. Thus, from the fits, which yield λ and η_1 (both of which are functions of K and B), the value for the membrane bending rigidity $\kappa = Kd$ and compression modulus B may be derived for each sample. For the samples, κ varies between $\approx 0.5 k_B T$ and $\approx 2 k_B T$, which is the driving force for large height fluctuations of the layers.

The ability of this membrane system to swell to interlayer spacings much larger than the membrane thickness, upon addition of dodecane, implies that a long-range repulsive force is present. Because the electric field is equal to zero in the oil region between the layers, one can rule out long-range Coulomb repulsions as a source of the observed large d spacings. On the other hand, the flexibility of the membrane and the observed massive increase in thermal diffuse scattering with dilution (i.e., increase in tail-to-peak scattering intensity seen in Figures 2.12 and 2.13c) suggest that the stability of the dilute membranes results from long-range repulsive Helfrich undulation forces described in Section 2.4.1.

To compare the measured power-law exponent $\eta_1(d)$ to Helfrich's prediction for the exponent, for each sample along the oil dilution line, we calculate the layer compressional modulus B as a function of d using Eq. (2.19): $B_{undulation} = d^2 \left(\partial^2 (F_{undulation}/V)/\partial d^2 \right)$ at constant layer density. Thus, for $\eta_1(d) = k_B T (2\pi/d)^2 / \left(8\pi \sqrt{KB_{undulation}}\right)$ with $K = \kappa/d$, Helfrich's model of undulation forces predicts

$$\eta_1(d) = 1.33 \times \left(1 - \delta/d\right)^2 \quad (2.22)$$

Figure 2.14 shows a plot of experimentally obtained values of $\eta_1(d)$ from fits of the first harmonic in the scattering profiles to the structure factor (Eq. 2.21) as a function of d (i.e., samples at distinct dodecane weight fractions along the dilution line). The solid line, which agrees well with the experimental data, is a plot of the predicted value of the Helfrich theory given by Eq. (2.22). The effective membrane thickness δ was taken to be 29 Å, which includes excluded volume effects of the surfactant tails in the oil

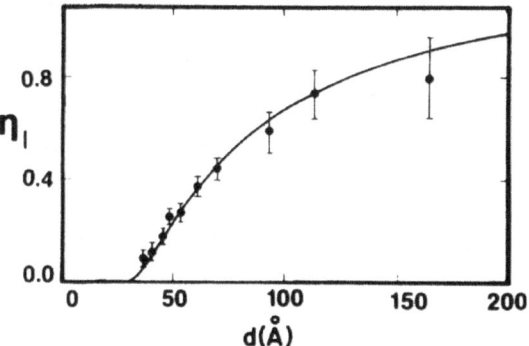

FIGURE 2.14 Plot of the experimentally measured power-law exponent $\eta_1(d)$ as a function of the intermembrane distance for mixtures along the dodecane dilution path. The solid line is the prediction of the model of Helfrich of entropically driven undulation interactions given by Eq. (2.22). (Reprinted with permission from Safinya et al. 1986. Copyright 1986 by the American Physical Society.)

(see Figure 2.3). Thus, we see that in this SDS/pentanol/water/dodecane quaternary mixture, the interlayer force stabilizing the dilute lamellar L_α phase swollen by dodecane is the entropically driven Helfrich undulation force.

2.5 CONCLUDING REMARKS

In this chapter, we have described studies of ordered (gel-like) and dilute fluid multilayer membranes. The structure of the $L_{\beta'}$ gel phase of the biological lipid DMPC membranes was elucidated in highly oriented freely suspended films under controlled temperature and relative humidity. Remarkably, the $L_{\beta'}$ gel phase, which occupies a large fraction of the phase diagram, consists of three lamellar $L_{\beta F}$, $L_{\beta L}$, and $L_{\beta I}$ phases where the distinction between the phases arises from the direction of tilt of the lipid chains in the distorted 2D hexagonal lattice. The elucidation of the precise nature of these phases relied entirely on the ability to study hydrated DMPC samples that are extremely well oriented, in the case described, due to freely suspended film preparations.

We further described the properties of dilute lyotropic lamellar L_α phases, which exhibit a pronounced Landau–Peierls effect arising from the low bending rigidity of the membranes comparable with thermal energies. The X-ray structure factor was seen to exhibit power-law behaviour indicative of quasi-long-range positional order. We showed that the long-range repulsive interaction in the dilute stack of membranes originated from entropically driven Helfrich undulation forces and depended only on geometric and elastic parameters such as the layer spacing and the membrane bending rigidity. We note that undulation forces have also been demonstrated in biological membranes of DMPC bilayers when the bending rigidity is softened (Safinya et al. 1989). The softening of the bending rigidity from $\kappa \approx 20~k_B T$ to $\kappa \approx 1-2~k_B T$ was achieved by inclusion of shorter-chain pentanol cosurfactants, which thins the membrane. In this case, the undulation forces were found to overwhelm the van der Waals attraction, precipitating a transition from the classical regime for "bound membranes", which occurs for fluid-phase DMPC bilayers, to the "nearly unbound" regime for soft DMPC/pentanol membranes.

Finally, a message of this chapter is that X-ray measurements may be used to obtain information beyond the average structure of the system studied. In special cases where thermal diffuse scattering is especially enhanced, for example, for the L_α phase of dilute multilayers discussed in this chapter, thermodynamic properties such as the bulk moduli of compression (derived from interlayer forces) can be obtained reliably from the analysis of the shape of the structure factor.

ACKNOWLEDGEMENTS

We want to acknowledge our collaborators involved in the original studies described in this chapter, in particular, Didier Roux, Eric Sirota, Noel Clark, and Sunil Sinha.

We acknowledge funding by the US Department of Energy (DOE), Office of Basic Energy Sciences, Division of Materials Sciences and Engineering, under Award DE-FG02-06ER46314 (self- and directed assembly in biomimetic-charged biomolecular materials systems studied by synchrotron X-ray scattering), the US National Institutes of Health under Awards R01GM130769 (structures of biological lipids), and the US National Science Foundation (NSF) under Award DMR-1807327 (membrane phase behaviour). Synchrotron X-ray scattering studies were carried out at the Stanford Synchrotron Radiation Lightsource, a Directorate of SLAC National Accelerator Laboratory and an Office of Science User Facility operated for the US DOE Office of Science by Stanford University.

REFERENCES

Als-Nielsen, J., J. D. Litster, R. J. Birgeneau, M. Kaplan, C.R. Safinya, A. Lindegaard-Andersen, and S. Mathiesen, 1980. Observation of algebraic decay of positional order in a smectic liquid crystal. *Phys. Rev. B* 22:312–20.

Andelman, D., M. E. Cates, D. Roux, and S. A. Safran. 1987. Structure and phase equilibria of microemulsions. *J. Chem. Phys.* 87:7229–41.

Avanti Polar Lipids Website. Accessed Dec 14, 2020. https://avantilipids.com/.

Bassereau, P., J. Marignan, and G. Porte. 1987. An X-Ray study of brine swollen lyotropic lamellar phases. *J. Phys. (Paris)* 48:673–8.

Bear, R. S., K. J. Palmer, and F. O. Schmitt. 1941. X-ray diffraction studies of nerve lipides. *J. Cell.Comp. Physiol.* 17:355–67.

Bivas, I., P. Hanusse, P. Bothorel, J. Lalanne, and O. Aguerre-Chariol. 1987. An application of the optical microscopy to the determination of the curvature elastic modulus of biological and model membranes. *J. Phys. (Paris)* 48:855–67.

Born, M. and E. Wolf. 1965. *Principles of Optics*. Pergamon Press, Oxford.

Caillé, A. 1972. Remarques sur la Diffusion des Rayons X dans les Smectiques. *C. R. Seances Acad. Sci., Ser. B* 274:891–3.

Cates, M. E., D. Roux, D. Andelman, S. T. Milner, and S. A. Safran. 1988. Random surface model for the L_3-phase of dilute surfactant solutions. *Europhys. Lett.* 5:733–9.

Chaikin, P. M. and T. C. Lubensky. 1995. *Principles of Condensed Matter Physics*. Cambridge University Press, Cambridge.

Chen, S. H. 1986. Small angle neutron scattering of the stucture and interaction in micellar and microemulsion systems. *Annu. Rev. Phys. Chem.* 37:351–99.

De Gennes, P. G. and J. Prost. 1993. *The Physics of Liquid Crystals*. Clarendon Press, Oxford.

De Gennes, P. G. and C. Taupin. 1982. Microemulsions and the flexibility of oil/water interfaces. *J. Phys. Chem.* 86:2294–304.

Di Meglio, J.-M., M. Dvolaitzky, L. Leger, and C. Taupin. 1985a. First observation of the undulation mode in birefringent microemulsions by quasielastic light scattering. *Phys. Rev. Lett.* 54:1686–9.

Di Meglio, J. M., M. Dvolaitzky, and C. Taupin. 1985b. Determination of the rigidity constant of the amphiphilic film in "birefringent microemulsions"; the role of the cosurfactant. *J. Phys. Chem.* 89:871–4.

Gruner, S. M. 1981. Controlled humidity gas circulators. *Rev. Sci. Instrum.* 52:134–6.

Guinier, A. 1994. *X-Ray Diffraction in Crystals, Imperfect Crystals, and Amorphous Bodies*. Dover Publication, Inc., Mineola, NY

Helfrich, W. 1978. Steric interaction of fluid membranes in multilayer systems. *Z. Naturforsch. A* 33:305–15.

Helfrich, W. and R. M. Servuss. 1984. Undulations, steric interaction and cohesion of fluid membranes. *Nuovo Cimento D* 3:137–51.

Helfrich, W. Z. 1973. Elastic properties of lipid bilayers: Theory and possible experiments. *Z. Naturforsch., C: J. Biosci.* 28:693–703.

Israelachvili, J. N. 2011. *Intermolecular and Surface Forces*. Elsevier, Amsterdam.

Israelachvili, J. N., D. J. Mitchell, and B. W. Ninham. 1977. Theory of self-assembly of lipid bilayers and vesicles. *Biochim. Biophys. Acta, Biomembr.* 470:185–201.

Janiak, M. J., D. M. Small, and G. G. Shipley. 1976. Nature of the thermal pretransition of synthetic phospholipids: Dimyristolyl- and dipalmitoyllecithin. *Biochemistry* 15:4575–80.

Janiak, M. J., D. M. Small, and G. G. Shipley. 1979. Temperature and compositional dependence of the structure of hydrated dimyristoyl lecithin. *J. Biol. Chem.* 254.

Katsaras, J. 1995. Structure of the subgel ($L_{c'}$) and Gel ($L_{\beta'}$) phases of oriented dipalmitoylphosphatidylcholine multibilayers. *J. Phys. Chem.* 99:4141–7.

Katsaras, J. 1998. Adsorbed to a rigid substrate, dimyristoylphosphatidylcholine multibilayers attain full hydration in all mesophases. *Biophys. J.* 75:2157–62.

Landau, L. D. 1937a. Zur Theorie der Phasenumwandlungen I. *Phys. Z. Sowjetunion* 11:26–47.

Landau, L. D. 1937b. Zur Theorie der Phasenumwandlungen II. *Phys. Z. Sowjetunion* 11:545–55.

Landau, L. D. 1965. On the theory of phase transitions. In *Collected Papers of L.D. Landau*. D. ter Haar, editor. Elsevier, New York, London, Paris, pp. 193–216.

Larche, F. C., J. Appell, G. Porte, P. Bassereau, and J. Marignan. 1986. Extreme swelling of a lyotropic lamellar liquid crystal. *Phys. Rev. Lett.* 56:1700–3.

Luzzati, V. 1968. X-ray diffraction studies of lipid-water systems. In *Biological Membranes; Physical Fact and Function*. D. Chapman, editor. Academic Press, London and New York, pp. 71–123.

Mahanty, J. and B. W. Ninham. 1976. *Dispersion Forces*. Academic Press, London.

Merzbacher, E. 1970. *Quantum Mechanics*. Wiley, New York.

Miller, C. E., J. Majewski, E. B. Watkins, D. J. Mulder, T. Gog, and T. L. Kuhl. 2008. Probing the local order of single phospholipid membranes using grazing incidence X-ray diffraction. *Phys. Rev. Lett.* 100:058103.

Parsegian, V. A., N. Fuller, and R. P. Rand. 1979. Measured work of deformation and repulsion of lecithin bilayers. *Proc. Natl. Acad. Sci. U. S. A.* 76:2750–4.

Peierls, R. 1934. Remarks on transition temperatures. *Helv. Phys. Acta, Suppl.* 7:81–3.

Peierls, R. 1935. Quelques Propriétés Typiques des Corps Solides. *Ann. Inst. Henri Poincare* 5:177–222.

Porte, G., P. Bassereau, J. Marignan, and R. May. 1987. Stability of brine swollen lamellar phases. In *Physics of Amphiphilic Layers*. D. Langevin, J. Meunier, and N. Boccara, editors. Springer, Heidelberg.

Rand, R. P. 1981. Interacting phospholipid bilayers: Measured forces and induced structural changes. *Annu. Rev. Biophys. Bioeng.* 10:277–314.

Roux, D. and A. M. Bellocq. 1985. In *Physics of Amphiphiles: Micelles, Vesicles, and Microemulsions*. V. Degiorgio and M. Corti, editors. North Holland, Amsterdam.

Roux, D. and C. R. Safinya. 1988. A synchrotron X-Ray study of competing undulation and electrostatic interlayer interactions in fluid multimembrane lyotropic phases. *J. Phys. (Paris)* 49:307–18.

Russell, T. P. 1991. The characterization of polymer interfaces. *Annu. Rev. Mater. Sci.* 21:249–68.

Safinya, C. R. 1989. Rigid and fluctuating surfaces: A series of synchrotron X-ray scattering studies of interacting stacked membranes. In *Phase Transitions in Soft Condensed Matter*. T. Riste and D. Sherrington, editors. Springer, Boston, pp. 249–70.

Safinya, C. R. 2006. Biophysics and biomolecular materials. In *The New Physics: For the Twenty-First Century*. G. Fraser, editor. Cambridge University Press, Cambridge and New York.

Safinya, C. R., D. Roux, G. S. Smith, S.K. Sinha, P. Dimon, N.A. Clark, and A.M. Bellocq 1986. Steric interactions in a model multimembrane system: A synchrotron X-ray study. *Phys. Rev. Lett.* 57:2718–21.

Safinya, C. R., E. B. Sirota, D. Roux, and G. S. Smith. 1989. Universality in interacting membranes: The effect of cosurfactants on the interfacial rigidity. *Phys. Rev. Lett.* 62:1134–7.

Schneider, M. B., J. T. Jenkins, and W. W. Webb. 1984. Thermal fluctuations of large quasi-spherical bimolecular phospholipid vesicles. *J. Phys. (Paris)* 45:1457–72.

Singer, S. J. and G. L. Nicolson. 1972. The fluid mosaic model of the structure of cell membranes. *Science* 175:720–31.

Sirota, E. B., G. S. Smith, C. R. Safinya, R. J. Plano, and N. A. Clark. 1988. X-Ray scattering studies of aligned, stacked surfactant membranes. *Science* 242:1406–9.

Smith, G. S. and C. F. Majkrzak. 2006. Neutron Reflectometry. In *International Tables for Crystallography, Volume C: Mathematical, Physical and Chemical Tables*. E. Prince, editor. International Union of Crystallography, Chester, pp. 126–46.

Smith, G. S., C. R. Safinya, D. Roux, and N. A. Clark. 1987. X-ray study of freely suspended films of a multilamellar lipid system. *Mol. Cryst. Liq. Cryst.* 144:235–55.

Smith, G. S., E. B. Sirota, C. R. Safinya, and N. A. Clark. 1988. Structure of the $L_{\beta'}$ phases in a hydrated phosphatidylcholine multimembrane. *Phys. Rev. Lett.* 60:813–6.

Smith, G. S., E. B. Sirota, C. R. Safinya, R. J. Plano, and N. A. Clark. 1990. X-ray structural studies of freely suspended ordered hydrated DMPC multimembrane films. *J. Chem. Phys.* 92:4519–29.

Tardieu, A., V. Luzzati, and F. C. Reman. 1973. Structure and polymorphism of the hydrocarbon chains of lipids: A study of lecithin-water phases. *J. Mol. Biol.* 75:711–33.

Torbet, J. and M. H. F. Wilkins. 1976. X-Ray diffraction studies of lecithin bilayers. *J. Theor. Biol.* 62:447–58.

Tristram-Nagle, S., Y. Liu, J. Legleiter, and J. F. Nagle. 2002. Structure of gel phase DMPC determined by X-ray diffraction. *Biophys. J.* 83:3324–35.

Wack, D. C. and W. W. Webb. 1989. Synchrotron X-ray study of the modulated lamellar phase $P_{\beta'}$ in the lecithin-water system. *Phys. Rev. A* 40:2712–30.

Watkins, E. B., C. E. Miller, W. P. Liao, and T. L. Kuhl. 2014. Equilibrium or quenched: fundamental differences between lipid monolayers, supported bilayers, and membranes. *ACS Nano* 8:3181–91.

Zasadzinski, J., J. Schneir, J. Gurley, V. Elings, and P. Hansma. 1988. Scanning tunneling microscopy of freeze-fracture replicas of biomembranes. *Science* 239:1013–5.

3 Structures of Lipid Membranes: Cubic and Inverse Hexagonal Phases

Charlotte E. Conn
RMIT University, Melbourne, Australia

John M. Seddon
Imperial College London, London, UK

CONTENTS

3.1 Introduction 49
3.2 Self-Assembly and Lipid Packing Parameter 50
3.3 Interfacial Curvature and Generic Phase Diagram 50
3.4 Curvature Elastic Energy and Chain Packing Frustration 51
3.5 Inverse Hexagonal Phases 52
3.6 Bicontinuous Cubic Phases 53
3.7 Discontinuous Inverse Cubic Phases 53
3.8 Mesophase Engineering: Bicontinuous Cubic and Hexagonal Phases by Design 54
3.9 Cubosomes and Hexosomes 55
3.10 Effects of Hydrostatic Pressure 55
3.11 Bicontinuous Cubic Phases *In Vivo*: Nanostructured Cubic Membranes 57
3.12 Observation of Bicontinuous Cubic and Hexagonal Phases During Lipid Digestion 58
3.13 Applications of Cubic and Hexagonal Phases: Drug Delivery 59
3.14 Bicontinuous Cubic Phases for Structural Studies of Membrane Proteins 60
References 60

3.1 INTRODUCTION

In addition to forming crystalline, gel, and fluid lamellar phases, many phospholipids and glycolipids can spontaneously self-assemble into a wide range of ordered lyotropic liquid–crystalline phases based on *curved* interfaces. Such phases occur in two variants, either oil-in-water (type I), where the interfaces have net mean curvature towards the lipid hydrocarbon chain regions, or water-in-oil (inverse, type II), where the interfaces curve towards the water regions and away from the hydrocarbon chains (see Figure 3.1). Type I curved mesophases generally break up into disordered micellar solutions upon high dilution in water. Type II ('inverse') mesophases, on the other hand, are usually stable in the presence of an excess aqueous phase, and are therefore of more direct relevance to biology, and are more suitable for any applications where stable ordered self-assembled structures are required.

Examples of lipids with strong tendencies to form such 'non-lamellar' phases are shown in Figure 3.2. Typically, such lipids have relatively small, weakly hydrated head groups and/or bulky hydrocarbon chain regions.

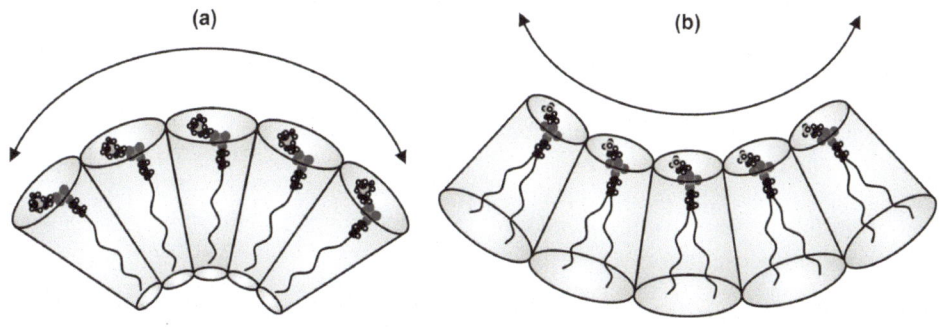

FIGURE 3.1 Type I lyotropic mesophases have *positive* interfacial mean curvature (a), whereas type II (inverse) phases have negative interfacial mean curvature (b). (Reproduced with permission from Shearman, Ces et al. 2006.)

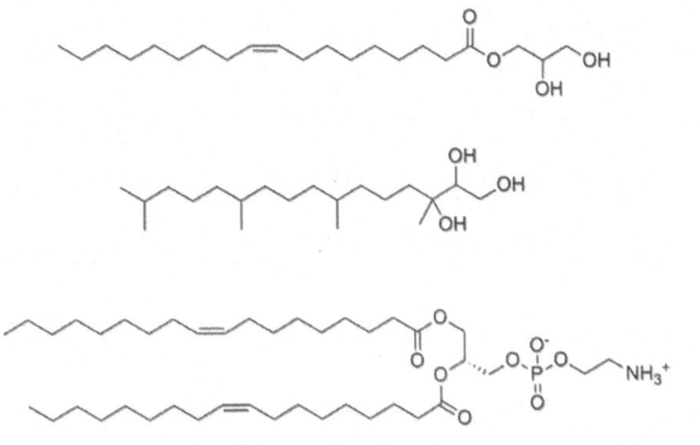

FIGURE 3.2 Example of nonlamellar lipids: (a) monoolein, (b) phytantriol, and (c) dioleoyl phosphatidylethanolamine (DOPE).

The structural ordering within nonlamellar phases can extend from the nanoscale (approximately 5 nm), up to much larger values (>50 nm), although locally the lipid molecules are liquid-like and undergo rapid rotation and lateral diffusion in the plane of the interface. Such lyotropic phases, in addition to their great relevance to biological membrane structure, dynamics, and function, also have great potential in applications such as nanoencapsulation, nanoreactors, drug delivery, gene transfer, and delivery of si-RNA (small-interfering RNA) to cells.

3.2 SELF-ASSEMBLY AND LIPID PACKING PARAMETER

Amphiphilic molecules such as lipids strongly tend to aggregate in water due to the hydrophobic effect, which drives them to shield their hydrocarbon chain regions away from contact with the aqueous regions, while simultaneously solvating their polar head groups. This leads to the formation of interfaces between the hydrophobic regions and the hydrophilic regions, with the lipid polar head groups sitting close to the interface. At low concentrations below the critical micelle concentration (CMC), amphiphiles exist as monomers in solution. The amphiphiles will tend to arrange themselves along the surface of the water with their head groups solvated and their tails pointing away from the surface into the air. At the CMC, the entropic penalty of assembling the amphiphilic molecules becomes less than that associated with solvating the hydrocarbon chains, and the hydrophobic chain regions aggregate together with the exclusion of water, forming micelles. Micelles are often spherical but may adopt a variety of shapes including rod- or disc-like assemblies having typical aggregation numbers of between 30 and 200 monomers. Diffusion of molecules around the micelle is a rapid process, given the typical values of amphiphile self-diffusion constants of the order of $10^{-10} m^2 s^{-1}$. Micelles may be divided into two types: type I (normal) micelles are those where the interfacial curvature is towards the chain region, and type II (inverse), where the curvature is towards the polar head groups and the surrounding water. The Krafft boundary is the temperature below which the amphiphile will separate out from solution as a crystal phase. The Krafft boundary varies with concentration for a particular system and at the CMC is known as the Krafft point.

The addition of more surfactant will result in an increase in micelle concentration with often a concomitant increase in micellar size. Eventually interactions between neighbouring micelles become significant, and aggregation of micelles into translationally ordered mesophases occurs. Such mesophases are often liquid crystalline in that they possess a degree of orientational and/or translational order of the interfaces, but only short-range positional order of the amphiphilic molecules.

The preferred structure of the aggregates – and hence of the lyotropic phase formed – is determined by the packing geometry of the polar and nonpolar regions. A highly simplified yet useful way to think about this is to associate a given lipid molecule, under specific thermodynamic conditions (temperature, pressure, hydration…), with a preferred molecular 'shape'. We can define a packing parameter P (Israelachvili 2011) by

$$P = \frac{v}{a_0 \ell_c}$$

where v is the volume of the hydrocarbon chain(s), a_o is the optimum head group area, and ℓ_c is the maximum length of the hydrocarbon chain(s). When $P<1$, type I phases are favoured, with a net positive interfacial mean curvature towards the hydrocarbon chain region. When $P=1$, the cross-sectional packing of the head groups is the same as that of the chains, and lamellar phases are preferred. For systems with relatively small, weakly hydrated head groups, $P>1$, and inverse phases are preferred. For such inverse systems, the packing parameter P is frequently quite strongly dependent on temperature, since the conformational disorder of the chains, and hence the net chain splay, increases with heating. Most lipid systems are lamellar (gel and/or fluid) at low temperature and form more highly curved phases such as the inverse bicontinuous cubic and inverse hexagonal phases as the temperature is increased.

3.3 INTERFACIAL CURVATURE AND GENERIC PHASE DIAGRAM

Two principal curvatures, c_1 and c_2, can be located at any point P on a surface by extending a normal to the surface as shown in Figure 3.3. c_1 and c_1, which are at right angles to each other, are defined as the maximum and minimum values of the curvature, respectively. The mean and Gaussian curvatures at any point on the surface are defined in terms of these two principal curvatures:

$$H = \frac{1}{2}(c_1 + c_2)$$

$$K = c_1 c_2$$

Structures of Lipid Membranes

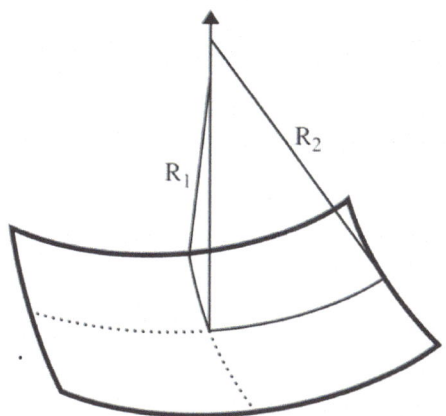

FIGURE 3.3 The principal curvatures at a point on a surface are given by $c_1 = 1/R_1$ and $c_2 = 1/R_2$. (Reproduced with permission from Shearman, Ces et al. 2006.)

The sign of the monolayer mean curvature is generally taken to be positive for bending towards the hydrocarbon chain region, although some authors adopt the opposite convention. Using the former convention, the mean curvature of the inverse bicontinuous cubic and inverse hexagonal phases is therefore negative.

Inverse lyotropic mesophases based on curved interfaces may be divided into three structural classes, depending on the sign of the Gaussian curvature K of the interface:

a. Packings of cylinders ($K=0$): inverse hexagonal H_{II} phase
b. Packings of saddle surfaces ($K<0$): inverse bicontinuous cubic phases
c. Packings of spheres/ellipsoids ($K>0$): inverse ordered micellar phases

A generic lyotropic phase diagram may be obtained by arranging the mesophases according to their average interfacial mean curvature, H (Figure 3.4). The centre of this phase diagram is occupied by the fluid lamellar phase, which has zero mean and Gaussian curvature. The lateral pressure, $\pi(z)$ varies across the lipid bilayer giving rise to a typical lateral pressure profile (Figure 3.5), which is positive in the head group and chain regions and strongly negative at the polar–apolar interface. This pressure can be

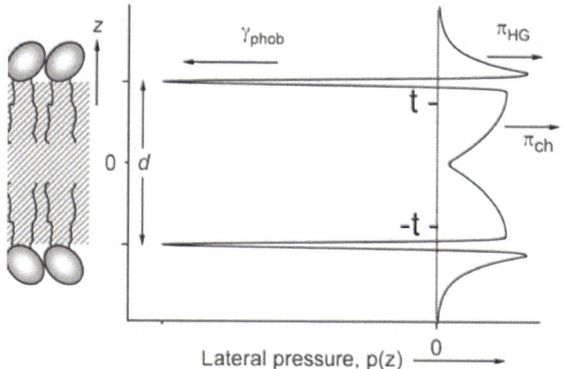

FIGURE 3.5 Typical lateral pressure profile across a fluid lipid bilayer. (Adapted from Marsh 2007 and reproduced with permission.)

high enough (>500 atm) to cause conformational changes in proteins embedded within the bilayer. Recent studies have shown that branched isoprenoid-type hydrocarbon chains can have a lateral pressure high enough to prevent proteins and polymers from being encapsulated within the bilayer (Meikle, Zabara et al. 2017).

3.4 CURVATURE ELASTIC ENERGY AND CHAIN PACKING FRUSTRATION

The total free energy of any amphiphile–water system (g_{tot}) is the sum of three different energy terms: the curvature elastic energy (g_C), which is the energy required to change the curvature of the membrane, the packing frustration (g_P), an energy term relating to the hydrocarbon chain packing, and a third energy term associated with various interaction forces including hydration and electrostatic forces (g_{inter}).

$$g_{tot} = g_C + g_P + g_{inter}$$

As g_{inter} is generally assumed to be negligible, this section will focus on curvature elastic energy and packing frustration.

The curvature elastic energy of a membrane is the energy required to bend the membrane away from its preferred curvature. The curvature elastic energy of a lipid bilayer (per unit area) is given by the Helfrich ansatz:

$$g_c = 2\kappa(H - H_0)^2 + \kappa_G K$$

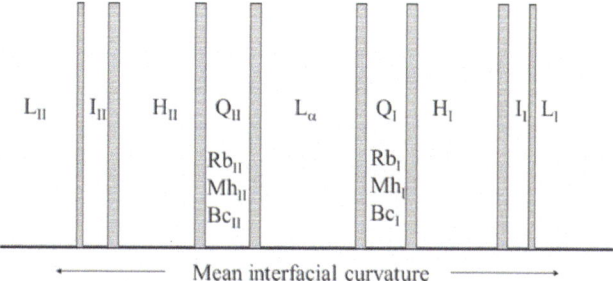

FIGURE 3.4 A schematic lyotropic 'phase diagram', showing the natural sequence of phases in terms of the average interfacial mean curvature. (Adapted from Seddon and Templer 1993 and reproduced with permission.)

where H_0 is the spontaneous mean curvature, i.e., the curvature of a fully relaxed membrane and H and K are the mean and Gaussian curvatures, respectively. κ, the mean curvature (bending) modulus, and κ_G, the Gaussian curvature modulus, describe the energy costs per unit area of changing the mean and Gaussian curvature. κ is always positive and has a monolayer value in the region of 4×10^{-20} J for most common lipids (although single chained lipids have a value approximately five times smaller). The value of κ for a bilayer is twice that of the corresponding lipid monolayer ($\kappa^b = 2\kappa^m$) as the energy required to bend two membranes is twice that for a single membrane. For a monolayer, a negative value of κ_G, with a magnitude somewhat less than that of κ (typically $-0.8\ \kappa^m$), has been predicted for systems with inverse curvature. The relationship between κ_G^b for a bilayer and that for a monolayer κ_G^m is more complex than that for κ and is given by

$$\kappa_G^b = 2\kappa_G^m - 4t\kappa^m H_o^m$$

where t is the distance from the end of the hydrocarbon chains to the neutral surface of the monolayer, i.e., the surface where the area does not change upon bending. This surface generally lies just below the polar–apolar interface.

An unusual outcome of this relationship is that κ_G^b can be positive, even though κ_G is typically negative for a monolayer. For κ_G^b to be positive, the value of $4t\kappa^m H_0$ must be sufficiently negative. As κ^m and t are both positive, then this only occurs when H_0 is sufficiently negative, which is the case for amphiphiles with small weakly hydrated head groups, with a larger lateral pressure in the hydrocarbon chain region than in the head group region. The Gaussian curvature contribution to the elastic energy is only relevant when the membrane undergoes a topological change (this is a consequence of the Gauss–Bonnet equation).

The curvature elastic energy across the whole lipid membrane (G_c) is given by the integral of g_c across the membrane surface:

$$G_c = \int g_c\, dA$$

The spontaneous mean curvature H_0 is proportional to the first moment of the lateral stress profile across the bilayer:

$$\kappa^b H_o = \int \pi(z) z\, dz$$

For a symmetric bilayer, where the lipid composition is the same in the inner and outer leaflet, the first moment of the lateral stress profile must be zero (by symmetry) and H_0 must therefore be zero (as κ^b is positive). For an asymmetric bilayer, however, as is the case for most cell membranes, the lateral pressure profile is not constrained to be zero, and H_0 will in general be nonzero.

The second moment of the lateral pressure profile across a monolayer gives the monolayer Gaussian curvature modulus (Helfrich, 1981):

$$\kappa_G^m = -\int_{-t}^{d_m-t} \pi(z) z^2\, dz$$

where t is the distance from the end of the hydrocarbon chains to the neutral surface of the monolayer, i.e., the surface where the area does not change upon bending. This surface generally lies just below the polar–apolar interface.

The packing frustration for a lipid mesophase may be simply understood by using the concept of 'packing fraction' which calculates the volume occupied by the elements within a system, compared with that of potential 'void space' within the structure.

3.5 INVERSE HEXAGONAL PHASES

The structure and stability of the inverse hexagonal phase has been extensively reviewed (see, for example, Seddon, 1990) and will not be repeated here. The packing fraction for the inverse hexagonal phase, due to the inability of close-packed circular cylinders to fill space, has a value of 0.91. This means that 9% of the volume consists of potential voids in the structure, as seen in Figure 3.6. This void volume must be filled, and this can only occur via stretching of the hydrocarbon chains away from their preferred conformation, or by the interface deforming away from circular in cross section. There is thus a significant degree of packing frustration within this phase, whereby both a uniform chain length and a uniform interfacial mean curvature are not geometrically possible, and a compromise solution ensues instead.

For cylindrical structures such as in the inverse hexagonal H_{II} phase, only a limited swelling can occur in water, due to the severe packing constraints in the hydrocarbon chain region. To a limited extent, swelling can be enhanced by the addition of nonpolar solutes such as alkanes, which can relax

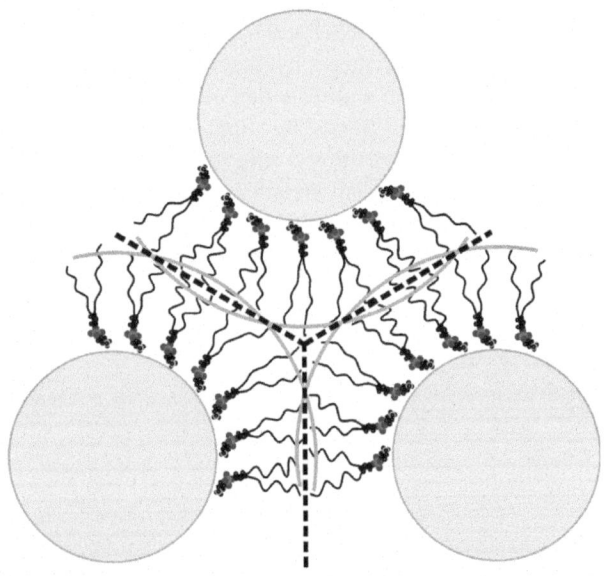

FIGURE 3.6 Packing frustration in the inverse hexagonal phase. (Reproduced with permission from Shearman, Ces et al. 2006.)

Structures of Lipid Membranes

such constraints. However, there is a risk that such solutes will destroy the ordered lattice, inducing a transition to an inverse micellar solution (possibly of inverse cylinders).

3.6 BICONTINUOUS CUBIC PHASES

For inverse bicontinuous phases (which consist of a lipid bilayer draped over an infinitely periodic minimal surface of zero mean curvature), the packing constraints are quite different. Three types of inverse bicontinuous phases, denoted Q_{II}^G, Q_{II}^D, and Q_{II}^P phases, have been observed in lipid–water systems (Figure 3.7). These are based on the Schoen gyroid (G), and the Schwarz diamond (D) and primitive (P) minimal surfaces, and have crystallographic space groups Ia3d, Pn3m, and Im3m, respectively. In all of these, the lipid bilayer acts as a membrane, subdividing space into two interpenetrating water networks. Inverse bicontinuous phases have only been observed for three of the multiple minimal surfaces which exist. This has been explained by a consideration of the Gaussian inhomogeneity which must exist over such a surface. It has been suggested that Gaussian inhomogeneity $<G^2>/<G>^2$ is minimized for the gyroid, diamond, and primitive surfaces, with a corresponding decrease in packing frustration for cubic phases based on them (Schwarz and Gompper 2000, Barriga, Tyler et al. 2015, Tyler, Barriga et al. 2015).

In principle, such phases could swell in water, without incurring a large chain packing cost. Calculations done some time ago (Bruinsma, 1992) suggested that thermal fluctuations should destroy such phases for lattice parameters larger than 30 nm. However, it now appears that this prediction is overconservative and that much larger unit cell sizes can be attained. There are a number of ways by which bicontinuous phases could be swollen, for example, by tuning the interfacial curvature with amphiphilic solutes (such as 'lamellar-forming' lipids), or by adding charged lipids to cause electrostatic swelling. This latter approach has been recently used to swell cubic phases to lattice parameters approaching 500 Å (Barriga, Tyler et al. 2015, Tyler, Barriga, et al. 2015). The swelling should in principle be controllable by varying the pH and/or the salt concentration, thereby varying the surface charge density and/or the Debye length.

The inverse bicontinuous cubic phases typically display the phase sequence $Q_{II}^G - Q_{II}^D - Q_{II}^P$ with increasing water content, which is consistent with a decreasing order of compactness of the associated minimal surfaces (assuming constant curvature). In other words, at the same curvature, the water required to fill the cubic phases increases in the order $G > D > P$. This order appears to be universally adopted in most phase diagrams, although it is rare for all three cubic phase to be adopted for a single lipid–water system.

A counterintuitive finding is that for a number of amphiphiles, for example, the monoacylglycerols, isothermal transitions from a flat fluid lamellar L_α phase, to inverse bicontinuous cubic phases, can be induced by increasing the water content. A theoretical attempt to explain this intriguing behaviour has attributed it to a competition between water–head group and head group–head group hydrogen bonding (Lee, 2008).

An interesting question is whether *asymmetric* curved lyotropic phases can be produced (having different lipid compositions in the two sides of the bilayer), and if so, what novel properties such structures would have. For example, the gyroid inverse bicontinuous cubic phase of space group Ia3d consists of two interwoven water networks of opposite chirality, separated by a continuous fluid bilayer. If the bilayer could be engineered to be asymmetric, then the two water channels would no longer be equivalent in diameter, and such a material might form the basis of a 'chiral mesoporous sieve', which might discriminate between enantiomeric solutes.

3.7 DISCONTINUOUS INVERSE CUBIC PHASES

For systems where the preferred interfacial mean curvature is extremely negative, ordered phases based upon 3-D packings of quasi-spherical inverse aggregates may be observed.

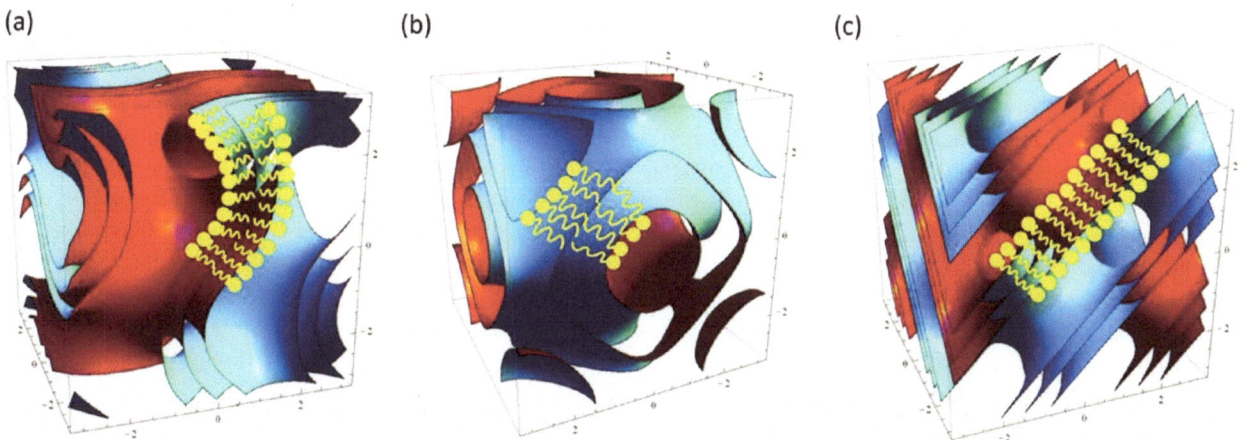

FIGURE 3.7 A schematic representation of the (a) Q_{II}^G (b) Q_{II}^P and (c) Q_{II}^D bicontinuous cubic phases. (Reproduced with permission from Conn, Darmanin et al. 2010.)

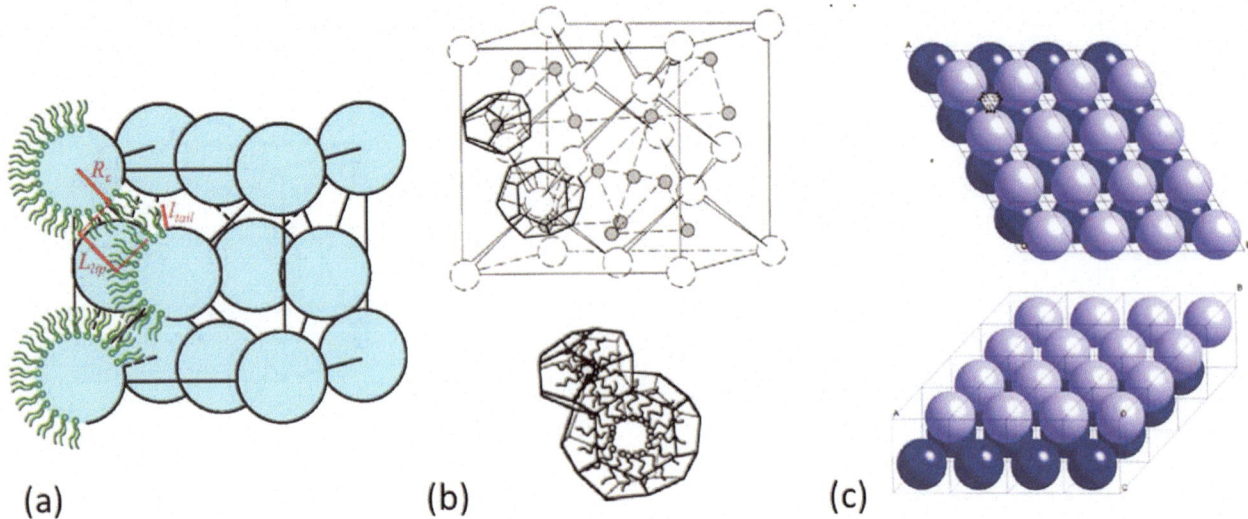

FIGURE 3.8 Ordered inverse micellar phases. (a) Cubic Fm3m (from Martiel, Sagalowicz et al., 2013); (b) Cubic Fd3m (from Seddon, Robins et al. 2000); (c) 3-D hexagonal P6$_3$/mmc (from Shearman, Tyler et al. 2009).

Three distinct phases have so far been reported (Figure 3.8): an Fd3m cubic phase (Seddon 1990, Luzzati, Vargas et al. 1992), a 3-D hexagonal (*hcp*) phase of space group P6$_3$/mmc (Shearman, Tyler et al. 2009), and an Fm3m (*fcc*) cubic phase (Martiel, Sagalowicz et al. 2013). The lattice parameters are in the region of 10–15 nm but can rise to over 30 nm for the Fm3m phase. The packing fractions for close-packed hard spheres in these three space groups are 0.71, 0.74, and 0.74, respectively. This means that the degree of chain packing frustration in all three of these phases is very large: more than 25% of the volume of the unit cell is hydrophobic potential 'void' volume, which has to be filled either by hydrocarbon chains stretching away from their preferred conformational state, or the interface becoming facetted, or by solute molecules relieving the packing frustration by partitioning into the hydrophobic regions. The Fd3m packing minimizes the degree of chain stretching that is required (Duesing, Templer et al. 1997, Rappolt, Cacho-Nerin et al. 2013), but at the expense of requiring the lipids to self-assemble into two types of quasi-spherical inverse micelle: 8 larger ones and 16 smaller ones (Seddon, Robins et al. 2000). This process is facilitated by using a second weakly amphiphilic molecule such as a long-chain fatty acid, fatty alcohol, or diacylglycerol to tune the preferred curvature to strongly negative values: this can in principle allow the two types of inverse micelles to have different lipid compositions, with the more polar lipid preferentially located in the larger aggregates, and the less polar one preferentially in the smaller, more strongly curved ones. However, this is not an absolute requirement: the Fd3m phase has been observed in two purely binary lipid/water systems (Seddon, Zeb et al. 1996, Minamikawa and Hato 1998), although both examples have been glycolipids rather than phospholipids. Certain lipid systems forming the Fd3m phase can be transformed to the P6$_3$/mmc phase by addition of cholesterol (Shearman, Tyler et al. 2009). It appears that the cholesterol is able to relieve the chain packing frustration because it is only weakly bound to the interface and can spend a certain amount of time partitioned into the hydrophobic region between the inverse micelles. A similar role is played by organic solvent molecules such as isooctane, cyclohexane, or limonene in stabilizing the Fm3m cubic phase (Martiel, Sagalowicz et al. 2013). In these cases, there is only a relatively narrow compositional window of stability: if too much organic solvent is added, the inverse micelles disorder into an inverse micellar (L_2) solution, with no long-range order.

3.8 MESOPHASE ENGINEERING: BICONTINUOUS CUBIC AND HEXAGONAL PHASES BY DESIGN

Simplistically, to engineer inverse phases such as the bicontinuous cubic and hexagonal phases, a characteristic 'wedge' shape of the amphiphilic molecule is required, i.e., that the effective hydrocarbon chain volume is much larger than the hydrophilic head group area. This may be simply achieved via the introduction of unsaturation or branching into the hydrocarbon chain. This rule has been applied successfully in the generation of bicontinuous cubic and hexagonal mesophases for a number of amphiphile systems including monoethanolamide (Sagnella, Conn et al. 2010, 2011, Sagnella, Gong et al. 2011), glycolipid (Zahid, Conn et al. 2013), ethylene oxide (Fong, Weerawardena et al. 2010, 2011), and urea-based (Fong, Wells et al. 2007) amphiphiles.

In a recent review, Fong et al. elucidated a set of design principles *a priori* to reproducibly engineer inverse phases of amphiphile self-assembly materials, including bicontinuous cubic and hexagonal mesophases (Fong, Le et al. 2012).

These selection criteria were obtained from a combination of experimental observations and QSPR-MLREM (quantitative structure–property relations multiple linear regression with expectation maximization) modelling. Specifically, the hydrocarbon chain region (which they term the hydrophobe) must contain either unsaturation (at least one cis unsaturated bond in a carbon chain of at least 14 carbons) or branching (a hydrocarbon chain of at least 12 carbons of which at least three are branched). For unsaturated chains, the greatest chain splay occurs when the cis double bond is positioned close to the middle of the chain. The hydrophobe must be sufficiently long (at least 200 amu), and the temperature must be above the chain melting temperature for cubic and hexagonal phases to form. While parameters relating to the hydrophobe were of significantly more importance than head group parameters in determining the ability of a particular amphiphile to form inverse phases, they determined three additional selection rules relating to the head group region. Firstly, the head group size must be small relative to the effective hydrocarbon chain volume. It must contain a minimum of three hydrophilic functional groups, e.g., hydroxyl groups. Thus phytantriol, which contains three hydroxyl groups in the head group, adopts a Q_{II}^G (gyroid) bicontinuous cubic phase, a Q_{II}^D (diamond) bicontinuous cubic phase, and a hexagonal (H_{II}) phase over a reasonably broad temperature range (20°C–50°C) (Barauskas and Landh 2003). In contrast, similar mono- and di-alcohols do not adopt liquid crystalline phases. Hydrogen bonding also plays a role, particularly for amphiphiles with amide- and urea-based head groups. For self-assembly to occur, the head group must be sufficiently hydrophilic for head group–head group interactions to be disrupted and head group–water H-bond networks to form.

The use of these rules in designing fit-for-purpose liquid crystalline materials has been facilitated by recent developments in the use of high-throughput formulation and structural characterization of lyotropic liquid crystalline materials (Mulet, Conn et al. 2013), allowing systematic testing of the effect of molecular architecture on phase behaviour (Feast, Hutt et al. 2014, Feast, Lepitre et al. 2014).

3.9 CUBOSOMES AND HEXOSOMES

Cubosomes and hexosomes (Figure 3.9) are nanostructured lyotropic liquid crystalline colloidal particles of cubic or hexagonal symmetry, respectively (Yaghmur and Glatter 2009). These particles, typically 200–300 nm in diameter, can in theory be prepared from any lipid that forms the cubic or hexagonal phases. The bulk lipidic material is dispersed into submicron particles via sonication or high-pressure homogenization (Barauskas, Johnsson et al. 2005). If the payload is fragile, then an alternative production method must be sought. Recently, a low energy route to produce cubosomes via the addition of phosphate-buffered saline (PBS) to cationic liposomes has been developed (Muir, Zhen et al. 2012). Cubosomes have also been produced by direct addition of a soluble protein, immunoglobulin, to a bulk cubic phase (Angelov, Angelov et al. 2005). A steric stabilizer is typically used to prevent aggregation of these particles. Traditionally, the triblock polyethylene oxide–polypropylene oxide–polyethylene oxide copolymer F127 was used to stabilize cubosomes over prolonged time periods, although Chong et al. have recently used high-throughput methodologies to determine the efficacy of a wide range of block copolymers in providing steric stabilization to cubosome dispersions (Chong, Mulet et al. 2011).

Cubosome and hexosome dispersions are typically characterized using small-angle X-ray scattering (SAXS), cryo-transmission electron microscopy (cryo-TEM), and dynamic light scattering (DLS) techniques (Yaghmur and Glatter 2009). SAXS provides information on the internal nanostructure of the particles, while DLS gives information on the particle size and polydispersity. Cryo-TEM allows for direct visualization of the particles. However, there are some challenges in phase identification using cryo-TEM; depending on the orientation of the particle, cubic phases can display either a cubic or hexagonal symmetry (Sagalowicz, Michel et al. 2006), while hexagonal phases display either hexagonal symmetry or a characteristic fingerprint pattern (Gustafsson, Ljusberg-Wahren et al. 1997). A combination of all three techniques is therefore typically used.

To date, the predominant use of cubosomes and hexosomes has been in the field of nanomedicine, including drug delivery and medical imaging (Mulet, Boyd et al. 2013). However, they have also demonstrated use in the food industry, including the delivery of nutraceuticals and food additives such as flavouring agents.

Recently, Janus nanoparticles, which contain two different lipidic mesophases within a single nanoparticle, have been discovered for the lipid system monoolein–capric acid, stabilized with Pluronic F127 (Figure 3.10; Tran, Mulet et al. 2015). These multicomponent lipid nanoparticles contain coexisting cubic and hexagonal phases, or two different cubic phases coexisting within the same particle. The introduction of the Janus bi-feature opens up additional opportunities for lipid nanoparticles including encapsulating drugs of varying size and release profile.

3.10 EFFECTS OF HYDROSTATIC PRESSURE

The effect of hydrostatic pressure on lipid mesophases is of fundamental biological importance in understanding the cell membrane behaviour of deep-sea organisms, which can exist at pressures of up to 1 kbar (Bartlett 2002). It also has considerable biotechnological relevance. However, experimental difficulties in working with elevated pressures mean that hydrostatic pressure has been less extensively studied than other variables such as temperature or composition. A major issue is the design of a suitable sample cell, capable of holding high hydrostatic pressure but also capable of transmitting X-rays. Recently, a high pressure cell, capable of

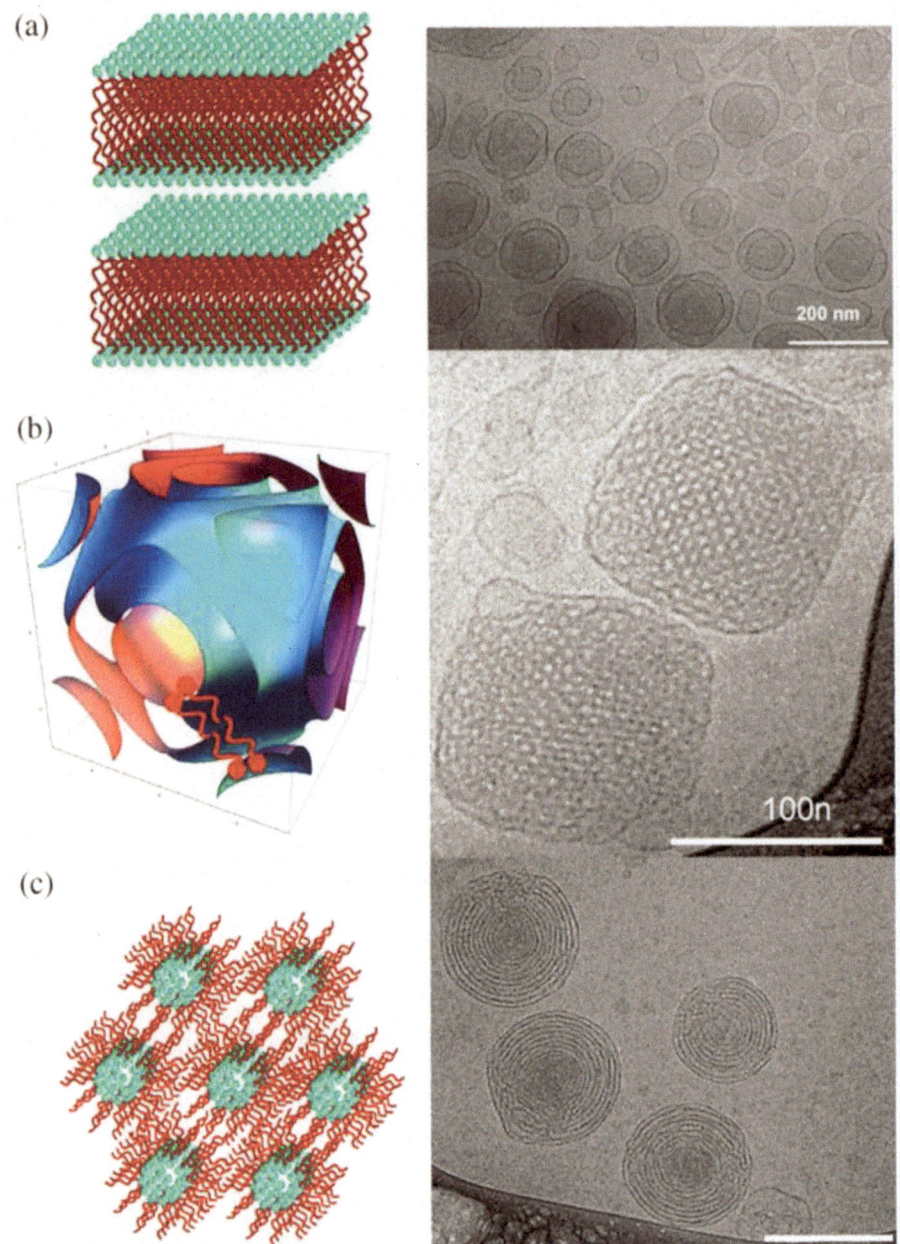

FIGURE 3.9 The bulk lamellar, bicontinuous cubic and hexagonal phases and their corresponding nanoparticles: (a) vesicles, (b) cubosomes, and (c) hexosomes, respectively. (Reproduced with permission from Fong, Le et al. 2011.)

pressure jump experiments in the millisecond regime, has been designed specifically for use at beamline I22, Diamond Light Source UK (Brooks, Gauthe et al. 2010). In addition, by coupling a fluorescent microscope to a high hydrostatic pressure system, McCarthy et al. have imaged small, highly dynamic microdomains in lipid membranes immediately after their formation (McCarthy, Ces et al. 2015).

The effect of increasing hydrostatic pressure in some systems tends to increase the lattice parameter of the bicontinuous cubic and hexagonal phases in excess water, which is the opposite effect to that of temperature (Seddon, Squires et al. 2006). The effect is due to the anisotropic compressibility of the lipid bilayer/monolayer. Upon increasing hydrostatic pressure, an increase in order within the hydrocarbon chain region is observed. The effect of pressure on the head group region is less than on the chain region with the result of increasing the desire for curvature towards the chain region. The concomitant reduction in (negative) curvature leads to an increase in lattice parameter. Extensive swelling of the hexagonal phase is prevented due to unfavourable chain packing frustration (Figure 3.11). In contrast, pressure-induced swelling of the bicontinuous cubic phases can be as high as 80 Å/kbar (Duesing, Seddon et al. 1997) provided enough water is available to fill the large aqueous pores in the swollen cubic phases.

Structures of Lipid Membranes

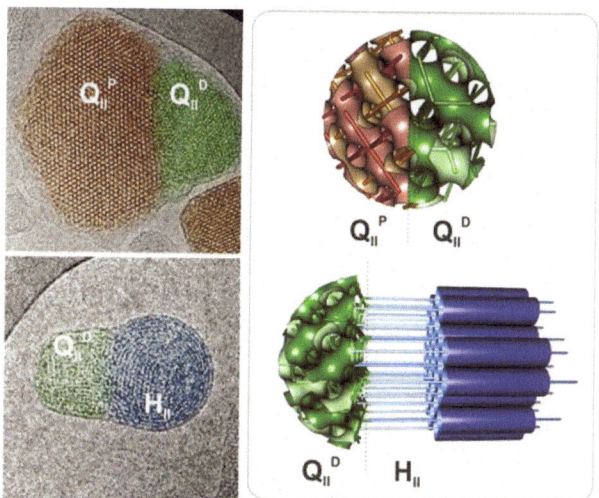

FIGURE 3.10 Cryo-TEM images (left) and models (right) of mixed phase nanoparticles. (top) Q_{II}^P/Q_{II}^D mixed cubic phase nanoparticle and (bottom) Q_{II}^D/H_{II} mixed cubic/hexagonal phase nanoparticle. TEM, transmission electron microscopy. (Reproduced with permission from Tran, Mulet et al. 2015.)

The pressure dependence of lipid phase transition temperatures T_t has been found to lie between 20 and 30°C/kbar for several phase transitions, including for transitions from the fluid lamellar to either the bicontinuous cubic or hexagonal phases (Duesing, Seddon et al. 1997).

A rapid change in pressure (a pressure-jump) is a useful means of studying transitions between different lipid mesophases, including the bicontinuous cubic and hexagonal mesophases. The use of pressure offers numerous advantages over the more commonly used temperature jumps (Seddon, Squires et al. 2006). Firstly, equilibrium is achieved quickly due to the rapid rate at which pressure propagates. In addition, pressure jumps can be in either the pressurization or depressurization direction, and the change in pressure does not have a significant effect on the solvent. The pressure jump technique has been used to monitor transitions between different bicontinuous cubic phases (Conn, Ces et al. 2008), between the cubic and hexagonal phases, and from the lamellar phase to a bicontinuous cubic or hexagonal phase (Conn, Ces et al. 2006). The rate of transition, and any structural intermediates formed during the transition, may be monitored as a function of the pressure jump amplitude. For the lamellar to cubic transition, and for intercubic transitions, the rate of transition is strongly dependent on the pressure jump amplitude, specifically the difference between the final pressure and the phase boundary pressure.

3.11 BICONTINUOUS CUBIC PHASES *IN VIVO*: NANOSTRUCTURED CUBIC MEMBRANES

The traditional view of the cell membrane as a flat phospholipid bilayer subdividing the cell into multiple organelles has been subverted by multiple studies confirming the existence of nonlamellar arrangements, such as cubic and hexagonal symmetries, in cellular membranes (Almsherqi, Kohlwein et al. 2006). Similarly to the bicontinuous cubic phase, cubic membranes consist of a folded bilayer arranged on a surface mathematically defined as a triply periodic minimal surface. They have been observed for numerous cell membrane assemblies and are particularly associated with viral infection, cellular stress, starvation, and deregulated protein synthesis (Almsherqi, Landh et al. 2009). A periodic lipid bilayer structure has also been observed at the air/water interface of the mammalian lung in rabbits (Figure 3.12; Larsson and Larsson 2014). Virus-induced cubic membranes have been observed in the presence of numerous viruses including the herpes simplex virus in human cerebral tissue (Baringer and Swovelan 1972), the HIV virus in human Kaposi's

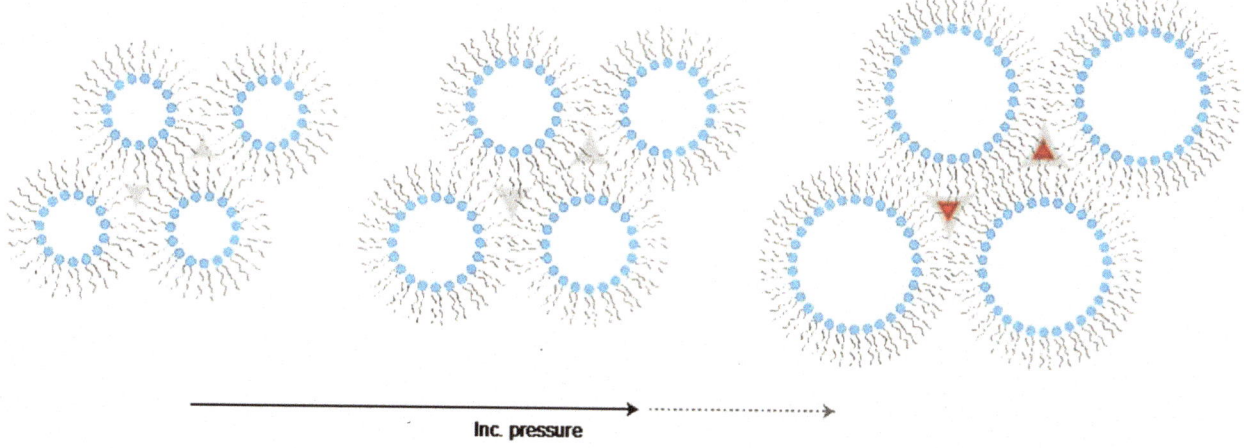

FIGURE 3.11 Pressure-induced swelling in an inverse hexagonal H_{II} lipid phase. Chain packing frustration increases as the diameter of the cylindrical inverse micelles increases. This can be accommodated up to a certain point, but voids (shown as dark red areas in the right hand structure) cannot be formed, thereby limiting the extent of the pressure-induced swelling. (Reproduced with permission from Brooks, Ces et al. 2011.)

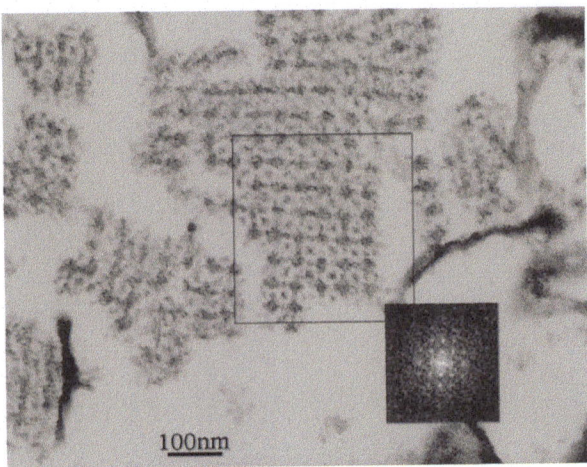

FIGURE 3.12 A sample of 'tubular myelin' (TM) from rabbit lung observed by transmission electron microscopy. The tetragonal symmetry of TM is illustrated by the Fourier transform of the indicated region. (Reproduced with permission from Larsson and Larsson 2014.)

sarcoma tissues (Marquart 2005), and poliovirus in monkey kidney (CMK) cells (Hashimoto, Hagiwara et al. 1984). Electron tomography studies have confirmed that Dengue virus–infected cells contain a nanostructured cubic membrane (Welsch, Miller et al. 2009). It has been proposed that the structural rearrangement to the membrane is induced via changes to the lipid composition, specifically focussed on sterol biosynthesis, induced by the virus itself (Deng, Almsherqi et al. 2010). As the cubic phases must be associated with a symmetric membrane, it has also been suggested that the cubic membrane occurs when energy production is too low to maintain the normal membrane asymmetry in lipid composition between the outer and inner membrane leaflets (Larsson and Larsson 2014).

Research in this area has been hampered by experimental limitations in identifying cubic and hexagonal symmetries *in vivo*. Techniques such as SAXS and NMR, which are commonly used to identify these structures in pure lipid systems, are problematic when studying whole cells, due to significant experimental issues with resolution and background noise. To date, the main technique used to characterize these structures has been TEM. TEM images from thin sections (70–90 nm) may be matched to computer-generated projections of the relevant cubic symmetry via a technique known as direct template matching (Almsherqi, Kohlwein et al. 2006). However, this technique can be challenging and may lead to misinterpretation of the data. EM (electron microscopy) tomography has proved successful in definitively assigning cubic membranes, for example, with the mitochondrial membrane of *Cytaea carolinensis* under starvation conditions (Deng, Marko et al. 1999). By utilizing multiple tilted angles, the technique can be used to assess much thicker samples (up to 400 nm) than can be characterized using EM. The large number of projections generated must be computationally reconstructed into a 3-D reconstruction of the object.

3.12 OBSERVATION OF BICONTINUOUS CUBIC AND HEXAGONAL PHASES DURING LIPID DIGESTION

To aid absorption though the epithelia, lipid (triglyceride) molecules must be hydrolyzed by lipases including gastric lipase and pancreatic lipase (Salentinig, Sagalowicz et al. 2011). Triglycerides may be hydrolyzed at the sn1 or sn3 position resulting in a diglyceride and corresponding fatty acid. It is believed that the fatty acid produced in this way acts to emulsify fat molecules in the stomach and thus aid digestion. The resulting emulsion droplets are partially stabilized by bile salts (BSs), which are found in the gall bladder and the intestine. The formation of fatty acid also modifies the curvature of the oil/water interface, promoting the formation of nanostructured phases. While vesicles have been found to be the predominant structures adopted during the digestion process (Salentinig, Sagalowicz et al. 2011), more complex nanostructures including the inverse bicontinuous cubic and hexagonal phases have also been observed (Salentinig, Phan et al. 2013, 2015).

Digestion of lipids by lipases was originally studied using microscopy techniques (Patton and Carey 1979). Characteristic textures observed during lipid digestion suggested the presence of self-assembled nanostructures such as the bicontinuous cubic and hexagonal phases. However, a definitive structure assignment was not possible. SAXS allows for the unambiguous identification of liquid crystalline structures during lipid digestion. Both lab-based SAXS (Salentinig, Sagalowicz et al. 2011) and synchrotron SAXS (Warren, Anby et al. 2011) have been used to study the evolving lipid nanostructure during digestion. The advantage of synchrotron SAXS is that data may be acquired very rapidly during the *in vitro* digestion process allowing the kinetics of lipid digestion to be tracked. Boyd et al. have designed a bespoke lipid digestion apparatus consisting of a quartz capillary coupled to a synchrotron SAXS flow-through cell (Warren, Anby et al. 2011). The sample is contained within a thermostated glass vessel, fitted with a pH stat auto-titrator, and held at a constant temperature of 37°C. The digestion process is initiated via the addition of simulated intestinal fluid consisting of BSs, pancreatin, and phospholipid in a digestion buffer adjusted to pH 6.5. Cryo-TEM is often used in combination with SAXS analysis, allowing the morphology of the evolving lipid nanostructure to be directly visualized.

For medium chain triglycerides (MCT), the evolution of nanostructure was monitored in real time using a combination of synchrotron SAXS and cryo-TEM. Digestion of MCT was found to lead to the formation of a vesicular phase, whose formation was linked to the ratio of lipid to BSs in the digestion mixture. Recently, the digestion of more complex milk-based products including cow's milk (Salentinig, Phan et al. 2013) and human breast milk (Salentinig, Phan et al. 2015) have been investigated. The standard milk oil-in-water emulsion mainly consists of fats (triglycerides), carbohydrates, proteins, salts, and water- and oil-soluble vitamins, contained within a protein-stabilized emulsion. Digestion of the milk triglycerides by pancreatin leads to the subsequent formation of diglycerides and fatty acids. Changes in lipid architecture associated with the formation of diglycerides and fatty acids modify the effective packing parameter of the lipids and drive the formation of nanostructured phases including the inverse bicontinuous cubic and hexagonal phases. The observed phase sequence (oil – emulsified microemulsion – Fd3m inverse micellar cubic – inverse hexagonal phase – inverse bicontinuous cubic phase) is consistent with an overall decrease in interfacial curvature of the system (Salentinig, Phan et al. 2013). A similar sequence of phases was observed during the *in vitro* digestion of human breast milk (Salentinig, Phan et al. 2015). The sequence and kinetics of transformation depended on a number of factors including pH, BS concentration, and time of digestion. In both cases corresponding cryo-TEM images confirmed the existence of a complex mixture of particles containing numerous self-assembly structures. It is suggested that the nanostructured materials act to function as carriers for poorly water-soluble molecules in the digestive tract.

3.13 APPLICATIONS OF CUBIC AND HEXAGONAL PHASES: DRUG DELIVERY

Liposomal and polymeric drug delivery vehicles currently dominate the nanomedicine research field with both cubosomes and hexosomes demonstrating a relatively slow uptake into clinical use. Nevertheless, these particles (Figure 3.13) offer a wide range of advantages over the ubiquitous liposome, including a high internal surface area, protection of the payload from chemical and physical degradation, and the ability to encapsulate both lipophilic and hydrophilic drugs of a range of sizes (Mulet, Boyd et al. 2013). Administration routes for cubosome- and hexosome-based drug preparations include oral, dermal, percutaneous,

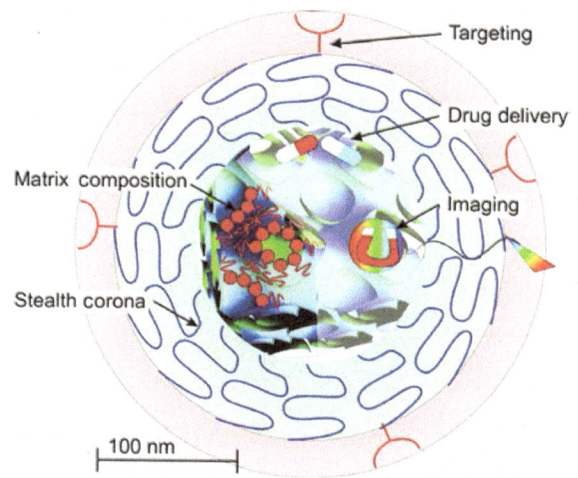

FIGURE 3.13 A schematic of a multifunctional lipidic nanoparticle drug delivery vehicle. (Reproduced with permission from Mulet, Boyd et al. 2013.)

and ocular (Mulet, Boyd et al. 2013). Controlled release has been demonstrated from the bicontinuous cubic phases (Shah, Sadhale et al. 2001). They are of particular advantage in the controlled delivery of therapeutic proteins and peptides, which represent a new generation of biopharmaceutical agents (Leader, Baca et al. 2008). The increasing use of cubosomes and hexosomes as drug delivery vehicles will depend on a fundamental understanding of the relationship between the encapsulated drug and the lipid matrix. Specifically, the impact of the encapsulated drug on mesophase geometry, and conversely, the impact of lipid microenvironment on drug loading and release rates must be considered.

In vitro and *in vivo* studies have demonstrated that cubosomes and hexosomes can be used to encapsulate drugs including vitamins such as vitamin K (Lopes, Speretta et al. 2007), the nonsteroidal anti-inflammatory drugs indomethacin (Esposito, Cortesi et al. 2005) and flurbiprofen (Han, Shen et al. 2010), the anti-histamine cinnarizine (Nguyen, Hanley et al. 2011), the sex hormone progesterone (Swarnakar, Jain et al. 2007), and irinotecan (Boyd, Whittaker et al. 2006) which is used in the treatment of colon cancer. The drug may be encapsulated in the aqueous, lipidic, or interfacial regions of the nanoparticle, with the exact location dependent on the lipophilicity and size of the drug. The encapsulated drug may impact the internal geometry of the nanoparticle, and the nanoparticle integrity, depending on the size and charge characteristics of the drug, and the physicochemical properties of the nanoparticle. Mulet et al. have shown, using high-throughput methodologies, that the octanol–water partition coefficient of the drug may be used to successfully predict the phase transition of the nanoparticle after drug encapsulation (Mulet, Kennedy et al. 2010). Specifically, lipophilic drugs such as progesterone, which have a high octanol–water partition coefficient, induce a phase transition to more curved phases. The maximum loading of each drug (before a phase transition occurs) will vary for individual systems. Drug loading may be increased via a variety of methods including charge functionalization of the membrane (Lynch, Ofori-Boateng et al. 2003) or by exploiting fundamental drug–lipid interactions (Conn, Mulet et al. 2011).

Most research on drug release from bicontinuous cubic and hexagonal materials has focussed on the bulk material, with relatively few studies looking at drug release from the dispersed, nanoparticulate, form. The release rate from the bulk material, which is typically diffusion controlled, has been shown to depend strongly on the material nanostructure. Phan et al. have compared drug release rates from cubic, hexagonal, and inverse micellar materials and shown that release rates (of glucose) were fastest from the cubic phase and relatively similar for the hexagonal and inverse micellar phases (Phan, Fong et al. 2011). While studies on release rates from dispersions are limited, it has been suggested that the 200,000-fold increase in surface area for the dispersed materials may lead to rapid, uncontrolled drug release from these systems (Boyd 2003).

The *in vitro* toxicity of cubosomes can depend significantly on the chemical structure of the lipid (Hinton, Grusche et al. 2014). For example, the *in vitro* toxicity of phytantriol cubosomes is much greater than that of monoolein cubosomes and is ascribed to the greater ability of phytantriol to disrupt the cell membrane (Hinton, Grusche et al. 2014). Nevertheless, the *in vitro* and *in vivo* toxicity of these systems has not been well established.

3.14 BICONTINUOUS CUBIC PHASES FOR STRUCTURAL STUDIES OF MEMBRANE PROTEINS

The bicontinuous cubic phase provides a unique, biomimetic matrix for the growth of protein crystals (Landau and Rosenbusch 1996), particularly hydrophobic and amphiphilic proteins and peptides, which are generally incompatible with solution phase crystallization (Figure 3.14). The fundamental bilayer structure of the bicontinuous cubic phase mimics the viscoelastic properties of the native cell membrane environment, which may promote retention of the active conformation of the protein (Conn and Drummond 2013). In addition, the protein can diffuse in three dimensions across the tortuous bilayer network, facilitating the growth of 3-D crystals of membrane proteins. While the percentage of membrane protein structures obtained using in meso crystallization is relatively small (12%), recent developments, particularly in high-throughput set-up of crystallization trials, have improved success rates for this technique and, in recent years, approximately 25% of all membrane proteins structures have been solved using in meso crystallization. Crystals grown in meso may be superior to detergent-grown crystals, as they are generally more robust and of higher quality. The technique has proved particularly useful in solving the structures of the G-protein-coupled receptors, including the adenosine A_{2A} receptor (Jaakola, Griffith et al. 2008), the dopamine D3 receptor (Chien, Liu et al. 2010), and the β_2-adrenergic receptor (Rasmussen, Choi et al. 2007), which are important drug targets. In meso crystallization is not limited to proteins containing a hydrophobic domain; small hydrophilic proteins have also been successfully crystallized from both cubic and inverse hexagonal phases (Mezzenga, Zabara et al. 2011).

While less successful than the lipidic cubic phase, the lipidic sponge mesophase has also been used for successful crystal growth. The lipidic sponge phase may be thought of as a 'melted' cubic phase; while the 3-D periodicity of the cubic phase is lost, the sponge phase locally retains the three-dimensional network of bilayers essential to crystal growth. Aqueous pore size in the sponge mesophase also tends to be larger which may be of use in crystallizing amphiphilic proteins with large extra- and/or intracellular domains, extending the range of proteins which may be investigated.

To date, the vast majority of in meso crystallization trials have used monoolein, or a similar monoacylglycerol lipid. The addition of cholesterol, which stiffens the membrane and may provide a more biomimetic environment for the protein, has proved successful in many crystallization trials. Crystal growth is induced by addition of a precipitant, typically some combination of polyethylene glycol, a salt, and buffer. Addition of the crystallization screen components is believed to trigger phase separation, with protein nucleation occurring in the protein-rich separated phase. The existence of a local lamellar phase, which is postulated to act as a conduit between the bulk lipidic cubic phase and the growing crystal, has been observed using SAXS and AFM (Caffrey 2008).

Recent developments in high-throughput formulation of lipid mixtures, and structural analysis using synchrotron SAXS, have allowed the evolving structure of the lipidic mesophase to be monitored during crystal growth (Conn, Darmanin et al. 2012). The multiple components within a crystallization trial (proteins, associated detergents and buffers, and the components of the crystallization screen) have been shown to destabilize the cubic phase matrix in many cases. Depending on the particular combination of lipid and protein, the system has been shown to transition to numerous different lipid mesophases, including hexagonal, fluid isotropic, and lamellar phases. A large discrepancy in size between the hydrophobic and hydrophilic domain sizes of the membrane protein, and the bilayer thickness and water channel diameter of the cubic mesophase was shown to promote loss of the cubic phase in many instances. This may render some lipids unsuitable for use with in meso crystallization trials. Despite significant research effort, the factors that promote in meso crystallization remain poorly understood and most crystallization trials therefore require screening across a very large compositional space.

FIGURE 3.14 Bicontinuous cubic phases for crystallizing membrane proteins. (Reproduced with permission from Caffrey 2008.)

REFERENCES

Almsherqi, Z. A., S. D. Kohlwein and Y. Deng (2006). "Cubic membranes: A legend beyond the Flatland of cell membrane organization." *Journal of Cell Biology* **173**(6): 839–844.

Almsherqi, Z. A., T. Landh, S. D. Kohlwein and Y. Deng (2009). Cubic membranes: The missing dimension of cell membrane organization. *International Review of Cell and Molecular Biology*, Elsevier. **274**: 275–342.

Angelova, A., B. Angelov, B. Papahadjopoulos-Sternberg, M. Ollivon and C. Bourgaux (2005). "Proteocubosomes: Nanoporous vehicles with tertiary organized fluid interfaces." *Langmuir* **21**(9): 4138–4143.

Barauskas, J., M. Johnsson, F. Johnson and F. Tiberg (2005). "Cubic phase nanoparticles (Cubosome): Principles for controlling size, structure, and stability." *Langmuir* **21**(6): 2569–2577.

Barauskas, J. and T. Landh (2003). "Phase behavior of the phytantriol/water system." *Langmuir* **19**(23): 9562–9565.

Baringer, J. R. and Swovelan, P. (1972). "Tubular aggregates in endoplasmic-reticulum - evidence against their viral nature." *Journal of Ultrastructure Research* **41**(3–4): 270–276.

Barriga, H., A. Tyler, N. L. McCarthy, E. Parsons, J. M. Seddon, R. V. Law, O. Ces and N. J. Brooks (2015). "Engineering swollen bicontinuous cubic phases and cubosomes; a new generation of artificial cells." *European Biophysics Journal with Biophysics Letters* **44**: S109-S109.

Bartlett, D. H. (2002). "Pressure effects on *in vivo* microbial processes." *Biochimica Et Biophysica Acta-Protein Structure and Molecular Enzymology* **1595**(1–2): 367–381.

Boyd, B. J. (2003). "Characterisation of drug release from cubosomes using the pressure ultrafiltration method." *International Journal of Pharmaceutics* **260**(2): 239–247.

Boyd, B. J., D. V. Whittaker, S. M. Khoo and G. Davey (2006). "Hexosomes formed from glycerate surfactants - Formulation as a colloidal carrier for irinotecan." *International Journal of Pharmaceutics* **318**(1–2): 154–162.

Brooks, N. J., O. Ces, R. H. Templer and J. M. Seddon (2011). "Pressure effects on lipid membrane structure and dynamics." *Chemistry and Physics of Lipids* **164**(2): 89–98.

Brooks, N. J., B. L. L. E. Gauthe, N. J. Terrill, S. E. Rogers, R. H. Templer, O. Ces and J. M. Seddon (2010). "Automated high pressure cell for pressure jump X-ray diffraction." *Review of Scientific Instruments* **81**(6): 064103.

Bruinsma, R. (1992). "Elasticity and excitations of minimal crystals." *Journal de Physique II France* **2**: 425–451.

Caffrey, M. (2008). "On the mechanism of membrane protein crystallization in lipidic mesophases." *Crystal Growth & Design* **8**(12): 4244–4254.

Chien, E. Y. T., W. Liu, Q. A. Zhao, V. Katritch, G. W. Han, M. A. Hanson, L. Shi, A. H. Newman, J. A. Javitch, V. Cherezov and R. C. Stevens (2010). "Structure of the human dopamine D3 receptor in complex with a D2/D3 selective antagonist." *Science* **330**(6007): 1091–1095.

Chong, J. Y. T., X. Mulet, L. J. Waddington, B. J. Boyd and C. J. Drummond (2011). "Steric stabilisation of self-assembled cubic lyotropic liquid crystalline nanoparticles: High throughput evaluation of triblock polyethylene oxide-polypropylene oxide-polyethylene oxide copolymers." *Soft Matter* **7**(10): 4768–4777.

Conn, C. E., O. Ces, X. Mulet, S. Finet, R. Winter, J. M. Seddon and R. H. Templer (2006). "Dynamics of structural transformations between lamellar and inverse bicontinuous cubic lyotropic phases." *Physical Review Letters* **96**(10): 108102.

Conn, C. E., O. Ces, A. M. Squires, X. Mulet, R. Winter, S. M. Finet, R. H. Templer and J. M. Seddon (2008). "A pressure-jump time-resolved x-ray diffraction study of cubic-cubic transition kinetics in monoolein." *Langmuir* **24**(6): 2331–2340.

Conn, C. E., C. Darmanin, X. Mulet, S. Le Cann, N. Kirby and C. J. Drummond (2012). "High-throughput analysis of the structural evolution of the monoolein cubic phase in situ under crystallogenesis conditions." *Soft Matter* **8**(7): 2310–2321.

Conn, C. E., C. Darmanin, S. M. Sagnella, X. Mulet, T. L. Greaves, J. N. Varghese and C. J. Drummond (2010). "Incorporation of the dopamine D2L receptor and bacteriorhodopsin within bicontinuous cubic lipid phases. 2. Relevance to in meso crystallization of integral membrane proteins in novel lipid systems." *Soft Matter* **6**(19): 4838–4846.

Conn, C. E. and C. J. Drummond (2013). "Nanostructured bicontinuous cubic lipid self-assembly materials as matrices for protein encapsulation." *Soft Matter* **9**(13): 3449–3464.

Conn, C. E., X. Mulet, M. J. Moghaddam, C. Darmanin, L. J. Waddington, S. M. Sagnella, N. Kirby, J. N. Varghese and C. J. Drummond (2011). "Enhanced uptake of an integral membrane protein, the dopamine D2L receptor, by cubic nanostructured lipid nanoparticles doped with Ni(II) chelated EDTA amphiphiles." *Soft Matter* **7**(2): 567–578.

Deng, Y., Z. A. Almsherqi, M. M. L. Ng and S. D. Kohlwein (2010). "Do viruses subvert cholesterol homeostasis to induce host cubic membranes?" *Trends in Cell Biology* **20**(7): 371–379.

Deng, Y., M. Marko, K. Buttle, A. Leith, C. A. Mannella and A. Mannella (1999). "Electron tomography of cubic membrane in amoeba mitochondria." *Biophysical Journal* **76**(1): A231-A231.

Duesing, P. M., J. M. Seddon, R. H. Templer and D. A. Mannock (1997). "Pressure effects on lamellar and inverse curved phases of fully hydrated dialkyl phosphatidylethanolamines and beta-D-xylopyranosyl-sn-glycerols." *Langmuir* **13**(10): 2655–2664.

Duesing, P. M., R. H. Templer and J. M. Seddon (1997). "Quantifying packing frustration energy in inverse lyotropic mesophases." *Langmuir* **13**(2): 351–359.

Esposito, E., R. Cortesi, M. Drechsler, L. Paccamiccio, P. Mariani, C. Contado, E. Stellin, E. Menegatti, F. Bonina and C. Puglia (2005). "Cubosome dispersions as delivery systems for percutaneous administration of indomethacin." *Pharmaceutical Research* **22**(12): 2163–2173.

Feast, G. C., O. E. Hutt, X. Mulet, C. E. Conn, C. J. Drummond and G. P. Savage (2014). "The high-throughput synthesis and phase characterisation of amphiphiles: A sweet case study." *Chemistry-a European Journal* **20**(10): 2783–2792.

Feast, G. C., T. Lepitre, X. Mulet, C. E. Conn, O. E. Hutt, G. P. Savage and C. J. Drummond (2014). "The search for new amphiphiles: Synthesis of a modular, high-throughput library." *Beilstein Journal of Organic Chemistry* **10**: 1578–1588.

Fong, C., T. Le and C. J. Drummond (2012). "Lyotropic liquid crystal engineering-ordered nanostructured small molecule amphiphile self-assembly materials by design." *Chemical Society Reviews* **41**(3): 1297–1322.

Fong, C., A. Weerawardena, S. M. Sagnella, X. Mulet, I. Krodkiewska, J. Chong and C. J. Drummond (2011). "Monodisperse nonionic isoprenoid-type hexahydrofarnesyl ethylene oxide surfactants: High throughput lyotropic liquid crystalline phase determination." *Langmuir* **27**(6): 2317–2326.

Fong, C., A. Weerawardena, S. M. Sagnella, X. Mulet, L. Waddington, I. Krodkiewska and C. J. Drummond (2010). "Monodisperse nonionic phytanyl ethylene oxide surfactants: High throughput lyotropic liquid crystalline phase determination and the formation of liposomes, hexosomes and cubosomes." *Soft Matter* **6**(19): 4727–4741.

Fong, C., D. Wells, I. Krodkiewska, A. Weerawardeena, J. Booth, P. G. Hartley and C. J. Drummond (2007). "Diversifying the solid state and lyotropic phase behavior of nonionic urea-based surfactants." *Journal of Physical Chemistry B* **111**(36): 10713–10722.

Gustafsson, J., H. Ljusberg-Wahren, M. Almgren and K. Larsson (1997). "Submicron particles of reversed lipid phases in water stabilized by a nonionic amphiphilic polymer." *Langmuir* **13**(26): 6964–6971.

Han, S., J. Q. Shen, Y. Gan, H. M. Geng, X. X. Zhang, C. L. Zhu and L. Gan (2010). "Novel vehicle based on cubosomes for ophthalmic delivery of flurbiprofen with low irritancy and high bioavailability." *Acta Pharmacologica Sinica* **31**(8): 990–998.

Hashimoto, I., A. Hagiwara and T. Komatsu (1984). "Ultrastructural studies on the pathogenesis of poliomyelitis in monkeys infected with poliovirus." *Acta Neuropathologica* **64**(1): 53–60.

Helfrich, W. (1981). *Physics of Defects*, Les Houches Summer School Proceedings. R. Ballan, M. Kléman and J. P. Poirier, Editors, 35, 715. North-Holland Publ. Co., Amsterdam

Hinton, T. M., F. Grusche, D. Acharya, R. Shukla, V. Bansal, L. J. Waddington, P. Monaghan and B. W. Muir (2014). "Bicontinuous cubic phase nanoparticle lipid chemistry affects toxicity in cultured cells." *Toxicology Research* **3**(1): 11–22.

Israelachvili, J. N. (2011). "Intermolecular and surface forces preface to the first edition." *Intermolecular and Surface Forces*, 3rd Edition: Xxi–Xxii. Academic Press, London.

Jaakola, V. P., M. T. Griffith, M. A. Hanson, V. Cherezov, E. Y. Chien, J. R. Lane, A. P. Ijzerman and R. C. Stevens (2008). "The 2.6 angstrom crystal structure of a human A2A adenosine receptor bound to an antagonist." *Science* **322**(5905): 1211–1217.

Landau, E. M. and J. P. Rosenbusch (1996). "Lipidic cubic phases: A novel concept for the crystallization of membrane proteins." *Proceedings of the National Academy of Sciences of the United States of America* **93**(25): 14532–14535.

Larsson, M. and K. Larsson (2014). "Periodic minimal surface organizations of the lipid bilayer at the lung surface and in cubic cytomembrane assemblies." *Advances in Colloid and Interface Science* **205**: 68–73.

Leader, B., Q. J. Baca and D. E. Golan (2008). "Protein therapeutics: A summary and pharmacological classification." *Nature Reviews Drug Discovery* **7**(1): 21–39.

Lee, W.B., R. Mezzenga and G.H. Fredrickson (2008). "Self-consistent field theory for lipid-based liquid crystals: Hydrogen bonding effects." *The Journal of Chemical Physics* **128**: 074504.

Lopes, L. B., F. F. F. Speretta, M. Vitoria and L. B. Bentley (2007). "Enhancement of skin penetration of vitamin K using monoolein-based liquid crystalline systems." *European Journal of Pharmaceutical Sciences* **32**(3): 209–215.

Luzzati, V., R. Vargas, A. Gulik, P. Mariani, J. M. Seddon and E. Rivas (1992). "Lipid polymorphism - a correction - the structure of the cubic phase of extinction symbol Fd—consists of 2 types of disjointed reverse micelles embedded in a 3-dimensional hydrocarbon matrix." *Biochemistry* **31**(1): 279–285.

Lynch, M. L., A. Ofori-Boateng, A. Hippe, K. Kochvar and P. T. Spicer (2003). "Enhanced loading of water-soluble actives into bicontinuous cubic phase liquid crystals using cationic surfactants." *Journal of Colloid and Interface Science* **260**(2): 404–413.

Marquart, K. H. (2005). "Occurrence of tuboloreticular structures and intracisternal paracrystalline inclusions in endothelial cells of tissue from different epidemiological types of Kaposi,s sarcoma." *Ultrastructural Pathology* **29**(2): 85–93.

Marsh, D. (2007). "Lateral pressure profile, spontaneous curvature frustration, and the incorporation and conformation of proteins in membranes." *Biophysical Journal* **93**(11): 3884–3899.

Martiel, I., L. Sagalowicz and R. Mezzenga (2013). "A reverse micellar mesophase of face-centered cubic Fm(3)over-barm symmetry in phosphatidylcholine/water/organic solvent ternary systems." *Langmuir* **29**(51): 15805–15812.

McCarthy, N. L. C., O. Ces, R. V. Law, J. M. Seddon and N. J. Brooks (2015). "Separation of liquid domains in model membranes induced with high hydrostatic pressure." *Chemical Communications* **51**(41): 8675–8678.

Meikle, T. G., A. Zabara, L. J. Waddington, F. Separovic, C. J. Drummond and C. E. Conn (2017). "Incorporation of antimicrobial peptides in nanostructured lipid membrane mimetic bilayer cubosomes." *Colloids and Surfaces B-Biointerfaces* **152**: 143–151.

Mezzenga, R., A. Zabara and I. Amar-Yuli (2011). "Tuning in-meso-crystallized lysozyme polymorphism by lyotropic liquid crystal symmetry." *Langmuir* **27**(10): 6418–6425.

Minamikawa, H. and M. Hato (1998). "Reverse micellar cubic phase in a phytanyl-chained glucolipid/water system." *Langmuir* **14**(16): 4503–4509.

Muir, B. W., G. L. Zhen, P. Gunatilake and P. G. Hartley (2012). "Salt induced lamellar to bicontinuous cubic phase transitions in cationic nanoparticles." *Journal of Physical Chemistry B* **116**(11): 3551–3556.

Mulet, X., B. J. Boyd and C. J. Drummond (2013). "Advances in drug delivery and medical imaging using colloidal lyotropic liquid crystalline dispersions." *Journal of Colloid and Interface Science* **393**: 1–20.

Mulet, X., C. E. Conn, C. Fong, D. F. Kennedy, M. J. Moghaddam and C. J. Drummond (2013). "High-throughput development of amphiphile self-assembly materials: Fast-tracking synthesis, characterization, formulation, application, and understanding." *Accounts of Chemical Research* **46**(7): 1497–1505.

Mulet, X., D. F. Kennedy, C. E. Conn, A. Hawley and C. J. Drummond (2010). "High throughput preparation and characterisation of amphiphilic nanostructured nanoparticulate drug delivery vehicles." *International Journal of Pharmaceutics* **395**: 290–297.

Nguyen, T. H., T. Hanley, C. J. H. Porter and B. J. Boyd (2011). "Nanostructured reverse hexagonal liquid crystals sustain plasma concentrations for a poorly water-soluble drug after oral administration." *Drug Delivery and Translational Research* **1**(6): 429–438.

Patton, J. S. and M. C. Carey (1979). "Watching fat digestion." *Science* **204**(4389): 145–148.

Phan, S., W.-K. Fong, N. Kirby, T. Hanley and B. J. Boyd (2011). "Evaluating the link between self-assembled mesophase structure and drug release." *International Journal of Pharmaceutics* **421**(1): 176–182.

Rappolt, M., F. Cacho-Nerin, C. Morello and A. Yaghmur (2013). "How the chain configuration governs the packing of inverted micelles in the cubic Fd3m-phase." *Soft Matter* **9**(27): 6291–6300.

Rasmussen, S. G., H. J. Choi, D. M. Rosenbaum, T. S. Kobilka, F. S. Thian, P. C. Edwards, M. Burghammer, V. R. Ratnala, R. Sanishvili, R. F. Fischetti, G. F. Schertler, W. I. Weis and B. K. Kobilka (2007). "Crystal structure of the human beta2 adrenergic G-protein-coupled receptor." *Nature* **450**(7168): 383–387.

Sagalowicz, L., M. Michel, M. Adrian, P. Frossard, M. Rouvet, H. J. Watzke, A. Yaghmur, L. De Campo, O. Glatter and M. E. Leser (2006). "Crystallography of dispersed liquid crystalline phases studded by cryo-transmission on electron microscopy." *Journal of Microscopy-Oxford* **221**: 110–121.

Sagnella, S. M., C. E. Conn, I. Krodkiewska, M. Moghaddam, J. M. Seddon and C. J. Drummond (2010). "Ordered nanostructured amphiphile self-assembly materials from endogenous nonionic unsaturated monoethanolamide lipids in water." *Langmuir* **26**(5): 3084–3094.

Sagnella, S. M., C. E. Conn, I. Krodkiewska, X. Mulet and C. J. Drummond (2011). "Anandamide and analogous endocannabinoids: A lipid self-assembly study." *Soft Matter* **7**(11): 5319–5328.

Sagnella, S. M., X. J. Gong, M. J. Moghaddam, C. E. Conn, K. Kimpton, L. J. Waddington, I. Krodkiewska and C. J. Drummond (2011). "Nanostructured nanoparticles of self-assembled lipid pro-drugs as a route to improved chemotherapeutic agents." *Nanoscale* **3**(3): 919–924.

Salentinig, S., S. Phan, A. Hawley and B. J. Boyd (2015). "Self-assembly structure formation during the digestion of human breast milk." *Angewandte Chemie-International Edition* **54**(5): 1600–1603.

Salentinig, S., S. Phan, J. Khan, A. Hawley and B. J. Boyd (2013). "Formation of highly organized nanostructures during the digestion of milk." *Acs Nano* **7**(12): 10904–10911.

Salentinig, S., L. Sagalowicz, M. E. Leser, C. Tedeschi and O. Glatter (2011). "Transitions in the internal structure of lipid droplets during fat digestion." *Soft Matter* **7**(2): 650–661.

Schwarz, U. S. and G. Gompper (2000). "Stability of inverse bicontinuous cubic phases in lipid-water mixtures." *Physical Review Letters* **85**(7): 1472–1475.

Seddon, J. M. (1990). "An inverse face-centered cubic phase formed by diacylglycerol phosphatidylcholine mixtures." *Biochemistry* **29**(34): 7997–8002.

Seddon, J. M., J. Robins, T. Gulik-Krzywicki and H. Delacroix (2000). "Inverse micellar phases of phospholipids and glycolipids." *Physical Chemistry Chemical Physics* **2**(20): 4485–4493.

Seddon, J. M., A. M. Squires, C. E. Conn, O. Ces, A. J. Heron, X. Mulet, G. C. Shearman and R. H. Templer (2006). "Pressure-jump X-ray studies of liquid crystal transitions in lipids." *Philosophical Transactions of the Royal Society a-Mathematical Physical and Engineering Sciences* **364**(1847): 2635–2655.

Seddon, J. M. and R. H. Templer (1993). "Cubic phases of self-assembled amphiphilic aggregates." *Philosophical Transactions of the Royal Society of London Series a-Mathematical Physical and Engineering Sciences* **344**(1672): 377–401.

Seddon, J. M., N. Zeb, R. H. Templer, R. N. McElhaney and D. A. Mannock (1996). "An Fd3m lyotropic cubic phase in a binary glycolipid/water system." *Langmuir* **12**(22): 5250–5253.

Shah, J. C., Y. Sadhale and D. M. Chilukuri (2001). "Cubic phase gels as drug delivery systems." *Advanced Drug Delivery Reviews* **47**(2–3): 229–250.

Shearman, G. C., O. Ces, R. H. Templer and J. M. Seddon (2006). "Inverse lyotropic phases of lipids and membrane curvature." *Journal of Physics-Condensed Matter* **18**(28): S1105-S1124.

Shearman, G. C., A. I. I. Tyler, N. J. Brooks, R. H. Templer, O. Ces, R. V. Law and J. M. Seddon (2009). "A 3-D hexagonal inverse micellar lyotropic phase." *Journal of the American Chemical Society* **131**(5): 1678-+.

Swarnakar, N. K., V. Jain, V. Dubey, D. Mishra and N. K. Jain (2007). "Enhanced oromucosal delivery of progesterone via hexosomes." *Pharmaceutical Research* **24**(12): 2223–2230.

Tran, N., X. Mulet, A. M. Hawley, C. E. Conn, J. L. Zhai, L. J. Waddington and C. J. Drummond (2015). "First direct observation of stable internally ordered janus nanoparticles created by lipid self-assembly." *Nano Letters*.

Tyler, A. I. I., H. M. G. Barriga, E. S. Parsons, N. L. C. McCarthy, O. Ces, R. V. Law, J. M. Seddon and N. J. Brooks (2015). "Electrostatic swelling of bicontinuous cubic lipid phases." *Soft Matter* **11**(16): 3279–3286.

Warren, D. B., M. U. Anby, A. Hawley and B. J. Boyd (2011). "Real time evolution of liquid crystalline nanostructure during the digestion of formulation lipids using synchrotron small-angle X-ray scattering." *Langmuir* **27**(15): 9528–9534.

Welsch, S., S. Miller, I. Romero-Brey, A. Merz, C. K. E. Bleck, P. Walther, S. D. Fuller, C. Antony, J. Krijnse-Locker and R. Bartenschlager (2009). "Composition and three-dimensional architecture of the dengue virus replication and assembly sites." *Cell Host & Microbe* **5**(4): 365–375.

Yaghmur, A. and O. Glatter (2009). "Characterization and potential applications of nanostructured aqueous dispersions." *Advances in Colloid and Interface Science* **147–48**: 333–342.

Zahid, I. N., C. E. Conn, N. J. Brooks, J. M. Seddon and R. Hashim (2013). "Effects of sugar stereochemistry on lyotropic mesophases of branched-chain synthetic glycolipids." *European Biophysics Journal with Biophysics Letters* **42**: S132-S132.

4 Structure of Lipid Membranes by Advanced X-Ray Scattering and Imaging

Tim Salditt
Universität Göttingen, Gottingen, Germany

CONTENTS

4.1 Introduction to X-Ray Analysis of Lipids 65
4.2 X-Ray Analysis of Lipid Membranes and Assemblies: Some Fundamentals 66
4.3 Nonequilibrium Dynamics of Model Membranes by Time-Resolved Diffraction 68
4.4 Nonlamellar Lipid Phases and Membrane Fusion Intermediate Structures 70
4.5 X-Ray Analysis of Biological Membranes: The Example of Synaptic Vesicles 73
4.6 Myelinated Axons Studied by Scanning Small-Angle X-Ray Scattering and Phase Contrast Tomography 75
4.7 Future Directions: Membrane Structure Analysis by Coherent X-Ray Imaging 77
Acknowledgements 78
References 79

4.1 INTRODUCTION TO X-RAY ANALYSIS OF LIPIDS

Biological membranes are often denoted as the nature's most important interface. Without doubt, they enable essential biological functions from compartmentalization to transport, from enzymatic activity to signal transduction. Their unique properties rely on the specific macromolecular and noncovalent association of its lipid and protein constituents. To unravel the underlying structures, we have powerful tools at hand concerning the protein part, notably macromolecular X-ray crystallography. Contrarily, the 'structural biology' of the lipids is less well developed, owing to a lack of resolution in noncrystallographic states. Nevertheless, X-ray diffraction and in particular small-angle X-ray scattering (SAXS) of lipid or lipid–protein assemblies in the fluid and hydrated state have provided important structural insights. As a classical technique of membrane biophysics, X-ray diffraction has received much attention and treatment on the advanced textbook level (Als-Nielsen, 2001). Rather than re-iterating these fundamentals, we concentrate in this chapter on novel approaches to X-ray analysis of membranes and lipid assemblies. While the progress in the field is certainly a global effort with merits of many active groups, this chapter is primarily based on our own recent work, in particular in the choice of the examples and graphical items. This is as much for the reason of being most familiar with the selected examples, as it is also a matter of perspective, representing a choice, not a merit.

The main questions addressed are the following: To which extent does X-ray diffraction still offer high (molecular) resolution in the noncrystallographic state, considering a fluid membrane assembly? How can oriented lipid bilayers help to overcome the loss of information associated with powder average of solution SAXS? At which resolution and when can electron density reconstructions be obtained in three dimensions (3D)? As is long known, 3D reconstructions are indeed possible for certain liquid–crystalline lipid mesophases at low water content, in particular for nonlamellar phases of cubic or hexagonal symmetry. These have been studied as early as in the 1960s in seminal work by Luzzati and Reiss-Husson (1966) and Luzzati et al. (1968). More recently, further mesophases have been discovered, and today, several high resolution electron density reconstructions have been acquired from oriented lipid mesophases studied with highly brilliant undulator radiation, as discussed below. Further, we address the question, how the dynamics of membrane systems can be probed, in particular in view of nonequilibrium effects. We also discuss how X-ray analysis can be extended from model membranes to biological membranes and how X-ray diffraction can be complemented by (lensless) X-ray microscopy. To date, molecular resolution is only achieved by X-ray diffraction, but not X-ray microscopy. In other words, high resolution is achieved in reciprocal (Fourier) space by diffracting from an ensemble of constituents, not in real space as in microscopic imaging. With recent progress in X-ray optics, however, interesting combinations of real-space and reciprocal-space information become possible, such as nanodiffraction or coherent diffractive imaging (CDI). We include a brief discussion of applications in membrane structure analysis.

The chapter is organized as follows: After this introduction, we first briefly consider structure analysis of oriented lipid membrane stacks. Section 4.2 presents time-resolved X-ray diffraction studies of lipid membranes under non-equilibrium conditions. In Section 4.3, we review studies of membrane fusion in model systems, followed by studies of synaptic vesicles and their interactions with lipid membranes in Section 4.4. In Section 4.5, we turn to myelin as an example for a biological membrane structure, studied in nerve tissues by scanning diffraction with nano-focused beams, and by full-field phase contrast tomography. We close the chapter by discussing X-ray structure analysis of lipid membranes based on CDI in the optical near field.

4.2 X-RAY ANALYSIS OF LIPID MEMBRANES AND ASSEMBLIES: SOME FUNDAMENTALS

To study the lipid bilayer profile function $\rho(z)$ by diffraction, one can choose between different experimental approaches: (i) SAXS from unilamellar vesicles (liposomes), (ii) SAXS from multilamellar vesicle, (iii) reflectivity from a single bilayer, or (iv) from multilamellar stacks deposited on a solid substrate. If the bilayer structure is unaffected by its neighbourhood, the different configurations share the same bilayer form factor $F(q_z) = |f(q_z)|^2$, with $f(q_z) = \int dz \exp(iq_z z)\rho(z)$, but differ in the structure factor $S(q_z) = |s(q_z)|^2$. In multilamellar phases, $S(q_z)$ takes into account the effects of the interbilayer interference. This not only adds information, for example, on elastic properties and interaction forces but also adds complexity to the analysis.

Let's briefly address each of the configurations. (i) SAXS from unilamellar vesicle is fairly simple, in particular if the radius is large enough and if the vesicles are radially symmetric. For small vesicles, the polydispersity has to be incorporated by an integral over the vesicle radius R. For larger liposomes with considerable shape and size fluctuations, interference between opposing bilayer patches averages out, and the simpler bilayer form factor of a quasiplanar, but powder-averaged system describes the data sufficiently well (Brzustowicz and Brunger, 2005, Pabst et al., 2000, Székely et al., 2010, Chappa et al., 2021). (ii) Multilamellar vesicles and more generally smectic liquid crystalline phases of lipids exhibit different thermal modes (undulations, compressional waves) which affect the structure factor leading to very characteristic lineshapes (Caille, 1972, Safinya et al., 1986). While this is an interesting problem in itself, decoding valuable information on the elastic properties of the membranes, it may complicate the structure analysis (if one is only interested in this!). (iii) Single bilayer reflectivity enables a robust and straightforward measurement of the bilayer structure, i.e., of its density profile $\rho(z)$. Importantly, $\rho(z)$ is obtained on an absolute scale (Novakova et al., 2006). The structure of the inorganic substrate and the corresponding reflectivity amplitude which 'interferes' with $f(q_z)$ are easily kept fixed from a control experiment of the bare surface. If the density step to the substrate is much larger than the density changes in the bilayer, unique inversion of reflectivity becomes possible, and the weak bilayer signal is amplified by the reflectivity of the substrate (Hohage et al., 2008). However, the substrate may induce structural alterations, and even denaturing of membrane proteins. Adding a soft cushion or tethering can overcome this problem, but at the expense of added complexity and additional structural parameters. If one is worried about these effects, it may therefore be a solution to switch to a multilamellar stack of membranes with thousands of bilayers. In this case, any eventual change in the direct vicinity of the substrate is irrelevant, since the signal is dominated by the many unaffected bilayers. However, again the same $S(q_z)$-related complications as for SAXS from multilamellar vesicles also apply to this case, even if the boundary conditions are different (Constantin et al., 2005). Again much can be learned about elasticity and interaction potentials from the thermal fluctuations (Salditt, 2005), and these effects are particularly pronounced when at full hydration (Mennicke et al., 2006). Full q-range fits which are the only really quantitative approach can be quite complicated. Therefore, one often resorts to a simplified Fourier analysis based on the integrated peak intensities. Here, only the form factor, and not structure factor, is taken into account. This approach must be regarded with caution since it falsely assumes that the peak intensities only depend on $F(q_z)$. In fact, the peak lineshape is significantly affected by the structure factor $S(q_z)$, accounting for translational order and the thermal fluctuations, which are significant for lower dimensional system such as smectic liquid crystalline systems. For these reasons, while fitting to a full model is better justified (Mennicke et al., 2006), the Fourier synthesis is a simple and robust alternative, but only at partial hydration (Table 4.1).

Figure 4.1 illustrates the technique of *X-ray reflectivity* from solid-supported lipid bilayers and the principal steps in the preparation of multilamellar lipid film by spreading from solvent. It is this combination of sample preparation and X-ray diffraction of oriented bilayers, which is the basis of many of the examples presented here. X-ray reflectivity alone is not sufficient, since the structure of a lipid bilayer is not fully contained in the one-dimensional profile function $\rho(z)$. Importantly, oriented bilayers offer the opportunity to record a two-dimensional plane in reciprocal space and to distinguish the momentum transfer parallel $q_\parallel$ and perpendicular to the membrane plane q_z. Height and density fluctuations, as well as correlations between membrane additives, peptides, pores, or simply the acyl chain packing are clearly and cleanly resolved by avoiding 'power averaging'. While synchrotron radiation is brilliant enough to pick up the diffraction signal of a single monolayer or bilayer, the diffraction signal increase in proportion

TABLE 4.1
To Study the Density Profile of a Bilayer by Diffraction, One Can Choose from Four Different Configurations

Configuration	Form Factor	Further Parameters	In-Plane Signal	Substrate Effects
(i)	Relative*	Size distribution	Powder averaged	None
(ii)	Relative*	Structure factor	Powder averaged	None
(iii)	Absolute	Inorganic surface	Weak	Likely
(iv)	Relative*	Structure factor	Strong	Negligible

(i) SAXS from unilamellar vesicles (liposome), (ii) SAXS from multilamellar vesicle, (iii) reflectivity from a single bilayer, or (iv) a multilamellar stack deposited on a solid substrate. * If SAXS is measured with calibrated intensity, the form factor can in principle be determined on absolute scale, but polydispersity and structure factor effects often hamper precise control of prefactors. This is also true of oriented stacks, which typically exhibit an uneven coverage with bilayers.

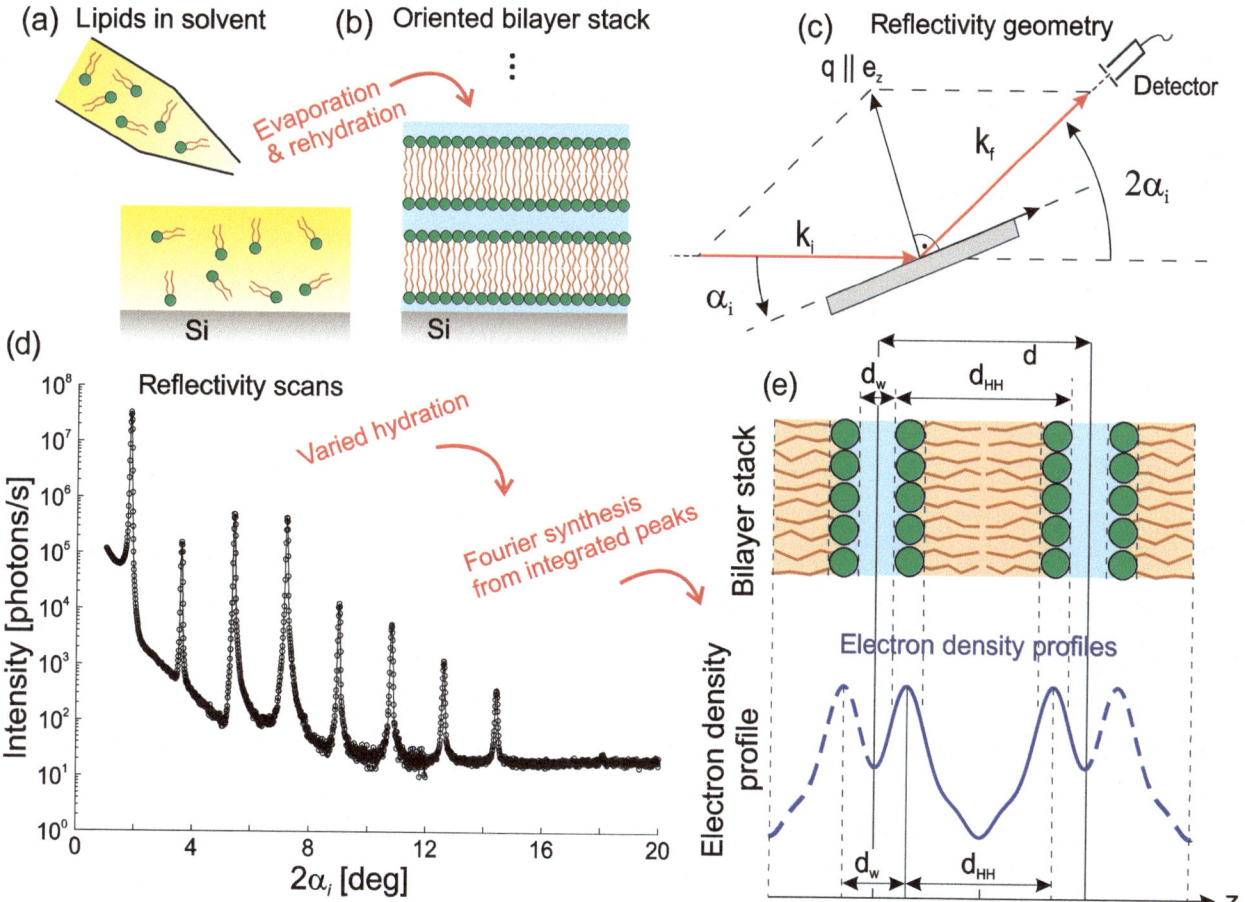

FIGURE 4.1 Preparation of highly oriented multilamellar lipid bilayers for X-ray reflectivity studies. (a, b) Sample preparation by spreading of lipids from organic solvent, subsequent evaporation, and nucleation of bilayers. (c) The reflectivity geometry. (d) Typical reflectivity curves of multilamellar (multibilayer) membrane stacks, exhibiting sharp Bragg reflection with characteristic lineshapes. (e) The resulting electron density profile $\rho(z)$. (Adapted and reproduced from Aeffner 2012, with permission.

to the number of bilayers N which for thick lipid stacks is easily in the range of thousands. In this case, secondary structure of membrane peptides, positional correlations between membrane peptides, and the tilt distribution of the hydrocarbon chains become accessible (Salditt, 2003).

Figure 4.2 illustrates the scattering geometries of *grazing incidence small-angle X-ray scattering* (GISAXS) and *grazing incidence diffraction* (GID) from lipid films, and the associated signals which can be observed in a mapping of reciprocal space.

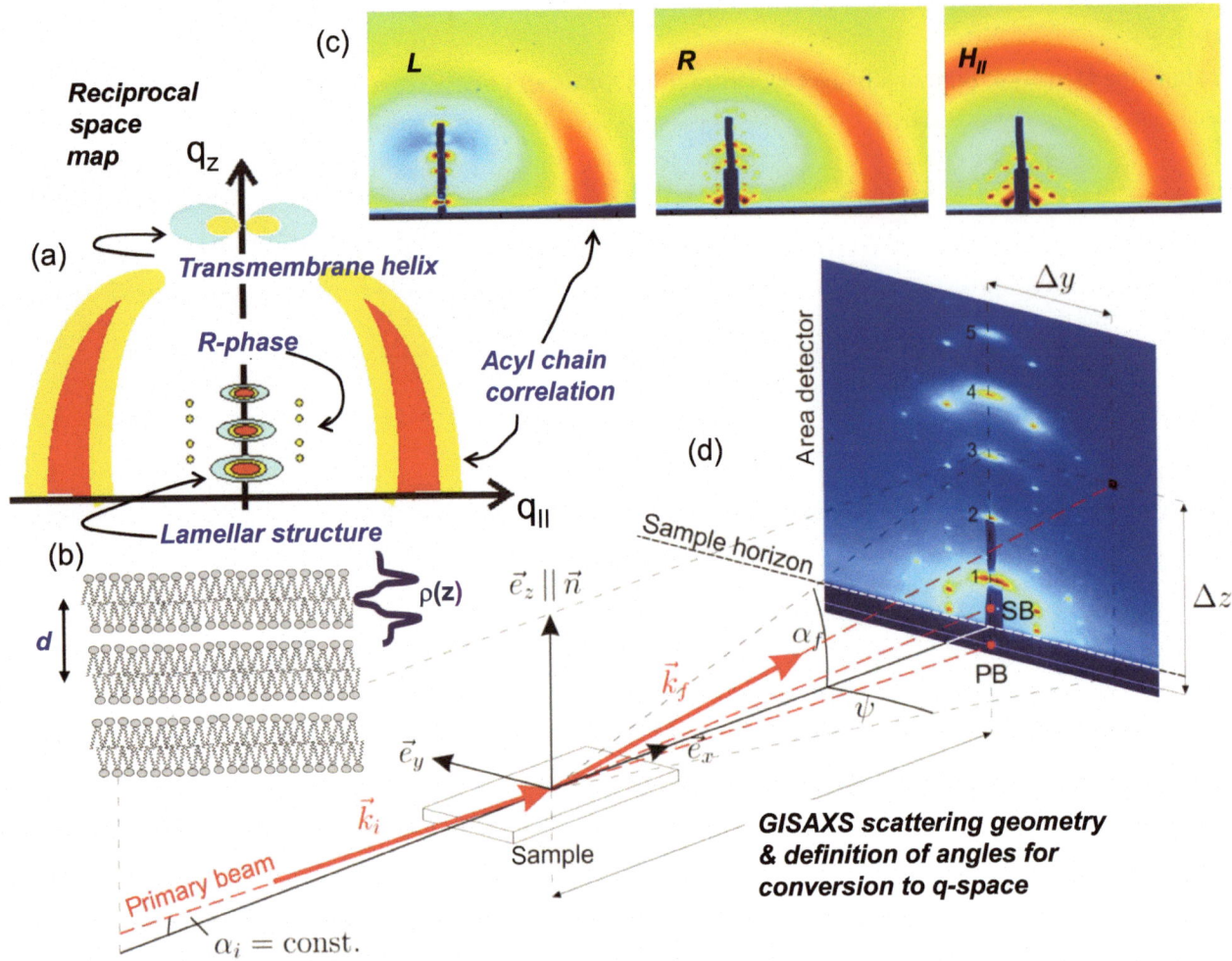

FIGURE 4.2 (a) Schematic of reciprocal space mapping (RSM) of oriented bilayers as a function of momentum parallel $q_{\parallel}$ and perpendicular q_z to the plane of the membranes, and (b) sketch of the multilamellar configuration. In the SAXS range, the lamellar peaks geometry and the superlattice peaks of the rhombohedral (R) phase (Section 4.3) are recorded. In the space WAXS range, the correlation peak of the lipid tails is observed with a characteristic banana shape. In some systems, signals of transmembrane helices can be detected. (c) Typical intensity distribution for the lamellar (L), the rhombohedral phase (R), and the inverse hexagonal (H_{II}) phase. From (Weinhausen et al., 2012), reproduced with permission. (d) Sketch of GISAXS geometry and definition of angles required for proper conversion of reciprocal space (Weinhausen et al., 2012). GISAXS, grazing incidence small-angle X-ray scattering; SAXS, small-angle X-ray scattering; WAXS, wide-angle X-ray scattering. (From Aeffner 2011, reproduced with permission.)

4.3 NONEQUILIBRIUM DYNAMICS OF MODEL MEMBRANES BY TIME-RESOLVED DIFFRACTION

Membranes in living systems are intrinsically out-of-thermal equilibrium. The metabolism results in a multitude of local force centres acting on the membrane on molecular scales. Biological membranes are driven out of equilibrium, for example, by osmotic pressure gradients, by cytoskeleton motors, by active transport or by ion channels and pumps. Shape transformations, for example, in membrane fusion can also be understood as extreme deviations from a 'ground-state' configuration. From a fundamental point of view, one may ask how a local energy release changes the properties of a lipid assembly? How does a membrane relax back to equilibrium state after it has been driven out of equilibrium by an external force? How do long-range, noncovalent interactions and collective modes of a membrane affect the energy dissipation pathway? A number of theoretical studies have first addressed 'active membranes' (Prost and Bruinsma, 1996, Sankararaman et al., 2002, Ramaswamy et al., 2000, Chen, 2004, Gov, 2004, Lin et al., 2006), focusing mainly on nonequilibrium fluctuations driven by active force centres such as ion pumps or other proteins. Knowing that on a fundamental level, nonequilibrium effects can change structural, dynamical, and mechanical properties of membranes, such effects are still extremely challenging to track down experimentally. Video microscopy has provided first insight (Manneville et al., 1999, 2001, Girard et al., 2005), but the most relevant length scales are below the resolution, at least of conventional optical techniques. Contrarily, inelastic neutron scattering would in principle be a very suitable tool to probe nonequilibrium dynamics of membranes. However, the sample volume typically required by the technique is prohibitive in view of most excitation mechanism. Time-resolved X-ray diffraction, on the other hand, offers both the required spatial and temporal resolution to probe generic nonequilibrium effects in membranes. To this end, one of the major experimental challenges is synchronization between excitation and probing pulses, as well as the excitation of a significantly large sample volume. Both are required to prevent that the desired effects are washed out by temporal and spatial averaging. Since lipid membranes are constantly subject to thermal motion, the various molecular and collective degrees of freedom, including rotations, vibrations, density waves, protrusions, diffusion, and undulations are constantly well populated even without external forces.

In view of the basic experimental principles and main results, we review two proof-of-concept experiments of time-resolved diffraction from lipid membranes under nonequilibrium conditions. The first experiment (Reusch et al., 2013a) used a laser pump and X-ray probe scheme for photo excitation of fluorophores in a labelled lipid membrane to drive the membranes out of equilibrium. The structural dynamics recorded at controlled time delay after excitation has revealed a collective response on mesoscopic scales, distinct from a thermal population of states, with a relaxation pathway involving specific undulation modes. In the second experiment (Reusch et al., 2014), acoustic excitation via continuous-wave (cw) surface acoustic waves (SAWs) was used to drive a stack of lipid membranes. In this case, periodic changes of the bilayer structure on molecular scales were evidenced by stroboscopic X-ray illumination. The nonequilibrium effects of the first study showed up in the structure factor of the bilayer stack. In the second study, they manifested themselves mainly in the form factor of lipid bilayers.

We first address the pump–probe diffraction study of fluorescently labelled phospholipid multibilayers by time-resolved X-ray scattering after optical excitation (Reusch et al., 2013a). Bilayer shape fluctuations were monitored in a stack of DOPC (1,2-dioleoyl-sn-glycero-3-phosphatidylcholine) bilayers, deposited on a quartz surface in solution. The membranes were excited by a nanosecond laser pulse, matched in wavelength to the absorption band of Texas-red labelled lipids, which were mixed into the membranes, as routinely done for optical fluorescence microscopy. After energy uptake, the system response was probed by well-controlled picosecond X-ray pulses, covering a broad range of time and length scales from near-molecular to the mesoscopic range. The characteristic diffraction pattern of the membrane stack (nonspecular diffuse scattering) was then recorded under grazing incidence small-angle X-ray scattering (GISAXS) geometry as a function of time delay τ, from a few picoseconds to several microseconds. The laser excitation with 1 kHz repetition rate was synchronized to the synchrotron pulses selected by a high-speed chopper system at beam line ID09 of the European Synchrotron Radiation Facility (ESRF) in Grenoble. From the diffuse (nonspecular) Bragg peaks of the multilamellar system, the evolution of membrane height–height correlation functions was monitored. Pronounced deviations of the collective undulation spectra from thermal equilibrium were observed. In particular, it was found that pulsed laser illumination even at quite moderate peak intensities of about $10^5 W\ cm^{-2}$ leads to significant changes of the in-plane membrane correlation length by up to 50% as well as the excitation of transient conformal undulation modes of a well-defined lateral wavelength. The observed phenomena evolved on nano- to microsecond timescales after optical excitation and could be modelled in terms of a modulation instability in the lipid multilamellar stack. The energy uptake at molecular level first leads to nonthermalized local vibrations ("local heat"), while the long-range undulation modes have not yet taken up any heat. In such a transient state, the lipid molecules tend to expand their intermolecular distances, but on short time scales, this motion is quenched. For compensation, the membranes buckle, opening up a faster relaxation mechanism than the thermal expansion of the entire system. In short, the experiments showed that local molecular energy uptake by photon absorption can lead to peculiar long-range mesoscopic changes in the membrane, relaxing over microseconds before the sample has reached equilibrium again, with an energy dissipation pathway involving a characteristic sequence of modes at length scales of a few hundred nanometres. While this setting of optical excitation is entirely artificial, there is a generic aspect of the experiment. In fact, it shows that nonequilibrium effects in membranes must not directly manifest themselves on molecular length scales, but rather on mesoscopic scales coupled by collective dynamics. To which physiological situations this may apply, deserves further investigation.

The second experiment reviewed here used an alternative excitation scheme, based on coupling an SAW with frequencies in the MHz range into a film of lipid membranes. The stack of lipid membranes was thus forced into controlled vibration by exciting a standing wave of ultrasound in the piezoelectric substrate, on which the lipid bilayers were deposited. The SAW standing wave can also be regarded as a coherent phonon, which is emitted into the lipid stack, superimposed on the multiple equilibrium phonon states. In fact, SAWs had previously been found to induce a number of phenomenal dynamic effects in soft matter films and samples deposited on top of the piezoelectric substrate, such as phase separation (Hennig et al., 2011). Of course, it can be questioned to which extent such a SAW excitation could mimic any real effects of external force membranes in a biological environment, but they clearly provide an excitation mechanism with mechanical displacement of well-controlled temporal and spatial scales (Figures 4.3 and 4.4).

The lipid stack was then probed by X-ray reflectivity in stroboscopic mode, by phase-locked synchronization of the SAW frequency generator (Reusch et al., 2013b), capturing the structure of the membranes during different

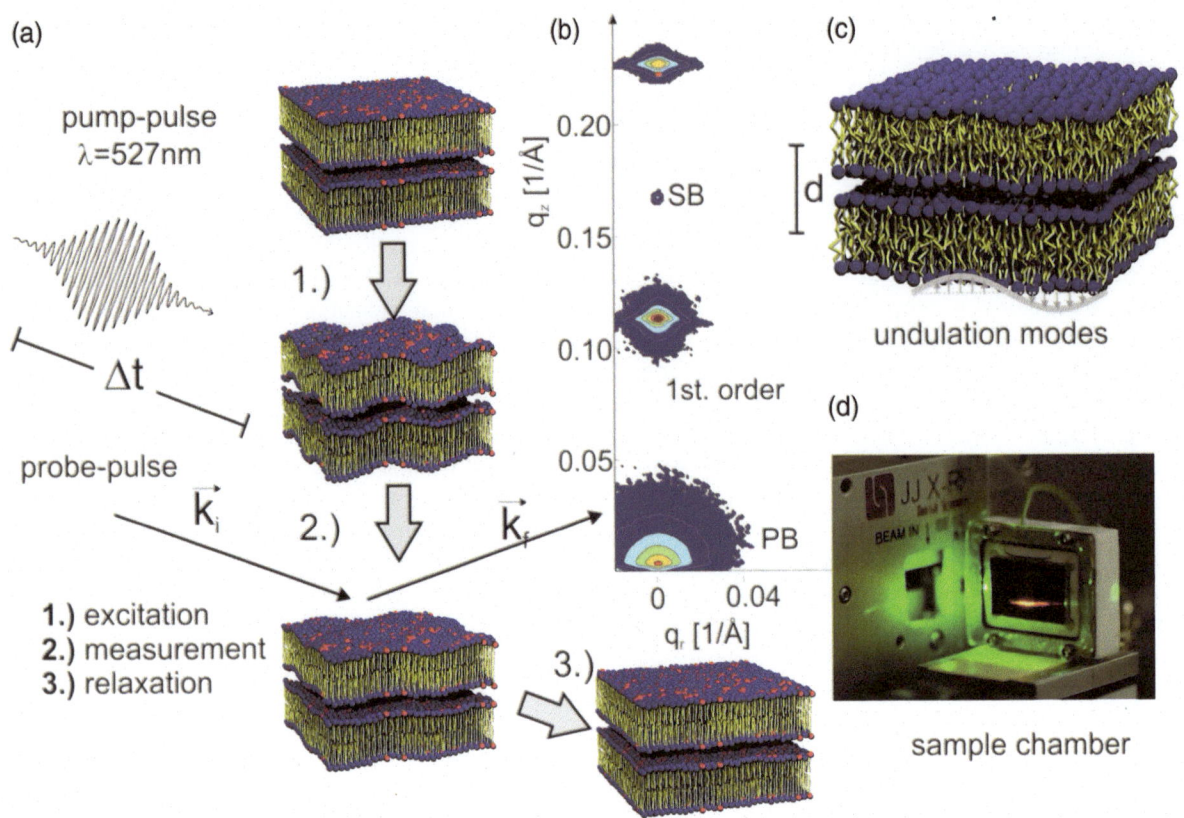

FIGURE 4.3 (a) Schematic of a laser pump/X-ray probe experiment on a stack of multilamellar lipid membranes in the fluid phase. The temporal evolution of membrane undulations is studied in response to a short pulse excitation, by recording the diffuse scattering pattern and lineshape analysis of the individual lamellar reflections. (b) Example of a lamellar diffraction pattern with primary beam (PB), specular beam (SB), and the two first lamellar diffraction orders. (c) From the two-dimensional distribution, the temporal evolution of the lateral and vertical correlation functions is extracted. (d) Sample chamber for oriented Texas-Red labelled lipid multilayers fully immersed in solution. (Adapted from Reusch et al. (2013a), copyright with the author.)

phases of the oscillation cycle. Using this stroboscopic approach, the response of the lipid bilayer structure to the controlled acoustic excitation was investigated (Reusch et al., 2014; see Figure 4.5). The experiment was performed using the favourable conditions of the 40 bunch mode of the PETRAIII storage ring at DESY, where the exceptionally long bunch spacing of 192 ns eliminates the need of a chopper. It was found that the forced oscillation is accompanied by unexpected changes in the density profile $\rho(x)$ of the bilayer. Instead of only a simple harmonic displacement of the bilayer (e.g., its centre of mass), the internal structure oscillates as well. In other words, the bilayers do not only move up and down with the ultrasound wave, but the bilayer thickness and density also changed periodically under the influence of the externally forced motion, as observed with sub-nm scale spatial and sub-100 ps temporal resolution. From data as shown in Figure 4.6, one can infer how the SAW couples to the arrangement and conformations of lipids and water. The periodic changes in thickness and density can be explained by collective stretching and compression of the lipid molecules. Estimations show that inertia effects can account for the amplitude of the response, given the experimental velocities and accelerations on the order of 1 nm ns^{-1} and 1 nm ns^{-2}, respectively. These values are close to what can be expected for a biological system and show that collective conformational changes on molecular scales can be an important component in the response to externally applied forces.

4.4 NONLAMELLAR LIPID PHASES AND MEMBRANE FUSION INTERMEDIATE STRUCTURES

Membrane fusion is well known as an essential process in exocytosis, neurotransmission, fertilization, or viral entry. Fusogenic proteins such as the protein family commonly denoted as *soluble N-ethylmaleimide-sensitive factor attachment protein receptors* (SNAREs) control membrane fusion and provide the required energy to bring the bilayers to close apposition. For the past two decades, SNARE-mediated fusion has been unravelled as an essential process in vesicular transport, recognized by the 2013 Noble Prize for medicine and physiology. While macromolecular crystallography (MX) has provided the structural basis to understand the role of the proteins, the role of the lipids has been experimentally more elusive. This is

X-Ray Scattering and Imaging of Membranes

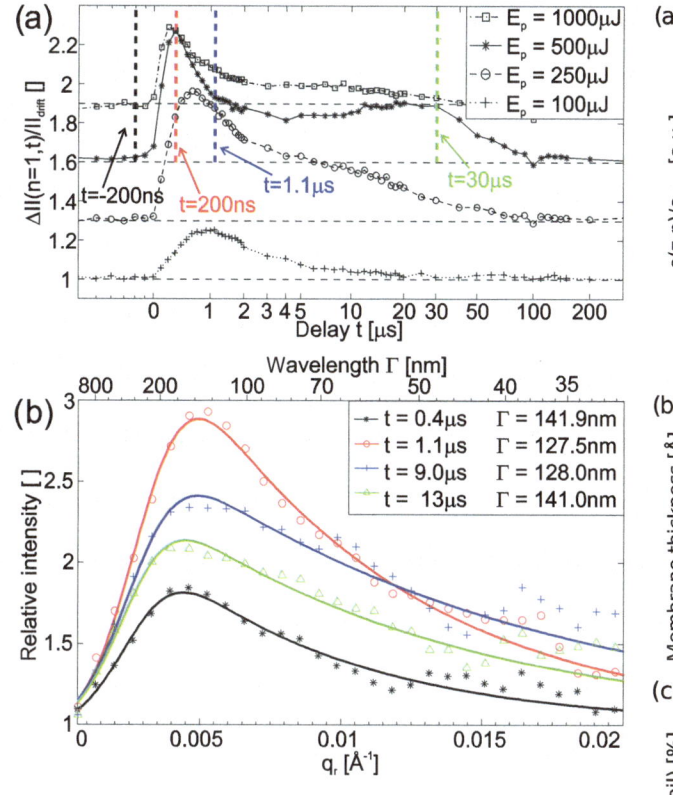

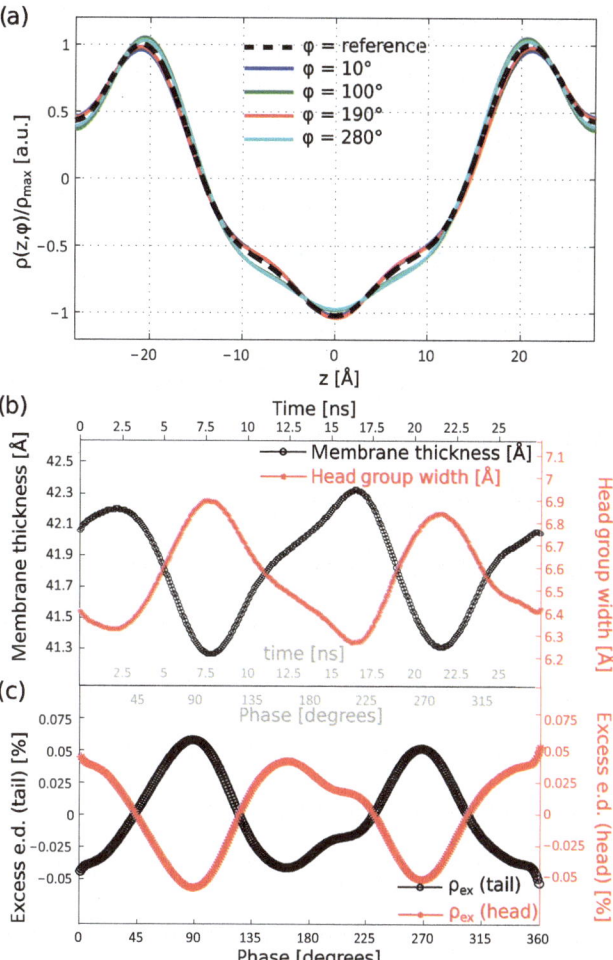

FIGURE 4.4 (a) Evolution of the diffuse scattering intensity as a function of time after excitation. The increase reflects the transient increase in membrane undulations. (b) Characteristic undulation modes appear with a wavelength band centred around 130 nm reaching its maximum about 1 μs after excitation. (Adapted from Reusch et al. 2013a, copyright with the author.)

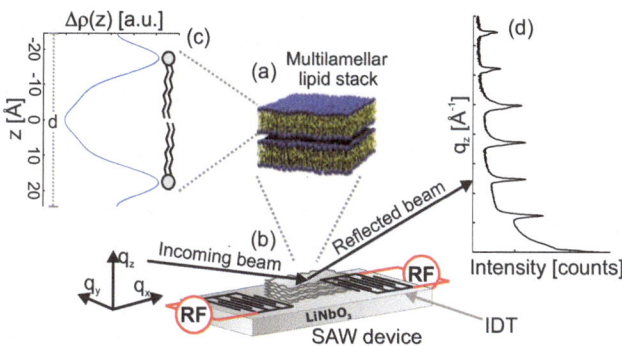

FIGURE 4.5 (a) Schematic of the experiment on acoustically driven membranes. Multilamellar stacks of lipid bilayers are deposited on the piezoelectric $LiNbO_3$ substrate of a surface acoustic wave (SAW) device. (b) By application of a RF signal to the interdigital transducers (IDTs), standing waves or propagating SAW pulses are induced. The ultrafast response of (c) the averaged bilayer electron density profile is measured by (d) time-resolved (phase-locked) X-ray reflectivity experiments. (Adapted from Reusch et al. 2014. Copyright with the author.)

for the simple reason that the relevant length scales of few nm and the highly transient structural rearrangements are inaccessible by existing microscopy techniques, nor by MX. Understanding the structural rearrangements during

FIGURE 4.6 (a) The lipid bilayers' response to the acoustic excitation is quantified by the time evolution of the electron density profile, shown here for selected phase angles along with the equilibrium reference profile (dashed line). Periodic vibrations of the resulting relative electron density are observed. The density variations are accompanied by a variation of membrane thickness. (b) Characteristic bilayer parameters (membrane thickness and head group width) extracted from the profile. The maxima in the head group width occur when the membrane thickness exhibits a minimum, and (c) a reduction of the electron density in the head group region (negative excess density) is accompanied by a density increase in the tail region (positive excess density). (Adapted from Reusch et al. 2014. Copyright with the author.)

membrane fusion from the perspective of the lipids, which actually undergo the merger, thus presented a particular challenge. The first hemifusion intermediate, in which the proximal lipid monolayer leaflets of adjacent membranes have merged, is the so-called stalk shown in Figure 4.7. In fact, stalks have been postulated nearly 30 years ago and have since then received much attention, starting with analytical models (see, for example, Kozlovsky and Kozlov, 2002, Markin and Albanesi, 2002, Chernomordik and Kozlov, 2005, 2008, and references therein). Subsequently, numerical simulations extended the continuum models by taking into account the molecular details of lipids and proteins, as well as a broader perspective of the entire fusion

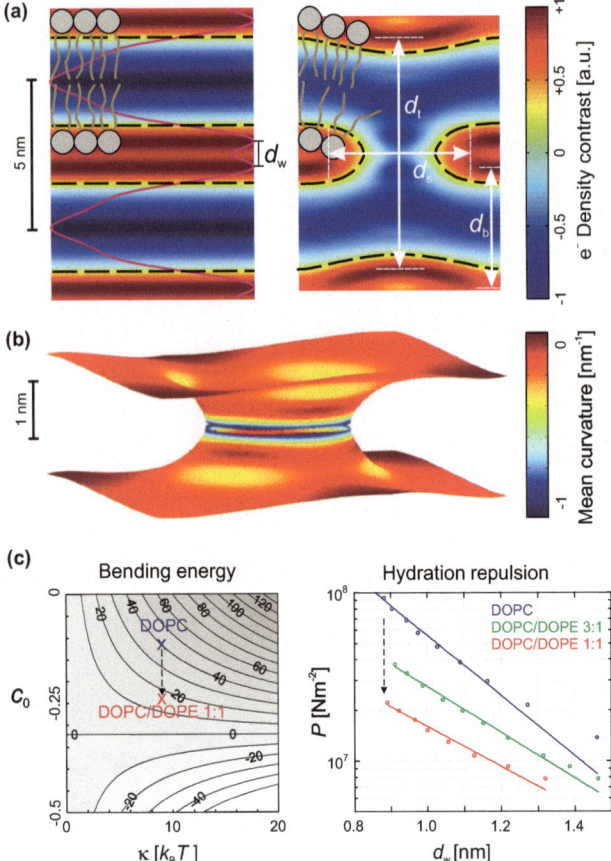

FIGURE 4.7 (a) 2d Electron density representations (DOPC/DOPE 1:1) of two lipid bilayers (left) and of a slice through a stalk (right). Irregardless of the molar fraction of the nonbilayer lipids DOPE or cholesterol, stalk phase formation occurred at a critical interbilayer distance $d_w^* = 9.0 \pm 0.5$ Å. Dashed black lines indicate contours of constant electron density. (b) Corresponding 3d electron density isosurface and the local distribution of mean curvature H. (c) Energy landscape of the bending energy $\kappa/2 \int dA(2H - c_0)^2$ (left). Literature values of the elastic coefficients lead to a minimum of about 15 k_BT. Pressure–distance curves $P(d_w)$ in the lamellar phase allow to determine the energy required for dehydration to d_w^*. (From Aeffner et al. 2012. Copyright with the author.)

pathway (Schick, 2011, Risselada and Grubmüller, 2012, Fuhrmans et al., 2015). Stalk formation is now widely accepted as a common step in all membrane fusion reactions. It is therefore important to know the associated energy barrier of this intermediate state, its structure, and curvature properties, as well as the variation for different types of lipids. At present, the only experimental method to provide this information is X-ray diffraction from lipid mesophases, as discussed in this section.

The field opened up when H.W. Huang and coworkers discovered that stalks form under conditions of strong dehydration in multilamellar stacks of many unsaturated lipids (Yang and Huang, 2002). The stalks form a superlattice of rhombohedral symmetry in the matrix of fluid membranes. Electron density reconstructions of stalks based on X-ray diffraction from such phases have provided an experimental basis for a field, which was previously restricted to theory and numerical simulations. This approach was subsequently extended, now covering a wide range of model lipid bilayers with high resolution and analysis of the energetics, yielding useful insights into the physical principles governing the merger (Aeffner et al., 2012). Of major biological relevance is the decisive role of the lipid composition, as the phase diagram and the stalk energy strongly depend on the lipid mixture. By ways of a structural assay, one can now classify lipids into stalk promoters and stalk inhibitors (Khattari et al., 2015, Xu et al., 2018). In search for 'magic' fusogenic mixtures, several ternary and quaternary lipid systems have been screened, including phosphatidylcholines, phosphatidylethanolamines, sphingomyelin, cholesterol, diacylglycerol, and phosphatidylinositol (Khattari et al., 2015). In most cases, the starting point for the formulation of more complex systems was a mixture of dioleoylphosphatidylcholine (DOPC) with dioleoylphosphatidylethanolamine (DOPE). Several fusion promoting lipids, i.e., lipids that reduce the osmotic pressure required for stalk formation, have been identified. Adding several of these at the same time may efficiently destabilize the lamellar phase. However, the result is not always a clean stalk phase but often leads to phase separation. Recently, a correlated stalk fluid was observed just above the phase transition to the stalk phase (Scheu et al. 2021).

Figure 4.7 shows the electron density $\rho(r)$ of a fusion stalk, as derived from the measured peak intensities, recorded by X-ray diffraction from the stalk (rhombohedral) phase. The Bragg reflection was phased by the swelling method, assuming inversion symmetry of the unit cell. The electron density maps are visualized by cuts through the 3D distribution $\rho(r)$. With $\rho(r)$ given as a Fourier series, analytical and numerical analysis is significantly facilitated. For example, the Gaussian and mean curvature of the stalk can be computed as an integral over the implicit electron density isosurface chosen to represent the hydrophobic–hydrophilic interface. This interface has a strongly peaked mean curvature at the waist of the stalk ($z=0$ plane). Furthermore, it was found that the stalk structure is highly conserved, i.e., the shapes become very similar, when characteristic dimensions such as the diameter of the stalk 'neck' d_s and the transversal extent d_t are scaled to the bilayer thickness d_b (Aeffner et al., 2012). The ratio d_t/d_b was found to be around 1.2, indicating a more stretched conformation of the acyl chains in this region, which is in line with wide-angle diffraction covering the chain correlation peak in stalk phases (Weinhausen et al., 2012). The experimental stalk shape is in excellent agreement with numerical simulations (Smirnova et al., 2010, Fuhrmans et al., 2015).

The structural variation of the stalk with lipid composition is surprisingly small. Furthermore, the onset of stalk formation occurred within a very narrow range of interbilayer (critical) bilayer distance Indeed, from the electron density profiles, a critical bilayer separation of $d_w = 9.0 \pm 0.5$ Å was found. This value varied only very slightly with lipid composition (Aeffner et al., 2012, Khattari et al., 2015). Contrarily, the differences in the energetics are considerable, since (i) the work performed in dehydrating the membranes and (ii) the coefficients (moduli) of mean and Gaussian curvature vary significantly with lipid composition. To compute the curvature energy, the mean and Gaussian curvatures are integrated over the entire density isosurface representing the neutral surface of the stalk. Interestingly, the experimental shape of the neutral surface is found in between two functional forms proposed in literature, namely that postulated in Kozlovsky and Kozlov (2002) with a kink in the neutral surface and that of Markin and Albanesi (2002) which is obtained by setting the mean curvature constant everywhere on the neutral surface. Given literature values for the elastic coefficients and within the limits of the Helfrich continuum model, the curvature energy of a stalk can then be computed, separately for mean and Gaussian curvature contributions. Together with the dehydration energy, this accounts for the entire energy barrier. To quantify the dehydration energies, the pressure–distance curves $d(p_{osm})$ (equation-of-state) were measured for a large variety of lipid mixtures (Aeffner et al., 2012, Khattari et al., 2015). By integration of the hydration repulsion $P(d_W)$ from full hydration to the critical value of the interbilayer water distance d_w, at which the stalk phase forms, the work per unit area required to reach the stalk phase can be computed, corresponding to the minimum energy which has to be released by fusogenic proteins. Considering both curvature and the dehydration energies, we illustrate the role of lipid composition with the example of an equimolar DOPC/DOPE mixture with mean curvature modulus 9 k_BT, Gaussian curvature modulus 8 k_BT, and spontaneous curvature value of 0.24 nm^{-1} (Siegel, 2008). This yields a total of 115 k_BT for the curvature energy per stalk (Aeffner et al., 2012), already significantly less than for pure DOPC. At the same time, the work performed in dehydration of the membranes from full hydration to the onset of stalk formation amounts to another 89.17 k_BT per stalk for the DOPC/DOPE mixture, while it is almost twice as high for pure DOPC, namely 17.317 k_BT (Aeffner et al., 2012). This shows that the high content of the nonbilayer lipids phosphatidylethanolamine and cholesterol lowers the energy barrier of the stalk significantly. Since 1–3 SNARE complexes were shown to be sufficient to induce membrane fusion (van den Bogaart et al., 2010, Mohrmann et al., 2010), and with an energy release of 35k_BT per complex (Li et al., 2007), the bilayer merger would become impossible for an example for the extreme case of pure DOPC, while addition of DOPE, cholesterol, or PIP$_2$ can sufficiently reduce this barrier. Possibly, the local lipid composition can also be changed in a controlled fashion at the synaptic membrane, for example, by Ca^{2+} influx. Altogether, the experimental model systems are highly indicative for a significant modulation of the fusion barrier by lipid composition. This and further conclusions from X-ray diffraction and quantitative analysis complement the protein-centred view of membrane fusion. However, to bring both aspects together, future work should be directed at reconstitution of SNARE proteins in stalk phases, followed by similar X-ray analysis (phase diagram, structure, and energetics) as discussed here.

4.5 X-RAY ANALYSIS OF BIOLOGICAL MEMBRANES: THE EXAMPLE OF SYNAPTIC VESICLES

Membrane fusion was discussed in the previous section solely from a perspective of lipid model bilayers. To extend beyond pure model lipids, we review in this section recent work on the structure and interaction of synaptic vesicles (SVs). SVs are small organelles consisting of a proteoliposome, designed by nature for controlled fusion with the synaptic membrane. They encapsulate the neurotransmitters for release into the synaptic cleft at the nerve terminals. Triggered by an action potential, SVs undergo a highly controlled fusion event with the synaptic membrane, catalysed by fusogenic SNARE proteins (Takamori et al., 2006). The composition and anatomy of SVs was mainly established based on biochemical characterization (Takamori et al., 2006). Given an average radius around $R = 20$ nm, structural analysis of SVs presents a challenge regarding the spatial resolution required to identify its lipid and protein constituents. By cryogenic electron microscopy (cryo-EM), a dense outer and inner layer of proteins around the vesicular membrane becomes visible, but the EM images do not allow the quantification of size or density, neither of the vesicular membrane itself nor of the associated protein layers.

We have therefore used small-angle X-ray scattering (SAXS) to study SV structure in solution, using the protocols to isolate and purify the SVs as established by the group of R. Jahn (Takamori et al., 2006). SAXS provides a structural probe with high (near molecular) resolution under physiological conditions (buffers, pH, osmotic pressure, salt, etc.) but had previously not been applied to the level of entire organelles. With the advent of highly brilliant undulator radiation suitable for small volumes, short recording times, and low concentrations, this has become feasible, for the example at ESRF's high brilliance SAXS beamline ID02. As illustrated in Figure 4.8, the following structural properties have been deduced from the least-square fit of the SAXS curves (Castorph et al., 2010b): (i) radial density profile $\rho(r)$ including the lipid bilayers and the inner and outer protein layers, (ii) the polydispersity (i.e. the distribution $p(R)$), and (iii) the 'graininess' of the protein layers. Since the scattering intensity was measured on absolute scale, the size distribution $p(R)$ could also be deduced without free parameter on an absolute scale. From the fully modelled SAXS results, the polydispersity $p(R)$ obtained by more readily accessible methods such as dynamic light scattering (DLS) could be properly 'calibrated' (Castorph et al., 2011). As it turned out, the SAXS curve could not be fitted to a model based on strictly radially symmetric profiles, but instead required distinct patches of density in the protein layers, simulated as Gaussian polymer chains (Castorph et al., 2010b). The fitting approach was checked

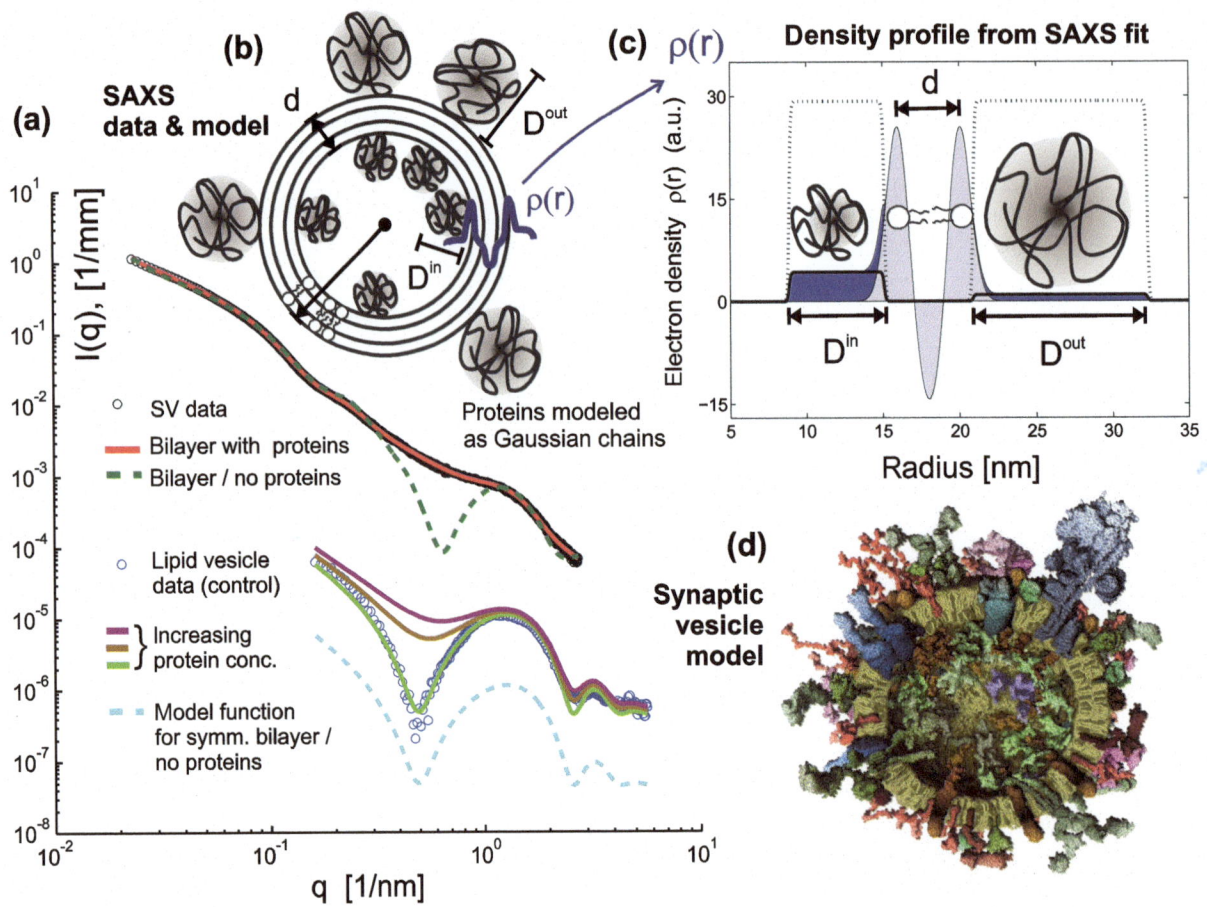

FIGURE 4.8 SAXS analysis of synaptic vesicles (SVs), adapted from Castorph et al. (2010a, b). (a) SAXS curve of a SV suspension (black open circles, shifted for clarity) along with the least-square fit (solid red line) to a model of the sketched protein-covered vesicle. The structural model parameterizes the protein layers with protein radii of gyration, effective number of protein patches, and densities as free fitting parameters. Models without patchy proteins cannot reproduce the data (dashed green curve). The lower curve (blue open circles) corresponds to a control sample of pure lipid, along with the model curves shown for increasing protein content. (c) The density profile of the SV membrane with the adjacent protein layers as obtained from the least-square fit of the SAXS curve shown in (a). (d) Visualization of a SV with the densely packed layer of proteins, from Takamori et al. (2006). (Reproduced with permission.)

by verifying that control samples of pure lipid vesicles did not yield any density in the protein layers. Overall, the SAXS study provided a quantitative confirmation for the model in Takamori et al. (2006), which was put forward earlier based on the crystal structures of the constituent proteins and stoichiometric analysis. The study is included in this chapter as a demonstration that the structure of entire membranous organelles can be studied by SAXS. Given the further developments in particular of size exclusion chromatography integrated into SAXS instruments, more interesting results may follow on quasi-monodisperse suspensions requiring less free parameters for the fits.

As a next step, the interaction of SVs with lipid membranes bilayers and monolayers was probed (Castorph et al., 2010b, Ghosh et al., 2010, 2012). By using mixed suspensions of SV and proteoliposomes, fusion and interactions of SV can be evidenced by SAXS (Castorph et al., 2010b, Ghosh et al., 2010). An obvious approach is to first prepare a solid-supported bilayer or a lipid monolayer at the air–water interface of a Langmuir film balance (trough), as a simple model of the presynaptic membrane. SVs are then added, and the interaction is probed by surface-sensitive scattering, i.e., by X-ray reflectivity, SAXS in grazing incidence (GISAXS), and grazing incidence (wide angle) diffraction (GID) (see Figure 4.9). The vesicle–membrane interface can thus be studied simultaneously by the diffraction and the Langmuir film balance read-out, with lipid monolayers of controlled composition, adjustable pH, and salt concentration. Since the interaction is known to be Ca^{2+} dependent, and controlled by the vesicular protein synaptotagmin in combination with phosphatidylinositol 4,5-bisphosphate (PIP_2), the phospholipid model bilayers and monolayers were prepared both with and without PIP_2. This highly negatively charged lipid of the plasma membrane is believed to act as a Ca^{2+} sensor and is also known for its propensity to induce nonlamellar phases in model bilayers, even at comparatively low concentrations (Ghosh et al., 2010). To probe the interactions with SVs, X-ray reflectivity and

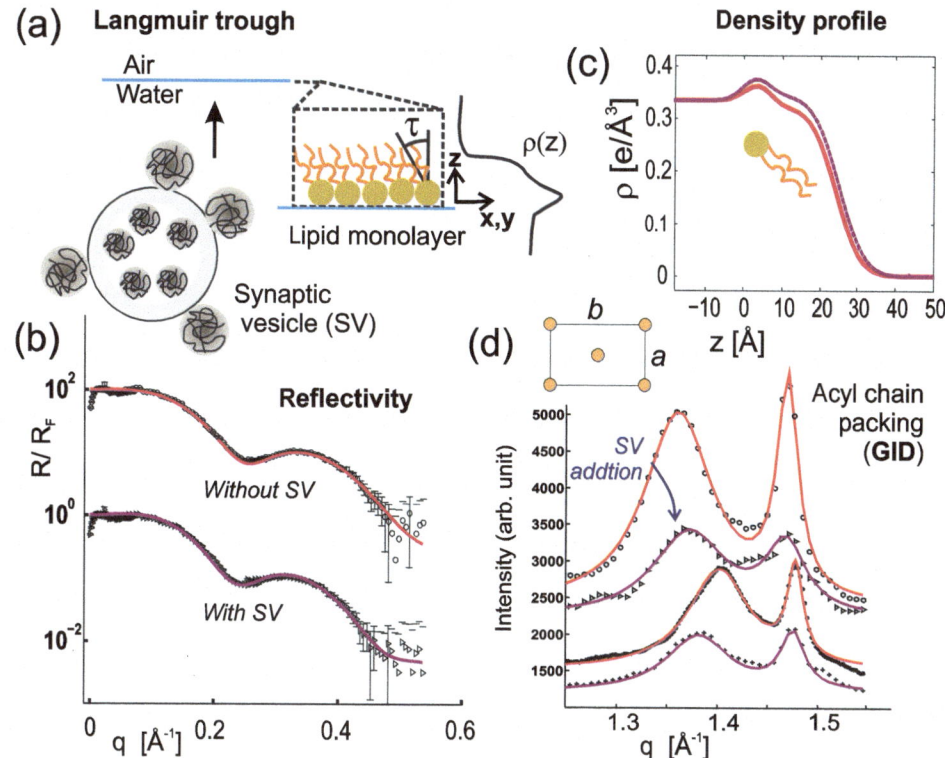

FIGURE 4.9 X-ray diffraction from lipid bilayers or monolayers prepared by a Langmuir film balance, with added synaptic vesicles (SVs) from solution. The film balance experiment serves as a model for the interaction of SVs with the inner leaflet of the synaptic membrane. (a) Schematic of the Langmuir monolayer experiments. (b) Reflectivity curves of the monolayer, with and without added SVs, and (c) corresponding density profiles, deduced from the reflectivity fits (solid red and purple lines, respectively). The results indicate structural changes between (a) DPPC monolayer before (top) and after (bottom) injection of SVs into the subphase, which are larger for DPPC/PIP2 than for the pure DPPC film. (d) Grazing incidence diffraction curves (shifted for clarity) indicating the acyl chain ordering in the film, for (from top to bottom): pure DPPC monolayer, DPPC after injection of SVs, DPPC/PIP$_2$, and DPPC/PIP$_2$ with SVs. The Bragg peaks are fitted by two Lorentzians (solid lines). The centred rectangular unit cell is illustrated in the inset. (Adapted from Ghosh et al. (2010) Ghosh et al. 2012. Copyright with the author.)

GID experiments were carried out on a monolayer of dipalmitoyl-sn-glycero-3-phosphatidylcoline (DPPC) with and without addition of PIP$_2$. Using a Langmuir trough installed at the ID10B beamline of ESRF, systematic changes in the lipid film were observed after SV injection in the subphase. A small decrease in the area per lipid and the tilt angle (from acyl chain diffraction) in the DPPC film were induced by SV association, accompanied by corresponding changes in the density profile. A collective reorganization of the lipid film was thus induced by local binding of a SVs. At the given concentration, the results are only plausible when assuming a nonlocal response of acyl chain tilt and film density to SV association. Furthermore, the relative changes were much more pronounced in the presence of PIP$_2$. Finally, it could be shown that the association was intensified by a physiologically relevant amount of Ca^{2+} ions in the subphase of the monolayer. The results led to the conclusion that the collective structural changes in the lipid induced by only 5% mol. concentration of PIP$_2$ may well modulate vesicle fusion in vivo and that collective degrees of freedom of the lipid membrane must be taken into account in addition to the specific interactions of fusion proteins.

4.6 MYELINATED AXONS STUDIED BY SCANNING SMALL-ANGLE X-RAY SCATTERING AND PHASE CONTRAST TOMOGRAPHY

As a further example of diffraction from biological membranes, we consider the example of the multilamellar myelin sheath, the well-known membranous structure in nerve tissue formed by extensions of the plasma membrane of the myelinating glial cells, the oligodendrocytes in the CNS, and the Schwann cells in the PNS (Quarles et al., 2006). Myelin ensures the electrical insulation of axons, and its unique segmental structure enables the saltatory conduction of nerve impulses necessary for fast nerve conduction in the thin fibres. The importance of myelin and its structure for signal conduction is further evidenced by its role in different neurological diseases, including multiple sclerosis. The periodicity of the lamellae is around 16 nm, depending on species and neuron type, and the typical length of each myelin sheath segment or internode is in the range of 150–200 μm in the CNS and up to 1 mm in the PNS (Kirschner and Blaurock, 1992, Siegel, 2006). Internodes are separated by spaces where myelin is lacking, the nodes of Ranvier (RN). In more or less regular distances, the compact myelin structure is interrupted by clefts, the so-called Schmidt-Lanterman (SL) incisures.

The basic structure of myelin was established by classical diffraction experiments which average over the entire nerve, and by EM which shows the individual configuration (width of the myelin sheath, compactness, number of lamellae, shape of the axon). Given the intrinsic heterogeneity of biomolecular assemblies, averages over large ensembles, which are typical for most diffraction experiments, are of limited use when considering cells or tissues. Therefore, we have recently tried to extend classical nerve diffraction to the scale of isolated axons. By using nano-focused beams, interesting regions of single axons, for example, at the node of Ranvier can be studied, complementing the reciprocal space resolution with beam-size-limited resolution in real space. This can, for example, be useful to study pathologies of myelin structure in neuronal diseases. As a proof-of-concept, we have studied myelin structure in sciatic neurons, isolated from wild-type mouse by a teased fibre preparation, by recording local SAXS diffraction patterns with a nano-focused beam scanning the axon, first in Ducic et al. (2011), and later with improved X-ray optics in Salditt et al. (2015). Important structural parameters such as the lamellar periodicity d (from the radial peak position), membrane orientation (from the angular orientation of diffraction peaks), and possibly parameters of myelin order/disorder (from the peak width and intensity variations) can thus be accessed on a local scale (Salditt et al., 2015). The requirements regarding beam preparation and instrumental settings (preparation of clean tails in focusing, detector read-out rate and dynamic range, protection from radiation damage by cryogenic conditions) have been addressed in Salditt et al. (2015). To this end, an optimized nano-focusing set-up based on a high-gain Kirkpatrick-Baez mirror system combined with different cleaning apertures was used in order to suppress spurious tails of the beam and to obtain clean myelin diffraction signals from sub-μm spot sizes.

Figure 4.10 shows the typical diffraction signals, recorded on freeze-dried single fibre preparations (Salditt et al., 2015),

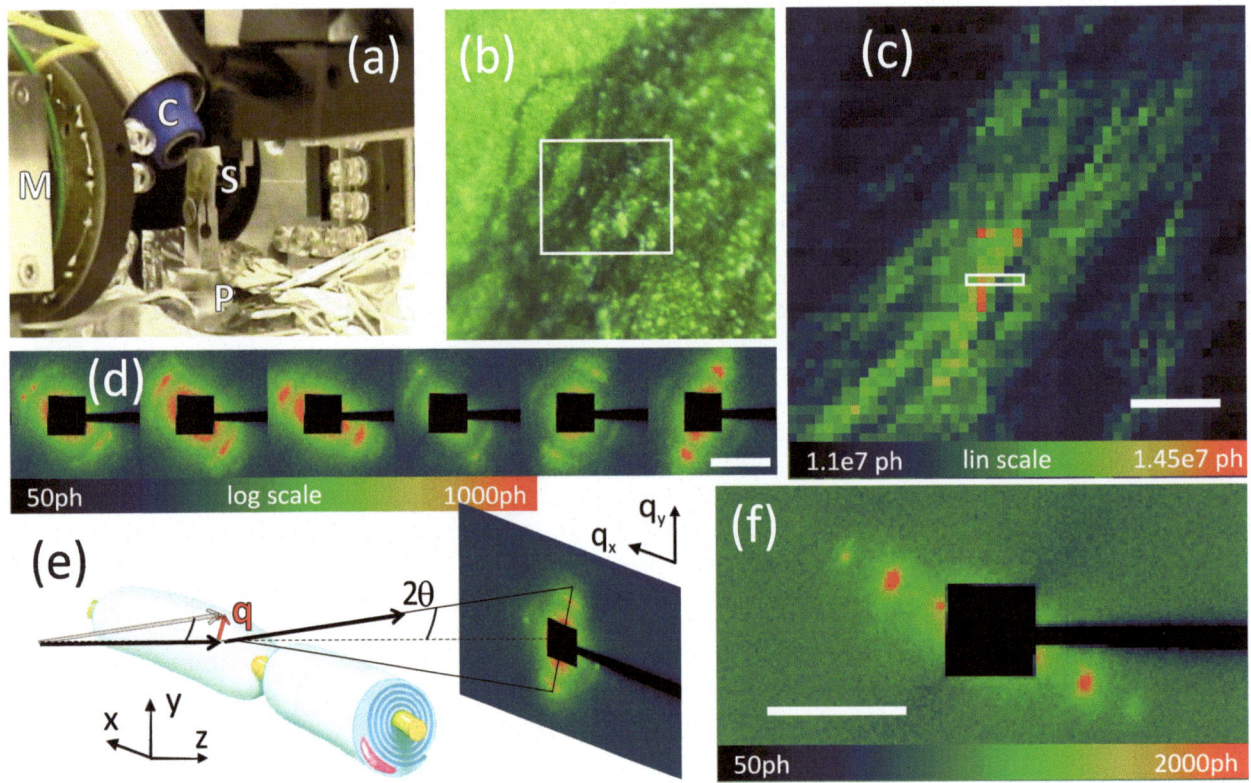

FIGURE 4.10 Nanobeam diffraction from myelinated nerves. (a) Photograph of the sample (S) mounted on the piezo table (P) in the cryo-stream (C), with the in situ on-axis visible light microscope (M). (b) View of the sample in the optical microscope (M), used for alignment and inspection during the X-ray exposure. The indicated rectangle (white lines) is scanned in (c). (c) Myelin scattering intensity map (darkfield) for scan field of 100 μm × 100 μm, in units of photon (ph) number per 50 (milli- s)$^{-1}$ acquisition time and detector pixel (region of interest around the lamellar diffraction). The real space pixel size in is 2 μm and hence significantly larger than the beam size for this overview image, showing several parallel nerve fibres as deposited on a polyimide foil. At each pixel, a complete diffraction pattern is recorded, as exemplified for six pixels in (d), which underline the surprisingly large diversity of local structures. The black square is a shadow of the beamstop. (e) Sketch of a diffraction experiment, illustrating the scattering geometry and a typical myelin diffraction intensity distribution. (f) In some spots, the order is high enough to yield pronounced higher harmonic reflections. Scale bars: (c) 20 μm; (d, f) 1 nm^{-1}. From Salditt et al. 2015. (Reproduced with permission of the International Union of Crystallography.)

by scanning at 7.9 keV photon energy with a beam size of 280 nm (vertical) and 360 nm (horizontal). To minimize damage, the sample was kept in a cryogenic nitrogen gas jet (Oxford CryoSystems). A photo of the sample mounted on a thin foil in the cryostream is shown in (a); an on-axis microscope image used for alignment is shown in (b). The corresponding X-ray darkfield map (c) shows the contour of axon fibres, which lead to strong diffraction, exemplified for several scan points in (d). For each scan point, a diffraction image was recorded, using a Pilatus 100 K pixel detector (Dectris), positioned 329 mm behind the sample. In some locations, the signal and lamellar ordering was found to be high enough to record higher lamellar reflection orders (see, for example, (f)). More recently, myelin diffraction signals were also recorded by scanning SAXS in histological sections of human brain tissue (Carboni et al., 2017).

After considering the diffraction from a single myelinated axon, as obtained by a complex and invasive preparation from the full nerve with its thousands of axons oriented in parallel along the nerve axis, we now address structure analysis on the scale of an entire unsliced nerve. In order to probe its three-dimensional (3D) anatomical structure underlying the biological function, it is necessary to go beyond diffraction and to reconstruct the entire volume by high-resolution phase contrast X-ray tomography. To date, not much quantitative data are available on the spatial organization of the axon bundles within the nerve, mainly since 3D capability is required for a relatively large volume. Some of the questions that can only be addressed by a 3D structural probe are, for example (Bartels et al., 2015): To what extent are axons organized in form of a rigid and regularly spaced bundle of nerve fibres? What is the size polydispersity in a given nerve and are the positions of RN and SL correlated between different axons? How does the amount and compactness of myelin vary within the nerve? Apart from the structure of the myelin sheath and axons, questions regarding the neuronal connectivity could be answered based on tomographic reconstructions. This would go well beyond the two-dimensional histological sections.

We have recently demonstrated a 3D study of different nerves by phase contrast X-ray tomography, which is intermediate in resolution between optical microscopy and EM, but compatible with high throughput and large field of view (Bartels et al., 2015). Phase contrast is based on the phase shifts originating from different indices of refraction, as governed by the local electron density, and transforms to measurable intensities by free space propagation of the wave field between the sample and the detector. In combination with geometric (cone beam) magnification, it is well suited to cover entire nerves with an average diameter in the range of 100–300 μm and voxel sizes in the range of 100–300 nm. By osmium staining, the myelin structure can be labelled with high contrast. As shown in Figure 4.11, the structures of thousands of axons in unsectioned nerve tissue were visualized simultaneously.

At voxel sizes down to 100 nm, the position and sizes of the neurons, including functional organelles such as the node of Ranvier and SL incisures were determined from segmentation of the 3D density distribution. The method was shown to yield data which is very consistent with histology sections, and EM micrographs, but offered the decisive advantage that much larger volumes can be covered at isotropic resolution.

4.7 FUTURE DIRECTIONS: MEMBRANE STRUCTURE ANALYSIS BY COHERENT X-RAY IMAGING

We close the chapter by considering X-ray analysis of lipid membranes beyond the paradigm of X-ray diffraction and the new opportunities opened up by the advent of coherent X-ray imaging. Diffraction typically yields the averaged structure of a large ensemble of identical scattering objects, for example, of many thousand membranes in the diffraction volume. By reflectivity, it is possible to reduce this to a single bilayer or lipid monolayer, but still at the prize of a global average over macroscopic membrane cross sections. The local membrane structure cannot be probed by conventional diffraction. To overcome this limitation, we have recently demonstrated a dose-efficient X-ray propagation imaging (near-field diffraction) approach to locally resolve thickness, density, and more generally the density profile of membranes along the membrane normal (Beerlink et al., 2009, 2012). By locally averaging the structure in the plane of the membrane over a length scale on the order of 5–20 μm, the molecular structure along the bilayer normal could be resolved in a hydrated sample environment (Beerlink et al., 2009). Importantly, local deviations and profiles become accessible. The scheme allows to probe membranes under a large variety of external control parameters such as exerted forces, out-of-equilibrium transport, or local fields.

Figure 4.12 presents the principle of the near-field diffractive imaging technique of lipid bilayers, and some main results obtained in Beerlink et al. (2012). For the experiments, single freely suspended lipid bilayer were prepared using a well-known set-up of membrane electrophysiology, the so-called black lipid membranes (BLMs), as an established model system in membrane biophysics. The bilayer was spanned in between two separated, aqueous compartments, allowing for studies of functional transport across the bilayer at controlled compositional and environmental parameters. The membrane was bulged by hydrostatic pressure and placed a few millimetres behind the focal plane of a highly efficient and partially coherent elliptical mirror system. This quasi-spherical illumination led to a magnified hologram of the sample which was then recorded a few metres behind the sample by a fast read-out low noise detector system. The Fresnel oscillations in this hologram corresponding to the native lipid bilayers with a thickness in the range of $d = 5$ nm thickness were recorded and analysed with respect to the local structure at molecular length scales.

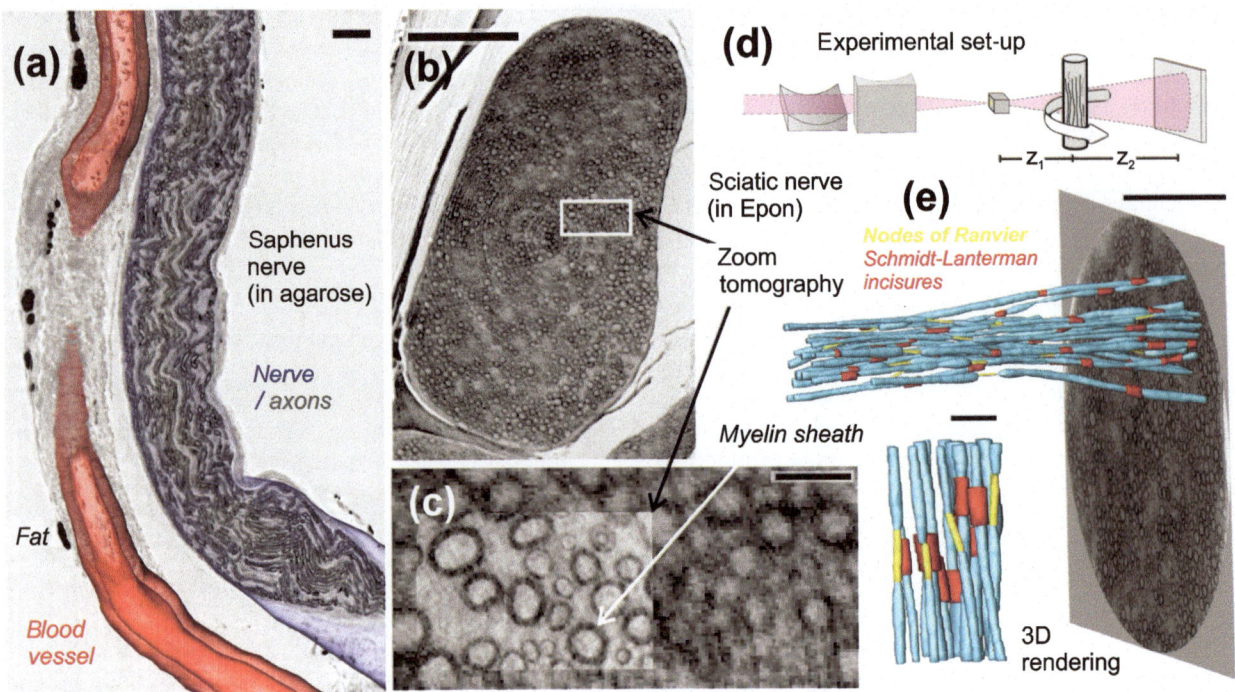

FIGURE 4.11 Phase contrast tomography of nerves (mouse). (a) Three-dimensional rendering of a mouse saphenous nerve (blue) with adjacent blood vessel (red) along with a longitudinal virtual slice. The nerve was stained with osmium tetroxide. Scale bar 50 μm. (b) Virtual slice through an EPON-embedded osmium-stained sciatic nerve, voxel size 430 nm. Scale bar 100 μm. (c) Magnified view of the region marked in (b) along with data obtained from a zoom-tomogram of the same nerve with 50 nm voxel size. Scale bar 10 μm. (d) Experimental set-up: The undulator beam is focused by two elliptically shaped mirrors (KB) and (optionally) filtered by an X-ray waveguide system. The nerve is placed at various distances z_1 from the focus within the divergent (partially) coherent beam and magnified Fresnel diffraction patterns (holograms) are recorded. Subsequent phase retrieval and tomographic reconstruction allows quantitative 3D electron density determination. (e) 3D rendering of a sciatic nerve probed by zoom tomography: (top) view of a 100 nm voxel size dataset is shown along with a virtual slice through the reconstructed volume. Nodes of Ranvier are rendered yellow, SL incisures red. Scale bars 100 μm. (bottom) Rendering of 13 axons, suggesting a correlation between positions of nodes and incisures of neighbouring axons. Scale bar 10 μm. (Adapted from Bartels et al. 2015. Reproduced with permission.)

Next, the formation of a bilayer from two monolayers was imaged by this near-field imaging technique (see Figure 4.12). To this end, the macroscopic sample chamber was replaced by a microfluidic device designed to observe the formation of a reconstituted lipid bilayer by fusion of two surfactant monolayers using controlled flow of oil and water in microfluidic channels and mixers. The throughput of such a system is extremely high, regarding supply of new material (lipids, solvent, and buffer) and change of parameters (pressure, concentration). In both systems, the membrane can be electrically excited by the implementation of electrodes into the aqueous channels. In future, near-field propagation imaging in microfluidic sample environments could be used to probe (hydro)dynamic processes in membranes, such as the formation of bilayers, thinning or bulging, and also membrane fusion, down to the length scale of a few nanometres.

The particular scheme of near-field imaging of membranes discussed here is just one of several examples, how progress in X-ray optics and generation of highly brilliant X-ray radiation enable entirely new approaches in X-ray structure analysis. Ultrashort fully coherent X-ray pulses at X-ray free electron lasers (FELs), for example, as proposed for single particle imaging (Gaffney and Chapman, 2007), could find applications in the analysis of vesicles and membranous assemblies. Furthermore, the prospect of probing not only the structure but also dynamics of membranes appealing, also in view of nonequilibrium effects by pump–probe techniques. However, first FEL data from oriented multibilayers have also indicated considerable technical challenges associated with the stochastic nature of single FEL pulses generated by self-amplified stimulated emission (Mai et al., 2013).

ACKNOWLEDGEMENTS

I thank all of our collaborators on the original work reviewed here, in particular Sebastian Aeffner, Tobias Reusch, Matthias Bartels, André Beerlink, Simon Castorph, Sajal Ghosh, Ziad Khattari, Sebastian Köhler, and Yihui Xu. We are thankful to the European Synchrotron Radiation Facility (ESRF, Grenoble, France), DESY Photon Science (Hamburg, Germany), and the Swiss Light Source (SLS, Villigen, Switzerland) for generous beam time allocation.

X-Ray Scattering and Imaging of Membranes

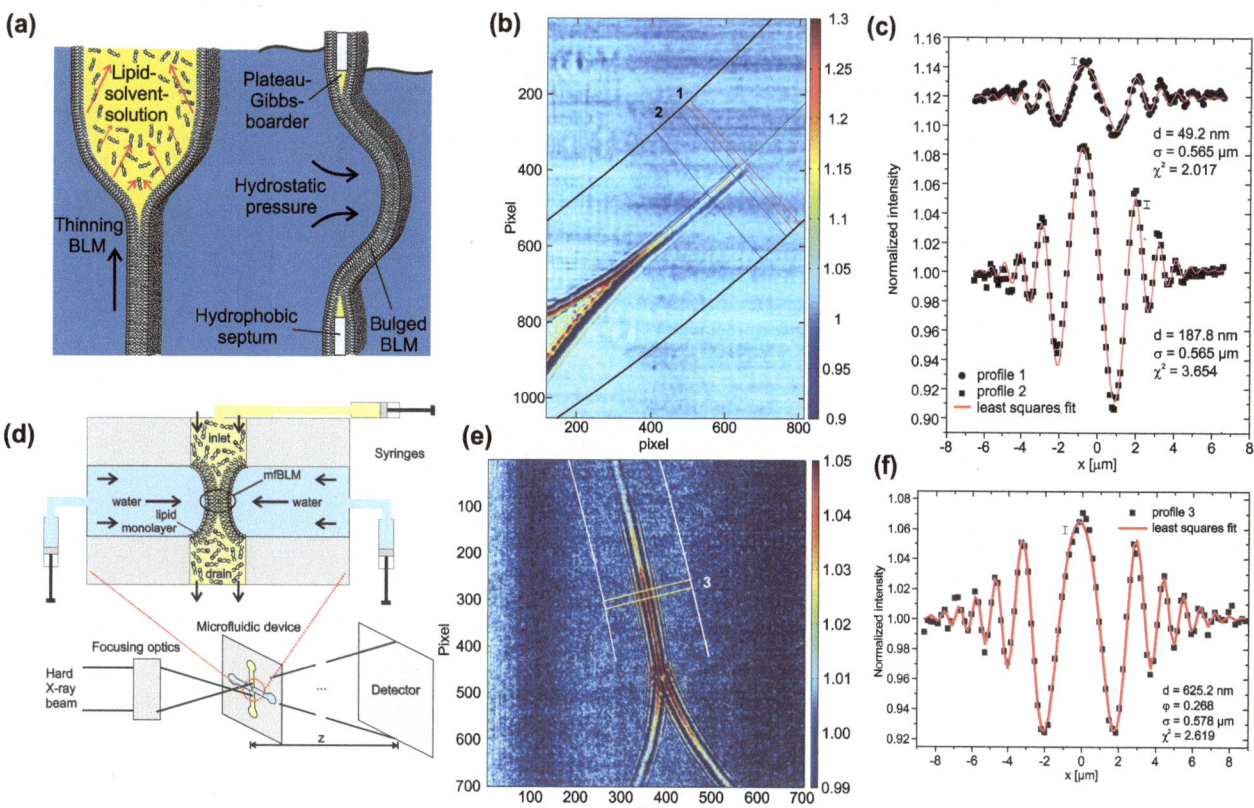

FIGURE 4.12 Phase contrast imaging of lipid membranes using an adapted set-up of black lipid membranes (BLMs). The goal of the experiment was to combine such a set-up equipped for electrophysiology with an in situ structural X-ray probe. To this end, a highly focused X-ray beam (17.5 keV) coherently illuminates a spherically BLM, spanned over a micro-machined hole and located downstream the focal plane of a KB-mirrors system. The image is formed by free propagation of the wave field between the sample and the detector, where the intensity profile is recorded. (a) Schematic of the formation and bulging process of a BLM. Organic solvent, used to dissolve the lipid molecules, diffuses towards the outer rim of the aperture supporting the membrane. The BLM starts to thin until the two monolayers at the oil–water interfaces approach to finally form a bilayer lipid membrane. Application of hydrostatic pressure to one side of the BLM leads to bulging of the interface. (b) Typical data set showing the BLM contour and Fresnel oscillations, shown in (c) for the indicated cuts, which are then fitted to a model of the bilayer profile, yielding the local bilayer structure. Analysis of the Fresnel fringes of a completely thinned native bilayer indicating a thickness of around 3 nm (not shown). (d) Microfluidic chamber for monolayer fusion experiments, and positioning in the X-ray beam. (e) Image of two adhering monolayers, along with (f) the associated analysis of the profile. In this way, the onset of bilayer formation, which proceeds via 'zippering effect', was monitored. (Adapted from Beerlink et al. 2012. Copyright with the author.)

Finally, we are grateful for very enjoyable collaborations with Reinhard Jahn, Matthew Holt, Wiebke Moebius, Achim Wixforth, Marcus Müller, and Yuliya Smirnova. Financial support by the German Science Foundation through grants SFB 1286/Project A2 and SFB 803/Project B1 is gratefully acknowledged.

REFERENCES

Aeffner, S. (2011). Stalk structures in lipid bilayer fusion studied by x-ray diffraction. PhD thesis, Institute for X-ray Physics, Georg-August University Goettingen.

Aeffner, S. (2012). *Phospholipids Bilayers: Physical Principles and Models*. Göttingen: Universitätsverlag Göttingen.

Aeffner, S., Reusch, T., Weinhausen, B., and Salditt, T. (2012). Energetics of stalk intermediates in membrane fusion are controlled by lipid composition. *PNAS*, 109(25):E1609–E1618.

Als-Nielsen, J. (2001). *Elements of Modern X-Ray Physics*. Hoboken, NJ: John Wiley & Sons.

Bartels, M., Krenkel, M., Cloetens, P., Möbius, W., and Salditt, T. (2015). Myelinated mouse nerves studied by X-ray phase contrast zoom tomography. *Journal of Structural Biology*, 192:561–568.

Beerlink, A., Mell, M., Tolkiehn, M., and Salditt, T. (2009). Hard X-ray phase contrast imaging of black lipid membranes. *Applied Physics Letters*, 95(20):203703.

Beerlink, A., Thutupalli, S., Mell, M., Bartels, M., Cloetens, P., Herminghaus, S., and Salditt, T. (2012). X-ray propagation imaging of a lipid bilayer in solution. *Soft Matter*, 8:4595–4601.

Brzustowicz, M. R. and Brunger, A. T. (2005). X-ray scattering from unilamellar lipid vesicles. *Journal of Applied Crystallography*, 38(1):126–131.

Caille, A. (1972). Remarques sur la diffusion des rayons X dans les smectiques. *Comptes Rendus de l Académie des Sciences. Series B*, 274:891–893.

Carboni, E., Nicolas, J.D., Topperwien, M., Stadelmann, C., Lingor, P., and Salditt, T. (2017). Imaging of neuronal tissues by x-ray diffraction and x-ray fluorescence microscopy: evaluation of contrast and biomarkers for neurodegenerative diseases. *Biomed. Opt. Express* 8, 4331–4347

Castorph, S., Arleth, L., Sztucki, M., Vainio, U., Ghosh, S. K., Holt, M., Jahn, R., and Salditt, T. (2010a). Synaptic vesicles studied by SAXS: Derivation and validation of a model form factor. *Journal of Physics: Conference Series*, 247(1):012015.

Castorph, S., Riedel, D., Arleth, L., Sztucki, M., Jahn, R., Holt, M., and Salditt, T. (2010b). Structure parameters of synaptic vesicles quantified by small-angle X-ray scattering. *Biophysical Journal*, 98(7):1200–1208.

Castorph, S., Schwarz Henriques, S., Holt, M., Riedel, D., Jahn, R., and Salditt, T. (2011). Synaptic vesicles studied by dynamic light scattering. *The European Physical Journal*, 34:1–11. Doi: 10.1140/epje/i2011-11063-2.

Chappa, V., Smirnova, Y.G., Komorowski, K., Müller, M., and Salditt, T. (2021). The effect of polydispersity, shape fluctuations, and local curvature on small unilamellar vesicle SAXS curves. *Journal of Applied Crystallography*, 54:557-568.

Chen, H.-Y. (2004). Internal states of active inclusions and the dynamics of an active membrane. *Physical Review Letters*, 92:168101.

Chernomordik, L. V. and Kozlov, M. M. (2005). Membrane hemifusion: Crossing a chasm in two leaps. *Cell*, 123(3):375–382.

Chernomordik, L. V. and Kozlov, M. M. (2008). Mechanics of membrane fusion. *Nature Structural & Molecular Biology*, 15(7):675–683.

Constantin, D., Ollinger, C., Vogel, M., and Salditt, T. (2005). Electric field unbinding of solid-supported lipid multilayers. *The European Physical Journal E: Soft Matter and Biological Physics*, 18(3):273–278.

Ducic, T., Quintes, S., Nave, K.-A., Susini, J., Rak, M., Tucoulou, R., Alevra, M., Guttmann, P., and Salditt, T. (2011). Structure and composition of myelinated axons: A multimodal synchrotron spectromicroscopy study. *Journal of Structural Biology*, 173(2):202–212.

Fuhrmans, M., Marelli, G., Smirnova, Y. G., and Müller, M. (2015). Mechanics of membrane fusion/pore formation. *Chemistry and Physics of Lipids*, 185:109–128.

Gaffney, K. J. and Chapman, H. N. (2007). Imaging atomic structure and dynamics with ultrafast x-ray scattering. *Science*, 316(5830):1444–1448.

Ghosh, S. K., Castorph, S., Konovalov, O., Jahn, R., Holt, M., and Salditt, T. (2010). In vitro study of interaction of synaptic vesicles with lipid membranes. *New Journal of Physics*, 12(10):105004.

Ghosh, S. K., Castorph, S., Konovalov, O., Salditt, T., Jahn, R., and Holt, M. (2012). Measuring Ca2+-induced structural changes in lipid monolayers: Implications for synaptic vesicle exocytosis. *Biophysical Journal*, 102(6):1394–1402.

Girard, P., Prost, J., and Bassereau, P. (2005). Passive or active fluctuations in membranes containing proteins. *Physical Review Letters*, 94:088102.

Gov, N. (2004). Membrane undulations driven by force fluctuations of active proteins. *Physical Review Letters*, 93(26):268104.

Hennig, M., Wolff, M., Neumann, J., Wixforth, A., Schneider, M. F., and Raedler, J. O. (2011). Dna concentration modulation on supported lipid bilayers switched by surface acoustic waves. *Langmuir*, 27(24):14721–14725.

Hohage, T., Giewekemeyer, K., and Salditt, T. (2008). Iterative reconstruction of a refractive-index profile from x-ray or neutron reflectivity measurements. *Physical Review E*, 77(5):051604–9.

Khattari, Z., Köhler, S., Xu, Y., Aeffner, S., and Salditt, T. (2015). Stalk formation as a function of lipid composition studied by X-ray reflectivity. *Biochimica et Biophysica Acta - Biomembranes*, 1848(1):41–50.

Kirschner, D. and Blaurock, A. (1992). *Myelin: Biology and Chemistry, Chapter Organization, Phylogenetic Variations, and Dynamic Transitions of Myelin*, 3–80. Boca Raton, FL: CRC Press.

Kozlovsky, Y. and Kozlov, M. M. (2002). Stalk model of membrane fusion: Solution of energy crisis. *Biophysical Journal*, 82(2):882–895.

Li, F., Pincet, F., Perez, E., Eng, W. S., Melia, T. J., Rothman, J. E., and Tareste, D. (2007). Energetics and dynamics of snarepin folding across lipid bilayers. *Nature Structural & Molecular Biology*, 14:890–896.

Lin, L. C.-L., Gov, N., and Brown, F. L. H. (2006). Nonequilibrium membrane fluctuations driven by active proteins. *The Journal of Chemical Physics*, 124(7):074903.

Luzzati, V., Gulik-Krzywicki, T., and Tardieu, A. (1968). Polymorphism of lecithins. *Nature*, 218:1031–1034.

Luzzati, V. and Reiss-Husson, F. (1966). Structure of the cubic phase of lipid-water systems. *Nature*, 210:1351–1352.

Mai, D., Hallmann, J., Reusch, T., Osterhoff, M., Düsterer, S., Treusch, R., Singer, A., Beckers, M., Gorniak, T., and Senkbeil, T. (2013). Single pulse coherence measurements in the water window at the free-electron laser flash. *Optics Express*, 21(11):13005–13017.

Manneville, J.-B., Bassereau, P., Lévy, D., and Prost, J. (1999). Activity of transmembrane proteins induces magnification of shape fluctuations of lipid membranes. *Physical Review Letters*, 82:4356–4359.

Manneville, J.-B., Bassereau, P., Ramaswamy, S., and Prost, J. (2001). Active membrane fluctuations studied by micropipette aspiration. *Physical Review E*, 64:021908.

Markin, V. S. and Albanesi, J. P. (2002). Membrane Fusion: Stalk Model Revisited. *Biophysical Journal*, 82:693–712.

Mennicke, U., Constantin, D., and Salditt, T. (2006). Structure and interaction potentials in solid-supported lipid membranes studied by x-ray reflectivity at varied osmotic pressure. *The European Physical Journal E: Soft Matter and Biological Physics*, 20(2):221–230.

Mohrmann, R., de Wit, H., Verhage, M., Neher, E., and Sørensen, J. B. (2010). Fast vesicle fusion in living cells requires at least three SNARE complexes. *Science*, 330(6003):502–505.

Novakova, E., Giewekemeyer, K., and Salditt, T. (2006). Structure of two-component lipid membranes on solid support: An x-ray reflectivity study. *Physical Review E*, 74:051911.

Pabst, G., Rappolt, M., Amenitsch, H., and Laggner, P. (2000). Structural information from multilamellar liposomes at full hydration: Full q-range fitting with high quality x-ray data. *Physical Review E*, 62(3):4000–4009.

Prost, J. and Bruinsma, R. (1996). Shape fluctuations of active membranes. *EPL (Europhysics Letters)*, 33(4):321.

Quarles, R., Macklin, W., and Morell, P. (2006). *Myelin Formation, Structure, and Biochemistry*. London: Academic Press.

Ramaswamy, S., Toner, J., and Prost, J. (2000). Nonequilibrium fluctuations, traveling waves, and instabilities in active membranes. *Physical Review Letters*, 84:3494–3497.

Reusch, T., Mai, D., Osterhoff, M., Khakhulin, D., Wulff, M., and Salditt, T. (2013a). Non-equilibrium collective dynamics in photo-excited lipid multilayers by time resolved diffuse x-ray scattering. *Physical Review Letters*, 11:268101.

Reusch, T., Schülein, F., Bömer, C., Osterhoff, M., Beerlink, A., Krenner, H. J., Wixforth, A., and Salditt, T. (2013b). Standing surface acoustic waves in linbo3 studied by time resolved x-ray diffraction at petra iii. *AIP Advances*, 3(7):072127.

Reusch, T., Schülein, F., Nicolas, J., Osterhoff, M., Beerlink, A., Krenner, H., M¨uller, M., Wixforth, A., and Salditt, T. (2014). Collective lipid bilayer dynamics excited by surface acoustic waves. *Physical Review Letters*, 113:118102.

Risselada, H. J. and Grubmüller, H. (2012). How SNARE molecules mediate membrane fusion: Recent insights from molecular simulations. *Current Opinion in Structural Biology*, 22:187–196.

Safinya, C., Roux, D., Smith, G., Sinha, S., Dimon, P., Clark, N., and Bellocq, A. (1986). Steric interactions in a model multimembrane system: A synchrotron study. *Physical Review Letters*, 57:2718.

Salditt, T. (2003). Lipid-peptide interaction in oriented bilayers probed by interface-sensitive scattering methods. *Current Opinion in Structural Biology*, 13:467.

Salditt, T. (2005). Thermal fluctuations and stability of solidsupported lipid membranes. *Journal of Physics: Condensed Matter*, 17(6):R287–.

Salditt, T., Osterhoff, M., Krenkel, M., Wilke, R. N., Priebe, M., Bartels, M., Kalbfleisch, S., and Sprung, M. (2015). Compound focusing mirror and X-ray waveguide optics for coherent imaging and nanodiffraction. *Journal of Synchrotron Radiation*, 22(4):867–878.

Sankararaman, S., Menon, G. I., and Kumar, P. S. (2002). Two-component fluid membranes near repulsive walls: Linearized hydrodynamics of equilibrium and nonequilibrium states. *Physical Review E*, 66(3):031914.

Scheu, M., Komorowski, K., Shen, C. and Salditt, T. (2021). A stalk fluid forming above the transition from the lamellar to the rhombohedral phase of lipid membranes. *European Biophysics Journal*, 50: 1–14.

Schick, M. (2011). Membrane fusion: The emergence of a new paradigm. *Journal of Statistical Physics*, 142:1317–1323.

Siegel, D. P. (2008). The Gaussian curvature elastic energy of intermediates in membrane fusion. *Biophysical Journal*, 95(11):5200–5215.

Siegel, G. (2006). *Basic Neurochemistry: Molecular, Cellular and Medical Aspects*. London: Elsevier Academic Press.

Smirnova, Y. G., Marrink, S.-J., Lipowsky, R., and Knecht, V. (2010). Solvent-Exposed Tails as Prestalk Transition States for Membrane Fusion at Low Hydration. *Journal of the American Chemical Society*, 132(19):6710–6718.

Székely, P., Ginsburg, A., Ben-Nun, T., and Raviv, U. (2010). Solution X-ray scattering form factors of supramolecular selfassembled structures. *Langmuir*, 26(16):13110–13129.

Takamori, S., Holt, M., Stenius, K., Lemke, E. A., Grønborg, M., Riedel, D., Urlaub, H., Schenck, S., Br¨ugger, B., Ringler, P., M¨uller, S. A., Rammner, B., Gr¨ater, F., Hub, J. S., Groot, B. L. D., Mieskes, G., Moriyama, Y., Klingauf, J., Grubm¨uller, H., Heuser, J., Wieland, F., and Jahn, R. (2006). Molecular anatomy of a trafficking organelle. *Cell*, 127(4):831–846.

van den Bogaart, G., Holt, M. G., Bunt, G., Riedel, D., Wouters, F. S., and Jahn, R. (2010). One snare complex is sufficient for membrane fusion. *Nature Structural & Molecular Biology*, 17:358–364.

Weinhausen, B., Aeffner, S., Reusch, T., and Salditt, T. (2012). Acyl-chain correlation in membrane fusion intermediates: X-ray diffraction from the rhombohedral lipid phase. *Biophysical Journal*, 102(9):2121–2129.

Xu, Y., Kuhlmann, J., Brennich, M., Komorowski, K., Jahn, R., Steinem, C., and Salditt, T. (2018). Reconstitution of SNARE proteins into solid-supported lipid bilayer stacks and X-ray structure analysis. *Biochimica et Biophysica Acta (BBA) – Biomembranes* 1860: 566–578.

Yang, L. and Huang, H. W. (2002). Observation of a membrane fusion intermediate structure. *Science*, 297(5588): 1877–1879.

5 Adhesion Protein Architecture and Intermembrane Potentials: Force Measurements and Biological Significance

Deborah E. Leckband
University of Illinois at Urbana-Champaign, Urbana, Illinois

CONTENTS

5.1 Introduction 83
5.2 Surface Forces Apparatus Measurements of Forces between Membrane-Bound Proteins 84
5.3 Lectin Structures and Intermembrane Binding Potentials 85
    5.3.1 Conformational Flexibility and Protein Recognition 85
    5.3.2 Genetic Length Variants and Receptor Function 86
5.4 Intrinsically Disordered Proteins (IDP) 87
5.5 Immunoglobulin Family Cell Adhesion Molecules 88
    5.5.1 Neural Cell Adhesion Molecule 88
    5.5.2 Posttranslational Modification and Neural Cell Adhesion Molecule Adhesion 90
5.6 Cadherins 90
    5.6.1 Cadherins form Multiple Bonds with Different Mechanical and Kinetic Properties 90
5.7 Intermembrane Confinement Alters Protein (Cadherin) Interactions 92
    5.7.1 Binding Kinetics from Intercellular Adhesion Frequency Measurements 93
    5.7.2 Confinement Reveals Protein Interactions Not Predicted by Models or Detected in Solution 94
5.8 Summary 94
References 95

5.1 INTRODUCTION

Membrane-associated cell adhesion proteins are crucial for a wide range of biological processes ranging from pathogen infectivity to neutrophil adhesion to embryogenesis. Adhesion proteins are essential for forming mechanical connections between cells and the surrounding environment. Yet their structural diversity affects not only their ability to form adhesive contacts but also their adhesion strength, receptor accessibility, avidity, and protein organization within adhesion zones. How protein structures contribute to those functions is not easily assessed from structure determinations or binding energies alone. Intriguingly, the extracellular (EC) regions of many adhesion proteins consist of multiple tandem repeats of similar domains that can extend 100 nm from the cell membrane. These structures are assumed to project binding sites away from membranes towards target surfaces and in some cases may regulate intermembrane spacing. In other cases, proteins are intrinsically unstructured and can bind ligands at much larger distances than more compact, folded structures.

The complex architectures, dimensions, posttranslational modifications, and polymorphisms impact the functions of these membrane proteins, but how? Studies of binding mechanisms typically rely on thermodynamic, kinetic, and structural studies of isolated, protein fragments. However, these approaches cannot determine how protein organization or flexibility affects binding distances, adhesion energies, or binding kinetics. Moreover, investigations of soluble fragments do not reveal the potential effects of confining proteins to two-dimensional membrane surfaces.

This chapter focuses on novel relationships between the architectures of adhesion proteins and their impact on cell surface and intermembrane interactions. We also discuss experimental approaches used to address these issues. The primary focus is on adhesion proteins with complex EC regions. We also consider intrinsically disordered proteins (IDPs), as an example of molecular strategies that enable protein interactions over large distances. This chapter emphasizes the elucidation of functional implications of protein structures that are not easily determined with standard biochemical measurements.

Case studies illustrate how protein lengths, oligomerization, flexibility, and posttranslational modifications affect intermembrane interactions, protein recognition, and adhesion. We also consider how confinement and crowding can alter membrane protein interactions. The physiological significance of these findings is illustrated in different examples.

5.2 SURFACE FORCES APPARATUS MEASUREMENTS OF FORCES BETWEEN MEMBRANE-BOUND PROTEINS

Protein structures affect the distance dependence of forces between membranes or cell surfaces. Quantifying the range of interactions between soft, deformable biomolecules requires both sensitive force sensors and the ability to accurately measure separation distances with nanometre spatial resolution. The surface forces apparatus (SFA) is currently the only approach with these capabilities. This instrument quantifies the normalized force between two, curved, macroscopic surfaces in liquid (or vapour) as a function of their separation distance (Figure 5.1) (Israelachvili 1973, Israelachvili and McGuiggan 1990). Surface separations are determined within ± 0.1 nm by interferometry (Israelachvili 1973). The net force between crossed, hemicylinders F_c normalized by the geometric average radius of the cylinders F_c/R is directly proportional to the energy per area E_A between equivalent flat plates, according to the Derjaguin approximation: $F_c/R = 2\pi R$ (Hunter 1989, Israelachvili 1992), which is valid when $R \gg D$. Because $R \sim 1$ cm and the range of molecular forces $D < 100$ nm, SFA measurements give intersurface energies. The effective contact area is ~300 µm², and the force typically reflects ~10^4 protein interactions. The net force is determined with a resolution of ±1 nN, but the normalized force sensitivity $\Delta F/R$ is ± 0.1 mJ m^{-2}, enabling quantification of interactions with energies on the order of the thermal energy $k_B T$ (Leckband and Israelachvili 2001). Here, k_B is Boltzmann's constant, and T is the absolute temperature. The SFA differs from single-bond rupture measurements such as atomic force microscopy, for example, because it quantifies near equilibrium energies between surfaces displaying large numbers of molecules, rather than the nonequilibrium strengths of single intermolecular bonds (Leckband and Israelachvili 2001). Importantly, the SFA is uniquely capable of quantifying the absolute distances between soft materials with subnanometre resolution.

The distance-dependent, normalized forces between two surfaces depend on the structure, composition, and density of molecules on those surfaces. In the case of SFA measurements of proteins, the protein dimensions (lengths) estimated from the range of protein interactions agreed quantitatively with the crystallographic dimensions of the proteins and/or complexes (Leckband et al. 1994, Leckband, Muller, et al. 1995, Yeung et al. 1999, Leckband, Kuhl, et al. 1995, Bayas et al. 2007, Zhu et al. 2002). The following sections illustrate additional, functionally important features that are revealed by such studies.

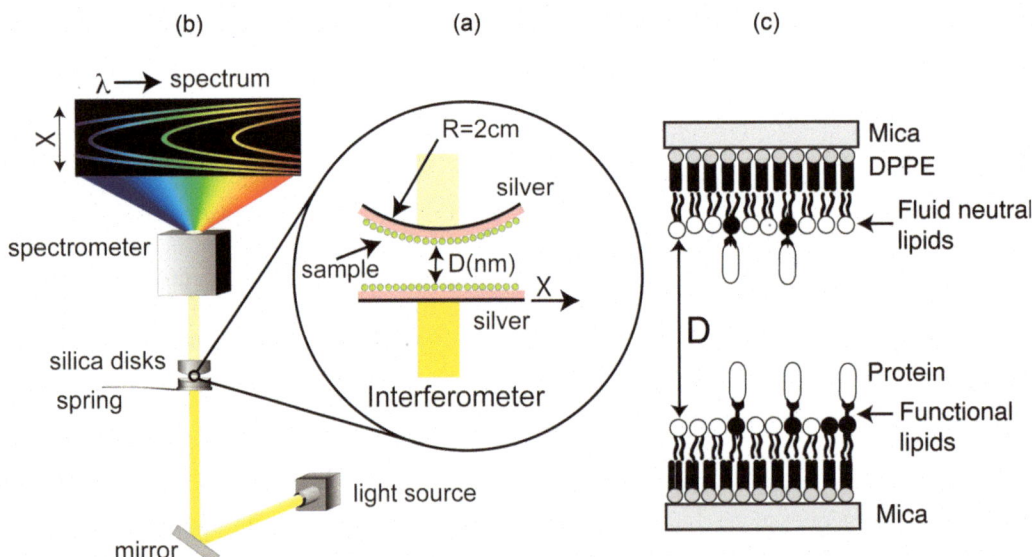

FIGURE 5.1 Surface force apparatus measurements of membrane anchored proteins. (a) Schematic of the samples in the surface force apparatus showing the two curved disks (equivalent to sphere interacting with flat surface) with the sample overlayers and the reflective silver surfaces of the interferometer. Here, X is the radial distance from the centre of the contact region and D is the surface separation. (b) Schematic of the interferometer showing the light source, the interference fringes used for distance measurements, the two opposing silica disks that support the samples, and the force-measuring spring. (c) Illustration of the sample configuration in the surface force apparatus experiment. Lipids used are gel phase 1,2-dipalmitoyl-sn-glycero-3-phosphoethanolamine (DPPE), neutral lipid 1,2-ditridecanoyl-sn-glycero-3-phosphocholine (DTPC), and lipid with reactive head groups used to immobilize and orient the proteins via specific amino acids or epitope tags. Here, D is the absolute distance between the bilayer surfaces.

5.3 LECTIN STRUCTURES AND INTERMEMBRANE BINDING POTENTIALS

5.3.1 Conformational Flexibility and Protein Recognition

Lectins are a broad class of carbohydrate recognition proteins expressed in both animals and plants. In many animal lectins, carbohydrate recognition domains (CRDs) are at the C-termini of long polypeptide tethers or necks (Drickamer and Taylor 1993) (see Figure 5.2a, for example). In some lectins, such as selectins, the polypeptide necks do not oligomerize into functional aggregates, but in others the collagenous neck domains oligomerize into coiled coils that cluster multiple CRDs at the C-termini (Drickamer and Taylor 1993). When such lectins are expressed on cell membranes, these extended neck structures are assumed to present single or clustered binding sites to pathogens or other cells. However, the neck extension and flexibility, its impact on the interaction range, and the consequences of oligomerization for recognition and adhesion are not easily determined from binding affinities or structures alone.

SFA measurements of the animal lectins dendritic cell-specific intercellular adhesion molecule-3-grabbing nonintegrin (DC-SIGN) and DC-SIGN-related protein (DC-SIGNR)—proteins of the innate immune system—revealed unique, functionally important structural features of these proteins. DC-SIGN is expressed on dendritic cells in the immune system and recognizes high mannose carbohydrate ligands on a variety of pathogens (Geijtenbeek, Engering, and Van Kooyk 2002). Most pathogens that bind DC-SIGN are internalized by cells and neutralized, but others such as HIV exploit DC-SIGN to infect cells (Geijtenbeek et al. 2000, van Kooyk and Geijtenbeek 2003). Knowledge of the mechanism of pathogen recognition could therefore guide the design of therapeutics that block infection (Borrok and Kiessling 2007, Kitov et al. 2000).

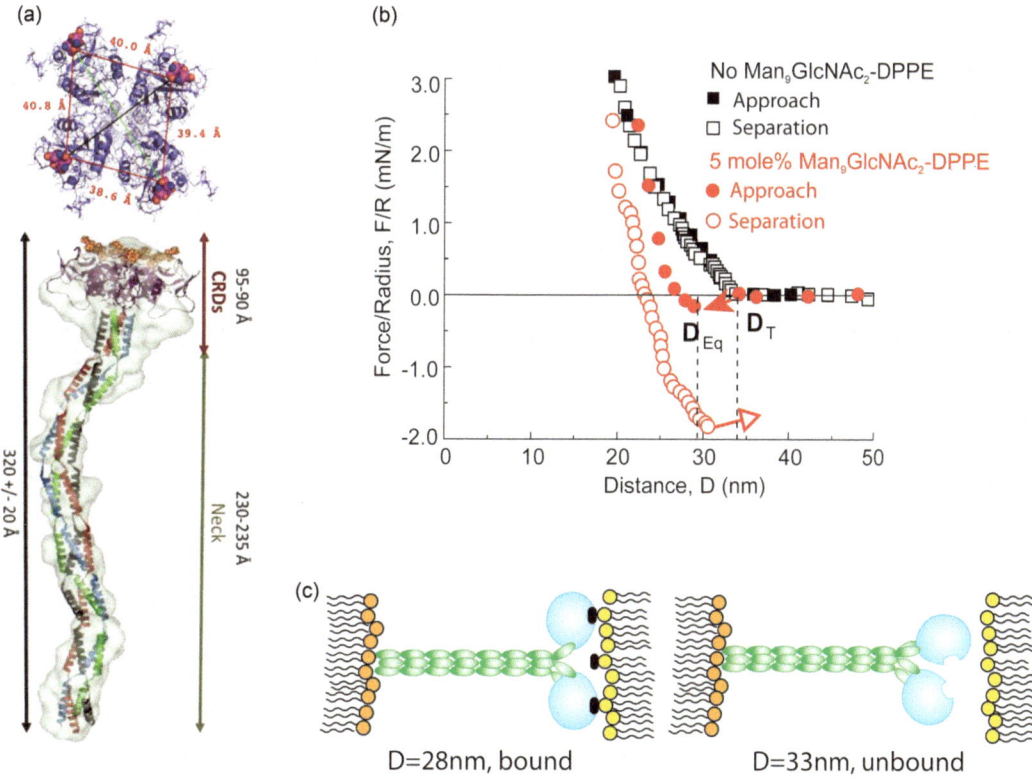

FIGURE 5.2 Lectin binding to membranes presenting neoglycolipid ligands. (a) Model of the tetrameric DC-SIGN based on small-angle X-ray scattering measurements showing the coiled-coil neck and cluster of C-terminal CRDs. (b) Normalized force (F/R) versus the distance, D between DC-SIGN and supported lipid bilayers with and without neoglycolipid ligand in the target membrane. The sample configuration and postulated protein orientations are shown in Figure 5.1c. The black squares show the forces measured during intersurface approach (open squares) and separation (filled squares) in the absence of neoglycolipid on the target membrane. The red circles show the force curves measured during approach (filled red) and separation (open red) when neoglycolipid is on the target membrane. The arrows show the distances at which the surfaces jump into (solid arrow) or out of (open arrow) adhesive contact. The vertical dashed lines indicated the onset of the steric repulsion D_T and the equilibrium position to which the surfaces jump into contact D_{Eq}. (c) Postulated DC-SIGN configurations in the absence (right) and presence (left) of neoglycolipid (black spheres, right) on the target membrane. CRDs, carbohydrate recognition domains; DC-SIGN, dendritic cell-specific intercellular adhesion molecule-3-grabbing nonintegrin; DPPE, 1,2-dipalmitoyl-sn-glycero-3-phosphoethanolamine. (Reproduced with permission from Menon et al. 2009.)

DC-SIGN contains a C-terminal CRD, a neck region comprising 7.5 repeats of a 23 amino acid, collagen-type domain, a transmembrane region, and a cytoplasmic domain. The EC domains oligomerize into tetramers, to form an α-helical, coiled-coil neck terminated by four, clustered CRDs (Figure 5.2a) (Feinberg et al. 2005). The neck is thought to direct the CRDs towards the pathogen, while the terminal CRD clusters enhance binding avidity. The estimated neck length based on hydrodynamic models was 20–30 nm (Feinberg et al. 2005), but such measurements could not determine the actual extension or rigidity of the membrane bound protein.

Force measurements between membrane-anchored, oriented DC-SIGN EC domains and bilayers displaying membrane-bound carbohydrate ligands quantified the overall dimensions of the DC-SIGN EC domains and the adhesion energy. They also revealed a novel binding-dependent conformational change that enhances adhesion between DC-SIGN and surface-bound ligands. Figure 5.2b shows the normalized force, F/R versus the distance D between lipid bilayers displaying oriented, immobilized DC-SIGN monolayers, and (i) a bare lipid membrane (Figure 5.2c, right) or (ii) a glycan-presenting bilayer (Figure 5.2c, left). The distance D_T (Figure 5.2a) at the onset of steric repulsion ($F > 0$) between DC-SIGN and the bare membrane indicates the steric thickness of the protein. The data indicate that the neck extends 28 nm from the surface, after accounting for the thickness of the CRD region. The steep increase in force at $D < D_T$ indicates that the neck is relatively rigid.

The force profiles measured with glycolipid (ligand) on the target membrane (Figure 5.2b and c) differ from measurements with bare membranes in two important ways. First, during approach, instead of the increase in repulsion at ~D_T, the surfaces jumped into contact (Figure 5.2b, solid red arrow). The surfaces came to rest at the equilibrium distance D_{Eq} of 28 nm, which corresponds to the equilibrium thickness of the DC-SIGN/carbohydrate complex. In force measurements, surfaces jump to contact when the gradient of the attractive intersurface potential exceeds the spring constant. Second, during separation, due to the protein–glycan adhesion, the force drops below zero ($F < 0$), and the adhesion energy is determined from the pull-off force, F_{po} at the minimum in the curve (Figure 5.2b, open red arrow).

Two features of the force–distance measurements revealed that DC-SIGN undergoes a binding-dependent conformational change: (i) the difference between the range of the steric repulsion and the equilibrium length of the DC-SIGN/glycan complex, $D_T - D_{Eq}$ and (ii) the jump into contact of DC-SIGN and glycolipid monolayers. The final equilibrium distance D_{Eq} measured with glycolipid membranes was less than D_T, even though the target membrane was coated with bulky glycans (Figure 5.2c, left). A binding-dependent conformation change would account for the difference in the range of steric repulsion between DC-SIGN and the opposing membranes, in the absence and presence of glycan. The difference $\Delta D_c = D_T - D_{Eq}$ indicates the magnitude of the change, which increased with the glycan surface density and hence with the fraction of bound CRDs, up to a limiting value of 4.8 nm (Menon et al. 2009).

The DC-SIGN conformation change was attributed to flexible linkers in the neck region adjacent to the CRDs. They appear to enable the CRDs to rearrange, in order to optimize ligand docking (Figure 5.2c). Although crystal structures of the isolated CRD region suggested linker flexibility, the distance resolution of the SFA measurements confirmed the flexibility and revealed a possible functional role. One postulate is that the flexible linkers allow the CRDs to adapt to different carbohydrate ligand arrangements (patterns) and for DC-SIGN to therefore recognize a broader range of pathogens than proteins with rigidly arranged CRDs (Menon et al. 2009).

5.3.2 Genetic Length Variants and Receptor Function

SFA measurements also revealed architectural features of related lectins that may contribute to differences in susceptibility to viral infections. DC-SIGN and the DC-SIGN-related protein (DC-SIGNR) are structurally very similar, but they are expressed in different tissues and recognize different pathogens (Pohlmann, Baribaud, and Doms 2001, Pohlmann et al. 2001, Simmons et al. 2003). Both proteins bind envelope glycoproteins on the human immunodeficiency virus to facilitate thymus cell (T cell) infection (Pohlmann, Baribaud, and Doms 2001). However, DC-SIGN modulates the host response to schistosomal parasites, whereas DC-SIGNR facilitates hepatitis C infection of hepatocytes (Gardner et al. 2003) and is associated with the maternal-foetal transmission of human immunodeficiency virus (Soilleux et al. 2001).

The diverse functions of these proteins arise in part to sequence differences in the carbohydrate recognition sites (Feinberg et al. 2001), but their lengths also differ. The neck region of DC-SIGN comprises 7.5 repeats of the 23-amino acid module, but the human population exhibits naturally occurring polymorphisms in the DC-SIGNR neck that vary between 4.5 and 9.5 repeats (Lichterfeld et al. 2003). In some cases, these neck length variants correlate with differences in viral infection (Nattermann et al. 2006).

SFA measurements addressed whether the functional differences are due to length-dependent binding energies, despite the autonomous folding of the CRDs and neck regions (Feinberg et al. 2005, Feinberg et al. 2009, Yu et al. 2009). The DC-SIGNR EC domain also exhibited a binding-dependent conformational change, and the equilibrium thickness of the length-equivalent 7-repeat form of DC-SIGNR is ~2 nm shorter than DC-SIGN (Leckband, Menon, et al. 2011). This may arise from differences in the neck sequences adjacent to the CRDs, as suggested by packing differences in structures of isolated CRDs of DC-SIGN and DC-SIGNR (Feinberg et al. 2005).

Force measurements of DC-SIGNR variants with five, six, and seven neck length repeats as observed in the human population revealed quantized decreases in the equilibrium binding distances D_{Eq} with the loss of each 23-amino acid neck repeat. The adhesion energy did not depend on the neck length, suggesting that differences in infection may be due to receptor accessibility. The 3.8 nm per neck repeat estimated from a plot of D_{Eq} versus the number of repeats agreed remarkably well with the 3.7 nm determined from a crystal structure of the neck region alone (Feinberg et al. 2009). Importantly, SFA measurements revealed the overall effect of length variations on the protein projection from the membrane, the neck rigidity, its interaction with membrane-bound ligands, and the length independence of the adhesion energy (Leckband, Menon, et al. 2011).

5.4 INTRINSICALLY DISORDERED PROTEINS (IDP)

The flexible, unstructured linkers in DC-SIGN influence pathogen recognition. In more extreme cases, IDPs are largely unstructured, but ligand interactions trigger folding (Dyson and Wright 2005). In the absence of ligand, IDPs can undergo large structural fluctuations, which increase their reaction cross section, analogous to fly-casting (Levy, Onuchic, and Wolynes 2007, Shoemaker, Portman, and Wolynes 2000). The distance dependence of ligand interactions and the onset of protein folding are central to understanding the physical chemistry underlying IDP functions.

The 'fly casting' behaviour was demonstrated in SFA measurements of membrane-anchored SNARE proteins, which facilitate fusion between cargo-carrying vesicles and target membranes in processes associated with protein transport and intercellular communication (Li et al. 2014, Wang et al. 2016, Zorman et al. 2014, Li et al. 2007). SNAREs are prevalent in the nervous system where they trigger rapid fusion of synaptic vesicles with the plasma membrane in neurotransmission (Brunger 2005). In classical SNARE-mediated vesicle fusion at neuronal synapses, two types of proteins on the vesicle and plasma membranes zipper together, and assemble a 'SNARE-pin', which triggers lipid bilayer fusion. The proteins include the synaptic vesicle SNARE (v-SNARE) and a dimer of two different alpha helical proteins (t-SNARE) on the target plasma membrane. The zippered 'SNAREpins' form highly stable coiled coils comprising four helix bundles (Brunger 2005).

As monomers, the proteins are mostly disordered (Hazzard, Sudhof, and Rizo 1999, Fasshauer et al. 1997). SFA studies were thus used to determine the range of SNARE interactions and the distance of the onset of zippering. Specifically, Wang et al. tested whether the unstructured v-SNARE and an alpha helical t-SNARE dimer actuate folding and zippering through initial contacts between their N-terminal domains (Wang et al. 2016, Li et al. 2007). However, to confirm protein binding as a function of membrane separation, they combined SFA measurements with fluorescence resonance energy transfer (FRET) between v- and t-SNAREs (Wang et al. 2016). The FRET measurements were necessary, because protein binding was not accompanied by a distinctive 'jump to contact', such as observed with DC-SIGN (see Figure 5.2b).

In force measurements between supported bilayers with immobilized, opposing v- and t-SNARES (Figure 5.3a), the proteins repelled at intermembrane distances of $D < 30$ nm (Li et al. 2007, Wang et al. 2016). The soft, exponential increase in force with decreasing distance to ~8 nm was similar to steric/osmotic repulsion between end-grafted, unstructured polymers, and the steep increase at $D < \sim 8$ nm suggested the compression of a more rigid, folded protein complex. The proteins adhered and jumped out of adhesive contact at ~8 nm, with an adhesion energy of ~30 kT (Figure 5.3a). Interestingly, the measured adhesion energy increased slowly with contact time, up to a plateau at $t > 60$ minutes, suggesting relatively slow SNAREpin assembly kinetics (Li et al. 2007).

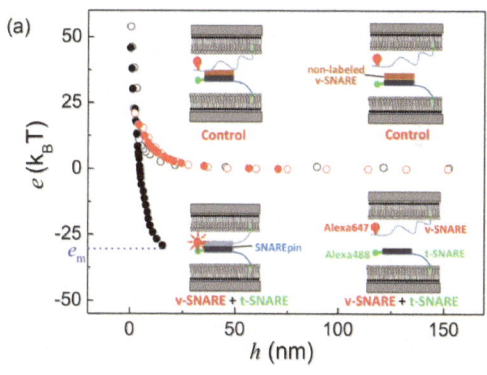

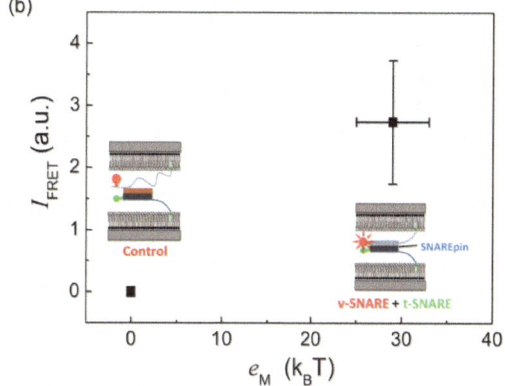

FIGURE 5.3 Interactions between v- and t-SNAREs on opposing supported lipid bilayers. (a) Energy (kT) between immobilized t-SNARE dimer (bottom surface) and v-SNARE (top surface) as a function of the distance between the supporting lipid bilayers. The v-SNARE and t-SNAREs were labelled with Alexa647 and Alexa488 fluorescent dyes, respectively. The black open and black filled symbols indicate the energy measured during approach and separation, respectively, between v- and t-SNAREs. The dotted line indicates the maximum attractive energy, e_M between the proteins. The red open and red filled circles indicate control measurements with inactive t-SNARE. (b) FRET intensity as a function of the adhesion energy between v-SNARE and t-SNARE monolayers. The initial FRET signal at large separations increases with increasing adhesion energy, and hence with increasing numbers of assembled SNAREpins. FRET, fluorescence resonance energy transfer.

FRET measurements provided greater molecular level detail regarding the distance at which the N-termini of the SNAREs bind (Figure 5.3b). Initial protein binding was detected at $D < 20$ nm, with the FRET signal increasing with decreasing distance to reach a plateau at $D < 8$ nm, which corresponds to the SNAREpin dimensions (Wang et al. 2016). Although some proteins bound at 20 nm, the initial protein encounters did not trigger rapid, extensive folding that might cause the surfaces to jump to 8 nm which corresponds to the folded complex dimensions. FRET data confirmed this slow folding transition.

It is interesting to compare the SNARE interactions with force measurements between monolayers of streptavidin and biotin-terminated, end-grafted polyethylene glycol chains (Jeppesen et al. 2001, Wong et al. 1997). Polymer fluctuations resulted in biotin–streptavidin binding at large membrane separations, indicative of large chain excursions beyond the radius of gyration. Upon receptor–ligand binding, the elastic restoring force of the stretched chains pulled the surfaces to equilibrium contact within seconds. In this case, the equilibrium membrane separation D_{Eq} was determined by the sum of the polymer radius of gyration and streptavidin thickness. The difference in the SNAREpin behaviour may be due to much weaker initial SNARE binding energy (<5 kT) (Wang et al. 2016) relative to biotin–streptavidin bonds (35 kT). The much slower streptavidin–biotin dissociation rate would result in the rapid accumulation of long-lived streptavidin bonds and corresponding rapid buildup of force from stretched chains that in turn would pull the surfaces to D_{Eq} (Jeppesen et al. 2001). By contrast, rapid dissociation of initial, weak SNARE bonds would slow the increase in force. The elastic restoring force of extended v-SNARE chains (which increases with the chain extension) might also be sufficient to break the weak bonds. These scenarios would impede the rate of SNARE binding and the increase in intersurface attraction.

The range of interactions (20 nm) relative to the thickness of the folded SNAREpin (8 nm) demonstrated the 'fly-casting' behaviour of the unstructured v-SNARE. Comparisons with polymer-tethered ligands also suggest how differences in binding energies and molecular flexibility may contribute to folding dynamics and intersurface potentials. Current theoretical models are unable to access long timescales of large IDP fluctuations, so we currently can only speculate on the physical–chemical basis of the slow v-SNARE binding and folding.

5.5 IMMUNOGLOBULIN FAMILY CELL ADHESION MOLECULES

5.5.1 Neural Cell Adhesion Molecule

Force measurements revealed how adhesion protein architectures determine intermembrane binding distances. The immunoglobulin (Ig) superfamily of proteins includes a large number of adhesion proteins whose EC, adhesive regions contain multiple immunoglobulin-type beta sandwich domains. An example is the neural cell adhesion molecule (NCAM). In the nervous system, NCAM forms homophilic bonds with identical proteins on adjacent cells to mediate cell–cell adhesion (Walsh and Doherty 1997). As one of the most abundant adhesion proteins in the brain, NCAM is associated with long-term memory formation and circadian rhythms (Walsh and Doherty 1997). The EC region consists of seven, structurally similar domains (Figure 5.4) (Chothia and Jones 1997). The first five domains are immunoglobulin-type (Ig) domains and are numbered 1–5 (Ig1-5), and the two juxtamembrane domains are fibronectin type III (FNIII) repeats. Electron microscopy also identified a distinct bend of ~138° in the ectodomain structure, presumably between Ig5 and the FNIII domains (Becker 1989, Hall and Rutishauser 1987) (Figure 5.4).

Three different NCAM binding models were proposed, based on structural data and equilibrium binding measurements (Figure 5.4a–c) (Atkins 2001, Cunningham et al. 1987, Jenson et al. 1999, Kasper 2000, Kiselyov 1997, Ranheim et al. 1996, Rao et al. 1992, Soroka et al. 2002, Soroka et al. 2003). Removing Ig3 eliminated cell adhesion, suggesting binding between Ig3 domains (Figure 5.4b) (Rao et al. 1992), although isolated Ig3 domains do not associate in solution (Atkins 2001). X-ray and NMR structures suggested that anti-parallel Ig12 domains adhere through intermolecular ion pairs. Alternatively, a structure of the Ig1–3 fragment (Soroka et al. 2003) suggested that the NCAM binds through antiparallel Ig1 to Ig3 contacts (Figure 5.4c) and that Ig12 domains form lateral bonds with proteins on the same membrane (Figure 5.4a).

Force–distance profiles measured between membrane-bound NCAM EC domain fragments distinguished between these different models (Johnson et al. 2004). Because of the bend in the protein (see Figure 5.4a), force measurements were done with engineered Ig1–5 fragments ~19 nm in length that were immobilized and oriented on supported bilayers via C-terminal hexahistidine tags (Figure 5.5b). In normalized force versus distance profiles measured between oriented Ig1–5 fragments (Figure 5.5a), the increase in repulsion at $D_T < 38$ nm indicates the steric thickness of the two Ig1-5 NCAM monolayers. Upon separation, the opposing NCAMs adhered at two, distinct intermembrane distances of 18.0 ± 0.5 and 29.0 ± 0.5 nm (Figure 5.5a) (Johnson et al. 2004).

In comparisons of the force profiles with the protein dimensions, adhesion at 18 nm is consistent with antiparallel binding between Ig3 domains (Figure 5.5b). Adhesion at 29.0 ± 0.5 nm is consistent with a double-reciprocal bond between Ig1 and Ig2 (Figure 5.5b). Parallel alignment of Ig12 domains cannot be ruled out from distances alone. The domains involved in the two adhesive configurations were also confirmed by measurements with proteins lacking different Ig domains (domain deletion mutants), (Wieland, Gewirth, and Leckband 2005, Johnson et al. 2004).

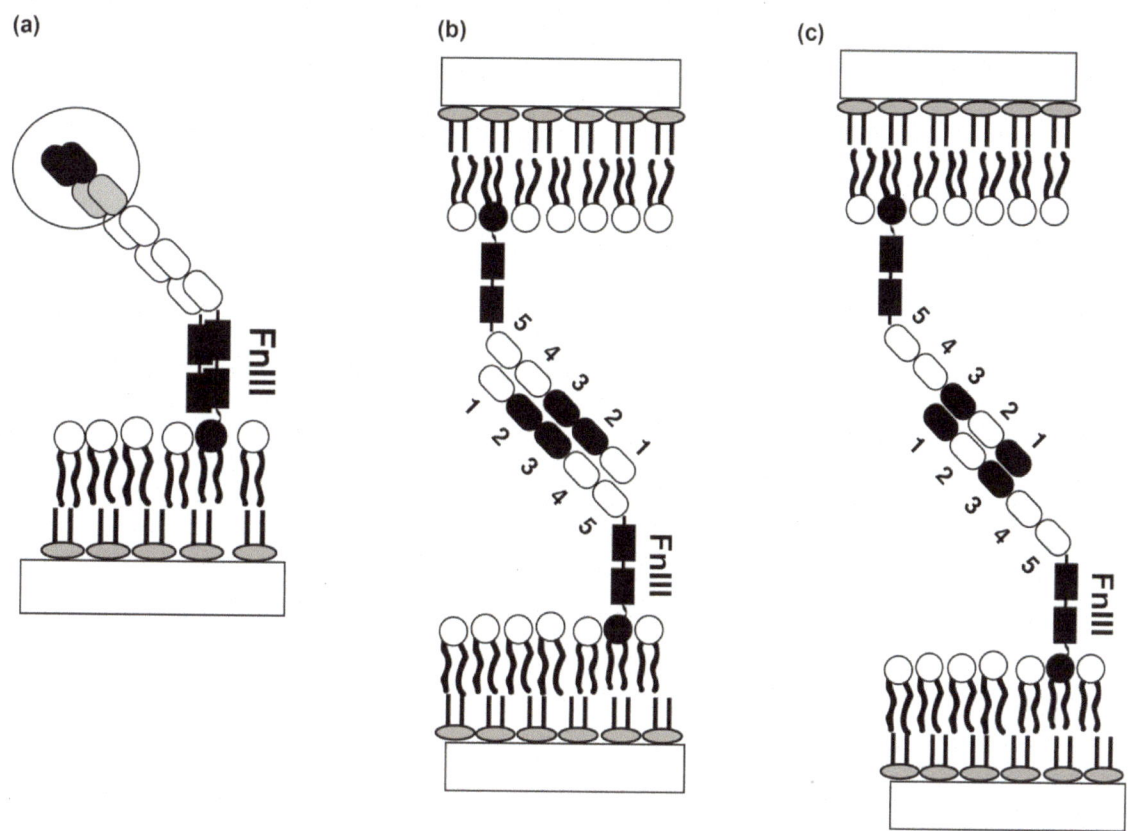

FIGURE 5.4 Proposed models for neural cell adhesion molecule (NCAM)-mediated intercellular adhesion based on crystallographic data. (a) The outer immunoglobulin Ig12 domains are postulated to associate laterally. (b) Opposing NCAM ectodomains were postulated to bind via antiparallel Ig3 contacts, by antiparallel Ig2–3 contacts, or by (c) antiparallel Ig1–Ig3 contacts.

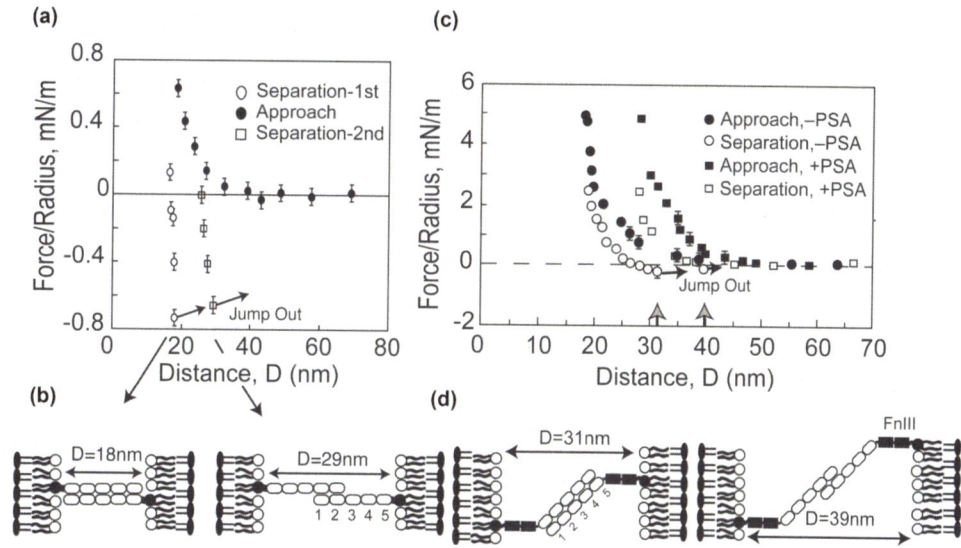

FIGURE 5.5 NCAM adhesion and posttranslational modification. (a) Normalized force, F/R versus distance, D between supported lipid bilayers with (b) immobilized, oriented NCAM Ig1–5 fragments. The filled, black circles indicate forces measured during approach, and open symbols indicate forces measured during separation from $D < 38$ nm (first bond at $D = 18$ nm) and from $18 < D < 29$ nm (second bond at 29 nm) The outward directed arrows indicate the distances at which interprotein bonds fail and the surfaces jump out of adhesive contact. (c) Normalized force versus distance between bilayers with (d) immobilized, oriented NCAM extracellular domains with and without polysialic acid (PSA) modification. The filled, black circles and squares indicate forces measured during approach, and open symbols indicate forces measured during separation. The unmodified proteins (–PSA) adhered at membrane separations of 31 and 39 nm, indicated by the gray arrows. With PSA-modified NCAM (+PSA), the onset of the intermembrane repulsive force is shifted to larger separations due to the steric repulsion between PSA chains. In parts (a) and (c), the arrows indicate the distances at which NCAMs jump out of contact at the pull-off force F_{po}. NCAM, neural cell adhesion molecule.

The full-length extracellular region, which include the FNIII repeats and the bend, adhered at 31 ± 0.5 nm and at 39 ± 0.5 nm (Figure 5.5c). Both of binding distances correspond to the same putative Ig1–5 overlaps shown in Figure 5.5b, assuming a 138° bend at the hinge between Ig5 and the first FNIII domain (Figure 5.5d). This is the maximum bend angle seen in electron micrographs and the maximum expected extension of the folded proteins under tension. The NCAM measurements thus identified two binding interactions, identified the Ig domains required for binding, and ruled out some postulated binding models. These findings illustrate the importance of both adhesion and distance information for elucidating binding mechanisms of more complex proteins.

5.5.2 Posttranslational Modification and Neural Cell Adhesion Molecule Adhesion

One of the more biologically important features of NCAM is that it has both adhesive and anti-adhesive forms. In the anti-adhesive form, two high-molecular-weight, linear polysialic acid (PSA) chains are covalently attached to Ig4 (Nelson 1995). The adhesive form is unmodified. The PSA modification is linked to neural plasticity during brain development, circadian rhythms, tumour progression, and spinal cord regeneration (Daston et al. 1996, El Maarouf et al. 2005, Rutishauser et al. 1988, Shen et al. 1997). High levels of PSA in the nervous system are associated with neural plasticity (Shen et al. 1997, Daston et al. 1996), and PSA increases intermembrane spacing (Yang, Yin, and Rutishauser 1992).

SFA measurements directly demonstrated that PSA perturbs protein-mediated intermembrane adhesion, by increasing the nonspecific intermembrane repulsion (Johnson, Fujimoto, et al. 2005). The net intermembrane force is a superposition of attractive and repulsive forces. Thus, in order for PSA to nonspecifically block NCAM adhesion, the magnitude and range of the polymer-dependent, intermembrane repulsion would have to exceed all NCAM–NCAM attraction.

SFA measurements carried out with both the modified and unmodified full NCAM ectodomains demonstrated directly that the PSA modification increases the range and magnitude of the intermembrane repulsive force (Figure 5.5c). Under physiological conditions, the range of the repulsion between PSA-modified NCAM monolayers ($D < 45$ nm) extends slightly beyond that of the unmodified NCAM ($D < 40$ nm). The PSA-dependent repulsion abolished NCAM-mediated adhesion at 31 and 39 nm (Figure 5.5c). However, the enzymatic removal of PSA restored NCAM binding (Johnson, Fujimoto, et al. 2005).

The ionic strength dependence of the steric repulsion also confirmed that PSA abrogates NCAM-mediated adhesion through nonspecific repulsion. PSA is a polyelectrolyte, and increasing ionic strength decreases the hydrodynamic radii of polyelectrolytes (Pincus 1991). Consistent with this, the range and magnitude of long-ranged repulsion between PSA–NCAM monolayers decreased at high ionic strength (1M NaCl) (Johnson, Fujimoto, et al. 2005). X-ray and neutron reflectivity confirmed the salt-dependent reduction in the thickness of monolayers of membrane-anchored PSA-modified NCAM (Johnson, Fragneto, et al. 2005). Concurrent with the reduced steric repulsion at high ionic strength, NCAM binding re-emerged at the same two distances that unmodified NCAM ectodomains adhered (see Figure 5.5c) (Johnson, Fujimoto, et al. 2005).

Importantly, the ability of SFA measurements to quantify both the range and magnitude of attractive and repulsive forces between membranes displaying NCAM monolayers enabled the direct demonstration, at the molecular level, that posttranslational modification regulates intercellular adhesion, by modulating nonspecific intermembrane repulsion. This mechanism contrasts with typical biochemical regulation of adhesion (Alberts 1983). These findings also revealed the mechanism by which PSA alters intercellular spacing in tissues and the origin of ionic strength dependent adhesion between cells expressing NCAM-PSA (Yang, Yin, and Rutishauser 1992, Rutishauser 1992, Rutishauser et al. 1988).

5.6 CADHERINS

5.6.1 Cadherins form Multiple Bonds with Different Mechanical and Kinetic Properties

Several members of the cadherin family of intercellular adhesion proteins also have large adhesive EC regions. Classical cadherins are essential calcium-dependent, transmembrane glycoproteins that mediate intercellular cohesion in all tissues, by binding cadherins on adjacent cells. Cadherin structures consist of an adhesive EC region, which folds into five structurally similar EC domains numbered 1–5 from the N-terminus (EC1-5), a transmembrane domain, and an intracellular domain, which binds actin and other cytosolic proteins (Figure 5.6a).

Crystal structures and biochemical data revealed that cadherins bind opposing cadherins via their N-terminal EC1 domains. A widely held view for many years was that the sole binding interaction involved the docking of tryptophan at position 2 (Trp2) into a hydrophobic pocket on the EC1 domain of the opposing protein (Figure 5.6d). This complex is referred to as a 'strand dimer'. Biophysical studies and adhesion-based measurements confirmed that the strand dimer is crucial for cadherin-mediated cell adhesion (Leckband, le Duc, et al. 2011). Nevertheless, other experimental evidence argued for a more complex binding mechanism. The EC12 fragment was shown to be the minimal functional unit required for cell adhesion (Shan et al. 2004). Cadherin mutations associated with diffuse gastric cancer are also distributed over the entire EC region, and a mutation at the EC2-EC3 junction abolished cell adhesion (Handschuh et al. 2001).

Force measurements confirmed greater cadherin binding complexity than strand dimerization alone. In SFA measurements between oriented, cadherin ectodomains immobilized on supported bilayers (Figure 5.6b), the proteins bound at

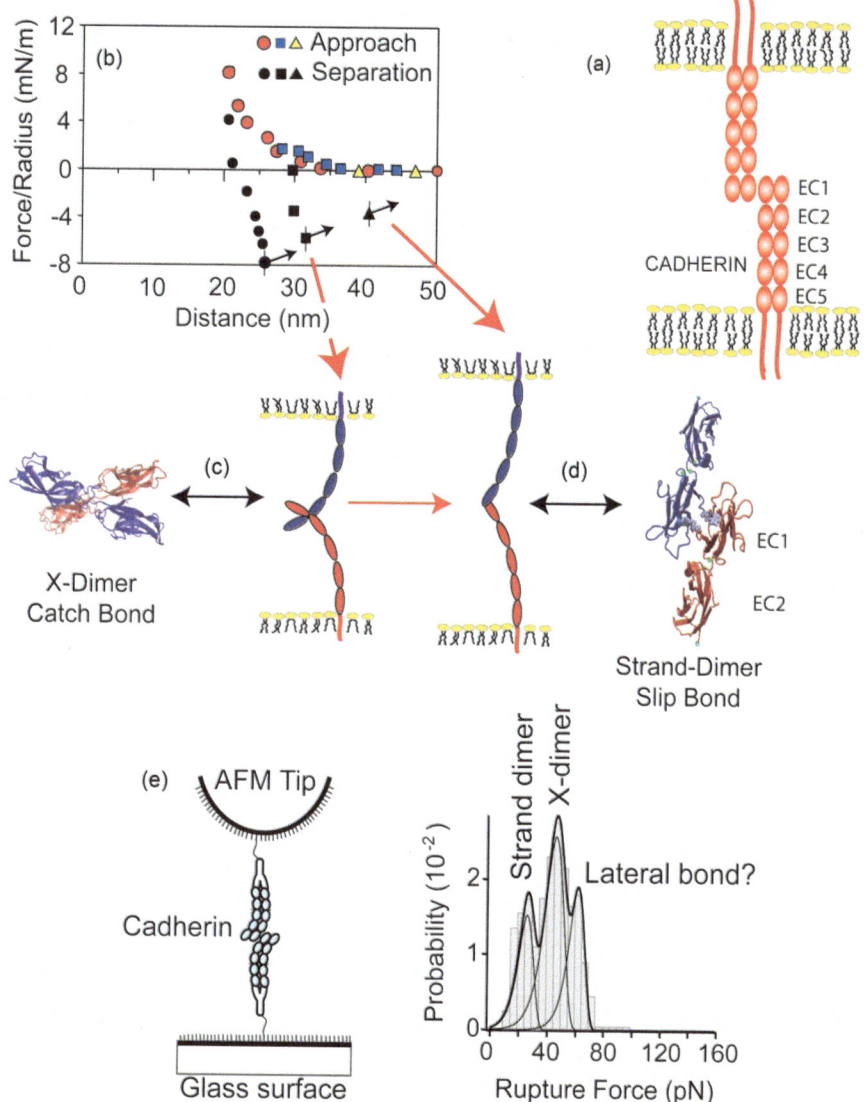

FIGURE 5.6 Cadherin structure and proposed binding interactions. (a) The extracellular region of the transmembrane protein consists of five tandemly arranged extracellular (EC) domains numbered 1–5 from the N-terminus. Proteins on the cell membrane also have transmembrane and intracellular regions. (b) Force versus the distance, D, between membranes displaying immobilized, oriented cadherin extracellular fragments (EC1–5). During approach, the intermembrane repulsion increases, but upon separation, the protein-coated membranes adhere at separations of 26, 32, and 39 nm. (c) Structure of the X-dimer formed upon initial contact (encounter complex) between EC1–2 domains and (d) the structure of the strand dimer formed by exchange of tryptophan residues (shown as van der Waals' spheres) between opposing EC1 domains. (e) (left) Configuration of cadherin extracellular domains immobilized to the AFM tip and substrate and (right) histograms of measured forces between single cadherin extracellular domains (left). The histograms exhibit three distinct peaks, which were fit by three Gaussian distributions that correspond to three distinct cadherin bonds. The weakest bond, rupturing at 20 pN, corresponds to the strand dimer interaction. The intermediate bond rupturing at ~40 pN corresponds to the X-dimer, and the strongest bond rupturing at 75 pN requires EC domains 1–3.

three distinct membrane separations of 26, 32, and 39 nm (Leckband and Prakasam 2006, Leckband and Sivasankar 2012a). The bonds with the lowest and highest adhesion energies formed at 39 and 26 nm, respectively. Studies of cadherin EC domain deletion mutants identified the protein regions required for each of the adhesive interactions (Leckband and Prakasam 2006). Adhesion at 39 nm required EC1 and was due to strand dimerization (see Figure 6.6d). Adhesion at 32 nm required EC2 and was at a distance consistent with the antiparallel alignment of EC12 domains (Figure 5.6c).

The EC3-dependent, strongest adhesion at 26 nm was at a distance consistent with antiparallel or parallel EC3–EC3 contacts, depending on the ectodomain flexibility.

Apart from the strand dimer, structures corresponding to adhesion at 26 and 32 nm were determined after SFA measurements revealed their existence (Leckband and Sivasankar 2012b, Rakshit et al. 2012, Ciatto et al. 2010, Harrison et al. 2010, Leckband and Prakasam 2006). In crystals, the EC12 fragments form an antiparallel complex, mediated by interprotein contacts at the EC1–EC2 junction (Figure 5.6c).

This crossed configuration is referred to as the 'X-dimer'. Kinetic and fluorescence measurements suggested that the X-dimer is an initial encounter complex, which rearranges to form the strand dimer (Harrison et al. 2010, Sivasankar et al. 2009). In SFA measurements, adhesion at 32 nm agreed quantitatively with the predicted end-to-end dimensions of an X-dimer complex (Figure 5.6c). A puzzle was how a postulated transient intermediate could sustain greater force than the strand dimer.

Single bond rupture measurements identified unique properties of cadherin bonds underlying the different adhesive properties of strand- and X-dimers. Analyses of histograms of the rupture forces of single cadherin bonds subjected to a steadily increasing force (force ramp) detected three, distinct force peaks (Figure 5.6e) (Shi, Maruthamuthu, and Leckband 2010, Bayas et al. 2006). These peaks were due to different cadherin bonds with distinct strengths and dissociation rates. Use of cadherin domain deletion mutants determined different structural regions required to form each of these bonds (Shi, Maruthamuthu, and Leckband 2010, Bayas et al. 2006). The strand dimer formed the weakest, EC1-dependent bond, which ruptured at ~20 pN (Figure 5.6e). The strongest bond (~60 pN) required the full-length EC domain and EC3 in particular. The peak at 40 pN required EC2 and was due to the X-dimer. Importantly, atomic force microscopy measurements also showed that the X-dimer forms a 'catch bond' and the strand dimer forms a 'slip' bond (Rakshit et al. 2012).

This distinction of 'catch' and 'slip bonds' was first proposed by Dembo et al. (1988) in a model based on Bell's postulate that force increases the rate of bond failure, according to the phenomenological equation: $k_{off} = k_0 \exp(-(E_{act} - \gamma x)/RT)$ (Bell 1978). Here, γ is the reactive compliance, x is the postulated distance from the ground state to the transition state along the unbinding coordinate, E_{act} is the activation energy for unbinding, k_0 is the intrinsic dissociation rate, R is the gas constant, and T is the absolute temperature. 'Slip bonds' are bonds for which k_{off} increases with increasing force, and $\gamma > 1$. However, 'catch bonds' result when $\gamma < 1$, such that the dissociation rate decreases with increasing force. Experimentally, catch and slip bond behaviour is distinguished by the dependence of bond survival time (lifetime), under a constant force (Dembo et al. 1988). The bond lifetime is $1/k_{off}$ such that the lifetime increased or decreased with force, for catch and slip bonds, respectively (Rakshit et al. 2012). Several proteins reportedly form catch bonds, based on these criteria (Marshall et al. 2003, Thomas 2008).

Measured cadherin bond lifetimes as a function of applied force demonstrated that X- and strand dimers are, respectively, 'catch–slip' and 'slip' bonds (Rakshit et al. 2012). The lifetime of the strand dimer decreased with increasing force. However, the lifetime of the X-dimer increased with increasing force (catch bond) to a peak value, after which the lifetime fell with increasing applied force (slip bond). These different mechanical responses account for greater X-dimer adhesion, relative to strand dimers, in both SFA and single bond rupture measurements (see Figure 5.6b and e). There is no contradiction between adhesion and equilibrium binding measurements, because the bond strength is a nonequilibrium parameter that depends on how the bond is ruptured. The X-dimer could thus exhibit no strength under equilibrium (no force) conditions but appear strong when force-loaded.

The Dembo model (Dembo et al. 1988) predicted catch bonds, but not their mechanisms. One postulate is that catch bonds arise from force-activated switching between conformational states, which have different unbinding reaction coordinates (Prezhdo and Pereverzev 2009, Pereverzev et al. 2005, Gunnerson, Pereverzev, and Prezhdo 2009, Thomas 2008, Barsegov and Thirumalai 2005, Evans et al. 2004). To form a catch bond, the conformational change would direct unbinding along a trajectory with a higher activation barrier, thus increasing the lifetime.

This postulate appears to be supported by steered molecular dynamics simulations of forced cadherin dissociation (Manibog et al. 2014). In simulations, the strand dimer failed under a steadily increasing force, without any conformational change impeding complex detachment. By contrast, in the simulated X-dimer detachment, force caused the EC1–2 complex to form new long-lived, force-dependent hydrogen bonds that appeared to lock X-dimers into tighter contact. Consistent with this postulate, a mutation that prevented H-bond formation abolished the catch bond signature (Manibog et al. 2014).

Together, structures, solution binding data, and force measurements revealed that cadherins form at least two distinct adhesive bonds that involved different domains and have different mechanical properties. The functional role of the X-dimer has yet to be established but may contribute to cadherin complex stability under tension. However, there is no similar experimental consensus regarding the strongest cadherin bond. Recent findings suggest that the EC3-dependent interaction only forms between confined or immobilized cadherins.

5.7 INTERMEMBRANE CONFINEMENT ALTERS PROTEIN (CADHERIN) INTERACTIONS

Binding mechanisms are typically based on studies of soluble fragments, but adhesion proteins function in confined, crowded gaps between cell membranes and EC matrix or adjacent cells. Recent theories and experimental findings are now revealing that confinement on or between membranes may alter protein interactions (Hu, Lipowsky, and Weikl 2013, Chesla, Selvaraj, and Zhu 1998, Dustin et al. 2001, Huang et al. 2010, Williams et al. 2001, Zhang et al. 2005, Wu et al. 2010, Wu et al. 2011, Dustin et al. 1996, Wu et al. 2008). Cadherins assemble into crowded adhesion plaques within intercellular gaps (He, Cowin, and Stokes 2003, Al-Amoudi et al. 2007, Wu, Kanchanawong, and Zaidel-Bar 2015) that are typically ~35 nm wide (Al-Amoudi et al. 2007, He, Cowin, and Stokes 2003). In crystal structures, an observed interface between adjacent proteins suggested that cadherins might also associate laterally with cadherins on the same membrane (see Figure 5.7c) (Harrison et al. 2011). However, solution binding studies failed to detect this proposed lateral interaction, even at high protein concentrations (Haussinger et al. 2002, Tomschy et al. 1996, Harrison et al. 2010).

Molecular dynamics and Monte Carlo simulations suggested that apparent discrepancies between force measurements, solution binding data, and crystal structures might be due to confinement (Wu et al. 2010, 2011). Simulations suggested that binding between flexible cadherin ectodomains on opposing cells reduced their configurational entropy. This would decrease entropic repulsion between proteins and enable them to associate through interactions that were otherwise masked by the fluctuations. Simulations suggested that cadherins on the same membrane cluster within adhesion zones (Figure 5.7c) but that lateral association first requires binding between opposed proteins, in order to suppress fluctuations (Wu et al. 2011).

Superresolution imaging of cadherins at cell–cell junctions and the effect of mutations on their organization suggested that postulated lateral interactions between confined, cadherin ectodomains may influence cadherin organization in adhesion zones (Wu, Kanchanawong, and Zaidel-Bar 2015). However, cadherin connections to actin, actin assembly/disassembly dynamics, and endocytosis complicated efforts to assess the influence of lateral ectodomain interactions on cadherin organization (Hong, Troyanovsky, and Troyanovsky 2013, Cavey et al. 2008, Truong Quang et al. 2013, Biswas et al. 2015). The force measurements revealed additional interactions (see Figure 5.6b and e), but the artificial configurations used might not reflect behaviour at *bona fide* intercellular adhesions. Instead, cadherin-mediated intercellular binding kinetics provided experimental evidence for confinement-enhanced cadherin interactions (Shashikanth, Kisting, and Leckband 2016).

5.7.1 Protein Binding Kinetics from Intercellular Adhesion Frequency Measurements

Measurements of intercellular binding kinetics instead revealed possible effects of confinement on cadherin binding. In particular, cell adhesion frequency measurements have been used to develop and test mechanistic models of protein-mediated cell adhesion (Chien et al. 2008, Chesla et al. 2000, Chesla, Selvaraj, and Zhu 1998, Huang et al. 2010, Williams et al. 2001, Zhang et al. 2005). In adhesion frequency measurements, two cells held by opposing micropipettes are repetitively brought to contact for defined intervals (Figure 5.7a) (Chesla, Selvaraj, and Zhu 1998). Adhesion events are detected from the membrane distortion as the cells are pulled apart, and the abrupt recoil upon bond rupture. The measured binding probability $P(t)$ is the number of detected binding events n_{adh} divided by the total number N_{total} of cell–cell contacts: $P(t) = n_{adh}/N_{total}$, and is related to the number of intercellular bonds. The dependence of $P(t)$ on intercellular contact time depends on the

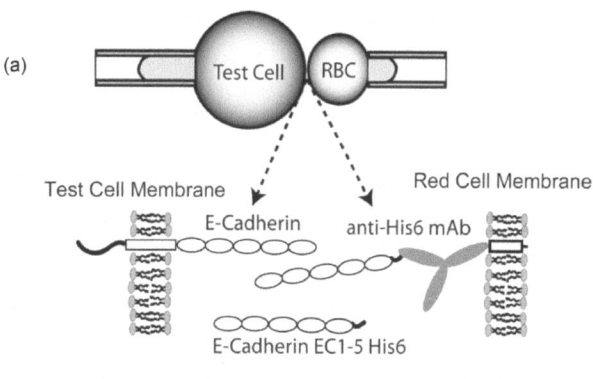

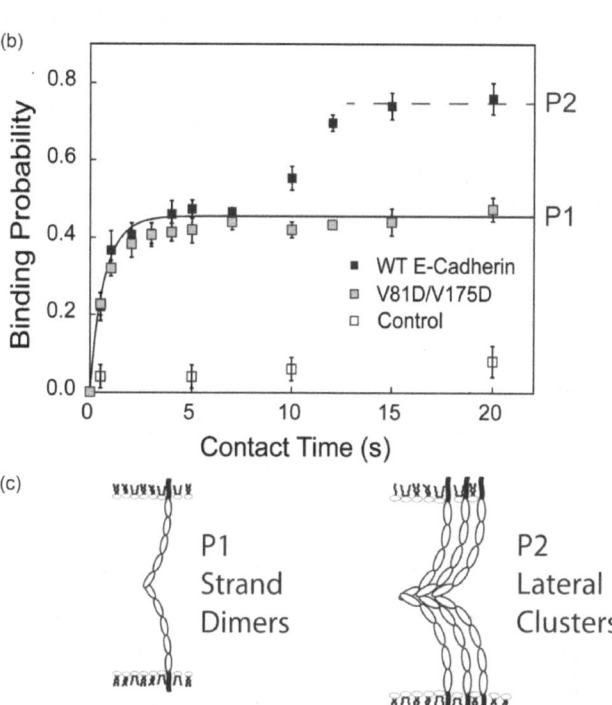

FIGURE 5.7 Adhesion frequency measurements of cadherin-mediated intercellular binding kinetics. (a) Typical experimental configuration in a micropipette measurement showing the 'test cell' and a modified red blood cell. The test cell, such as a Chinese Hamster Ovary (CHO) cell that expresses cadherin (left), is aspirated into a 7 micron inner diameter glass pipette. The red blood cell (right) is aspirated into a 1.3 inner diameter pipette. (b) Illustration of test cell surface and modified red blood cell in part A with surface-immobilized cadherin. The RBC surface is covalently modified with 'capture' monoclonal antibody. The capture–antibody binds an epitope tag on cadherin such as hexahistidine (shown here). (c) Binding probability versus contact time between a red blood cell modified with wild-type epithelial cadherin (E-cadherin) extracellular domains and a test cell expressing either wild type E-cadherin or the E-cadherin mutant V81D/L175D. P1 indicates the initial plateau corresponding to strand dimerization at steady state, and P2 indicates the final steady state binding probability measured with wild type E-cadherin. (d) Postulated binding interactions corresponding to P1 and P2 in part c. The initial kinetic step (increase to P1) is due to strand dimerization, and the second kinetic step (increase to P2) is attributed to lateral interactions (clustering) between cadherins on the same membrane.

kinetic mechanism and on the associated kinetic rates (Chesla, Selvaraj, and Zhu 1998). This approach is analogous to determining enzyme reaction mechanisms, by comparing the rate of product formation with kinetic model predictions. Such measurements were used to study the binding kinetics of several different cell surface receptors (Zhang et al. 2005, Chesla, Selvaraj, and Zhu 1998, Chesla et al. 2000, Long et al. 2001, Williams, Selvaraj, and Zhu 2000, Williams et al. 2001).

Typically, one of the cells is a soft red blood cell (RBC), and the second 'test cell' expresses the adhesion molecule of interest (Figure 5.7a and b). The RBC membrane is soft enough to enable single bond detection, although kinetic measurements typically involve tens to hundreds of molecules. The kinetics depends on the receptor and ligand surface densities and on the contact area. In contrast to force measurements, binding rates and affinities determined from adhesion frequency measurements do not depend on the force applied to bonds.

5.7.2 Confinement Reveals Protein Interactions Not Predicted by Models or Detected in Solution

The time-dependent binding probability measured between an RBC modified with cadherin EC domains and a cell expressing the full length cadherin deviated from that predicted for strand dimerization. The strand dimer model (see Figure 5.6d) predicts that the binding probability would increase monotonically to a steady-state plateau, according to $P = 1 - \exp\{-A_c m_R m_L K_a (1 - \exp(-k_r t)]\}$ where A_c is the contact area, m_R and m_L are the receptor and ligand surface densities on the cells, K_a is the 'two-dimensional' binding affinity, and k_r is the dissociation rate (Chesla, Selvaraj, and Zhu 1998). Instead, the measured time course exhibits two kinetic processes: a rapid increase to an initial plateau at P1, followed by a 2–5s lag, and then a second increase to a final steady-state plateau at P2 (Figure 5.7b).

The use of domain deletion mutants mapped the first kinetic step to strand dimer formation (Figure 5.7c) (Chien et al. 2008). The time course was well described by the rate expression $P(t)$ for the trans dimer model; hence, strand dimerization affinities and dissociation rates could be determined, by fitting data for the first kinetic step to $P(t)$ (solid line, Figure 5.7b) (Shashikanth, Kisting, and Leckband 2016). Mutants lacking either EC3–5 or EC3 can form strand dimers, but they only exhibited the initial rise to the first plateau (P1, strand dimerization) in adhesion frequency measurements (Chien et al. 2008). Importantly, the two-stage kinetic signature is independent of the cadherin transmembrane and intracellular domains, indicating that the underlying mechanism reflects EC domain interactions. Therefore, the second EC3-dependent, kinetic step revealed ectodomain interactions at intercellular contacts that are not predicted by the strand dimerization model or observed in solution binding measurements.

Investigations with mutants that disrupted the putative lateral cadherin binding interface observed in crystals (Harrison et al. 2011) suggested that the two-stage kinetic profile reflects initial strand dimerization followed by lateral cadherin interactions (Figure 5.7c) (Shashikanth, Kisting, and Leckband 2016). Mutations in the proposed lateral binding interface disrupted cadherin organization at artificial cadherin adhesions between giant vesicles and altered cadherin organization at cell–cell contacts (Harrison et al. 2011, Shashikanth, Kisting, and Leckband 2016). In adhesion frequency measurements, fast, initial binding was similar for all 'lateral interface mutants', and their strand dimerization affinities were quantitatively similar. However, the lateral interface mutations all perturbed the second kinetic step, but to different extents. A double mutant V81D/L175D completely eliminated the second kinetic step (Figure 5.7b) (Shashikanth, Kisting, and Leckband 2016). Intriguingly, in solution, cadherin EC domains do not associate through this putative (lateral) binding interface, even in concentrated solutions. Nevertheless, perturbing residues in this region altered cadherin-mediated cell binding kinetics and cadherin organization within adhesion zones (Shashikanth, Kisting, and Leckband 2016). Together, these results suggest that this interaction only manifests between confined cadherins at adhesive contacts. Aspects of these findings are consistent with EC3-dependent adhesion detected in the SFA and single bond rupture studies (Figure 5.6b and e) because the putative binding interface includes amino acids in EC3 (Harrison et al. 2011).

The postulated confinement-enhanced cadherin interactions are also functionally relevant. The lateral interface mutations and corresponding kinetic perturbations correlated with altered cadherin organization at intercellular adhesions and with increased macromolecular transport across cell monolayers (due to leakier junctions). Although the latter observations could reflect other biochemical processes, the kinetic data link cadherin ectodomain interactions to the perturbed cell behaviour.

5.8 SUMMARY

The examples described in this chapter illustrate how complex EC domains of membrane-bound adhesion proteins, posttranslational modifications, and protein confinement can alter binding mechanisms and intermembrane potentials. This emphasis on the biophysics of EC domain interactions reveals novel ways in which structure governs function, beyond the mere formation of mechanical connections. The focus on EC domains is somewhat oversimplified because inside–out signalling, cytoskeletal anchorage, or possible interactions with other membrane proteins can also modulate adhesion. Nevertheless, the findings described in this chapter illustrate the variety of ways that the ectodomain architectures can influence intermolecular potentials and adhesion protein functions. Conversely, they also identified functionally important protein interactions—beyond simple receptor–ligand docking—that could be regulated by intracellular signals or by association with other proteins, as well as by posttranslational modifications or confinement.

REFERENCES

Al-Amoudi, A., D. C. Diez, M. J. Betts, and A. S. Frangakis. 2007. "The molecular architecture of cadherins in native epidermal desmosomes." *Nature* 450 (7171):832–7. doi: 10.1038/nature05994.

Alberts, B., D. Bray, J. Lewis, M. Raff, K. Roberts, and J. jd. Watson. 1983. *The Molecular Biology of the Cell*. Garland, NY.

Atkins, A. R., J. Chung, D. Songpon, E. Little, G. M. Edelman, P. E. Wright, B. A. Cunningham, and H. J. Dyson. 2001. "Solution structure of the third immunoglobulin domain of the neural cell adhesion molecule NCAM: can solution studies define the mechanism of homophilic binding?" *J Mol Biol* 311:161–172.

Barsegov, V., and D. Thirumalai. 2005. "Dynamics of unbinding of cell adhesion molecules: transition from catch to slip bonds." *Proc Natl Acad Sci USA* 102 (6):1835–9.

Bayas, M. V., A. Kearney, A. Avramovic, P. A. van der Merwe, and D. E. Leckband. 2007. "Impact of salt bridges on the equilibrium binding and adhesion of human CD2 and CD58." *J. Biol. Chem.* 282 (8):5589–96. doi: 10.1074/jbc.M607968200.

Bayas, M. V., A. Leung, E. Evans, and D. Leckband. 2006. "Lifetime measurements reveal kinetic differences between homophilic cadherin bonds." *Biophys J* 90 (4):1385–95.

Becker, J. W., H. P. Erickson, S. Hoffmann, B. A. Cunningham, and G. M. Edelman. 1989. "Topology of cell adhesion molecules." *Proc Natl Acad Sci USA* 86:1088–1092.

Bell, G. I. 1978. "Models for the specific adhesion of cells to cells." *Science* 200 (4342):618–27.

Biswas, K. H., K. L. Hartman, C. H. Yu, O. J. Harrison, H. Song, A. W. Smith, W. Y. Huang, W. C. Lin, Z. Guo, A. Padmanabhan, S. M. Troyanovsky, M. L. Dustin, L. Shapiro, B. Honig, R. Zaidel-Bar, and J. T. Groves. 2015. "E-cadherin junction formation involves an active kinetic nucleation process." *Proc Natl Acad Sci USA* 112 (35):10932–7. doi: 10.1073/pnas.1513775112.

Borrok, M. J., and L. L. Kiessling. 2007. "Non-carbohydrate inhibitors of the lectin DC-SIGN." *J Am Chem Soc* 129 (42):12780–5.

Brunger, A. T. 2005. "Structure and function of SNARE and SNARE-interacting proteins." *Q Rev Biophys* 38 (1):1–47. doi: 10.1017/S0033583505004051.

Cavey, M., M. Rauzi, P. F. Lenne, and T. Lecuit. 2008. "A two-tiered mechanism for stabilization and immobilization of E-cadherin." *Nature* 453 (7196):751–6. doi: 10.1038/nature06953.

Chesla, S. E., P. Li, S. Nagarajan, P. Selvaraj, and C. Zhu. 2000. "The membrane anchor influences ligand binding two-dimensional kinetic rates and three-dimensional affinity of FcgammaRIII (CD16)." *J Biol Chem* 275 (14):10235–46.

Chesla, S. E., P. Selvaraj, and C. Zhu. 1998. "Measuring two-dimensional receptor-ligand binding kinetics by micropipette." *Biophys J* 75 (3):1553–72. doi: 10.1016/S0006-3495(98)74074-3.

Chien, Y. H., N. Jiang, F. Li, F. Zhang, C. Zhu, and D. Leckband. 2008. "Two stage cadherin kinetics require multiple extracellular domains but not the cytoplasmic region." *J Biol Chem* 283 (4):1848–56. doi: 10.1074/jbc.M708044200.

Chothia, C., and E. Y. Jones. 1997. "The molecular structure of cell adhesion molecules." *Annu Rev Biochem* 66:823–62.

Ciatto, C., F. Bahna, N. Zampieri, H. C. VanSteenhouse, P. S. Katsamba, G. Ahlsen, O. J. Harrison, J. Brasch, X. Jin, S. Posy, J. Vendome, B. Ranscht, T. M. Jessell, B. Honig, and L. Shapiro. 2010. "T-cadherin structures reveal a novel adhesive binding mechanism." *Nat Struct Mol Biol* 17 (3):339–47. doi: 10.1038/nsmb.1781.

Cunningham, B. A., J. J. Hemperly, B. A. Murray, E. A. Prediger, R. Brackenbury, and G. M. Edelman. 1987. "Neural cell adhesion molecule: structure, immunoglobulin-like domains, cell surface modulation, and alternative RNA splicing." *Science* 236:799–806.

Daston, M. M., M. Bastmeyer, U. Rutishauser, and D. D. O'Leary. 1996. "Spatially restricted increase in polysialic acid enhances corticospinal axon branching related to target recognition and innervation." *J Neurosci* 16 (17):5488–97.

Dembo, M., D. C. Torney, K. Saxman, and D. Hammer. 1988. "The reaction-limited kinetics of membrane-to-surface adhesion and detachment." *Proc R Soc Lond B Biol Sci* 234 (1274):55–83.

Drickamer, K., and M. E. Taylor. 1993. "Biology of animal lectins." *Annu Rev Cell Biol* 9:237–64. doi: 10.1146/annurev.cb.09.110193.001321.

Dustin, M. L., S. K. Bromley, M. M. Davis, and C. Zhu. 2001. "Identification of self through two-dimensional chemistry and synapses." *Annu Rev Cell Dev Biol* 17:133–57. doi: 10.1146/annurev.cellbio.17.1.133.

Dustin, M. L., L. M. Ferguson, P. Y. Chan, T. A. Springer, and D. E. Golan. 1996. "Visualization of CD2 interaction with LFA-3 and determination of the two-dimensional dissociation constant for adhesion receptors in a contact area." *J Cell Biol* 132 (3):465–74.

Dyson, H. J., and P. E. Wright. 2005. "Intrinsically unstructured proteins and their functions." *Nat Rev Mol Cell Biol* 6 (3):197–208. doi: 10.1038/nrm1589.

El Maarouf, A., Y. Kolesnikov, G. Pasternak, and U. Rutishauser. 2005. "Polysialic acid-induced plasticity reduces neuropathic insult to the central nervous system." *Proc Natl Acad Sci U S A* 102 (32):11516–20. doi: 10.1073/pnas.0504718102.

Evans, E., A. Leung, V. Heinrich, and C. Zhu. 2004. "Mechanical switching and coupling between two dissociation pathways in a P-selectin adhesion bond." *Proc Natl Acad Sci U S A* 101 (31):11281–6.

Fasshauer, D., H. Otto, W. K. Eliason, R. Jahn, and A. T. Brunger. 1997. "Structural changes are associated with soluble N-ethylmaleimide-sensitive fusion protein attachment protein receptor complex formation." *J Biol Chem* 272 (44):28036–41.

Feinberg, H, Y Guo, DA Mitchell, K Drickamer, and W. I. Weis. 2005. "Extended neck regions stabilize tetramers of the receptors DC-SIGN and DC-SIGNR." *J Biol Chem* 280 (2):1327–1335.

Feinberg, H., D. A. Mitchell, K Drickamer, and W. I. Weis. 2001. "Structural basis for selective recognition of oligosaccharides by DC-SIGN and DC-SIGNR." *Science* 294 (5549):2163–2166.

Feinberg, H., C. K. Tso, M. E. Taylor, K. Drickamer, and W. I. Weis. 2009. "Segmented helical structure of the neck region of the glycan-binding receptor DC-SIGNR." *J. Mol. Biol.* 394 (4):613–20.

Gardner, J. P., R. J. Durso, R. R. Arrigale, G. P. Donovan, P. J. Maddon, T. Dragic, and W. C. Olson. 2003. "L-SIGN (CD 209L) is a liver-specific capture receptor for hepatitis C virus." *Proc Natl Acad Sci USA* 100 (8):4498–503.

Geijtenbeek, T. B., A. Engering, and Y. Van Kooyk. 2002. "DC-SIGN, a C-type lectin on dendritic cells that unveils many aspects of dendritic cell biology." *J Leukoc Biol* 71 (6):921–31.

Geijtenbeek, T. B., D. S. Kwon, R. Torensma, S. J. van Vliet, G. C. van Duijnhoven, J. Middel, I. L. Cornelissen, H. S. Nottet, V. N. KewalRamani, D. R. Littman, C. G. Figdor, and Y. van Kooyk. 2000. "DC-SIGN, a dendritic cell-specific HIV-1-binding protein that enhances trans-infection of T cells." *Cell* 100 (5):587–97.

Gunnerson, K. N., Y. V. Pereverzev, and O. V. Prezhdo. 2009. "Atomistic simulation combined with analytic theory to study the response of the P-selectin/PSGL-1 complex to an external force." *J Phys Chem B* 113 (7):2090–100.

Hall, A. K., and U. Rutishauser. 1987. "Visualization of neural cell adhesion molecule by electron microscopy." *J. Cell Biol.* 104:1579–86.

Handschuh, G., B. Luber, P. Hutzler, H. Hofler, and K. F. Becker. 2001. "Single amino acid substitutions in conserved extracellular domains of E-cadherin differ in their functional consequences." *J Mol Biol* 314 (3):445–54. doi: 10.1006/jmbi.2001.5143.

Harrison, O. J., F. Bahna, P. S. Katsamba, X. Jin, J. Brasch, J. Vendome, G. Ahlsen, K. J. Carroll, S. R. Price, B. Honig, and L. Shapiro. 2010. "Two-step adhesive binding by classical cadherins." *Nat Struct Mol Biol* 17 (3):348–57. doi: 10.1038/nsmb.1784.

Harrison, O. J., X. Jin, S. Hong, F. Bahna, G. Ahlsen, J. Brasch, Y. Wu, J. Vendome, K. Felsovalyi, C. M. Hampton, R. B. Troyanovsky, A. Ben-Shaul, J. Frank, S. M. Troyanovsky, L. Shapiro, and B. Honig. 2011. "The extracellular architecture of adherens junctions revealed by crystal structures of type I cadherins." *Structure* 19 (2):244–56. doi: 10.1016/j.str.2010.11.016.

Haussinger, D., T. Ahrens, H. J. Sass, O. Pertz, J. Engel, and S. Grzesiek. 2002. "Calcium-dependent homoassociation of E-cadherin by NMR spectroscopy: changes in mobility, conformation and mapping of contact regions." *J Mol Biol* 324 (4):823–39.

Hazzard, J., T. C. Sudhof, and J. Rizo. 1999. "NMR analysis of the structure of synaptobrevin and of its interaction with syntaxin." *J Biomol NMR* 14 (3):203–7.

He, W., P. Cowin, and D. L. Stokes. 2003. "Untangling desmosomal knots with electron tomography." *Science* 302 (5642):109–13.

Hong, S., R. B. Troyanovsky, and S. M. Troyanovsky. 2013. "Binding to F-actin guides cadherin cluster assembly, stability, and movement." *J Cell Biol* 201 (1):131–43. doi: 10.1083/jcb.201211054.

Hu, J., R. Lipowsky, and T. R. Weikl. 2013. "Binding constants of membrane-anchored receptors and ligands depend strongly on the nanoscale roughness of membranes." *Proc Natl Acad Sci USA* 110 (38):15283–8. doi: 10.1073/pnas.1305766110.

Huang, J., V. I. Zarnitsyna, B. Liu, L. J. Edwards, N. Jiang, B. D. Evavold, and C. Zhu. 2010. "The kinetics of two-dimensional TCR and pMHC interactions determine T-cell responsiveness." *Nature* 464 (7290):932–6. doi: 10.1038/nature08944.

Hunter, R. 1989. *Foundations of Colloid Science*. 2 vols. Vol. 1. Oxford: Oxford University Press.

Israelachvili, J. 1992. *Intermolecular and Surface Forces*. 2 ed. New York: Academic Press.

Israelachvili, J., and P. McGuiggan. 1990. "Adhesion and short-range forces between surfaces: new apparatus for surface force measurements." *J Mater Res* 5:2223–2231.

Israelachvili, J. N. 1973. "Thin film studies using multiple-beam interferometry." *J Coll Int Sci* 44:259–272.

Jenson, P., V. Soroka, N. K. Thompson, I. Ralets, V. Berezin, E. Bock, and F. M. Poulsen. 1999. "Structure and interactions of NCAM modules 1 and 2-basic elements in neural cell adhesion." *Nat Struct Biol* 6:486–493.

Jeppesen, C., J. Y. Wong, T. L. Kuhl, J. N. Israelachvili, N. Mullah, S. Zalipsky, and C. M. Marques. 2001. "Impact of polymer tether length on multiple ligand-receptor bond formation." *Science* 293 (5529):465–8. doi: 10.1126/science.293.5529.465293/5529/465 [pii].

Johnson, C. P., G. Fragneto, O. Konovalov, V. Duboscard, J. F. Legrand, and D. E. Leckband. 2005. "Structural studies of the neural-cell-adhesion molecule by X-ray and neutron reflectivity." *Biochemistry* 44 (2):546–54. doi: 10.1021/bi048263j.

Johnson, C. P., I. Fujimoto, C. Perrin-Tricaud, U. Rutishauser, and D. Leckband. 2004. "Mechanism of homophilic adhesion by the neural cell adhesion molecule: use of multiple domains and flexibility." *Proc Natl Acad Sci USA* 101 (18):6963–8.

Johnson, C. P., I. Fujimoto, U. Rutishauser, and D. E. Leckband. 2005. "Direct evidence that neural cell adhesion molecule (NCAM) polysialylation increases intermembrane repulsion and abrogates adhesion." *J Biol Chem* 280 (1):137–45.

Kasper, C., H. Rasmussen, J. S. Kastrup, S. Ikemizu, R. Y. Jones, V. Berezin, E. Bock, and I. K. Larsen. 2000. "Structural basis of cell-cell adhesion by NCAM." *Nat Struct Biol* 7:389–393.

Kiselyov, V., V. Berezin, T. E. Maar, V. Soroka, K. Edvardsen, A. Schousboe, and E. Bock. 1997. "The first immunoglobulin-like neural cell adhesion molecule (NCAM) domain is involved in double-reciprocal interaction with the second immunoglobulin-like NCAM domain and in heparin binding." *J Biol Chem* 272:10125–10134.

Kitov, P. I., J. M. Sadowska, G. Mulvey, G. D. Armstrong, H. Ling, N. S. Pannu, R. J. Read, and D. R. Bundle. 2000. "Shiga-like toxins are neutralized by tailored multivalent carbohydrate ligands." *Nature* 403 (6770):669–72.

Leckband, D., and J. Israelachvili. 2001. "Intermolecular forces in biology." *Q Rev Biophys* 34 (2):105–267.

Leckband, D., W. Muller, F. J. Schmitt, and H. Ringsdorf. 1995. "Molecular mechanisms determining the strength of receptor-mediated intermembrane adhesion." *Biophys J* 69 (3):1162–9.

Leckband, D., and A. Prakasam. 2006. "Mechanism and dynamics of cadherin adhesion." *Annu Rev Biomed Eng* 8:259–87.

Leckband, D., and S. Sivasankar. 2012a. "Biophysics of cadherin adhesion." *Subcell Biochem* 60:63–88. doi: 10.1007/978-94-007-4186-7_4.

Leckband, D., and S. Sivasankar. 2012b. "Cadherin recognition and adhesion." *Curr Opin Cell Biol.* doi: 10.1016/j.ceb.2012.05.014.

Leckband, D. E., T. Kuhl, H. K. Wang, J. Herron, W. Muller, and H. Ringsdorf. 1995. "4-4-20 anti-fluorescyl IgG Fab' recognition of membrane bound hapten: direct evidence for the role of protein and interfacial structure." *Biochemistry* 34 (36):11467–78.

Leckband, D. E., Q. le Duc, N. Wang, and J. de Rooij. 2011. "Mechanotransduction at cadherin-mediated adhesions." *Curr Op Cell Biol* 23 (5):523–30. doi: 10.1016/j.ceb.2011.08.003.

Leckband, D. E., S. Menon, K. Rosenberg, S. A. Graham, M. E. Taylor, and K. Drickamer. 2011. "Geometry and adhesion of extracellular domains of DC-SIGNR neck length variants analyzed by force-distance measurements." *Biochemistry* 50 (27):6125–32. doi: 10.1021/bi2003444.

Leckband, D. E., F. J. Schmitt, J. N. Israelachvili, and W. Knoll. 1994. "Direct force measurements of specific and nonspecific protein interactions." *Biochemistry* 33 (15):4611–24.

Levy, Y., J. N. Onuchic, and P. G. Wolynes. 2007. "Fly-casting in protein-DNA binding: frustration between protein folding and electrostatics facilitates target recognition." *J Am Chem Soc* 129 (4):738–9. doi: 10.1021/ja065531n.

Li, F., D. Kummel, J. Coleman, K. M. Reinisch, J. E. Rothman, and F. Pincet. 2014. "A half-zippered SNARE complex represents a functional intermediate in membrane fusion." *J Am Chem Soc* 136 (9):3456–64. doi: 10.1021/ja410690m.

Li, F., F. Pincet, E. Perez, W. S. Eng, T. J. Melia, J. E. Rothman, and D. Tareste. 2007. "Energetics and dynamics of SNAREpin folding across lipid bilayers." *Nat Struct Mol Biol* 14 (10):890–6. doi: 10.1038/nsmb1310.

Lichterfeld, M., H. D. Nischalke, J. van Lunzen, J. Sohne, N. Schmeisser, R. Woitas, T. Sauerbruch, J. K. Rockstroh, and U. Spengler. 2003. "The tandem-repeat polymorphism of the DC-SIGNR gene does not affect the susceptibility to HIV infection and the progression to AIDS." *Clin. Immunol.* 107 (1):55–9.

Long, M., H. Zhao, K. S. Huang, and C. Zhu. 2001. "Kinetic measurements of cell surface E-selectin/carbohydrate ligand interactions." *Annals Biomed Eng* 29 (11):935–46.

Manibog, K., H. Li, S. Rakshit, and S. Sivasankar. 2014. "Resolving the molecular mechanism of cadherin catch bond formation." *Nat Commun* 5:3941. doi: 10.1038/ncomms4941.

Marshall, B. T., M. Long, J. W. Piper, T. Yago, R. P. McEver, and C. Zhu. 2003. "Direct observation of catch bonds involving cell-adhesion molecules." *Nature* 423 (6936):190–3.

Menon, S., K. Rosenberg, S. A. Graham, E. M. Ward, M. E. Taylor, K. Drickamer, and D. E. Leckband. 2009. "Binding-site geometry and flexibility in DC-SIGN demonstrated with surface force measurements." *Proc Natl Acad Sci USA* 106 (28):11524–9. doi: 10.1073/pnas.0901783106.

Nattermann, J., G. Ahlenstiel, T. Berg, G. Feldmann, H. D. Nischalke, T. Muller, J. Rockstroh, R. Woitas, T. Sauerbruch, and U. Spengler. 2006. "The tandem-repeat polymorphism of the DC-SIGNR gene in HCV infection." *J. Viral Hepat* 13 (1):42–6.

Nelson, R. W., P. A. Bates, U. Rutishauser. 1995. "Protein determinants for specific polysialylation of the neural cell adhesion molecule." *J Biol Chem* 270:17171–17179.

Pereverzev, Y. V., O. V. Prezhdo, M. Forero, E. V. Sokurenko, and W. E. Thomas. 2005. "The two-pathway model for the catch-slip transition in biological adhesion." *Biophys J* 89 (3):1446–54.

Pincus, P. 1991. "Colloid stabilization with grafted polyelectrolytes." *Macromolecules* 24:2912–2919.

Pohlmann, S., F. Baribaud, and R. W. Doms. 2001. "DC-SIGN and DC-SIGNR: helping hands for HIV." *Trends Immunol* 22 (12):643–6.

Pohlmann, S., E. J. Soilleux, F. Baribaud, G. J. Leslie, L. S. Morris, J. Trowsdale, B. Lee, N. Coleman, and R. W. Doms. 2001. "DC-SIGNR, a DC-SIGN homologue expressed in endothelial cells, binds to human and simian immunodeficiency viruses and activates infection in trans." *Proc Natl Acad Sci USA* 98 (5):2670–5.

Prezhdo, O. V., and Y. V. Pereverzev. 2009. "Theoretical aspects of the biological catch bond." *Acc Chem Res* 42 (6):693–703.

Rakshit, S., Y. Zhang, K. Manibog, O. Shafraz, and S. Sivasankar. 2012. "Ideal, catch, and slip bonds in cadherin adhesion." *Proc Natl Acad Sci USA* 109 (46):18815–20. doi: 10.1073/pnas.1208349109.

Ranheim, T. S., G. M. Edelman, and B. A. Cunningham. 1996. "Homophilic adhesion mediated by the neural cell adhesion molecule involves multiple immunoglobulin domains." *Proc Natl Acad Sci* 93:4071–4075.

Rao, Y., X.-F. Wu, J. Gariepy, U. Rutishauser, and C.-H. Siu. 1992. "Identification of a peptide sequence involved in homophilic binding in the neural cell adhesion molecule NCAM." *J Cell Biol* 118:937–949.

Rutishauser, U. 1992. "NCAM and its polysialic acid moiety: a mechanism for pull/push regulation of cell interactions during development?" *Dev Suppl*:99–104.

Rutishauser, U., A. Acheson, A. K. Hall, D. M. Mann, and J. Sunshine. 1988. "The neural cell adhesion molecule (NCAM) as a regulator of cell-cell interactions." *Science* 240 (4848):53–7.

Shan, W., Y. Yagita, Z. Wang, A. Koch, A. Fex Svenningsen, E. Gruzglin, L. Pedraza, and D. R. Colman. 2004. "The minimal essential unit for cadherin-mediated intercellular adhesion comprises extracellular domains 1 and 2." *J Biol Chem* 279 (53):55914–23.

Shashikanth, N., M. A. Kisting, and D. E. Leckband. 2016. "Kinetic Measurements Reveal Enhanced Protein-Protein Interactions at Intercellular Junctions." *Sci Rep* 6:23623. doi: 10.1038/srep23623.

Shen, H., M. Watanabe, H. Tomasiewicz, U. Rutishauser, T. Magnuson, and J. D. Glass. 1997. "Role of neural cell adhesion molecule and polysialic acid in mouse circadian clock function." *J Neurosci* 17 (13):5221–9.

Shi, Q., V. Maruthamuthu, and D. E. Leckband. 2010. "Allosteric cross-talk between cadherin ectodomains." *Biophys J* 99:95–104.

Shoemaker, B. A., J. J. Portman, and P. G. Wolynes. 2000. "Speeding molecular recognition by using the folding funnel: the fly-casting mechanism." *Proc Natl Acad Sci U S A* 97 (16):8868–73. doi: 10.1073/pnas.160259697.

Simmons, G., J. D. Reeves, C. C. Grogan, L. H. Vandenberghe, F. Baribaud, J. C. Whitbeck, E. Burke, M. J. Buchmeier, E. J. Soilleux, J. L. Riley, R. W. Doms, P. Bates, and S. Pohlmann. 2003. "DC-SIGN and DC-SIGNR bind Ebola glycoproteins and enhance infection of macrophages and endothelial cells." *Virology* 305 (1):115–123.

Sivasankar, S., Y. Zhang, W. J. Nelson, and S. Chu. 2009. "Characterizing the initial encounter complex in cadherin adhesion." *Structure* 17 (8):1075–81. doi: 10.1016/j.str.2009.06.012.

Soilleux, E. J., L. S. Morris, B. Lee, S. Pohlmann, J. Trowsdale, R. W. Doms, and N. Coleman. 2001. "Placental expression of DC-SIGN may mediate intrauterine vertical transmission of HIV." *J Pathol* 195 (5):586–92.

Soroka, V., D. Kiryushko, V. Novitskaya, C. B. Ronn, F. M. Poulson, A. Holm, E. Bock, and V. Berezin. 2002. "Induction of neuronal differentiation by a peptide corresponding to the homophilic binding site of the second Ig module of NCAM." *J Biol Chem* 277:24676–24683.

Soroka, V., K. Kolkova, J. S. Kastrup, K. Diederichs, J. Breed, V. V. Kiselyov, F. M. Poulsen, F. M. Poulsen, I. K. Larsen, W. Welte, V. Berezin, E. Bock, and C. Kasper. 2003. "Structure and interactions of NCAM Ig1-2-3 suggest a novel zipper mechanism for homophilic adhesion." *Structure* 10:1291–1301.

Thomas, W. 2008. "Catch bonds in adhesion." *Annu Rev Biomed Eng* 10:39–57.

Tomschy, A., C. Fauser, R. Landwehr, and J. Engel. 1996. "Homophilic adhesion of E-cadherin occurs by a cooperative two-step interaction of N-terminal domains." *EMBO J* 15 (14):3507–14.

Truong Quang, B. A., M. Mani, O. Markova, T. Lecuit, and P. F. Lenne. 2013. "Principles of E-cadherin supramolecular organization in vivo." *Curr Biol* 23 (22):2197–207. doi: 10.1016/j.cub.2013.09.015.

van Kooyk, Y., and T. B. Geijtenbeek. 2003. "DC-SIGN: escape mechanism for pathogens." *Nat Rev Immunol* 3 (9):697–709.

Walsh, F. S., and P. Doherty. 1997. "Neural cell adhesion molecules of the immunoglobulin superfamily: role in axon growth and guidance." *Annu Rev Cell Dev Biol* 13:425–56.

Wang, Y. J., F. Li, N. Rodriguez, X. Lafosse, C. Gourier, E. Perez, and F. Pincet. 2016. "Snapshot of sequential SNARE assembling states between membranes shows that N-terminal transient assembly initializes fusion." *Proc Natl Acad Sci USA* 113 (13):3533–8. doi: 10.1073/pnas.1518935113.

Wieland, J. A., A. A. Gewirth, and D. E. Leckband. 2005. "Single molecule adhesion measurements reveal two homophilic neural cell adhesion molecule bonds with mechanically distinct properties." *J Biol Chem* 280 (49):41037–46.

Williams, T. E., S. Nagarajan, P. Selvaraj, and C. Zhu. 2001. "Quantifying the impact of membrane microtopology on effective two-dimensional affinity." *J Biol Chem* 276 (16):13283–8. doi: 10.1074/jbc.M010427200.

Williams, T. E., P. Selvaraj, and C. Zhu. 2000. "Concurrent binding to multiple ligands: kinetic rates of CD16b for membrane-bound IgG1 and IgG2." *Biophys J* 79 (4):1858–66. doi: S0006-3495(00)76435-6 [pii]10.1016/S0006-3495(00)76435-6 [doi].

Wong, J. Y., T. L. Kuhl, J. N. Israelachvili, N. Mullah, and S. Zalipsky. 1997. "Direct measurement of a tethered ligand-receptor interaction potential." *Science* 275 (5301):820–2.

Wu, J., Y. Fang, V. I. Zarnitsyna, T. P. Tolentino, M. L. Dustin, and C. Zhu. 2008. "A coupled diffusion-kinetics model for analysis of contact-area FRAP experiment." *Biophys J* 95 (2):910–9. doi: 10.1529/biophysj.107.114439.

Wu, Y., X. Jin, O. Harrison, L. Shapiro, B. H. Honig, and A. Ben-Shaul. 2010. "Cooperativity between trans and cis interactions in cadherin-mediated junction formation." *Proc Natl Acad Sci USA* 107 (41):17592–7. doi: 10.1073/pnas.1011247107.

Wu, Y., P. Kanchanawong, and R. Zaidel-Bar. 2015. "Actin-delimited adhesion-independent clustering of e-cadherin forms the nanoscale building blocks of adherens junctions." *Dev Cell* 32 (2):139–54. doi: 10.1016/j.devcel.2014.12.003.

Wu, Y., J. Vendome, L. Shapiro, A. Ben-Shaul, and B. Honig. 2011. "Transforming binding affinities from three dimensions to two with application to cadherin clustering." *Nature* 475 (7357):510–3. doi: 10.1038/nature10183.

Yang, P. Yin, X., and Rutishauser, U. 1992. "Intercellular space is affected by polysialic acid content of NCAM." *J Cell Biol* 116:1487–1496.

Yeung, C., T. Purves, A. A. Kloss, T. L. Kuhl, S. Sligar, and D. Leckband. 1999. "Cytochrome c recognition of immobilized, orientational variants of cytochrome b5: direct force and equilibrium binding measurements." *Langmuir* 15 (20):6829–6836. doi: 10.1021/la990019j.

Yu, Q. D., A. P. Oldring, A. S. Powlesland, C. K. Tso, C. Yang, K. Drickamer, and M. E. Taylor. 2009. "Autonomous tetramerization domains in the glycan-binding receptors DC-SIGN and DC-SIGNR." *J Mol Biol* 387 (5):1075–80.

Zhang, F., W. D. Marcus, N. H. Goyal, P. Selvaraj, T. A. Springer, and C. Zhu. 2005. "Two-dimensional kinetics regulation of alphaLbeta2-ICAM-1 interaction by conformational changes of the alphaL-inserted domain." *J. Biol. Chem.* 280 (51):42207–18. doi: 10.1074/jbc.M510407200.

Zhu, B., E. A. Davies, P. A. van der Merwe, T. Calvert, and D. E. Leckband. 2002. "Direct measurements of heterotypic adhesion between the cell surface proteins CD2 and CD48." *Biochemistry* 41 (40):12163–70.

Zorman, S., A. A. Rebane, L. Ma, G. Yang, M. A. Molski, J. Coleman, F. Pincet, J. E. Rothman, and Y. Zhang. 2014. "Common intermediates and kinetics, but different energetics, in the assembly of SNARE proteins." *Elife* 3:e03348. doi: 10.7554/eLife.03348.

6 Charged Membranes: Poisson–Boltzmann Theory, The DLVO Paradigm, and Beyond

Tomer Markovich and David Andelman
Tel Aviv University, Tel Aviv, Israel

Rudolf Podgornik
University of Chinese Academy of Sciences, Beijing, China
Institute of Physics, Chinese Academy of Sciences, Beijing, China
Wenzhou Institute of the University of Chinese Academy of Sciences, Wenzhou, China

CONTENTS

6.1 Introduction ... 100
6.2 Poisson–Boltzmann Theory ... 101
    6.2.1 Debye–Hückel Approximation ... 103
6.3 One Planar Membrane ... 104
    6.3.1 Counterions Only ... 104
    6.3.2 Added Electrolyte ... 105
    6.3.3 The Grahame Equation ... 107
6.4 Modified Poisson–Boltzmann Theory ... 107
6.5 Two-Membrane System: Osmotic Pressure ... 109
6.6 Two Symmetric Membranes, $\sigma_1 = \sigma_2$ ... 111
    6.6.1 Counterions Only ... 111
    6.6.2 Added Electrolyte ... 112
    6.6.3 Debye–Hückel Regime ... 114
    6.6.4 Intermediate Regime ... 114
    6.6.5 Other Pressure Regimes ... 115
6.7 Two Asymmetric Membranes, $\sigma_1 \neq \sigma_2$ ... 115
    6.7.1 The Debye–Hückel Regime ... 116
    6.7.2 The Debye–Hückel Regime with Constant Surface Potential ... 116
    6.7.3 Counterions Only ... 117
    6.7.4 Attraction/Repulsion Crossover ... 118
6.8 Charge Regulation ... 118
    6.8.1 Charge Regulation via Free Energy ... 121
6.9 van der Waals' Interactions ... 122
    6.9.1 The Hamaker Pairwise Summation ... 122
    6.9.2 Macroscopic Theory of vdW Interactions ... 123
    6.9.3 The Derjaguin–Landau–Verwey–Overbeek Theory ... 124
6.10 Limitations and Generalizations ... 125
Notes ... 126
Bibliography ... 126

6.1 INTRODUCTION

It is of great importance to understand electrostatic interactions and their key role in soft and biological matter. These systems typically consist of aqueous environment in which charges tend to dissociate and affect a wide variety of functional, structural, and dynamical properties. Among the numerous effects of electrostatic interactions, it is instructive to mention their effect on elasticity of flexible charged polymers (polyelectrolytes) and cell membranes, formation of self-assembled charged micelles, and stabilization of charged colloidal suspensions that results from the competition between repulsive electrostatic interactions and attractive van der Waals' interactions (Verwey and Overbeek 1948, Andelman 1995, 2005, Holm, Kekicheff and Podgornik 2000, Dean et al. 2014, Churaev, Derjaguin and Muller 2014).

In this chapter, we focus on charged membranes. Biological membranes are complex heterogeneous two-dimensional interfaces separating the living cell from its extracellular surrounding. Membranes also surround intercellular organelles such as the cell nucleus, Golgi apparatus, mitochondria, endoplasmic reticulum, and ribosomes. Electrostatic interactions control many of the membrane structural properties and functions, e.g., rigidity, structural stability, lateral phase transitions, and their dynamics. Moreover, electric charges are a key player in processes involving more than one membrane such as membrane adhesion and cell–cell interaction, as well as the overall interactions of membranes with other intra- and extracellular proteins, biopolymers, and DNA.

How do membranes interact with their surrounding ionic solution? Charged membranes attract a cloud of oppositely charged mobile ions that forms a *diffuse electric double layer* (Gouy 1910, 1917, Chapman 1913, Debye and Hückel 1923, Verwey and Overbeek 1948, Israelachvili 2011). The system favours *local electroneutrality*, but while achieving it, entropy is lost. The competition between electrostatic interactions and entropy of ions in solution determines the exact distribution of mobile ions close to charged membranes. This last point shows the significance of temperature in determining the equilibrium properties, because temperature controls the strength of entropic effects as compared to electrostatic interactions. For soft materials, the thermal energy $k_B T$ where k_B is the Boltzmann constant and T is the temperature, is also comparable with other characteristic energy scales associated with elastic deformations and structural degrees of freedom.

It is convenient to introduce a length scale for which the thermal energy is equal to the Coulombic energy between two unit charges. This is called the *Bjerrum length*, defined as follows:

$$\ell_B = \frac{e^2}{4\pi\varepsilon_0\varepsilon_w k_B T}, \quad (6.1)$$

where e is the elementary charge, $\varepsilon_0 = 8.85 \cdot 10^{-12} [F/m]$ is the vacuum permittivity,[1] and the dimensionless dielectric constant of water is $\varepsilon_w = 80$ (for a complete list of symbols, see Table 6.1). The Bjerrum length is equal to about 0.7 nm at room temperatures, $T = 300$ K.

A related length is the *Gouy–Chapman (GC) length* defined as

$$\ell_{GC} = \frac{2\varepsilon_0\varepsilon_w k_B T}{e|\sigma|} = \frac{e}{2\pi\ell_B|\sigma|} \sim \sigma^{-1}. \quad (6.2)$$

At this length scale, the thermal energy $k_B T$ is equal to the Coulombic energy between a unit charge and a planar surface with a constant surface-charge density, σ. The Gouy–Chapman length ℓ_{GC} is inversely proportional to σ. For strongly charged membranes, ℓ_{GC} is rather small, on the order of a tenth of nanometre.

In their pioneering work of almost a century ago, Debye and Hückel introduced the important concept of screening of the electrostatic interactions between two charges in presence of all other cations and anions of the solution (Debye and Hückel 1923). This effectively limits the range of electrostatic interactions as will be further discussed below. The characteristic length for which the electrostatic interactions are screened is called the *Debye length*, λ_D, defined for monovalent 1:1 electrolyte, as

$$\lambda_D = \kappa_D^{-1} = (8\pi\ell_B n_b)^{-1/2} \simeq \frac{0.3[nm]}{\sqrt{n_b[M]}}, \quad (6.3)$$

with n_b being the salt concentration (in molar) and κ_D being the inverse Debye length. The Debye screening length for 1:1 monovalent salts varies from about 0.3 nm in strong ionic solutions of 1 M to about 1 μm in pure water, where the concentration of the dissociated OH^- and H^+ ions is 10^{-7} M.

The aim of this chapter is to review some of the basic considerations underlying the behaviour of charged membranes in aqueous solutions using the three important length scales introduced above. We will not account for the detailed structure of real biological membranes, which can add considerable complexity, but restrict ourselves to simple model systems, relying on several assumptions and simplifications. The membrane is treated as a flat interface with a continuum surface charge distribution or constant surface potential. The mobile charge distributions are continuous, and we disregard the discreteness of surface charges that can lead to multipolar charge distributions.

This chapter is focussed only on static properties in thermodynamic equilibrium, excluding the interesting phenomena of dynamical fluctuations and dynamical responses to external fields (such as in electrochemistry systems). We mainly treat the mean-field approximation of the electric double-layer problem and the solutions of the classical Poisson–Boltzmann (PB) equation. Nevertheless, some effects of fluctuations and correlations will be briefly discussed in Section 6.10. We will also discuss the ionic finite size in Section 6.4, where the 'modified PB equation' is introduced.

TABLE 6.1
Table of Symbols

Symbol	Interpretation
a	Microscopic molecular size
$\alpha = \ln(a^3 K_d)$	Surface interaction parameter
α_i	Polarizability of the ith molecule
$\beta = 1/k_B T$	Inverse thermal energy
C_{PB}	Differential capacitance of the PB model
C_{mPB}	Differential capacitance of the mPB model
D	Intermembrane separation
ε_w	Water dimensionless dielectric constant
ε_L	Lipid dimensionless dielectric constant
F	Helmholtz free energy
$\mathcal{F} = F(d) - F(d \to \infty)$	Excess Helmholtz free energy
$\phi_b = a^3 n_b$	Bulk volume fraction for monovalent electrolyte
$\phi_s = |\sigma| a^2 / e$	Surface area fraction
h	Membrane thickness
$\mathcal{H}$	Hamaker constant
$I = \frac{1}{2} \sum_{i=1}^{M} z_i^2 n_i^{(b)}$	Ionic strength
K_d	Kinetic constant
$\ell_B = e^2/(4\pi\varepsilon_0 \varepsilon_w k_B T)$	Bjerrum length
$\ell_{GC} = e/(2\pi\ell_B |\sigma|); \ell_{1,2} = e/(2\pi\ell_B |\sigma_{1,2}|)$	Gouy–Chapman length
$\lambda_D = \kappa_D^{-1} = (8\pi\ell_B n_b)^{-1/2}$	Debye length
μ_i	Intrinsic chemical potential
μ_i^{tot}	Total chemical potential
$n_i(\mathbf{r})$	Concentration of the ith ionic species
$n_i^{(b)}$	Bulk concentration of the ith ionic species
n_b	Bulk concentration of monovalent electrolyte
n_0	Reference density, taken at zero potential
$n_i^{(s)}; n_s$	Surface density
$n_i^{(m)}; n_m$	Midplane density (two membranes)
P	Pressure
$\Pi = P_{in} - P_{out}$	Osmotic pressure
$\psi(\mathbf{r})$	Electrostatic potential
$\Psi(\mathbf{r}) = \beta e \psi(\mathbf{r})$	Dimensionless electrostatic potential
$\Psi_s; \psi_s$	Surface potential
$\Psi_m; \psi_m$	Midplane potential
$q_i = z_i e$	Charge of the ith ionic species
$\rho(\mathbf{r})$	Charge density of mobile ions
$\rho_f(\mathbf{r})$	Charge density of fixed charges
$\rho_{tot}(\mathbf{r})$	Total charge density
$\sigma; \sigma_{1,2}$	Surface charge density
V	Volume
z_i	Valency of the ith ionic species

The classical reference for the electric double layer is the book of Verwey and Overbeek (1948), which explains the DLVO (Derjaguin–Landau–Verwey–Overbeek) theory for stabilization of charged colloidal systems. More recent treatments can be found in many textbooks and monographs on colloidal science and interfacial phenomena, such as Evans and Wennerström (1999), Israelachvili (2011), and in two reviews by one of the present authors, Andelman (1995, 2005).

6.2 POISSON–BOLTZMANN THEORY

In Figure 6.1, a schematic view of a charged amphiphilic (phospholipid) membrane is presented. A membrane of thickness $h \simeq 4$ nm is composed of two monomolecular leaflets packed in a back-to-back configuration. The constituting molecules are amphiphiles having a charged 'head' and a hydrocarbon hydrophobic 'tail'. For phospholipids, the

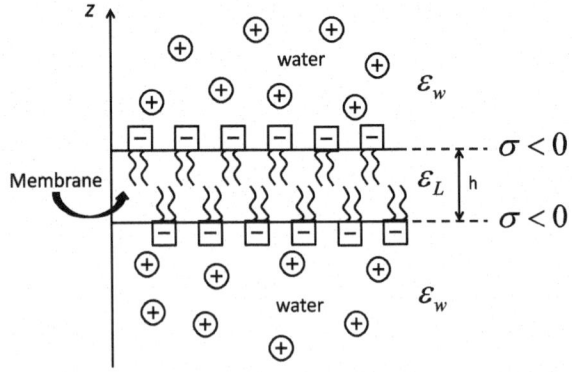

FIGURE 6.1 A bilayer membrane of thickness h composed of two monolayers (leaflets), each having a negative charge density, $\sigma<0$. The core membrane region (hydrocarbon tails) is modelled as a continuum medium with a dielectric constant ε_L, while the embedding medium (top and bottom) is water and has a dielectric constant, ε_w.

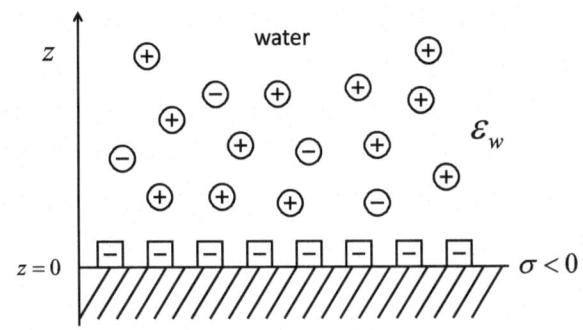

FIGURE 6.2 Schematic illustration of a charged membrane, located at $z=0$, with charge density σ. Without loss of generality, we take $\sigma<0$. For the counterion-only case, the surface charge is neutralized by the positive counterions. When monovalent (1:1) electrolyte is added to the reservoir, its bulk ionic density is $n_\pm^{(b)} = n_b$.

amphiphiles have a double tail. We model the membrane as a medium of thickness h having a dielectric constant, ε_L, coming essentially from the closely packed hydrocarbon ('oily') tails. The molecular heads contribute to the surface charges, and the entire membrane is immersed in an aqueous solution characterized by another dielectric constant, ε_w, assumed to be the water dielectric constant throughout the fluid.

The membrane charge can have two origins: either a charge group (*e.g.*, H$^+$) dissociates from the polar head group into the aqueous solution, leaving behind an oppositely charged group in the membrane; or, an ion from the solution (*e.g.*, Na$^+$) binds to a neutral site on the membrane and charges it (Borkovec, Jönsson, and Koper 2001). These association/dissociation processes are highly sensitive to the ionic strength and pH of the aqueous solution.

When the ionic association/dissociation is slow as compared with experimental times, the charges on the membrane can be considered as fixed and time independent, while for rapid association/dissociation, the surface charge can vary and is determined self-consistently from the thermodynamical equilibrium equations. We will further discuss the two processes of association/dissociation in Section 6.8. In many situations, the finite thickness of the membrane can be safely taken to be zero, with the membrane modelled as a planar surface as displayed in Figure 6.2. We will see later under what conditions this simplifying limit is valid.

Let us consider such an ideal membrane represented by a sharp boundary (located at $z=0$) that limits the ionic solution to the positive half space. The ionic solution contains, in general, the two species of mobile ions (anions and cations), and is modelled as a continuum dielectric medium as explained above. Thus, the boundary at $z=0$ marks the discontinuous jump of the dielectric constant between the ionic solution (ε_w) and the membrane (ε_L), which the ions cannot penetrate.

The PB equation can be obtained using two different approaches. The first is the one we present below combining the Poisson equation with the Boltzmann distribution, while the second one (presented later) is done through a minimization of the system free-energy functional. The PB equation is a mean-field (MF) equation, which can be derived from a field theoretical approach as the zeroth order in a systematic expansion of the grand-partition function (Podgornik and Žekš 1988, Borukhov, Andelman and Orland 1998, 2000, Netz and Orland 2000, Markovich, Andelman and Podgornik 2014, 2015, 2016, Markovich, Andelman and Orland 2016).

Consider M ionic species, each of them with charge q_i, where $q_i = ez_i$ and z_i is the valency of the ith ionic species. It is negative ($z_i < 0$) for anions and positive ($z_i > 0$) for cations. The mobile charge density (per unit volume) is defined as $\rho(\mathbf{r}) = \sum_{i=1}^{M} q_i n_i(\mathbf{r})$ with n_i being the number density (per unit volume), and both ρ and n_i are continuous functions of $\mathbf{r}$.

In the MF approximation, each of the ions sees a local environment constituting of all other ions, which dictates a local electrostatic potential $\psi(\mathbf{r})$. The potential $\psi(\mathbf{r})$ is a continuous function that depends on the total charge density through the Poisson equation:

$$\nabla^2 \psi(\mathbf{r}) = -\frac{\rho_{tot}(\mathbf{r})}{\varepsilon_0 \varepsilon_w} = -\frac{1}{\varepsilon_0 \varepsilon_w}\left[\sum_{i=1}^{M} q_i n_i(\mathbf{r}) + \rho_f(\mathbf{r})\right], \quad (6.4)$$

where $\rho_{tot} = \rho + \rho_f$ is the total charge density and $\rho_f(\mathbf{r})$ is a fixed external charge contribution.

As stated above, the aqueous solution (water) is modelled as a continuum featureless medium. This by itself represents an approximation because the ions themselves can change the local dielectric response of the medium (Ben-Yaakov, Andelman and Podgornik 2011, Levy, Andelman and Orland 2012, Adar et al. 2018) by inducing strong localized electric field. However, we will not include such refined local effects in this chapter.

Charged Membranes

The ions dispersed in solution are mobile and are allowed to adjust their positions. As each ionic species is in thermodynamic equilibrium, its density obeys the Boltzmann distribution:

$$n_i(\mathbf{r}) = n_i^{(b)} e^{-\beta q_i \psi(\mathbf{r})}, \quad (6.5)$$

where $\beta = 1/k_B T$, and $n_i^{(b)}$ is the bulk density of ith species taken at zero reference potential, $\psi = 0$.

> **BOLTZMANN DISTRIBUTION VIA ELECTROCHEMICAL POTENTIAL**
>
> A simple derivation of the Boltzmann distribution is obtained through the requirement that the *electrochemical potential* (total chemical potential) μ_i^{tot}, for each ionic species, is constant throughout the system:
>
> $$\mu_i^{\text{tot}} = \mu_i(\mathbf{r}) + q_i \psi(\mathbf{r}) = \text{const}, \quad (6.6)$$
>
> where $\mu_i(\mathbf{r})$ is the intrinsic chemical potential. For dilute ionic solutions, the ith ionic species entropy is taken as an ideal gas one, $\mu_i(\mathbf{r}) = k_B T \ln[n_i(\mathbf{r}) a^3]$. By substituting $a^3 n_i^{(b)} = \exp(\beta \mu_i^{\text{tot}})$ into Eq. (6.6), the Boltzmann distribution of Eq. (6.5) follows. This relation between the bulk ionic density and chemical potential is obtained by setting $\psi = 0$ in the bulk and shows that one can consider the chemical potential, μ_i^{tot}, as a Lagrange multiplier setting the bulk densities to be $n_i^{(b)}$. Note that we have introduced a microscopic length scale, a, defining a reference close packing density, $1/a^3$. Equation (6.6) assumes that the ions are point-like and have no other interactions in addition to their electrostatic one.

We now substitute Eq. (6.5) into Eq. (6.4) to obtain the *Poisson–Boltzmann equation*:

$$\nabla^2 \psi(\mathbf{r}) = -\frac{1}{\varepsilon_0 \varepsilon_w} \left[\sum_{i=1}^{M} q_i n_i^{(b)} e^{-\beta q_i \psi(\mathbf{r})} + \rho_f(\mathbf{r}) \right]. \quad (6.7)$$

For binary monovalent electrolytes (denoted as 1:1 electrolyte), $z_i = \pm 1$, the PB equation reads

$$\nabla^2 \psi(\mathbf{r}) = \frac{1}{\varepsilon_0 \varepsilon_w} \left[2 e n_b \sinh(\beta e \psi(\mathbf{r})) - \rho_f(\mathbf{r}) \right]. \quad (6.8)$$

Generally speaking, the PB theory is a very useful analytical approximation with many applications. It is a good approximation at physiological conditions (electrolyte strength of about 0.1 M) and for other dilute monovalent electrolytes and moderate surface potentials and surface charge. Although the PB theory produces good results in these situations, it misses some important features associated with charge correlations and fluctuations of multivalent ions.

Moreover, close to a charged membrane, the finite size of the surface ionic groups and that of the counterions lead to deviations from the PB results (see Sections 6.4 and 6.8 for further details).

As the PB equation is a nonlinear equation, it can be solved analytically only for a limited number of simple boundary conditions. On the other hand, by solving it numerically or within further approximations or limits, one can obtain ionic profiles and free energies of complex structures. For example, the free energy change for a charged globular protein that binds onto an oppositely charged lipid membrane.

In an alternative approach, the PB equation can also be obtained by a minimization of the system free-energy functional. One can assume that the internal energy, U_{el}, is purely electrostatic and that the Helmholtz free energy, $F = U_{el} - TS$, is composed of an internal energy and an ideal mixing entropy, S, of a dilute solution of mobile ions.

The electrostatic energy, U_{el}, is expressed in terms of the potential $\psi(\mathbf{r})$:

$$U_{el} = \frac{\varepsilon_0 \varepsilon_w}{2} \int_V d^3 r |\nabla \psi(\mathbf{r})|^2$$

$$= \frac{1}{2} \int_V d^3 r \left[\sum_{i=1}^{M} q_i n_i(\mathbf{r}) \psi(\mathbf{r}) + \rho_f(\mathbf{r}) \psi(\mathbf{r}) \right], \quad (6.9)$$

while the mixing entropy of ions is written in the dilute solution limit as

$$S = -k_B \sum_{i=1}^{M} \int_V d^3 r \left(n_i(\mathbf{r}) \ln[n_i(\mathbf{r}) a^3] - n_i(\mathbf{r}) \right). \quad (6.10)$$

Using Eqs. (6.9) and (6.10), the Helmholtz free energy can be written as

$$F = \int_V d^3 r \left[-\frac{\varepsilon_0 \varepsilon_w}{2} |\nabla \psi(\mathbf{r})|^2 + \left(\sum_{i=1}^{M} q_i n_i(\mathbf{r}) + \rho_f(\mathbf{r}) \right) \psi(\mathbf{r}) \right.$$

$$\left. + k_B T \sum_{i=1}^{M} \left(n_i(\mathbf{r}) \ln[n_i(\mathbf{r}) a^3] - n_i(\mathbf{r}) \right) \right], \quad (6.11)$$

where the sum of the first two terms is equal to U_{el} and the third one is $-TS$. The variation of this free energy with respect to $\psi(\mathbf{r})$, $\delta F/\delta \psi = 0$, gives the Poisson equation, Eq. (6.4), while from the variation with respect to $n_i(\mathbf{r})$, $\delta F/\delta n_i = \mu_i^{\text{tot}}$, we obtain the electrochemical potential of Eq. (6.6). As before, substituting the Boltzmann distribution obtained from Eq. (6.6), into the Poisson equation, Eq. (6.4), gives the PB equation, Eq. (6.7).

6.2.1 Debye–Hückel Approximation

A useful and quite tractable approximation to the nonlinear PB equation is its linearized version. For electrostatic potentials smaller than 25 mV at room temperature (or equivalently $e|\psi| < k_B T$, $T \simeq 300\text{K}$), this approximation can be justified and the well-known Debye–Hückel (DH) theory (Debye and Hückel 1923) is recovered. Linearization of Eq. (6.7) is obtained by expanding its right-hand side to the first order in ψ,

$$\nabla^2 \psi(\mathbf{r}) = -\frac{1}{\varepsilon_0 \varepsilon_w} \sum_{i=1}^{M} q_i n_i^{(b)} + 8\pi \ell_B I \psi(\mathbf{r}) - \frac{1}{\varepsilon_0 \varepsilon_w} \rho_f(\mathbf{r}), \quad (6.12)$$

where $I = \frac{1}{2} \sum_{i=1}^{M} z_i^2 n_i^{(b)}$ is the *ionic strength* of the solution. The first term on the right-hand side of Eq. (6.12) vanishes because of electroneutrality of the bulk reservoir

$$\sum_{i=1}^{M} q_i n_i^{(b)} = 0, \quad (6.13)$$

recovering the DH equation:

$$\nabla^2 \psi(\mathbf{r}) = \kappa_D^2 \psi(\mathbf{r}) - \frac{1}{\varepsilon_0 \varepsilon_w} \rho_f(\mathbf{r}), \quad (6.14)$$

with the inverse Debye length, κ_D, defined as

$$\kappa_D^2 = \lambda_D^{-2} = 8\pi \ell_B I = 4\pi \ell_B \sum_{i=1}^{M} z_i^2 n_i^{(b)}. \quad (6.15)$$

For monovalent electrolytes, $z_i = \pm 1$, $\kappa_D^2 = 8\pi \ell_B n_b$ with $n_i^{(b)} = n_b$, and Eq. (6.3) is recovered. Note that the Debye length, $\lambda_D = \kappa_D^{-1} \sim n_b^{-1/2}$, is a decreasing function of the salt concentration.

The DH treatment gives a simple tractable description of the pair interactions between ions. It is related to the Green function associated with the electrostatic potential around a point-like ion that can be calculated by using Eq. (6.14) for a point-like charge, q, placed at the origin, $\mathbf{r} = 0$, $\rho_f(\mathbf{r}) = q\delta(\mathbf{r})$,

$$(\nabla^2 - \kappa_D^2) \psi(\mathbf{r}) = -\frac{q}{\varepsilon_0 \varepsilon_w} \delta(\mathbf{r}), \quad (6.16)$$

where $\delta(\mathbf{r})$ is the Dirac δ-function. The solution to the above equation can be written in spherical coordinates as

$$\psi(r) = \frac{q}{4\pi \varepsilon_0 \varepsilon_w r} e^{-\kappa_D r}. \quad (6.17)$$

It manifests the exponential decay of the electrostatic potential with a characteristic length scale, $\lambda_D = 1/\kappa_D$. In a crude approximation, this exponential decay is replaced by a Coulombic interaction, which is only slightly screened for $r \leq \lambda_D$ and, thus, varies as $\sim r^{-1}$, while for $r > \lambda_D$, $\psi(r)$ is strongly screened and can sometimes be completely neglected.

6.3 ONE PLANAR MEMBRANE

We consider the PB equation for a single membrane, assumed to be planar and charged, and discuss separately two cases: (i) a charged membrane in contact with a solution containing only counterions, and (ii) a membrane in contact with a monovalent electrolyte reservoir.

As the membrane is taken to have an infinite extent in the lateral (x, y) directions, the PB equation is reduced to an effective one-dimensional equation, where all local quantities, such as the electrostatic potential, $\psi(\mathbf{r}) = \psi(z)$, and ionic densities, $n(\mathbf{r}) = n(z)$, depend only on the z-coordinate perpendicular to the planar membrane.

For a binary monovalent electrolyte (1:1 electrolyte, $z_i = \pm 1$), the PB equation from Eq. (6.7), reduces in its effective one-dimensional form to an ordinary differential equation depending only on the z-coordinate:

$$\Psi''(z) = \kappa_D^2 \sinh \Psi(z), \quad (6.18)$$

where $\Psi \equiv \beta e \psi$ is the rescaled dimensionless potential and we have assumed that the external charge, ρ_f, is restricted to the system boundaries and will only affect the boundary conditions.

We will consider two boundary conditions in this section. A fixed surface potential (Dirichlet boundary condition), $\Psi_s \equiv \Psi(z = 0) = \text{const}$, and constant surface charge (Neumann boundary condition), $\sigma \propto \Psi'_s = \text{const}$. A third and more specialized boundary condition of charge regulation will be treated in detail in Section 6.8. In the constant charge case, the membrane charge is modelled via a fixed surface charge density, $\rho_f = \sigma \delta(z)$ in Eq. (6.8). A variation of the Helmholtz free energy, F of Eq. (6.11) with respect to the surface potential, Ψ_s, $\delta F / \delta \Psi_s = 0$, is equivalent to constant surface charge boundary:

$$\left. \frac{d\Psi}{dz} \right|_{z=0} = -\frac{4\pi \ell_B \sigma}{e}. \quad (6.19)$$

Although we focus in the rest of the chapter on monovalent electrolytes, the extension to multivalent electrolytes is straightforward.

The boundary condition of Eq. (6.19) is valid if the electric field does not penetrate the 'oily' part of the membrane. This assumption can be justified (Kiometzis and Kleinert 1989, Winterhalter and Helfrich 1992), as long as $\varepsilon_L/\varepsilon_w \simeq 1/40 \ll h/\lambda_D$, where h is the membrane thickness (see Figure 6.1). All our results for one or two flat membranes, Sections 6.3–6.4 and 6.5–6.7, respectively, rely on this decoupled limit where the two sides (monolayers) of the membrane are completely decoupled and the electric field inside the membrane is negligible.

6.3.1 COUNTERIONS ONLY

A single charged membrane in contact with a cloud of counterions in solution is one of the simplest problems that has an analytical solution. It has been formulated and solved in

Charged Membranes

the beginning of the 20th century by Gouy (1910, 1917) and Chapman (1913). The aim is to find the profile of the counterion cloud forming a diffusive electric double layer close to a planar membrane (placed at $z = 0$) with fixed surface charge density (per unit area), σ, as in Figure 6.2.

Without loss of generality, the single-membrane problem is treated here for negative (anionic) surface charges ($\sigma < 0$) and positive monovalent counterions (cations) in the solution, $q_+ = e$ and $n(z) = n_+(z)$, such that the charge neutrality condition

$$\sigma = -e \int_0^\infty n(z) \, dz \qquad (6.20)$$

is fulfilled.

The PB equation for monovalent counterions is written as

$$\Psi''(z) = -4\pi \ell_B n_0 e^{-\Psi(z)}, \qquad (6.21)$$

where n_0 is the reference density, taken at zero potential in the absence of a salt reservoir. The PB equation, Eq. (6.21), with the boundary condition for one charged membrane, Eq. (6.19), and vanishing electric field at infinity, can be integrated analytically twice, yielding

$$\Psi(z) = 2\ln(z + \ell_{GC}) + \Psi_0, \qquad (6.22)$$

so that the density is

$$n(z) = \frac{1}{2\pi \ell_B} \frac{1}{(z + \ell_{GC})^2}, \qquad (6.23)$$

where Ψ_0 is a reference potential and ℓ_{GC} is the Gouy–Chapman length defined in Eq. (6.2). For example, for a choice of $\Psi_0 = -2\ln(\ell_{GC})$, the potential at $z = 0$ vanishes and Eq. (6.22) reads

$$\Psi(z) = 2\ln(1 + z/\ell_{GC}). \qquad (6.24)$$

Although the entire counterion profile is diffusive as it decays algebraically, half of the counterions ($|\sigma|/2$ per unit area) accumulate in a layer of thickness ℓ_{GC} close to the membrane,

$$e \int_0^{\ell_{GC}} n(z) \, dz = \frac{1}{2}|\sigma|. \qquad (6.25)$$

As an example, we present in Figure 6.3 the potential ψ (in mV) and ionic profile n (in M) for a surface density of $\sigma = -e/(2\,\mathrm{nm}^2)$, leading to a Gouy–Chapman length, $\ell_{GC} \simeq 0.46\,\mathrm{nm}$. The figure clearly shows the build-up of the diffusive layer of counterions attracted by the negatively charged membrane, reaching a limiting value of $n_s = n(0) \simeq 1.82$ M. Note that the potential has a weak logarithmic divergence as $z \to \infty$. This divergence is a consequence of the vanishing ionic reservoir (counterions only) with counterion density obeying the Boltzmann distribution. However, the physically measured electric field, $E = -d\psi/dz$, properly decays to zero as $\sim 1/z$, at $z \to \infty$.

6.3.2 Added Electrolyte

Another case of experimental interest is that of a single charged membrane at $z = 0$ in contact with an electrolyte reservoir. For a symmetric electrolyte, $n_+^{(b)} = n_-^{(b)} \equiv n_b$, and the same boundary condition of constant surface charge σ, Eq. (6.19), holds at the $z = 0$ surface. The negatively charged membrane attracts the counterions and repels the co-ions.

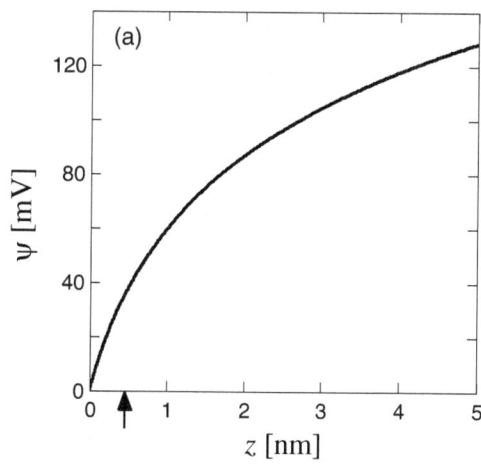
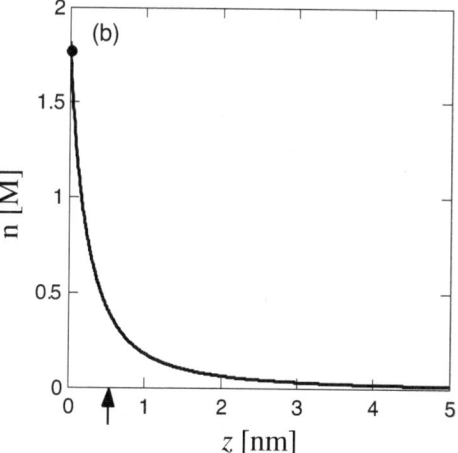

FIGURE 6.3 The electric double layer for a single charged membrane in contact with an aqueous solution of neutralizing monovalent counterions. In (a), the electrostatic potential $\psi(z)$ (in mV) is plotted as function of the distance from the membrane, z, Eq. (6.24). The charged membrane is placed at $z=0$ with $\sigma=-e/(2\,\mathrm{nm}^2)<0$. The zero of the potential is chosen to be at the membrane, $\psi(z=0)=0$. In (b), the density profile of the counterions, n (in M), is plotted as function of the distance z. Its value at the membrane is $n(z=0) = n_s \simeq 1.82$ M and the Gouy–Chapman length, $\ell_{GC} \simeq 0.46$ nm, is marked by an arrow.

As will be shown below, the potential decays to zero from below at large z; hence, it is always negative. Since the potential is a monotonic function, this also implies that $\Psi'(z)$ is always positive. At large z, where the potential decays to zero, the ionic profiles tend to their bulk (reservoir) densities, $n^{\pm}|_{\infty} = n_b$.

The PB equation for monovalent electrolyte, Eq. (6.18), with the boundary conditions as explained above can be solved analytically. The first integration of the PB equation for 1:1 electrolyte yields

$$\frac{d\Psi}{dz} = -2\kappa_D \sinh(\Psi/2), \qquad (6.26)$$

where we have used $d\Psi/dz(z \to \infty) = 0$ that is implied by the Gauss law and electroneutrality, and chose the bulk potential, $\Psi(z \to \infty) = 0$, as the reference potential. A further integration yields

$$\Psi = -4\tanh^{-1}\left(\gamma e^{-\kappa_D z}\right) = -2\ln\left(\frac{1+\gamma e^{-\kappa_D z}}{1-\gamma e^{-\kappa_D z}}\right), \qquad (6.27)$$

where γ is an integration constant, $0 < \gamma < 1$. Its value is determined by the boundary condition at $z = 0$.

The two ionic profiles, $n_{\pm}(z)$, are calculated from the Boltzmann distribution, Eq. (6.5), and from Eq. (6.27), yielding:

$$n_{\pm}(z) = n_b \left(\frac{1 \pm \gamma e^{-\kappa_D z}}{1 \mp \gamma e^{-\kappa_D z}}\right)^2. \qquad (6.28)$$

For constant surface charge, the parameter γ is obtained by substituting the potential from Eq. (6.27) into the boundary condition at $z = 0$, Eq. (6.19). This yields a quadratic equation, $\gamma^2 + 2\kappa_D \ell_{GC} \gamma - 1 = 0$, with γ as its positive root:

$$\gamma = -\kappa_D \ell_{GC} + \sqrt{(\kappa_D \ell_{GC})^2 + 1}. \qquad (6.29)$$

For constant surface potential, the parameter γ can be obtained by setting $z = 0$ in Eq. (6.27),

$$\Psi_s = e\psi_s/k_B T = -4\tanh^{-1}\gamma. \qquad (6.30)$$

We use the fact that the surface potential Ψ_s is uniquely determined by the two lengths, $\ell_{GC} \sim \sigma^{-1}$ and λ_D, and write the electrostatic potential as

$$\Psi(z) = -2\ln\left[\frac{1 - \tanh(\Psi_s/4)e^{-\kappa_D z}}{1 + \tanh(\Psi_s/4)e^{-\kappa_D z}}\right], \qquad (6.31)$$

where $\Psi_s < 0$, in accord with our choice of $\sigma < 0$. In Figure 6.4, we show typical profiles for the electrostatic potential and ionic densities, for $\sigma = -5e/\text{nm}^2$ ($\ell_{GC} \simeq 0.046\,\text{nm}$). Note that this surface charge density is ten times larger than σ of Figure 6.3. For electrolyte bulk density of $n_b = 0.1\,\text{M}$, the Debye screening length is $\lambda_D \simeq 0.97\,\text{nm}$.

The DH (linearized) limit of the PB equation, Eq. (6.14), is obtained for small surface charge and/or high electrolyte strength, $\kappa_D \ell_{GC} \gg 1$. This limit yields $\gamma \simeq (2\kappa_D \ell_{GC})^{-1}$, and the potential can be approximated as

$$\Psi \simeq \Psi_s e^{-\kappa_D z} \simeq -\frac{2}{\kappa_D \ell_{GC}} e^{-\kappa_D z}. \qquad (6.32)$$

As expected for the DH limit, the solution is exponentially screened and falls off to zero for $z \gg \kappa_D^{-1} = \lambda_D$.

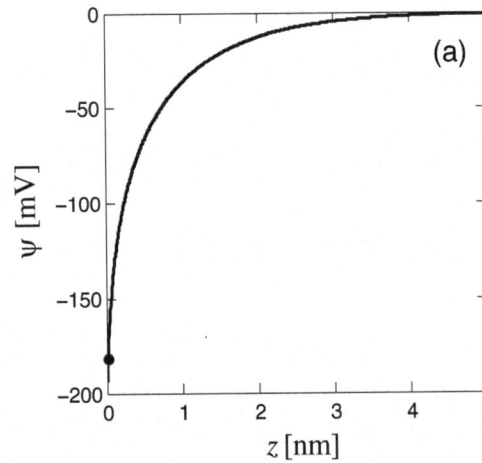 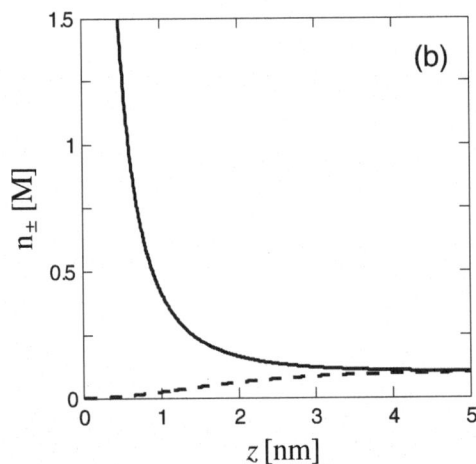

FIGURE 6.4 The electric double layer for a single charged membrane in contact with a 1:1 monovalent electrolyte reservoir of concentration $n_b = 0.1\,\text{M}$, corresponding to $\lambda_D \simeq 0.97\,\text{nm}$. The membrane located at $z=0$ is negatively charged with $\sigma = -5e/\text{nm}^2$, yielding $\ell_{GC} \simeq 0.046\,\text{nm}$. Note that the value of σ is ten times larger than the value used in Figure 6.3. In (a), we plot the electrostatic potential, $\psi(z)$ as function of z, the distance from the membrane. The value of the surface potential is $\psi_s \simeq -194$ mV. In (b), the density profiles of counterions (solid line) and co-ions (dashed line), $n_{\pm}$ (in M), are plotted as function of the distance from the membrane, z. The positive counterion density at the membrane is $n_+(z=0) \simeq 182\,\text{M}$ (not shown in the figure).

The counterion-only case, considered earlier in Section 6.3.1, is obtained by formally taking the $n_b \to 0$ limit in Eqs. (6.27)–(6.29) or, equivalently, $\kappa_D \ell_{GC} \ll 1$. This means that $\gamma \simeq 1 - \kappa_D \ell_{GC}$, and from Eq. (6.27), we recover Eq. (6.23) for the counterion density, $n(z) = n_+(z)$, while the co-ion density, $n_-(z)$, vanishes.

For a system in contact with an electrolyte reservoir, the potential always has an exponentially screened form in the distal region (far from the membrane). This can be seen by taking $z \to \infty$ while keeping $\kappa_D \ell_{GC}$ finite in Eq. (6.27):

$$\Psi(z) \simeq -4\gamma e^{-\kappa_D z}. \quad (6.33)$$

Moreover, it is possible to extract from the distal form an effective surface charge density, σ_{eff}, by comparing the coefficient 4γ of Eq. (6.33) with an effective coefficient $2/(\kappa_D \ell_{GC})$ from the DH form, Eq. (6.32),

$$|\sigma_{\text{eff}}| = 2\gamma \kappa_D \ell_{GC} |\sigma| = \frac{e \kappa_D}{\pi \ell_B} \gamma. \quad (6.34)$$

Note that $\gamma = \gamma(\kappa_D \ell_{GC})$ is calculated for the nominal parameter values in Eq. (6.29). The same concept of an effective σ is useful in several situations other than the simple planar geometry considered here.

6.3.3 The Grahame Equation

In the planar geometry, for any amount of salt, the nonlinear PB equation can be integrated analytically, resulting in a useful relation known as the Grahame equation (Grahame 1947). This equation is a relation between the surface charge density, σ, and the limiting value of the ionic density profile at the membrane, $n_\pm^{(s)} \equiv n_\pm(z=0)$. The first integration of the PB equation for a 1:1 electrolyte yields, Eq. (6.26), $d\Psi/dz = -2\kappa_D \sinh(\Psi/2)$. Using the boundary condition, Eq. (6.19), and simple hyperbolic function identities gives a relation between σ and Ψ_s

$$\pi \ell_B \left(\frac{\sigma}{e}\right)^2 = n_b \left(\cosh(\Psi_s) - 1\right), \quad (6.35)$$

and via the Boltzmann distribution of $n_\pm$, the Grahame equation is obtained:

$$\sigma^2 = \frac{e^2}{2\pi \ell_B} \left(n_+^{(s)} + n_-^{(s)} - 2n_b\right). \quad (6.36)$$

This equation implies a balance of stresses on the surface, with the Maxwell stress of the electric field compensating the van 't Hoff ideal pressure of the ions.

For large and negative surface potential, $|\Psi_s| \gg 1$, the co-ion density, $n_-^{(s)} \sim \exp(-|\Psi_s|)$, can be neglected, and Eq. (6.36) becomes

$$\sigma^2 = \frac{e^2}{2\pi \ell_B} \left(n_+^{(s)} - 2n_b\right). \quad (6.37)$$

For example, for a surface charge density of $\sigma = -5e/\text{nm}^2$ (as in Figure 6.4) and an ionic strength of $n_b = 0.1\,\text{M}$, the limiting value of the counterion density at the membrane is $n_+^{(s)} \simeq 182\,\text{M}$ and that of the co-ions is $n_-^{(s)} \simeq 5 \times 10^{-5}\,\text{M}$. The very high and unphysical value of $n_+^{(s)}$ should be understood as an artefact of the continuum PB theory. In physical situations, the ions accumulate in the membrane vicinity till their concentration saturates due to the finite ionic size and other ion–surface interactions. We will further explore this point in Sections 6.4 and 6.8.

The differential capacitance is another useful quantity to calculate, and it gives a physical measurable surface property. By using Eq. (6.35), we obtain

$$C_{\text{PB}} = \frac{d\sigma}{d\psi_s} = \frac{e}{k_B T} \frac{d\sigma}{d\Psi_s} = \varepsilon_0 \varepsilon_w \kappa_D \cosh(\Psi_s/2). \quad (6.38)$$

As shown later in Figure 6.6, the PB differential capacitance has a minimum at the potential of zero charge, $\Psi_s = 0$, and increases exponentially for $|\Psi_s| \gg 1$.

6.4 MODIFIED POISSON–BOLTZMANN THEORY

Within the PB theory, the density of accumulated counterions at the membrane might reach unphysical high values (see Figure 6.4). This unphysical situation is avoided on the MF level by accounting for the solvent entropy. Including this additional term yields a modified free-energy and PB equation (mPB). Taking this entropy into account yields a modified free energy, written here for monovalent electrolyte:

$$\beta F = \int_V d^3 r \left[-\frac{1}{8\pi \ell_B} |\nabla \Psi(\mathbf{r})|^2 + [n_+(\mathbf{r}) - n_-(\mathbf{r})]\Psi(\mathbf{r}) \right.$$
$$+ n_+ \ln(n_+ a^3) + n_- \ln(n_- a^3)$$
$$\left. + \frac{1}{a^3}(1 - a^3 n_+ - a^3 n_-)\ln(1 - a^3 n_+ - a^3 n_-) \right].$$

$$(6.39)$$

This is the free energy of a Coulomb lattice gas (Borukhov, Andelman and Orland 1997, 2000, Kilic, Bazant and Ajdari 2007, Adar et al 2018). Taking the variation of the above free energy with respect to $n_\pm$, $\delta F/\delta n_\pm = \mu_\pm$, gives the ionic profiles

$$n_\pm(z) = \frac{n_b e^{\mp \Psi}}{1 - 2\phi_b + 2\phi_b \cosh\Psi}, \quad (6.40)$$

with $\phi_b = n_b a^3$ being the bulk volume fraction of the ions. For simplicity, a is taken to be the same molecular size of all ionic species and the solvent.

ENTROPY DERIVATION OF THE MPB

Let us start with a homogenous system containing an ionic solution inside a volume v, with N_+ cations, N_- anions and N_w water molecules, such that $N_+ + N_- + N_w = N$. The number of different combinations of cations, anions, and water molecules is $N!/(N_+!N_-!N_w!)$. Therefore, the entropy is

$$S_v = -k_B \log\left(\frac{N!}{N_+!N_-!N_w!}\right) \simeq -k_B\left[N_+\ln\left(\frac{N_+}{N}\right) + N_-\ln\left(\frac{N_-}{N}\right) + (N - N_+ - N_-)\ln\left(1 - \frac{N_+}{N} - \frac{N_-}{N}\right)\right], \quad (6.41)$$

where we have used Stirling's formula for $N_\pm, N_w \gg 1$.

We now consider a system of volume $V \gg v$. The entropy of such system can be written in the continuum limit as

$$S_V = \int \frac{d^3r}{v} S_v = k_B \int d^3r \left[n_+\ln(n_+a^3) + n_-\ln(n_-a^3) + \frac{1}{a^3}(1 - a^3n_+ - a^3n_-)\ln(1 - a^3n_+ - a^3n_-)\right]. \quad (6.42)$$

where $n_\pm = N_\pm/v$ and $n_w = N_w/v$ are the densities of the cations, anions, and water molecules, respectively, and $N = v/a^3$ is the total number of molecules in the volume v. In this last equation, we have used the lattice gas formulation, in which the solution is modelled as a cubic lattice with unit cell of size $a \times a \times a$. Each unit cell contains only one molecule, $a^3(n_+ + n_- + n_w) = 1$.

In the above equation, we have also used the equilibrium relation

$$e^{\beta\mu_\pm} = \frac{n_b a^3}{1 - 2n_b a^3} = \frac{\phi_b}{1 - 2\phi_b}, \quad (6.43)$$

valid in the bulk where $\Psi = 0$. Variation with respect to Ψ, $\delta F/\delta\Psi = 0$, yields the mPB equation for 1:1 electrolyte:

$$\nabla^2 \Psi(\mathbf{r}) = -4\pi\ell_B[n_+(\mathbf{r}) - n_-(\mathbf{r})] = \frac{\kappa_D^2 \sinh\Psi}{1 - 2\phi_b + 2\phi_b \cosh\Psi}. \quad (6.44)$$

For small electrostatic potentials, $|\Psi| \ll 1$, the ionic distribution, Eq. (6.40), reduces to the usual Boltzmann distribution, but for large electrostatic potentials, $|\Psi| \gg 1$, this model gives very different results with respect to the PB theory. In particular, the ionic concentration is unbound in the standard PB theory, whereas it is bound for the mPB by the close packing density, $1/a^3$. This effect is important close to strongly charged membranes immersed in an electrolyte solution, while the regular PB equation is recovered in the dilute bulk limit, $n_b a^3 \ll 1$, for which the solvent entropy can be neglected.

For large electrostatic potentials, the contribution of the co-ions is negligible, and the counterion concentration follows a distribution reminiscent of the Fermi–Dirac distribution

$$n_-(\mathbf{r}) \simeq \frac{1}{a^3} \frac{1}{1 + \exp(-\Psi - \beta\mu)}, \quad (6.45)$$

where electroneutrality dictates $\mu = \mu_\pm$. In Figure 6.5, we show for comparison the modified and regular PB profiles for a 1:1 electrolyte. To emphasize the saturation effect of the mPB theory, we chose in the figure a large ion size, $a = 0.8$ nm.

The mPB theory also implies a modified Grahame equation that relates the surface charge density to the ion surface density, $n_\pm^{(s)}$. First, we find the relation between σ and the surface potential, Ψ_s,

$$\left(\frac{\sigma}{e}\right)^2 = \frac{1}{2\pi a^3 \ell_B} \ln\left[1 + 2\phi_b(\cosh(\Psi_s) - 1)\right]. \quad (6.46)$$

This equation represents a balance of stresses on the surface, where the Maxwell stress of the electric field is equal to the lattice-gas pressure of the ions. The surface potential can also be calculated:

$$\Psi_s = \cosh^{-1}\left(\frac{e^\xi - 1 + 2\phi_b}{2\phi_b}\right), \quad (6.47)$$

with the dimensionless parameter $\xi = a^3/(2\pi\ell_B\ell_{GC}^2)$.

For large surface charge or large surface potential, the co-ions concentration at the membrane is negligible, $n_-^{(s)} \ll 1$, and the surface potential, Eq. (6.47) is approximated by

$$\Psi_s \simeq \ln(e^\xi - 1 + 2\phi_b) - \ln(\phi_b), \quad (6.48)$$

and from Eq. (6.46), we obtain the Grahame equation,

$$\left(\frac{\sigma}{e}\right)^2 \simeq \frac{1}{2\pi a^3 \ell_B} \ln\left(\frac{1 - 2\phi_b}{1 - a^3 n_+^{(s)}}\right). \quad (6.49)$$

Note that in the dilute limit $\phi_b \ll 1$, the Grahame equation reduces to the regular PB case (Eq. 6.36).

Charged Membranes

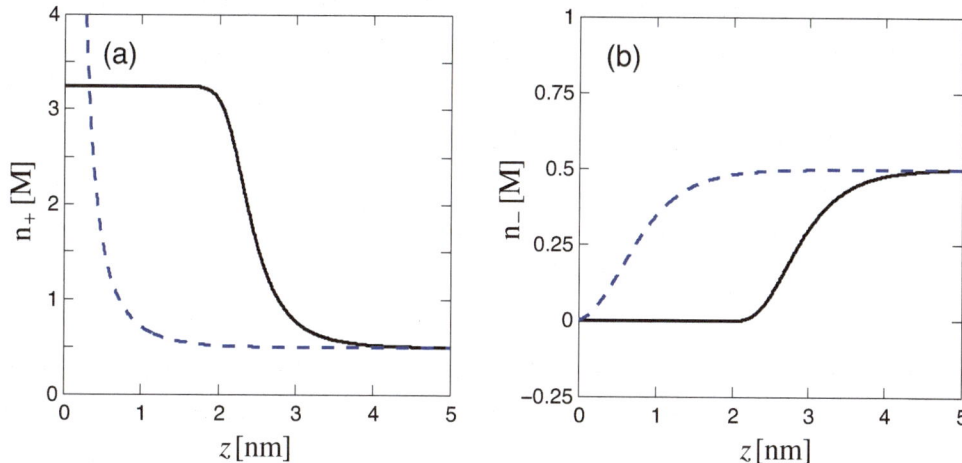

FIGURE 6.5 Comparison of the modified PB (mPB) profiles (black solid lines) with those of the regular PB one (dashed blue lines). In (a), we show the counterion profile, and in (b), the co-ion profile. The parameters used are ion size $a=0.8$ nm, surface charge density $\sigma=-5e/\text{nm}^2$, and 1:1 electrolyte with ionic strength $n_b=0.5$ M. Note that while the PB value at the membrane is $n_s^+ \simeq 182$ M, the mPB density saturates at $n_s^+ \simeq 3.2$ M.

It is also straightforward but more cumbersome to calculate the differential capacitance, $C = d\sigma/d\psi_s$, for the mPB theory. From Eq. (6.46), we obtain (Kornyshev 2007, Nakayama and Andleman 2015)

$$C_{\text{mPB}} = \frac{C_{\text{PB}}}{1+4\phi_b \sinh^2(\Psi_s/2)}\sqrt{\frac{4\phi_b \sinh^2(\Psi_s/2)}{\ln\left[1+4\phi_b \sinh^2(\Psi_s/2)\right]}}.$$

(6.50)

Although it can be shown that for $\phi_b \to 0$ the mPB differential capacitance reduces to the standard PB result, the resulting C_{mPB} is quite different for any finite value of ϕ_b. The main difference is that instead of an exponential divergence of C_{PB} at large potentials, C_{mPB} decreases for high-bias $|\Psi_s| \gg 1$. For rather small bulk densities, $\phi_b < 1/6$, the C_{mPB} shows a behaviour called *camel-shape* or *double-hump*. This behaviour is also observed in experiments at relatively low salt concentrations. As shown in Figure 6.6, the double-hump C_{mPB} has a minimum at $\Psi_s = 0$ and two maxima. The peak positions can roughly be estimated by substituting the close packing concentration, $n = 1/a^3$, into the Boltzmann distribution (Eq. 6.5), yielding $\Psi_s^{\max} \simeq \mp\ln(\phi_b)$. Using parameter values as in Figure 6.6, $\Psi_s^{\max}$ is estimated as ± 4.6 as compared with the exact values, $\Psi_s^{\max} = \pm 5.5$.

Furthermore, it can be shown that for high salt densities, $\phi_b > 1/6$, C_{mPB} exhibits (see also Figure 6.6) a unimodal maximum close to the potential of zero charge, rather than a minimum as for C_{PB}. Such results that take into account finite ion size for the differential capacitance are of importance in the theory of confined ionic liquids (Kornyshev 2007, Nakayama and Andleman 2015).

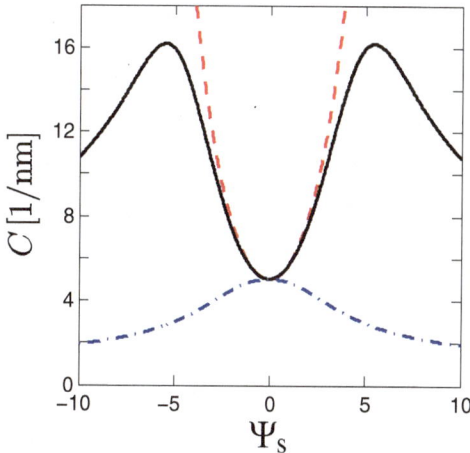

FIGURE 6.6 Comparison of the differential capacitance, C, calculated from the regular PB theory (dashed red line), Eq. (6.38), with $n_b \simeq 0.4$ mM (chosen so that it corresponds to $\phi_b=0.01$ and $a=0.3$ nm), and from the mPB theory, Eq. (6.50). The mPB differential capacitance is calculated for $a=0.3$ nm. For low $\phi_b=0.01$, it shows a camel shape (black solid line), while for high $\phi_b=0.2$, it shows a unimodal (dash-dotted blue line).

6.5 TWO-MEMBRANE SYSTEM: OSMOTIC PRESSURE

We now consider the PB theory of two charged membranes as shown in Figure 6.7. The two membranes can, in general, have different surface charge densities: σ_1 at $z=-d/2$ and σ_2 at $z=d/2$. The boundary conditions of the two-membrane system are written as $\rho_f = \sigma_1\delta(z+d/2)+\sigma_2\delta(z-d/2)$, and using the variation of the free energy, $\delta F/\delta \Psi_s = 0$:

$$\Psi'\big|_{-d/2} = -4\pi\ell_B \frac{\sigma_1}{e},$$
$$\Psi'\big|_{d/2} = 4\pi\ell_B \frac{\sigma_2}{e}.$$

(6.51)

It is of interest to calculate the force (or the osmotic pressure) between two membranes interacting across the ionic solution. The osmotic pressure is defined as $\Pi = P_{in} - P_{out}$, where P_{in} is the inner pressure and P_{out} is the pressure exerted by the reservoir that is in contact with the two-membrane system. Sometimes the osmotic pressure is referred to as the disjoining pressure, introduced first by Derjaguin (Churaev, Derjaguin and Muller 2014).

Let us start by calculating the inner and outer pressures from the Helmholtz free energy. The pressure (P_{in} or P_{out}) is the variation of the free energy with the volume:

$$P = -\frac{\delta F}{\delta V} = -\frac{1}{A}\frac{\delta F}{\delta d}, \quad (6.52)$$

with $V = Ad$ being the system volume, A the lateral membrane area, and d the intermembrane distance. As the interaction between the two membranes can be either attractive ($\Pi < 0$) or repulsive ($\Pi > 0$), we will analyse the criterion for the crossover ($\Pi = 0$) between these two regimes as function of the surface charge asymmetry and intermembrane distance.

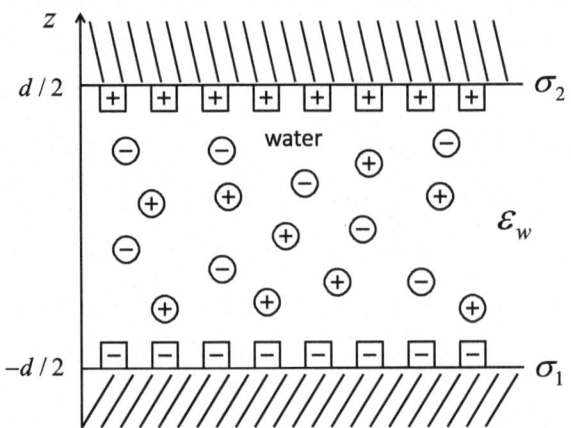

FIGURE 6.7 Schematic drawing of two asymmetric membranes. The planar membrane located at $z=-d/2$ carries a charge density σ_1, while the membrane at $z=d/2$ has a charge density of σ_2. The antisymmetric membrane set-up is a special case with $\sigma_1 = -\sigma_2$, while in the symmetric case, $\sigma_1 = \sigma_2$.

GENERAL DERIVATION OF THE PRESSURE

The Helmholtz free-energy obtained from Eq. (6.11) can be written in a general form as $F = A\int f[\Psi(z), \Psi'(z)]dz$, where we use the Poisson equation to obtain the relation, $n_{\pm} = n_{\pm}(\Psi)$. As the integrand f depends only implicitly on the z coordinate through $\Psi(z)$, one can obtain from the Euler–Lagrange equations the following relation (Ben-Yaakov et al. 2009a):

$$f - \sum_i \mu_i^{tot} n_i - \frac{\partial f}{\partial \Psi'}\Psi' = \text{const}$$

$$= \frac{k_B T}{8\pi \ell_B}(\Psi')^2 + k_B T \sum_{i=1}^{M} z_i n_i \Psi + k_B T \sum_{i=1}^{M}\left[n_i \ln(n_i a^3) - n_i\right], \quad (6.53)$$

where the sum is over $i = 1,...,M$ ionic species and the total chemical potential defined as before, $\delta F / \delta n_i = \partial f / \partial n_i = \mu_i^{tot}$. Let us understand the meaning of the constant on the right-hand side of the above equation. For uncharged solutions, the Helmholtz free energy per unit volume contains only the entropy term, $f = k_B T\sum_i\left[n_i \ln(n_i a^3) - n_i\right]$, and from Eq. (6.53), we obtain $f - \sum_i \mu_i^{tot} n_i = \text{const}$. A known thermodynamic relation is $P = \sum_i \mu_i^{tot} n_i - f$, implying that the right-hand side constant is $-P$. However, even for charged liquid mixtures, the electrostatic potential vanishes in the bulk, away from the boundaries, and reduces to the same value as for uncharged solutions. Therefore, we conclude that the right-hand side constant is $-P$, yielding

$$P = -\frac{k_B T}{8\pi \ell_B}(\Psi')^2 + k_B T \sum_{i=1}^{M} n_i. \quad (6.54)$$

If the electric field and ionic densities are calculated right at the surface, we obtain the *contact theorem* that gives the osmotic pressure acting on the surface. Another and more straightforward way to calculate the pressure is to calculate the incremental difference in free energy, F, for an intermembrane separation d, i.e., $\left[F(d+\delta d)-F(d)\right]/\delta d$. The calculation of $F(d+\delta d)$ can be done by including an additional slab of width δd in the space between the two membranes at an arbitrary position. We remark that the validity of the contact theorem itself is not limited to the PB theory but is an exact theorem of statistical mechanics (Henderson and Blum 1981, Evans and Wennerström 1999, Dean and Horgan 2003).

Charged Membranes

We are interested in the osmotic pressure, Π. For an ionic reservoir in the dilute limit, Eq. (6.54) gives $P_{\text{out}} = k_B T \sum_i n_i^{(b)}$, where $n_i^{(b)}$ is the ith ionic species bulk density. Thus, the osmotic pressure can be written as

$$\Pi = -\frac{k_B T}{8\pi \ell_B}(\Psi')^2 + k_B T \sum_{i=1}^{M}\left(n_i(z) - n_i^{(b)}\right)$$

$$= \left(\cosh[\Psi(z)] - 1\right), \quad (6.55)$$

and for monovalent 1:1 ions:

$$\Pi = -\frac{k_B T}{8\pi \ell_B}(\Psi')^2 + 2k_B T n_b\left(\cosh[\Psi(z)] - 1\right) = \text{const.}$$

$$(6.56)$$

At any position z between the membranes, the osmotic pressure has two contributions. The first is a negative Maxwell electrostatic pressure proportional to $(\Psi')^2$. The second is due to the entropy of mobile ions and measures the local entropy change (at an arbitrary position, z) with respect to the ion entropy in the reservoir.

6.6 TWO SYMMETRIC MEMBRANES, $\sigma_1 = \sigma_2$

For two symmetrically charged membranes, $\sigma_1 = \sigma_2 \equiv \sigma$ at $z = \pm d/2$, the electrostatic potential is symmetric about the midplane yielding a zero electric field, $E = 0$ at $z = 0$. It is then sufficient to consider the interval $[0, d/2]$ with the boundary conditions,

$$\Psi'\big|_{z=d/2} = \Psi'_s = 4\pi\ell_B\sigma/e, \quad \Psi'\big|_{z=0} = \Psi'_m = 0. \quad (6.57)$$

As Π is constant (independent of z) between the membranes, one can calculate the disjoining pressure, Π, from Eq. (6.55), at any position z, between the membranes. A simple choice will be to evaluate it at $z = 0$ (the midplane), where the electric field vanishes for the symmetric $\sigma_1 = \sigma_2$ case,

$$\Pi = k_B T \sum_{i=1}^{M}\left(n_i^{(m)} - n_i^{(b)}\right) = k_B T \sum_{i=1}^{M} n_i^{(b)}\left(e^{-z_i\Psi_m} - 1\right) > 0,$$

$$(6.58)$$

and for monovalent ions, $z_i = \pm 1$, we get

$$\Pi = 4k_B T n_b \sinh^2(\Psi_m/2) > 0, \quad (6.59)$$

where $n_i^{(m)} = n_i(z=0)$ is the midplane concentration of the ith species. It can be shown that the electroneutrality condition implies that the osmotic pressure is always *repulsive* for any shape of boundaries (Sader and Chan 1999, Neu 1999) as long as we have two symmetric membranes $(\sigma_1 = \sigma_2)$.

Note that the Grahame equation can be derived also for the two-membrane case with added electrolyte. One way of doing it is by comparing the pressure of Eq. (6.55) evaluated at one of the membranes, $z = \pm d/2$, and at the midplane, $z = 0$. The pressure is constant between the two membranes; thus, by equating these two pressure expressions, the Grahame equation emerges

$$\left(\frac{\sigma}{e}\right)^2 = \frac{1}{2\pi\ell_B}\sum_{i=1}^{M}\left(n_i^{(s)} - n_i^{(m)}\right). \quad (6.60)$$

By taking the limit of infinite separation between the two membranes and $n_i^{(m)} \to n_i^{(b)}$, the Grahame equation for a single membrane (Eq. 6.36) is recovered.

6.6.1 COUNTERIONS ONLY

In the absence of an external salt reservoir, the only ions in the solution for a symmetric two-membrane system are positive monovalent ($z = +1$) counterions with density $n(z)$ that neutralizes the negative surface charge,

$$2\sigma = -e\int_{-d/2}^{d/2} n(z)\,dz. \quad (6.61)$$

The PB equation has an analytical solution for this case. Integrating twice the PB equation (Eq. 6.18) with the appropriate boundary conditions (Eq. 6.57) yields an analytical expression for the electrostatic potential:

$$\Psi(z) = \ln\left(\cos^2 Kz\right), \quad (6.62)$$

and consequently the counterion density is

$$n(z) = n_m e^{-\Psi(z)} = \frac{n_m}{\cos^2(Kz)}. \quad (6.63)$$

In the above, we have defined $n_m = n(z=0)$ and chose arbitrarily $\Psi_m = 0$. We also introduced a new length scale, K^{-1}, related to n_m by

$$K^2 = 2\pi\ell_B n_m. \quad (6.64)$$

Notice that K plays a role similar to the inverse Debye length $\kappa_D = \sqrt{8\pi\ell_B n_b}$, with the midplane density replacing the bulk density, $n_b \to n_m$. Using the boundary condition at $z = d/2$, we get a transcendental relation for K

$$Kd\tan(Kd/2) = \frac{d}{\ell_{GC}}. \quad (6.65)$$

In Figure 6.8, we show a typical counterion profile with its corresponding electrostatic potential for $\sigma = -e/(7\,\text{nm}^2)$ and $d = 4\,\text{nm}$.

The osmotic pressure (Eq. 6.55), calculated for the counterion-only case, is

$$\Pi = \frac{k_B T}{2\pi\ell_B}K^2. \quad (6.66)$$

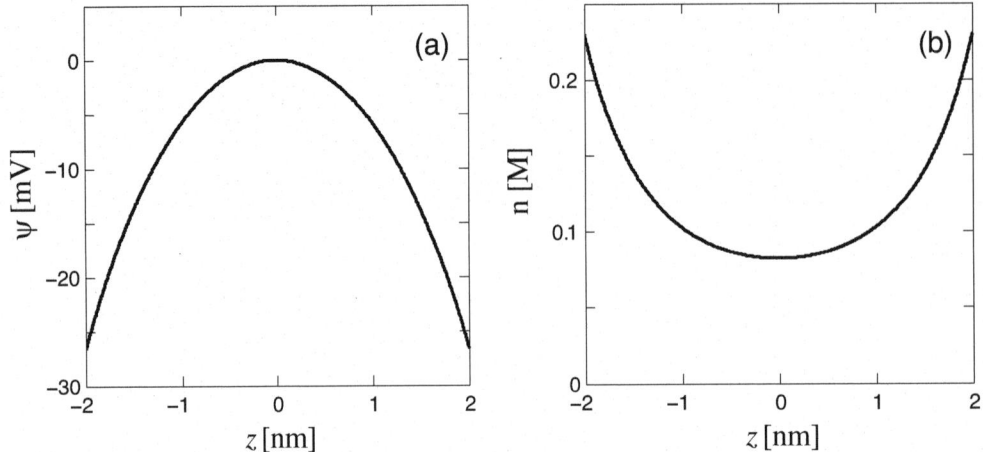

FIGURE 6.8 The counterion-only case for two identically charged membranes located at $z = \pm d/2$ with $d = 4$ nm and $\sigma = -e/(7\,\text{nm}^2)$ on each membrane ($\ell_{GC} \simeq 1.6$ nm). In (a), we plot the electrostatic potential, ψ, and in (b), the counterion density profile, n. The plots are obtained from Eqs. (6.62)–(6.65).

For weak surface charge, $d/\ell_{GC} \ll 1$, one can approximate $(Kd)^2 \simeq 2d/\ell_{GC} \ll 1$, and the pressure is given by

$$\Pi \simeq -\frac{2k_B T \sigma}{e}\frac{1}{d} = \frac{k_B T}{\pi \ell_B \ell_{GC}}\frac{1}{d} \sim \frac{1}{d}. \qquad (6.67)$$

The $\Pi \sim 1/d$ behaviour is similar to an ideal-gas equation of state, $P = Nk_B T/V$ with $V = Ad$ and N the total number of counterions. The density (per unit volume) of the counterions is almost constant between the two membranes and is equal to $2|\sigma|/(ed)$. This density neutralizes the surface charge density, σ, on the two membranes. The main contribution to the pressure comes from an ideal-gas like pressure of the counterion cloud. This regime can be reached experimentally for small intermembrane separation, $d < \ell_{GC}$. For example, for e/σ in the range of $1-100\,\text{nm}^2$, $\ell_{GC} \sim 1/\sigma$ varies between 0.2 nm and 20 nm.

For the opposite case of strong surface charge, $d/\ell_{GC} \gg 1$, one gets $Kd \simeq \pi$ from Eq. (6.65). This is the Gouy–Chapman regime. It is very different from the weak surface charge, as the density profile between the two membranes varies substantially leading to $n_s \gg n_m$, and to a pressure

$$\Pi \simeq \frac{\pi k_B T}{2\ell_B d^2} \sim \frac{1}{d^2}. \qquad (6.68)$$

It is interesting to note that the above pressure expression does not depend explicitly on the surface charge density. This can be rationalized as follows. Counterions are accumulated close to the surface, at an average separation $\ell_{GC} \sim 1/|\sigma|$. Therefore, creating a surface dipole density of $|\sigma|\ell_{GC}$. The interaction energy per unit area is proportional to the electrostatic energy between two such planar dipolar layers, which scales as $1/d$ for the free energy density and d^{-2} for the pressure. The surface charge density dependence itself vanishes because the effective dipolar-moment surface density, $|\sigma|\ell_{GC}$, is charge independent. In the Gouy–Chapman regime, the electrostatic interactions are most dominant as they are long-ranged and unscreened. Of course, even in pure water, the effective Debye screening length is about 1 μm, and the electrostatic interactions will be screened for larger distances.

6.6.2 Added Electrolyte

When two charged membranes are placed in contact with an electrolyte reservoir, the co-ions and counterions between the membranes have a nonhomogenous density profile. The PB equation does not have a closed-form analytical solution for two (or more) ionic species, even when we restrict ourselves to a 1:1 symmetric and monovalent electrolyte. Instead, the solution can be expressed in terms of elliptic functions.

The PB equation for a monovalent 1:1 electrolyte (Eq. 6.18) is $\Psi''(z) = \kappa_D^2 \sinh \Psi$, while the same boundary conditions as in Eq. (6.57) are satisfied. The first integration from the midplane ($z = 0$) to an arbitrary point between the membranes, $z \in [-d/2, d/2]$, gives

$$\frac{d\Psi}{dz} = -\kappa_D \sqrt{2\cosh \Psi(z) - 2\cosh \Psi_m}. \qquad (6.69)$$

As explained in the beginning of Section 6.6, $\Psi'_m = 0$ for two symmetric membranes and the second integration leads to an elliptic integral (see box below)

$$z = -\lambda_D \int_{\Psi_m}^{\Psi} \frac{d\eta}{\sqrt{2\cosh \eta - 2\cosh \Psi_m}}. \qquad (6.70)$$

Inverting the relation $z = z(\Psi)$ leads to the expression for the profile, $\Psi(z)$.

Charged Membranes

THE ELECTROSTATIC POTENTIAL VIA JACOBI ELLIPTIC FUNCTIONS

It is possible to write Eq. (6.70) in terms of an incomplete elliptic integral of the first kind

$$F(\theta \mid a^2) \equiv \int_0^\theta \frac{\mathrm{d}\eta}{\sqrt{1-a^2\sin^2\eta}}. \tag{6.71}$$

After change of variables and some algebra, we write Eq. (6.70) with the help of Eq. (6.71) as

$$z = 2\lambda_\mathrm{D}\sqrt{m}\left[F\left(\frac{\pi}{2}\bigg|m^2\right) - F\left(\varphi\big|m^2\right)\right], \tag{6.72}$$

with $m = \exp(\Psi_\mathrm{m})$ and $\varphi = \sin^{-1}\left[\exp([\Psi - \Psi_\mathrm{m}]/2)\right]$.

The electrostatic potential, which is the inverse relation of Eq. (6.72), can then be written in terms of the Jacobi elliptic function, $\mathrm{cd}(u \mid a^2)$,

$$\Psi = \Psi_\mathrm{m} + 2\ln\left[\mathrm{cd}\left(\frac{z}{2\lambda_\mathrm{D}\sqrt{m}}\bigg|m^2\right)\right]. \tag{6.73}$$

In writing this equation, we have used the definition of the Jacobi elliptic functions:

$$\mathrm{sn}(u \mid a^2) = \sin\alpha, \tag{6.74}$$

$$\mathrm{cn}(u \mid a^2) = \cos\alpha = \sqrt{1 - \mathrm{sn}^2(u \mid a^2)}, \tag{6.75}$$

$$\mathrm{dn}(u \mid a^2) = \sqrt{1 - a^2\,\mathrm{sn}^2(u \mid a^2)}, \tag{6.76}$$

$$\mathrm{cd}(u \mid a^2) = \frac{\mathrm{cn}(u \mid a^2)}{\mathrm{dn}(u \mid a^2)}, \tag{6.77}$$

with $u \equiv F(\alpha \mid a^2)$.

Using one of the boundary conditions (Eq. 6.57) with the first integration (Eq. 6.69) yields

$$\cosh(\Psi_\mathrm{s}) = \cosh(\Psi_\mathrm{m}) + 2\left(\frac{\lambda_\mathrm{D}}{\ell_\mathrm{GC}}\right)^2. \tag{6.78}$$

The above equation also gives a relation between σ and the midplane potential, Ψ_m, in terms of Jacobi elliptic functions (see box above),

$$\frac{\sigma}{e} = \frac{\kappa_\mathrm{D}}{4\pi\ell_\mathrm{B}}\frac{m^2-1}{\sqrt{m}}\frac{\mathrm{sn}(u_\mathrm{s} \mid m^2)}{\mathrm{cn}(u_\mathrm{s} \mid m^2)\,\mathrm{dn}(u_\mathrm{s} \mid m^2)}, \tag{6.79}$$

with $u_\mathrm{s} \equiv d/(4\lambda_\mathrm{D}\sqrt{m})$ and $m = \exp(\Psi_\mathrm{m})$ as defined after Eq. (6.72). For fixed surface charge, this relation gives the midplane potential, Ψ_m, and the osmotic pressure can then be calculated from Eq. (6.58). The other boundary condition can also be expressed as an elliptic integral

$$\frac{d}{2\lambda_\mathrm{D}} = -\int_{\Psi_\mathrm{m}}^{\Psi_\mathrm{s}} \frac{\mathrm{d}\eta}{\sqrt{2\cosh\eta - 2\cosh\Psi_\mathrm{m}}}$$

$$= 2\sqrt{m}\left[F\left(\frac{\pi}{2}\bigg|m^2\right) - F\left(\varphi_\mathrm{s}\big|m^2\right)\right], \tag{6.80}$$

where $\varphi_\mathrm{s} = \sin^{-1}\left[\exp([\Psi_\mathrm{s} - \Psi_\mathrm{m}]/2)\right]$.

The three equations, Eqs. (6.70), (6.78), and (6.80), completely determine the potential $\Psi(z)$, the two species density profiles, $n_\pm(z) = n_\mathrm{b}\exp(\mp\Psi)$, and their midplane values $n_\pm^{(\mathrm{m})} = n_\mathrm{b}\exp(\mp\Psi_\mathrm{m})$, as function of the three parameters: the intermembrane spacing d, the surface charge density σ (or equivalently ℓ_GC), and the electrolyte bulk ionic strength n_b (or equivalently λ_D).

The exact form of the profiles and pressure can be obtained either from the numerical solution of Eqs. (6.70), (6.78) and (6.80) or by the usage of the elliptic functions. For example, we calculate numerically the counterion,

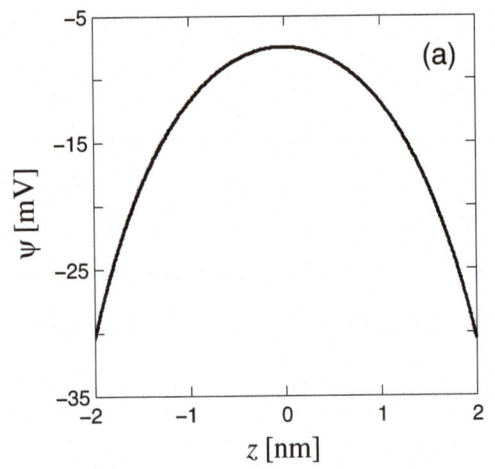
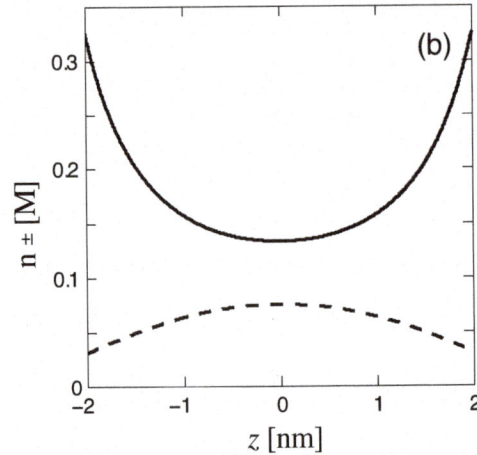

FIGURE 6.9 Monovalent 1:1 electrolyte with $n_b = 0.1\,\text{M}$ ($\lambda_D \simeq 0.97\,\text{nm}$) between two identically charged membranes with $\sigma = -e/(7\,\text{nm}^2)$ each ($\ell_{GC} \simeq 1.6\,\text{nm}$) located at $z = \pm d/2$ with $d = 4\,\text{nm}$. In (a), we plot the electrostatic potential, ψ, and in (b), we show the co-ion (dashed line) and counterion (solid line) density profiles, $n_\pm$. The plots are obtained from Eqs. (6.70), (6.78), and (6.80).

co-ion, and potential profiles as shown in Figure 6.9, where the three relevant lengths are $d = 4\,\text{nm}$, $\ell_{GC} \simeq 0.4d$, and $\lambda_D \simeq 0.25d$.

6.6.3 Debye–Hückel Regime

Broadly speaking (see Section 6.2.1), the PB equation can be linearized when the surface potential is small, $|\Psi_s| \ll 1$. In this case, the potential is small everywhere because it is a monotonous function that vanishes in the bulk. The DH solution for monovalent electrolytes has the general form

$$\Psi(z) = A\cosh(\kappa_D z) + B\sinh(\kappa_D z), \quad (6.81)$$

and the boundary conditions of Eq. (6.57) dictate the specific solution

$$\Psi(z) = -\frac{2}{\kappa_D \ell_{GC}}\frac{\cosh(\kappa_D z)}{\sinh(\kappa_D d/2)}. \quad (6.82)$$

In the DH regime, the potential is small, and the disjoining pressure (Eq. 6.58) can be expanded to second order in Ψ_m. As the first order vanishes from electroneutrality, we obtain

$$\Pi \simeq k_B T n_b \Psi_m^2 = \frac{k_B T}{2\pi \ell_B \ell_{GC}^2}\frac{1}{\sinh^2(\kappa_D d/2)}. \quad (6.83)$$

The DH regime can be further divided into two subcases: DH_1 and DH_2. For large separations, $d \gg \lambda_D$, the above expression reduces to

$$\Pi \simeq \frac{2k_B T}{\pi \ell_B \ell_{GC}^2} e^{-\kappa_D d}. \quad (6.84)$$

This DH_1 subregime is valid for $d \gg \lambda_D$ and $\ell_{GC} \gg \lambda_D$.

In the other limit of small separations, the pressure is approximated by

$$\Pi \simeq \frac{2k_B T}{\pi \ell_B (\kappa_D \ell_{GC})^2}\frac{1}{d^2}. \quad (6.85)$$

The limits of validity for this DH_2 subregime are $d \ll \lambda_D$ and $\ell_{GC} \gg \lambda_D^2/d$ (see Table 6.2).

6.6.4 Intermediate Regime

When d is the largest length scale in the system, $d \gg \lambda_D$ and $d \gg \ell_{GC}$, the interaction between the membranes is weak, and one can use the superposition principle. This defines the *distal* region, where the midplane potential is obtained by adding the contributions from two identical charged single surfaces, located at $z = \pm d/2$.

In the distal region, the midplane potential is obtained from Eq. (6.33) by the above-mentioned superposition:

$$\Psi_m = -8\gamma e^{-\kappa_D d/2}. \quad (6.86)$$

Since Ψ_m is small, the pressure expression (Eq. 6.58) can be expanded to second order in Ψ_m, as was done in Eq. (6.83), giving

$$\Pi \simeq k_B T n_b \Psi_m^2 = 64\, k_B T \gamma^2 n_b e^{-\kappa_D d}. \quad (6.87)$$

This osmotic pressure expression is valid for large distances, $d \gg \lambda_D$ and $d \gg \ell_{GC}$, and partially holds for the DH_1 regime.

The *intermediate* regime is obtained by further assuming strongly charged surfaces, $\lambda_D \gg \ell_{GC}$. In this limit, $\gamma = \tanh(-\Psi_s/4) \simeq 1$, and the osmotic pressure is written as

$$\Pi \simeq \frac{8k_B T \kappa_D^2}{\pi \ell_B} e^{-\kappa_D d}. \quad (6.88)$$

Charged Membranes

TABLE 6.2
The Five Pressure Regimes of the Symmetric Two-Membrane System

Pressure Regime	Pressure (Π)	Range of Validity
Ideal-Gas	$\dfrac{k_B T}{\pi \ell_B \ell_{GC}} \dfrac{1}{d}$	$\lambda_D/d \gg \ell_{GC}/\lambda_D \gg d/\lambda_D$
Gouy–Chapman	$\dfrac{\pi k_B T}{2 \ell_B} \dfrac{1}{d^2}$	$1 \gg d/\lambda_D \gg \ell_{GC}/\lambda_D$
Intermediate	$\dfrac{8 k_B T}{\pi \ell_B \lambda_D^2} e^{-d/\lambda_D}$	$d/\lambda_D \gg 1 \gg \ell_{GC}/\lambda_D$
Debye–Hückel (DH$_1$)	$\dfrac{2 k_B T}{\pi \ell_B \ell_{GC}^2} e^{-d/\lambda_D}$	$d/\lambda_D \gg 1;\ \ell_{GC}/\lambda_D \gg 1$
Debye–Hückel (DH$_2$)	$\dfrac{2 k_B T \lambda_D^2}{\pi \ell_B \ell_{GC}^2} \dfrac{1}{d^2}$	$1 \gg d/\lambda_D \gg \lambda_D/\ell_{GC}$

The intermediate regime is valid for $d \gg \lambda_D \gg \ell_{GC}$ (see Table 6.2).

6.6.5 Other Pressure Regimes

The pressure expression can be derived analytically in two other limits, which represent the two regimes obtained for the counterions-only case: the Ideal-Gas (IG) regime (Eq. 6.67),

$$\Pi \simeq \frac{k_B T}{\pi \ell_B \ell_{GC}} \frac{1}{d}, \qquad (6.89)$$

valid for $\lambda_D^2/d \gg \ell_{GC} \gg d$, and the Gouy–Chapman (GC) regime, (Eq. 6.68),

$$\Pi \simeq \frac{\pi k_B T}{2 \ell_B} \frac{1}{d^2}, \qquad (6.90)$$

whose range of validity is $\lambda_D \gg d \gg \ell_{GC}$.

The five pressure regimes complete the discussion of the various limits as function of the two ratios: ℓ_{GC}/λ_D and d/λ_D. They are summarized in Table 6.2 and plotted in Figure 6.10.

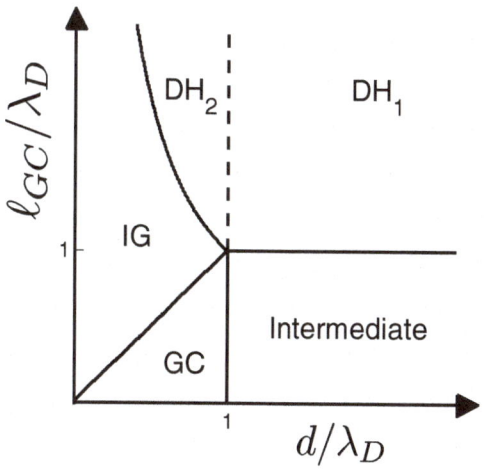

FIGURE 6.10 Schematic representation of the various regimes of the PB equation for two flat and equally charged membranes at separation d. We plot the four different pressure regimes: Ideal-Gas (IG), Gouy–Chapman (GC), intermediate, and Debye–Hückel (DH). The two independent variables are the dimensionless ratios d/λ_D and ℓ_{GC}/λ_D. The four regimes are detailed in Table 6.2. The DH regime is further divided into two subregimes: DH$_1$ for large d/λ_D and DH$_2$ for small d/λ_D.

6.7 TWO ASYMMETRIC MEMBRANES, $\sigma_1 \neq \sigma_2$

For asymmetrically charged membranes, $\sigma_1 \neq \sigma_2$, the interacting membranes impose different boundary conditions. Such a system can model, for example, two surfaces that are coated with two different polyelectrolytes or two lipid membranes with different charge/neutral lipid compositions.

It is possible to have an overall attractive interaction between two asymmetric membranes, unlike the symmetric $\sigma_1 = \sigma_2$ case. When σ_1 and σ_2 have the same sign, the boundary condition of Eq. (6.51) implies that $\Psi'(d/2)$ has the opposite sign of $\Psi'(-d/2)$. Since Ψ' is monotonous, it means that there is a point between the plates for which $\Psi' = 0$. The osmotic pressure, Π of Eq. (6.55), calculated at this special point, has only an entropic contribution and is positive for any intermembrane separation, d, just as in the $\sigma_1 = \sigma_2$ case.

However, when σ_1 and σ_2 have opposite signs, Ψ' is always negative between the two membranes, and the sign of the pressure can be either positive (repulsive) or negative (attractive). A crossover between repulsive and attractive pressure occurs when $\Pi = 0$ and depends on four system parameters: $\sigma_{1,2}$, λ_D, and d (see Figure 6.11).

Although the general expression for $\Pi(d)$ cannot be cast in an analytical form, a closed-form criterion exists for the crossover pressure, $\Pi = 0$, for any amount of salt (Ben-Yaakov et al. 2007). The crossover criterion has two rather simple limits: for the linearized DH (high salt) limit, the general criterion reduces to the well-known result of Parsegian and Gingell (1972), while in the counterion-only

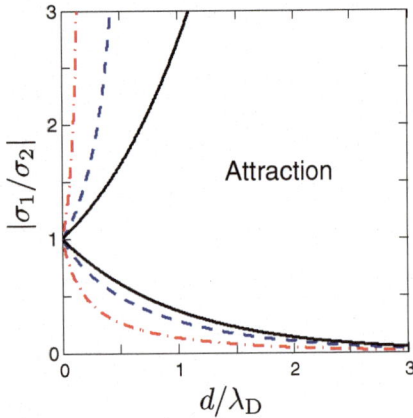

FIGURE 6.11 Crossover from attraction to repulsion, $\Pi(d)=0$, for two oppositely charged membrane, $\sigma_1 \cdot \sigma_2 < 0$, in the $(|\sigma_1/\sigma_2|, d/\lambda_D)$ plane. The solid lines show the crossover in the high-salt DH limit, Eq. (6.95). For lower salinity, derived from the full PB theory, Eq. (6.107), the attractive region is increased as is seen in the examples we choose: $\ell_2/\lambda_D = 0.2$ (red dash-dotted line) and $\ell_2/\lambda_D = 0.7$ (blue dashed line).

limit, another analytical expression has been derived more recently by Lau and Pincus (1999).

6.7.1 The Debye–Hückel Regime

The crossover criterion between attraction and repulsion has an analytical limit for high salinity, (Parsegian and Gingell 1972). We repeat here the well-known argument (Ben-Yaakov and Andelman 2010) where the starting point is the linear DH limit (Eq. 6.14) of the full PB equation.

The DH equation for monovalent electrolyte in planar geometries has a solution (Eq. 6.81), for which the boundary conditions of Eq. (6.51) yield

$$\Psi(z) = \frac{2\pi \ell_B}{\kappa_D e} \left[\frac{\sigma_1 + \sigma_2}{\sinh(\kappa_D d/2)} \cosh(\kappa_D z) \right.$$
$$\left. + \frac{\sigma_2 - \sigma_1}{\cosh(\kappa_D d/2)} \sinh(\kappa_D z) \right]. \quad (6.91)$$

In the DH regime, the pressure expression (Eq. 6.55) can be expanded in powers of the electrostatic potential, Ψ. Keeping only terms of order Ψ^2, the pressure can be written as

$$\Pi \simeq \frac{k_B T}{2\pi \ell_B \sinh^2(\kappa_D d)} \left[\frac{1}{\ell_1^2} + \frac{1}{\ell_2^2} \pm \frac{2}{\ell_1 \ell_2} \cosh(\kappa_D d) \right], \quad (6.92)$$

where $\ell_{1,2} = e/(2\pi \ell_B |\sigma_{1,2}|)$ are the two Gouy–Chapman lengths corresponding to the two membranes with σ_1 and σ_2, respectively. The $\pm$ sign of the last term corresponds to the two situations: $\sigma_1 \cdot \sigma_2 > 0$ and $\sigma_1 \cdot \sigma_2 < 0$, respectively.

This Π expression can be simplified in two limits. For small separation, $d \ll \lambda_D$, the expansion of the hyperbolic functions yields a power law divergence $\sim d^{-2}$ for $d \to 0$,

$$\Pi \simeq \frac{k_B T}{2\pi \ell_B} \left[\left(\frac{1}{\kappa_D \ell_1} \pm \frac{1}{\kappa_D \ell_2} \right)^2 \frac{1}{d^2} \pm \frac{1}{\ell_1 \ell_2} \right] > 0. \quad (6.93)$$

Clearly, it is positive definite (hence repulsive) for both ± signs. However, when $\sigma_1 = -\sigma_2$ (the antisymmetric case with $\sigma_1 \cdot \sigma_2 < 0$), the pressure goes to a negative constant (independent of d), $\Pi = -k_B T/(2\pi \ell_B \ell_1 \ell_2)$.

For the opposite limit of large separation $d \gg \lambda_D$, the pressure decays exponentially, while its sign depends on the sign of $\sigma_1 \cdot \sigma_2$,

$$\Pi \simeq \pm \frac{k_B T}{\pi \ell_B \ell_1 \ell_2} e^{-\kappa_D d}, \quad (6.94)$$

and thus, it is attractive for $\sigma_1 \cdot \sigma_2 < 0$.

The attraction/repulsion crossover is calculated from the zero-pressure condition of Eq. (6.92), while keeping in mind that attraction is possible only for oppositely charged membranes, $\sigma_1 \cdot \sigma_2 < 0$ (see the beginning of this section):

$$e^{-\kappa_D d} < \left| \frac{\sigma_1}{\sigma_2} \right| < e^{\kappa_D d}. \quad (6.95)$$

This is exactly the result obtained by Parsegian and Gingell (1972). Interestingly, in the linear DH case, the crossover depends only on the ratio of the two surface charges $|\sigma_1/\sigma_2|$ and not on their separate values, as can be seen in Figure 6.11. For comparison, we plot (with dashed and dash-dotted lines on the same figure) two examples of low-salt crossovers, as calculated from the general criterion presented below for the full PB theory (Section 6.7.4). The low-salt line has a smaller repulsive region. Increasing the salt concentration increases the repulsion region due to screening of electrostatic interactions. The repulsive region increases till it reaches the Parsegian–Gingell result for the DH limit (solid line in Figure 6.11).

6.7.2 The Debye–Hückel Regime with Constant Surface Potential

So far we have solved the PB equation using the constant charge boundary conditions. However, constant potential boundary conditions are appropriate when the surfaces are metal electrodes, and it is important to understand this case as well.

Let us examine the effect of constant surface potential on the pressure. For simplicity, we will focus on the linearized PB equation (DH) in the asymmetric membrane case for monovalent electrolyte. We still refer to the set-up as in Figure 6.7. The two membranes at $z = \mp d/2$ are held at different values of constant surface potential, $\Psi_{1,2}$:

$$\Psi\big|_{z=-d/2} = \Psi_1,$$
$$\Psi\big|_{z=d/2} = \Psi_2. \quad (6.96)$$

Charged Membranes

Applying the DH solution of Eq. (6.81) with the boundary conditions of Eq. (6.96) leads to

$$\Psi(z) = \frac{\Psi_1 + \Psi_2}{2\cosh(\kappa_D d/2)} \cosh(\kappa_D z) + \frac{\Psi_1 - \Psi_2}{2\sinh(\kappa_D d/2)} \sinh(\kappa_D z), \quad (6.97)$$

and expanding the pressure Π from Eq. (6.56) to the second order in powers of $\Psi_{1,2}$ yields

$$\Pi \simeq \frac{k_B T n_b}{\sinh^2(\kappa_D d)} \left(2\Psi_2 \Psi_1 \cosh(\kappa_D d) - \Psi_2^2 - \Psi_1^2 \right). \quad (6.98)$$

This expression is similar to the one obtained for constant surface charge (Eq. 6.92). Indeed, for large separations, $d \gg \lambda_D$, the relative sign of Ψ_2 and Ψ_1 determines the sign of the pressure

$$\Pi \simeq 2 k_B T n_b \Psi_2 \Psi_1 \, e^{-\kappa_D d}, \quad (6.99)$$

as for the large-separation behaviour of the constant-charge case.

However, for small separations, $d \ll \lambda_D$, the pressure is different than for the constant surface-charge case,

$$\Pi \simeq -\frac{k_B T (\Psi_2 - \Psi_1)^2}{8 \pi \ell_B} \frac{1}{d^2} + k_B T n_b \Psi_2 \Psi_1. \quad (6.100)$$

It yields a pure attractive (negative) pressure that diverges as $\sim 1/d^2$, and does not depend on n_b. For the special symmetric case $\Psi_2 = \Psi_1$, at those small d, the pressure does not diverge and reaches a positive constant, $\Pi > 0$, proportional to n_b. Unlike the constant-charge case, here the counterion concentration remains constant near each of the membranes, because it depends only on the surface potential through the Boltzmann factor (Ben-Yaakov and Andelman 2010). However, the induced surface charge ($\sigma \propto \Psi_s'$, Eq. (6.97)) diverges when the membranes are brought closer together, resulting in a diverging electrostatic attraction.

Note that the crossover from repulsive to attractive pressure is obtained for zero pressure in Eq. (6.98) and is possible only for potentials of the same sign, $\Psi_2 \cdot \Psi_1 > 0$ including $\Psi_1 = \Psi_2$. The condition for attraction reads

$$e^{-\kappa_D d} < \frac{\Psi_2}{\Psi_1} < e^{\kappa_D d}. \quad (6.101)$$

For potentials of opposite sign, $\Psi_2 \cdot \Psi_1 < 0$, the pressure is purely attractive.

6.7.3 Counterions Only

In the absence of an external salt reservoir, the only mobile ions in the solution are monovalent counterions with density $n(z)$, such that the system is charge neutral,

$$\sigma_1 + \sigma_2 = -e \int_{-d/2}^{d/2} n(z) \, dz. \quad (6.102)$$

For the assumed overall negative charge on the two membranes, $\sigma_1 + \sigma_2 < 0$, the counterions are positive, $z_+ = 1$.

The PB equation for the two-membrane system is the same as for the single membrane (Eq. 6.18), with the boundary condition as in Eq. (6.51). The osmotic pressure (Eq. 6.55) reduces here to

$$\Pi = -\frac{k_B T}{8 \pi \ell_B} (\Psi')^2 + k_B T n_0 \, e^{-\Psi(z)}, \quad (6.103)$$

where n_0 is defined as the reference density for which $\Psi = 0$. This equation is a first-order ordinary differential equation and can be integrated. Nevertheless, its solution depends on the sign of the osmotic pressure. We will not present here the solution of the PB equation but rather discuss the crossover between attractive and repulsive pressures. This crossover is obtained by solving Eq. (6.103) with $\Pi = 0$. As the total surface charge is chosen to be negative, $\sigma_1 + \sigma_2 \leq 0$, and attraction occurs only for $\sigma_1 \cdot \sigma_2 < 0$, we choose σ_1 to be negative and σ_2 to be positive.

Integrating this equation and using the boundary condition at $z = -d/2$, we obtain the same electrostatic potential profile as in the single membrane case (Eq. 6.22), with a shifted z-axis origin: $z \to z + d/2$ and $\ell_{GC} \to \ell_1$ (with ℓ_1 defined as before):

$$\Psi = \Psi_0 + 2 \ln(z + \ell_1 + d/2). \quad (6.104)$$

The second boundary condition at $z = d/2$ gives a relation between d and $\sigma_{1,2}$. The condition for attraction can be expressed in terms of the surface densities (Kanduč et al. 2008):

$$|\sigma_1| - |\sigma_2| < \frac{|\sigma_2 \sigma_1|}{\sigma_d}, \quad (6.105)$$

where $\sigma_d \equiv e/(2\pi \ell_B d)$.

The crossover between attraction and repulsion is plotted in Figure 6.12. Two crossover lines separate the central attraction region from two repulsion ones. The upper one lies above the diagonal, $|\sigma_1| > |\sigma_2|$, and corresponds directly to the condition of Eq. (6.105). A second crossover line lies in the lower wedge below the diagonal, $|\sigma_1| < |\sigma_2|$. It corresponds to the crossover of the complementary problem of an overall positive surface charge, $\sigma_1 + \sigma_2 > 0$, and negative counterions. Note that the figure is symmetric about the principal diagonal, $|\sigma_1| \leftrightarrow |\sigma_2|$, as expected. The condition of attraction, irrespectively of the sign of $\sigma_1 + \sigma_2$, can be written as

$$\left| |\sigma_1| - |\sigma_2| \right| < \frac{|\sigma_1 \sigma_2|}{\sigma_d}, \quad (6.106)$$

as was obtained by Lau and Pincus (1999).

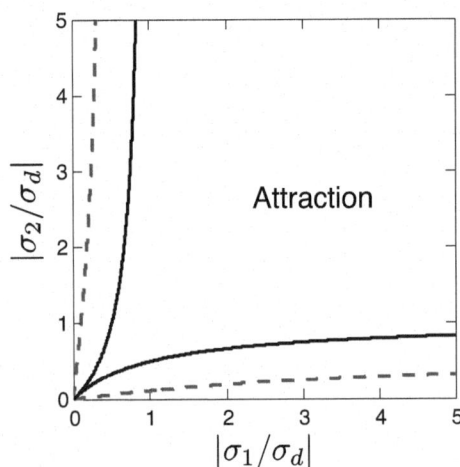

FIGURE 6.12 Regions of attraction ($\Pi<0$) and repulsion ($\Pi>0$) for the counterion-only case, plotted in terms of the two rescaled charge densities, $|\sigma_1/\sigma_d|$ and $|\sigma_2/\sigma_d|$, where $\sigma_d = e/2\pi\ell_B d$. The figure is plotted for oppositely charged membranes, $\sigma_1 \cdot \sigma_1 < 0$, and is symmetric about the diagonal $|\sigma_1|=|\sigma_2|$. The two solid lines delimit the boundary between repulsion and attraction in the no-salt limit, $n_b \to 0$, Eq. (6.106). For comparison, we also plot the crossover between attraction and repulsion for finite n_b from Eq. (6.107) for $d/\lambda_D = 2.2$ (blue dashed line). For the case of $\sigma_1 \cdot \sigma_2 > 0$, Π is always repulsive and there is no crossover.

6.7.4 Attraction/Repulsion Crossover

We now calculate the general criterion of the attractive-to-repulsive crossover. The $\Pi = 0$ pressure between two membranes located at $z = \pm d/2$ can be mapped exactly into the problem of a single membrane at $z = 0$ in contact with the same electrolyte reservoir. The only difference is that beside the boundary at $z = 0$, there is another boundary at $z = d$. The mapping to the single-membrane case is possible as the osmotic pressure of a single membrane is zero. This equivalence can be checked by substituting $\Pi = 0$ in Eq. (6.55) to recover the first integration of the PB equation for the single membrane system with added 1:1 electrolyte (Eq. 6.26).

The potential can be written as in Eq. (6.27) with $\gamma \to \gamma_1 = \sqrt{1+(\kappa_D\ell_1)^2} - \kappa_D\ell_1$. This solution already satisfies the boundary condition, $\Psi'(0) = -4\pi\ell_B\sigma_1/e$, but another boundary condition at $z = d$ needs to be satisfied as well, $\Psi'(d) = 4\pi\ell_B\sigma_2/e$.

Attraction will occur only for charged membranes of different sign, $\sigma_1 \cdot \sigma_2 < 0$. Using the boundary condition at $z = d$ for the two cases, $\sigma_1 < 0$ and $\sigma_1 > 0$, determines the region of attraction, $\Pi < 0$, by the inequalities (Ben-Yaakov et al. 2007)

$$e^{-\kappa_D d} < \frac{\gamma_2}{\gamma_1} < e^{\kappa_D d}, \qquad (6.107)$$

with $\gamma_2 = \sqrt{1+(\kappa_D\ell_2)^2} - \kappa_D\ell_2$. It can be shown that the above general expression (Eq. 6.107) reduces to the expression of Eq. (6.106), in the limit of counterion only and to that of Eq. (6.95) in the high-salt (DH) limit.

A similar general crossover criterion can also be obtained for constant potential boundary conditions. As we explained above, the crossover condition maps to the single membrane problem, yielding the generalized relation of Eq. (6.30), $\gamma_{1,2} = \pm\tanh(\Psi_{1,2}/4)$. The $\pm$ sign is chosen such that $\gamma_{1,2}$ is positive. For opposite surface potentials, $\Psi_1 \cdot \Psi_2 < 0$, there is a point between the membranes in which the potential vanishes. The osmotic pressure of Eq. (6.55), calculated at this point, has only the negative Maxwell stress contribution, and therefore, it is always *attractive*. On the other hand, for $\Psi_1 \cdot \Psi_2 > 0$, the following condition on Ψ_1 and Ψ_2 results in an attraction:

$$e^{-\kappa_D d} < \frac{\tanh(\Psi_2/4)}{\tanh(\Psi_1/4)} < e^{\kappa_D d}. \qquad (6.108)$$

In Figure 6.13, we show the osmotic pressure, Π, in units of $k_B T / (4\pi\ell_B\lambda_D^2)$, as a function of the (dimensionless) intermembrane separation, d/λ_D. The pressure is calculated for several values of constant charge and constant potential boundary conditions. Three types of pressure profiles are seen in the figure: attractive, repulsive, and the crossover between attraction and repulsion.

6.8 CHARGE REGULATION

As discussed in Section 6.7.2, the difference between constant surface potential, Ψ_s, and constant surface charge density, σ, is large when the distance between the two membranes is of order of the Debye screening length, λ_D, or smaller. Ninham and Parsegian (1971) considered an interesting intermediate case of great practical importance. Membranes with ionizable groups that can release ions into the aqueous solution or trap them – a situation intermediate between a constant σ, describing inert ionic groups on the membrane, and constant Ψ_s, relevant for a surface (an electrode or a membrane) held at a constant potential by an external potential source.

Let us consider a system where the membrane is composed of ionizable groups (lipids), and each of them can release a counterion into the solution. This surface dissociation/association (see Figure 6.14) is described by the reaction:

$$A^+ + B^- \rightleftharpoons AB, \qquad (6.109)$$

where A denotes a surface site that can be either ionized (A^+) or neutral (AB). The process of membrane association/dissociation is characterized by a kinetic constant K_d through the law of mass action:

$$K_d = \frac{[A^+][B^-]_s}{[AB]}, \qquad (6.110)$$

Charged Membranes

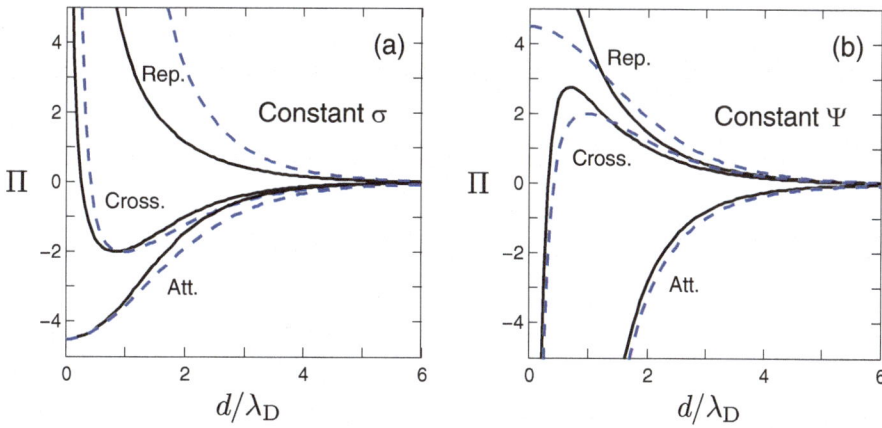

FIGURE 6.13 Osmotic pressure, Π, in units of $k_BT/(4\pi\ell_B\lambda_D^2)$, as function of the dimensionless intermembrane separation, d/λ_D. We present the solution of the nonlinear (solid black lines) PB equation and the linear (blue dashed lines) DH equation for two boundary conditions: (a) constant surface charge, and (b) constant surface potential. In each of the figure parts we show three profiles: repulsive, crossover, and attractive. In (a), the boundary conditions for the repulsive, crossover, and attractive profiles are $\sigma_1=\sigma_2=3$; $\sigma_1=3$ and $\sigma_2=-2$; and $\sigma_1=3$ and $\sigma_2=-3$, respectively, where σ is given in units of $e/(4\pi\ell_B\lambda_D)$. In (b), the boundary conditions for the repulsive, crossover, and attractive profiles are $\Psi_1=\Psi_2=3$; $\Psi_1=3$ and $\Psi_2=2$; and $\Psi_1=3$ and $\Psi_2=-3$, respectively. Note that the nonlinear PB solution of Π for the symmetric (repulsive) osmotic pressure for constant potential reaches a constant value as $d \to 0$, like in the DH case, but with a different value, $\Pi(d\to 0) \simeq 9.07$ in units of $k_BT/(4\pi\ell_B\lambda_D^2)$ (not shown in the figure).

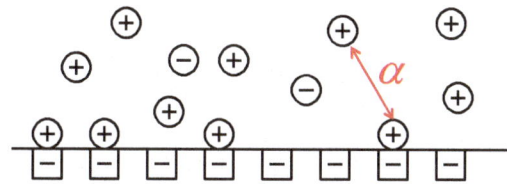

FIGURE 6.14 Illustration of the charge regulation boundary condition for cation association/dissociation energy α.

where $[A^+]$, $[B^-]_s$ and $[AB]$ denote the three corresponding surface concentrations (per unit volume). We define ϕ_s to be the area fraction of the A^+ ions, related to $\sigma > 0$, the membrane charge density by $\phi_s = \sigma a^2/e \sim [A^+]$ and $1-\phi_s \sim [AB]$, where a^2 is the surface area per charge. The equilibrium condition of Eq. (6.110) is then written as

$$K_d = \frac{\phi_s}{1-\phi_s}[B_s^-]. \quad (6.111)$$

As the counterions are released into the ionic solution, the relation between surface and bulk B^- concentrations, $[B^-]_s$ and $[B^-]_\infty$, is obtained via the Boltzmann distribution, $[B^-]_s = [B^-]_\infty \exp(\Psi_s)$ where $[B^-]_\infty = n_b$ is the bulk salt concentration. We note that in this section we choose $\sigma > 0$ and it implies $\Psi_s > 0$, and the reference potential in the bulk is set to be $\Psi = 0$. The concentration of dissociated B^- ions at the surface is $[B^-]_s$ and it is equal to $n_s^- = n_-(z \to 0)$. Note that this is *not* the charge concentration of the membrane itself (which comprises the bound ionic groups and is proportional to ϕ_s) but is the concentration of mobile ions evaluated at the submembrane position (just as in Section 6.3.3), where the standard PB equation holds. One can then write the area fraction of the membrane ionized sites as

$$\phi_s = \frac{K_d}{K_d + n_b e^{\Psi_s}}. \quad (6.112)$$

It is useful to introduce a surface interaction parameter $\alpha = \ln(a^3 K_d)$ instead of using the kinetic constant (see Figure 6.14). This gives the adsorption isotherm for ϕ_s (the fraction of A^+ groups on the membrane):

$$\phi_s = \frac{1}{1+\phi_b e^{-\alpha+\Psi_s}}, \quad (6.113)$$

where $\phi_b = a^3 n_b$ as before. The typical shape of ϕ_s as function of Ψ_s is a sigmoid and is shown in Figure 6.15a for $a = 0.3$ nm, $n_b = 0.1$ M, and $\alpha = -6$ (pK $\simeq 0.82$). The fraction of charge groups, ϕ_s, varies between $\phi_s = 1$ (fully charged) and $\phi_s = 0$ (neutral) as Ψ_s varies from negative values to positive ones. For the half-filled surface charge, $\phi_s = 0.5$, the surface potential is $\Psi_s^* = \alpha - \ln\phi_b$. At this special point, the slope of $\phi_s(\Psi_s)$ is exactly -0.25. The differential capacitance as discussed in Section 6.4 can be cast into a simple form for the charge regulation case

$$C_{CR} = \frac{d\sigma}{d\psi_s} = \frac{e^2}{k_BTa^2}\frac{d\phi_s}{d\Psi_s} = \frac{e^2}{k_BTa^2}\phi_s(1-\phi_s). \quad (6.114)$$

It follows that for the sigmoid shape of $\phi_s(\Psi_s)$, the differential capacitance has a unimodal shape with a universal maximum whose value is $e^2/(4k_BTa^2)$.

Similarly, for the fraction of neutral AB groups on the membrane, $1-\phi_s$, we write

$$1-\phi_s = \frac{\phi_b}{\phi_b + e^{\alpha-\Psi_s}}. \quad (6.115)$$

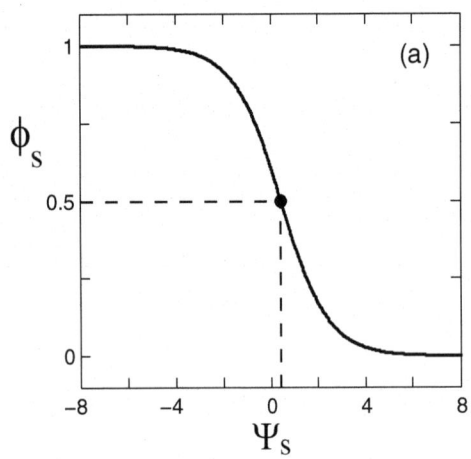
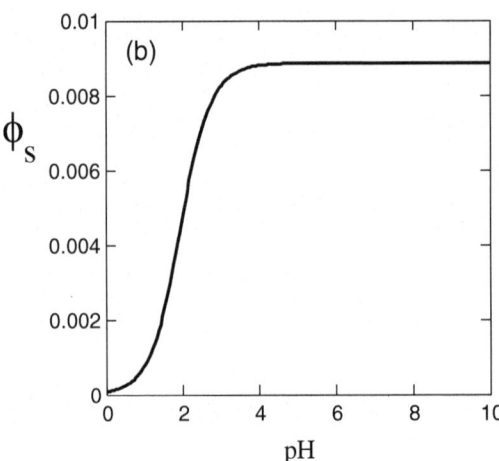

FIGURE 6.15 In (a), the area fraction of charge groups on the membrane, ϕ_s, is plotted as a function of the dimensionless electrostatic potential, Ψ_s, using Eq. (6.113). The parameters used are $a = 0.3$ nm, $n_b = 0.1$ M, and $\alpha = -6$ (pK $\simeq 0.82$). The symmetric point of $\phi_s = 0.5$ occurs at $\Psi_s^* \simeq 0.425$, and the slope there is exactly -0.25. In (b), we present the area fraction of charge groups on the membrane, ϕ_s, as a function of pH for surface binding $A^- + H^+ \rightleftharpoons AH$ (see "Surface pH and PK" box below). The parameters used are: $a = 0.5$ nm and $\alpha = -12$ (pK $\simeq 4.1$).

Equation (6.115) is the *Langmuir–Davies isotherm* (Davies 1958) and is an extension of the Langmuir adsorption isotherm (see, *e.g.*, Adamson and Gast 1997) for charged adsorbing particles. This can be understood because $1 - \phi_s$ is the fraction of the membrane AB neutral groups. Hence, it effectively describes the adsorption of B^- ions onto a charged membrane.

SURFACE pH AND pK

When an acidic reservoir exchanges H^+ ions with the membrane, the membrane chemical reaction is $A^- + H^+ \rightleftharpoons AH$, and the same local equilibrium of Eq. (6.110) can be expressed in terms of three logarithms: $pK \equiv -\log_{10} K_d = -\log_{10}(a^{-3} e^{\alpha})$ where the close packing density a^{-3} is measured in molar, $pH_s \equiv -\log_{10}[H^+]_s$, and $pH \equiv -\log_{10}[H^+]_\infty$. The regular pH measures the acidic strength in the reservoir, while the membrane pK (or K_d) is a fixed (and usually unknown) parameter that depends on the membrane as well as on the binding H^+ ions. Equation (6.111) can then be written as

$$\frac{\phi_s}{1 - \phi_s} = 10^{-pK + pH} e^{\Psi_s}. \tag{6.116}$$

When half of the membrane is charged, $\phi_s = 0.5$, the given pK relates the solution pH with the surface potential:

$$\Psi_s = (pH - pK)\ln(10). \tag{6.117}$$

In Figure 6.15b, we show the dependence of ϕ_s on pH for process of protonation/deprotonation for the case of one membrane with: $a = 0.5$ nm and $\alpha = -12$ (pK $\simeq 4.1$). The typical shape of ϕ_s(pH) is a sigmoid and is obtained by solving numerically Eq. (6.118) with the charge regulation boundary condition of Eq. (6.113).

The Langmuir–Davies isotherm (Eq. 6.115) relates the self-adjusting surface charge fraction ϕ_s and surface potential Ψ_s with K_d (or α) and the bulk density, n_b. In order to find Ψ_s and ϕ_s separately as function of K_d and n_b, one needs to find the electrostatic relation between them. This relation can be obtained from the Grahame equation (6.36), introduced in Section 6.3.3:

$$w\phi_s = \sinh(\Psi_s/2); \quad w \equiv \frac{2\pi \ell_B \lambda_D}{a^2}, \tag{6.118}$$

where we expressed the Grahame equation as $\phi_s = \phi_s(\Psi_s)$ in terms of the Bjerrum and Debye lengths via a dimensionless parameter, w. By inverting the relation, $\Psi_s = \Psi_s(\phi_s)$, we get

$$e^{\Psi_s/2} = w\phi_s + \sqrt{1 + (w\phi_s)^2}. \tag{6.119}$$

Recall that this Grahame equation (6.36) is obtained from the PB equation for one planar membrane and gives a relationship between the density σ at the membrane and $n_s^\pm$ at the subsurface layer. The Grahame equation depends only on the electrostatics properties and applies for any boundary condition: constant σ, constant Ψ_s, or the present case of charge regulation due to association/dissociation of the ionizable groups on the membrane. These two equations (Eqs. 6.115 and 6.118) determine completely the surface potential, $\Psi_s = \Psi_s(K_d, n_b)$, and the surface charge fraction, $\phi_s = \phi_s(K_d, n_b)$.

It is instructive to take the two limits of large and small ϕ_s. When the ionizable surface sites are almost fully dissociated, $\phi_s \simeq 1$, K_d is large enough so that $K_d \gg n_b \exp(\Psi_s)$. In this limit, Eqs. (6.112) and (6.119) reduce to

$$\phi_s \simeq 1 - \frac{n_b}{K_d} e^{\Psi_s},$$
$$\Psi_s \simeq 2\ln(w + \sqrt{1 + w^2}), \tag{6.120}$$

Charged Membranes

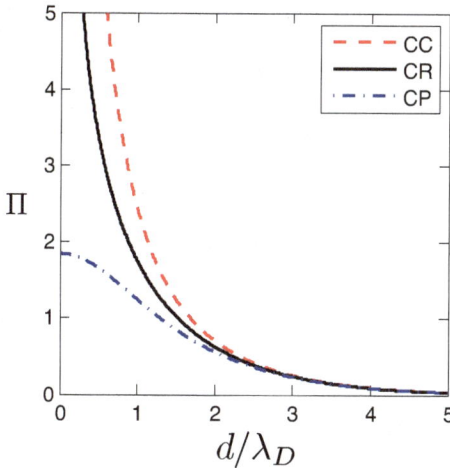

FIGURE 6.16 The pressure for two symmetric membranes in units of $k_B T/(4\pi \ell_B \lambda_D^2)$, in the presence of three different boundary conditions: constant charge (CC), constant potential (CP), and charge regulation (CR). The pressure inequality, seen in the figure, $\Pi_{CC} > \Pi_{CR} > \Pi_{CP}$, is an inequality that holds in general. The parameters used are $a = 0.5$ nm, $n_b = 0.1$ M, and $\alpha = -6$ (pK $\simeq 1.48$).

and for the additional requirement of large w (or $\lambda_D \gg a^2/\ell_B$)

$$\Psi_s \simeq 2\ln(2w) = 2\ln\left(\frac{8\pi \ell_B \lambda_D}{a^2}\right). \quad (6.121)$$

In the opposite limit, only a small fraction of the ionizable surface groups are dissociated, $\phi_s \ll 1$, which results in $K_d \ll n_b \exp(\Psi_s)$ and

$$\phi_s \simeq \frac{K_d}{n_b} e^{-\Psi_s} \ll 1. \quad (6.122)$$

If, in addition to small ϕ_s also $w\phi_s \ll 1$, expanding the right-hand side of Eq. (6.119) gives

$$\Psi_s \simeq 2w\phi_s \ll 1. \quad (6.123)$$

As stated in the beginning of this section, the charge regulation boundary condition lies between the constant charge and constant potential boundary conditions. In other words, the osmotic pressure obtained for constant charge (CC) boundary condition, Π_{CC}, will always be more repulsive than the pressure, Π_{CR}, of charge regulation (CR). Furthermore, the latter is more repulsive than Π_{CP}, the pressure for constant potential (CP), yielding $\Pi_{CC} > \Pi_{CR} > \Pi_{CP}$.

A calculated example manifests this fact and can be seen in Figure 6.16, where we show the osmotic pressure, Π, in units of $k_B T/(4\pi \ell_B \lambda_D^2)$ as a function of the dimensionless inter-membrane separation, d/λ_D. We use Eqs. (6.69) and (6.113) to obtain numerically the surface potential, Ψ_s, for the symmetric charge regulation boundary condition for any intermembrane separation, d/λ_D. The midplane potential is obtained from Eq. (6.79), and the pressure is calculated via Eq. (6.58). The difference in the osmotic pressure for the three boundary conditions arises only for small separations, $d \simeq \lambda_D$, while in the large separation limit, $d \gg \lambda_D$, the three pressures coincide. From the extrapolation of the CR surface potential at large separations, $\Psi_s(d \to \infty)$, we find the constant surface potential to be, $\Psi_s^{(const)} \simeq -1.7$, while the constant surface charge is $\sigma_{const} \simeq -e/4.35 \,\text{nm}^2$.

The behaviour of Π_{CP} for small distances, $d \ll \lambda_D$, is very different from the behaviour of Π_{CC} and of Π_{CR}, as Π_{CP} saturates at a value of $\Pi_{CP} \simeq 1.85$ (in units of $k_B T/(4\pi \ell_B \lambda_D^2)$). For $d/\lambda_D \to 0$ and finite ℓ_{GC}, Π_{CC} is always in the ideal gas regime (see Table 6.2), thus diverging as $\Pi_{CC} \sim 1/d$. Note that the osmotic pressure Π_{CR} also diverges when $d/\lambda_D \to 0$, but only as $\Pi_{CR} \sim 1/d^{1/2}$, which is weaker than for Π_{CC} (Markovich, Andelman and Podgornik 2016).

6.8.1 Charge Regulation via Free Energy

The Langmuir–Davies isotherm (Eq. 6.115) can also be derived from a free-energy minimization (Diamant and Andelman 1996). It is done by including a surface free energy, F_s, to account for the association/dissociation of ions onto/from the membrane. The total free energy is $F_t = F_v + F_s$, where the volume contribution, F_v, is the same as in Eq. (6.11) and the surface free energy is

$$a^2 F_s/(A k_B T) = \Psi_s \phi_s + \phi_s \ln \phi_s$$
$$+ (1-\phi_s)\ln(1-\phi_s) + \alpha(1-\phi_s), \quad (6.124)$$

where A is the lateral membrane area and a is its thickness. The first term describes the coupling between the surface charge density $\sigma = e\phi_s/a^2$ (taken as positive) and the surface potential $\Psi_s = e\psi_s/k_B T$. The second and third terms describe the mixing entropy of dissociated (charged) surface sites of fraction ϕ_s and associated (thus neutral) ones of fraction $1-\phi_s$. Finally, the fourth term is proportional to $1-\phi_s$, the amount of B$^-$ ions adsorbing onto the membrane. It accounts for the excess surface interaction as an ion binds onto the membrane creating a neutral AB group. The surface interaction parameter, α, is the same as the one defined after Eq. (6.112).

In order to obtain the surface isotherm, one needs to take the variation of F_t with respect to $\phi_s = \sigma a^2/e$, $\delta F_t/\delta \phi_s = \delta F_s/\delta \phi_s = \mu_s$, giving

$$\Psi_s + \ln\left(\frac{\phi_s}{1-\phi_s}\right) + (\beta \mu_s - \alpha) = 0. \quad (6.125)$$

For a dilute ionic solution, the chemical potential of the ions is related to their bulk density by $\mu_b = k_B T \ln \phi_b$, as explained after Eq. (6.6). In thermodynamical equilibrium, the chemical potential is equal throughout the solution, hence, $\mu_s = \mu_b$, and by rearranging Eq. (6.125) the Langmuir–Davies isotherm emerges,

$$1 - \phi_s = \frac{\phi_b}{\phi_b + e^{\alpha - \Psi_s}}. \quad (6.126)$$

The above equation is exactly the Langmuir–Davies isotherm of Eq. (6.115).

The advantage of the free-energy formulation presented in this section over the chemical equilibrium one presented earlier is that the former can be generalized to other cases of surface interaction, such as cooperativity between the surface sites modelled by adding a $b(1-\phi_s)^2$ term to F_s, adsorbing of several ion types with different ion–surface interactions, α, and other extensions of the simple charge regulation mechanism (Diamant and Andelman 1996, Ariel, Diamant and Andelman 1999). Recently, the free-energy formulation was also used to study the collective behaviour of mobile charge-regulation macromolecules (Markovich, Andelman and Podgornik 2017b, Avni et al. 2018, Hallett et al. 2018, Avni, Andelman and Podgornik 2019, Avni, Podgornik and Andelman 2020).

6.9 VAN DER WAALS' INTERACTIONS

Long-range van der Waals' (vdW) interactions between molecules are universal and result from the molecular dipolar fluctuations (for details see Parsegian 2005, Bordag et al. 2009). These fluctuations can have different origins. For polar molecules with permanent dipoles, their orientational fluctuations lead to Keesom interaction. When orientational fluctuations of a permanent dipole induce a dipole in another nonpolar but polarizable molecule, the induced dipole leads to Debye interaction. In all remaining cases, transient dipoles of nonpolar polarizable molecules induce other transient dipoles and lead to London dispersion interactions.

In general, one can write the total vdW interaction potential $V(\mathbf{r})$ between two molecules at positions $\mathbf{r}_1$ and $\mathbf{r}_2$ separated by $\mathbf{r} = \mathbf{r}_1 - \mathbf{r}_2$, in the form

$$V(\mathbf{r}) = -\frac{\mathcal{C}}{r^6} = -\frac{3k_BT\alpha_1\alpha_2}{(4\pi\varepsilon_0)^2}\frac{1}{r^6}. \quad (6.127)$$

The $1/r^6$ interaction reflects the dipolar nature of vdW interaction: dipolar field of the first molecule decays as $1/r^3$, interacts with the second molecule, and then propagates back to the first molecule, yielding a squared dipolar interaction, $1/r^6$. This argument does not take into account the relativistic corrections and, thus, corresponds to the nonretarded case.

The prefactor $\mathcal{C}$ gives the strength of the interaction and is proportional to the product of polarizabilities, $\alpha_1\alpha_2$, of the two molecules and, thus, also to the product of molecular volumes. If the molecules interact in a medium of dielectric constant ε_w, then the strength of the interaction is proportional to $(\alpha_1\alpha_2)/\varepsilon_w^2$.

6.9.1 THE HAMAKER PAIRWISE SUMMATION

The simplest way to account for the vdW interactions between two large macroscopic bodies is called the *Hamaker summation*. It results from the pairwise summation of the

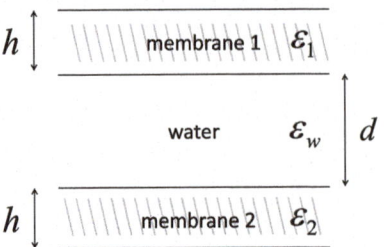

FIGURE 6.17 Illustration of two planar membranes of thickness h each and with dielectric constants ε_1 and ε_2 interacting across a water slab of thickness d of dielectric constant ε_w.

molecular vdW interactions, yielding for the interaction free energy:

$$\mathcal{F} = -\frac{3k_BT}{(4\pi\varepsilon_0)^2}\int_{V_1}d^3r_1\int_{V_2}d^3r_2\,\frac{\alpha_1 n_1(\mathbf{r}_1)\alpha_2 n_2(\mathbf{r}_2)}{|\mathbf{r}_1-\mathbf{r}_2|^6}, \quad (6.128)$$

where $n_{1,2}$ are the densities of the molecules in the two bodies, and the volume integrals go over the volumes of the two bodies, V_1 and V_2.

For two planar membranes of constant molecular density n_1 and n_2, each of finite thickness h at a separation d (see Figure 6.17), the above Hamaker integral yields an interaction energy per unit surface area A of the form

$$\frac{\mathcal{F}(d,h)}{A} = -\frac{3k_BT\alpha_1 n_1\alpha_2 n_2}{(4\pi\varepsilon_0\varepsilon_w)^2}$$
$$\times \int_{d/2}^{d/2+h}dz_1\int_{-d/2-h}^{-d/2}dz_2\int_0^\infty\frac{2\pi\rho d\rho}{\left[(z_1-z_2)^2+\rho^2\right]^3}, \quad (6.129)$$

where $\mathbf{r}=(z,\rho)$ in cylindrical coordinates. In this case of two planar dielectric media (membranes) interacting across a gap of dielectric constant ε_w (water), the excess polarizabilities are given by $n_{1,2}\alpha_{1,2} = 2\varepsilon_0\varepsilon_w(\varepsilon_{1,2}-\varepsilon_w)/(\varepsilon_{1,2}+\varepsilon_w)$ (Israelachvili 2011). The integrals over z_1 and z_2 can be evaluated analytically, yielding

$$\frac{\mathcal{F}(d,h)}{A} = -\frac{\mathcal{H}}{12\pi}\left(\frac{1}{d^2}-\frac{2}{(d+h)^2}+\frac{1}{(d+2h)^2}\right), \quad (6.130)$$

with a prefactor defined as the *Hamaker constant*

$$\mathcal{H} = \frac{3k_BT}{4}\left(\frac{\varepsilon_1-\varepsilon_w}{\varepsilon_1+\varepsilon_w}\right)\left(\frac{\varepsilon_2-\varepsilon_w}{\varepsilon_2+\varepsilon_w}\right). \quad (6.131)$$

From this general expression, one can derive two interesting scaling limits for small and large inter-membrane separation, d.

For small separations, $d \ll h$, corresponding to the vdW interaction between two semi-infinite media separated by distance d,

$$\frac{\mathcal{F}(d)}{A} \simeq -\frac{\mathcal{H}}{12\pi d^2} \sim \frac{1}{d^2}, \quad (6.132)$$

while for large separation, $d \gg h$, corresponding to the interaction of two thin sheets,

$$\frac{\mathcal{F}(d,h)}{A} \simeq -\frac{\mathcal{H}h^2}{2\pi d^4} \sim \frac{1}{d^4}. \qquad (6.133)$$

The d dependence obtained from the Hamaker summation is, to the lowest order, the same as obtained in more sophisticated approaches. However, the Hamaker constant, $\mathcal{H}$, can only be taken heuristically and, in fact, cannot be obtained from the simple pair-wise summation procedure.

6.9.2 Macroscopic Theory of vdW Interactions

A more sophisticated theory for vdW interactions between macroscopic bodies was developed by J. M. Lifshitz in the 1950s (see Parsegian 2005). In the Lifshitz theory, the vdW interactions are *electromagnetic fluctuation interactions*, and the Hamaker coefficient, $\mathcal{H}$, is a functional of the frequency-dependent dielectric permeabilities of the interacting media. It can be evaluated from either experimentally determined dispersion properties or calculated dispersion spectra of the interacting materials. It consistently includes the relativistic retardation effects due to the finite velocity of light propagation and finite temperature effects (Woods et al. 2016).

In the Lifshitz theory, the free energy of two planar semi-infinite bodies is typically cast into the Hamaker-type form (Safran 1994, Parsegian 2005)

$$\frac{\mathcal{F}(d,h)}{A} = -\frac{\mathcal{H}(d,h)}{12\pi d^2}, \qquad (6.134)$$

TABLE 6.3

Values of the Hamaker Constant for Different Materials Interacting across Water (from Parsegian 2005)

Material	Hamaker Constant × 10^{21} [J]
Polystyrene	13
Polycarbonate	3.5
Hydrocarbons	3.8
Polymethyl methacrylate	1.47
Proteins	5–9

as in Eq. (6.132), but with the important difference that the d-dependent $\mathcal{H}(d)$ can now be calculated explicitly via the Lifshitz formalism, when dielectric frequency-dependent properties of the interacting materials are available.

The most important characteristics of vdW interactions in the context of biomatter and, specifically, intermembrane interactions come from the presence of solvent, *i.e.*, water (Ninham and Parsegian 1970) and can be consistently taken into account within the Lifshitz theory. The calculated nonretarded Hamaker constant between lipid bilayers in water is found to be in the range of $10^{-20} - 10^{-21}$ J. The high static dielectric constant of water and the low static dielectric constant of hydrocarbons (consisting of the membrane core region) lead to an anomalously large contribution to the entropic part of the vdW free-energy, which remains unretarded at all separations. Some characteristic values of the Hamaker constant for interaction of different materials across a water layer are given in Table 6.3.

LIFSHITZ THEORY

The vdW interaction free energy is obtained in the Lifshitz theory as a sum, Eq. (6.136), over discrete imaginary Matsubara frequencies:

$$\zeta_n = \frac{2\pi k_B T n}{\hbar}, \; n = 0, 1, 2..., \qquad (6.135)$$

where $\hbar = h/2\pi$ and h is the Planck constant. The terms of this sum involve the dielectric response functions of the different media: the two lipid bilayers separated by a slab of aqueous medium. The Hamaker constant in Eq. (6.134) depends on the dielectric response functions and can be obtained quantitatively.

The dielectric response function at imaginary frequencies, $\varepsilon(i\zeta)$, is given formally by the Kramers–Kronig relation (Smith 1985)

$$\varepsilon(i\zeta) = 1 + \frac{2}{\pi}\int_0^\infty \frac{\omega \varepsilon''(\omega)}{\omega^2 + \zeta^2}d\omega, \qquad (6.136)$$

with $\varepsilon''(\omega)$ being the imaginary part of the complex frequency-dependent dielectric function, $\varepsilon(\omega) = \varepsilon'(\omega) + i\varepsilon''(\omega)$. Quite generally, $\varepsilon(i\zeta)$ is a real, monotonically decreasing function of its argument, ζ. The Kramers–Kronig relation also establishes the connection between the vdW interactions and the measurable dispersion part of the dielectric response functions, $\varepsilon''(\omega)$. For this reason, the vdW interactions are also referred to as the *dispersion interactions*. The imaginary frequencies can be rationalized intuitively as follows: just as $\varepsilon(\omega)$ characterizes the temporal response of a material to an external oscillating electric field $\sim \exp(i\omega t)$, $\varepsilon(i\zeta)$ characterizes the spontaneous time decaying fluctuation $\sim \exp(-\zeta t)$.

From the full Lifshitz formula for interacting lipid membranes, the limit of thick membranes, $h \gg d$, Eq. (6.134), without any retardation effects, yields the Hamaker constant in the form

$$\mathcal{H} = \frac{3}{2} k_B T \sum_{n=0}^{\infty}{}' \left[\Delta(i\zeta_n)\right]^2, \quad (6.137)$$

where the dielectric contrast is defined as

$$\Delta(i\zeta_n) = \frac{\varepsilon_L(i\zeta_n) - \varepsilon_w(i\zeta_n)}{\varepsilon_L(i\zeta_n) + \varepsilon_w(i\zeta_n)}, \quad (6.138)$$

in terms of the dielectric response function between the interacting media, i.e., the lipid ε_L and water ε_w, at imaginary frequencies. Note that the prime in the summation of Eq. (6.137), Σ', means that we have taken the lowest $n = 0$ term with weight 1/2.

Standard forms for these dielectric responses can be used (Mahanty and Ninham 1976, Dagastine, Prieve and White 2000), where the dielectric response of water is described by twelve different relaxation frequencies: one microwave relaxation frequency, five infrared relaxation frequencies, and six ultraviolet relaxation frequencies. Similarly, the hydrocarbon materials (lipid membrane) are modelled by four ζ ultraviolet relaxation frequencies (for details, see Parsegian 2005). In this limit of thick membranes, $h \gg d$, the Hamaker constant is indeed a constant, independent of the separation d but becomes d-dependent, $\mathcal{H} = \mathcal{H}(d)$, if the finite velocity of light is taken into account.

These retardation effects change the scaling of the interaction free energy from the $1/d^2$ into a $1/d^3$, at separations of $d \simeq 10-100$ nm, usually too large to be of practical importance for interacting lipid membranes. We further stress that the summation over the discrete frequencies set is something that cannot be derived from a simple Hamaker summation procedure of Section 6.9.1.

Using model expressions for the dielectric response of the hydrocarbon core of lipid bilayers and the aqueous medium, one ends up with the value of 4.3×10^{-21} J for the relevant Hamaker coefficient (Podgornik, French and Parsegian 2006). A comparable value has been obtained in experiments for small membrane spacings. For example, it was found that the values of the Hamaker constant for dimyristoyl phosphatidylcholine (DMPC) and for dipalmitoyl phosphatidylcholine (DPPC) membranes, which forms a multistack, are in the range of $2.87 - 9.19 \times 10^{-21}$ J (Petrache et al. 1998).

For thin membranes, $h \ll d$, the Lifshitz formula is valid in the nonretarded limit. The Hamaker constant of Eq. (6.137) has the same scaling of $1/d^4$ as in the Hamaker summation of Eq. (6.133). The same Hamaker constant is obtained also for membranes of finite thickness (see Eq. 6.130).

VdW interactions for symmetric bodies, e.g., two identical membranes interacting across a finite gap, are always attractive, just as electrostatic interactions between two symmetric bodies in the PB theory are always repulsive (Neu 1999). For asymmetric bodies, e.g., material '1' interacting with material '2' across water 'w', the Hamaker constant is given by a generalization of Eq. (6.137):

$$\mathcal{H} = \frac{3}{2} k_B T \sum_{n=0}^{\infty}{}' \Delta_{1w}(i\zeta_n) \Delta_{w2}(i\zeta_n). \quad (6.139)$$

This form of the Hamaker constant allows us to propose approximate *combining relations* that allow to extract unknown Hamaker constants from known ones (Israelachvili 2011). For the case of two media of material '1' interacting over material '3', the combining relation assumes the simple form $\mathcal{H}_{131} \simeq \mathcal{H}_{11} + \mathcal{H}_{33} - 2\mathcal{H}_{13}$, where $\mathcal{H}_{13}$ is for media '1' and '3' interacting across vacuum.[2]

The Lifshitz form of the Hamaker constant for asymmetric bodies also suggests that the vdW interaction can change sign, becoming repulsive. Sometimes this repulsive vdW interaction is referred to as quantum levitation (Munday, Capasso and Parsegian 2009).

6.9.3 The Derjaguin–Landau–Verwey–Overbeek Theory

In the Derjaguin–Landau–Verwey–Overbeek (DLVO) theory (Verwey and Overbeek 1948), the total interaction energy between charged bodies (colloidal particles or membranes) is assumed to be a simple sum of the electrostatic and vdW interactions. For interacting planar bilayer membranes, the total interaction free energy is

$$\mathcal{F}(d,h) = \mathcal{F}_{el}(d,h) + \mathcal{F}_{vdW}(d,h), \quad (6.140)$$

where the electrostatic part, $\mathcal{F}_{el}(d,h)$, is calculated in the PB framework as in Section 6.6. Since the PB osmotic pressure is easier to evaluate, the corresponding free-energy can be obtained via the integral

$$\frac{1}{A} \mathcal{F}_{el}(d,h) = \int_d^{\infty} d\ell \, \Pi(\ell), \quad (6.141)$$

with Π calculated as in Section 6.8. The vdW interaction free energy, $\mathcal{F}_{vdW}$, is calculated either from the Hamaker summation procedure or, more appropriately, from the Lifshitz theory.

Because of the $1/d^2$ dependence of the vdW free energy, the total interaction free energy exhibits a universal *primary minimum* at vanishing spacings. However, since the assumption of the continuum solvent is bound to break down in this small d limit, this minimum is more related to the continuum theory assumption inherent in the Lifshitz theory than to the physical interactions between membranes. In fact, a strong *hydration interaction* (related to the breakdown of the continuum model of the solvent) usually obliterates the primary minimum and results in a

Charged Membranes

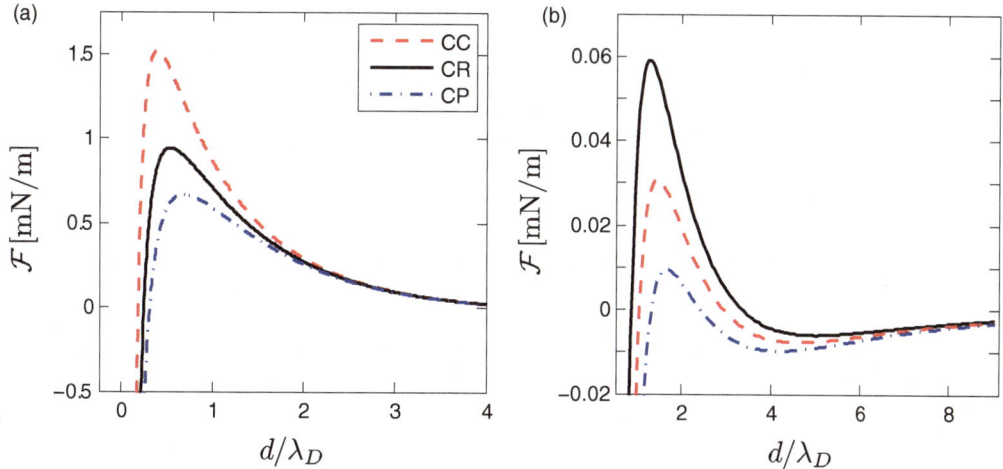

FIGURE 6.18 The interaction free energy density for two symmetric membranes, $\mathcal{F}(d,h)/A$, Eq. (6.140). In (a), the electrostatic part is calculated with three different boundary conditions: constant charge (CC), constant potential (CP), and charge regulation (CR). The inequality, $\mathcal{F}_{CC} > \mathcal{F}_{CR} > \mathcal{F}_{CP}$, seen here is a general relation. The parameters used are $a = 0.5$ nm, $n_b = 0.1$ M, and $\alpha = -6$ (pK $\simeq 1.48$). In (b), we calculate the electrostatic part of the free energy for constant charge boundary conditions with $\sigma = -e/(7 \text{ nm}^2)$ for three different bulk ionic concentrations, $n_b = 0.32$ M (solid black line), $n_b = 0.35$ M (dotted red line), and $n_b = 0.38$ M (dash-dotted blue line). The membrane width in (a) and (b) is $h = 4$ nm.

monotonic repulsive interaction even for very small inter-membrane separations (Parsegian, Fuller and Rand 1979).

Apart from the primary minimum at small separations, a secondary minimum can emerge at larger separations, depending on the system parameters. In Figure 6.18, we plot some of these scenarios for various assumptions on the electrostatic interactions. The curves shown in Figure 6.18 embody the essence of the DLVO theory. They can either show a monotonic repulsion extending over the whole ranges of separation, or attraction that is turned into repulsion or vice versa. In Figure 6.18a, the DLVO free energy is shown for the three different boundary conditions: constant charge (CC), constant potential (CP), and charge regulation (CR). Two minima are clearly seen: a minimum at $d/\lambda_D \to 0$, which is the nonphysical primary minimum mentioned earlier, and a minimum at $d/\lambda_D \to \infty$.

In Figure 6.18b, we show the DLVO free energy for CC boundary conditions with three different ionic concentrations: $n_b = 0.32$ M $(\lambda_D \simeq 0.54 \text{ nm})$, $n_b = 0.35$ M $(\lambda_D \simeq 0.52 \text{ nm})$, and $n_b = 0.38$ M $(\lambda_D \simeq 0.5 \text{ nm})$. As shown in the figure, for these bulk salt concentrations, a secondary, very shallow, minimum appears. Increasing n_b lowers the energy barrier and strengthens the shallow secondary minimum. The appearance of this secondary minimum is an essential ingredient in the explanation of the stability of colloidal particles and interacting membranes. They come into stable equilibrium at this secondary minimum. The high energy barrier between the secondary and the primary minimum makes it stable.

6.10 LIMITATIONS AND GENERALIZATIONS

The DLVO theory relies on approximations that have a finite range of validity (Naji et al. 2013). First, the vdW interaction is not really decoupled from the PB mean-field formulation and should be correspondingly modified. And second, one of the central results of PB theory that symmetric bodies always repel each other is *incorrect*. For physically interesting situations involving highly charged interfaces, or multivalent mobile ions, the electrostatic interaction can, in fact, be attractive.

These drawbacks of the classical DLVO theory, describing interactions between charged colloidal bodies or interacting membranes, can be amended. Recently, a new paradigm introduced a transparent systematization of the electrostatic interactions between charged bodies in terms of two useful regimes: *weak coupling* (WC) and *strong coupling* (SC) (Boroudjerdi et al. 2005). This allows a more accurate evaluation of the electrostatic interactions and the coupling between electrostatic and vdW interactions.

In order to introduce the WC and SC regimes, one needs to consider the relative strength of electrostatic interactions as compared with the background thermal energy. The thermal energy can be compared either with Coulomb interaction between two $q = ze$ charges of valency z giving rise to a modified Bjerrum length, $z^2 \ell_B$, as well as a modified Gouy–Chapman length, ℓ_{GC}/z, quantifying the strength of the electrostatic interaction between a point charge ($q = ze$) and a surface charge density, σ. Dividing the two lengths leads to a fundamental dimensionless *electrostatic coupling parameter* introduced by Netz (2001):

$$\Xi = z^2 \ell_B / (\ell_{GC}/z) = 2\pi z^3 \ell_B^2 |\sigma|. \quad (6.142)$$

For a system composed purely of counterions, the regimes of WC and SC can be understood in the following way. When the coupling parameter is small, *i.e.*, $\Xi \ll 1$, one goes back to the PB theory, with an addition of thermal fluctuations contributing a vdW-like interaction that scales linearly with Ξ and partially replaces the Lifshitz theory results.

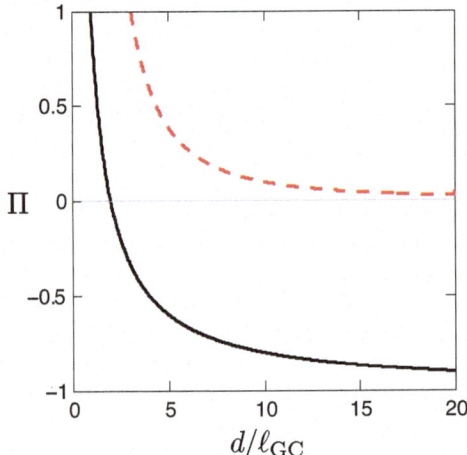

FIGURE 6.19 The pressure for two symmetric membranes in units of $k_B T/(2\pi \ell_B \ell_{GC}^2)$ as function of the dimensionless distance d/ℓ_{GC}. The calculation is done for the counterions-only case as obtained from the PB theory (dashed red line), Eqs. (6.65) and (6.66), in comparison with the strong coupling (SC) result (solid black line), $\Pi = -1 + 2\,\ell_{GC}/d$ in units of $k_B T/(2\pi \ell_B \ell_{GC}^2)$ (Ghodrat et al. 2015).

This clearly establishes a connection between electrostatic and vdW interactions.

In the opposite limit of large $\Xi \gg 1$, one observes a very different behaviour and important deviations from the standard DLVO paradigm. Here, the mobile ions become strongly correlated. It leads to a fundamental consequence that the interactions between nominally equally charged surfaces can become *attractive*. This is shown clearly in Figure 6.19 that compares the WC and SC results for the osmotic pressure between two equally charged membranes. Extensive Monte-Carlo simulations show that the WC to SC crossover is associated with a jump in the heat capacity and the appearance of short-range correlations between counterions for the coupling range $10 < \Xi < 100$ (Naji et al. 2013). For very high values of the coupling parameter, $\Xi \simeq 3 \times 10^4$, a transition to a Wigner crystalline phase (characterized by a diverging heat capacity) occurs. Therefore, the whole DLVO idea that the interaction is composed of repulsive electrostatic interaction and an attractive vdW, needs a serious revision for $\Xi \gg 1$ (SC regime).

The WC/SC paradigm is clear for the simple counterions-only case but becomes more complex in the presence of additional mobile charge components, *e.g.*, polyvalent plus monovalent salt, surface charge heterogeneity, mobile ion with multipolar structure, and polarizable mobile ions. These different components introduce new coupling parameters (similar to Ξ) that lead to additional features making the general form of the electrostatic interactions more complicated than in the simple PB and DLVO framework (Podgornik and Andelman 2020).

In this review, we have presented in detail the PB treatment for mobile ions in solutions for planar geometries as applicable to charged membranes. The PB equation is a mean-field equation that is obtained from the zeroth approximation in the WC regime (Podgornik and Žekš 1988, Borukhov, Andelman and Orland 1998, 2000, Netz and Orland 2000, Markovich, Andelman and Podgornik 2014, 2015).

Our chapter does not treat the fluctuations around the mean-field solution explicitly but only indirectly via the vdW interactions. Thermal fluctuations, which represent ion–ion correlations, have a key role in many interesting phenomena. Among them, we mention surface tension of electrolyte solutions (Onsager and Samaras 1934, Markovich, Andelman and Podgornik 2014, 2015, 2016, 2017a) and their dielectric decrement (Ben-Yaakov, Andelman and Podgornik 2011, Levy, Andelman and Orland 2012, Adar, Markovich and Andelman 2017, Adar et al. 2018). Furthermore, we do not cover the SC regime but only give some of its interesting results in this last section.

Apart from a more accurate treatment of electrostatic interactions, other simplifying assumptions were made at the base of the PB and DLVO theories presented in this chapter. As noted in the beginning of the chapter, the membrane structure was completely ignored. Therefore, membrane heterogeneities, curvature, and undulations were not discussed. The solvent (water) was treated as a featureless media with dielectric constant ε_w, and the effect of the mobile ions and the solvent structure (*e.g.*, water permanent dipole) on the decrement of the solvent dielectric constant was ignored (Ben-Yaakov, Andelman and Podgornik 2011, Levy, Andelman and Orland 2012, Adar et al. 2018). This effect also gives rise to a dielectrophoretic saturation of the counterions close to the membrane, similar to the steric mPB saturation in Section 6.4 (Nakayama and Andelman 2015).

Other extensions of the PB theory can be done by considering mixture of solvents (Ben-Yaakov et al. 2009a and 2009b) or nonelectrostatic interaction between the mobile ions themselves such as hydration interactions (Burak and Andelman 2000) or between the membrane and the mobile ions (Markovich, Andelman and Podgornik 2014, 2015, 2016, 2017a). For simplicity, the ions were treated as point-like particles which neglects their internal structure and polarizability (see, *e.g.*, Démery, Dean and Podgornik 2012 and references therein). Unfortunately, these interesting developments lie beyond the scope of this chapter and will be covered elsewhere (Andelman, Burak, and Orland, to be published).

NOTES

1. Throughout this chapter we use the SI unit system.
2. Note that the *combining relations* can be also obtained from the simpler Hamakar pair-wise summation of Section 6.9.1.

BIBLIOGRAPHY

Adamson, A. W. and A. P. Gast. *Physical Chemistry of Surfaces*, 6th ed. New York: Wiley, (1997).

Adar, R. M., T. Markovich and D. Andelman. "Bjerrum pairs in ionic solutions: A Poisson-Boltzmann approach". *J. Chem. Phys.* (2017): 146, 194904.

Adar, R. M., T. Markovich, A. Levy, H. Orland and D. Andelman. "Bjerrum pairs in ionic solutions: A Poisson-Boltzmann approach". *J. Chem. Phys.* (2018): 149, 054504.

Andelman, D., "Electrostatic properties of membranes: The Poisson-Boltzmann theory". In *Handbook of Physics of Biological Systems*, edited by Lipowsky R. and Sackman E., Vol. I. Amsterdam: Elsevier Science, (1995), Chap. 12.

Andelman, D., "Introduction to electrostatics in soft and biological matter", In *Soft Condensed Matter Physics in Molecular and Cell Biology*, Ed. by Poon W. and Andelman D., Scottish Graduate Series: SUSSP 59. New York: Taylor and Francis, (2005), pp. 97–122.

Andelman, D., Y. Burak and H. Orland. *Electrostatic Interactions in Soft and Biological Matter: Ions, Membranes, Polymers and Colloids*. New York: Cambridge University Press, to be published.

Ariel, G., H. Diamant and D. Andelman. "Kinetics of surfactant adsorption at fluid-fluid interfaces: Surfactant mixtures". *Langmuir* (1999): 15, 3574–3581.

Avni, Y., D. Andelman and R. Podgornik. "Charge regulation with fixed and mobile charged macromolecules". *Curr. Opin. Electrochem.* (2019): 13, 70–77.

Avni, Y., T. Markovich, R. Podgornik and D. Andelman. "Charge regulating macro-ions in salt solutions: Screening properties and electrostatic interactions". *Soft Matter.* (2018): 14, 6058–6069.

Avni, Y., R. Podgornik and D. Andelman. "Critical Behavior of Charged-regulated Macro-ion". *J. Chem. Phys.* (2020): 153, 024901

Ben-Yaakov, D. and D. Andelman. "Revisiting the Poisson-Boltzmann theory: Charge surfaces, multivalent ions and inter-plate forces". *Physica A* (2010): 389, 2956–2961.

Ben-Yaakov, D., D. Andelman, D. Harries and R. Podgornik. "Ions in mixed dielectric solvents: Density profiles and osmotic pressure between charged interfaces". *J. Phys. Chem. B* (2009a): 113, 6001–6011.

Ben-Yaakov, D., D. Andelman, D. Harries and R. Podgornik. "Beyond standard Poisson-Boltzmann Theory: Ion-specific interactions in aqueous solutions". *J. Phys.: Condens. Matter* (2009b): 21, 424106.

Ben-Yaakov, D., D. Andelman and R. Podgornik. "Dielectric decrement as a source of ion-specific effects". *J. Chem. Phys.* (2011): 134, 074705.

Ben-Yaakov, D., Y. Burak, D. Andelman and S. A. Safran. "Electrostatic interactions of asymmetrically charged membranes". *Europhys. Lett.* (2007): 79, 48002.

Bordag, M., G. L. Klimchitskaya, U. Mohideen, and V. M. Mostapanenko. *Advances in the Casimir Effect*. Oxford: Oxford University Press, 2009.

Borkovec, M., B. Jönsson and G. J. M. Koper. "Ionization processes and proton binding in polyprotic systems: Small molecules, proteins, interfaces and polyelectrolytes". *Surf. Colloid Sci.* (2001): 16, 99–339.

Boroudjerdi, H., Y. W. Kim, A. Naji, R. R. Netz, X. Schlagberger and A. Serr. "Statics and dynamics of strongly charged soft matter". *Phys. Rep.* (2005): 416 129–199.

Borukhov, I., D. Andelman and H. Orland. "Steric effects in electrolytes: A modified Poisson-Boltzmann equation". *Phys. Rev. Lett.* (1997): 79, 435–438.

Borukhov, I., D. Andelman and H. Orland. "Random polyelectrolytes and polyampholytes in solution". *Eur. Phys. J. B* (1998): 5, 869–880.

Borukhov, I., D. Andelman and H. Orland. "Adsorption of large ions from an electrolyte solution: A modified Poisson-Boltzmann equation". *Electrochimica Acta* (2000): 46, 221–229.

Burak, Y. and D. Andelman. "Hydration interactions: Aqueous solvent effects in electric double layers". *Phys. Rev. E* (2000): 62, 5296–5312.

Chapman, D. L. "A contribution to the theory of electrocapillarity". *Philos. Mag* (1913): 25, 475.

Churaev, N. V., B. V. Derjaguin and V. M. Muller. *Surface Forces*. New York: Springer, 2014.

Dagastine R. R., D. C. Prieve and L. R. White. "The dielectric function for water and its application to van der Waals forces". *J. Colloid Interface Sci.* (2000): 231, 351–358.

Davies, J. T. "Adsorption of long-chain ions. I". *Proc. R. Soc. A* (1958): 245, 417–428.

Dean D. S., J. Dobnikar, A. Naji, and R. Podgornik, eds. *Electrostatics of Soft and Disorder Matter*. CRC Press, 2014.

Dean, D. S. and R. R. Horgan. "Field theoretic derivation of the contact value theorem in planar geometries and its modification by the Casimir effect". *Phys. Rev. E* (2003): 68, 061106.

Debye, P. and E. Hückel. "Zur Theorie der Elektrolyte. Gefrierpunktserniedrigung und verwandte Erscheinungen. (The theory of electrolytes. Lowering of freezing point and related phenomena)". *Phyzik Z.* (1923): 24, 185–206.

Démery, V., D. S. Dean and R. Podgornik. "Electrostatic interactions mediated by polarizable counterions: Weak and strong coupling limits". *J. Chem. Phys.* (2012): 137, 174903.

Diamant, H. and D. Andelman. "Kinetics of surfactant adsorption at fluid-fluid interfaces". *J. Phys. Chem.* (1996): 100, 13732–13742.

Evans, D. F. and H. Wennerström, 1999. *The Colloidal Domain*, 2nd edition. New York: VCH Publishers, 1999.

Ghodrat, M., A. Naji, H. Komaie-Moghaddam and R. Podgornik. "Strong coupling electrostatics for randomly charged surfaces: Antifragility and effective interactions". *Soft Matter.* (2015): 11, 3441–3459.

Gouy, G. "Sur la constitution de la charge électrique à la surface d'un électrolyte". *J. Phys. (France)* (1910): 9, 457–468.

Gouy, G. "Sur la fonction électrocapillaire". *Ann. Phys. (Leipzig)* (1917): 7, 129-184.

Grahame, D. C. "The electrical double layer and the theory of ectrocapillarity". *Chem. Rev.* (1947): 41, 441–501.

Hallett, J. E., D. A. Gillespie, R. M. Richardson and P. Bartlett. "Charge regulation of nonpolar colloids". *Soft Matter* (2018): 14, 331–343.

Henderson, D. and L. Blum. "Some comments regarding the pressure tensor and contact theorem in a nonhomogeneous electrolyte". *J. Chem. Phys.* (1981): 75, 2025–2026.

Holm, C., P. Kékicheff and R. Podgornik, eds. *Electrostatic Effects in Soft Matter and Biophysics*. Amsterdam: Kluwer Academic Press, 2000.

Israelachvili, J. N. *Intermolecular and Surface Forces, 3rd edition*. London: Academic Press, 2011.

Kanduč, M., M. Trulsson, A. Naji, Y. Burak, J. Forsman and R. Podgornik. "Weak- and strong-coupling electrostatic interactions between asymmetrically charged planar surfaces". *Phys. Rev. E* (2008): 78, 061105.

Kilic, M. S., M. Z. Bazant and A. Ajdari. "Steric effects in the dynamics of electrolytes at large applied voltages: I. Double-layer charging". *Phys. Rev. E* (2007): 75, 021502.

Kiometzis, M. and H. Kleinert. "Electrostatic stiffness properties of charged bilayers". *Phys. Lett. A* (1989): 140, 520–524.

Kornyshev, A. A. "Double layer in ionic liquids: Paradigm change?" *J. Phys. Chem. B* (2007): 111, 5545–5557.

Lau, A. W. C. and P. Pincus. "Binding of oppositely charged membranes and membrane reorganization". *Eur. Phys. J. B* (1999): 10, 175–180.

Levy, A., D. Andelman and H. Orland. "Dielectric constant of ionic solutions: A field-theory approach". *Phys. Rev. Lett.* (2012): 108, 227801.

Mahanty, J. and B. W. Ninham. *Dispersion Forces.* London: Academic Press, 1976.

Markovich, T., D. Andelman and H. Orland. "Ionic profiles close to dielectric discontinuities: Specific ion-surface interactions". *J. Chem. Phys.* (2016): 145, 134704.

Markovich, T., D. Andelman and R. Podgornik. "Surface tension of electrolyte solutions: A self-consistent theory". *Europhys. Lett.* (2014): 106, 16002.

Markovich, T., D. Andelman and R. Podgornik. "Surface tension of electrolyte interfaces: Ionic specificity within a field-theory approach". *J. Chem. Phys.* (2015): 142, 044702.

Markovich, T., D. Andelman and R. Podgornik. "Charge regulation: A generalized boundary condition?" *EPL* (2016): 113, 26004.

Markovich, T., D. Andelman and R. Podgornik. "Surface tension of acid solutions: Fluctuations beyond the nonlinear Poisson-Boltzmann theory". *Langmuir* (2017a): 33, 34–44.

Markovich, T., D. Andelman and R. Podgornik. "Complex fluids with mobile charge-regulating macro-ions". EPL (2017b): 120, 26001.

Munday, J. N., F. Capasso and V. A. Parsegian. "Measured long-range repulsive Casimir-lifshitz forces." *Nature* (2009): 457, 170–173.

Naji, A., M. Kanduč, J. Forsman and R. Podgornik. "Perspective: Coulomb fluids - weak coupling, strong coupling, in between and beyond". *J. Chem. Phys.* (2013): 139, 150901.

Nakayama, Y. and D. Andelman. "Differential capacitance of the electric double layer: The interplay between ion finite size and dielectric decrement". *J. Chem. Phys.* (2015): 142, 044706.

Netz, R. R. "Electrostatics of counter-ions at and between planar charged walls: From Poisson-Boltzmann to the strong-coupling theory". *Eur. Phys. J. E* (2001): 5, 557–574.

Netz, R. R. and H. Orland. "Beyond Poisson-Boltzmann: Fluctuation effects and correlation functions". *Eur. Phys. J. E* (2000): 1, 203–214.

Neu, J. C. "Wall-mediated forces between like-charged bodies in an electrolyte". *Phys. Rev. Lett.* (1999): 82, 1072–1074.

Ninham, B. W. and V. A. Parsegian. "Van der Waals forces: Special characteristics in lipid-water systems and a general method of calculation based on the lifshitz theory". *Biophys. J.* (1970): 10, 646–663.

Ninham, B. W. and V. A. Parsegian. "Electrostatic potential between surfaces bearing ionizable groups in ionic equilibrium with physiologic saline solution". *J. Theor. Biol.* (1971): 31, 405–428.

Onsager, L. and N. N. T. Samaras. "The surface tension of Debye-Hückle electrolytes". *J. Chem. Phys.* (1934): 2, 528–536.

Parsegian V. A. *Van der Waals Forces.* New York: Cambridge University Press, 2005.

Parsegian V. A., N. Fuller and R. P. Rand. "Measured work of deformation and repulsion of lecithin bilayers". *Proc. Natl. Acad. Sci. USA* (1979): 76, 2750–2754.

Parsegian V. A. and D. Gingell. "On the electrostatic interaction across a salt solution between two bodies bearing unequal charges". *Biophys. J.* (1972): 12, 1192–1204.

Petrache, H. I., N. Gouliaev, S. Tristram-Nagle, R. Zhang, R. M. Suter, and J. F. Nagle. "Interbilayer interactions from high-resolution X-Ray scattering". *Phys. Rev. E* (1998): 57, 7014–7024.

Podgornik, R. and D. Andelman. "Embarras de Richesses in non-DLVO Colloidal Interactions". *J. Club Cond. Matt. Phys.* https://arxiv.org/abs/2101.00187 (2020).

Podgornik, R., R. H. French and V. A. Parsegian. "Nonadditivity in van der Waals interactions within multilayers". *J. Chem. Phys.* (2006): 124, 044709.

Podgornik, R. and B. Zeks. "Inhomogeneous coulomb fluid: A functional integral approach". *J. Chem. Soc., Faraday Trans.* 2(1988): 84, 611–631.

Sader, J. E. and D. Y. C. Chan. "Electrical double layer interaction between charged particles near surfaces and in confined geometries". *J. Colloid Interface Sci.* (1999): 213, 268–269.

Safran, S. A. *Statistical Thermodynamics of Surfaces, Interfaces, and Membranes.* Boston, MA: Addison-Wesley, 1994.

Verwey, E. J. W. and J. T. G. Overbeek. *Theory of the Stability of Lyophobic Colloids.* New York: Elsevier, 1948.

Winterhalter, M. and W. Helfrich. "Bending elasticity of electrically charged bilayers: Coupled monolayers, neutral durfaces, and balancing stresses". *J. Phys. Chem.* (1992): 96, 327–330.

Woods, L. M., D. A. R. Dalvit, A. Tkatchenko, P. Rodriguez-Lopez, A. W. Rodriguez and R. Podgornik. "Materials perspective on Casimir and van der Waals interactions," *Rev. Mod. Phys.* (2016): 88, 045003.

7 Membrane Shape Evolution *In Vitro*

Alexandra Zidovska
New York University, New York, New York

CONTENTS

- 7.1 Introduction 129
- 7.2 From Sphere to Starfish: Experimental Observations 130
  - 7.2.1 Membrane Shape Evolution in Scientific Discovery 130
  - 7.2.2 Membrane Shape Fluctuations and Transitions 131
- 7.3 Physical Concepts Underlying Membrane Shape Evolution 132
  - 7.3.1 Shape of Lipid Molecules 132
  - 7.3.2 Membrane Composition 133
  - 7.3.3 Molecular Interactions 134
- 7.4 Continuum Models of Membrane Shape Evolution 134
  - 7.4.1 The Spontaneous Curvature Model 134
  - 7.4.2 The Bilayer Coupling Model 135
  - 7.4.3 The Area Difference Elasticity Model 136
- 7.5 From *In Vitro* to *In Vivo*: Biological Membranes 136
  - 7.5.1 Membrane Curvature Generation and Stabilization *In Vivo* 136
- 7.6 Future Directions 138
- Acknowledgements 138
- Bibliography 138

7.1 INTRODUCTION

Liposomes, the self-assemblies of lipids in aqueous solutions, have attracted much interest since their initial discovery by Bangham et al. (1974). And rightly so, as they proved to be a powerful system and a versatile tool for basic science as well as numerous applications in medicine, pharmacology, diagnostics, cosmetics, and food industry. As the first part of this book discusses in great detail, in basic science liposomes facilitated in-depth studies of physicochemical properties of lipids and membranes, their biochemistry, structure, and molecular interactions (Lipowsky and Sackmann, 1995, Israelachvili, 2011). Furthermore, their similarities with biological membranes make them an invaluable model system for mimicking cell membranes and their physiological functions such as cellular adhesion, fusion, or signalling (Alberts, 2015, Sackmann and Merkel, 2010). Liposomes also play a prominent role in biomedical applications due to their ability to encapsulate other molecules, such as drugs and nucleic acids (Lasic, 1993). In particular, in gene therapy, the goal of which is the delivery of a therapeutic gene inside a cell *in vivo*, liposomes present a highly promising nonviral delivery vector (Huang et al., 2005, Yin et al., 2014). In fact, cationic liposome–nucleic acid complexes are currently the focus of about 10% of the ongoing human clinical trials in gene therapy (Ginn et al., 2013).

Our understanding of the shape evolution of liposomes, i.e., their ability to assume complex morphologies, has progressed substantially over the past four decades through fruitful interaction between experiment and theory. Development of optical techniques such as interference contrast and fluorescence imaging combined with digital video recording (Murphy and Davidson, 2012) opened new possibilities to study the membrane morphologies and their shape fluctuations as a function of their lipid composition, lipid phase, temperature, osmotic pressure, and chemical environment (e.g., pH, salt content). Such studies have mainly focused on the shape transitions and bending fluctuations of a freely suspended vesicle (a closed lipid bilayer structure), interacting membranes, membranes exposed to external forces, and heterogeneous membranes with respect to bilayer asymmetry, intramembrane domains, and anchored polymers (Lipowsky, 1991, Sackmann, 1994, Lipowsky, 1995, Seifert, 1997).

In this chapter, we will review the liposomal shapes found by experiments *in vitro*, with our main focus on membranes, i.e., the lipid bilayers, in their closed form, a vesicle. The main emphasis is put on the underlying physical principles such as regulation of the vesicle shape by lipid composition and solvent-dependent molecular interactions, which dictate the shape of membranes. A comprehensive overview of current continuum models of shape transitions of vesicles will be presented and compared with experimental observations. Finally, we will leap from *in vitro* to *in vivo* and compare shape transitions of model membranes and natural membranes by discussing concrete examples of complex membrane morphologies occurring in a living cell, their differences, and agreements in the context of the framework introduced in this chapter.

7.2 FROM SPHERE TO STARFISH: EXPERIMENTAL OBSERVATIONS

Here, we provide a survey of the membrane shape evolution in two ways: shape evolution from the simplest to the most complex shapes in the historical context of their scientific discovery, and shape evolution, when a membrane undergoes shape transitions or shape fluctuations in time.

7.2.1 Membrane Shape Evolution in Scientific Discovery

Lipid molecules in aqueous environment assemble spontaneously into liposomes, i.e., lipid aggregates. Numerous techniques have been developed to fabricate large closed lipid bilayer shells, called giant vesicles. A simplest and most common shape of a vesicle is a sphere, a simple lipid bilayer 'bubble'. But due to the fluidity of the lipid bilayer, and thus high mobility of the lipids in 2D, a variety of fascinating morphologies arises, which as we shall see is governed by the curvature and bending elasticity of the lipid bilayer.

Fluid bilayers comprised of a single lipid component generally form vesicles of spherical topology, such as ellipsoids, prolates and oblates, pears and dumbbells, discocytes and stomatocytes (Yager and Schoen, 1984, Lipowsky and Sackmann, 1995, Seifert, 1997), tubules (Chiruvolu et al., 1994), and starfish (Wintz et al., 1996). Topologically invariant, these morphologies have a genus of zero and therefore could be derived by deforming a sphere. Higher genus topologies followed shortly after, when morphologies like tori (Michalet and Bensimon, 1995b) and buttons (Michalet and Bensimon, 1995a) were observed. Figure 7.1 shows examples of complex vesicle morphologies such as a torus (Figure 7.1a and b) (Michalet and Bensimon, 1995b), a button (Figure 7.1c–e) (Michalet and Bensimon, 1995a), a pear (Figure 7.1f), a dumbbell (Figure 7.1g) (Käs and Sackmann, 1991), and tubules (Figure 7.1h). In addition, Figure 7.1i and j shows vesicles shaped like starfish (Wintz et al., 1996).

All of the above shapes can be found in systems containing a single lipid component. Shape transitions are induced by varying the area to volume ratio through temperature changes, e.g., ellipsoid–discocyte–stomatocyte transition or dumbbell–pear transition (Käs and Sackmann, 1991). Even more complex shape behaviour is observed for multicomponent membranes. As we shall see later, the differences in the structure and melting temperature T_m of different lipid components give rise to rich shape phase diagrams with changing relative concentrations of the lipid components in the mixture. Tubular morphology is in this aspect particularly interesting, as solid-phase lipid tubules can be obtained in single-lipid systems, while at least two lipid

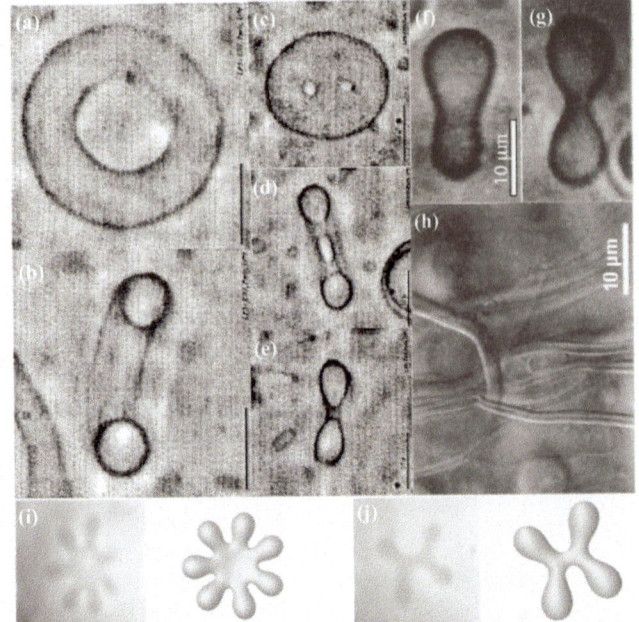

FIGURE 7.1 Complex vesicle morphologies. (a, b) Micrographs of a torus vesicle shown from different angles. (Reprinted in part from Michalet and Bensimon 1995b, with permission. Copyright 1995, EDP Sciences.) (c–e) Different views of a button-shaped vesicle. (Reprinted in part from Michalet and Bensimon 1995a, with permission. Copyright 1995, American Association for the Advancement of Science.) (f, g) Images of a pear-shaped (f) and a dumbbell-shaped (g) lipid vesicle. (Reprinted in part from Käs and Sackmann 1991 with permission. Copyright 1991, Biophysical Society.) (h) A micrograph of micrometre-scale lipid tubules. (Reprinted in part from Chiruvolu et al. 1994 with permission. Copyright 1994, American Association for the Advancement of Science.) (i–j) Micrographs of starfish vesicles and their theoretically calculated shapes. (Reprinted in part from Wintz et al. 1996 with permission. Copyright 1996, Europhysics Letters.)

components are required to generate liquid-phase tubular vesicles (consisting of chain-melted lipids above T_m). One way to generate tubules in a binary lipid mixture is to use two lipids with a cylindrical and a conical molecular shape, respectively. A spontaneous breaking of symmetry in lipid composition between the outer and inner layers occurs, which generates a membrane curvature leading to a cylindrical geometry (Safran et al., 1990). Liquid-phase vesicles are particularly attractive for biomedical applications, i.e., delivery of drugs and nucleic acids, since upon incorporation *in vivo* they assimilate into the cell membrane without disrupting its biological activity.

Another step in the evolution towards more complex shapes was the discovery of block liposomes (BLs), which are vesicles consisting of several connected, yet distinctly shaped liposomes such as spheres, pears, tubes, and cylindrical micelles (rods) (Zidovska et al., 2008). BLs were shown to form by two-component lipid mixtures containing neutral and multivalent lipids (Zidovska et al., 2011a). Figure 7.2 shows an overview of BLs, e.g., pear–tube–pear triblock (Figure 7.2a), pear–tube diblock (Figure 7.2c and d),

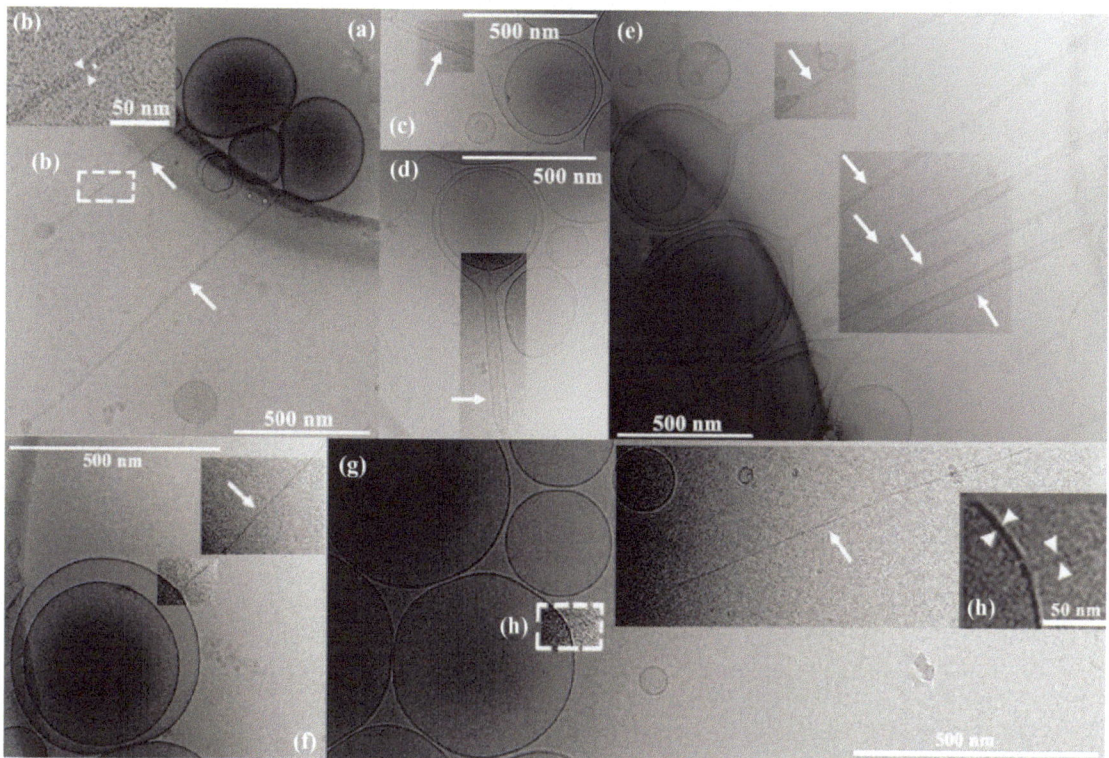

FIGURE 7.2 Cryo-TEM images of block liposomes formed by mixture of the hexadecavalent cationic lipid MVLBG2 and neutral 1,2-dioleoyl-sn-glycero-3-phosphatidylcholine (DOPC) in water. (a) Pear–tube–pear triblock liposome. (b) An inset of (a) highlighting the tubular structure (white arrowheads and white bar point out the bilayer thickness of 4 nm). (c) A pear–tube triblock liposome. (d) One block liposome encapsulated within another one. (e) A group of block liposomes. The block liposomes shown in (a–e) are comprised of liquid-phase lipid nanotube segments capped by spherical vesicles with diameters of a few hundred nm. The nanotubes (white arrows) are 10–50 nm in diameter and 1 mm in length. (f–g) Diblock liposomes comprised of lipid nanorods (white arrows) connected to spherical vesicles. Lipid nanorods are stiff cylindrical micelles with an aspect ratio ~1000. Their diameter equals the thickness of a lipid bilayer (~4 nm), and their length can reach up to several millimetres with a persistence length at the order of millimetres. (Reprinted in part from Zidovska et al. 2008 with permission. Copyright 2008, American Chemical Society.)

and sphere–nanorod diblock (Figure 7.2f and g). While the morphology of BLs is highly sensitive to ionic strength and pH of the solution (Zidovska et al., 2009b), BLs can be dried and rehydrated, resuming their original shape, which may be useful for potential drug storage and drug delivery applications.

For student of the subject, it is important to realize the difference between shapes that a membrane can assume while in and out of the thermodynamic equilibrium. All the shapes discussed above are equilibrium shapes and thus stable at given temperature and environmental conditions. However, there are a variety of transient shapes that a membrane assumes while in a process of seeking a thermodynamic equilibrium, which can be often quite obscure and not trivial. Figure 7.3 shows two fascinating examples of such metastable nonequilibrium shapes: the Rayleigh pearling instability occurring in tubular vesicles (Figure 7.3a–d, (Tsafrir et al., 2001)) and budding or tubulation in oblate vesicles (Figure 7.3e–h, (Tsafrir et al., 2003)) upon anchoring polymers into the lipid bilayer. Such instability occurs due to an increase of the membrane lateral tension, caused in this specific example by anchoring of a polymer into the lipid bilayer. In fact, one can find a plethora of nonequilibrium shapes inside a living cell, and we shall discuss later in this chapter the mechanisms that nature invented to stabilize such metastable membrane shapes.

7.2.2 Membrane Shape Fluctuations and Transitions

For a given area to volume ratio, the morphology of a vesicle is determined by the spontaneous curvature of the membrane and its bending elasticity. Spontaneous curvature of the membrane arises due to the asymmetry between the inner and outer leaflets of the lipid bilayer, while the bending rigidity is a material property dependent on the lipid composition of the membrane. Since the bending rigidity of a typical lipid bilayer is ~10 k_BT, thermal fluctuations cause pronounced bending undulations of the membrane (Helfrich, 1973). In the case of giant unilamellar vesicles (10–20 μm in diameter), such undulations are visible with light microscopy and thus can be measured and analysed. Specifically, if one carries out Fourier analysis of the vesicle contour bending excitations and applies the equipartition theorem, the bending modulus κ of the membrane can

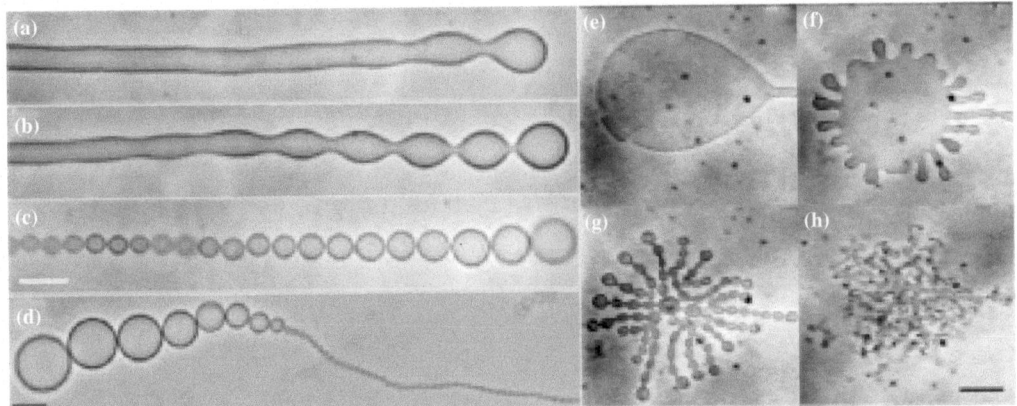

FIGURE 7.3 Nonequilibrium vesicle shapes. (a–d) Pearling instability occurring in tubular vesicles upon anchoring polymers into the lipid bilayer. (Reprinted in part from Tsafrir et al. (2001) with permission. Copyright 2001, American Physical Society.) (e–h) Budding and tubulation in oblate vesicles upon anchoring polymers into lipid bilayer. (Reprinted in part from Tsafrir et al. 2003 with permission. Copyright 2003, American Physical Society.)

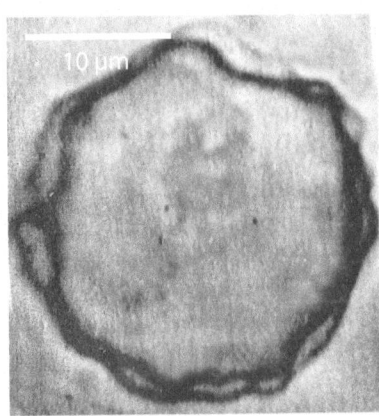

FIGURE 7.4 Thermal fluctuations of a vesicle shape. Contour of fluctuating vesicle made of plant lipid di-galactosyl-diglyceride. (Reprinted in part from Duwe and Sackmann 1990 with permission. Copyright 1990, American Physical Society.)

7.3 PHYSICAL CONCEPTS UNDERLYING MEMBRANE SHAPE EVOLUTION

The formation of liposomes is governed mainly by the hydrophobic effect and molecular shape of the lipids in a given solvent. The hydrophobic effect has been discussed in detail in the first part of this book; therefore, here we will focus mainly on the lipids, their molecular shape, and interactions within the membrane, responsible for the membrane polymorphism.

7.3.1 Shape of Lipid Molecules

There are two sets of parameters that can be used to describe molecular shape of a lipid, when considering its preferred packing geometry: *microscopic*, or molecular, and *macroscopic*, referring to the final preferred membrane shape.

The packing geometry of the lipid molecules within a lipid aggregate is strongly dictated by their molecular shape, which can be cylindrical, conical, or inverse conical. The molecular shape of a lipid can be described by following characteristics: head group area a_0, volume of the hydrocarbon chains v, and the length of the hydrocarbon chains l_c. In case of l_c, one refers to the maximum extension length of the hydrocarbon chain, which sets the geometrical limit for the molecular shape considerations. Both a_0 and v depend also on the saturation of the hydrocarbon chains. All three molecular characteristics, a_0, v, and l_c, commonly referred to as *microscopic* parameters, can be measured experimentally. Thus, for lipids with known a_0, v, and l_c, we can predict their favoured packing geometry in a lipid aggregate. In addition to the geometric constraints given by the molecular shape of lipids, a formed lipid aggregate (liposome) must also satisfy the system's tendency to maximize its entropy. Therefore, a large number of small lipid aggregates are preferred over a small number of large lipid aggregates (Israelachvili, 2011).

be directly obtained (Engelhardt et al., 1985). Figure 7.4 shows a fluctuating contour of a giant unilamellar vesicle made of plant lipid di-galactosyl-diglyceride, which has exceptionally low bending rigidity κ, thus exhibiting pronounced thermally driven bending fluctuations (Duwe and Sackmann, 1990).

Furthermore, as mentioned in 7.2.1, the membrane shape can undergo transitions from one shape to another in response to change in temperature, pH, salt content, or other environmental factors, which lead to change of the physical properties of the membrane such as spontaneous curvature and bending elasticity, which dictate its shape. For example, changes in temperature can lead to changes in the area to volume ratio causing ellipsoid–discocyte–stomatocyte transition or dumbbell–pear transition (Käs and Sackmann, 1991). In the following section, we will explain the molecular picture behind the macroscopic membrane shape transition.

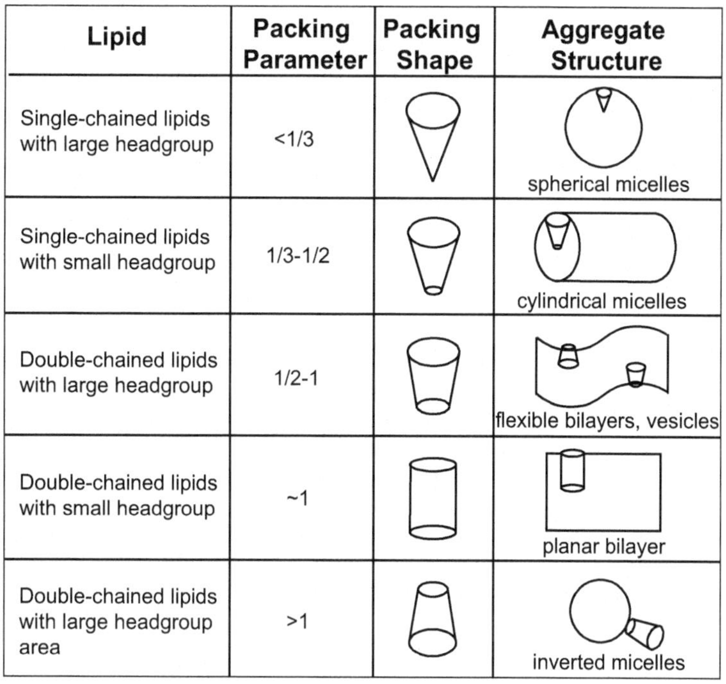

Lipid	Packing Parameter	Packing Shape	Aggregate Structure
Single-chained lipids with large headgroup	<1/3	(cone)	spherical micelles
Single-chained lipids with small headgroup	1/3–1/2	(truncated cone)	cylindrical micelles
Double-chained lipids with large headgroup	1/2–1	(truncated cone)	flexible bilayers, vesicles
Double-chained lipids with small headgroup	~1	(cylinder)	planar bilayer
Double-chained lipids with large headgroup area	>1	(inverted truncated cone)	inverted micelles

FIGURE 7.5 Summary of lipid packing shapes and the formed aggregates redrawn according to Israelachvili (2011).

Furthermore, to compare packing properties of different lipids, one can evaluate their *packing parameter* $v/(a_0 l_c)$, which is characteristic of every lipid in a given solution. The packing parameter is sensitive to the molecular structure of a lipid (e.g., chain saturation/unsaturation, head group area), as well as to the changes in the lipid's biochemical environment (e.g., pH, salt content) or temperature. Figure 7.5 provides an overview of the liposome structures corresponding to different values of the packing parameter $v/(a_0 l_c)$ as described in Israelachvili (2011).

Another set of parameters used to describe the liposome aggregate structure are the *macroscopic* parameters related to the properties of the membrane: spontaneous curvature C_0, bending rigidity κ, and Gaussian modulus κ_G. Spontaneous curvature C_0 is defined as the inverse radius of the curvature R_0, $C_0 = 1/R_0$, accounting for the asymmetry between the inner and outer leaflets of the bilayer, thereby carrying indirect information about the molecular shape of participating lipid molecules. In the case that $C_0 \approx 0$, lipid molecules have a cylindrical shape, and they form mostly flat bilayers. If $C_0 > 0$, lipid molecules have a conical shape forming cylindrical or spherical micelles, while when $C_0 < 0$, lipid molecules have an inverse cone shape assembling into inverse cylindrical or spherical micelles. Bending rigidity κ is a material parameter providing a measure for the energy necessary to bend the membrane of given lipid composition into a shape described by principal curvatures C_1 and C_2, thus directly influencing the dynamic roughness of the membrane by governing the amplitude of the thermal fluctuations. Gaussian modulus κ_G is often referred to as the saddle splay elastic modulus and depends on the membrane topology. The local elastic energy of such a system can be expressed as

$$f_{\text{elastic}} = \frac{1}{2}\kappa (C_1 + C_2 - C_0)^2 + \kappa_G C_1 C_2$$

where C_1 and C_2 are the principal curvatures of the membrane; $C_1 C_2$ is referred to as the Gaussian curvature; and $1/2(C_1 + C_2)$ is referred to as the mean curvature of the membrane (Helfrich, 1973).

7.3.2 Membrane Composition

In multicomponent lipid mixtures, the molecular shape of each lipid species and their relative concentration determine the final shape of the lipid aggregate. Figure 7.6 illustrates this point in great detail showing a phase diagram of binary lipid mixture of hexadecavalent cationic lipid MVLBG2 and neutral 1,2-dioleoyl-sn-glycero-3-phosphatidylcholine (DOPC) in water as a function of the lipid stoichiometry (Zidovska et al., 2008). MVLBG2 has a conical shape, while DOPC can be represented by a cylinder (Figure 7.6a). Therefore, mixing these two components at different ratios leads to formation of different liposome morphologies: at low content of MVLBG2, multilamellar spherical vesicles (onions) form (0–8 mol% MVLBG2, Figure 7.6a). In a narrow composition range of 8–10 mol% MVLBG2, BLs can be found (Figure 7.6b), while with further increase of MVLBG2 content, onions reenter the phase diagram (11–50 mol% MVLBG2, Figure 7.6d, e). Yet further increase of the content of conical MVLBG2 leads to a coexistence of vesicles and micelles (~50 mol% MVLBG2, Figure 7.6f), until only micelles (75–100 mol% MVLBG2, Figure 7.6g–i) can be found.

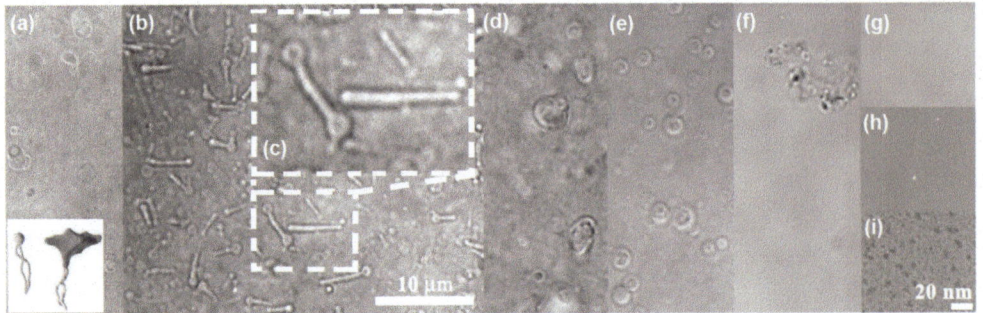

FIGURE 7.6 Phase behaviour of the MVLBG2/DOPC/water system. (a–g) Differential interference contrast (DIC) microscopy images of vesicle shapes as a function of their composition: multilamellar spherical vesicles (onions) (0–8 mol% MVLBG2, a), block liposomes (8–10 mol% MVLBG2, b), reentrant onions (11–50 mol% MVLBG2, d, e), macroscopic coexistence of vesicles and micelles (~50 mol% MVLBG2, f), and micelles (75–100 mol% MVLBG2, g). An inset in (a) shows a cartoon of the MVLBG2 and DOPC schematically depicting their molecular shapes. (h) Fluorescence microscopy image, demonstrating the existence of micelles. (i) Cryo-TEM image, showing micelle size and morphology. (c) An inset of (b), showing the block liposome morphology in detail: long (~5 μm) cylindrical cores of diameter ~0.5 μm capped at both ends with spherical vesicles of a few μm diameter. (Reprinted in part from Zidovska et al. 2008 with permission. Copyright 2008, American Chemical Society.)

7.3.3 Molecular Interactions

When we discuss the molecular shape of lipids, we must always carefully specify 'in a given solvent'. This refers to the fact that the molecular shape will be dependent on the solvent mediating the molecular interactions within the membrane. While the hydrophobic effect drives the assembly of the amphiphilic lipids in aqueous environment, their shape will be dependent on their physicochemical properties in the given environment. Temperature, which can be below or above their melting temperature T_m, will determine whether the hydrocarbon chain is in an ordered all-trans or in a disordered conformation, often referred to as solid and fluid state. In the former case, the lipids tend to condense the local environment, while it expands the membrane in the latter case. The number of the double bonds in the hydrocarbon chains plays an important role in the membrane state as well as membrane bending rigidity. A major effect on the lipid packing is played by cholesterol, a steroid molecule, which strongly condenses fluid lipid membranes, thus increasing the membrane stiffness and prevents the formation of solid phases, thus acting as a fluidizer. The solution pH influences the number of charges that lipid molecules will effectively carry, while the salt content screens the electrostatic interaction in between the molecules. Taken together, the packing parameter $v/(a_0 l_c)$ will depend on all of these influences and will lead to different aggregate shapes under different conditions.

To illustrate on our example from 7.3.2, the phase diagram of MVLBG2/DOPC changes dramatically when exposed to different pH or ionic strength, when MVLBG2 changes its molecular shape from conical to less conical to cylindrical in response to screening the electrostatic interaction between its 16 charges (salt content) or in response to effective change in the number of its charges (pH) (Zidovska et al., 2009b). For an interested reader, we would like to point out the special case of zwitterionic lipids, which while neutral, possess an electric dipole, which can change its orientation in response to electric fields caused by charged lipids in the bilayer and thus change its effective head group area (Gurtovenko et al., 2004, Zidovska et al., 2009a). Every dependence of a shape on an environmental condition represents an opportunity to possibly control the shape and its potential transition to other shapes.

7.4 CONTINUUM MODELS OF MEMBRANE SHAPE EVOLUTION

The manifold of shapes described in 7.2 can be well understood on the basis of bending elasticity concept of closed shells. It is based on the assumption that the shapes are determined by the mean curvature $H = 1/2(C_1+C_2)$ and the Gaussian curvature $G = C_1 C_2$, where C_1 and C_2 are the local principal curvatures. The asymmetry of the membrane can be accounted for by a spontaneous curvature C_0. Current theoretical treatments are based on the elastic free energy models of membranes described by the membrane bending and Gaussian moduli as well as the spontaneous curvature, at a constant total surface area of the vesicle (Lipowsky and Sackmann, 1995, Safran, 1994, Seifert, 1997). In the following, we will review the main principles of the three continuum models developed over past 40 years to describe membrane shape evolution.

7.4.1 The Spontaneous Curvature Model

In the early 1970s, it was shown by Helfrich (1973) that lipid bilayers can be very effectively described by their local bending elastic energy, and thus vesicles, composed of fluid lipid bilayers, can be regarded as elastic shells. A thin shell model is especially appropriate, considering the thickness of lipid bilayer is ~4 nm, while the diameter of a giant unilamellar vesicle can be 10–20 μm, i.e., several orders of magnitude larger. This property allows to describe a membrane as a 2D surface in a 3D space characterized by its local principal curvatures, C_1 and C_2, which are the inverse principal radii $R_1=1/C_1$ and $R_2=1/C_2$, respectively. C_0 is the

Membrane Shape Evolution *In Vitro*

spontaneous curvature of the membrane accounting for its inherent local asymmetry between the inner and outer leaflets of the membrane. Such asymmetry can occur due to different lipid composition or due to the impact of different physicochemical environment to the outside and inside of the vesicle. As defined in 7.3.1, $C_1 C_2$ is the Gaussian curvature, and $1/2(C_1 + C_2)$ is the mean curvature, with the elastic energy of the 2D surface being (Helfrich, 1973)

$$F_{elastic} = \oint_A \frac{1}{2} \kappa (C_1 + C_2 - C_0)^2 \, dA + \oint_A \kappa_G C_1 C_2 \, dA.$$

While the first term of this expression strongly depends on the surface A, i.e., vesicle shape, the second term is topologically invariant and therefore, as per the Gauss–Bonnet theorem remains constant for the surfaces of the same genus and can be formally neglected.

The vesicle shapes are calculated by minimizing the elastic energy. Helfrich's elastic energy has successfully predicted a number of vesicle shapes, including the discocyte–stomatocyte transition. However, it could not explain vesicle budding or branching of tubular vesicles. Further adjustments to the model were needed.

7.4.2 The Bilayer Coupling Model

Helfrich's model assumes a homogeneous membrane, not taking into account the difference in the lipid packing density between the inner and outer leaflets of the lipid bilayer (Helfrich, 1973). As pointed out by Evans in 1974, when a vesicle is formed, the lipids in the inner leaflet will be more compressed than the lipids in the outer leaflet; thus, an asymmetry arises within the membrane, which is more pronounced the longer the hydrocarbon tails of the lipids (Evans, 1974). To resolve this discrepancy, in 1989, Svetina and Zeks introduced the bilayer coupling model, which was able to predict shapes of red blood cells, such as discocyte and stomatocyte, and phospholipid vesicle shapes (Svetina and Žekš, 1989).

In the bilayer coupling model, membrane is considered having two laterally incompressible leaflets, an inner and outer one, which are decoupled. As in the previous model, vesicle shapes can be calculated by minimizing the elastic energy, while the volume V and area A are kept constant. Importantly, the difference of area ΔA between the inner and outer leaflets is being considered. The bilayer coupling model provides a phase diagram, which shows that different vesicle shapes occur for different $v/\Delta a$, with v and Δa being the volume and leaflet area difference relative to those of a sphere with volume V and area A.

Figure 7.7a shows a phase diagram of minimum bending energy vesicle shapes calculated by the bilayer coupling model (Käs and Sackmann, 1991). It denotes the regimes where certain shapes such as dumbbells, pears, stomatocytes, and discocytes exist as a function of the reduced volume v and area difference Δa. C^{pear} and C^{sto} represent phase boundaries corresponding to two continuous transitions at which the up-down symmetry is broken, while the outer solid lines L^{dumb}, L^{pear}, and L^{sto} denote the limiting shapes. Furthermore, large v values in the dumbbell regime contain

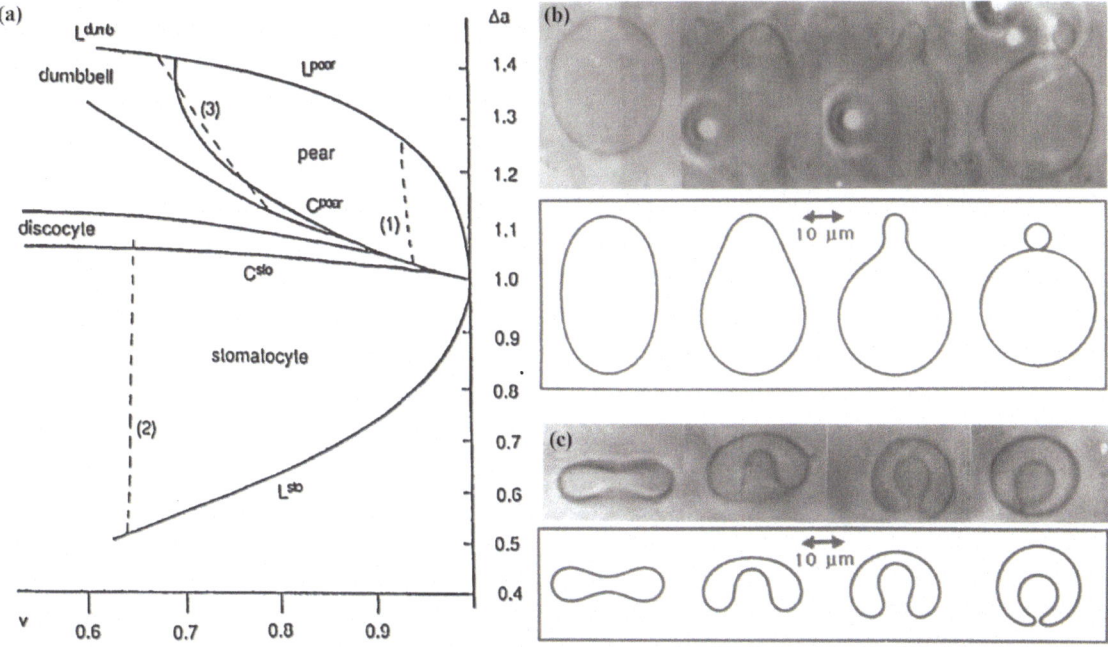

FIGURE 7.7 Phase behaviour of lipid vesicles. (a) Phase diagram of minimum bending energy vesicle shapes calculated by the bilayer coupling approach as a function of the reduced volume, v, and area difference, Δa, between inner and outer leaflets. (b) Comparison between the experimental and theoretical shapes of dimyristoylphosphatidylcholine (DMPC) vesicles in pure water according to the phase trajectories shown in (a) for budding transition belonging to the trajectory (1) and (c) for discocyte–stomatocyte transition described by trajectory (2). (Reprinted in part from Käs and Sackmann, 1991 with permission. Copyright 1991, Biophysical Society.)

also prolate ellipsoids, while ellipsoids can be found also in the discocyte regime. This phase diagram is in good agreement with experiments for dimyristoylphosphatidylcholine (DMPC) and palmitoyloleylphosphatidylcholine (POPC) vesicles in pure water (Käs and Sackmann, 1991).

7.4.3 THE AREA DIFFERENCE ELASTICITY MODEL

In the early 1990s, a next generation of vesicle shape predicting models was developed (Seifert et al., 1992, Wiese et al., 1992, Bozic et al., 1992). This model is called the area–difference–elasticity (ADE) model and represents an interpolation between the spontaneous curvature model (Section 7.4.1) and the bilayer coupling model (Section 7.4.2). ADE accounts for tight coupling between the inner and outer leaflets of the lipid bilayer, while not only considering the spontaneous curvature C_0, but also introducing an effective spontaneous curvature, which considers two previously described contributing phenomena, i.e., local spontaneous curvature and the intrinsic area difference between the outer and inner leaflet (Miao et al., 1994). Thus, ADE model unifies the two classes of previous approaches, the spontaneous curvature model and the bilayer coupling model, and provides one generalized approach. While its phase diagram is more complex than those of the two previous models, no new shapes were found. The ADE model proved to be particularly useful for studies of the budding regime (Seifert et al., 1992, Miao et al., 1994).

7.5 FROM *IN VITRO* TO *IN VIVO*: BIOLOGICAL MEMBRANES

It is quite remarkable that all the shapes that we have discussed up to now *in vitro* can be found inside a single living cell, i.e., *in vivo*. They occur naturally as an integral part of the cellular physiology due to the fact that membraneous compartments are omnipresent inside a cell. Most of the cellular organelles are membrane bound in general; beside the compartmentalization function, they serve also a very specific and specialized purpose for the given organelle. For example, inner membranes of mitochondria are folded in a systematic way to increase their surface area but also to pose a physical barrier allowing the build-up of a proton gradient, which is used by F_0F_1-ATPases as a mechanical driving force for the ATP synthesis (Alberts, 2015). Endoplasmic reticulum (ER) alone exhibits a variety of shapes, from tubules to flat sheets to large sacks called the cisternae (Figure 7.8a). One of the most prominent examples is during cytokinesis (Figure 7.8b), the final step in cell division, which presents sphere–oblate–dumbbell transitions followed by fission, when one cell becomes two. Such shape transitions have long fascinated biologists, physicists, and mathematicians alike, attempting to describe this elegant shape series during cytokinesis, especially since it does resemble the Poincare conjecture, a mathematical theorem about a sphere homeomorphism, quite strikingly

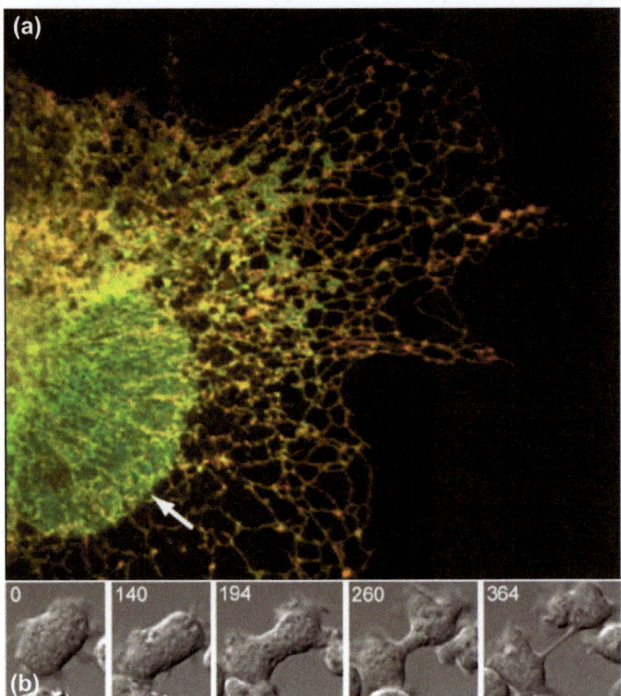

FIGURE 7.8 Examples of biological membranes and their morphologies. (a) Endoplasmic reticulum (ER) in a eukaryotic cell. Fluorescent micrograph shows the low-curvature domains of the nuclear envelope (arrow) and the cisternal sheets (green), as well as the high-curvature domains of the peripheral tubules (red). (Reprinted in part from Shibata et al., 2006 with permission. Copyright 2006, Elsevier Inc.) (b) A *Dictyostelium discoideum* cell undergoing cytokinesis. Time passed indicated in seconds. First, an oblate–prolate–dumbbell transition is observed followed by cell fission of the mother cell into two daughter cells. (Reprinted in part from Reichl et al., 2008 with permission. Copyright 2008, Elsevier Ltd.)

(Mackenzie, 2006). However, most of the membrane shapes in a cell are nonequilibrium shapes, observed only as metastable or transient states *in vitro*. How does nature manage to make use of these nonequilibrium shapes then and moreover stabilize them over long periods of time? The secret ingredients in all of these situations are proteins, which through a plethora of resourceful mechanisms can form, stabilize, and reshape the lipid membranes.

7.5.1 MEMBRANE CURVATURE GENERATION AND STABILIZATION *IN VIVO*

Figure 7.9 shows an overview of the processes leading to membrane deformation. In Section 7.3, we described how membrane curvature arises by changing the lipid composition appropriately (Figure 7.9a). As discussed in great detail in McMahon and Gallop (2005), there are five pathways for introduction of curvature into a lipid membrane, of which four pathways involve proteins.

Influence of Integral Membrane Proteins. Integral membrane proteins of conical shape possess their own intrinsic curvature; thus upon insertion into the membrane, they

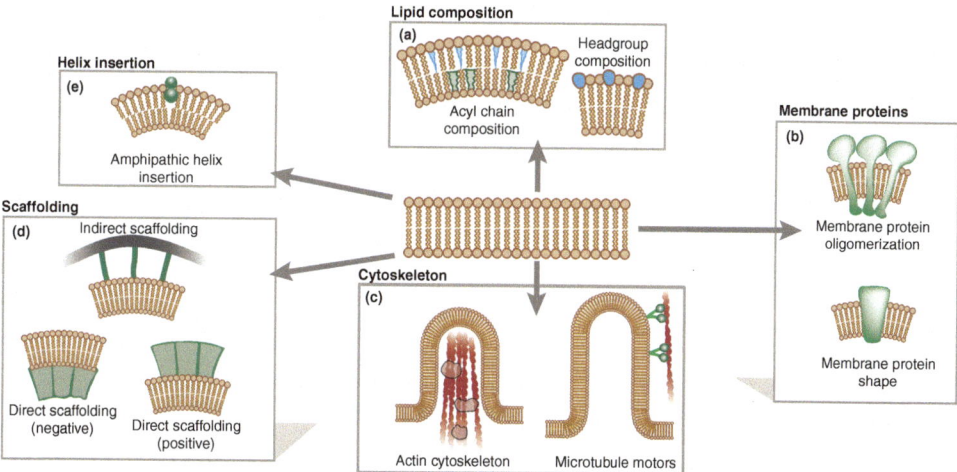

FIGURE 7.9 Mechanisms of membrane deformation. The phospholipid bilayer can be deformed causing positive or negative membrane curvature. There are five main categories: (a) changes in lipid composition, (b) influence of integral membrane proteins that have intrinsic curvature or have curvature upon oligomerization, (c) changes in cytoskeletal polymerization and pulling of tubules by motor proteins, (d) direct and indirect scaffolding of the bilayer, and (e) active amphipathic helix insertion into one leaflet of the bilayer. (Adapted from McMahon and Gallop (2005) with permission. Copyright 2005, Nature Publishing Group.)

locally introduce curvature (Figure 7.9b). Examples of such proteins are reticulins, acetylcholine receptors, and K^+ channels. Some proteins can in addition oligomerize and thus achieve an even larger effect of induced curvature.

Cytoskeletal and Molecular Motor Activity. Polymerization of the cytoskeleton can also cause strong deformation of the membrane (Figure 7.9c). Actin cytoskeleton is known to reorganize easily to facilitate processes such as cell migration, change in the cellular shape, or cell division. Membrane protrusions such as filopodia or lamellipodia form during the phagocytosis or cell migration, respectively. Filopodia are long membrane protrusions filled with parallel actin bundles, and their elongation is attributed to the growth of actin bundles (Mogilner and Rubinstein, 2005, Lan and Papoian, 2008, Zidovska et al., 2011b). In addition, molecular motors such as kinesins are known to pull on membranes of ER or Golgi apparatus and are thought to be able to generate tubules.

Scaffolding of the Membrane. Another means to bending membrane and stabilizing its generated curvature is to build a scaffold on or underneath the membrane which serves as a brace and holds the membrane in the appropriate shape (Figure 7.9d). Prominent representatives from this category are dynamins, which can provide a helical scaffold leading to a formation of membrane tubules. Other examples involve clathrin, generating the clathrin-coated pits, caveolin, facilitating the shape of caveolae, and BAR domains, banana-shaped proteins bending the membrane upon their oligomerization. Such scaffolding can cause both positive and negative curvature (Figure 7.9d).

Insertion of Amphiphatic Helices. Insertion of an amphipathic helix into the lipid bilayer acts as inserting a wedge into it, thus immediately generating a positive curvature (Figure 7.9e). A prominent example of this type of membrane shape regulating protein is epsin. Interestingly, a number of scaffolding proteins, e.g., some of the BAR domains, contain epsin or epsin-like amphipathic helix, which they insert into membrane, therefore generating membrane curvature, while the BAR domain itself then serves for the curvature stabilization.

For all membrane shape regulating proteins, the interaction with the lipid molecules within the membranes is of critical importance. Specialized lipids such as phosphatidylinositol 4,5-bisphosphate (PIP2) or phosphatidylinositol 3,4,5-trisphosphate (PIP3) are required for many proteins to interact with membranes and thus play a key role in number of cellular processes, e.g., organization and polymerization of the cytoskeletal filaments near the membrane, endocytosis and exocytosis, activity of some ion channels, and cell signalling (Alberts, 2015).

Membrane curvature generation and stabilization processes are present across the cell: from organelle shaping, protein sorting, membrane scission, membrane fusion, or enzyme activation. Most of the organelles have very specific shapes to accommodate the function and/or the machinery that they house, e.g., mitochondria and ER. Receptor and channel proteins are often sorted by either affinity to or exclusion from areas of high curvature. Dynamins can lead to membrane scission, while SNARE proteins promote membrane fusion in cells. Enzymes, such as synaptojanins, which play an important role in the recovery of the synaptic vesicles at the neuronal synapse, become activated upon sensing higher curvature (smaller vesicle), while they become deactivated when the curvature decreases (large vesicle). These are only a few examples of the cellular processes where membrane curvature is of critical importance.

In the context of the continuum models discussed in 7.4, for live cell shape prediction, the contribution of all these factors above needs to be considered. In particular, the role of the cytoskeleton in the cellular shape determination has been discussed since early days of the membrane continuum models (Zarda et al., 1977, Evans and Skalak, 1980). While initially cytoskeleton was thought to cause a perturbative modification of basic shapes, later continuum-mechanics treatments were developed for the composite plasma membrane consisting of lipid bilayer supported by cytoskeletal filaments (Mukhopadhyay et al., 2002). Specifically, the shape of red blood cells has received much attention over decades, with many morphological details successfully captured by these models (Canham, 1970, Peterson, 1992, Wong, 1994, Mukhopadhyay et al., 2002).

7.6 FUTURE DIRECTIONS

Liposomes have played an important role in both basic and applied science in the past 40 years. The basic research has shown their enormous potential in variety of applications thanks to their ability to encapsulate other molecules. One of the most promising directions exploiting the liposomes remains the gene and drug delivery. Specifically, with respect to the focus of this chapter, the membrane shape, and its evolution, the knowledge and control of the liposome shape and structure could have far-reaching impact on drug storage, delivery, and its controlled release. It is conceivable that such goal might be soon reached, considering recent advances in, e.g., gene therapy, where controlled release of nucleic acids was achieved using lipid digestion (Shirazi et al., 2012). The advantages of such nonviral delivery vehicles for drugs and genes are numerous, the main advantage being the lack of the immune response of the human body to the nonviral lipid-based delivery vector.

Another direction worthwhile of future explorations is the use of a liposome as an artificial cell. Such efforts started about 20 years ago, when first vesicles were filled with actin filaments to simulate the actin cytoskeleton under the cell membrane (Limozin and Sackmann, 2002). Such studies were further developed to mimic very particular cellular situations such as filopodia (Pronk et al., 2008), which are cell membrane protrusions filled with parallel actin bundles, formed by, e.g., macrophages as a part of the immune response.

Highly decorated membranes were used for studies of glycocalyx and the cell–cell interactions (Tanaka and Sackmann, 2005). The use of liposomes as an artificial cell to mimic the physiological process or cellular property of choice is far from explored. In fact, liposomes can serve as a model system not only for studies of the cell membrane, but also for those of specific organelles (Liu and Fletcher, 2009). For example, the peculiar shapes and shape transformations of the ER or the ones occurring during cytokinesis were explored in such model systems *in vitro* (Shibata et al., 2009). In combination with proteins, lipid vesicles provide a treasure chest for a modern biophysicist.

One of the current frontiers in the soft condensed matter physics is the understanding of the active and smart matter. While the shape and shape evolution of the membranes in the thermal equilibrium are relatively well understood, both experiments and theories are needed to understand their complex nonequilibrium behaviour. Experimental observations of the membrane out-of-equilibrium shape transitions could guide new theories and simulations of such transitions, providing further insight into these complex phenomena. Biological membranes *in vivo* exhibit a plethora of such situations inside a living cell. While understanding of such systems is without a doubt critical for our knowledge about cell physiology in health and disease, they also provide an endless inspiration for man-made reconstituted systems *in vitro*, which are mimicking their active nonequilibrium behaviour.

ACKNOWLEDGEMENTS

A.Z. would like to thank Erich Sackmann, Leif Ristroph, and Janka Zidovska for critical reading of the manuscript, Fang-Yi Chu and Christina Caragine for technical assistance with the manuscript, and the Center for Soft Matter Research at New York University for their support. A.Z. is grateful for the support from the National Institutes of Health Grant R00-GM104152, National Science Foundation CAREER Grant PHY-1554880, and New York University Whitehead Fellowship for Junior Faculty in Biomedical and Biological Sciences.

BIBLIOGRAPHY

B Alberts. *Molecular Biology of the Cell*. Garland Science, 2015.

AD Bangham, MW Hill, and NGA Miller. *Preparation and Use of Liposomes as Models of Biological Membranes*. Springer, 1974.

B Bozic, S Svetina, B Zeks, and RE Waugh. Role of lamellar membrane structure in tether formation from bilayer vesicles. Biophysical journal, 61(4): 963, 1992.

PB Canham. The minimum energy of bending as a possible explanation of the biconcave shape of the human red blood cell. *Journal of theoretical biology*, 26(1):61IN777–76IN881, 1970.

S Chiruvolu, HE Warriner, E Naranjo, SH Idziak, JO Radler, RJ Plano, JA Zasadzinski, and CR Safinya. A phase of liposomes with entangled tubular vesicles. *Science*, 266(5188): 1222–1225, 1994.

HP de Duwe and E Sackmann. Bending elasticity and thermal excitations of lipid bilayer vesicles: modulation by solutes. *Physica A: Statistical Mechanics and its Applications*, 163(1): 410–428, 1990.

H Engelhardt, HP Duwe, and E Sackmann. Bilayer bending elasticity measured by Fourier analysis of thermally excited surface undulations of flaccid vesicles. *Journal de Physique Lettres*, 46(8): 395–400, 1985.

EA Evans. Bending resistance and chemically induced moments in membrane bilayers. *Biophysical Journal*, 14(12): 923–931, 1974.

EA Evans and R Skalak. *Mechanics and Thermodynamics of Biomembranes*. CRC Press, 1980.

SL Ginn, IE Alexander, ML Edelstein, MR Abedi, and J Wixon. Gene therapy clinical trials worldwide to 2012–an update. *The Journal of Gene Medicine*, 15(2): 65–77, 2013.

AA Gurtovenko, M Patra, M Karttunen, and I Vattulainen. Cationic DMPC/DMTAP lipid bilayers: molecular dynamics study. *Biophysical Journal*, 86(6): 3461–3472, 2004.

W Helfrich. Elastic properties of lipid bilayers: theory and possible experiments. *Zeitschrift für Naturforschung C*, 28(11–12):693–703, 1973.

L Huang, M-C Hung, and E Wagner. *Nonviral vectors for gene therapy*, volume 54. Academic Press, 2005.

JN Israelachvili. *Intermolecular and Surface Forces: Revised Third Edition*. Academic Press, 2011.

J Käs and E Sackmann. Shape transitions and shape stability of giant phospholipid vesicles in pure water induced by area-to-volume changes. *Biophysical Journal*, 60(4): 825, 1991.

Y Lan and GA Papoian. The stochastic dynamics of filopodial growth. *Biophysical Journal*, 94(10): 3839–3852, 2008.

DD Lasic. *Liposomes: From Physics to Applications*. Elsevier Science Ltd, 1993.

L Limozin and E Sackmann. Polymorphism of cross-linked actin networks in giant vesicles. *Physical Review Letters*, 89(16): 168103, 2002.

R Lipowsky. The conformation of membranes. *Nature*, 349(6309): 475–481, 1991.

R Lipowsky. The morphology of lipid membranes. *Current Opinion in Structural Biology*, 5(4): 531–540, 1995.

R Lipowsky and E Sackmann. *Structure and Dynamics of Membranes: I. From Cells to Vesicles/II. Generic and Specific Interactions*, volume 1. Elsevier, 1995.

AP Liu and DA Fletcher. Biology under construction: in vitro reconstitution of cellular function. *Nature Reviews Molecular Cell Biology*, 10(9): 644–650, 2009.

D Mackenzie. The Poincaré conjecture–proved. *Science*, 314(5807): 1848–1849, 2006.

HT McMahon and JL Gallop. Membrane curvature and mechanisms of dynamic cell membrane remodelling. *Nature*, 438(7068): 590–596, 2005.

L Miao, U Seifert, M Wortis, and H-G Döbereiner. Budding transitions of fluid-bilayer vesicles: the effect of area-difference elasticity. *Physical Review E*, 49(6): 5389, 1994.

X Michalet and D Bensimon. Observation of stable shapes and conformal diffusion in genus 2 vesicles. *Science*, 269(5224): 666–668, 1995.

X Michalet and D Bensimon. Vesicles of toroidal topology: observed morphology and shape transformations. *Journal de Physique II*, 5(2): 263–287, 1995.

A Mogilner and B Rubinstein. The physics of filopodial protrusion. *Biophysical Journal*, 89(2): 782–795, 2005.

R Mukhopadhyay, HWG Lim, and M Wortis. Echinocyte shapes: bending, stretching, and shear determine spicule shape and spacing. *Biophysical Journal*, 82(4): 1756–1772, 2002.

DB Murphy and MW Davidson. *Fundamentals of Light Microscopy and Electronic Imaging, Second Edition*. Wiley Online Library, 2012.

MA Peterson. Linear response of the human erythrocyte to mechanical stress. *Physical Review A*, 45(6): 4116, 1992.

S Pronk, PL Geissler, and DA Fletcher. Limits of filopodium stability. *Physical Review Letters*, 100(25): 258102, 2008.

EM Reichl, Y Ren, MK Morphew, M Delannoy, JC Effler, KD Girard, S Divi, PA Iglesias, SC Kuo, and DN Robinson. Interactions between myosin and actin crosslinkers control cytokinesis contractility dynamics and mechanics. *Current Biology*, 18(7): 471–480, 2008.

E Sackmann. Membrane bending energy concept of vesicle-and cell-shapes and shape-transitions. *FEBS Letters*, 346(1): 3–16, 1994.

Erich Sackmann and R Merkel. *Lehrbuch der Biophysik*. Wiley-VCH, 2010.

SA Safran. *Statistical Thermodynamics of Surfaces, Interfaces, and Membranes*. Westview Press, 1994.

SA Safran, P Pincus, and D Andelman. Theory of spontaneous vesicle formation in surfactant mixtures. *Science*, 248(4953): 354–356, 1990.

U Seifert. Configurations of fluid membranes and vesicles. *Advances in Physics*, 46(1): 13–137, 1997.

U Seifert, L Miao, H-G Döbereiner, and Michael Wortis. Budding transition for bilayer fluid vesicles with area-difference elasticity. In *The Structure and Conformation of Amphiphilic Membranes*, pp.93–96. Springer, 1992.

Y Shibata, J Hu, MM Kozlov, and TA Rapoport. Mechanisms shaping the membranes of cellular organelles. *Annual Review of Cell and Developmental*, 25:329–354, 2009.

Y Shibata, GK Voeltz, and TA Rapoport. Rough sheets and smooth tubules. *Cell*, 126(3): 435–439, 2006.

RS Shirazi, KK Ewert, C Leal, and CR Safinya. Environmentally responsive cationic liposome-DNA complexes for cell delivery. *Biophysical Journal*, 102(3):638a, 2012.

S Svetina and B Žekš. Membrane bending energy and shape determination of phospholipid vesicles and red blood cells. *European Biophysics Journal*, 17(2): 101–111, 1989.

M Tanaka and E Sackmann. Polymer-supported membranes as models of the cell surface. *Nature*, 437(7059): 656–663, 2005.

I Tsafrir, Y Caspi, M-A Guedeau-Boudeville, T Arzi, and J Stavans. Budding and tubulation in highly oblate vesicles by anchored amphiphilic molecules. *Physical Review Letters*, 91(13): 138102, 2003.

I Tsafrir, D Sagi, T Arzi, M-A Guedeau-Boudeville, V Frette, D Kandel, and J Stavans. Pearling instabilities of membrane tubes with anchored polymers. *Physical Review Letters*, 86(6): 1138, 2001.

W Wiese, W Harbich, and W Helfrich. Budding of lipid bilayer vesicles and flat membranes. *Journal of Physics: Condensed Matter*, 4(7): 1647, 1992.

W Wintz, H-G Döbereiner, and U Seifert. Starfish vesicles. *EPL (Europhysics Letters)*, 33(5): 403, 1996.

P Wong. Mechanism of control of erythrocyte shape: a possible relationship to band 3. *Journal of Theoretical Biology*, 171(2): 197–205, 1994.

P Yager and PE Schoen. Formation of tubules by a polymerizable surfactant. *Molecular Crystals and Liquid Crystals*, 106(3–4):371–381, 1984.

H Yin, RL Kanasty, AA Eltoukhy, AJ Vegas, JR Dorkin, and DG Anderson. Non-viral vectors for gene-based therapy. *Nature Reviews Genetics*, 15(8): 541–555, 2014.

PR Zarda, S Chien, and R Skalak. Elastic deformations of red blood cells. *Journal of Biomechanics*, 10(4): 211–221, 1977.

A Zidovska, HM Evans, A Ahmad, KK Ewert, and CR Safinya. The role of cholesterol and structurally related molecules in enhancing transfection of cationic liposome- DNA complexes. *The Journal of Physical Chemistry B*, 113(15): 5208–5216, 2009a.

A Zidovska, KK Ewert, J Quispe, B Carragher, CS Potter, and CR Safinya. Block liposomes from curvature-stabilizing lipids: connected nanotubes, -rods, or -spheres. *Langmuir*, 25(5): 2979–2985, 2008.

A Zidovska, KK Ewert, J Quispe, B Carragher, CS Potter, and CR Safinya. The effect of salt and pH on block liposomes studied by cryogenic transmission electron microscopy. *Biochimica et Biophysica Acta (BBA)-Biomembranes*, 1788(9): 1869–1876, 2009b.

A Zidovska, KK Ewert, J Quispe, B Carragher, CS Potter, and CR Safinya. Block liposome and nanotube formation is a general phenomenon of two-component membranes containing multivalent lipids. *Soft Matter*, 7(18): 8363–8369, 2011.

A Zidovska and E Sackmann. On the Mechanical Stabilization of Filopodia. *Biophysical Journal*, 100(6):1428-1437, 2011b.

8 Mechanisms of Membrane Curvature Generation by Peptides and Proteins: A Unified Perspective on Antimicrobial Peptides

Michelle W. Lee, Nathan W. Schmidt, and Gerard C. L. Wong
University of California Los Angeles, Los Angeles, California

CONTENTS

8.1 Introduction 141
8.2 Definition of Terms 142
8.3 Survey of Basic Membrane Curvature Generation Mechanisms 143
    8.3.1 Membrane Partitioning and Insertion 143
    8.3.2 Membrane Scaffolding 145
    8.3.3 Curvature Sensing, Curvature-Mediated Attraction 147
    8.3.4 Molecular Crowding 147
    8.3.5 Membrane Wrapping 148
    8.3.6 Composite Mechanism 148
8.4 A Case Study on Antimicrobial Peptides 148
    8.4.1 Background and History 148
    8.4.2 Induced Curvature and the Phase Behaviour of Antimicrobial Peptide–Lipid Systems 150
    8.4.3 Why Membrane-Active Antimicrobial Peptides Exhibit Selective Action against Bacterial Membranes 151
    8.4.4 Composite Mechanisms of Curvature Generation in Antimicrobial Peptides 152
    8.4.5 Amino Acid Requirements for Peptides That Permeate Membranes 153
References 154

8.1 INTRODUCTION

Adsorption of peptides or proteins onto a membrane can alter the membrane curvature and remodel the membrane. This phenomenon has been observed in a broad range of biological processes. Examples include the action of antimicrobial peptides (AMPs) on bacterial membranes, of cell-penetrating peptides (CPPs) in drug delivery, of viral budding peptides on the budding and release of enveloped viruses, of dynamin-related proteins in tubulation and fission, of Bcl2 proteins in the mitochondrial apoptosis pathway, of coat proteins such as clathrin, caveolin, and the endosomal sorting complex required for transport (ESCRT) machinery in active cellular transport. Here, we review the membrane curvature generation in these processes. The goal is not only to appreciate the diversity of outcomes but also to begin to organize our knowledge of simple, constitutive curvature generation mechanisms into a unified taxonomic framework that can engage the induction of complex curvatures via composite mechanisms.

We start with a definition of terms, in the form of a review of the conceptual language that we will use. Next, we survey the "simple" peptide- or protein-induced membrane curvature generation mechanisms that are currently known in various fields of biology. These will form the vocabulary and grammar with which we will use to help parse through new problems in induced membrane curvature. Finally, we end with a case study on AMPs, which shows how different peptide-induced effects combine to form complex curvatures. Central to our approach is the idea that bulk phase diagrams can be immensely informative in identifying structural tendencies in peptide–membrane interactions and resultant induced curvature. We illustrate the approach by showing that we can identify some basic sequence principles for peptides that permeate membranes based on induced curvature considerations and apply these principles to impart membrane permeability to non–membrane-active molecules.

8.2 DEFINITION OF TERMS

Biological membranes can be thought of as a fluid bilayer consisting of two leaflets. Membrane shape and deformations can be described using geometric concepts. There are many good reviews and books on this venerable topic. Here, we include a cursory treatment, with the purpose of acquainting the reader with the basic relevant concepts useful in thinking about peptide- or protein-induced membrane curvature, rather than a unified and rigorous treatment. On a curved 2-D membrane surface that exists in 3-D space, the curvature at any point on the surface can be defined by a tangent plane at that point. Planes that are normal to this tangent plane intersect the surface as a normal section, and each normal section is associated with a curvature at the tangent point, defined as $c = 1/R$, with R being the radius of curvature. Thus, one can see that there are in principle many different possible normal sections and therefore many different curvatures that can be defined through the point in question. It turns out that the extrema of these diverse curvatures, the maximum and minimum curvatures that go through the point, correspond to normal sections that are orthogonal to one another. These directions are referred to as the two *principal directions*, and their corresponding curvatures are called the *principal curvatures*. The two principal curvatures can therefore be defined as $c_1 = 1/R_{max}$ and $c_2 = 1/R_{min}$ (with R_{max} and R_{min} being the principal radii of curvature) (Figure 8.1). Once the principal directions and their corresponding curvatures are established, every directional curvature at that point can be calculated using a formula derived by Euler (Kreyszig 1991):

$$c(\alpha) = c_1 \cos^2 \alpha + c_2 \sin^2 \alpha$$

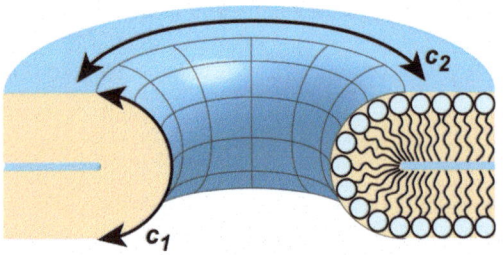

FIGURE 8.1 The membrane of a pore features a curved surface with a saddle shape, which is characterized by principal curvatures c_1 and c_2 (directions indicated by arrows) that are opposite in sign and together create negative Gaussian curvature (NGC).

where α is the angle between the selected direction and the principal direction of curvature c_1. In combination, the principal curvatures can be used to describe the shape of a surface with two general expressions:

$$H = \frac{1}{2}(c_1 + c_2)$$

$$K = c_1 c_2$$

Their arithmetic mean defines the mean curvature, H, and their product is the Gaussian curvature, K.

It is important to note that curvature can be positive or negative. By convention, a membrane monolayer that bends to form a convex hydrophilic surface exhibits positive curvature. Conversely, a monolayer that bends in the opposite direction to form a concave hydrophilic surface results in negative curvature. When referring to cell membranes, the distinction between the sides of the membrane depends on the intracellular and extracellular space. Positive curvature is generally defined as the cell membrane bulging convexly towards the extracellular medium.

The lipid molecules that comprise a bilayer each have a specific "shape" and can thereby contribute to the membrane curvature. (Here, it is worthwhile to point out that "molecular shape" can be determined by different effects. We often think in terms of steric shapes, but patterns of hydrogen bonding can also be important.) In a pioneering paper, Israelachvili et al. have described how the properties of a specific lipid molecule can influence packing and the spontaneous morphology of aggregated lipids (Israelachvili et al. 1980). This model introduces an average geometric shape for the lipid molecule that is described using a dimensionless packing parameter, S, which depends on the repulsive steric and electrostatic interactions between the polar lipid head groups as well as the attractive hydrophobic interactions and repulsive steric forces experienced by the lipid tails:

$$S = \frac{V}{A_0 L_c}$$

where V is the molecular volume of the lipid tails, A_0 is the optimal area occupied by the lipid head group, and L_c is the length of the lipid tails. Lipids tend to form aggregates of certain geometries based on their packing parameters. Generally, $S < \frac{1}{3}$ forms spherical aggregates, $\frac{1}{3} < S < \frac{1}{2}$ forms cylindrical aggregates, $\frac{1}{2} < S < 1$ forms planar aggregates, and $S > 1$ forms inverted aggregates. Correspondingly, the packing behaviour of lipid molecules dictates their intrinsic curvature, which is a reflection of their molecular shape. For instance, a lipid with $S < \frac{1}{2}$ tends to be cone- or wedge-shaped and takes on positive intrinsic curvature, whereas a lipid with $S > 1$ is shaped like an inverted cone and tends to take on negative intrinsic curvature. In the context of this review, it can already be noted that insertion of objects of different shapes into the membrane is one method of inducing different curvatures.

Let us say we want to use peptide–membrane interactions to induce curvature. Shaping or deforming a membrane away from its unperturbed native state is associated with an energetic cost determined by the structure and elasticity of the membrane. The energetics of membrane shape changes have been generally described using the approach by Helfrich (1973). In this formalism, deformation of the membrane would have an energetic cost that is only dependent on curvature changes of the surface, which are defined by the mean and Gaussian curvatures. The curvature elastic energy per unit area of bending a membrane is

$$f = 2\kappa(H - c_0)^2 + \kappa_G K$$

where c_0 is the spontaneous mean curvature, also referred to as the intrinsic curvature. The energetic cost per unit area is therefore given by the bending modulus, κ, and the Gaussian modulus, κ_G, which are elastic constants that describe the stiffness of the membrane and the resistance of the membrane to topological transitions, respectively. To arrive at the total elastic energy of a symmetric membrane, in which case $c_0 = 0$, the Helfrich curvature elastic energy density is integrated over the surface of the membrane, with dA being the area element on the membrane:

$$F = \int dA \, (2\kappa H^2 + \kappa_G K)$$

The first term accounts for the energetic cost of bending the membrane in a manner reminiscent of Hooke's law, while the second term accounts for the energetic cost of distortions related to topological complexity. To see this qualitatively, one can use the Gauss–Bonnet theorem, which states that the integral of the Gaussian curvature over a closed surface is a topological constant:

$$\int K \, dA = 4\pi(1 - g)$$

where the integer g is the genus of the surface, a topological property characterizing the connectivity of a surface and often described as the number of "holes" or "handles". For instance, a sphere has $g=0$, whereas a torus, which can be viewed as a sphere with one handle, has $g=1$ (Siegel and Kozlov 2004). For each additional handle, the genus increases by one. One can see that the more negative the total Gaussian curvature is on the surface, the "holier" the surface is. In contrast, closed surfaces with no holes have a total Gaussian curvature on the surface that is positive.

While the Helfrich formalism has been used extensively, it is not the only way to parameterize a curvature energy functional, since the Gauss–Bonnet theorem allows some latitude (Deserno 2009). Moreover, it is also important to observe that Helfrich theory is based on the initial assumption of small membrane curvatures. A curvature is typically considered small if its corresponding radius of curvature is significantly larger than the membrane thickness, which is approximately 4 nm. Hence, the applicability of the elastic theory approaches its limits as the radius of curvature becomes of a similar magnitude to the membrane thickness (Shearman et al. 2006, Zimmerberg and Kozlov 2006). This incompatibility is exemplified when considering a saddle-like membrane shape that is found in inverse bicontinuous cubic phases, in which the infinitely thin membrane traces the minimal surface. Because in this case, the mean curvature and the intrinsic curvature are both zero, the Helfrich elastic energy per unit area is then simplified to be the product of the Gaussian modulus and the Gaussian curvature, which is positive for minimal surfaces (Siegel and Kozlov 2004). As Gaussian curvature depends on lattice dimensions, the curvature elastic energy can be presumably reduced by decreasing the lattice size. Accordingly, minimization of the curvature elastic energy would thus minimize the lattice parameter to an arbitrarily small value, which in reality cannot occur (Shearman et al. 2006). Such nonphysical behaviour can be corrected by including higher-order terms in the expression for curvature elastic energy, which predict cubic phase stability (Ljunggren and Eriksson 1992, Seddon and Templer 1993). The Helfrich approach is almost always used as a starting point in the context of curvature induced by protein–membrane interactions.

Now that we have a biophysical description of membranes, we can examine what happens when peptides or proteins interact with membranes. We begin with a survey of membrane curvature generating mechanisms for different systems.

8.3 SURVEY OF BASIC MEMBRANE CURVATURE GENERATION MECHANISMS

8.3.1 Membrane Partitioning and Insertion

Many peptides that insert into membranes are amphiphilic. A natural question to ask is how hydrophobic interactions contribute to the self-assembly between the peptide and the membrane. Here, the corpus of work on AMPs provides useful examples.

AMPs are a group of peptides that exhibit diverse secondary structures but share two common motifs: cationic charge and amphiphilicity (Zasloff 2002). A large class of AMPs adopts amphipathic α-helical structures at the membrane interface, in which one face is polar (charged) and the other is nonpolar (hydrophobic). The strong electrostatic attraction between an anionic membrane and a cationic peptide is driven by the entropic gain of counterion release (discussed below), resulting in electrostatic binding and membrane deformation, with some trade-off between the two. However, many AMPs are unstructured in solution, before binding to membranes. The binding or embedding of an amphiphilic peptide into a membrane can involve a competition between mutually antagonistic events. For example, insertion of such a peptide into a membrane can create favourable hydrophobic interactions between nonpolar amino acids and the hydrocarbon lipid tails of the membrane, but at the cost of partitioning polar amino acids and peptide bonds into the membrane (Shai 1999). By forming an amphipathic α-helical structure, which facilitates electrostatic interactions between the polar groups of the peptide and the membrane and maintains hydrogen bonding along the peptide, the energetic cost of inserting the peptide into the membrane is reduced (White and Wimley 1999). Thus, an AMP can adsorb onto a membrane surface with its long axis parallel to the membrane surface and adopt a stable amphipathic helix with the hydrophobic and charged residues segregated on opposite faces. In this arrangement, the amphipathic AMP helix is oriented with the hydrophobic domains of the peptide contacting the hydrocarbon chains in the interior of the membrane, while the exposed cationic residues are able to interact with anionic lipid head groups of the membrane. In doing so, the helix acts like a rod-like inclusion embedded in one leaflet of the membrane, promoting anisotropic disruption of lipid packing and generating anisotropic positive curvature as a result of compressive effects from adding hydrophobic volume to one monolayer. As a result, the maximal disruption of lipid packing is experienced in the direction along the axis of the helix.

Previous work has suggested that a deeper penetration of amphipathic helical peptides into a bilayer increases the perturbation of membrane-stabilizing hydrophobic lipid interactions. This membrane-disruptive process depends on the hydrophobic interactions, and thus, the hydrophobic content of the peptide. Indeed, studies have found that the reduction of membrane activity of cationic amphipathic α-helices is correlated with decreased hydrophobicity (Bechinger 2009, Dathe et al. 1996), although the actual impact on activity from curvature generation is likely complex and multifactorial. In a similar manner, the geometric properties of the hydrophobic and polar faces also influence the curvature induced by an amphipathic helical peptide. As we have described, AMPs can create positive curvature through membrane insertion; however, the generation of negative curvature is also possible. These differences in curvature effects have been attributed to a phenomenological "wedge shape" specific to an individual peptide. For example, Tytler et al. examined the membrane effects of two classes of amphipathic helices (identified by Segrest et al. 1990) with contrasting residue distributions, one featuring wide polar faces of charged residues, while the other class having a small cluster of cationic residues to form a narrow polar face (Tytler et al. 1993). These characteristics cause the two classes of helices to form different cross-sectional shapes, with the first described as an upright wedge with a polar base and a hydrophobic apex, and the other viewed as an inverted wedge with a polar apex and a hydrophobic base. It is hypothesized that these wedge-shaped or inverted wedge-shaped amphipathic helices can produce membrane effects analogous to those of wedge-shaped and inverted wedge-shaped lipids (Campelo et al. 2008, Israelachvili et al. 1980, Zimmerberg and Gawrisch 2006). Indeed, wedge-shaped helices were found to induce positive curvature, which stabilized micellar and bilayer structures in both model and biological membranes (Epand et al. 1995, Segrest 1990, Shai 1999, Zemel et al. 2008). In contrast, inverted wedge-shaped helices induced negative curvature, which destabilized bilayers and promoted inverted lipid structures in model membranes (Segrest 1990, Shai 1999, Tytler et al. 1993, Zemel et al. 2008). These results, when taken together with the body of work on electrostatics, suggest a unified way of qualitatively understanding membrane curvature generation by facially amphiphilic cationic peptides, one in which the roles of cationic charge and hydrophobicity combine synergistically to create something new. We note that most peptide insertion scenarios will involve a fair amount of membrane thinning, due to the disturbance to lipid packing by the embedded peptide (Chen et al. 2003, Herce and Garcia 2007, Tristram-Nagle et al. 2010). Such thinning has been experimentally observed in a number of systems and can cause membranes to become more amenable to curvature deformations. (For example, it is known that the bending modulus of a membrane varies roughly as the third power of the membrane thickness (Safinya et al. 1989, Szleifer et al. 1988).) There exists a large collection of literature on correlating the relative sizes of the polar and hydrophobic faces of amphipathic helices with downstream cellular outcomes such as lysis (Drin and Antonny 2010, Epand et al. 1995, Segrest 1990, Tytler et al. 1993, Wieprecht et al. 1997, Zemel et al. 2008), which itself depends on a broad range of effects such as the form of lysis and the efficiency of peptide–membrane binding. In the context of curvature generation, we can look at it in the following way: For peptides with a sufficiently small circumferential sector of cationic residues and large circumferential sector of hydrophobicity (Figure 8.2), the

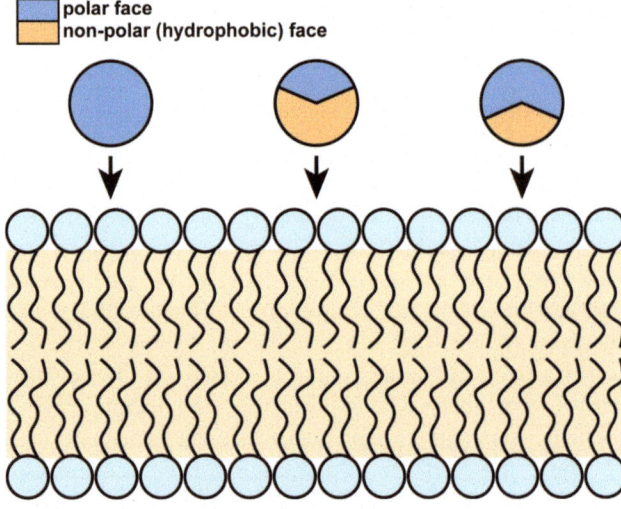

FIGURE 8.2 Studies on amphipathic α-helical peptides have correlated their membrane effects with structural properties, such as the widths of polar and hydrophobic sectors.

excluded volume interactions from hydrophobic insertion will dominate, and anisotropic positive curvature will result, with a larger hydrophobic volume perturbation along the helical axis than perpendicular to it. Since this interaction is based mostly on hydrophobic insertion, we expect these peptides to generate strong positive curvature for a broad range of membranes. However, for peptides with a sufficiently large circumferential sector of cationic residues and small circumferential sector of hydrophobicity, electrostatic wrapping of the membrane to optimize contact between cationic and anionic surfaces will be more important, and the resultant induction of negative curvature will dominate over the positive curvature generated from the small amount of hydrophobic insertion. When this induced negative curvature perpendicular to the helix is combined with the induced positive curvature along the helical axis, the result is negative Gaussian curvature (NGC), which is the type of curvature topologically necessary for pore formation, and other membrane destabilization mechanisms such as blebbing and budding.

Transmembrane (TM) proteins play important roles as channels, receptors, and enzymes in cells. Interactions between these integral proteins and their membrane environment can influence their conformations and distributions, and consequently, their functions and activities (Botelho et al. 2006, Cornea and Thomas 1994, de Planque et al. 2001, Johannsson et al. 1981, Kik et al. 2010, Lundbæk et al. 1997, Michelangeli et al. 1991, Montecucco et al. 1982). It is energetically favourable for the hydrophobic domain length of a membrane protein to match the thickness of the hydrophobic interior of a lipid bilayer due to the high energetic cost of exposing hydrophobic surfaces to an aqueous environment. Hydrophobic mismatches can occur, as proteins can have different sizes of hydrophobic domains and bilayers can vary in thickness. Hydrophobic mismatch is one mechanism by which TM proteins can deform their surrounding membranes. If the hydrophobic length of the protein is greater than the thickness of the hydrophobic region of the membrane, it is described as being "positively mismatched", while the opposite case is called "negatively mismatched". In the case of mismatch, the protein–lipid system adapts its structure to minimize contact between polar and hydrophobic regions, which involves modification in protein or membrane structure, or both. With respect to the membrane, lipid chains surrounding the protein adjust their lengths to accommodate the hydrophobic portion of the protein. They can stretch to thicken a bilayer for a positive mismatch, or they can compress to provide a thinner bilayer for a negative mismatch (Killian et al. 1998, Lee 2004). Experiments with gramicidin have demonstrated this modulation of lipid chain length (de Planque et al. 1998, Killian 1998, Watnick et al. 1990). Moreover, the stretching of lipids in the vicinity of a membrane protein in response to positive mismatch will change the effective shape of lipid molecules and thereby generate membrane curvature. Such induced curvatures are experimentally observed in the formation of nonlamellar (inverted hexagonal and cubic) lipid phases (Killian 1998, Lee 2004, Morein et al. 2000). It is interesting to note that the above effects can work synergistically with protein shape, as in the case of membrane-spanning "wedges". An example of these effects can be observed in the influenza M2 protein (Schmidt et al. 2013).

8.3.2 Membrane Scaffolding

Membrane scaffolding by curved proteins or complexes is a mechanism by which proteins can directly force membrane remodelling and curvature. Protein complexes, which can be found covering the surface of buds and invaginations, are able to act as scaffolds and impart their geometric shape on the membrane. This can be observed in the budding and release of enveloped viruses, as well as other fission processes that involve membrane neck formation and scission. To effectively drive membrane curvature, the intrinsic shape of the protein or protein network must expose a curved surface to interact with the lipid bilayer while having sufficient affinity for the polar lipid head groups. Furthermore, the rigidity of the protein coat must be able to counteract both the membrane intrinsic curvature and resistance to mechanical bending, as described by its elastic modulus, to allow for membrane deformation (Farsad and Camilli 2003, Zimmerberg and Kozlov 2006). In general, the protein–membrane interaction energy needs to be greater than the membrane bending energy. As can be seen below, recent work has shown that the scaffolding mechanism often coexists with other curvature-generating mechanisms and cannot be idealized as a purely mechanical effect. We illustrate these concepts using several protein families: dynamin and BAR (Bin, Amphiphysin, Rvs)-domain-containing proteins can wrap around membranes to create scaffolds for cylindrical curvature, while COPI and COPII complexes and clathrin–adaptor-protein complexes can form scaffolds for spherical curvature (Baumgart et al. 2011, McMahon and Gallop 2005, Zimmerberg and Kozlov 2006).

Dynamin, a GTPase, is a primary member of a large family of proteins called the dynamin-related or dynamin-like proteins (Morlot and Roux 2013). It self-assembles into rigid helical oligomers on membrane surfaces to drive bilayer tubulation and fission (Farsad and Camilli 2003, Hinshaw and Schmid 1995, Marks et al. 2001, McMahon and Gallop 2005, Morlot and Roux 2013, Sweitzer and Hinshaw 1998, Takei et al. 1998, 1995, 1999, Zimmerberg and Kozlov 2006). Studies have implicated dynamin in organelle division, necking and scission of clathrin-coated vesicles, and other membrane-trafficking events (Damke et al. 1994, Hinshaw 2000, McNiven et al. 2000). Dynamin oligomerizes to form helical rings around the neck of endocytic buds, which scaffolds the membrane into a cylindrical shape. Upon GTP hydrolysis, dynamin undergoes conformational changes that trigger the constriction of the helical ring, reducing the neck radius to mediate membrane fission (Hinshaw 2000, Kozlov 2010, Marks et al. 2001, McMahon and Gallop 2005, Zhang and Hinshaw 2001). Membrane tension and rigidity, which describe membrane elasticity, control the shape of the neck and its elastic energy. Through this mechanism, it is currently believed that dynamin constricts the membrane down to radii approximately the thickness of a bilayer (Morlot and Roux 2013). Because dynamin works against membrane elasticity in this process, creating such high curvatures requires large forces. While the dynamin helix is rigid enough to constrain a membrane, studies have shown that the rotational force (torque) produced during its conformational change can further drive membrane deformation (Roux et al. 2010). This constriction has been found to be necessary, though not sufficient for complete dynamin-mediated fission (Morlot and Roux 2013). Indeed, complementary to the mechanochemical action of dynamin, recent work has further suggested that the hydrophobic insertion mechanism also plays an important role. This is evidenced by the membrane insertion of the hydrophobic loops of the dynamin PH domains and interactions of dynamin with other proteins, amphiphysin and endophilin, which contain amphipathic helices (Farsad and Camilli 2003, Graham and Kozlov 2010, Kozlov 2010, Ramachandran et al. 2009).

BAR domains are present in a wide variety of proteins, including amphiphysins, arfaptins, endophilins, nadrins, and oligophrenins, and can also be classified into several different types based on their structural characteristics, such as BAR, N-BAR, F-BAR, I-BAR, and PX-BAR (Baumgart et al. 2011). The BAR domain is a crescent-shaped dimeric α-helical bundle, with each monomer having three kinked α-helices to yield a six-helix bundle upon dimerization. The BAR domain binds to membranes through electrostatic interactions at its concave surface, which features a high concentration of positively charged residues, lysines and arginines, allowing for preferential interaction with the negatively charged polar head groups of the membrane lipids (Gallop and McMahon 2005). This dependence on electrostatics has been demonstrated with amphiphysins, in which lysine and arginine residues were mutated to anionic residue glutamate and resulted in reduced binding to membranes and inhibited membrane tubulation (Peter et al. 2004). For this reason, BAR domains are believed to induce membrane curvature along these interaction surfaces. To initiate membrane tubulation, a high surface density of BAR domains is required, suggesting a cooperative effect among individual domains. Moreover, the curvature of membrane tubes covered by BAR domains has been found to be close to that of the concave domain surface, which indicates that the domains have greater rigidity than membrane bilayers. Interestingly, BAR domains are frequently found in combination with N-terminal amphipathic helices (together called N-BAR domains), as observed in amphiphysins, endophilins, BRAP (BRCA1-associated protein), and nadrins. These amphipathic N-terminal helices are believed to stabilize and further promote curvature generation via the hydrophobic insertion mechanism (Boucrot et al. 2012, Gallop and McMahon 2005, Kozlov 2010, McMahon and Gallop 2005).

Coat proteins such as clathrin, COPI, and COPII can also be considered protein complexes that influence membrane curvature through curved scaffolding structures. The polymerization of coat proteins was previously believed to drive curvature formation for membrane vesicle trafficking (Mashl and Bruinsma 1998), but it is now understood that the process involves direct membrane–protein interactions. Studies have revealed that coat proteins work cooperatively with other proteins to bend membranes and form vesicles. For example, this can be observed with clathrin-mediated endocytosis. In the absence of membranes, clathrin complexes with adaptor proteins to self-assemble into cages that take on spheroid polyhedral geometries and have curvatures comparable with the curvatures observed in clathrin-coated vesicles (Fotin et al. 2004, Schmid 1997, Zimmerberg and McLaughlin 2004). This indicates that clathrin complexes feature an intrinsic curvature, which can potentially be applied towards scaffolding. On the contrary, the rigidity of clathrin–adaptor-protein complexes is on the same order of magnitude as that of membranes (Graham and Kozlov 2010, Nossal 2001), which then suggests that polymerization of a clathrin coat on a membrane itself does not lead to effective membrane bending and vesicle budding. Therefore, additional mechanisms alongside a coat protein lattice are likely to be involved in vesicle formation. Indeed, clathrin–adaptor-protein complexes interact with epsin, a protein that contains the ENTH (epsin N-terminal homology) domain. An amphipathic helix, the ENTH domain drives membrane curvature by inserting into the membrane bilayer. Epsin binds to phosphatidylinositol-4,5-bisphosphate on a membrane and subsequently bends the membrane and attracts clathrin–adaptor-protein complexes, which polymerize into a coat that can stabilize the curvature (Zimmerberg and Kozlov 2006).

Like clathrin, coat proteins COPI and COPII complexes are not expected to form a scaffold with sufficient rigidity to effectively bend membranes (Lee et al. 2005, McMahon and Mills 2004). Thus, cooperation between coat proteins and amphipathic helices to help drive membrane curvature may have a similar role in COPI- and COPII-coated vesicles. Analogous to epsin, Arf and Sar1 proteins have N-terminal amphipathic helices and are predicted to function in curvature generation in association with stabilization by COPI and COPII, respectively. For instance, studies have demonstrated the importance of Sar1 membrane insertion in COPII vesicle budding (Lee et al. 2005). Mutations of Sar1 have been shown to uncouple its membrane-deforming and coat-recruitment activities and also impair membrane bending and COPII vesicle budding (Lee et al. 2005). Furthermore, the observed curvature of the COPII coat surface matches that of a vesicle membrane, which suggests a role in facilitating membrane deformation and stabilizing the final curvature of COPII-coated vesicles (Bi et al. 2002).

Additional coat proteins recognized to promote membrane curvature generation include caveolin and ESCRT. Although their precise molecular mechanisms are not fully understood, these proteins are likely to bend membranes using mechanisms different from those discussed above. Caveolin oligomerizes into a coat that produces membrane curvature to form caveoli, flask-shaped membrane invaginations. In contrast with COP and clathrin-coated vesicles, caveolin can directly interact with membranes, and this interaction is thought to mediate membrane bending (Drab et al. 2001, Fernandez et al. 2002, Fra et al. 1995, McMahon and Gallop 2005, Rothberg et al. 1992). While clathrin, COPI, and COPII complexes polymerize on the external side of budding vesicles, ESCRT complexes are located on the internal side of the endosomes and generate opposite curvature. ESCRT is characterized by four complexes that together enable membrane remodelling to form intraluminal vesicles in the endosomal pathway (Bassereau 2010, Baumgart et al. 2011, Farsad and Camilli 2003, Wollert and Hurley 2010, Wollert et al. 2009). There is still no clear mechanism to explain how ESCRT generates curvature, as understanding how vesicle scission can be triggered by protein binding inside the neck still remains a technical challenge (Bassereau 2010, Fabrikant et al. 2009).

Although protein coats are predominantly discussed in relation to vesicle budding and fission, similar interconnected coats of proteins on membranes have also been found to drive fusion (Kozlov and Chernomordik 2002). While many assumptions of the fusion coat hypothesis are still open to question, experimental studies have substantiated its prediction that fusion proteins located far from the fusion site can contribute to the fusion reaction through long-range membrane forces (Leikina et al. 2004, Lenz et al. 2005).

8.3.3 Curvature Sensing, Curvature-Mediated Attraction

Recent work has identified proteins with membrane-binding affinities that are influenced by the curvature of the membrane (Antonny 2011). Proteins that can induce membrane curvature necessarily imply that they can also sense membrane curvature (Baumgart et al. 2011, Zimmerberg and Kozlov 2006). Curvature sensing by proteins is based on the energy of protein–membrane interaction and the energy of membrane deformation caused by the protein. Together, these energetic contributions make up an effective binding energy. A protein with a shape that matches the membrane curvature requires little or no membrane deformation for membrane–protein binding, which results in a minimal effective binding energy. Conversely, increasing mismatch between the protein shape and membrane curvature increases the effective binding energy (Zimmerberg and Kozlov 2006). Flexible proteins that cannot effectively bend a membrane can still sense curvature in an analogous manner. In this case, the protein experiences deformation upon binding to the membrane, and this energy contributes to the effective binding energy. Similarly, the effective binding energy is minimized when the protein shape and membrane curvature match and becomes greater with increasing mismatch.

One well-characterized example of a curvature sensor is the aforementioned BAR domain. In its dimeric form, the membrane-binding region is a concave surface, which imparts its ability to preferentially bind to curved membranes (Gallop and McMahon 2005, McMahon and Gallop 2005). The curvature-sensing ability of the BAR domain has been demonstrated by curvature partitioning and tighter binding to liposomes that have curvatures similar to its intrinsic curvature (Baumgart et al. 2011, Peter et al. 2004). Experiments have shown the curvature-partitioning behaviour for a range of proteins, including dynamin (Bashkirov et al. 2008, Pucadyil and Schmid 2008, Roux et al. 2010), ENTH (Capraro et al. 2010), Arf (Ambroggio et al. 2010), Sar1 (Baumgart et al. 2011), and caveolin (Sens and Turner 2004).

Curvature sensing can also result in collective behaviour between membrane-bound proteins. It has been found that if one protein induces a given membrane curvature, it attracts additional proteins that favour a similar curvature (Koltover et al. 1999, Leibler 1986, Sens et al. 2008). While entropic membrane shape fluctuations can cause long-range attractive interactions between proteins (Goulian et al. 1993), curvature-inducing proteins adsorbed onto membranes also experience attractive short-range interactions that are a consequence of membrane curvature (Reynwar et al. 2007). For the latter, it is believed that the local elastic membrane perturbation that occurs in the vicinity of an adsorbed protein leads to oscillations in the membrane profile, which mediate attractive short-range interactions between proteins (Dan et al. 1993, Koltover et al. 1999). Because curvature-mediated attraction can occur between proteins that do not have any specific interactions, this mechanism may provide a force to facilitate aggregation of proteins (Reynwar et al. 2007). In fact, simulations have shown that once a minimal local bending is achieved, curvature-mediated interactions cause attraction between membrane-adsorbed proteins and drive protein clustering (Reynwar et al. 2007). The formation of protein clusters and aggregation further demonstrate the ability of membrane curvature to induce lateral phase segregation (Koltover et al. 1999).

8.3.4 Molecular Crowding

Protein adsorption on membrane surfaces can result in local crowding of proteins, which has been demonstrated by recent studies to promote the generation of high membrane curvature (Baumgart et al. 2011, Sens et al. 2008, Stachowiak et al. 2010, 2012). This protein–protein crowding mechanism, by which protein molecules concentrate to form dense clusters, creates lateral steric pressure that leads to membrane bending and aids in the formation of lipid buds and tubules (Farsad and Camilli 2003, Stachowiak et al. 2010, 2012). Experiments on proteins involved in tubulation and proteins that are not associated with membrane-bending events have revealed a strong correlation between the frequency of membrane tubulation and the percentage of membrane surface covered by proteins (Romer et al. 2007,

Stachowiak et al. 2010, 2012). More specifically, increasing membrane coverage results in increasing tubule formation, independent of the membrane-binding chemistry. This is believed to stem from collisions among crowded membrane-bound proteins, which create steric pressure that can drive membrane bending. Accordingly, this crowding mechanism is sufficient to generate membrane curvature on its own without involving other processes such as hydrophobic insertion.

Together, these observations suggest a mechanism for inducing and amplifying membrane curvature through steric congestion of proteins bound to membrane surfaces such that lateral pressures from local asymmetries in protein densities can cause drastic membrane shape changes. Furthermore, by concentrating protein binding to spatially confined regions of the membrane, such as lipid domains, the membrane deformation effects of protein–protein crowding may be amplified due to increased protein density on the surface, and thus, increased steric interactions (Stachowiak et al. 2010). Membrane-mediated interactions can bring proteins together, such as a protein having higher affinity for one lipid species or a certain lipid conformation (Callan-Jones and Bassereau 2013, Gil et al. 1998, Reynwar and Deserno 2008). For instance, lipid rafts are able to concentrate protein-binding regions on a membrane (Stachowiak et al. 2010) and is the process hypothesized to cause the aggregation of caveolin (Sens and Turner 2004). However, it should be noted that aggregation of proteins on membrane surfaces is not necessarily coupled to appreciable membrane deformation (Sens et al. 2008).

8.3.5 Membrane Wrapping

It is possible for a membrane to wrap partially around a peptide or protein if attractive interactions exist. For example, the interaction between an anionic membrane and a cationic peptide is driven by the entropic gain from counterion release, as the peptide and the membrane charge compensate one another. Since counterion release will be maximized when the cationic moieties of the peptide are closely associated with the anionic and polar head groups of the membrane, there is a tendency for the membrane to wrap around the peptide to maximize contact between the charged portions of the peptide and membrane (Bruinsma 1998, Harries et al. 1998, Manning 1969, Wong and Pollack 2010). We refer the reader to the following studies for more information on electrostatic interactions in membrane-based systems (Koltover et al. 1998, Raviv et al. 2005, 2006, Safinya et al. 2006, Wong et al. 2000).

8.3.6 Composite Mechanism

With the survey above, we see that proteins can employ insertion, scaffolding, curvature sensing, curvature-mediated protein aggregation, molecular crowding, and membrane wrapping mechanisms to amplify and promote effective membrane curvature generation (Ambroggio et al. 2010, Bashkirov et al. 2008, Baumgart et al. 2011, Callan-Jones and Bassereau 2013, Capraro et al. 2010, Farsad and Camilli 2003, Gallop and McMahon 2005, Gil et al. 1998, McMahon and Gallop 2005, Peter et al. 2004, Pucadyil and Schmid 2008, Reynwar and Deserno 2008, Romer et al. 2007, Roux et al. 2010, Sens et al. 2008, Sens and Turner 2004, Stachowiak et al. 2010, 2012, Zimmerberg and Kozlov 2006). In addition, one can see that curvature generation in membranes is often achieved through the combination of different constitutive mechanisms, sometimes between proteins, or even between motifs or domains within a protein. This suggests that the effectiveness and reliability of cellular membrane deformation is attained through a concerted action of multiple mechanisms. The rich interplay between mechanisms can be illustrated by proteins involved in vesicle budding, such as dynamin, which interacts with amphiphysin and endophilin. Similarly, coat protein clathrin binds to epsin. In both combinations of proteins, the former employs the scaffolding mechanism, while the latter features amphipathic helices that apply the hydrophobic insertion mechanism.

From our discussion, it can be seen that even though we can describe idealized curvature-generating mechanisms that are conceptually distinct, nature does not always respect these demarcations. In fact, from recent work, it would appear that compound mechanisms may be the rule rather than the exception. Generation of membrane curvature involves a complex interplay between lipids, proteins, and membranes, and experimental data show that robust and effective membrane curvature generation is often a result from combining different mechanisms. We present an illustrative case study that highlights these effects.

8.4 A CASE STUDY ON ANTIMICROBIAL PEPTIDES

8.4.1 Background and History

Cationic AMPs comprise an important part of innate host defence (Brogden 2005, Hancock and Lehrer 1998, Hancock and Sahl 2006, Shai 1999, Yeaman and Yount 2003, Zasloff 2002). To date, well over 1000 AMPs have been discovered. Collectively, AMPs display broad-spectrum antimicrobial activity (Brogden 2005, Hancock and Sahl 2006, Zasloff 2002). Whereas many commonly used antibiotics such as β-lactams, quinolones, macrolides, and tetracyclines have core structural features that are responsible for their antibacterial activities, AMP sequences are highly diverse and do not have a common core structure. However, we know that most AMPs tend to be short (<50 amino acids) and share two fundamental features: net cationic charge (+2 to +9) and amphiphilicity (Brogden 2005, Hancock and Lehrer 1998, Hancock and Sahl 2006, Shai 1999, Yeaman and Yount 2003, Zasloff 2002). AMPs are often classified according to their secondary structures. The α-helical AMPs include magainins (Zasloff 1987) and cathelicidins (Dürr et al. 2006), and the β-sheet AMPs include protegrins

(Kokryakov et al. 1993) and defensins (Ganz 2003, Lehrer 2004, Selsted and Ouellette 2005). A third group is categorized by extended linear peptides with sequences dominated by a few amino acid species, like the tryptophan-rich indolicidin (Selsted 1992) and the arginine- and proline-rich PR-39 (Agerberth et al. 1991).

From *in vitro* studies, the general mechanism of AMP activity involves the selective disruption and permeabilization of microbial membranes. This in principle can lead to depolarization, dissipation of electrochemical gradients, leakage, and cell death. Evidence that AMPs can target lipids in membranes comes from *in vitro* studies using a broad range of techniques including X-ray, nuclear magnetic resonance (NMR), atomic force microscopy (AFM), Raman spectroscopy, circular dichroism, differential scanning calorimetry, and variations of fluorescence experiments on vesicles and dye leakage (Brogden 2005, Epand and Vogel 1999) (and references therein). Clearly AMPs interact strongly with lipid bilayers and can alter their structure. For completeness, it should be noted that increasing evidence indicates that some AMPs employ alternative modes of action or act upon multiple bacterial cell targets. For instance, observations suggest that translocated AMPs can interact with intracellular targets to inhibit cell wall synthesis, nucleic acid synthesis, protein synthesis, and enzymatic activity (Andreu and Rivas 1998, Brogden 2005, Brötz et al. 1998, Otvos et al. 2000, Park et al. 1998, Patrzykat et al. 2002, Subbalakshmi and Sitaram 1998, Yonezawa et al. 1992). While membrane permeabilization alone may not be sufficient to achieve maximum antimicrobial activity in such cases, it is nonetheless required for their activity, as these peptides must still cross the cell membrane in order to reach their internal targets. Therefore, regardless of the precise antimicrobial mechanism, AMP activity is typically dependent on interaction with bacterial membranes (Hancock and Rozek 2002). Furthermore, studies have established that the role of mammalian AMPs in host defense is not restricted to acting as direct microbicides and that many AMPs can modulate the immune response through interactions with components of the innate and adaptive immune systems (Bowdish et al. 2005, Schmidt et al. 2015). Previous work has even shown that the AMP LL-37 plays a role in autoimmune disorders such as lupus (Lande et al. 2011) and psoriasis (Lande et al. 2007).

The selective action of AMPs against bacterial membranes and not against eukaryotic membranes is thought to be rooted in the compositional differences between bacterial cell membranes and eukaryotic cell membranes (Brogden 2005, Hancock and Sahl 2006, Huang 2000, Matsuzaki 1999, Matsuzaki et al. 1998, Shai 1999, Zasloff 2002). Microbial cell surfaces are decorated with polyanionic molecules such as lipopolysaccharides in Gram-negative bacteria, and lipoteichoic acids in Gram-positive bacteria (Hancock and Sahl 2006). Additionally, the outer leaflet of bacterial plasma membranes contains large amounts of anionic lipids such as those with phosphatidylglycerol (PG) and cardiolipin (CL) head groups. The membranes of the Gram-positive bacteria *Staphylococcus aureus* and *Streptococcus pneumoniae* are composed primarily of PG and CL lipids. Phosphatidylethanolamine (PE), a negative intrinsic curvature amphiphile, is the principle zwitterionic phospholipid found in Gram-negative bacteria such as *Escherichia coli*, *Pseudomonas aeruginosa*, and *Salmonella typhimurium* (Epand and Epand 2011). The lipid compositions of animal cell membranes differ from bacterial plasma membranes in a number of ways (van Meer et al. 2008, Zachowski 1993, Zasloff 2002). Animal cell membranes have more lipids with neutral zwitterionic head groups such as phosphatidylcholine (PC) and sphingomyelin (SM). Moreover, their lipid compositions are asymmetrically distributed between the inner and outer bilayer leaflets. For example, PC and SM are found in the outer leaflet of human erythrocytes, while PE and the anionic lipids, phosphatidylserine (PS) and phosphatidylinositol (PI), are found on the inner leaflet. Finally, sterols such as cholesterol (~30% by mole) constitute a major component of animal plasma membranes. Consistent with the above differences, *in vitro* studies on both natural and synthetic cationic membrane-active antimicrobials have shown that the presence of anionic lipids increases membrane disruption and permeabilization (Shai 1999).

To interact with bacterial cells, AMPs must first be attracted to their membrane surfaces, which can occur through nonspecific electrostatic binding between the cationic region of the peptide and anionic components on the surface. These peptides adsorb and orient parallel to the surface, followed by partitioning into the amphiphilic interface (between the hydrophilic and hydrophobic regions of the membrane). This process is driven by the hydrophobic interactions that govern the hydrophobic insertion mechanism. The amphipathic nature of AMPs is essential for this behaviour, as the hydrophobic domain of the peptide allows direct interaction with the lipid components of the membrane. Furthermore, because AMPs embed themselves near the lipid head group region of the membrane, they will naturally perturb the packing of lipid chains in the hydrophobic membrane core. By adding more volume near the hydrophilic–hydrophobic interface, the peptides create space that neighbouring acyl chains must fill, thereby effectively thinning the membrane. Local stretching or thinning of the membrane has been experimentally observed and is dependent on the specific AMP and its concentration (Bechinger and Lohner 2006, Buffy et al. 2004, Chen et al. 2003, Mecke et al. 2005, Münster et al. 2000, Weiss et al. 2002). Moreover, peptide-induced membrane thinning has been associated with reduced membrane bending moduli, which indicates softening of the membrane, thereby lowering the free energy required to induce membrane curvature (Safinya et al. 1989, Szleifer et al. 1988, Tristram-Nagle et al. 2010).

Building on the picture presented above, three early models have been proposed to describe the insertion of AMPs into a lipid bilayer, based on experimental work on membranes with single or mixed lipids: the "carpet" model, the "barrel-stave" model, and the "toroidal pore" model. In the carpet model, AMPs adsorb onto a bilayer in an orientation parallel to the membrane plane and accumulate in a carpet-like manner until a threshold concentration is reached (Bechinger 1991, Brogden 2005, Gazit et al. 1994, Jenssen et al. 2006, Matsuzaki et al. 1994, Pouny et al. 1992, Shai 1999), at which point the adsorbed peptides disintegrate the membrane in a detergent-like manner via the formation of micelles (Shai 1999). From the structure of micelles (a convex hydrophilic surface surrounding a hydrophobic interior), one can see that this model amounts to the induction of positive curvature, which enables the formation of composite peptide–lipid micelles. Thus, for the carpet model, membrane disruption does not involve pore formation, and these peptides are not inserted into the hydrophobic membrane core. This model has been proposed for the action of ovispirin (Yamaguchi et al. 2001) and cecropins (Gazit et al. 1995, Shai 1995, Yeaman and Yount 2003). In the barrel-stave model, amphipathic α-helical peptides form a bundle with a central lumen, which orients perpendicular to the plane of the membrane. This arrangement can be described as a "barrel" composed of helical peptides as the "staves" (Ehrenstein and Lecar 1977, Yang et al. 2001). The hydrophobic regions of each component peptide align and associate with the acyl chains of the bilayer membrane, while the hydrophilic regions face inwards towards the central lumen of the formed aqueous pore. Experiments using oriented circular dichroism (Lee et al. 2004), neutron scattering (Yang et al. 2001), and X-ray scattering (Spaar et al. 2004) have suggested that α-helical alamethicin induces this type of TM pore (Jenssen et al. 2006, Shai 1999, Spaar et al. 2004, Yeaman and Yount 2003). In the toroidal pore model, α-helical AMPs insert perpendicular into the plane of the membrane to form a pore, in which the membrane contributes to its lining (Bechinger 2009, Jenssen et al. 2006, Matsuzaki et al. 1996). More specifically, the lipids from the outer leaflet connect with lipids from the inner to form a continuous membrane that curves through the toroidal pore. In this model, the TM pore is lined with both the inserted peptides and lipid head groups, with the polar faces of the peptides associating with the polar head groups of the lipids. It has been suggested that the hydrophobic portions of the peptides displace the polar lipid head groups to induce positive curvature strain in the membrane (Dathe and Wieprecht 1999, Hara et al. 2001, Jenssen et al. 2006), via the hydrophobic insertion mechanism, which manifests as the membrane bending around the toroidal pore. AMPs that are proposed to form this type of TM pore include magainin (Ludtke et al. 1996, Matsuzaki et al. 1996, 1998, 1997), melittin (Yang et al. 2001), and protegrins (Tang et al. 2007).

There are many ways in which AMPs can act against bacterial membranes beyond these three basic models. Scanning electron microscopy (SEM) micrographs of *P. aeruginosa* showed the formation of large blebs on the bacterial membrane after exposure to sheep cathelicidin SMAP29 (Kalfa et al. 2001, Saiman et al. 2001). Similarly, electron micrographs of polymyxin B-treated *E. coli* showed significant surface vesicularization (Falagas et al. 2005), thin finger-like membrane protrusions that bead off into small vesicles. These protrusions are similar to EM images of microvilli emanating from the membranes of *E. coli* treated with protegrin-1 (Gidalevitz et al. 2003), an effect which was recapitulated in AFM studies on supported lipid bilayers treated with the peptide (Lam et al. 2006). Likewise, recent work shows that the pore-forming AMP melittin can induce budding in vesicles by promoting lipid-phase separation (Yu et al. 2010). Together, this diversity of structural outcomes suggests that in addition to pore formation, AMPs can permeate membranes and compromise their barrier function through multiple processes including blebbing, budding, and vesicularization. Furthermore, the resultant membrane destabilization mechanism is due to interplay between the unique properties of *both* the AMP and membrane. Rather than debate whether a specific AMP uses a particular mechanism, an interesting alternate question to ask is what these different membrane permeation mechanisms (pore formation, blebbing, budding, and vesicularization) have in common.

From the survey on membrane curvature generation in the first half of this review, it is evident that the physical chemistry of AMPs and cell membranes leads to a whole taxonomy of local membrane distortions, specific combinations of which are topologically active and can lead to membrane destabilization. Induction of positive mean curvature and positive Gaussian curvature can be seen in the generation of micelles from flat membranes. Negative Gaussian curvature (NGC, or equivalently, saddle-splay curvature) seen in lyotropic cubic phases is expected to be especially disruptive since this type of curvature is geometrically necessary (Gelbart et al. 1994) for pore formation and a broad range of known AMP-induced membrane destabilizing processes (Schmidt et al. 2010, Schmidt and Wong 2013) (Figure 8.3). For example, NGC can be seen in the lining of a TM pore, the base of a bleb, and the neck of a bud.

8.4.2 Induced Curvature and the Phase Behaviour of Antimicrobial Peptide–Lipid Systems

Defensins are a potent class of membrane-disruptive AMPs. Since the biology of defensins has been extensively studied, it is an ideal prototypical family of AMPs to investigate AMP-induced membrane curvature. Synchrotron small-angle X-ray scattering (SAXS) is used to map out the curvature deformation modes induced in model cell membranes by these AMPs and elucidate the relationships between induced curvature deformations and membrane disruption processes. In particular, SAXS was used to characterize the structures and phase behaviour of the peptide–lipid complexes generated by representative defensins from each of the three defensin families, α-defensins (Crp-4), β-defensins (HBD-2, HBD-3), and θ-defensins (RTD-1,

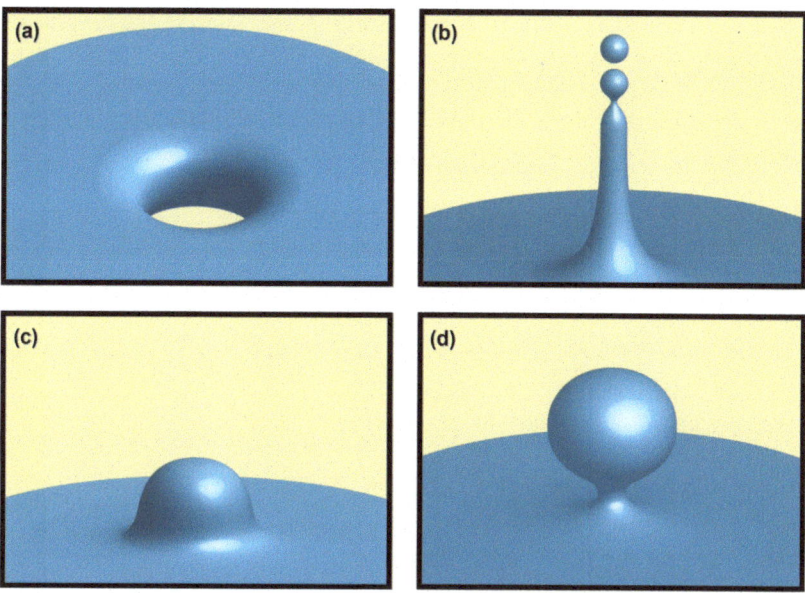

FIGURE 8.3 Saddle-splay curvature manifests in different membrane destabilizing processes, such as in the interior of a transmembrane pore (a), at the bases of tubules (b) and blebs (c), and at the necks of budding vesicles (b, d).

BTD-7). The membrane curvature deformations induced by these AMPs were probed by modifying the biophysical properties of the membrane via changes in the membrane lipid composition. Specifically, we assayed the effects of negative intrinsic curvature lipids on phase behaviour, as bacterial plasma membranes are known to have higher concentrations of negative intrinsic curvature lipids (PE, CL) compared with eukaryotic cell membranes. We see that defensins typically generate NGC, as evidenced by the presence of cubic phases, in model bacterial membranes rich in negative intrinsic curvature lipid PE. Moreover, when the concentration of PE is decreased, so that the target lipid composition is closer to those of mammalian membranes, this tendency to generate NGC is drastically suppressed. These observations are consistent with vesicle leakage assays and with the known activity profile of defensins, which permeate bacterial membranes but not eukaryotic membranes. A mechanism of action based on saddle-splay membrane curvature generation is broadly enabling since it is a necessary condition for processes such as pore formation, blebbing, budding, and vesicularization, all of which destabilize the barrier function of cell membranes. This trend of selective NGC generation and permeation in bacterial membranes, but not in eukaryotic membranes, is robust and can be observed in a broad range of other membrane-permeating molecules, natural AMPs, AMP mutants, and synthetic mimics of AMPs (Hu et al. 2013, Lee et al. 2014, Mishra et al. 2011b, Schmidt et al. 2011, 2012a,b, Schmidt and Wong 2013, Lee et al. 2017, Dishman et al. 2020, Lee et al. 2020). Generation of NGC is also observed in a wide range of CPPs and transporter sequences (Mishra et al. 2008, 2011a, Schmidt et al. 2010, Zhao et al. 2012). In fact, in systems where the peptide activity is strongly determined by its membrane activity, the correlation between NGC generation and activity is striking. For example, AMP defensin mutants (Schmidt et al. 2012b) with decreased activity are correlated with reduced NGC generation and shifted instances of NGC generation to different lipid compositions. Conversely, circular transporter sequences with increased activity over their linear counterparts have increased NGC generation (Zhao et al. 2012).

8.4.3 Why Membrane-Active Antimicrobial Peptides Exhibit Selective Action against Bacterial Membranes

One interesting question is why is it easier for AMPs to generate NGC in bacterial membranes than in eukaryotic membranes. We can sketch out an answer related to membrane composition, as bacterial membranes have higher concentrations of negative intrinsic curvature lipids ("traffic cone" -shaped lipids with small head groups and large hydrocarbon tails).

From the Helfrich formulation, bending modulus, κ, and Gaussian modulus, κ_G, for a symmetric bilayer membrane can be expressed in terms of the monolayer parameters (indicated by superscript m) through the following relations (Helfrich and Rennschuh 1990, Kozlov and Winterhalter 1991, Siegel and Kozlov 2004):

$$\kappa = 2\kappa^m$$
$$\kappa_G = 2\left(\kappa_G^m - 4\kappa^m c_0^m d\right)$$

where d is the distance between the midsurface of the bilayer and the neutral surface of either monolayer (this neutral surface can be approximated by the pivotal plane

(Kozlov 2007)). Here, we see that the bilayer Gaussian modulus, κ_G, depends on the monolayer Gaussian modulus κ_G^m, but it also depends on the monolayer bending modulus, κ^m, the monolayer intrinsic curvature, c_0^m, and d. The formation energy of a single pore can be estimated as follows. The Gauss–Bonnet theorem shows that the formation of a toroidal pore in a membrane alters the Gaussian curvature K in the membrane by $\Delta \int K\, dA = -4\pi$. The estimated energy change from pore formation is then $\Delta F = \kappa_G \int K\, dA = -4\pi \kappa_G$. Pore formation is generally not energetically favourable since empirically $\kappa_G < 0$ (Siegel and Kozlov 2004, Zimmerberg and Kozlov 2006). In fact, the negative value of the monolayer Gaussian modulus ($\kappa_G^m < 0$, from the lateral stress profile of the monolayer) (Seddon and Templer 1995) ensures that κ_G remains negative. Considering $\kappa_G < 0$, to lower the energy cost of forming a pore, we want to minimize the magnitude of κ_G. Since both κ^m and d must be positive-definite quantities, the sign of the second term in the expression for the bilayer Gaussian modulus depends on the monolayer intrinsic curvature, $\kappa_G \propto -c_0^m$. (Similar expressions have been derived for block copolymers (Wang 1992).) Therefore, enriching membranes with negative intrinsic curvature lipids decreases the energetic barrier to pore formation (Chernomordik and Kozlov 2003, Zimmerberg and Gawrisch 2006, Zimmerberg and Kozlov 2006), and membranes with elevated concentrations of these lipids are expected to be more susceptible to membrane disruption by pore-forming peptides such as AMPs (Schmidt et al. 2011, Yang et al. 2008). Indeed, *E. coli* mutants that are PE deficient (knock-out of the enzyme for making PE from PS lipids) show orders of magnitude lower susceptibility to membrane-active antimicrobials (Yang et al. 2008). Similarly, *P. aeruginosa* strains resistant to the antibiotic polymyxin have also been found to have reduced PE content (Conrad and Gilleland 1981).

8.4.4 Composite Mechanisms of Curvature Generation in Antimicrobial Peptides

Another interesting question to ask is how AMPs generate NGC. This is a deep question that requires much additional rigorous work to answer fully. However, in the context of this review, we can sketch out a qualitative answer based on recent studies.

AMPs are essentially oligomers that are cationic and hydrophobic, while target bacterial membranes are anionic, as previously mentioned. Therefore, electrostatic interactions between a cationic AMP and an anionic membrane will generally result in membrane wrapping from counter-ion release and negative membrane curvature generation. Embedding the hydrophobic portions of an AMP into the membrane requires displacing lipids in order to accommodate the peptide. The amount of displacement will be related to the amount of hydrophobicity and its presentation, but the displacement itself will lead to two consequences. One is membrane thinning and the other is an increase of hydrophobic volume in the perturbed outer monolayer in a manner dependent on the hydrophobic properties of the AMP (Campelo et al. 2008, Drin and Antonny 2010, McMahon and Gallop 2005). This gain in hydrophobic volume has the effect of bending the membrane towards its interior. Thus, hydrophobic interactions between an amphipathic AMP and a membrane will generally result in positive membrane curvature generation. The ability of AMPs to generate both negative and positive membrane curvatures at the same nanoscopic location, which together yield NGC, is important for its ability to destabilize membranes.

Combinations of cationic amino acids in AMPs have additional subtleties in the context of membrane curvature generation. AMPs often contain positively charged residues arginine and lysine, which play a crucial role in the activity of AMPs due to their ability to facilitate electrostatic interactions with negatively charged bacterial membranes. Through these interactions, AMPs are able to generate negative curvature via membrane wrapping. To understand the differences between arginine and lysine in the context of membrane curvature generation, we used quantum mechanical calculations to investigate interactions between phosphates of lipid head groups and guanidinium (the side chain of arginine) in comparison with the behaviour between phosphates and amines (the side chain of lysine) (Schmidt et al. 2012a). Differences between these two cationic side chains become apparent when multiple side chains are packed close to one another, as they are in AMPs. The planar Y-shape of the guanidinium group combined with its bidentate hydrogen-bonding ability allows it to rigidly arrange two phosphates. By stacking their guanidinium groups "face-to-face", nearby arginine side chains can maintain a diphosphate coordination. In contrast, the I-shape of the amine group, along with its monodentate hydrogen-bonding ability, cannot organize phosphates in such a structured manner, which results in a significant energetic penalty when amine groups are placed at separations characteristic of polylysine. These results suggest that arginines in AMPs can undergo stable bidentate hydrogen bonding with lipid head groups, thereby "cross-linking" lipid head groups into a composite object with an effectively larger head group area, thereby generating positive curvature along the peptide chain. (This effect is analogous to the molecular crowding mechanism summarized in the survey.) Molecular dynamics simulations on the interactions of arginine and lysine homopolymers with lipid membranes displayed the same behaviour qualitatively (Wu et al. 2013). This positive curvature from lipid head crowding is induced in a direction perpendicular to the negative curvature from electrostatic membrane wrapping. Therefore, arginines produce NGC, in agreement with experiments (Mishra et al. 2008, 2011a). In contrast, lysines in AMPs can only undergo monodentate hydrogen bonding with lipid head groups and cannot generate this additional curvature through lipid head crowding. Lysines therefore only produce negative mean curvature, which is also in agreement with experiments.

Consistent with the above model, we find that increasing the spacing between side chains in guanidinium homopolymers decreases the amount of saddle-splay curvature that can be generated, since increasing the average distance between guanidinium groups relieves the stress from molecular crowding interactions that lead to positive curvature generation (Mishra et al. 2011a, Schmidt et al. 2012a). Furthermore, a critical number of arginines are needed in polyarginine before the peptide can generate saddle-splay curvature; tetra-arginine (R4) produced inverted hexagonal phases with negative mean curvature only, and a minimum number of five residues are necessary before polyarginine can generate cubic phases (Mishra et al. 2011a). Mechanisms of curvature generation based on lipid head group reorganization are inherently sensitive to lipid head group chemistry; therefore, such mechanisms will be conducive to mediating specific peptide–lipid interactions.

8.4.5 Amino Acid Requirements for Peptides That Permeate Membranes

Now that we see how cationic and hydrophobic residues generate membrane curvature, we can recast these fundamental structural motifs of AMPs in terms of outcomes in membrane curvature generation. Since a general mechanism of action for AMPs is membrane disruption, the topological requirement to generate NGC constrains their arginine, lysine, and hydrophobic content. There will be a compositional trade-off between the relative amounts of arginine and lysine-plus-hydrophobicity used in an AMP. Arginine can generate NGC (both positive and negative curvatures in perpendicular directions), lysine generates negative curvature only, and hydrophobicity generates positive curvature only. These observations imply that we should be able to mimic the curvature-generating abilities of arginine using a combination of lysine and hydrophobic amino acids. Therefore, AMP sequences should exhibit this type of compositional trade-off between arginines on the one hand, and lysines plus hydrophobes on the other: a decrease in arginine content in an AMP can be offset by an increase in both lysine and hydrophobic content. This saddle-splay curvature selection rule is indeed observed when we analyze the sequences of 1080 cationic AMPs in the AMP database (Schmidt et al. 2011, Wang and Wang 2004). In our analysis, AMP hydrophobicity was determined by using three distinct, widely used hydrophobicity scales: the Kyte–Doolittle scale (Kyte and Doolittle 1982), the Eisenberg consensus scale (Eisenberg et al. 1982), and the Wimley–White biological scale (Hessa et al. 2005). Consistent with the saddle-splay curvature selection rule, all three scales show a strong positive trend between the average hydrophobicity and lysine content in AMPs (Schmidt et al. 2011). The good correspondence between the trends in AMP cationicity and hydrophobicity with those predicted by the selection rule supports our model that AMP sequences enforce the selective generation of NGC in bacterial membranes. Interestingly, a similar selection rule was found for CPPs (Mishra et al. 2011a). For CPPs, the same qualitative trade-off is observed, but with significantly less hydrophobicity (Figure 8.4). We hypothesize that this is related to the shorter residence times of CPPs on membranes, since they cross bilayers via transient interactions rather than remain on the membrane and kill the cell.

The saddle-splay curvature selection rule shows that to impart a peptide membrane-permeating ability, constraints on the average amounts of arginine, lysine, and hydrophobicity need to be placed in the peptide sequence in order to generate the required NGC. (Other principles for optimizing activity exist of course.) It must be noted that this saddle-splay curvature selection rule underdetermines a full AMP sequence, and therefore, leaves plenty of sequence diversity to allow for additional functions besides membrane activity. One demonstration of this is our recent work on engineering antibiotics for persister bacterial strains using tobramycin, an aminoglycoside antibiotic that binds to bacterial ribosomes to disrupt protein synthesis but is

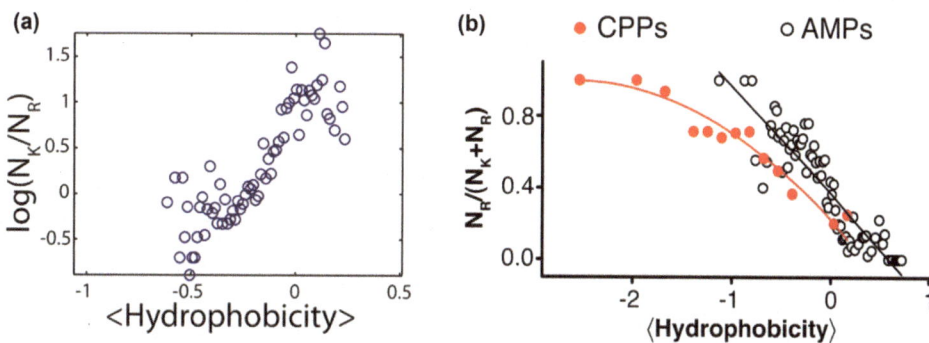

FIGURE 8.4 The trade-off between arginine content and lysine-plus-hydrophobicity content is observed among the sequences of cationic AMPs (a, b) and CPPs (b). Generally, as the average hydrophobicity (<Hydrophobicity>) of an AMP or CPP increases, the number of arginines (N_R) decreases relative to the number of lysines (N_K). AMPs, antimicrobial peptides; CPPs, cell-penetrating peptides. (Figure (a) reprinted with permission from Schmidt et al. 2011. Copyright © 2011 American Chemical Society. Figure (b) reprinted with permission from Mishra et al. 2011a.)

normally ineffective against persisters due to their reduced metabolic and uptake activity. By providing tobramycin with the capability of generating NGC via the addition of 12 amino acids, we designed a tobramycin–AMP hybrid. The resulting molecule spontaneously permeates bacterial membranes, recovers the high antibiotic activity of aminoglycosides, and kills *E. coli* and *S. aureus* persisters 4–6 logs better than tobramycin, while remaining noncytotoxic to eukaryotes (Schmidt et al. 2014). These results suggest that we can still learn a lot from AMPs and that adapting the AMP design to other contexts may allow us to renovate other traditional antibiotics.

REFERENCES

Agerberth, B., Lee, J.-Y., Bergman, T. et al. 1991. Amino acid sequence of PR-39. *Eur. J. Biochem.* 202: 849–854.

Ambroggio, E., Sorre, B., Bassereau, P. et al. 2010. ArfGAP1 generates an Arf1 gradient on continuous lipid membranes displaying flat and curved regions. *EMBO J.* 29: 292–303.

Andreu, D. and Rivas, L. 1998. Animal antimicrobial peptides: An overview. *Pept. Sci.* 47: 415–433.

Antonny, B. 2011. Mechanisms of membrane curvature sensing. *Annu. Rev. Biochem.* 80: 101–123.

Bashkirov, P. V., Akimov, S. A., Evseev, A. I. et al. 2008. GTPase cycle of dynamin is coupled to membrane squeeze and release, leading to spontaneous fission. *Cell* 135: 1276–1286.

Bassereau, P. 2010. Division of labour in ESCRT complexes. *Nat. Cell Biol.* 12: 422–423.

Baumgart, T., Capraro, B. R., Zhu, C., and Das, S. L. 2011. Thermodynamics and mechanics of membrane curvature generation and sensing by proteins and lipids. *Annu. Rev. Phys. Chem.* 62: 483–506.

Bechinger, B. 2009. Rationalizing the membrane interactions of cationic amphipathic antimicrobial peptides by their molecular shape. *Curr. Opin. Colloid Interface Sci.* 14: 349–355.

Bechinger, B., Kim, Y., Chirlian, L. E. et al. 1991. Orientations of amphipathic helical peptides in membrane bilayers determined by solid-state NMR spectroscopy. *J Biomol NMR* 1: 167–173.

Bechinger, B. and Lohner, K. 2006. Detergent-like actions of linear amphipathic cationic antimicrobial peptides. *Biochim. Biophys. Acta, Biomembr.* 1758: 1529–1539.

Bi, X., Corpina, R. A., and Goldberg, J. 2002. Structure of the Sec23/24-Sar1 pre-budding complex of the COPII vesicle coat. *Nature* 419: 271–277.

Botelho, A. V., Huber, T., Sakmar, T. P., and Brown, M. F. 2006. Curvature and hydrophobic forces drive oligomerization and modulate activity of rhodopsin in membranes. *Biophys. J.* 91: 4464–4477.

Boucrot, E., Pick, A., Çamdere, G. et al. 2012. Membrane fission is promoted by insertion of amphipathic helices and is restricted by crescent BAR domains. *Cell* 149: 124–136.

Bowdish, D. M. E., Davidson, D. J., and Hancock, R. E. W. 2005. A re-evaluation of the role of host defence peptides in mammalian immunity. *Curr Protein Pept Sci* 6: 35–51.

Brogden, K. A. 2005. Antimicrobial peptides: Pore formers or metabolic inhibitors in bacteria? *Nat. Rev. Microbiol.* 3: 238–250.

Brötz, H., Bierbaum, G., Leopold, K., Reynolds, P. E., and Sahl, H.-G. 1998. The lantibiotic mersacidin inhibits peptidoglycan synthesis by targeting lipid II. *Antimicrob. Agents Chemother.* 42: 154–160.

Bruinsma, R. 1998. Electrostatics of DNA–cationic lipid complexes: Isoelectric instability. *Eur. Phys. J. B* 4: 75–88.

Buffy, J. J., McCormick, M. J., Wi, S. et al. 2004. Solid-state NMR investigation of the selective perturbation of lipid bilayers by the cyclic antimicrobial peptide RTD-1. *Biochemistry* 43: 9800–9812.

Callan-Jones, A. and Bassereau, P. 2013. Curvature-driven membrane lipid and protein distribution. *Curr. Opin. Solid State Mater. Sci.* 17: 143–150.

Campelo, F., McMahon, H. T., and Kozlov, M. M. 2008. The hydrophobic insertion mechanism of membrane curvature generation by proteins. *Biophys. J.* 95: 2325–2339.

Capraro, B. R., Yoon, Y., Cho, W., and Baumgart, T. 2010. Curvature sensing by the epsin N-terminal homology domain measured on cylindrical lipid membrane tethers. *J. Am. Chem. Soc.* 132: 1200–1201.

Chen, F.-Y., Lee, M.-T., and Huang, H. W. 2003. Evidence for membrane thinning effect as the mechanism for peptide-induced pore formation. *Biophys. J.* 84: 3751–3758.

Chernomordik, L. V. and Kozlov, M. M. 2003. Protein–lipid interplay in fusion and fission of biological membranes. *Annu. Rev. Biochem.* 72: 175–207.

Conrad, R. S. and Gilleland, H. E. 1981. Lipid alterations in cell envelopes of polymyxin-resistant *Pseudomonas aeruginosa* isolates. *J. Bacteriol.* 148: 487–497.

Cornea, R. L. and Thomas, D. D. 1994. Effects of membrane thickness on the molecular dynamics and enzymic activity of reconstituted Ca-ATPase. *Biochemistry* 33: 2912–2920.

Damke, H., Baba, T., Warnock, D. E., and Schmid, S. L. 1994. Induction of mutant dynamin specifically blocks endocytic coated vesicle formation. *J. Cell Biol.* 127: 915–934.

Dan, N., Pincus, P., and Safran, S. A. 1993. Membrane-induced interactions between inclusions. *Langmuir* 9: 2768–2771.

Dathe, M., Schümann, M., Wieprecht, T. et al. 1996. Peptide helicity and membrane surface charge modulate the balance of electrostatic and hydrophobic interactions with lipid bilayers and biological membranes. *Biochemistry* 35: 12612–12622.

Dathe, M. and Wieprecht, T. 1999. Structural features of helical antimicrobial peptides: Their potential to modulate activity on model membranes and biological cells. *Biochim. Biophys. Acta, Biomembr.* 1462: 71–87.

de Planque, M. R. R., Goormaghtigh, E., Greathouse, D. V. et al. 2001. Sensitivity of single membrane-spanning α-helical peptides to hydrophobic mismatch with a lipid bilayer: Effects on backbone structure, orientation, and extent of membrane incorporation. *Biochemistry* 40: 5000–5010.

de Planque, M. R. R., Greathouse, D. V., Koeppe, R. E. et al. 1998. Influence of lipid/peptide hydrophobic mismatch on the thickness of diacylphosphatidylcholine bilayers. A 2H NMR and ESR study using designed transmembrane α-helical peptides and gramicidin A. *Biochemistry* 37: 9333–9345.

Deserno, M. 2009. Membrane elasticity and mediated interactions in continuum theory: A differential geometric approach. In *Biomembrane Frontiers*, eds. Faller, R., Longo, M. L., Risbud, S. H., and Jue, T., 41–74. Humana Press.

Dishman, A. F., Lee, M. W., de Anda, J., et al. 2020. Switchable membrane remodeling and antifungal defense by metamorphic chemokine XCL1. *ACS Infect. Dis.* 6: 1204–1213.

Drab, M., Verkade, P., Elger, M. et al. 2001. Loss of caveolae, vascular dysfunction, and pulmonary defects in caveolin-1 gene-disrupted mice. *Science* 293: 2449–2452.

Drin, G. and Antonny, B. 2010. Amphipathic helices and membrane curvature. *FEBS Lett.* 584: 1840–1847.

Dürr, U. H. N., Sudheendra, U. S., and Ramamoorthy, A. 2006. LL-37, the only human member of the cathelicidin family of antimicrobial peptides. *Biochim. Biophys. Acta, Biomembr.* 1758: 1408–1425.

Ehrenstein, G. and Lecar, H. 1977. Electrically gated ionic channels in lipid bilayers. *Q. Rev. Biophys.* 10: 1–34.

Eisenberg, D., Weiss, R. M., Terwilliger, T. C., and Wilcox, W. 1982. Hydrophobic moments and protein structure. *Faraday Symp. Chem. Soc.* 17: 109–120.

Epand, R. M. and Epand, R. F. 2011. Bacterial membrane lipids in the action of antimicrobial agents. *J. Pept. Sci.* 17: 298–305.

Epand, R. M., Shai, Y., Segrest, J. P., and Anantharamiah, G. M. 1995. Mechanisms for the modulation of membrane bilayer properties by amphipathic helical peptides. *Biopolymers* 37: 319–338.

Epand, R. M. and Vogel, H. J. 1999. Diversity of antimicrobial peptides and their mechanisms of action. *Biochim. Biophys. Acta, Biomembr.* 1462: 11–28.

Fabrikant, G., Lata, S., Riches, J. D. et al. 2009. Computational model of membrane fission catalyzed by ESCRT-III. *PLoS Comput. Biol.* 5: e1000575.

Falagas, M. E., Kasiakou, S. K., and Saravolatz, L. D. 2005. Colistin: The revival of polymyxins for the management of multidrug-resistant gram-negative bacterial infections. *Clin. Infect. Dis.* 40: 1333–1341.

Farsad, K. and Camilli, P. D. 2003. Mechanisms of membrane deformation. *Curr. Opin. Cell Biol.* 15: 372–381.

Fernandez, I., Ying, Y., Albanesi, J., and Anderson, R. G. W. 2002. Mechanism of caveolin filament assembly. *Proc. Natl. Acad. Sci. U. S. A.* 99: 11193–11198.

Fotin, A., Cheng, Y., Sliz, P. et al. 2004. Molecular model for a complete clathrin lattice from electron cryomicroscopy. *Nature* 432: 573–579.

Fra, A. M., Williamson, E., Simons, K., and Parton, R. G. 1995. De novo formation of caveolae in lymphocytes by expression of VIP21-caveolin. *Proc. Natl. Acad. Sci. U. S. A.* 92: 8655–8659.

Gallop, J. L. and McMahon, H. T. 2005. BAR domains and membrane curvature: Bringing your curves to the BAR. *Biochem. Soc. Symp.* 72: 223–231.

Ganz, T. 2003. Defensins: Antimicrobial peptides of innate immunity. *Nat. Rev. Immunol.* 3: 710–720.

Gazit, E., Boman, A., Boman, H. G., and Shai, Y. 1995. Interaction of the mammalian antibacterial peptide cecropin P1 with phospholipid vesicles. *Biochemistry* 34: 11479–11488.

Gazit, E., Lee, W.-J., Brey, P. T., and Shai, Y. 1994. Mode of action of the antibacterial cecropin B2: A spectrofluorometric study. *Biochemistry* 33: 10681–10692.

Gelbart, W. M., Ben-Shaul, A., and Roux, D., eds. 1994. *Micelles, Membranes, Microemulsions, and Monolayers.* New York: Springer-Verlag.

Gidalevitz, D., Ishitsuka, Y., Muresan, A. S. et al. 2003. Interaction of antimicrobial peptide protegrin with biomembranes. *Proc. Natl. Acad. Sci. U. S. A.* 100: 6302–6307.

Gil, T., Ipsen, J. H., Mouritsen, O. G. et al. 1998. Theoretical analysis of protein organization in lipid membranes. *Biochim. Biophys. Acta, Rev. Biomembr.* 1376: 245–266.

Goulian, M., Bruinsma, R., and Pincus, P. 1993. Long-range forces in heterogeneous fluid membranes. *Europhys. Lett.* 22: 145.

Graham, T. R. and Kozlov, M. M. 2010. Interplay of proteins and lipids in generating membrane curvature. *Curr. Opin. Cell Biol.* 22: 430–436.

Hancock, R. E. W. and Lehrer, R. 1998. Cationic peptides: A new source of antibiotics. *Trends Biotechnol.* 16: 82–88.

Hancock, R. E. W. and Rozek, A. 2002. Role of membranes in the activities of antimicrobial cationic peptides. *FEMS Microbiol. Lett.* 206: 143–149.

Hancock, R. E. W. and Sahl, H.-G. 2006. Antimicrobial and host-defense peptides as new anti-infective therapeutic strategies. *Nat. Biotechnol.* 24: 1551–1557.

Hara, T., Kodama, H., Kondo, M. et al. 2001. Effects of peptide dimerization on pore formation: Antiparallel disulfide-dimerized magainin 2 analogue. *Biopolymers* 58: 437–446.

Harries, D., May, S., Gelbart, W. M., and Ben-Shaul, A. 1998. Structure, Stability, and Thermodynamics of Lamellar DNA–Lipid Complexes. *Biophys. J.* 75: 159–173.

Helfrich, W. 1973. Elastic properties of lipid bilayers: Theory and possible experiments. *Z. Naturforsch. C* 28: 693–703.

Helfrich, W. and Rennschuh, H. 1990. Landau theory of the lamellar-to-cubic phase transition. *J. Phys. Colloques* 51: C7-189-C7-195.

Herce, H. D. and Garcia, A. E. 2007. Molecular dynamics simulations suggest a mechanism for translocation of the HIV-1 TAT peptide across lipid membranes. *Proc. Natl. Acad. Sci. U. S. A.* 104: 20805–20810.

Hessa, T., Kim, H., Bihlmaier, K. et al. 2005. Recognition of transmembrane helices by the endoplasmic reticulum translocon. *Nature* 433: 377–381.

Hinshaw, J. E. 2000. Dynamin and its role in membrane fission. *Annu. Rev. Cell Dev. Biol.* 16: 483–519.

Hinshaw, J. E. and Schmid, S. L. 1995. Dynamin self-assembles into rings suggesting a mechanism for coated vesicle budding. *Nature* 374: 190–192.

Hu, K., Schmidt, N. W., Zhu, R. et al. 2013. A Critical Evaluation of Random Copolymer Mimesis of Homogeneous Antimicrobial Peptides. *Macromolecules* 46: 1908–1915.

Huang, H. W. 2000. Action of Antimicrobial Peptides: Two-State Model. *Biochemistry* 39: 8347–8352.

Israelachvili, J. N., Marčelja, S., and Horn, R. G. 1980. Physical principles of membrane organization. *Q. Rev. Biophys.* 13: 121–200.

Jenssen, H., Hamill, P., and Hancock, R. E. W. 2006. Peptide antimicrobial agents. *Clin. Microbiol. Rev.* 19: 491–511.

Johannsson, A., Keightley, C. A., Smith, G. A. et al. 1981. The effect of bilayer thickness and n-alkanes on the activity of the (Ca^{2+}+Mg^{2+})-dependent ATPase of sarcoplasmic reticulum. *J. Biol. Chem.* 256: 1643–1650.

Kalfa, V. C., Jia, H. P., Kunkle, R. A. et al. 2001. Congeners of SMAP29 kill ovine pathogens and induce ultrastructural damage in bacterial cells. *Antimicrob. Agents Chemother.* 45: 3256–3261.

Kik, R. A., Leermakers, F. A. M., and Kleijn, J. M. 2010. Molecular modeling of proteinlike inclusions in lipid bilayers: Lipid-mediated interactions. *Phys. Rev. E* 81: 021915.

Killian, J. A. 1998. Hydrophobic mismatch between proteins and lipids in membranes. *Biochim. Biophys. Acta, Rev. Biomembr.* 1376: 401–416.

Kokryakov, V. N., Harwig, S. S. L., Panyutich, E. A. et al. 1993. Protegrins: Leukocyte antimicrobial peptides that combine features of corticostatic defensins and tachyplesins. *FEBS Lett.* 327: 231–236.

Koltover, I., Rädler, J. O., and Safinya, C. R. 1999. Membrane mediated attraction and ordered aggregation of colloidal particles bound to giant phospholipid vesicles. *Phys. Rev. Lett.* 82: 1991–1994.

Koltover, I., Salditt, T., Rädler, J. O., and Safinya, C. R. 1998. An inverted hexagonal phase of cationic liposome–DNA complexes related to DNA release and delivery. *Science* 281: 78–81.

Kozlov, M. M. 2007. Determination of lipid spontaneous curvature from X-ray examinations of inverted hexagonal phases. In *Methods in Membrane Lipids*, ed. Dopico, A. M., 355–366. Humana Press.

Kozlov, M. M. and Chernomordik, L. V. 2002. The protein coat in membrane fusion: Lessons from fission. *Traffic* 3: 256–267.

Kozlov, M. M., McMahon, H. T., and Chernomordik, L. V. 2010. Protein-driven membrane stresses in fusion and fission. *Trends Biochem. Sci.* 35: 699–706.

Kozlov, M. M. and Winterhalter, M. 1991. Elastic moduli for strongly curved monoplayers. Position of the neutral surface. *J. Phys. II France* 1: 1077–1084.

Kreyszig, E. 1991. *Differential Geometry*. New York: Dover Publications.

Kyte, J. and Doolittle, R. F. 1982. A simple method for displaying the hydropathic character of a protein. *J. Mol. Biol.* 157: 105–132.

Lam, K. L. H., Ishitsuka, Y., Cheng, Y. et al. 2006. Mechanism of supported membrane disruption by antimicrobial peptide protegrin-1. *J. Phys. Chem. B* 110: 21282–21286.

Lande, R., Ganguly, D., Facchinetti, V. et al. 2011. Neutrophils activate plasmacytoid dendritic cells by releasing self-DNA–peptide complexes in systemic lupus erythematosus. *Sci. Transl. Med.* 3: 73ra19.

Lande, R., Gregorio, J., Facchinetti, V. et al. 2007. Plasmacytoid dendritic cells sense self-DNA coupled with antimicrobial peptide. *Nature* 449: 564–569.

Lee, A. G. 2004. How lipids affect the activities of integral membrane proteins. *Biochim. Biophys. Acta, Biomembr.* 1666: 62–87.

Lee, M. C. S., Orci, L., Hamamoto, S. et al. 2005. Sar1p N-terminal helix initiates membrane curvature and completes the fission of a COPII vesicle. *Cell* 122: 605–617.

Lee, M.-T., Chen, F.-Y., and Huang, H. W. 2004. Energetics of pore formation induced by membrane active peptides. *Biochemistry* 43: 3590–3599.

Lee, M. W., Chakraborty, S., Schmidt, N. W. et al. 2014. Two interdependent mechanisms of antimicrobial activity allow for efficient killing in nylon-3-based polymeric mimics of innate immunity peptides. *Biochim. Biophys. Acta, Biomembr.* 1838: 2269–2279.

Lee, M. W., de Anda, J., Kroll, C., et al. 2020. How do cyclic antibiotics with activity against Gram-negative bacteria permeate membranes? A machine learning informed experimental study. *Biochim. Biophys. Acta, Biomembr.* 1862: 183302.

Lee, M. W., Han, M., Bossa, G. V. et al. 2017. Interactions between Membranes and "Metaphilic" Polypeptide Architectures with Diverse Side-Chain Populations. *ACS Nano* 11: 2858–2871.

Lehrer, R. I. 2004. Primate defensins. *Nat. Rev. Microbiol.* 2: 727–738.

Leibler, S. 1986. Curvature instability in membranes. *J. Phys. France* 47: 507–516.

Leikina, E., Mittal, A., Cho, M.-S. et al. 2004. Influenza hemagglutinins outside of the contact zone are necessary for fusion pore expansion. *J. Biol. Chem.* 279: 26526–26532.

Lenz, O., Dittmar, M. T., Wagner, A. et al. 2005. Trimeric Membrane-anchored gp41 Inhibits HIV Membrane Fusion. *J. Biol. Chem.* 280: 4095–4101.

Ljunggren, S. and Eriksson, J. C. 1992. Minimal surfaces and Winsor III microemulsions. *Langmuir* 8: 1300–1306.

Ludtke, S. J., He, K., Heller, W. T. et al. 1996. Membrane pores induced by magainin. *Biochemistry* 35: 13723–13728.

Lundbæk, J. A., Maer, A. M., and Andersen, O. S. 1997. Lipid bilayer electrostatic energy, curvature stress, and assembly of gramicidin channels. *Biochemistry* 36: 5695–5701.

Manning, G. S. 1969. Limiting laws and counterion condensation in polyelectrolyte solutions I. Colligative properties. *J. Chem. Phys.* 51: 924–933.

Marks, B., Stowell, M. H. B., Vallis, Y. et al. 2001. GTPase activity of dynamin and resulting conformation change are essential for endocytosis. *Nature* 410: 231–235.

Mashl, R. J. and Bruinsma, R. F. 1998. Spontaneous-curvature theory of clathrin-coated membranes. *Biophys. J.* 74: 2862–2875.

Matsuzaki, K. 1999. Why and how are peptide–lipid interactions utilized for self-defense? Magainins and tachyplesins as archetypes. *Biochim. Biophys. Acta, Biomembr.* 1462: 1–10.

Matsuzaki, K., Murase, O., Fujii, N., and Miyajima, K. 1996. An antimicrobial peptide, magainin 2, induced rapid flip-flop of phospholipids coupled with pore formation and peptide translocation. *Biochemistry* 35: 11361–11368.

Matsuzaki, K., Murase, O., Tokuda, H. et al. 1994. Orientational and aggregational states of magainin 2 in phospholipid bilayers. *Biochemistry* 33: 3342–3349.

Matsuzaki, K., Sugishita, K., Harada, M., Fujii, N., and Miyajima, K. 1997. Interactions of an antimicrobial peptide, magainin 2, with outer and inner membranes of Gram-negative bacteria. *Biochim. Biophys. Acta, Biomembr.* 1327: 119–130.

Matsuzaki, K., Sugishita, K.-i., Ishibe, N. et al. 1998. Relationship of membrane curvature to the formation of pores by magainin 2. *Biochemistry* 37: 11856–11863.

McMahon, H. T. and Gallop, J. L. 2005. Membrane curvature and mechanisms of dynamic cell membrane remodelling. *Nature* 438: 590–596.

McMahon, H. T. and Mills, I. G. 2004. COP and clathrin-coated vesicle budding: different pathways, common approaches. *Curr. Opin. Cell Biol.* 16: 379–391.

McNiven, M. A., Cao, H., Pitts, K. R., and Yoon, Y. 2000. The dynamin family of mechanoenzymes: Pinching in new places. *Trends Biochem. Sci.* 25: 115–120.

Mecke, A., Lee, D.-K., Ramamoorthy, A., Orr, B. G., and Banaszak Holl, M. M. 2005. membrane thinning due to antimicrobial peptide binding: An atomic force microscopy study of MSI-78 in lipid bilayers. *Biophys. J.* 89: 4043–4050.

Michelangeli, F., Grimes, E. A., East, J. M., and Lee, A. G. 1991. Effects of phospholipids on the function of the calcium-magnesium ATPase. *Biochemistry* 30: 342–351.

Mishra, A., Gordon, V. D., Yang, L., Coridan, R., and Wong, G. C. L. 2008. HIV TAT forms pores in membranes by inducing saddle-splay curvature: Potential role of bidentate hydrogen bonding. *Angew. Chem., Int. Ed.* 47: 2986–2989.

Mishra, A., Lai, G. H., Schmidt, N. W. et al. 2011a. Translocation of HIV TAT peptide and analogues induced by multiplexed membrane and cytoskeletal interactions. *Proc. Natl. Acad. Sci. U. S. A.* 108: 16883–16888.

Mishra, A., Tai, K. P., Schmidt, N. W., Ouellette, A. J., and Wong, G. C. L. 2011b. Chapter four - small-angle X-ray scattering studies of peptide–lipid interactions using the mouse paneth cell α-defensin cryptdin-4. In *Methods in Enzymology*, eds. Michael L. Johnson, J. M. H. and Gary, K. A., 127–149. Academic Press.

Montecucco, C., Smith, G. A., Dabbeni-sala, F. et al. 1982. Bilayer thickness and enzymatic activity in the mitochondrial cytochrome c oxidase and ATPase complex. *FEBS Lett.* 144: 145–148.

Morein, S., Koeppeli, R.E., Lindblom, G., de Kruijff, B., and Antoinette Killian, J. 2000. The effect of peptide/lipid hydrophobic mismatch on the phase behavior of model membranes mimicking the lipid composition in *Escherichia coli*membranes. *Biophys. J.* 78: 2475–2485.

Morlot, S. and Roux, A.2013. Mechanics of dynamin-mediated membrane fission. *Annu. Rev. Biophys.* 42: 629–649.

Münster, C., Lu, J., Bechinger, B., and Salditt, T.2000. Grazing incidence X-ray diffraction of highly aligned phospholipid membranes containing the antimicrobial peptide magainin 2. *Eur Biophys J* 28: 683–688.

Nossal, R. 2001. Energetics of clathrin basket assembly. *Traffic*2: 138–147.

Otvos, L., O, I., Rogers, M. E. et al. 2000. Interaction between heat shock proteins and antimicrobial peptides. *Biochemistry* 39: 14150–14159.

Park, C. B., Kim, H. S., and Kim, S. C. 1998. Mechanism of action of the antimicrobial peptide Buforin II: Buforin II kills microorganisms by penetrating the cell membrane and inhibiting cellular functions. *Biochem. Biophys. Res. Commun.* 244: 253–257.

Patrzykat, A., Friedrich, C. L., Zhang, L., Mendoza, V., and Hancock, R. E. W. 2002. Sublethal concentrations of pleurocidin-derived antimicrobial peptides inhibit macromolecular synthesis in *Escherichia coli*. *Antimicrob. Agents Chemother.* 46: 605–614.

Peter, B. J., Kent, H. M., Mills, I. G. et al. 2004. BAR domains as sensors of membrane curvature: The amphiphysin BAR structure. *Science* 303: 495–499.

Pouny, Y., Rapaport, D., Mor, A., Nicolas, P., and Shai, Y. 1992. Interaction of antimicrobial dermaseptin and its fluorescently labeled analogs with phospholipid membranes. *Biochemistry* 31: 12416–12423.

Pucadyil, T. J. and Schmid, S. L. 2008. Real-time visualization of dynamin-catalyzed membrane fission and vesicle release. *Cell*135: 1263–1275.

Ramachandran, R., Pucadyil, T. J., Liu, Y.-W. et al. 2009. Membrane insertion of the pleckstrin homology domain variable loop 1 is critical for dynamin-catalyzed vesicle scission. *Mol. Biol. Cell* 20: 4630–4639.

Raviv, U., Needleman, D. J., Li, Y. et al. 2005. Cationic liposome–microtubule complexes: Pathways to the formation of two-state lipid–protein nanotubes with open or closed ends. *Proc. Natl. Acad. Sci. U. S. A.* 102: 11167–11172.

Raviv, U., Needleman, D. J., and Safinya, C. R. 2006. Cationic membranes complexed with oppositely charged microtubules: Hierarchical self-assembly leading to bio-nanotubes. *J. Phys.: Condens. Matter* 18: S1271.

Reynwar, B. J. and Deserno, M. 2008. Membrane composition-mediated protein–protein interactions. *Biointerphases* 3: FA117-FA124.

Reynwar, B. J., Illya, G., Harmandaris, V. A. et al. 2007. Aggregation and vesiculation of membrane proteins by curvature-mediated interactions. *Nature* 447: 461–464.

Romer, W., Berland, L., Chambon, V. et al. 2007. Shiga toxin induces tubular membrane invaginations for its uptake into cells. *Nature* 450: 670–675.

Rothberg, K. G., Heuser, J. E., Donzell, W. C. et al. 1992. Caveolin, a protein component of caveolae membrane coats. *Cell* 68: 673–682.

Roux, A., Koster, G., Lenz, M. et al. 2010. Membrane curvature controls dynamin polymerization. *Proc. Natl. Acad. Sci. U. S. A.* 107: 4141–4146.

Safinya, C. R., Ewert, K., Ahmad, A. et al. 2006. Cationic liposome–DNA complexes: From liquid crystal science to gene delivery applications. *Philos. Trans. R. Soc., A* 364: 2573–2596.

Safinya, C. R., Sirota, E. B., Roux, D., and Smith, G. S. 1989. Universality in interacting membranes: The effect of cosurfactants on the interfacial rigidity. *Phys. Rev. Lett.* 62: 1134–1137.

Saiman, L., Tabibi, S., Starner, T. D. et al. 2001. Cathelicidin peptides inhibit multiply antibiotic-resistant pathogens from patients with cystic fibrosis. *Antimicrob. Agents Chemother.* 45: 2838–2844.

Schmid, S. L. 1997. Clathrin-coated vesicle formation and protein sorting:An integrated process. *Annu. Rev. Biochem.* 66: 511–548.

Schmidt, N., Mishra, A., Lai, G. H., and Wong, G. C. L. 2010. Arginine-rich cell-penetrating peptides. *FEBS Lett.* 584: 1806–1813.

Schmidt, N. W., Deshayes, S., Hawker, S. et al. 2014. Engineering persister-specific antibiotics with synergistic antimicrobial functions. *ACS Nano* 8: 8786–8793.

Schmidt, N. W., Jin, F., Lande, R. et al. 2015. Liquid-crystalline ordering of antimicrobial peptide–DNA complexes controls TLR9 activation. *Nat. Mater.* 14: 696–700.

Schmidt, N. W., Lis, M., Zhao, K. et al. 2012a. Molecular basis for nanoscopic membrane curvature generation from quantum mechanical models and synthetic transporter sequences. *J. Am. Chem. Soc.* 134: 19207–19216.

Schmidt, N. W., Mishra, A., Lai, G. H. et al. 2011. Criterion for amino acid composition of defensins and antimicrobial peptides based on geometry of membrane destabilization. *J. Am. Chem. Soc.*133: 6720–6727.

Schmidt, N. W., Mishra, A., Wang, J., DeGrado, W. F., and Wong, G. C. L. 2013. Influenza virus A M2 protein generates negative Gaussian membrane curvature necessary for budding and scission. *J. Am. Chem. Soc.*135: 13710–13719.

Schmidt, N. W., Tai, K. P., Kamdar, K. et al. 2012b. Arginine in α-Defensins: Differential effects on bactericidal activity correspond to geometry of membrane curvature generation and peptide-lipid phase behavior. *J. Biol. Chem.* 287: 21866–21872.

Schmidt, N. W. and Wong, G. C. L.2013. Antimicrobial peptides and induced membrane curvature: Geometry, coordination chemistry, and molecular engineering. *Curr. Opin. Solid State Mater. Sci.*17: 151–163.

Seddon, J. M. and Templer, R. H. 1993. Cubic phases of self-assembled amphiphilic aggregates. *Philos. Trans. R. Soc., A*344: 377–401.

Seddon, J. M. and Templer, R. H. 1995. Chapter 3- polymorphism of lipid-water systems. In*Handbook of Biological Physics*, eds. Lipowsky, R. and Sackmann, E., 97–160. Amsterdam: Elsevier Science B.V.

Segrest, J. P., De Loof, H., Dohlman, J. G., Brouillette, C. G., and Anantharamaiah, G. M. 1990. Amphipathic helix motif: Classes and properties. *Proteins: Struct. Funct. Bioinf.*8: 103–117.

Selsted, M. E., Novotny, M. J., Morris, W. L. et al. 1992. Indolicidin, a novel bactericidal tridecapeptide amide from neutrophils. *J. Biol. Chem.* 267: 4292–4295.

Selsted, M. E. and Ouellette, A. J. 2005. Mammalian defensins in the antimicrobial immune response. *Nat. Immunol.*6: 551–557.

Sens, P., Johannes, L., and Bassereau, P. 2008. Biophysical approaches to protein-induced membrane deformations in trafficking. *Curr. Opin. Cell Biol.* 20: 476–482.

Sens, P. and Turner, M. S. 2004. Theoretical model for the formation of caveolae and similar membrane invaginations. *Biophys. J.* 86: 2049–2057.

Shai, Y. 1995. Molecular recognition between membrane-spanning polypeptides. *Trends Biochem. Sci.* 20: 460–464.

Shai, Y. 1999. Mechanism of the binding, insertion and destabilization of phospholipid bilayer membranes by α-helical antimicrobial and cell non-selective membrane-lytic peptides. *Biochim. Biophys. Acta, Biomembr.* 1462: 55–70.

Shearman, G. C., Ces, O., Templer, R. H., and Seddon, J. M. 2006. Inverse lyotropic phases of lipids and membrane curvature. *J. Phys.: Condens. Matter* 18: S1105.

Siegel, D. P. and Kozlov, M. M. 2004. The Gaussian curvature elastic modulus of n-monomethylated dioleoylphosphatidylethanolamine: Relevance to membrane fusion and lipid phase behavior. *Biophys. J.* 87: 366–374.

Spaar, A., Münster, C., and Salditt, T. 2004. Conformation of peptides in lipid membranes studied by X-ray grazing incidence scattering. *Biophys. J.* 87: 396–407.

Stachowiak, J. C., Hayden, C. C., and Sasaki, D. Y. 2010. Steric confinement of proteins on lipid membranes can drive curvature and tubulation. *Proc. Natl. Acad. Sci. U. S. A.*107: 7781–7786.

Stachowiak, J. C., Schmid, E. M., Ryan, C. J. et al. 2012. Membrane bending by protein–protein crowding. *Nat. Cell Biol.* 14: 944–949.

Subbalakshmi, C. and Sitaram, N.1998. Mechanism of antimicrobial action of indolicidin. *FEMS Microbiol. Lett.* 160: 91–96.

Sweitzer, S. M. and Hinshaw, J. E. 1998. Dynamin undergoes a GTP-dependent conformational change causing vesiculation. *Cell* 93: 1021–1029.

Szleifer, I., Kramer, D., Ben-Shaul, A., Roux, D., and Gelbart, W. M. 1988. Curvature elasticity of pure and mixed surfactant films. *Phys. Rev. Lett.* 60: 1966–1969.

Takei, K., Haucke, V., Slepnev, V. et al. 1998. Generation of coated intermediates of clathrin-mediated endocytosis on protein-free liposomes. *Cell* 94: 131–141.

Takei, K., McPherson, P. S., Schmid, S. L., and Camilli, P. D. 1995. Tubular membrane invaginations coated by dynamin rings are induced by GTP-γS in nerve terminals. *Nature* 374: 186–190.

Takei, K., Slepnev, V. I., Haucke, V., and De Camilli, P. 1999. Functional partnership between amphiphysin and dynamin in clathrin-mediated endocytosis. *Nat. Cell Biol.* 1: 33–39.

Tang, M., Waring, A. J., and Hong, M. 2007. Phosphate-mediated arginine insertion into lipid membranes and pore formation by a cationic membrane peptide from solid-state NMR. *J. Am. Chem. Soc.* 129: 11438–11446.

Tristram-Nagle, S., Chan, R., Kooijman, E. et al. 2010. HIV fusion peptide penetrates, disorders, and softens T-cell membrane mimics. *J. Mol. Biol.*402: 139–153.

Tytler, E. M., Segrest, J. P., Epand, R. M. et al. 1993. Reciprocal effects of apolipoprotein and lytic peptide analogs on membranes. Cross-sectional molecular shapes of amphipathic alpha helixes control membrane stability. *J. Biol. Chem.* 268: 22112–22118.

van Meer, G., Voelker, D. R., and Feigenson, G. W. 2008. Membrane lipids: Where they are and how they behave. *Nat. Rev. Mol. Cell Biol.* 9: 112–124.

Wang, Z. and Wang, G. 2004. APD: The antimicrobial peptide database. *Nucleic Acids Res.* 32: D590–D592.

Wang, Z. G. 1992. Curvature instability of diblock copolymer bilayers. *Macromolecules* 25: 3702–3705.

Watnick, P. I., Chan, S. I., and Dea, P. 1990. Hydrophobic mismatch in gramicidin A'/lecithin systems. *Biochemistry* 29: 6215–6221.

Weiss, T. M., Yang, L., Ding, L. et al. 2002. Two states of cyclic antimicrobial peptide RTD-1 in lipid bilayers. *Biochemistry* 41: 10070–10076.

White, S. H. and Wimley, W. C. 1999. Membrane protein folding and stability: Physical principles. *Annu. Rev. Biophys. Biomol. Struct.* 28: 319–365.

Wieprecht, T., Dathe, M., Epand, R. M. et al. 1997. Influence of the angle subtended by the positively charged helix face on the membrane activity of amphipathic, antibacterial peptides. *Biochemistry* 36: 12869–12880.

Wollert, T. and Hurley, J. H. 2010. Molecular mechanism of multivesicular body biogenesis by ESCRT complexes. *Nature* 464: 864–869.

Wollert, T., Wunder, C., Lippincott-Schwartz, J., and Hurley, J. H. 2009. Membrane scission by the ESCRT-III complex. *Nature* 458: 172–177.

Wong, G. C. L. and Pollack, L. 2010. Electrostatics of strongly charged biological polymers: Ion-mediated interactions and self-organization in nucleic acids and proteins. *Annu. Rev. Phys. Chem.* 61: 171–189.

Wong, G. C. L., Tang, J. X., Lin, A. et al. 2000. Hierarchical self-assembly of F-actin and cationic lipid complexes: Stacked three-layer tubule networks. *Science* 288: 2035–2039.

Wu, Z., Cui, Q., and Yethiraj, A. 2013. Why do arginine and lysine organize lipids differently? Insights from coarse-grained and atomistic simulations. *J. Phys. Chem. B*117: 12145–12156.

Yamaguchi, S., Huster, D., Waring, A. et al. 2001. Orientation and dynamics of an antimicrobial peptide in the lipid bilayer by solid-state NMR spectroscopy. *Biophys. J.* 81: 2203–2214.

Yang, L., Gordon, V. D., Trinkle, D. R. et al. 2008. Mechanism of a prototypical synthetic membrane-active antimicrobial: Efficient hole-punching via interaction with negative intrinsic curvature lipids. *Proc. Natl. Acad. Sci. U. S. A.* 105: 20595–20600.

Yang, L., Harroun, T. A., Weiss, T. M., Ding, L., and Huang, H. W. 2001. Barrel-stave model or toroidal model? A case study on melittin pores. *Biophys. J.* 81: 1475–1485.

Yeaman, M. R. and Yount, N. Y. 2003. Mechanisms of antimicrobial peptide action and resistance. *Pharmacol. Rev.* 55: 27–55.

Yonezawa, A., Kuwahara, J., Fujii, N., and Sugiura, Y. 1992. Binding of tachyplesin I to DNA revealed by footprinting analysis: Significant contribution of secondary structure to DNA binding and implication for biological action. *Biochemistry* 31: 2998–3004.

Yu, Y., Vroman, J. A., Bae, S. C., and Granick, S. 2010. Vesicle budding induced by a pore-forming peptide. *J. Am. Chem. Soc.* 132: 195–201.

Zachowski, A. 1993. Phospholipids in animal eukaryotic membranes: Transverse asymmetry and movement. *Biochem. J.* 294: 1–14.

Zasloff, M. 1987. Magainins, a class of antimicrobial peptides from *Xenopus* skin: Isolation, characterization of two active forms, and partial cDNA sequence of a precursor. *Proc. Natl. Acad. Sci. U. S. A.* 84: 5449–5453.

Zasloff, M. 2002. Antimicrobial peptides of multicellular organisms. *Nature* 415: 389–395.

Zemel, A., Ben-Shaul, A., and May, S. 2008. Modulation of the spontaneous curvature and bending rigidity of lipid membranes by interfacially adsorbed amphipathic peptides. *J. Phys. Chem. B* 112: 6988–6996.

Zhang, P. and Hinshaw, J. E. 2001. Three-dimensional reconstruction of dynamin in the constricted state. *Nat. Cell Biol.* 3: 922–926.

Zhao, K., Choe, U.-J., Kamei, D. T., and Wong, G. C. L. 2012. Enhanced activity of cyclic transporter sequences driven by phase behavior of peptide-lipid complexes. *Soft Matter* 8: 6430–6433.

Zimmerberg, J. and Gawrisch, K. 2006. The physical chemistry of biological membranes. *Nat. Chem. Biol.* 2: 564–567.

Zimmerberg, J. and Kozlov, M. M. 2006. How proteins produce cellular membrane curvature. *Nat. Rev. Mol. Cell Biol.* 7: 9–19.

Zimmerberg, J. and McLaughlin, S. 2004. Membrane curvature: How BAR domains bend bilayers. *Curr. Biol.* 14: R250–R252.

9 Lipid Membrane Shape Evolution and the Actin Cytoskeleton

David R. Slochower
University of California San Diego, La Jolla, California

Yu-Hsiu Wang
The University of Texas Medical Branch, Galveston, Texas

Ravi Radhakrishnan and Paul A. Janmey
University of Pennsylvania, Philadelphia, Pennsylvania

CONTENTS

9.1 Introduction 161
9.2 Mechanisms of Producing Membrane Curvature Driven by the Cytoskeleton 162
9.3 Phosphoinositides Regulate the Membrane–Cytoskeletal Interface 162
    9.3.1 Phosphoinositides Bind Many Curvature-Generating Proteins 163
    9.3.2 Structure of Phosphoinositides in Eukaryotic Cells 163
    9.3.3 Nonrandom Distribution of PPIs in the Cell Membrane 163
9.4 Regulation of Actin Cytoskeleton Remodelling Depends on Lipid Lateral Organization 164
    9.4.1 Regulation of Submembrane Actin Assembly by Membrane PPIs 164
    9.4.2 The Cortical Cytoskeleton Perturbs the Lateral Structure of Membranes 165
    9.4.3 Actomyosin Dynamics Drives Transient Membrane Reorganization 165
    9.4.4 Membrane Curvature Driven by Actin Polymerization 165
9.5 Phosphoinositide Effects on Curvature-Dependent Vesicle Trafficking 166
    9.5.1 Modelling Cell Membrane Curvature 168
    9.5.2 Modelling the Actin–Membrane Linkage 169
9.6 What Can Molecular Dynamics Simulations Tell Us about How Phosphoinositides and Other Phospholipids Lead to Membrane Binding and Shape Changes? 169
    9.6.1 Molecular Dynamics Simulations of Phosphoinositides in Membranes 169
    9.6.2 Molecular Dynamics Simulations of Phospholipid–Protein Interactions 170
9.7 Conclusions 171
Acknowledgements 171
Note 171
References 171

9.1 INTRODUCTION

The two-dimensional fluid lipid bilayer of the cell membrane is coupled chemically and physically to a two- or three-dimensional protein meshwork that is best characterized as a viscoelastic solid on the timescale at which processes that regulate membrane and cell shape occur. The lipid bilayer and its complement of transmembrane proteins provide docking sites for cytoskeletal elements, both actin filaments and intermediate filaments, and generate signals that control cytoskeletal assembly. In turn, the cytoskeleton controls many features of the cell membrane including curvature, area, and lipid lateral distribution within the bilayer that under some conditions can lead to phase demixing. In addition to immediate contacts between the cytoskeleton and the membrane, the cytoskeleton also helps determine the total plasma membrane surface area by means of regulating the microtubule- and actin-motor-dependent transport of vesicles that fuse with the plasma membrane or that mediate invagination and fission of endocytic and phagocytic vesicles.

The membrane–cytoskeleton interface controls a multitude of cellular processes including chemical and mechanical signal transduction, motility/migration, membrane traffic, as well as cell–cell and cell–matrix adhesion. Central to nearly all of these processes is a small group

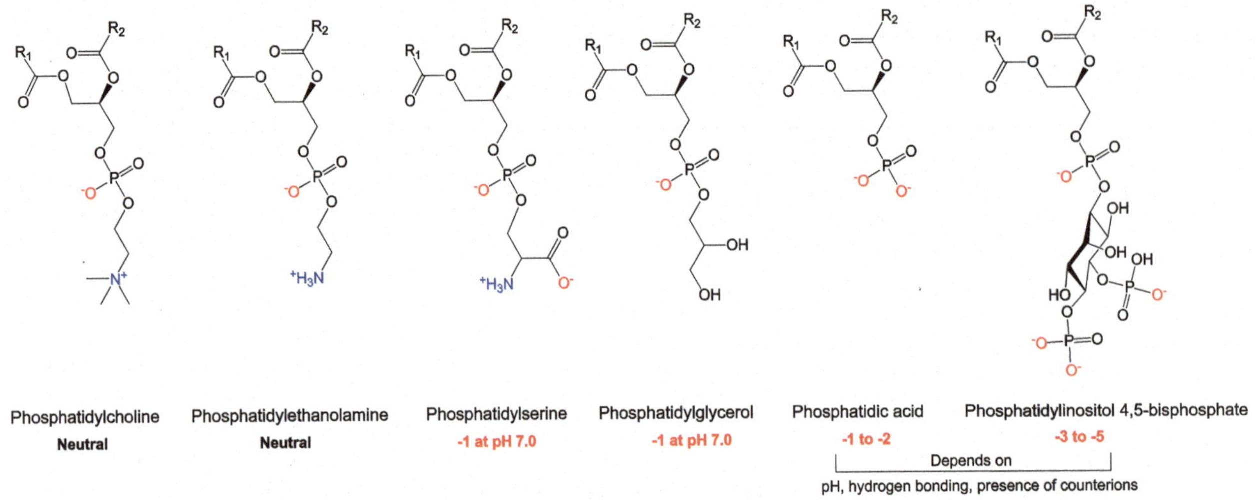

Phosphatidylcholine — Neutral
Phosphatidylethanolamine — Neutral
Phosphatidylserine — -1 at pH 7.0
Phosphatidylglycerol — -1 at pH 7.0
Phosphatidic acid — -1 to -2
Phosphatidylinositol 4,5-bisphosphate — -3 to -5
Depends on pH, hydrogen bonding, presence of counterions

FIGURE 9.1 Structures and electrostatic charges of PtdIns(4,5)P_2 and PtdIns(3,4,5)P_3 in comparison with the more abundant inner leaflet phospholipids phosphatidylserine and phosphatidylethanolamine.

of highly negatively charge phospholipids, the polyphosphoinositides (PPIs). These lipids, most prominently phosphatidylinositol 4,5 bisphosphate (PIP2), but also the less abundant PI3,4,5P3, PI3,5P2, and PI4P,[1] are each essential for different specific cellular processes (Figure 9.1). Many processes such as endo-, exo-, or phagocytosis as well as membrane protrusion during cell motility involve changes in membrane curvature. The aim of this chapter is to provide an overview of the role of phosphoinositides in coupling between the membrane and the cytoskeleton, and the potential that these lipids alone or in concert with proteins are key regulators of curvature.

9.2 MECHANISMS OF PRODUCING MEMBRANE CURVATURE DRIVEN BY THE CYTOSKELETON

On the surface, it might seem that the presence of a solid three-dimensional cytoskeleton might inhibit or retard the rapid changes in membrane curvature associated with many cell processes, since the rates of actin polymerization and depolymerization occur on timescales of seconds, whereas the thermally driven fluctuations of the lipid bilayer on a subcellular length scale are typically much faster. However, the nonthermal motions generated as cytoskeletal polymers assemble, cross-links form, motor proteins activate, and network dissolving proteins function play a large role in shaping the dynamic interface between the lipid bilayer in the cytoskeleton. For cases in which lipid bilayers are firmly attached to an underlying cytoskeleton, mismatches between the total surface area of the lipid bilayer and the underlying cortical cytoskeleton create mechanical tensions and allow for mechanisms of energy storage not possible for lipid bilayers themselves. One example of this effect is seen in the flickering of red blood cells in which the red cell bilayer is coupled to an underlying two-dimensional spectrin–actin network of unequal effective surface area, and in which the rates and amplitudes of membrane fluctuations depend on ATP hydrolysis as the strength of coupling between the membrane and the cytoskeleton is actively altered (Rodriguez-Garcia et al. 2015, Zeman et al. 1990).

9.3 PHOSPHOINOSITIDES REGULATE THE MEMBRANE–CYTOSKELETAL INTERFACE

A comprehensive summary of the many proteins that link the cytoskeleton to the membrane and how these linkages are regulated is beyond the scope of this chapter and has been considered in other recent publications (Bezanilla et al. 2015, Kapus and Janmey 2013). Here, we focus on one specific aspect of this linkage, mainly the role of polyphosphoinositides that are rapidly created, modified, and destroyed at the inner leaflet of the cell plasma membrane by the activities of an array of lipid kinases, phosphatases, and phospholipases. These lipids, which constitute less than 1% of the total phospholipid content of the typical eukaryotic cell (Bunney and Katan 2010, Funakoshi et al. 2011, Kwiatkowska 2010, Mao and Yin 2007, Miled et al. 2007, Sun et al. 2010, van den Bout and Divecha 2009, Yin and Janmey 2003), interact with scores or hundreds of soluble and membrane-bound proteins (Cadmel et al. 2008), including many proteins that regulate actin assembly and that link the actin cytoskeleton to cell–matrix, or cell–cell contact sites, proteins that induce lipid bilayer curvature, and transmembrane ion channels that can produce osmotic pressures at sites of activation and that trigger the activities of cytoskeletal remodelling proteins. The power of PPIs to drive membrane curvature is illustrated by a recent report that perturbation of local concentrations of PIP2 and PIP3 by optogenetic activation of PPI phosphatases is sufficient to create membrane waves driven by the actin cytoskeleton (Xiong et al. 2016).

9.3.1 Phosphoinositides Bind Many Curvature-Generating Proteins

At least ten different classes of proteins across several species, from yeast to humans, interact with PPIs in the plasma membrane with different degrees of specificity and affinity (Lemmon 2007, 2008, Moravcevic et al. 2010). The first domain discovered to localize to membranes in a target-specific manner was the pleckstrin homology (PH) domain (Lemmon and Ferguson 2000), and since then, the 100-amino acid phospholipase Cδ_1 PH (PLCδ-PH) domain has been shown to recognize PPIs – either PtdIns(4,5)P_2 or the cleaved head group Ins(1,4,5)P_3 – in an isomer-specific manner. The PLCδ-PH domain binds Ins(1,4,5)P_3 even more effectively than PtdIns(4,5)P_2 through a set of four or more positively charged residues without significant membrane penetration of the protein. There are several mechanisms through which PH domains localize to membranes (Lumb and Sansom 2012). The selectivity of various PH domains has been used to mark the location of PPI isomers in different cellular compartments.

Other domains – PX, C1, and FYVE – bind PPIs and concomitantly insert small hydrophobic side chains into the membrane. The ENTH domain of epsin induces membrane curvature by inserting an entire amphipathic alpha helix into one leaflet of the membrane upon binding PtdIns(4,5)P_2 (Campelo et al. 2008). The insertion increases the surface area of one leaflet, and the mismatch in area of the two bilayer leaflets produces membrane curvature and can even lead to extended membrane tubules; mutations that abolish insertion of the helix also abolish formation of membrane curvature (Zimmerberg and Kozlov 2006). In addition to the insertion of amphipathic helices, membranes can be deformed when proteins act as scaffolds (for positive or negative curvature) or by polymerization of the cytoskeleton at the membrane interface (McMahon and Gallop 2005). BAR domains have an intrinsic curvature and bend membranes to match their shape. Clathrin and clathrin–adapter–protein complexes have a strongly concave surface and can stabilize regions of high curvature by forming a polyhedral lattice surrounding vesicles. Lipid composition and the inclusion of integral membrane proteins have additional effects on membrane morphology.

9.3.2 Structure of Phosphoinositides in Eukaryotic Cells

In addition to their specific recruitment of curvature-generating or curvature-sensing proteins, PPIs themselves can promote curvature of the bilayer due to their unusually large head group size, their high electrostatic charge, and the frequent incorporation of the large and polyunsaturated fatty acid arachidonate as one of its two acyl chains (Figure 9.1). Both steric and electrostatic effects of PIP2 head groups tend to increase the surface area of the leaflet in which these lipids are enriched (Levental et al. 2008a, b), thereby promoting positive curvature. Their rapid hydrolysis by phospholipase C would leave diacyl glycerol, with a much smaller surface area, leading to negative curvature. The degree to which these purely lipid-driven mechanical effects control the cell membrane would depend on the local concentration of PIP2, and direct evidence that the lipids rather than the protein they recruit dominate their effects on cell membrane curvature is unclear (Baumgart et al. 2011, Shi and Baumgart 2015).

9.3.3 Nonrandom Distribution of PPIs in the Cell Membrane

One of the major unresolved issues in understanding how phosphoinositides regulate cytoskeletal assembly and membrane curvature is how these scarce lipids, with total concentrations between 10 and 100 micromolar calculated as total lipid divided by cell volume, nevertheless affect the activity of many proteins (Cadmel et al. 2008), some of which, like profilin, the first discovered PIP2 binding protein, alone have concentrations exceeding 100 micromolar. Selective binding of PPIs to their target proteins under specific cellular states appears to depend on local production (Kwiatkowska 2010), delivery (Chierico et al. 2014) or hydrolysis of the lipids (Xiong et al. 2016), their lateral distribution within the bilayer (Picas et al. 2016), how their head groups are exposed to the cytosol (Slochower et al. 2013), and the functions of proteins that appear to electrostatically sequester and inactivate PPIs at the membrane surface (Gambhir et al. 2004). Measurement of PIP2 diffusion shows that most of the plasma membrane PIP2 pool is bound or sequestered to some extent (Golebiewska et al. 2008). A major unresolved question is how PIP2 distributes laterally within the plasma membrane (Slochower et al. 2015, Wang et al. 2014) and whether all PIP2 molecules are equally effective at binding their targets.

Several recent studies show the relevance of nanoscale PIP2 clusters to critical PIP2-triggered cellular functions (Chierico et al. 2014, Honigmann et al. 2013, van den Bogaart et al. 2011, Wang and Richards 2012). Clusters of the SNAP receptor protein syntaxin-1A on the plasma membrane (Sieber et al. 2007), which control neuronal exocytosis, require formation of lipid domains with approximately 73 nm diameter, which are 80% PIP2 (Figure 9.2a and d) (van den Bogaart et al. 2011, Wang and Richards 2012), and similar PIP2 clusters were detected in cleavage furrows of intact cells (Abe et al. 2012). Strikingly, μM Ca^{2+} induces PIP2 clustering with similar size distribution in PIP2-containing model membranes in the absence of proteins (Figure 9.2d and f) (Wang et al. 2012). PIP2-dependent clustering of syntaxin 1 in the presynaptic plasma membrane is reduced when cholesterol is depleted (Dason et al. 2014), but significant PIP2 clustering is not reduced when cortical actin assembly is disrupted with

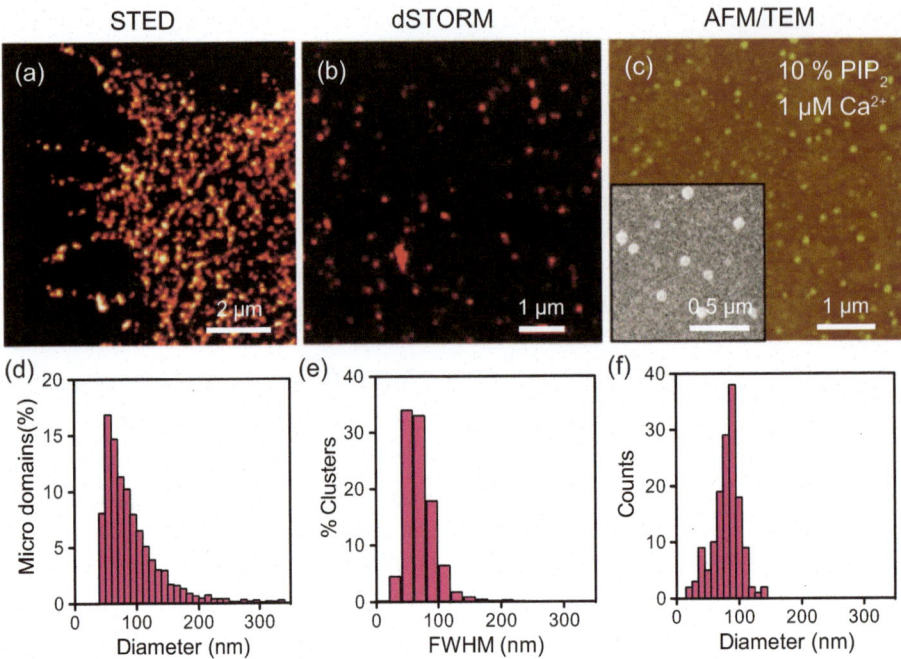

FIGURE 9.2 Submicron domains of PIP2. (a) STED image of PC12 cell membrane (van den Bogaart et al. 2011) and (b) dSTORM image of PC12 cell stained with anti-PIP2 antibody (Wang and Richards 2012). (c) PIP2 clusters in supported lipid monolayers with 10 mol% PIP2 (Wang et al. 2012). (d–f) Size distribution of PIP2 clusters from (a) to (c).

inhibitors of ARP2.3-dependent actin nucleation, suggesting that PIP2 clusters precede and do not result from actin nucleation.

The interactions between actin assembly and the formation of lipid rafts (Byrum and Rodgers 2015, Caroni 2001, Chichili and Rodgers 2009, Koushik et al. 2013) or PIP2 nanoclusters (Arumugam and Bassereau 2015, Brown 2015, Chierico et al. 2014, Honigmann et al. 2013, Masters et al. 2016, Wang et al. 2012) have elicited much attention and are discussed in detail by several recent reviews (Arumugam and Bassereau 2015, Brown 2015, Chichili and Rodgers 2009, Kusumi et al. 2012, Slochower et al. 2014, Wang et al. 2014). In the following section, we focus on how actin cytoskeleton remodelling might be regulated by the phase segregation of membrane lipids, and conversely, how actin cytoskeleton remodelling perturbs the lateral structure of membrane constituents.

These structural features in turn depend on the distribution of PPIs with other lipids that constitute most of the lipid bilayer.

The membrane bilayer segregates into different phases including the L_d (liquid disordered), the L_o (liquid ordered), and the L_β (gel) phases when the membrane is composed of cholesterol, saturated and unsaturated lipids (Veatch and Keller 2002, 2003), and PIP_2 partitions preferentially into the liquid disordered phase (Leventhal et al. 2009). As many PIP2-binding proteins interact with PIP2 through an unstructured polybasic stretch, their interactions with the membrane depend on its surface potential and are therefore sensitive to the local density of PIP2, which is the most highly charge lipid in the bilayer. Increased protein–PIP2 binding can occur after inducing L_o/L_d phase segregation, and this effect can promote protein crowding on model membranes (Stachowiak et al. 2012).

9.4 REGULATION OF ACTIN CYTOSKELETON REMODELLING DEPENDS ON LIPID LATERAL ORGANIZATION

PPIs bind many cytoskeleton and membrane-associated proteins by mechanisms that include both electrostatic interactions with areas of local membrane charge density and insertion of protein domains into the lipid bilayer. Both of these features depend on how the PPIs are distributed within the lipid bilayer and how accessible their head groups as well as their acyl chains are to the target protein.

9.4.1 Regulation of Submembrane Actin Assembly by Membrane PPIs

PIP2-mediated effects on the assembly and dynamics of the cortical cytoskeleton include interaction with nucleation promoting factors that initiate actin assembly, as well as proteins that mediate actin monomer sequestering, actin severing, and barbed-end capping. The interaction between PIP2 and the actin cytoskeleton has been reviewed in detail elsewhere (Janmey 1994, Kapus and Janmey 2013, Yin and Janmey 2003).

9.4.2 THE CORTICAL CYTOSKELETON PERTURBS THE LATERAL STRUCTURE OF MEMBRANES

Membrane reorganization triggered by actin polymerization was demonstrated by a study showing that the polymerized actin outside of a phase-demixed giant unilamellar vesicle (GUV) was associated with the L_d domain and led to a positive shift in the miscibility transition temperature (Liu and Fletcher 2006), suggesting that PIP2-mediated actin polymerization, but not the binding of N-WASP alone, promotes phase segregation by stabilizing PIP2-containing microdomains. More surprisingly, it was found that a pre-existing actin network spatially directs the formation of L_d microdomains under the actin meshwork patch. Such spatial biasing may be explained by a preferential localization of PIP2-N-WASP with actin networks that in turn localizes associated lipids in the L_d phase.

However, large-scale phase segregation evident by standard fluorescence microscopy is not generally seen on the plasma membrane of a live cell, although the liquid/liquid phase segregation does occur in cell-derived membranes when they are detached from the cytoskeleton and at 25°C or below (Baumgart et al. 2007). This finding suggests that the membrane compositions of mammalian cells at physiological temperature are close to a miscibility critical point at which compositional fluctuations allow the formation of domains smaller than 50 nm, as shown in giant plasma membrane vesicles at 37°C (Veatch et al. 2008).

Another factor that prevents lipids from phase separating in plasma membranes is likely imposed by the cortical actin cytoskeleton, by a mechanism called cytoskeletal pinning (Ehrig et al. 2011, Machta et al. 2011). The pinning effect can be understood as the coupling between membrane and the underlying actin cytoskeleton that helps preserve the heterogeneous nature of membranes at 37°C, while preventing macroscopic domains from forming through the dense meshwork of polymerized actin (Ehrig et al. 2011). Experimental evidence for the pinning effect in model membranes was reported by super-resolution microscopy (Honigmann et al. 2014). Pinning-mediated perturbation is not limited to actin filaments but can also be triggered by cytoskeletal filaments composed of the prokaryotic tubulin homolog, FtsZ (Arumugam et al. 2015).

The ezrin, radixin, moesin (ERM) protein family is one potential mediator of pinning, which interacts with actin directly through its actin-binding domain (ABD) and with PIP2 through its FERM domain (for 4.1, ezrin, radixin, and moesin) (Fehon et al. 2010, Pearson et al. 2000). The dual structural functionality allows ERM family protein to serve as a membrane anchor for actin filaments. Talin is also a potential membrane anchor for actin filaments that also possesses a FERM domain and binds both PIP2 and integrins (Elliott et al. 2010).

The actin cytoskeleton might also impose other effects on the plasma membrane of a cell. Cellular studies revealed that the attachment of actin cytoskeleton at the plasma membrane induces the formation of L_o domains (Dinic et al. 2013), which is in line with an *in vitro* finding that the adhesion between vesicles promotes the formation of L_o domains at the contact site (Gordon et al. 2008). Moreover, actin-driven formation of cholesterol-dependent microdomains in a tubular membrane invagination is part of the scission mechanism for clathrin-independent endocytosis (Romer et al. 2010).

9.4.3 ACTOMYOSIN DYNAMICS DRIVES TRANSIENT MEMBRANE REORGANIZATION

Another aspect of cytoskeleton-induced perturbation in membranes is not mediated by actin itself but through its actomyosin contractility. A growing number of observations appear inconsistent with thermodynamic equilibrium. As a result, these studies lead to an increasing interest in actomyosin contractility-mediated perturbation in model membranes, and a membrane-associated component that is capable of linking actin cytoskeleton to the membrane has become a straightforward target for such investigation.

The concept for an active cytoskeletal remodelling-mediated membrane perturbation has been described in a coarse-grained simulation and corroborated by experiment in live cells (Gowrishankar et al. 2012). The underlying mechanism of this phenomenon was demonstrated with a recombinant peptide derived from the actin-binding domain of ezrin (EzrABD) (Saldanha et al. 2016). The complex membrane-binding capability of EzrABD was replaced by a direct interaction with Ni:NTA lipids through a recombinant His tag for the sake of simplicity. These studies showed that transient aggregation of EzrABD, and presumably its associated lipids, in a lipid bilayer happens in the presence of an active actomyosin complex, but not in the absence of type II myosin or in a jammed system. These results suggest that a contractile flow of actin could lead to a transient accumulation of membrane components.

9.4.4 MEMBRANE CURVATURE DRIVEN BY ACTIN POLYMERIZATION

Actin assembly at the leading edge of cells is well documented to provide the force required for cell protrusion, and similar physical principles of actin polymerization at a membrane boundary also power intracellular motions such as movement of endosomes, bacteria, viruses, and other membrane-bound organelles or pathogens (Boulant et al. 2011, Tsujita et al. 2015). A large fraction of the actin regulatory factors that initiate actin assembly at these interfaces are activated by PPIs. The importance of membrane lipids for organizing and controlling force generation by actin assembly is illustrated by the finding that introducing purified phospholipid vesicles containing PtdIns(4,5)P_2 or PtdIns(3,4,5)P_3 to a cell extract is sufficient to generate a tail of polymerizing actin that propels these vesicles through the extract (Ma et al. 1998). Despite numerous reviews on the topic (Anitei and Hoflack 2011, Takenawa and Suetsugu 2007), a major unrealized goal is to understand the effect of force coupling, cytoskeletal membrane attachment, and actin recruitment on the membrane signalosome assembly.

9.5 PHOSPHOINOSITIDE EFFECTS ON CURVATURE-DEPENDENT VESICLE TRAFFICKING

Spatiotemporal orchestration of functional protein–lipid assemblies transduces outside-in as well as inside-out signalling and ultimately links (often governs) intracellular signalling to cell function. Integrative approaches have been pursued to unravel the long-standing question of how phosphoinositides regulate such processes. These can be classified into several categories as briefly described below. The physical chemistry of polymers and phospholipids in the context of their effect on cell behaviour can be influenced by the effects of PPIs on protein function, as well as intracellular signalling related to cytoskeletal assembly (Kapus and Janmey 2013). Modular domains in signalling and trafficking proteins that recognize specific membrane phospholipids (Lemmon 2007, Moravcevic et al. 2012) have been identified, which have led to a clearer picture of mechanisms surrounding several membrane association domains such as PH domains and to how these domains can play effector roles by altering phospholipid distribution and/or membrane morphology (Bethoney et al. 2009). As a complement to such studies focused on the structural aspects, other investigations have focused on intracellular membrane trafficking events that underlie endosomal processing events and the formation of lysosome-related organelles (Marks et al. 2013). The use of quantitative experimental biophysical techniques, theoretical methods, and simulation methodologies has contributed significantly to the understanding of membrane curvature/shape, and how this is coupled to membrane composition (Baumgart et al. 2011, Ramakrishnan et al. 2014b). The biophysical approaches are beginning to synergize with cellular approaches studying mechanisms of trafficking, motility, and cytoskeletal dynamics to provide an integrated picture of function (Zhao et al. 2013). The use of mouse genetics to study PPI-dependent processes and their essential roles in cytoskeletal dynamics and PPI metabolism have also been reported (Suzuki et al. 2013, Wang et al. 2013).

The formation of membrane-bound functional assemblies (membrane signalosomes) is often linked to recruitment and reorganization of actin filaments, which provides the force required for multiple biological processes. Most such processes are coupled to deformation of the cell membrane. Examples of these membrane–cytoskeleton-coupled processes include formation of filopodia, lamellipodia, and podosomes for cell movement or cancer cell invasion, endocytosis, phagocytosis, exocytosis, and various membrane-trafficking events (Gallop et al. 2013, Kapus and Janmey 2013). The central interactions of the various components of membrane and intracellular trafficking are summarized in Table 9.1.

TABLE 9.1
Membrane Signalosome and the Interactions within Membrane and Intracellular Trafficking

	Endocytosis	Exocytosis	Endosomal Sorting and Recycling
Biochemical Signal: Proteins	Clathrin, AP-2, ENTH, BAR, dynamin (PH) (Kutateladze 2010, Lemmon 2008)	Exocyst: Exo70, Sec3 (PH), Snare (FYVE, C2) (Kutateladze 2010, Lemmon 2008, Liu and Guo 2012, Liu et al. 2009b, Orlando and Guo 2009, Orlando et al. 2011)	KIF3 (PH), AP-3, SNX (PX), ESCRT (GLUE) (Grant and Donaldson 2009, Kutateladze 2010, Lemmon 2008, Williams and Urb 2007)
Biochemical Signal: Lipids	PtdIns(3,4)P_2, PtdIns(4,5)P_2 PtdIns(3,4,5)P_3, PtdIns4P, DAG, other PPIs? (Posor et al. 2013, Vicinanza et al. 2008a, b)	PtdIns(3,5)P_2, other PPIs? (Lee et al. 2010, Liu et al. 2007a, Vicinanza et al. 2008a, b)	PtdIns(3,5)P_2, PtdIns(3,4,5)P_2, PtdIns(4,5)P_2, DAG, other PPIs? (Cullen and Korswagen 2011, Vicinanza et al. 2008a, b, Zoncu et al. 2009)
Mechanical Signal	Curvature, tension, force, adhesion (cargo size), line tension (Kapus and Janmey 2013)	Curvature, Force, tension (Kapus and Janmey 2013, Lee et al. 2010)	Curvature, tension, force, adhesion (cargo size), line tension (Gallop et al. 2013, Kapus and Janmey 2013)
Mechanobiochemical Coupling	FCHo, clathrin, N-WASP, WAVE, Arp2/3, ARF, GTPases (PH), actin, myosin (Kapus and Janmey 2013, Papayannopoulos et al. 2005)	Exo-70, WASP, WAVE, ARP2/3, GTPases (PH), actin, myosin (Liu et al. 2012b, Papayannopoulos et al. 2005, Ren and Guo 2012, Zuo et al. 2006)	SNX, dynein, AP-3, kinesin, microtubule, annexin, Arp2/3, GTPases (PH), actin (Cullen and Korswagen 2011, Kapus and Janmey 2013, Papayannopoulos et al. 2005, Puthenveedu et al. 2010, Williams and Urb 2007)

The studies summarized in this table discuss how membrane properties such as lipid composition, and membrane tension, modulate and lead to the induction of membrane shape transitions that promote formation of highly bent membrane structures, with the goal to link clustering of membrane components functionally with the onset of membrane shape transitions. One of the major functions of the membrane signalosome is to orchestrate membrane remodelling and actin assembly, where the coupling is through curvature inducing/sensing proteins.

Formation of membrane signalosomes can be understood by spatiotemporal studies (using live cell imaging approaches and *in vitro* analyses of functional protein–lipid assemblies) of endocytosis, exocytosis, and endosomal sorting (Table 9.1). Published studies have defined and elucidated the underlying lipid heterogeneity, specificity, and cooperativity in protein recruitment (e.g., through PH, ENTH, BAR domain, and other proteins).

The recruitment of effector proteins and specific lipids (e.g., phosphoinositides) into functional nanodomains (see Table 9.1), and their defined spatial organization, is key to the orchestration of trafficking processes. Self-assembly of these protein/lipid complexes and membrane nanodomains is driven by electrostatic interactions (van den Bogaart et al. 2011), line tension (Liu et al. 2006), restricted diffusion (Golebiewska et al. 2011), and membrane curvature effects (Zhao et al. 2013). In considering the dynamics of signalosome assembly in vesicle biogenesis at the plasma membrane (PM) and in the *trans*-Golgi network (TGN), it is important to focus on lipid clustering (especially for phosphoinositides) and signalosome assembly at the initial step of endocytosis, induced by clathrin, ENTH and BAR domains (Zhu et al. 2012), and the PH domain-containing GTPase dynamin that catalyses the force generation step in vesicle scission – possibly by modulating lipid distribution (Bethoney et al. 2009). Other settings in which PPIs are intimately associated with membrane curvature include the lipid and protein distribution in the exocyst complex (Liu and Guo 2012) involved in exocytosis, and the protein/lipid assemblies involved in endosomal sorting and recycling – where it is hypothesized, for example, that the phosphoinositide PtdIns(3,4)P_2 plays a special role (Boucrot et al. 2015).

Membrane curvature has recently been identified as a biological signal for membrane recruitment of signalosome components (Gallop and McMahon 2005, Hatzakis et al. 2009), adding functional importance to long-known local plasma membrane heterogeneities in morphology and composition (Figure 9.3). The membrane curvature-generating role of BAR and ENTH domains is well documented through electron microscopy studies (Farsad et al. 2001, Mim et al. 2012), typically using high concentrations of individual pure proteins combined with highly

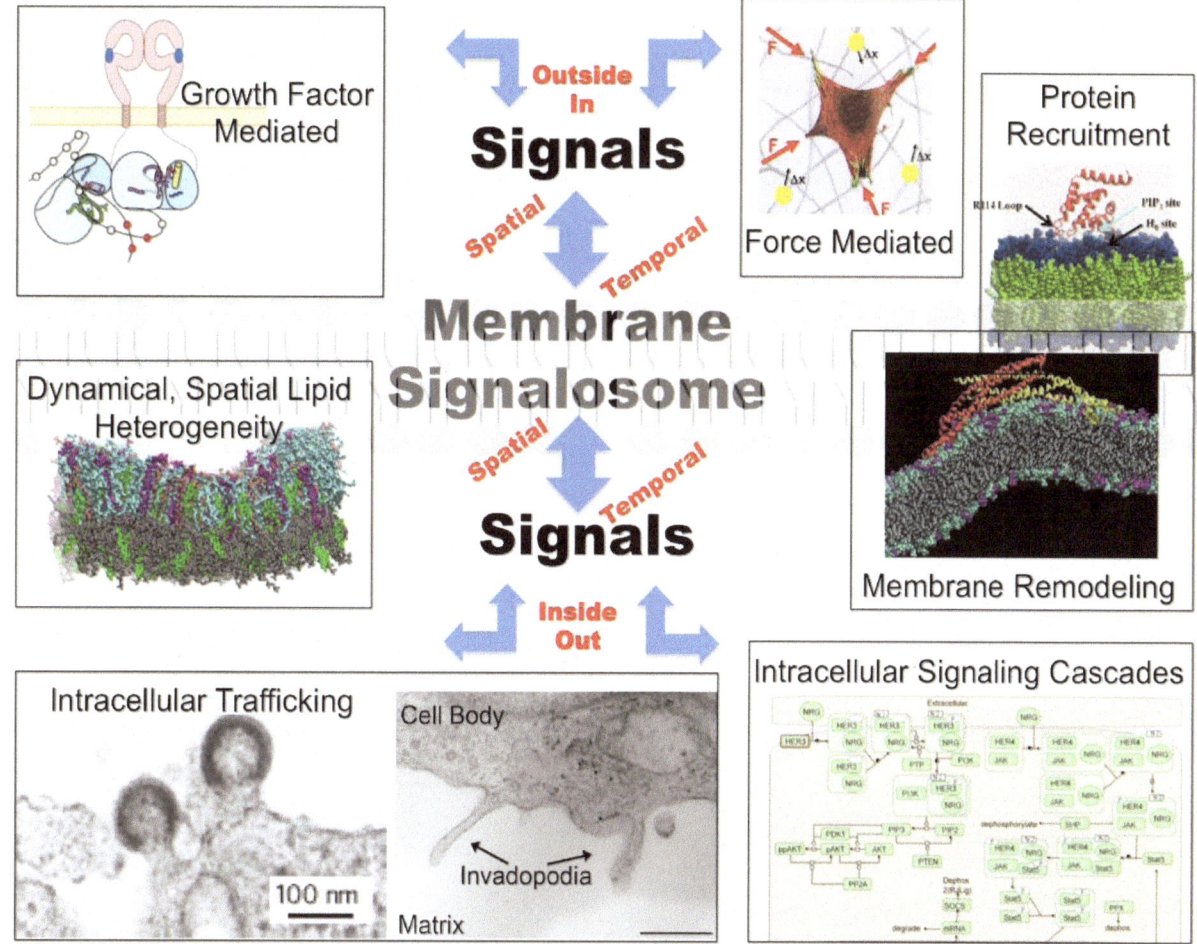

FIGURE 9.3 Schematic overview of membrane recognition by proteins leading to functional modules in several cell functions.

charged membranes. Very recently, techniques have been developed that allow characterization at equilibrium of membrane curvature generation by (and curvature-mediated membrane binding of) lipid-binding proteins (Bhatia et al. 2009, Sorre et al. 2012, Zhu et al. 2012). Much less is known about the dynamics of the formation of membrane elements with curvature, such as buds, tubes, and vesicles that may involve signalosome assembly and function.

9.5.1 Modelling Cell Membrane Curvature

Modelling studies have investigated how membrane properties such as lipid composition, and membrane tension, modulate (i) the recruitment of peripheral proteins (Diz-Munoz et al. 2013, Tourdot et al. 2014, 2015), and (ii) the induction of membrane shape transitions that lead to the formation of highly bent membrane structures (Ramakrishnan et al. 2014a, 2015, Tourdot et al. 2014) (Figure 9.3). Experiments as well as simulations have suggested that transitions between homogeneous lateral distributions of membrane

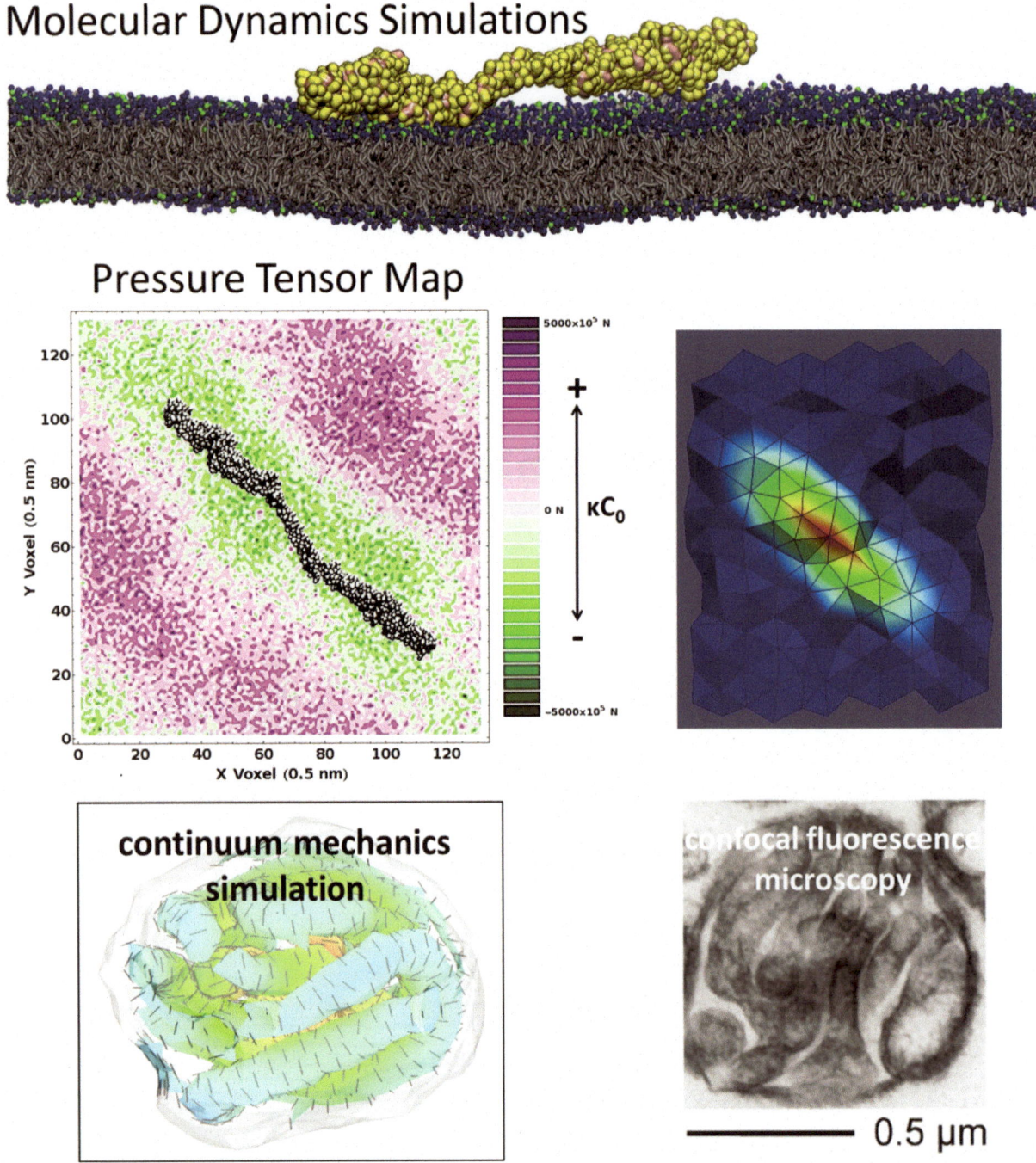

FIGURE 9.4 Multiscale simulation linking molecular interactions to emergent membrane morphology.

components and inhomogeneous localization on bent membrane elements occur at well-defined membrane tensions and protein densities on the membrane (Shi and Baumgart 2015, Tourdot et al. 2015). The onset of membrane shape transitions while varying lipid compositions by changing the mole fraction of charged lipids, and/or varying levels of PtdIns(4,5)P_2, PtdIns(3,4)P_2, and PtdIns(3,5)P_2 has not yet been investigated.

Models and simulations at multiple resolutions (Figure 9.4) have provided valuable insight into phosphoinositide properties at the electronic structure scale, into their protein–lipid interactions, into membrane remodelling processes at both atomic and nano scales (Arkhipov et al. 2009, Ayton and Voth 2010b, Bradley and Radhakrishnan 2013, Cui et al. 2009, 2013, Lai et al. 2012, Mim et al. 2012, Simunovic et al. 2013, Voth 2013, Zhao et al. 2013), and into the energetics of carrier biogenesis at the cellular scale (Agrawal et al. 2010, Liu et al. 2006, 2009a, 2012a, Ramanan et al. 2011, Zhao et al. 2013). Although the treatment of larger length-scale processes in models at a subcellular scale necessitates additional approximations, such models provide a unique opportunity to connect with cellular experiments. Recently, several studies have explored a multiscale approach that links information from the molecular scale modelling to the subcellular-scale morphology modelling (Ayton and Voth 2010a, Bradley and Radhakrishnan 2013, Liu et al. 2007b, Ramanan et al. 2011, Shih et al. 2008, Simunovic et al. 2013, Telesco and Radhakrishnan 2012, Zhao et al. 2013). The key to such a model is the bridging of scales, which is achieved using a method called atom-field mapping. In brief, by analysing the molecular dynamics (MD) simulations for the deformation of the bilayer midplane, computing the pressure tensor in the bilayer, and analyzing the fluctuation spectrum, one can determine the curvature field induced by a protein on the bilayer. This computed curvature field is then passed to a continuum model for the membrane and is used to compute the membrane morphology as well as the free energy as a function of protein concentration (Agrawal et al. 2010, Agrawal and Radhakrishnan 2009, Bradley and Radhakrishnan 2013, Liu et al. 2012b, Zhao et al. 2013). The structural basis for membrane binding by PH, PX, FYVE, ENTH, BAR, and other domains is well studied (Lemmon 2008, Moravcevic et al. 2012) and provides a rich set of tools for future modelling studies.

9.5.2 Modelling the Actin–Membrane Linkage

Adhesion of mammalian cells mediated by the binding of cell surface integrin receptors has been studied by specifically probing the effect the ligand affinity on the assembly of focal adhesions (Kato and Mrksich 2004). Beyond the focal adhesions (which are generally regarded as representing a canonical mechanism of coupling membrane cytoskeletal interactions), there can be several noncanonical interactions defining the coupling (Diz-Munoz et al. 2013, Tourdot et al. 2014, 2015). A conceptual framework as to how the actin assembly process and cytoskeletal interactions can be bidirectionally coupled to the recruitment of curvature inducing and sensing proteins is through the coupling of curvature generation and membrane tension. A stated hypothesis that is worth pursuing in the cellular context is that the recruitment of curvature-sensing proteins can be influenced by the physical/mechanical state of the cell membrane thermal fluctuations relaxing the membrane undulations, the fluid nature of the membrane, and membrane physical environment variables such as tension, osmotic stress, stiffness, and excess membrane area.

9.6 WHAT CAN MOLECULAR DYNAMICS SIMULATIONS TELL US ABOUT HOW PHOSPHOINOSITIDES AND OTHER PHOSPHOLIPIDS LEAD TO MEMBRANE BINDING AND SHAPE CHANGES?

MD simulations allow us to investigate the behaviour of PPIs over a range of timescales and length scales – from picoseconds to microseconds and from Angstroms to microns – to study the dynamics of biological membranes. Additionally, MD simulation methods allow the inspection and manipulation of PPI properties – such as their net charge and partial atomic charge distribution – that are intimately related to their biological function by mediating intermolecular contacts with other lipids and ions. Many *in vitro* and *in vivo* experimental characterizations of PPIs have focused on their localization (by using fluorescently tagged proteins) and structure (X-ray and neutron scattering), and these measurements generally report static properties. MD simulations are indispensable for testing variable membrane compositions in different ionic environments in the absence and presence of proteins.

9.6.1 Molecular Dynamics Simulations of Phosphoinositides in Membranes

MD simulations of PPIs began in the early 2000s by characterizing how a small number of PPIs embedded in a phosphatidylcholine bilayer interacted with the protein gelsolin (Liepina et al. 2003). The phosphoinositides carried a charge of −5e and caused fluctuations in the thickness of the membrane, exposing the 4- and 5-phosphate groups to solvent, whereas the nonpolar side chains of gelsolin formed hydrophobic contacts with the acyl chains of the phosphoinositides. By the end of the decade, the number of MD simulations involving highly charged phospholipids was growing. In 2009, a set of MD simulations were carried out to determine the positions and orientations of phosphoinositides (here charged at −4e) at physiologically relevant concentrations in phosphatidylcholine bilayers with different levels of hydration ionic concentration (Li et al. 2009), simulating for 50 ns compared with just 3 ns in Liepina et al. (2003). These studies found that the phosphodiester of phosphoinositides was roughly in line with the phosphodiester of phosphatidylcholine – and owing to the much larger head group of PPIs relative to the zwitterionic phosphatidylcholine (PC) – extension of the inositol phosphate groups into solvent. This extension causes a negative electrostatic potential bulge at the location of PPIs in membranes contributing to nonspecific attraction with proteins.

In 2010, another MD simulation investigated neighbour packing between DPPC and PIP2 with new force field parameters and partial atomic charges (using a net charge of $-5e$) (Lupyan et al. 2010). The addition of PIP2 to a pure PC bilayer increases the order of the PC acyl chains and decreases the area per lipid of PC molecules. This study also concluded that the PIP2 head group extends further into solvent than PC and yet it requires much more work to extract a single PIP2 from the membrane than a single PC molecule. In the same year, the CHARMM all-atom force field was updated with new lipid parameters (C36) with the goal of alleviating the spurious positive surface tension in bilayers simulated with the previous iteration of the CHARMM force field (C27) (Hatcher et al. 2009, Klauda et al. 2010, Pastor and Mackerell 2011). The new C36 force field contained phosphate group patches carrying charges of $-1e$, $-2e$, or $-3e$, allowing PIP2 to be simulated at $-4e$ or $-5e$ using the same force field. Around the same time, a united-atom model based on C27 was released, removing explicit hydrogens atoms on the acyl chains of lipids (Henin et al. 2008). At a higher level of coarse graining, the popular Martini force field maps four heavy atoms and their associated hydrogens to a single bead – reducing the degrees of freedom in the system and allowing much longer simulation times – and has been used for a wide number of phospholipid simulations (Marrink et al. 2007, Marrink and Tieleman 2013).

In 2013, work using new quantum calculations incorporating experimental data on the pKa of inositol phosphate groups found that the most stable conformation of PIP2 is proton sharing between the 4- and 5-phosphate with a net charge of $-4e$. Placing the proton on the 5-phosphate group is a good compromise for classical simulations (Kooijman et al. 2009, Slochower et al. 2013, 2015). A comparison of four recent all-atom force fields – C36, GROMOS54a7, Lipid14, and Slipids – demonstrates that many of these force fields can accurately reproduce area per lipid and bilayer thickness to within 10% of experimental values but need work on their dynamic properties such as lipid diffusion coefficients and melting temperature (Pluhackova et al. 2016).

9.6.2 Molecular Dynamics Simulations of Phospholipid–Protein Interactions

Associated with the increased number and accuracy of lipid force fields, there has been a proliferation of simulations with phosphoinositides and proteins (Poyry and Vattulainen 2016). In 2008, the parameters of (Lupyan et al. 2010) were used in a simulation of the N-BAR domain binding to membranes with and without PIP2 (Blood et al. 2008). Mutating the positively charged residues on the concave surface of the N-BAR domain to glutamine reduced the ability of the protein to drive membrane curvature. Exchanging a few PS residues for PIP2 in the simulation resulted in the longest time of protein–membrane contacts, although curvature was not clearly induced in this case. In separate simulations, this study also showed that the insertion of amphipathic helices can induce a significant amount of curvature after just 28 ns.

The interaction between the PLCδ-PH domain and PIP2 was investigated in 2008 (Psachoulia and Sansom 2008) showing restriction in protein flexibility upon binding the head group and allowing a small amount of membrane penetration. Brownian dynamics simulations have shown how anionic membranes attract and align PH domains, effectively steering the protein towards PtdIns(3,4,5)P_3 (Lumb and Sansom 2012). The C2 domain, an eight-stranded antiparallel β-sandwich, is found in the tumour suppressor phosphatase and tensin homolog (PTEN) (Katan and Allen 1999) that associates with membranes in a calcium-dependent manner. PTEN catalyses the dephosphorylation of PtdIns(3,4,5)P_3 to PtdIns(4,5)P_2 on the inner leaflet of the plasma membrane in the PD domain, negatively regulating phosphatidylinositol 3-kinase and downregulating cell proliferation. Coarse-grained MD of PTEN (and a similar domain) demonstrated PtdIns(3,4,5)P_3 in proximity of the active site of PTEN in the PD domain and some attraction of PtdIns(3,4,5)P3 for the cationic C2 domain (Kalli et al. 2014).

Coarse-grained MD simulations of ENTH domains have shown them to stabilize tubular geometry (Lai et al. 2012), and the protein Exo70 has been shown to associate with negatively charged phospholipids and induce negative curvature by acting as a scaffold, similar to BAR domains but in the opposite direction as N-BAR (Zhao et al. 2013).

PIP2 has been found to bind membrane-spanning potassium ion channels in all-atom and coarse-grained MD over the course of 1.5 μs of simulation time with the PIP2 head group making several hydrogen bonds with arginine and lysine residues (Stansfeld et al. 2009). In a separate simulation, the binding of PIP2 appeared to stabilize the C-linker of the ion channel in a manner similar to its gating mechanism (Schmidt et al. 2013).

The highly basic region (HBR) of the human immunodeficiency virus-1 myristoylated matrix protein (MA protein) has a preference for negatively charged membranes that is enhanced by to PIP2 (Charlier et al. 2014). Coarse-grained and all-atom MD simulations confirmed the binding of MA to PIP2 and lateral aggregation of PIP2 around MA on the membrane surface. Additionally, MD simulations of PPIs have been used to investigate interactions with talin (Kalli et al. 2013), actin-capping protein (Pleskot et al. 2012), TRPV1 channels (Poblete et al. 2015), receptor tyrosine kinases (Hedger et al. 2015), epidermal growth factor receptor (Abd Halim et al. 2015), focal adhesion kinase (Zhou et al. 2015), human dopamine transporter (Khelashvili et al. 2015), amyloid Aβ42 (Lemkul and Bevan 2011), and many others.

9.7 CONCLUSIONS

A wealth of experimental data now show that interactions of PPIs with cytoskeletal-binding proteins play a central role in regulating how the lipid bilayer couples to the protein framework immediately beneath it in the cell membrane. The cytoskeletal proteins regulated by PPIs include both those that link the cytoskeleton on the cell membrane and those that regulate actin assembly and molecular motors that generate the forces that can bend the membrane. PPIs also bind many of the proteins that either cause or sense local curvature. The large size and anionic charge of PPIs might also directly affect the mechanics of the lipid bilayer and contribute to the energy needed to bend it (Levental et al. 2008a, b). Many of the interactions between PPIs and their protein ligands depend on the manner in which the lipids are laterally distributed in the bilayer. Advances in MDs and other simulation and theoretical studies have begun to incorporate the vast amount of data detailing PPI effects at the membrane–cytoskeletal interface to produce an integrated model for how these fascinating lipids perform their many cellular functions.

ACKNOWLEDGEMENTS

We thank the members of the Radhakrishnan and Janmey Laboratories, and Changsong Yang and Tatyana Svitkina for helpful discussions. We acknowledge financial support from the National Institutes of Health through grants CA227550, CA244660 and GM136259.

NOTE

1 Nomenclature for phosphoinositides is cumbersome when each species is named correctly and unambiguously, but for most purposes a simple shortened form suffices. For example, PIP2 nearly always stands for phosphatidylinositol 4,5 bisphosphate (also often denoted PtdIns(4,5)P_2) and not its other possible isomers PtdIns(3,5)P_2 or PtdIns(3,4)P_2, and PIP is usually PtdIns(4)P, although the less abundant PtdIns(3)P and PtdIns(5)P also have important functions. Unless another isomer is specifically meant, PIP2 here will stand for PtdIns(4,5)P_2.

REFERENCES

Abd Halim, K. B., H. Koldso, and M. S. P. Sansom. 2015. Interactions of the EGFR juxtamembrane domain with PIP2-containing lipid bilayers: Insights from multiscale molecular dynamics simulations. *Biochimica et Biophysica Acta* 1850:1017–1025.

Abe, M., A. Makino, F. Hullin-Matsuda et al. 2012. A role for sphingomyelin-rich lipid domains in the accumulation of phosphatidylinositol-4,5-bisphosphate to the cleavage furrow during cytokinesis. *Molecular and Cellular Biology* 32:1396–1407.

Agrawal, N. J., J. Nukpezah, and R. Radhakrishnan. 2010. Minimal mesoscale model for protein-mediated vesiculation in clathrin-dependent endocytosis. *PLoS Computational Biology* 6:e1000926.

Agrawal, N. J., and R. Radhakrishnan. 2009. Calculation of free energies in fluid membranes subject to heterogeneous curvature fields. *Physical Review E* 80:011925.

Anitei, M., and B. Hoflack. 2011. Bridging membrane and cytoskeleton dynamics in the secretory and endocytic pathways. *Nature Cell Biology* 14:11–19.

Arkhipov, A., Y. Yin, and K. Schulten. 2009. Membrane-bending mechanism of amphiphysin N-BAR domains. *Biophysical Journal* 97:2727–2735.

Arumugam, S., and P. Bassereau. 2015. Membrane nanodomains: contribution of curvature and interaction with proteins and cytoskeleton. In *Membrane Nanodomains*. 109–119.

Arumugam, S., E. P. Petrov, and P. Schwille. 2015. Cytoskeletal pinning controls phase separation in multicomponent lipid membranes. *Biophysical Journal* 108:1104–1113.

Ayton, G. S., and G. A. Voth. 2010a. Multiscale computer simulation of the immature HIV-1 virion. *Biophysical Journal* 99:2757–2765.

Ayton, G. S., and G. A. Voth. 2010b. Multiscale simulation of protein mediated membrane remodeling. *Seminars in Cell and Developmental Biology* 21:357–362.

Baumgart, T., B. R. Capraro, C. Zhu, and S. L. Das. 2011. Thermodynamics and mechanics of membrane curvature generation and sensing by proteins and lipids. *Annual Review of Physical Chemistry* 62:483–506.

Baumgart, T., A. T. Hammond, P. Sengupta et al. 2007. Large-scale fluid/fluid phase separation of proteins and lipids in giant plasma membrane vesicles. *Proceedings of the National Academy of Sciences of the United States of America* 104:3165–3170.

Bethoney, K. A., M. C. King, J. E. Hinshaw, E. M. Ostap, and M. A. Lemmon. 2009. A possible effector role for the pleckstrin homology (PH) domain of dynamin. *Proceedings of the National Academy of Sciences of the United States of America* 106:13359–13364.

Bezanilla, M., A. S. Gladfelter, D. R. Kovar, and W. L. Lee. 2015. Cytoskeletal dynamics: a view from the membrane. *The Journal of Cell Biology* 209:329–337.

Bhatia, V. K., K. L. Madsen, P. Y. Bolinger et al. 2009. Amphipathic motifs in BAR domains are essential for membrane curvature sensing. *The EMBO Journal* 28:3303–3314.

Blood, P. D., R. D. Swenson, and G. A. Voth. 2008. Factors influencing local membrane curvature induction by N-BAR domains as revealed by molecular dynamics simulations. *Biophysical Journal* 95:1866–1876.

Boucrot, E., A. P. Ferreira, L. Almeida-Souza et al. 2015. Endophilin marks and controls a clathrin-independent endocytic pathway. *Nature* 517:460–465.

Boulant, S., C. Kural, J. C. Zeeh, F. Ubelmann, and T. Kirchhausen. 2011. Actin dynamics counteract membrane tension during clathrin-mediated endocytosis. *Nature Cell Biology* 13:1124–1131.

Bradley, R., and R. Radhakrishnan. 2013. Coarse-grained models for protein-cell membrane interactions. *Polymers (Basel)* 5:890–936.

Brown, D. A. 2015. PIP2Clustering: from model membranes to cells. *Chemistry and Physics of Lipids* 192:33–40.

Bunney, T. D., and M. Katan. 2010. Phosphoinositide signalling in cancer: beyond PI3K and PTEN. *Nature Reviews Cancer* 10:342–352.

Byrum, J. N., and W. Rodgers. 2015. Membrane-cytoskeleton interactions in cholesterol-dependent domain formation. *Essays in Biochemistry* 57:177–187.

Cadmel, B., C. Schleber, M. Condron et al. 2008. The PI(3,5)P2 and PI(4,5)P2 interactomes. *Journal of Proteome Research* 7:5295–5313.

Campelo, F., H. T. McMahon, and M. M. Kozlov. 2008. The hydrophobic insertion mechanism of membrane curvature generation by proteins. *Biophysical Journal* 95:2325–2339.

Caroni, P. 2001. New EMBO members' review: actin cytoskeleton regulation through modulation of PI(4,5)P(2) rafts. *The EMBO Journal* 20:4332–4336.

Charlier, L., M. Louet, L. Chaloin et al. 2014. Coarse-grained simulations of the HIV-1 matrix protein anchoring: revisiting its assembly on membrane domains. *Biophysical Journal* 106:577–585.

Chichili, G. R., and W. Rodgers. 2009. Cytoskeleton-membrane interactions in membrane raft structure. *Cellular and Molecular Life Sciences* 66:2319–2328.

Chierico, L., A. S. Joseph, A. L. Lewis, and G. Battaglia. 2014. Live cell imaging of membrane/cytoskeleton interactions and membrane topology. *Scientific Reports* 4:6056.

Cui, H., G. S. Ayton, and G. A. Voth. 2009. Membrane binding by the endophilin N-BAR domain. *Biophysical Journal* 97:2746–2753.

Cui, H., C. Mim, F. X. Vazquez et al. 2013. Understanding the role of amphipathic helices in N-BAR domain driven membrane remodeling. *Biophysical Journal* 104:404–411.

Cullen, P. J., and H. C. Korswagen. 2011. Sorting nexins provide diversity for retromer-dependent trafficking events. *Nature Cell Biology* 14:29–37.

Dason, J. S., A. J. Smith, L. Marin, and M. P. Charlton. 2014. Cholesterol and F-actin are required for clustering of recycling synaptic vesicle proteins in the presynaptic plasma membrane. *The Journal of Physiology* 592:621–633.

Dinic, J., P. Ashrafzadeh, and I. Parmryd. 2013. Actin filaments attachment at the plasma membrane in live cells cause the formation of ordered lipid domains. *Biochimica et Biophysica Acta* 1828:1102–1111.

Diz-Munoz, A., D. A. Fletcher, and O. D. Weiner. 2013. Use the force: membrane tension as an organizer of cell shape and motility. *Trends in Cell Biology* 23:47–53.

Ehrig, J., E. P. Petrov, and P. Schwille. 2011. Near-critical fluctuations and cytoskeleton-assisted phase separation lead to subdiffusion in cell membranes. *Biophysical Journal* 100:80–89.

Elliott, P. R., B. T. Goult, P. M. Kopp et al. 2010. The Structure of the talin head reveals a novel extended conformation of the FERM domain. *Structure* 18:1289–1299.

Farsad, K., N. Ringstad, K. Takei et al. 2001. Generation of high curvature membranes mediated by direct endophilin bilayer interactions. *The Journal of Cell Biology* 155:193–200.

Fehon, R. G., A. I. McClatchey, and A. Bretscher. 2010. Organizing the cell cortex: the role of ERM proteins. *Nature Reviews. Molecular Cell Biology* 11:276–287.

Funakoshi, Y., H. Hasegawa, and Y. Kanaho. 2011. Regulation of PIP5K activity by Arf6 and its physiological significance. *Journal of Cellular Physiology* 226:888–895.

Gallop, J. L., and H. T. McMahon. 2005. BAR domains and membrane curvature: bringing your curves to the BAR. *Biochemical Society symposium* 72:223–231.

Gallop, J. L., A. Walrant, L. C. Cantley, and M. W. Kirschner. 2013. Phosphoinositides and membrane curvature switch the mode of actin polymerization via selective recruitment of toca-1 and Snx9. *Proceedings of the National Academy of Sciences of the United States of America* 110:7193–7198.

Gambhir, A., G. Hangyas-Mihalyne, I. Zaitseva et al. 2004. Electrostatic sequestration of PIP2 on phospholipid membranes by basic/aromatic regions of proteins. *Biophysical Journal* 86:2188–2207.

Golebiewska, U., J. G. Kay, T. Masters et al. 2011. Evidence for a fence that impedes the diffusion of phosphatidylinositol 4,5-bisphosphate out of the forming phagosomes of macrophages. *Molecular Biology of the Cell* 22:3498–3507.

Golebiewska, U., M. Nyako, W. Woturski, I. Zaitseva, and S. McLaughlin. 2008. Diffusion coefficient of fluorescent phosphatidylinositol 4,5-bisphosphate in the plasma membrane of cells. *Molecular Biology of the Cell* 19:1663–1669.

Gordon, V. D., M. Deserno, C. M. J. Andrew, S. U. Egelhaaf, and W. C. K. Poon. 2008. Adhesion promotes phase separation in mixed-lipid membranes. *EPL (Europhysics Letters)* 84:48003.

Gowrishankar, K., S. Ghosh, S. Saha et al. 2012. Active remodeling of cortical actin regulates spatiotemporal organization of cell surface molecules. *Cell* 149:1353–1367.

Grant, B. D., and J. G. Donaldson. 2009. Pathways and mechanisms of endocytic recycling. *Nature Reviews. Molecular Cell Biology* 10:597–608.

Hatcher, E., O. Guvench, and A. D. Mackerell, Jr. 2009. CHARMM additive all-atom force field for acyclic polyalcohols, acyclic carbohydrates and inositol. *Journal of Chemical Theory and Computation* 5:1315–1327.

Hatzakis, N. S., V. K. Bhatia, J. Larsen et al. 2009. How curved membranes recruit amphipathic helices and protein anchoring motifs. *Nature Chemical Biology* 5:835–841.

Hedger, G., M. S. Sansom, and H. Koldso. 2015. The juxtamembrane regions of human receptor tyrosine kinases exhibit conserved interaction sites with anionic lipids. *Scientific Reports* 5:9198.

Henin, J., W. Shinoda, and M. L. Klein. 2008. United-atom acyl chains for CHARMM phospholipids. *The Journal of Physical Chemistry. B* 112:7008–7015.

Honigmann, A., S. Sadeghi, J. Keller et al. 2014. A lipid bound actin meshwork organizes liquid phase separation in model membranes. *eLife* 3:e01671.

Honigmann, A., G. van den Bogaart, E. Iraheta et al. 2013. Phosphatidylinositol 4,5-bisphosphate clusters act as molecular beacons for vesicle recruitment. *Nature Structural & Molecular Biology* 20:679–686.

Janmey, P. A. 1994. Phosphoinositides and calcium as regulators of cellular actin assembly and disassembly. *Annual Review of Physiology* 56:169–191.

Kalli, A. C., I. D. Campbell, and M. S. Sansom. 2013. Conformational changes in talin on binding to anionic phospholipid membranes facilitate signaling by integrin transmembrane helices. *PLoS Computational Biology* 9:e1003316.

Kalli, A. C., I. Devaney, and M. S. Sansom. 2014. Interactions of phosphatase and tensin homologue (PTEN) proteins with phosphatidylinositol phosphates: insights from molecular dynamics simulations of PTEN and voltage sensitive phosphatase. *Biochemistry* 53:1724–1732.

Kapus, A., and P. Janmey. 2013. Plasma membrane--cortical cytoskeleton interactions: a cell biology approach with biophysical considerations. *Comprehensive Physiology* 3:1231–1281.

Katan, M., and V. L. Allen. 1999. Modular PH and C2 domains in membrane attachment and other functions. *FEBS Letters* 452:36–40.

Kato, M., and M. Mrksich. 2004. Using model substrates to study the dependence of focal adhesion formation on the affinity of integrin-ligand complexes. *Biochemistry* 43:2699–2707.

Khelashvili, G., M. Doktorova, M. A. Sahai et al. 2015. Computational modeling of the N-terminus of the human dopamine transporter and its interaction with PIP2-containing membranes. *Proteins* 83:952–969.

Klauda, J. B., R. M. Venable, J. A. Freites et al. 2010. Update of the CHARMM all-atom additive force field for lipids: validation on six lipid types. *The Journal of Physical Chemistry. B* 114:7830–7843.

Kooijman, E. E., K. E. King, M. Gangoda, and A. Gericke. 2009. Ionization properties of phosphatidylinositol polyphosphates in mixed model membranes. *Biochemistry* 48:9360–9371.

Koushik, A. B., R. R. Powell, and L. A. Temesvari. 2013. Localization of phosphatidylinositol 4,5-bisphosphate to lipid rafts and uroids in the human protozoan parasite Entamoeba histolytica. *Infection and Immunity* 81:2145–2155.

Kusumi, A., T. K. Fujiwara, R. Chadda et al. 2012. Dynamic organizing principles of the plasma membrane that regulate signal transduction: commemorating the fortieth anniversary of Singer and Nicolson's fluid-mosaic model. *Annual Review of Cell and Developmental Biology* 28: 215–250.

Kutateladze, T. G. 2010. Translation of the phosphoinositide code by PI effectors. *Nature Chemical Biology* 6:507–513.

Kwiatkowska, K. 2010. One lipid, multiple functions: how various pools of PI(4,5)P(2) are created in the plasma membrane. *Cellular and Molecular Life Sciences* 67: 3927–3946.

Lai, C. L., C. C. Jao, E. Lyman et al. 2012. Membrane binding and self-association of the epsin N-terminal homology domain. *Journal of Molecular Biology* 423:800–817.

Lee, K., J. L. Gallop, K. Rambani, and M. W. Kirschner. 2010. Self-assembly of filopodia-like structures on supported lipid bilayers. *Science* 329:1341–1345.

Lemkul, J. A., and D. R. Bevan. 2011. Lipid composition influences the release of Alzheimer's amyloid beta-peptide from membranes. *Protein Science: A Publication of the Protein Society* 20:1530–1545.

Lemmon, M. A. 2007. Pleckstrin homology (PH) domains and phosphoinositides. *Biochemical Society Symposium* 74:81–93.

Lemmon, M. A. 2008. Membrane recognition by phospholipid-binding domains. *Nature Reviews. Molecular Cell Biology* 9:99–111.

Lemmon, M. A., and K. M. Ferguson. 2000. Signal-dependent membrane targeting by pleckstrin homology (PH) domains. *The Biochemical Journal* 350 Pt 1:1–18.

Levental, I., A. Cebers, and P. A. Janmey. 2008a. Combined electrostatics and hydrogen bonding determine intermolecular interactions between polyphosphoinositides. *Journal of the American Chemical Society* 130:9025–9030.

Levental, I., D. A. Christian, Y. H. Wang et al. 2009. Calcium-dependent lateral organization in phosphatidylinositol 4,5-bisphosphate (PIP2)- and cholesterol-containing monolayers. *Biochemistry* 48:8241–8248.

Levental, I., P. A. Janmey, and A. Cebers. 2008b. Electrostatic contribution to the surface pressure of charged monolayers containing polyphosphoinositides. *Biophysical Journal* 95:1199–1205.

Li, Z., R. M. Venable, L. A. Rogers, D. Murray, and R. W. Pastor. 2009. Molecular dynamics simulations of PIP2 and PIP3 in lipid bilayers: determination of ring orientation, and the effects of surface roughness on a Poisson-Boltzmann description. *Biophysical Journal* 97:155–163.

Liepina, I., C. Czaplewski, P. Janmey, and A. Liwo. 2003. Molecular dynamics study of a gelsolin-derived peptide binding to a lipid bilayer containing phosphatidylinositol 4,5-bisphosphate. *Biopolymers* 71:49–70.

Liu, A. P., and D. A. Fletcher. 2006. Actin polymerization serves as a membrane domain switch in model lipid bilayers. *Biophysical Journal* 91:4064–4070.

Liu, J., and W. Guo. 2012. The exocyst complex in exocytosis and cell migration. *Protoplasma* 249:587–597.

Liu, J., M. Kaksonen, D. G. Drubin, and G. Oster. 2006. Endocytic vesicle scission by lipid phase boundary forces. *Proceedings of the National Academy of Sciences of the United States of America* 103:10277–10282.

Liu, J., Y. Sun, D. G. Drubin, and G. F. Oster. 2009a. The mechanochemistry of endocytosis. *PLoS biology* 7:e1000204.

Liu, J., R. Tourdot, V. Ramanan, N. J. Agrawal, and R. Radhakrishanan. 2012a. Mesoscale simulations of curvature-inducing protein partitioning on lipid bilayer membranes in the presence of mean curvature fields. *Molecular Physics* 110:1127–1137.

Liu, J., P. Yue, V. V. Artym, S. C. Mueller, and W. Guo. 2009b. The role of the exocyst in matrix metalloproteinase secretion and actin dynamics during tumor cell invadopodia formation. *Molecular Biology of the Cell* 20: 3763–3771.

Liu, J., Y. Zhao, Y. Sun et al. 2012b. Exo70 stimulates the Arp2/3 complex for lamellipodia formation and directional cell migration. *Current Biology* 22:1510–1515.

Liu, J., X. Zuo, P. Yue, and W. Guo. 2007a. Phosphatidylinositol 4,5-bisphosphate mediates the targeting of the exocyst to the plasma membrane for exocytosis in mammalian cells. *Molecular Biology of the Cell* 18:4483–4492.

Liu, Y., J. Purvis, A. Shih et al. 2007b. A multiscale computational approach to dissect early events in the Erb family receptor mediated activation, differential signaling, and relevance to oncogenic transformations. *Annals of Biomedical Engineering* 35:1012–1025.

Lumb, C. N., and M. S. Sansom. 2012. Finding a needle in a haystack: the role of electrostatics in target lipid recognition by PH domains. *PLoS Computational Biology* 8:e1002617.

Lupyan, D., M. Mezei, D. E. Logothetis, and R. Osman. 2010. A molecular dynamics investigation of lipid bilayer perturbation by PIP2. *Biophysical Journal* 98:240–247.

Ma, L., L. C. Cantley, P. A. Janmey, and M. W. Kirschner. 1998. Corequirement of specific phosphoinositides and small GTP-binding protein Cdc42 in inducing actin assembly in Xenopus egg extracts. *The Journal of Cell Biology* 140:1125–1136.

Machta, B. B., S. Papanikolaou, J. P. Sethna, and S. L. Veatch. 2011. Minimal model of plasma membrane heterogeneity requires coupling cortical actin to criticality. *Biophysical Journal* 100:1668–1677.

Mao, Y. S., and H. L. Yin. 2007. Regulation of the actin cytoskeleton by phosphatidylinositol 4-phosphate 5 kinases. *Pflügers Archiv* 455:5–18.

Marks, M. S., H. F. Heijnen, and G. Raposo. 2013. Lysosome-related organelles: unusual compartments become mainstream. *Current Opinion in Cell Biology* 25:495–505.

Marrink, S. J., H. J. Risselada, S. Yefimov, D. P. Tieleman, and A. H. de Vries. 2007. The Martini force field: coarse grained model for biomolecular simulations. *The Journal of Physical Chemistry. B* 111:7812–7824.

Marrink, S. J., and D. P. Tieleman. 2013. Perspective on the Martini model. *Chemical Society Reviews* 42:6801–6822.

Masters, T. A., M. P. Sheetz, and N. C. Gauthier. 2016. F-actin waves, actin cortex disassembly and focal exocytosis driven by actin-phosphoinositide positive feedback. *Cytoskeleton (Hoboken)* 73:180–196.

McMahon, H. T., and J. L. Gallop. 2005. Membrane curvature and mechanisms of dynamic cell membrane remodelling. *Nature* 438:590–596.

Miled, N., Y. Yan, W. C. Hon et al. 2007. Mechanism of two classes of cancer mutations in the phosphoinositide 3-kinase catalytic subunit. *Science* 317:239–242.

Mim, C., H. Cui, J. A. Gawronski-Salerno et al. 2012. Structural basis of membrane bending by the N-BAR protein endophilin. *Cell* 149:137–145.

Moravcevic, K., J. M. Mendrola, K. R. Schmitz et al. 2010. Kinase associated-1 domains drive MARK/PAR1 kinases to membrane targets by binding acidic phospholipids. *Cell* 143:966–977.

Moravcevic, K., C. L. Oxley, and M. A. Lemmon. 2012. Conditional peripheral membrane proteins: facing up to limited specificity. *Structure* 20:15–27.

Orlando, K., and W. Guo. 2009. Membrane organization and dynamics in cell polarity. *Cold Spring Harbor Perspectives in Biology* 1:a001321.

Orlando, K., X. Sun, J. Zhang et al. 2011. Exo-endocytic trafficking and the septin-based diffusion barrier are required for the maintenance of Cdc42p polarization during budding yeast asymmetric growth. *Molecular Biology of the Cell* 22:624–633.

Papayannopoulos, V., C. Co, K. E. Prehoda et al. 2005. A polybasic motif allows N-WASP to act as a sensor of PIP(2) density. *Molecular Cell* 17:181–191.

Pastor, R. W., and A. D. Mackerell, Jr. 2011. Development of the CHARMM force field for lipids. *The Journal of Physical Chemistry Letters* 2:1526–1532.

Pearson, M. A., D. Reczek, A. Bretscher, and P. A. Karplus. 2000. Structure of the ERM protein moesin reveals the FERM domain fold masked by an extended actin binding tail domain. *Cell* 101:259–270.

Picas, L., F. Gaits-Iacovoni, and B. Goud. 2016. The emerging role of phosphoinositide clustering in intracellular trafficking and signal transduction. *F1000Research* 5(F1000 Faculty Rev):422. doi:10.12688/f1000research.7537.1.

Pleskot, R., P. Pejchar, V. Zarsky, C. J. Staiger, and M. Potocky. 2012. Structural insights into the inhibition of actin-capping protein by interactions with phosphatidic acid and phosphatidylinositol (4,5)-bisphosphate. *PLoS Computational Biology* 8:e1002765.

Pluhackova, K., S. A. Kirsch, J. Han et al. 2016. A critical comparison of biomembrane force fields: structure and dynamics of model DMPC, POPC, and POPE bilayers. *The Journal of Physical Chemistry. B* 120:3888–3903.

Poblete, H., I. Oyarzun, P. Olivero et al. 2015. Molecular determinants of phosphatidylinositol 4,5-bisphosphate (PI(4,5)P2) binding to transient receptor potential V1 (TRPV1) channels. *The Journal of Biological Chemistry* 290:2086–2098.

Posor, Y., M. Eichhorn-Gruenig, D. Puchkov et al. 2013. Spatiotemporal control of endocytosis by phosphatidylinositol-3,4-bisphosphate. *Nature* 499:233–237.

Poyry, S., and I. Vattulainen. 2016. Role of charged lipids in membrane structures - Insight given by simulations. *Biochimica et Biophysica Acta* 1858:2322–2333.

Psachoulia, E., and M. S. Sansom. 2008. Interactions of the pleckstrin homology domain with phosphatidylinositol phosphate and membranes: characterization via molecular dynamics simulations. *Biochemistry* 47:4211–4220.

Puthenveedu, M. A., B. Lauffer, P. Temkin et al. 2010. Sequence-dependent sorting of recycling proteins by actin-stabilized endosomal microdomains. *Cell* 143:761–773.

Ramakrishnan, N., R. P. Bradley, R. Tourdot, and R. Radhakrishnan. 2014a. Cellular scale biophysical models of membrane sculpting by the proteins during endocytosis and exocytosis. *Biophysical Journal* 106:310a–310a.

Ramakrishnan, N., P. B. Sunil Kumar, and R. Radhakrishnan. 2014b. Mesoscale computational studies of membrane bilayer remodeling by curvature-inducing proteins. *Physics Reports* 543:1–60.

Ramakrishnan, N., R. Tourdot, and R. Radhakrishanan. 2015. Thermodynamic free energy methods to investigate shape transitions in bilayer membranes. *International Journal of Advances in Engineering Sciences and Applied Mathematics* 2015.

Ramanan, V., N. J. Agrawal, J. Liu et al. 2011. Systems biology and physical biology of clathrin-mediated endocytosis. *Integrative Biology: Quantitative Biosciences from Nano to Macro* 3:803–815.

Ren, J., and W. Guo. 2012. ERK1/2 regulate exocytosis through direct phosphorylation of the exocyst component Exo70. *Developmental Cell* 22:967–978.

Rodriguez-Garcia, R., I. Lopez-Montero, M. Mell et al. 2015. Direct cytoskeleton forces cause membrane softening in red blood cells. *Biophysical Journal* 108:2794–2806.

Romer, W., L. L. Pontani, B. Sorre et al. 2010. Actin dynamics drive membrane reorganization and scission in clathrin-independent endocytosis. *Cell* 140:540–553.

Saldanha, O., M. E. Brennich, M. Burghammer, H. Herrmann, and S. Koster. 2016. The filament forming reactions of vimentin tetramers studied in a serial-inlet microflow device by small angle X-ray scattering. *Biomicrofluidics* 10:024108.

Schmidt, M. R., P. J. Stansfeld, S. J. Tucker, and M. S. Sansom. 2013. Simulation-based prediction of phosphatidylinositol 4,5-bisphosphate binding to an ion channel. *Biochemistry* 52:279–281.

Shi, Z., and T. Baumgart. 2015. Membrane tension and peripheral protein density mediate membrane shape transitions. *Nature Communications* 6:5974.

Shih, A. J., J. Purvis, and R. Radhakrishnan. 2008. Molecular systems biology of ErbB1 signaling: bridging the gap through multiscale modeling and high-performance computing. *Molecular Biosystems* 4:1151–1159.

Sieber, J. J., K. I. Willig, C. Kutzner et al. 2007. Anatomy and dynamics of a supramolecular membrane protein cluster. *Science* 317:1072–1076.

Simunovic, M., C. Mim, T. C. Marlovits et al. 2013. Protein-mediated transformation of lipid vesicles into tubular networks. *Biophysical Journal* 105:711–719.

Slochower, D. R., P. J. Huwe, R. Radhakrishnan, and P. A. Janmey. 2013. Quantum and all-atom molecular dynamics simulations of protonation and divalent ion binding to phosphatidylinositol 4,5-bisphosphate (PIP2). *The Journal of Physical Chemistry. B* 117:8322–8329.

Slochower, D. R., Y. H. Wang, R. Radhakrishnan, and P. A. Janmey. 2015. Physical chemistry and membrane properties of two phosphatidylinositol bisphosphate isomers. *Physical Chemistry Chemical Physics* 17:12608–12615.

Slochower, D. R., Y. H. Wang, R. W. Tourdot, R. Radhakrishnan, and P. A. Janmey. 2014. Counterion-mediated pattern formation in membranes containing anionic lipids. *Advances in Colloid and Interface Science* 208:177–188.

Sorre, B., A. Callan-Jones, J. Manzi et al. 2012. Nature of curvature coupling of amphiphysin with membranes depends on its bound density. *Proceedings of the National Academy of Sciences of the United States of America* 109:173–178.

Stachowiak, J. C., E. M. Schmid, C. J. Ryan et al. 2012. Membrane bending by protein-protein crowding. *Nature Cell Biology* 14:944–949.

Stansfeld, P. J., R. Hopkinson, F. M. Ashcroft, and M. S. Sansom. 2009. PIP(2)-binding site in Kir channels: definition by multiscale biomolecular simulations. *Biochemistry* 48:10926–10933.

Sun, Y., D. A. Turbin, K. Ling et al. 2010. Type I gamma phosphatidylinositol phosphate kinase modulates invasion and proliferation and its expression correlates with poor prognosis in breast cancer. *Breast Cancer Research* 12:R6.

Suzuki, A., J. W. Shin, Y. Wang et al. 2013. RhoA is essential for maintaining normal megakaryocyte ploidy and platelet generation. *PloS one* 8:e69315.

Takenawa, T., and S. Suetsugu. 2007. The WASP-WAVE protein network: connecting the membrane to the cytoskeleton. *Nature Reviews. Molecular Cell Biology* 8:37–48.

Telesco, S. E., and R. Radhakrishnan. 2012. Structural systems biology and multiscale signaling models. *Annals of Biomedical Engineering* 40:2295–2306.

Tourdot, R. W., N. Ramakrishnan, T. Baumgart, and R. Radhakrishnan. 2015. Application of a free-energy-landscape approach to study tension-dependent bilayer tubulation mediated by curvature-inducing proteins. *Physical Review E* 92:042715.

Tourdot, R. W., N. Ramakrishnan, and R. Radhakrishnan. 2014. Defining the free-energy landscape of curvature-inducing proteins on membrane bilayers. *Physical Review E* 90:022717.

Tsujita, K., T. Takenawa, and T. Itoh. 2015. Feedback regulation between plasma membrane tension and membrane-bending proteins organizes cell polarity during leading edge formation. *Nature Cell Biology* 17:749–758.

van den Bogaart, G., K. Meyenberg, H. J. Risselada et al. 2011. Membrane protein sequestering by ionic protein-lipid interactions. *Nature* 479:552–555.

van den Bout, I., and N. Divecha. 2009. PIP5K-driven PtdIns(4,5)P2 synthesis: regulation and cellular functions. *Journal of Cell Science* 122:3837–3850.

Veatch, S. L., P. Cicuta, P. Sengupta et al. 2008. Critical fluctuations in plasma membrane vesicles. *ACS Chemical Biology* 3:287–293.

Veatch, S. L., and S. L. Keller. 2002. Organization in lipid membranes containing cholesterol. *Physical Review Letters* 89:268101.

Veatch, S. L., and S. L. Keller. 2003. Separation of liquid phases in giant vesicles of ternary mixtures of phospholipids and cholesterol. *Biophysical Journal* 85:3074–3083.

Vicinanza, M., G. D'Angelo, A. Di Campli, and M. A. De Matteis. 2008a. Function and dysfunction of the PI system in membrane trafficking. *The EMBO Journal* 27:2457–2470.

Vicinanza, M., G. D'Angelo, A. Di Campli, and M. A. De Matteis. 2008b. Phosphoinositides as regulators of membrane trafficking in health and disease. *Cellular and Molecular Life Sciences* 65:2833–2841.

Voth, G. A. 2013. New and notable: key new insights into membrane targeting by proteins. *Biophysical Journal* 104:517–519.

Wang, J., and D. a. Richards. 2012. Segregation of PIP2 and PIP3 into distinct nanoscale regions within the plasma membrane. *Biology Open* 1:857–862.

Wang, Y., L. Zhao, A. Suzuki et al. 2013. Platelets lacking PIP5KIgamma have normal integrin activation but impaired cytoskeletal-membrane integrity and adhesion. *Blood* 121:2743–2752.

Wang, Y. H., A. Collins, L. Guo et al. 2012. Divalent cation-induced cluster formation by polyphosphoinositides in model membranes. *Journal of the American Chemical Society* 134:3387–3395.

Wang, Y. H., D. R. Slochower, and P. A. Janmey. 2014. Counterion-mediated cluster formation by polyphosphoinositides. *Chemistry and Physics of Lipids* 182:38–51.

Williams, R. L., and S. Urb. 2007. The emerging shape of the ESCRT machinery. *Nature Reviews Molecular Cell Biology* 8: 355–368.

Xiong, D., S. Xiao, S. Guo et al. 2016. Frequency and amplitude control of cortical oscillations by phosphoinositide waves. *Nature Chemical Biology* 12:159–166.

Yin, H. L., and P. A. Janmey. 2003. Phosphoinositide regulation of the actin cytoskeleton. *Annual Review of Physiology* 65:761–789.

Zeman, K., H. Engelhard, and E. Sackmann. 1990. Bending undulations and elasticity of the erythrocyte membrane: effects of cell shape and membrane organization. *European Biophysics Journal* 18:203–219.

Zhao, Y., J. Liu, C. Yang et al. 2013. Exo70 generates membrane curvature for morphogenesis and cell migration. *Developmental Cell* 26:266–278.

Zhou, J., A. Bronowska, J. Le Coq, D. Lietha, and F. Grater. 2015. Allosteric regulation of focal adhesion kinase by PIP2 and ATP. *Biophysical Journal* 108:698–705.

Zhu, C., S. L. Das, and T. Baumgart. 2012. Nonlinear sorting, curvature generation, and crowding of endophilin N-BAR on tubular membranes. *Biophysical Journal* 102: 1837–1845.

Zimmerberg, J., and M. M. Kozlov. 2006. How proteins produce cellular membrane curvature. *Nature Reviews Molecular Cell Biology* 7:9–19.

Zoncu, R., R. M. Perera, D. M. Balkin et al. 2009. A phosphoinositide switch controls the maturation and signaling properties of APPL endosomes. *Cell* 136:1110–1121.

Zuo, X., J. Zhang, Y. Zhang et al. 2006. Exo70 interacts with the Arp2/3 complex and regulates cell migration. *Nature Cell Biology* 8:1383–1388.

10 Effects of Osmotic Stress on Topologically Closed Membrane Compartments

James C. S. Ho and Bo Liedberg
Nanyang Technological University, Singapore

Atul N. Parikh
University of California, Davis, California
Nanyang Technological University, Singapore

CONTENTS

10.1 Introduction 177
10.2 Model Membrane Compartments 179
10.3 Osmotic Pressure Gradient across Giant Unilamellar Vesicles 182
    10.3.1 Osmotic Upshift: Giant Unilamellar Vesicles in Hypertonic Media 183
    10.3.2 Osmotic Downshift: Giant Unilamellar Vesicles in Hypotonic Media 186
10.4 Outlook and Perspectives 189
Bibliography 190

10.1 INTRODUCTION

A solute – excluded from a spatial "compartment" within an aqueous continuum – exerts an osmotic pressure on the compartment by lowering the chemical activity of water on the "outside". This in turn triggers a withdrawal of water from the compartment, changing size, shape, and hydration of the compartment itself (Leneveu, Rand et al. 1976). This universal and nonspecific osmotic stress, in conjunction with accompanying activity of water, can drive conformational change in proteins (Rand 2004), gate membrane channels (Zimmerberg and Parsegian 1986), and mediate enzyme actions (Reid and Rand 1997), giving the cell a global mechanism for regulating protein activity (Figure 10.1a) simply by modulating physical interactions between the macromolecule, solute, and water (Rand, Parsegian et al. 2000).

Consequences of osmotic forces become quite drastic when the compartment itself is a topologically closed, structurally deformable, and compositionally complex structure (Figure 10.1b) such as a living cell. In the constant struggle with their environment, free-living cells (e.g., bacterial cells) experiencing a sudden spike or drop in the amount of dissolved molecules in water – such as due to high or fluctuating salinity, desiccation, or freezing – are confronted with significant osmotic stresses (Wood 1999). If left unchecked, these environmental perturbations would result in an instantaneous flow of water out of the cell as the ambient aqueous phase becomes more concentrated (an osmotic upshift) and into the cell as the surrounding aqueous phase becomes more dilute (an osmotic downshift). The former will cause the cell to shrink and dehydrate, while the latter will result in swelling, potentially inducing rupture and cell death.

To avoid these catastrophic outcomes, most bacterial, plant, and animal cells have evolved mechanisms – conserved and often highly sophisticated – allowing them to regulate their water content in response to osmotic assaults imposed by variations in their local microenvironments or global environmental conditions (Kung, Martinac et al. 2010, Wood 2011). Cells accomplish these osmoregulatory activities, as part of their normal cellular homeostasis, primarily by altering the compositions of the cytoplasm, such as by accumulating or releasing electrolytes and small organic solutes (Wood 2011) but also by altering chemical compositions and physical properties of their membrane shells and/or cytoplasmic cores (Hoekstra, Golovina et al. 2001). Faced with osmotic upshift, through increased concentrations of solute in the surrounding milieu, cells face plasmolysis and dehydration (Koch 1984). In response, they accumulate "compatible solutes", or osmoprotectants genetically or biochemically, which can attain high cytoplamic concentrations without disrupting normal cellular functions (Kempf and Bremer 1998). This adaptive strategy against osmotic stress is evolutionarily well conserved across bacteria, archaea, and eukarya. Typical examples of osmoprotectants include low-molecular-weight metabolic

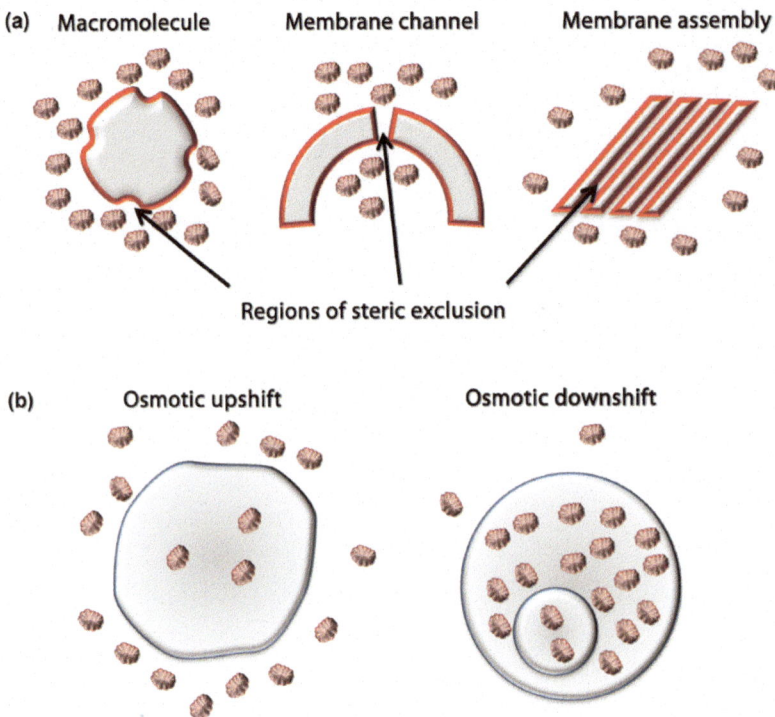

FIGURE 10.1 Universal effects of osmotic stress on topologically opened and closed systems. (a) Steric exclusion of solutes produces osmotic stress on water-filled cavities on macromolecular surface, membrane channels, or membrane assembly. Adapted from Parsegian et al. (2000). (b) Osmotic forces on topologically closed structures could cause membrane shape changes (in response to osmotic upshift) or swelling (in response to osmotic downshift).

products including polyhydric alcohols (polyols), such as glycerol, trehalose, and sucrose; free amino acids and amino acid derivatives (proline, taurine, and 13-alanine); and urea and methylamines, such as trimethylamine-N-oxide (TMAO), betaine, and sarcosine (Yancey, Clark et al. 1982). In the absence of the availability of osmoprotectants, such as in metabolically simple archeae halobacteria, cells rapidly accumulate inorganic (e.g., K^+) ions. In either case, the accumulation of osmolytes in the cytoplasm serves to reduce the osmotic imbalance allowing the cell to restore hydration and thereby normal functions (Kempf and Bremer 1998). Osmotic downshifts, on the other hand, tend to increase the cell volume through influx of water. This osmotic stress, as a result, strains the membrane and alters the cytoskeleton networks (Stewart, Helenius et al. 2011, Sachs and Sivaselvan 2015), converting the osmotic energy into mechanical energy, until the cell reaches a new mechanical equilibrium. By opening up channels of mechanosensitive kind (e.g., MscL), which allows efficient efflux of cytoplasmic solutes, the cells can then achieve cell volume regulation, avoid lysis, and regain normal functionality (Kung 2005, Kung, Martinac et al. 2010). Extensive reviews of the effects of osmotic pressure changes and cellular osmoregulatory mechanisms capture these issues in the literature (Hamill and Martinac 2001, Poolman, Spitzer et al. 2004, Spitzer and Poolman 2005, Booth, Edwards et al. 2007, Altendorf, Booth et al. 2009, Kramer 2009, Kung, Martinac et al. 2010, Wood 2010). A recent series of perspectives on how cells handle osmotic challenges is particularly noteworthy (Andersen 2015, Haswell and Verslues 2015, Sachs and Sivaselvan 2015, Wood 2015).

The movement of water accompanying an osmotic stress, involving rapid influx or efflux across cellular compartments, is intrinsically intertwined with the membrane's structural (i.e., elastic properties) and compositional (i.e., chemical heterogeneity) degrees of freedom (Dai, Sheetz et al. 1998). During osmotic upshift, increased area-to-volume ratio in topologically closed membrane compartments induces global shape transformations in flaccid membranes, whereas during osmotic downshift, increased membrane tension irons out thermal undulations, drives lateral redistributions of membrane components, and beyond a threshold tension, induces membrane portion. Moreover, in living cellular compartments, besides changes in the membrane area and mechanical tension, osmotic shifts tend to induce additional changes in physical properties including cell volume, turgor pressure, solute concentrations, the ionic strength, cytoplasmic macromolecular crowding, and changes in cytoskeleton–membrane interactions, all of which modulate membrane structure and organization (Wood 1999, 2011). Many of these membrane deformations and reorganizations have important implications for cellular osmoregulation, which are only beginning to be appreciated. Indeed, several bacterial mechanosensitive channels, which effect water efflux under osmotic downshifts, have been suggested to become "allosterically" activated (by conformational change) through the transduction of mechanical tension generated in the surrounding lipid matrix of the membrane (Perozo, Kloda et al. 2002, Kung 2005). Understanding how cellular capacities for osmosensing and osmoregulation affect (and is affected by) membrane properties continues to be an exciting area for enquiry and research (Wood 2015).

In what follows, we draw from a selection of efforts, which highlights how model membranes respond to osmotic stresses. Utilizing synthetic models for cellular compartments, designed with systematically varied degrees of structural and compositional complexities, these studies serve to isolate and provide a framework for understanding the roles that membrane bilayers play in sensing and regulating cellular response to osmotic assaults. Key membrane biophysical properties that are altered under osmotic stress include membrane's two-dimensional lateral fluidity, mechanical tension, lateral molecular distributions, and curvature – all dependent on membrane compositions. These studies thus delineate membrane-mediated biophysical mechanisms, which are activated during osmotic activity across topologically closed compartments. More generally, they may also shed light on how primitive protocells might have produced stable and functional compartmentalization against osmotic assaults prior to the advent of more complex, osmosensing and osmoregulating protein machineries (Szostak, Bartel et al. 2001, Chen, Roberts et al. 2004, Luisi, Ferri et al. 2006, Luisi, de Souza et al. 2008). The references and attributions here are merely illustrative and far from comprehensive but should serve as a guide to the literature.

10.2 MODEL MEMBRANE COMPARTMENTS

Perhaps, the simplest, topologically closed cell-like compartment allowing independent control of the properties of the encapsulated aqueous core and the surrounding aqueous phase – needed for introducing osmotic stresses – is a phospholipid vesicle. Formed through spontaneous self-assembly of amphiphilic lipids (Lipowsky 1991), vesicles are delimited by a 3- to 5-nm-thick spherical shell consisting of two apposed monomolecular leaflets of amphiphiles, isolating the encapsulated aqueous core from the surrounding bulk (Safinya and Ewert 2012). They were first discovered serendipitously by Bangham in 1964 during investigations of phospholipids (Bangham and Horne 1964). Although the bilayer shells are spontaneously formed through the hydrophobic interactions (Chandler 2005) between lipids and the aqueous phase, monodisperse population of vesicles is seldom obtained without an extraneous input of energy: spontaneously formed vesicles through uncontrolled hydration are invariably polydisperse, multilamellar dispersions. A variety of well-established techniques including mechanical extrusion (Hope, Bally et al. 1985), sonication, and electroformation (Angelova and Dimitrov 1986), through an energetic input, reproducibly transform the spontaneous multilamellar lipid dispersions in water into well-defined monodisperse vesicles, broadly categorized as small (50–100 nm), large (100–1000 nm), and giant (1–50 μm) unilamellar vesicles. Although water equilibrates over millisecond timescales across the vesicular shell with high permeability (10^{-2}–10^{-3} cm s^{-1}) (Fettiplace and Haydon 1980, Mathai, Tristram-Nagle et al. 2008), passive permeation of solutes (e.g., protons, ions, and neutral molecules) is strongly hindered (Deamer and Bramhall 1986). As a result, stable encapsulation of cargo (e.g., osmolytes, enzymes, drugs, and imaging agents) can be readily achieved in vesicular configurations (Gregoriadis 1995, Langer 1998).

Cell-sized giant unilamellar vesicles (GUVs) (Menger and Angelova 1998, Walde, Cosentino et al. 2010) (Figure 10.2) are ideally suited for unravelling roles that membrane play during cellular osmoregulation for a variety of important reasons. *First*, the ability to prepare vesicular compartments using purified natural/synthetic lipids as well as non-native amphiphilic block copolymers (Discher, Won et al. 1999, Mai and Eisenberg 2012) enables molecularly tailored selection of membrane components. The importance of this is obvious as the membranes of different organisms and organelles are composed of a highly diverse repertoire of lipids, which undergo significant compositional adaptations in response to osmotic stresses (Sajbidor 1997, Zhang and Rock 2008). This ability to tailor membrane chemical compositions in model membranes then allows one to map the relations between osmotic stress and membrane's chemical make-up towards appreciating membrane lipid homeostasis in free living cells (Zhang and Rock 2008). *Second*, many important osmosensory and osmoregulatory membrane proteins can be purified (or isolated in native membrane fragments) and subsequently reconstituted allowing to insert minimal cellular osmoregulatory machines in synthetic GUVs (Sukharev, Martinac et al. 1993, Wood 1999, Poolman, Spitzer et al. 2004, Kung, Martinac et al. 2010). This ability is important for developing an understanding of the effects of membrane properties on conformations and functions of osmoregulatory, mechanosensitive channels. *Third*, GUVs allow to isolate (and independently adjust) the properties of the encapsulated aqueous phase milieu (in addition to the external ambient phase), thus affording a control over the internal and external solute compositions and concentrations. *Fourth*, the semipermeable property also makes it possible to mimic the intracellular macromolecular crowding (Zimmerman and Minton 1993, Ellis 2001) such as by introducing the so-called crowding agents (e.g., polyethylene glycol, dextran), thus reconstituting the properties of the cytoplasmic aqueous phase (i.e., crowding, phase separation, viscosity, reduced diffusivities of solutes) in GUVs (Zimmerman and Minton 1993, Dix and Verkman 2008, Dominak and Keating 2008, Zhou, Rivas et al. 2008, Keighron and Keating 2011, Mika and Poolman 2011, Keating 2012). Taken together, the structural and compositional tailorability of the GUV membranes and

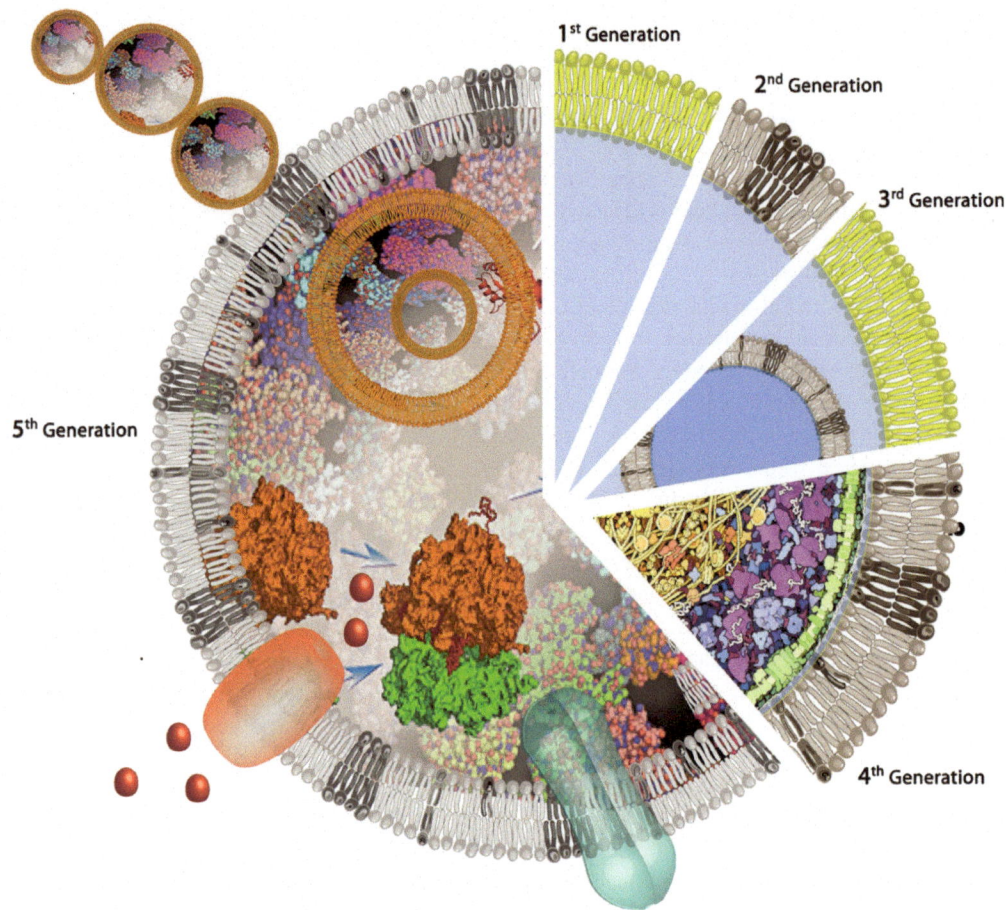

FIGURE 10.2 Evolution of an integrated, self-sufficient protocell from a simple vesicle, depicted using five hierarchies of complexity. 1st generation: Phospholipid vesicles of a single lipid component, isolating the encapsulated aqueous core from the surrounding bulk. 2nd generation: Compositionally more complex phospholipid vesicles. 3rd generation: Compositionally complex multivesicular phospholipid vesicles. 4th generation: Multivesicular vesicles, with complex lipid make-up, encapsulating solutes at high concentrations. 5th generation: Self-sufficient vesicles (protocells) consisting of a complex, integrated set of lipid composition and self-replicating machineries.

the encapsulated aqueous phase then make them elegant models of cellular compartments allowing to recapitulate the diverse conditions that characterize variations in the cytosolic milieu and membrane compositions such as that occurs during cellular osmoregulation.

GUVs are most commonly formed by the techniques of gentle hydration and electroformation (Figure 10.3). First established by Reeves and Dowben (1969), the simpler, gentle hydration method is driven by the electrostatic repulsion force between charged lipids bilayers. Vesicles form even under high ionic strength but often suffer from lamellarity issue and the necessity to include charged lipids in the vesicle composition. The latter method, introduced first by Angelova and Dimitrov (1986), consistently produces relatively more monodisperse populations of GUVs. Its less robust performance in forming vesicles under high ionic strength condition has been dealt with variants in the method (Estes and Mayer 2005, Montes, Alzonso et al. 2007, Peterlin and Arrigler 2008, Pott, Bouvrais et al. 2008). Mechanisms by which electroformation process produces monodisperse GUV population are complex. Qualitatively, it is thought to involve a balance of a variety of disparate forces, including van der Waals, electrostatic, hydration, osmotic pressure, the undulation forces, and resistance to motion (Dimitrov and Angelova 1988). Step-wise mechanisms of the swelling process induced by external electric field, involving these forces, have been proposed (Dimitrov and Angelova 1988) and substantiated (Angelova, Soléau et al. 1992, Shimanouchi, Umakoshi et al. 2009), which involve (i) gentle mechanical agitation of thin lipid bilayer films, driven by hydration and electrostatic interactions, as the electric field interacts with the dipole moments of lipid molecules, leading to curvature fluctuation (Heimburg 2007); (ii) differential membrane tension of the two monolayers, giving rise to instability of the lamellar membrane to bending deformation and swelling; and (iii) membrane closure as a result of the progressive bending instability, the kinetics of which depends on the edge energy, membrane viscosity, and bending elasticity, largely determining the size distribution of resulting vesicles. Excellent articles and

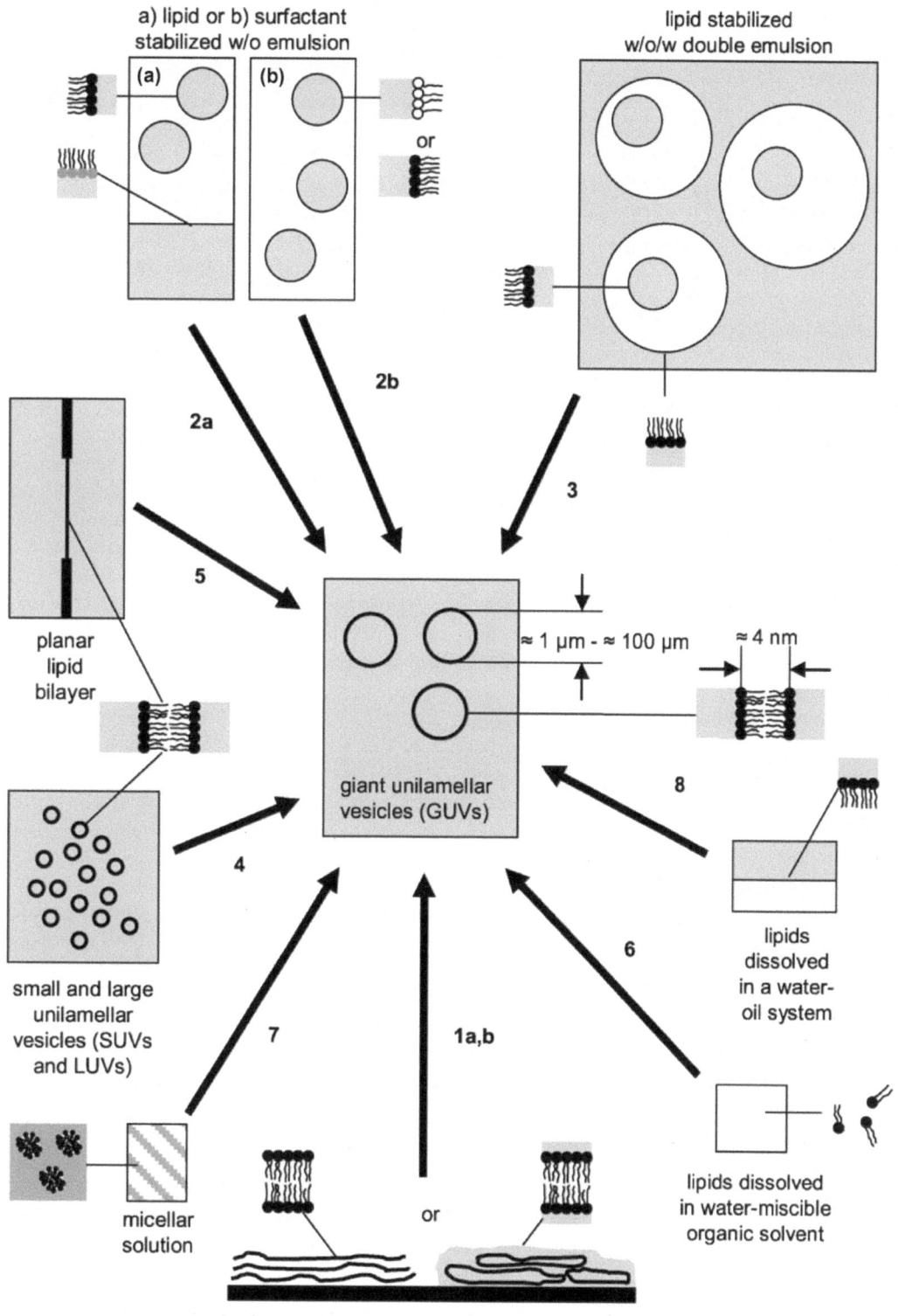

FIGURE 10.3 Key concepts of the main methods for the formation of giant unilamellar vesicles (GUVs). 1: Controlled hydration of dried films of lipids deposited on a solid surface, either in the absence of electric field (1a: spontaneous swelling or gentle hydration), or in the presence of an applied electric field (1b: electroformation or electroswelling). 2: Transformation of a lipid-stabilized w/o emulsion (2a) or a surfactant-stabilized w/o emulsion (method 2b) into giant vesicles. 3: Transformation of a lipid stabilized w/o/w double emulsion into giant vesicles. 4: Fusion of small unilamellar vesicles (SUVs) or large unilamellar vesicles (LUVs) into giant vesicles. 5: Formation of giant vesicles by jet-blowing onto a planar lipid bilayer maintained between two aqueous solutions. 6: Spontaneous formation of giant vesicles from lipids in a water-miscible organic solvent. 7: Giant vesicles formed from micellized bilayer-forming lipids. 8: Giant vesicles formed from bilayer-forming lipids that are initially present in a w/o two-phase system. Images reproduced with permission from Walde et al. 2010 (Copyright © 2010, John Wiley and Sons).

comprehensive reviews detailing the available approaches and challenges associated with the formation of GUV are available in literature (Morales-Penningston, Wu et al. 2010, Walde, Cosentino et al. 2010).

Despite the progress these accomplishments portend, there are still myriad challenges that must be overcome to recapitulate structural and compositional complexities of living cells and organelles using GUVs (Richmond, Schmid et al. 2011, Schmid, Richmond et al. 2015). Some of these challenges include lack of availability of general methods that afford (i) the use of variable ionic strength and pH conditions for the aqueous phase(s); (ii) control over size and lamellarity; (iii) control over lipid composition, surface charge, and concentrations of encapsulated components; (iv) reproducing compositional asymmetry of membranes; (v) incorporation of membrane proteins in controlled orientations and concentrations; (iv) integration of cytoskeleon- and glycocalyx elements; and (vii) an accurate reproduction of physical properties of the intravesicular cytosolic environment, including macromolecular crowding, diffusional complexities, and cytosolic viscosity. Although generic approaches to prepare GUVs satisfying all of the desired abilities to mimic cellular complexity are lacking, there has been promising development in recent years; key among them are highlighted below.

First, the formation of complex GUVs in challenging aqueous conditions, such as high salt concentrations, has been recently achieved using agarose- (Tsai, Stuhrmann et al. 2011) and polyvinyl alcohol-gel surfaces (Weinberger, Tsai et al. 2013), or by including a small concentration of PEG anchored lipids (Yamashita, Oka et al. 2002). Innovations using micropatterned polymer film or new electroformation apparatus that provides in-plane confinement to growing vesicles yield GUVs of narrow-sized distributions (Taylor, Xu et al. 2003, Bi, Yang et al. 2013, Kang, Wostein et al. 2013). In addition, an in situ Ca^{2+} (or $CaCO_3$) adsorption layer on dried lipid films (Tao and Yang 2015) has recently been shown to significantly reduce the polydispersity of GUVs.

Second, asymmetric membranes in specific cases can be produced by employing chemical methods, such as use of pH gradients, osmotic pressure, or lipid transfer proteins, or by using emulsion-based approaches, which allow to construct GUVs step-wise in a leaflet-by-leaflet manner. For instance, standard vesicle preparation techniques with pH gradients (Hope, Redelmeier et al. 1989, Eastman, Hope et al. 1991) has been reported to yield asymmetric lipid distribution across membrane leaflets, which is, however, limited and difficult to control. More promising methods involve asymmetric addition, removal, or transfer of lipids between the monolayer leaflets comprising the bilayer, here, the use of lipid transfer proteins; a more controlled lipid asymmetry could be achieved (Bloj and Zilversmit 1981, Everett, Zlotnick et al. 1986). Recently, using methyl β-cyclodextrin to add or remove cholesterol to or from the outer leaflet of the lipid bilayer, stable asymmetry in GUVs can be generated (Cheng, Megha et al. 2009, Chiantia, Schwille et al. 2011).

Third, there has been some success in reconstituting membrane proteins within GUVs. General procedures used to incorporate membrane proteins into GUVs utilize a two-step process. The first of which involves the delivery of the protein of interest into large unilamellar vesicles (LUVs) typically by detergent-mediated delivery (Helenius, Sarvas et al. 1981, Eytan 1982, Rigaud, Pitard et al. 1995, Gummadi, Hrafnsdóttir et al. 2003). The second step then transfers the protein by dehydration and rehydration (or electroformation) of the proteoliposomes to produce proteo-GUVs (Girard, Pécréaux et al. 2004). In some cases, anhydrobiotic sugars, such as trehalose or sucrose, have been used to confer structural stability to the protein during dehydration stage. Other methods involving direct incorporation using fusogenic peptides and detergent removal using biobeads have also proved successful (Doeven, Folgering et al. 2005). More recently, an integrated approach using microfluidic jetting has shown the potential to make GUVs with controlled lipid composition and internal content, coupled to oriented membrane proteins (Richmond, Schmid et al. 2011). Furthermore, added complexity involving the incorporation of membrane-binding actin (Limozin, Bärmann et al. 2003) and microtubule network (Roux, Cappello et al. 2002, Koster, VanDuijn et al. 2003, Herold, Leduc et al. 2012) have been recently reported.

Fourth, in addition to controlling membrane components and their distributions, control over the substances occupying the encapsulated volume within GUVs is also beginning to be demonstrated. Several species including polymers, nucleic acids, proteins, small molecules, and gels (Oberholzer, Albrizio et al. 1995, Oberholzer, Nierhaus et al. 1999, Walde and Ichikawa 2001, Fischer, Franco et al. 2002, Dominak and Keating 2007) have been shown to be successfully encapsulated into GUVs, providing in vitro tools to mimic physical–chemical properties of the cytoplasmic milieu. Insightful reviews covering experimental and theoretical bases are available (Zimmerman and Minton 1993, Ellis and Minton 2003, Minton 2007, Zhou, Rivas et al. 2008, Minton 2007, 2008; Ellis 2001; Zimmerman and Minton 1993). *Fifth*, recent demonstrations of successful encapsulation of complex transcription–translation machinery within GUVs (Luisi, Ferri et al. 2006), coupled to sustainable vesicle growth machinery, powered by simple metal catalytic systems (DeClue, Monnard et al. 2009) or an autocatalytic lipid self-synthesis (Hardy, Yang et al. 2015), suggest significant progress in building structurally complex GUVs or "minimal" protocells.

10.3 OSMOTIC PRESSURE GRADIENT ACROSS GIANT UNILAMELLAR VESICLES

Osmotic stress originates from the difference in osmotic pressures (ΔP) of the aqueous solution on the "inside" and the "outside" of the GUV compartments. It is perhaps best expressed in terms of the osmotic potential (Π, atm) of an aqueous solution, which is proportional to its water activity and is determined by the activities (but not the identities) of all its solutes (Castellan 1964):

$$\Pi = (RT/\overline{V_w}) \ln a_w$$

where R is the gas constant (0.082054 L atm mol^{-1} K^{-1}), T is the temperature (kelvin, K), and $\overline{V_w}$ is the partial molar volume of water (0.01801 L mol^{-1}). By definition, the activity (a_w) of pure water is 1.0, which corresponds to zero osmotic potential for pure water. As solutes are added, a_w generally falls towards zero, and therefore the osmotic potential associated with solutions becomes negative. Absent preferential interactions between the solute particles and those between the solute and the membrane bilayer in the limit of dilute solution exhibiting ideal behaviour, the osmotic potential reduces to $\Pi = -RTc_s$ where c_s represents the total molar concentration of the solute (Csonka 1989). Applied to vesicular compartments embedded in solutions containing osmotically active solutes (rather than pure water), the expression for the osmotic potential gradient generalizes to $\Delta\Pi = RT(c_{s-\text{inside}} - c_{s-\text{outside}})$. This then indicates that when a solution is separated from another by a vesicular membrane that is permeable to water but impermeable to the solutes, water will move into the vesicle from the region of higher to lower osmotic potential, down the activity (or concentration) gradient. When the internal vesicular volume cannot freely expand to accommodate the increased water content, such as in bacterial and plant cells bounded by cell walls, the influx of water in cells containing higher concentrations of osmolytes than their surroundings would necessarily result in a build-up of pressure, called the turgor pressure, experienced by the wall (Csonka 1989).

10.3.1 Osmotic Upshift: Giant Unilamellar Vesicles in Hypertonic Media

GUVs immersed in hypertonic bath experience osmotic upshift due to the osmotic potential gradient ($\Delta\Pi$) across the membrane boundary, which acts to expel water from the vesicular compartment. Assuming that the initial shape of the vesicle is spherical, the rapid expulsion of water (typically over millisecond timescales) causes the vesicle volume (V) to decrease, rendering it flaccid. A dominant effect of the availability of excess membrane area for enclosing the vesicle volume then is the transformation of vesicle shapes. This can be described in terms of a dimensionless parameter termed reduced volume, $v = V/[4\pi R_0^3/3]$ where R_0 corresponds to the radius of an equivalent sphere of area $S = [4\pi R_0^2]$. From geometrical considerations alone, when vesicle volume is reduced below the maximum, $V_o [= 4\pi R_o^3/3]$ at the constant surface area S, $v < 1$, the vesicles deviate from a spherical motif assuming irregular shapes: a practically unlimited number of different shapes are accessible for deflated vesicles with excess membrane area. Experimentally, however, a large but finite number of distinct shape types have been observed (Svetina and Zeks 2002). These include oblates, prolates, star-fish, dumb-bells, and pearls (Figure 10.4), often following well-defined pathways for shape changes (Hotani 1984). Two characteristic oblate shapes include cup (shapes 1–3) and disc (shape 4) shapes. Two typical prolate shapes are the cigar (shape 5) and pear shapes (shapes 6–8). For vesicles with lower volumes, shapes 9–12 are most frequently observed and shapes 13–16 are those characterized by narrow necks, namely dumb-bells and pearls. In parallel sets of experiments in which area-to-volume ratios are altered by temperature variations, comparable morphologies have been observed (Kas and Sackmann 1991). The finite range of distinct vesicle shapes that are experimentally observed in experiments modulating area-to-volume ratios reflects selection of some symmetry characteristics. This in turn suggests that their formation may not be determined solely by the geometric criterion of reduced volume alone, but rather that the membrane material properties play a role in determining vesicle shapes.

Several different theoretical models have been advanced to predict the shape behaviour of vesicles, which incorporate materials elastic properties of the membrane bilayers. Two independent elastic deformational modes, namely in-plane elasticity and bending rigidity, are relevant. Because lipid molecules comprising the membrane are two-dimensionally fluid, membranes do not exhibit shear elasticity (Helfrich 1973). Moreover, the area expansion moduli (10^2–10^3 m Nm^{-1}) of typical membranes are large, and bending rigidities (10^{-19} Nm) are low. Thus the vesicular shells respond to changing vesicle volume during osmotic upshift by producing deformations and shape changes determined primarily by their bending rigidities and hence curvatures (Needham and Nunn 1990, Hallett, Marsh et al. 1993, Seifert 1997). Major models, which predict vesicle shapes under conditions of $v < 1$, consider minimal bending energy configurations of vesicles: An implicit assumption in these models is that the vesicles adopt shapes, which correspond to the smallest possible value of the membrane bending energy (Canham 1970). In these models, the membrane is regarded as a two-dimensional surface embedded in three-dimensional space. The spontaneous-curvature (SC) model (Canham 1970, Helfrich 1973) explicitly accounts for the *local* elastic energy penalty due to any deviation of the local mean curvature from the preferred curvature (C_o, where $2C_o$ is the spontaneous curvature) of the membrane. This is formally accounted for in terms of local bending rigidity (k) and reflects the membrane asymmetry due to differences in the chemical composition (or the immediate chemical environment) of the two constituent leaflets. The alternate bilayer-couple (BC) model (Sheetz and Singer 1974, Svetina, Brumen et al. 1985) considers the *global* elastic energy due to bending in terms of the departure, albeit

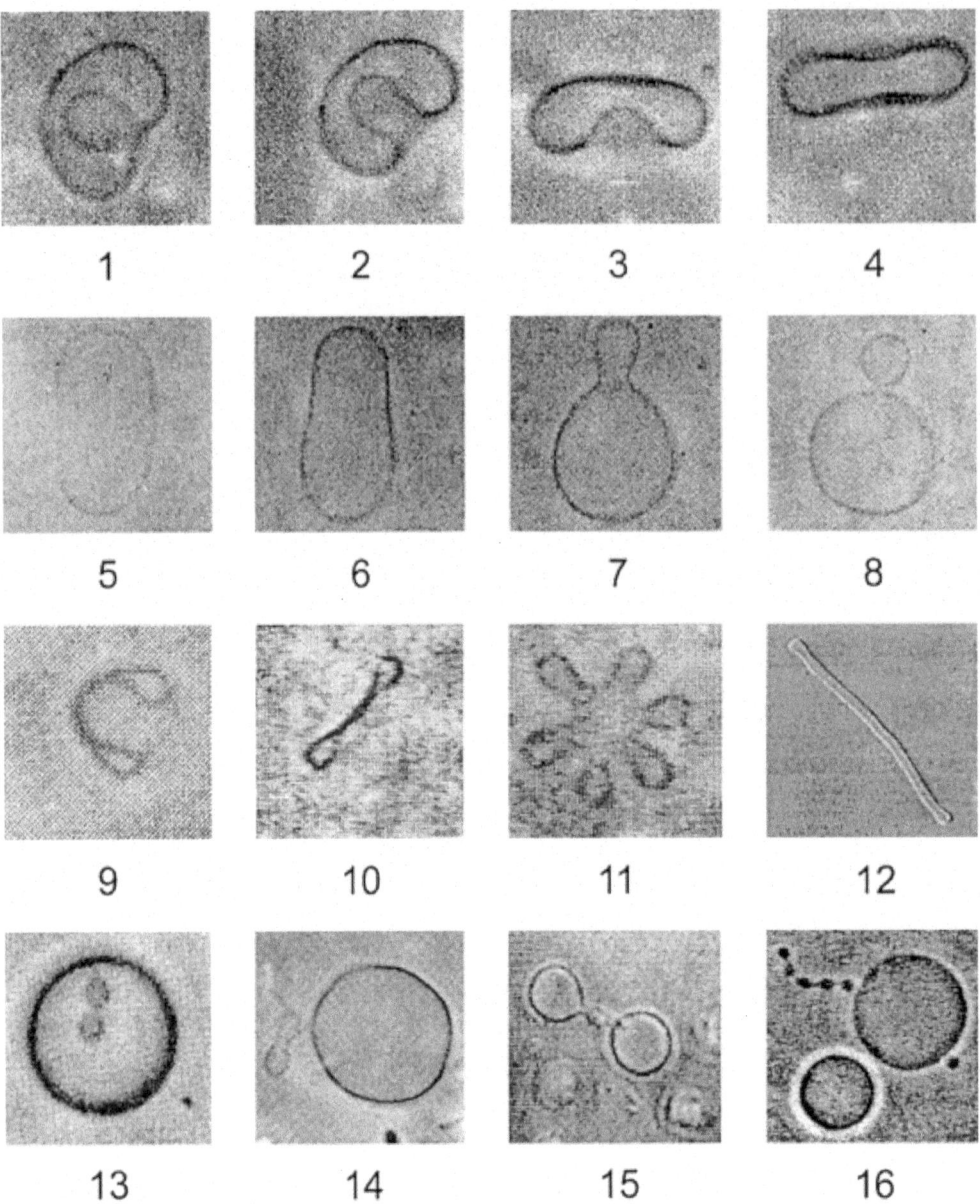

FIGURE 10.4 Flaccid vesicle shapes, formed as a result of vesicle formation process or in response to osmotic upshift. Phase contrast microscopic images. First row: Characteristic oblate shapes exemplified by the cup-shape class (1–3) and the disc-shape class (4). Second row: Characteristic prolate shapes exemplified by the cigar-shape (5) and pear-shape (6–8) classes. Third row: Codocyte (9), torocyte (10), starfish (11), and worm (12) shapes are characterized by their relatively small vesicle volume/membrane area ratios. The fourth row shows shapes characterized by narrow necks connecting nearly spherical vesicle parts (13–16). Images produced with permission from Svetina and Zeks 2002 (Copyright © 2002, John Wiley and Sons).

fixed, of the geometric area difference, ΔA, between the two monolayers from its equilibrium value ΔA_o. This then becomes basis for the nonlocal bending rigidity. In this approach, the two monolayers are considered to be coupled at a fixed distance, but they are not allowed to exchange lipid molecules. This then leads to an additional constraint which can be incorporated into the continuum model. In addition to the differences in chemical environment surrounding the two monolayers (See above), area difference can also originate from (i) asymmetric insertion of molecules (Sheetz and Singer 1974); (ii) differences in thermal expansivities of the two monolayers (Kas and Sackmann 1991); and (iii) increased flip-flop of molecules due to changes in lateral membrane tension (e.g., during osmotic deflation, flow of water through the membrane can lead to a greatly enhanced flip-flop rate) (Bruckner, Mansy et al. 2009). The generalized BC model or the area difference elasticity (ADE) model (Miao, Seifert et al. 1994, Mui, Dobereiner et al. 1995), which interpolates between the two, SC and BC, models, describes the shape-free-energy functional formally in terms of both the SC (first term) and the BC (second term) models:

Osmotic Stress on Membrane Compartments

$$F_{ADE}[S] = \frac{\kappa_b}{2} \oint_S dA(C_1 + C_2 - 2C_o)^2$$
$$+ \frac{\bar{\kappa}}{2} \frac{\pi}{AD^2}(\Delta A - \Delta A_0)^2$$

in which C_1 and C_2 are the local principle radii of curvature (m^{-1}), C_o represents the spontaneous curvature of the membrane surface, and k_b represents the mean local bending rigidity – a material property describes the SC term (first term). In the BC contribution (second term), A is the membrane area; D is the membrane thickness; and $\bar{k}$ is the nonlocal bending rigidity. Thus, minimal energy vesicle shape can be determined for given reduced volume V and by an effective reduced area difference, which is determined by the spontaneous curvature of the SC model and by a curvature-induced area difference between the inner and outer monolayers from the BC model.

By minimization of the total curvature energy of vesicular compartments under the imposed constraints of constant area and volume, minimal energy shapes can be obtained as functions of the enclosed volume and the area of the vesicle. Applying the SC model alone, Deuling and Helfrich obtained a rich catalog of shapes (Deuling and Helfrich 1976), which recapitulates the experimentally observed GUV shapes. Organizing the shapes in so-called "phase diagrams" as functions of reduced volume then allows one to predict shape transformations such as that occurs when an osmotic stress is applied (Seifert 1997). Svetina and Zeks (1989) combined bending elasticity of the membrane with the BC model and developed corresponding phase diagrams for vesicle shapes. The two models identify comparable shapes although the order of the shape transitions and the paths of transition differ for the two models when area-to-volume ratio (or osmotic pressure differences) is varied (Seifert, Berndl et al. 1991).

GUV response to osmotic deflation becomes more involved when additional structural and compositional complexities are introduced (Farge and Devaux 1992, Julicher and Lipowsky 1996). For instance, GUVs prepared using mixtures of lipids (e.g., egg–PC lecithin and phospholipid–cholesterol mixtures) have been observed to respond to osmotic upshift by gradually shrinking their size without exhibiting any measurable distortion from the nominally spherical shape (Boroske, Elwenspoek et al. 1981). The gradual shrinkage, in these cases, appears to be accompanied by an irreversible topological transition producing daughter vesicles, which detach from the parent GUV. In many instances, the daughter vesicles remain attached to the mother vesicle through thin tethers. These observations support the notion that budding and vesiculation might be driven by the generation of spontaneous membrane curvature through nonhomogeneous distribution of lipids. In another study, Döbereiner, Kas, and Sackman (1993) examined morphological behaviours of vesicles consisting of lipid–cholesterol mixtures under conditions of increasing area-to-volume ratio by heating and through osmotic effects. They too find budding and repeated fission, which they attribute to a combination of liquid/gel domain formation, asymmetric transverse distribution of lipids, and coupling of the spontaneous curvature of the membrane to the local lipid composition – all consequences of membrane's compositional heterogeneity. Using phase-separating ternary system, Yanagisawa and coworkers (2008) monitored shape deformations of ternary vesicles undergoing phase separation under an osmotic upshift. They observe that lateral phase separation became coupled with shape changes converging prolate, discocyte, and starfish vesicles to discocytes (Figure 10.5). Moreover, in late stages, they find that domains bud vectorially, inwards or outwards depending

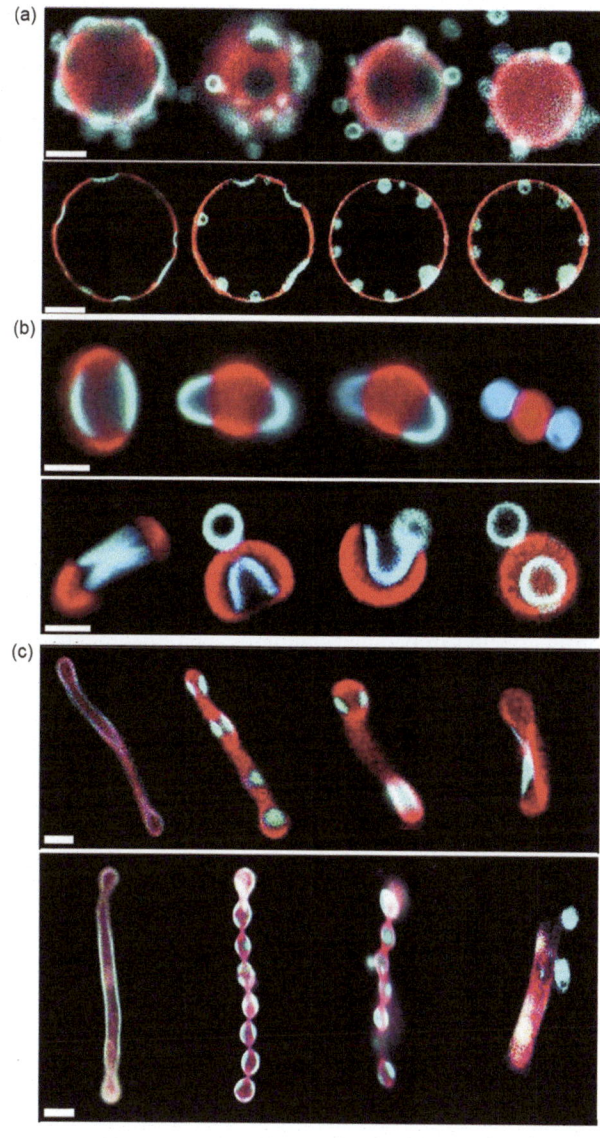

FIGURE 10.5 Time evolution of shape deformations induced by osmotic stress, coupled with phase separation, of multicomponent vesicles. Outside (a–c, top panels) and inside (a–c, bottom panels) budding of a sphere (a), a discocyte (b), and a tube (c) in response to osmotic stress. Scale bars are 5 μm. Images reproduced with permission from Yanagisawa et al. 2008 (Copyright © 2008, American Physical Society).

on the amount of available excess area. All of the observations above can be reconciled in terms of theoretical frameworks above for understanding equilibrium shapes for compositionally complex vesicles. Here, in addition to bending energy, the minimization of membrane energy requires consideration of additional contributions from the free energy for lateral phase separation and the line tension (Lipowsky 1992, Julicher and Lipowsky 1993, Wolff, Komura et al. 2015).

The next level of complexity introduced into synthetic vesicle involves recapitulation of internal structure of the vesicle volume. Unlike synthetic vesicles, which typically employ dilute aqueous solutions, properties of cellular cytoplasm, cytoskeleton network, and macromolecular crowding in the cellular interior all contribute to the cellular response to the osmotic challenge. Studies of osmotic response of complex GUVs, which incorporate these physical properties of cytoplasm are, however, sparse. In one such study, Viallat, Dalous, and Abkarian (2004) compared osmotically induced shape changes in vesicles embedding a dense, viscoelastic agarose gel with those containing sucrose. They find that the presence of gels within the vesicles, above a threshold concentration, suppresses large-scale deformations and produces unique sets of morphologies characterized by the appearance at the vesicle surface of facets, bumps, and spikes for low moduli gels or wrinkles for higher moduli gels – reminiscent of echinocytic shapes of red blood cells (Mukhopadhyay, Lim et al. 2002). The latter erythrocyte shape is well described by combining ADE model with the elasticity of the spectrin network anchored to the plasma membrane (Iglic 1997, Lim, Wortis et al. 2002, Mukhopadhyay, Lim et al. 2002). These findings then suggest a prospect for resolving a significant controversy surrounding how cell volume regulation is effected in cellular compartments. Specifically, by allowing to incorporate gel-like, poroelastic material, the lumen of the GUVs, these complex compartments promise to resolve whether the osmotic stress is distributed throughout the cell volume and not confined to the membrane cortex alone (Sachs and Sivaselvan 2015).

These and other recent advances summarized above in designing complex vesicles promise new insights towards a more complete understanding of how internal structure of the encapsulated phase in vesicles, which mimic cytoplasmic contents, affects vesicular osmotic responses.

10.3.2 Osmotic Downshift: Giant Unilamellar Vesicles in Hypotonic Media

Osmotic downshift is readily established between the compartmentalized volume and the surrounding free bath when the vesicles experience hypotonic environment. The relaxation process triggered by osmotic downshift then acts to reduce the osmotic pressure difference across the closed semipermeable membrane by influx of water, increasing its volume-to-surface area ratio (Ertel, Marangoni et al. 1993). An immediate consequence of the osmotic influx of water in vesicles embedded in hypotonic media is the build-up of lateral membrane tension due to changes in the balance of forces within the bilayer producing high energy states (compared with isotonic relaxes vesicles) (Needham and Nunn 1990). Beyond a threshold tension, rupture and pore formation become energetically favourable, lysing the GUVs at lateral tensions corresponding to ≈30–40 m N m^{-1} (Needham and Nunn 1990, Ertel, Marangoni et al. 1993, Mui, Cullis et al. 1993). Consistent with Laplace's law ($\sigma = \Delta\Pi\ r/2$, where σ is the lateral membrane tension, Π is the osmotic pressure difference or osmotic potential, and r is the vesicle radius), vesicle size is inversely proportional to the osmotic differential, with small vesicles able to tolerate greater residual osmotic differentials (Needham and Nunn 1990, Mui, Cullis et al. 1993).

A theoretical framework on osmoregulatory behaviour is advanced by the use of these vesicular compartments. Theories and experiments (Haleva and Diamant 2008, Peterlin, Arrigler et al. 2012) have shown that, consistent with early work on 100–200 nm large unilamellar vesicles (Ertel, Marangoni et al. 1993, Mui, Cullis et al. 1993), an osmotically tense vesicle does not lyse catastrophically; rather, it follows a step-wise sequence of events (Peterlin, Arrigler et al. 2012) (Figure 10.6). The vesicles first progressively swell, transforming themselves from an oblate shape to a spherical shape, which then undergo the transition, or crossover, to a stretching stage, at which the vesicles transform into an inflated sphere, prior to vesicle rupture (Peterlin, Arrigler et al. 2012). The action of this "pressure-release valve" is not all-or-nothing. During each membrane rupture event, the valve (or hole) opens for a very short period of time, releasing only a fraction of the intravesicular solute (and water) before the bilayer reseals leaving the vesicle hyperosmotic with a lower osmotic differential. This then prompts subsequent events of water influx, vesicle swelling, and rupture until sufficient intravesicular solute has been lost, and the membrane is able to withstand the residual sublytic osmotic pressure without collapsing (Wood 1999). Thus, GUVs in hypotonic media exhibit oscillations in their sizes – characterized by alternating modulations of vesicular volume, tension, and solute in efflux – prompted by repeated cycles of swelling and bursting (Sandre, Moreaux et al. 1999, Karatekin, Sandre et al. 2003, Popescu and Popescu 2008, Peterlin, Arrigler et al. 2012).

More recently, we and others have shown that the long-lived transient swell–burst cycles become coupled with membrane's compositional degrees of freedom. Using hypertonic giant vesicles, consisting of domain-forming lipid mixtures consisting of ternary mixtures of a saturated (sphingomyelin, SM), an unsaturated (1-palmitoyl-2-oleoyl-sn-1-glycerol, POPC), and cholesterol (Chol), we found that the swell-burst cycles are accompanied by periodic oscillations of an optically uniform swelling phase, formation of large microscopic L_o and L_d domains, and transient pore formation and closure (Oglęcka, Rangamani et al. 2014). The valve (or hole) opens for a very short period

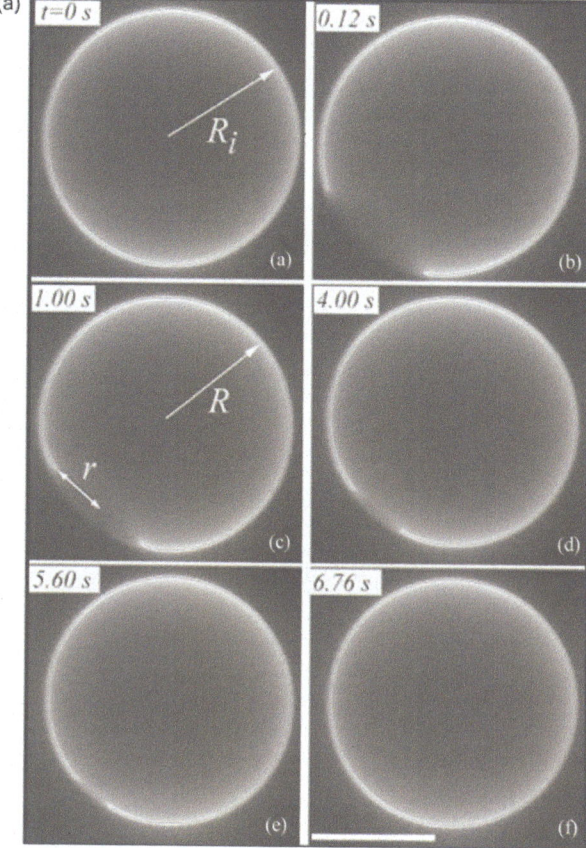

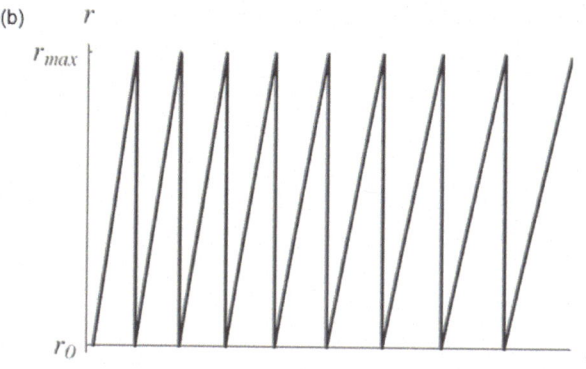

FIGURE 10.6 Transient pores in giant vesicles and the evolution of vesicle radius over time, in response to osmotic downshift. (a) Appearance of a transient pore in a DOPC vesicle of initial radius R_i. (a), a membrane pore forms when the membrane tension reaches a critical value. The pore size reaches its maximum very rapidly (b), thereafter decreasing slowly until complete resealing (c–f). The white bar corresponds to 10 μm. The vesicle contains 20 mol% cholesterol. (b) Theoretical model of the behaviour of a vesicle radius r as a function of time t upon a transfer from a sucrose/glucose solution into an iso-osmolar glycerol solution. Images reproduced with permission from Karatekin et al. 2003 (Copyright © 2003, Elsevier) and Peterlin and Arrigler 2008 (Copyright © 2008, Elsevier).

of time – typically less than 1 second – releases some of the internal solutes several times producing a pulsating, breathing pattern in size, and molecular texture of the vesicle (Figure 10.7). Specifically, a highly coordinated interplay between a number of distinct elementary physical mechanisms appears to regulate osmoregulation in synthetic vesicles. These are briefly described below.

First, osmotically triggered water influx. A trans-bilayer osmotic differential, imposed on osmotically equilibrated isotonic GUVs, such as by dilution of the extravesicular dispersion medium, triggers an osmotic relaxation process, which acts to reduce the pressure difference across the semipermeable membrane by an influx of water (Mui, Cullis et al. 1993). *Second, the retention of osmotic pressure and build-up of membrane tension ironing out thermal undulations.* In the initial isotonic environment, vesicles are flaccid, with thermally undulating surface topology caused by bending-dominated shape fluctuations (Seifert 1997). As water enters, the GUVs swell, ironing out the undulations and generating lateral membrane tension (Peterlin, Arrigler et al. 2012) and rendering the vesicular boundary tense (Haleva and Diamant 2008). At equilibrium, the lateral tension generated in the membrane then compensates for the osmotic pressure, consistent with Laplace's law.

Third, the appearance of microscopic domains in the membrane subject to osmotic pressure and lateral tension. Membranes formed from multiple components can partition laterally into binary coexisting microscopic liquid phases or domains. A ternary mixture, namely POPC, SM, and Chol, at a molar ratio of 1:1:1, at 23°C, exhibits a uniform phase behaviour, as outlined in its partial phase diagram (Veatch and Keller 2005). Below a miscibility transition temperature, these lipid components coexist as liquid-ordered (L_o), a liquid phase with short-range order populated by SM and Chol, and POPC-rich liquid disordered (L_d) phases (Veatch and Keller 2003, 2005). The use of POPC is highly relevant due to its abundance in the plasma membrane. SM and Chol, on the other hand, in addition to their presence on the outer leaflet of the plasma membrane, lie in the unique geometry of SM to hydrogen bond with 3-OH group of the cholesterol through ester linkage, amine, and hydroxyl groups (Barenholz and Thompson 1980, Scherfeld, Kahya et al. 2003, Veatch and Keller 2005). When subject to hypotonic conditions as described above (the first step above), the optically homogeneous vesicles break up into surface patterns consisting of large microscopic rounded liquid-disordered domains, bounded by spatially continuous liquid-ordered region. The stochastic nature of the phase transformation is evident from the heterogeneous landscape of optically homogeneous vesicles and those decorated with microscopic domains. The time-dependent process of appearance and disappearance of large microscopic domains repeats itself multiple times over several tens of minutes, ultimately producing a steady state characterized by a fixed microstructure of a rounded boundary. Importantly, the oscillatory phase separation behaviour is fully reproducible for a variety of neutral osmolytes (including glycerol, glucose, lactose, galactose, dextran, sorbitol, and sucrose), GUV sizes (5–50 μm), strength of osmotic gradients (20–2000 mM), and a wide range of lipid compositions within the phase coexistence window (Veatch and Keller 2005).

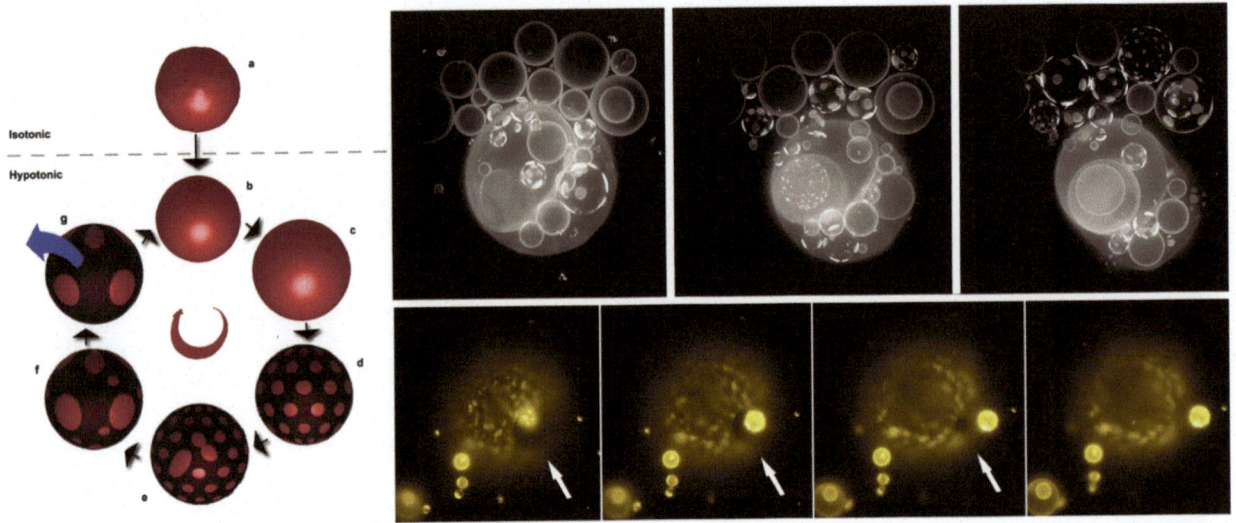

FIGURE 10.7 Schematic representations of physical mechanisms and changes in vesicular membrane properties during osmotic downshift. (Left panel) (a) GUV in isotonic medium exhibiting a flaccid morphology. (b,c) Immersion in a hypotonic bath initiates an osmotically triggered influx of water rendering the GUV tense. (d–f) The optically uniform vesicular surface breaks up into a pattern of microscopic domains, which grow by collision and coalescence. (g) Transient appearance of a microscopic pore (~0.3–0.5 s lifetime), enabling solute efflux and tension relaxation, which drives pore closure, producing closed GUVs with a reduced osmotic differential and homogeneous surface. Steps (b–g) repeat until the sub-lytic solute concentration differential is reached and the Laplace tension in the membrane is able to compensate for the residual osmotic pressure. (Right panel, top) Microscopic images of the oscillatory phase separation (corresponds to steps d–f on the left panel) and (Right penel, bottom) transient appearance of pore (corresponds to step g on the left panel). Images reproduced with permission from Oglecka et al. 2014 (Copyright © 2010, Elife Sciences Publications).

Fourth, coarsening of domains. Under lateral tension and pressure difference, (i) the domains coarsen through collision and coalescence and (ii) the appearance of phase-separated state invariably coincides with the swollen, tense state of GUV during the cyclical swell-burst processes. Although lateral tension and pressure difference influence membrane phase behaviour in our osmotically driven case, it is instructive to consider how each of the two factors individually affects membrane phase behaviour. A recent thermodynamic analysis and experiments examining the effects of mechanically generated tension reveal a lowering of miscibility phase transition temperature between the L_o and L_d phases with increase in tension ($dT/d\sigma$, ~−1 K [m Nm^{-1}]$^{-1}$) (Portet, Gordon et al. 2012, Uline, Schick et al. 2012). However, how this shift in transition temperature affects membrane phase behaviour and domain morphology is not obvious: a recent experimental study suggests that even tension alone can stabilize complex domain morphologies (Chen and Santore 2014). To add to the complexity, a theoretical model by Givli and Bhattacharya (Givli, Giang et al. 2012), explicitly introducing osmotic pressure contributions within the generalized Helfrich energy treatment, suggests that pressure can perturb isothermal phase diagram, driving domain formation primarily by affecting the interaction between geometry and composition. An alternate explanation involves separate theoretical arguments, which require preexisting phase-separated domains in the optically homogeneous state. It suggests that the lateral tension elevates line tension between coexisting phases (Akimov, Kuzmin et al. 2007). Therefore, although membrane tension disfavours nucleation of a new phase (by raising the energy barrier that must be met for the formation of critical nuclei), it can promote coalescence of small preexisting nanoscale domains driven by minimization of line tension between L_o and L_d phase. Clearly at variance with these predictions, current and earlier observations indicate that osmotic differentials induce phase separation (Hamada, Kishimoto et al. 2011). Additional experiments using ternary lipid mixtures (DOPC, DPPC, and Chol), which have been thought not to produce nanodomains at temperatures above 20°C (Hamada, Kishimoto et al. 2011), also produce oscillatory phase behaviour. This then suggests that the osmotically generated tension alone might be insufficient to explain the observed osmotically induced isothermal phase transition and that the nonideality in mixing is likely a consequence of a combined effect of the pressure and tension.

Fifth, the appearance of a short-lived transient pore, which enables partial solute efflux, reducing osmotic pressure and membrane tension. According to classical nucleation theory, the cost (E) of creating a pore in a tense membrane is determined by the competition between membrane tensional energy ($-\pi r^2 d$) and the line tension energy ($+2\pi r \Upsilon$) at the edge of the pore. Thus, under conditions of sufficient membrane tension ($dE/dr > 0$), pores nucleate and grow, enabling solute efflux (Sandre, Moreaux et al. 1999, Peterlin and Arrigler 2008). Membrane lysis proceeds via cascades of pores during each cycle of the swell–burst sequence (Karatekin, Sandre et al. 2003): during the swell segment of each oscillation cycle, a single microscopic pore, several micrometres across, becomes visible under conditions of maximum swelling and the largest domain size, typically for a period not exceeding 1.0 second. Although domain formation is not required for pore formation, the probability of pore nucleation might be enhanced by surface defects, such as present at the boundary between coexisting phases, since the energy required to open a pore (> 40 K_BT) is considerably higher than the thermal activation energy (Karatekin, Sandre et al. 2003). The long life spans (1.0 second) of the pores are likely supported by two opposing processes, namely osmotic influx of water and the leakage rate of solute through the pore (Koslov and Markin 1984).

Sixth, pore closure resulting in closed GUVs with reduced osmotic differential. Healing of the pore is promoted by the reduction in the net membrane area and partial loss of the encapsulated solutes, both of which reduce membrane tension, σ (Karatekin, Sandre et al. 2003). Thus, during each membrane rupture event, only a fraction of the intravesicular solute is released before the bilayer reseals, leaving the vesicle hyperosmotic, albeit with a reduced osmotic differential. This then prompts subsequent cycles of water influx, vesicle swelling, and rupture until sufficient intravesicular solute is lost and Laplace's tension in the membrane is able to compensate for the residual osmotic pressure (Ertel, Marangoni et al. 1993).

Several features are noteworthy in this extraordinary sequence of cyclical changes in membrane properties driving vesicular osmoregulation. The tension- and pressure-mediated appearance of the phase-separated state in osmotically swollen membrane does not account for the oscillations in domain pattern. The period of oscillation between optically uniform and phase-separated states increases with the passage of time. The cycle period – defined as the time elapsed between two consecutive instances of homogeneous fluorescence – increases 3- to 10-fold, before reaching a nonoscillating quiescent state, 60–120 minutes after the imposition of the osmotic differential. This temporal dynamics reveals that the osmotic differential and accompanying tension weaken with each cycle.

Taken together, this extraordinary and seemingly autonomous vesicle response – in which an external osmotic perturbation is managed by a coordinated and cyclical sequence of physical mechanisms allowing vesicles to sense (by domain forming) and regulate (by solute efflux) their local environment – suggests a primitive form of a quasi-homeostatic regulation in a synthetic material system (He, Aizenberg et al. 2012). The observations also illustrate the coupling between out-of-plane osmotic activity of water and membrane's in-plane compositional degrees of freedom producing an exquisite and complex response, thus underscoring the intrinsic coupling between membrane phase and mechanical tension.

This osmotically generated tension and poration can produce additional complex effects. In a recent study, Chen et al. (2004) have shown that the entrapment of high concentrations of osmolytes (e.g., 1 M sucrose) within nanometre-scale protocell compartments (e.g., ~100-nm fatty acid vesicles) – mimicking the physical effect of counterions that accumulate during RNA replication – produces an osmotic pressure difference across the compartmental boundary, which renders the vesicular membrane tense but does not lyse it. Subsequent uptake of additional membrane amphiphiles from surrounding relaxed vesicles, they show, then allows the osmolyte-loaded protocell to preferentially grow. This simple demonstration, thus, suggests how osmotic activity of membrane compartments can drive growth and in conjunction with budding and fission may pave for rudimentary constructs exhibiting self-reproducing behaviours.

10.4 OUTLOOK AND PERSPECTIVES

The ability to engineer ever-more complex giant unilamellar vesicles has led to tremendous progress in the understanding of the roles of membrane's structural and compositional degrees of freedom in aiding the fundamental processes of cellular homeostasis, such as osmoregulation. From a biological perspective, work at this interface of membrane biophysics and cell biology has yielded a clearer mechanistic picture of how membrane physical–chemical attributes play a role in (i) facilitating the sensing of osmotic irregularities, (ii) activating osmoregulatory mechanosensitive channels, and (iii) driving cellular shape changes that occur in response to an osmotic challenge. Viewed from a physical science perspective, these studies abstract fundamental principles and suggest practical means to develop cell-like, synthetic model compartments which exhibit complex shapes and molecular textures, undergo well-defined shape transformations and lateral compositional redistributions, and ultimately recapitulate living cell's extraordinary responsiveness to changes in the properties of their environment.

In this chapter, we have revisited the fundamental physical chemical processes that determine vesicular response to osmotic perturbations, most of which are already well understood within the framework of membrane mechanics. Contextualizing their relative roles in determining vesicle osmoregulation has allowed us to organize the cumulated evidence and progress in understanding osmotic responses of membrane vesicles. It has also offered a glimpse into challenges and opportunities for future work, which we have highlighted within the chapter. The abilities to engineer these complex vesicular compartments now promise clearer molecular-level mechanistic understanding of the relations between membrane deformations including changes in membrane shape, curvature, tension, fluidity, and lateral phase separation and functioning of osmoregulatory machines, such as mechanosensitive channels (Kung 2005). It should also help unravel mechanisms of more involved consequences of osmotic activities, including recent demonstrations that reveal how a polarized distribution of channels in the cell membrane – creating a net inflow of water (and ions) at the cell leading edge and a net outflow of water (and ions) at the trailing edge – propels the cell leading to migration of cells – a fundamental physiological process of broad relevance (Stroka, Jiang et al. 2014). In a parallel vein, an understanding of osmosensing and osmoregulating capacities of simple vesicular compartments made of amphiphiles alone may help understand how early evolution of cellular design of living systems might have used these simple and pervasive perturbations to enable stability, growth, fusion, division, and motility of early protocells at the dawn of life (Szostak, Bartel et al. 2001, Hanczyc, Fujikawa et al. 2003, Chen, Roberts et al. 2004, Zhu and Szostak 2009, Budin and Szostak 2011).

BIBLIOGRAPHY

Akimov, S. A., P. I. Kuzmin, J. Zimmerberg and F. S. Cohen (2007). "Lateral tension increases the line tension between two domains in a lipid bilayer membrane." *Physical Review E* **75**(1): 011919.

Altendorf, K., I. Booth, J. Gralla, J. Greie, A. Rosenthal and J. Wood (2009). "Osmotic Stress." *EcoSal Plus.* **3**(2). doi:10.1128/ecosalplus.5.4.5.

Andersen, O. S. (2015). "Perspectives on: The response to osmotic challenges." *Journal of General Physiology* **145**(5): 371–372.

Angelova, M. I. and D. S. Dimitrov (1986). "Liposome electroformation." *Faraday Discussions* **81**: 303–311.

Angelova, M. I., S. Soléau, P. Méléard, F. Faucon and P. Bothorel (1992). Preparation of giant vesicles by external AC electric fields. Kinetics and applications. *Trends in Colloid and Interface Science VI.***89**: 127–131.

Bangham, A. D. and R. W. Horne (1964). "Negative staining of phospholipids+their structural modification by-surface active agents as observed in electron microscope." *Journal of Molecular Biology* **8**(5): 660–668.

Barenholz, Y. and T. E. Thompson (1980). "Sphingomyelins in bilayers and biological membranes." *Biochimica et Biophysica Acta (BBA) - Reviews on Biomembranes* **604**(0): 129–158.

Bi, H., B. Yang, L. Wang, W. Cao and X. Han (2013). "Electroformation of giant unilamellar vesicles using interdigitated ITO electrodes." *Journal of Materials Chemistry A* **1**(24): 7125–7130.

Bloj, B. and D. B. Zilversmit (1981). "Lipid transfer proteins in the study of artificial and natural membranes." *Molecular and Cellular Biochemistry* **40**(3): 163–172.

Booth, I. R., M. D. Edwards, S. Black, U. Schumann and S. Miller (2007). "Mechanosensitive channels in bacteria: Signs of closure?" *Nature Reviews Microbiology* **5**(6): 431–440.

Boroske, E., M. Elwenspoek and W. Helfrich (1981). "Osmotic shrinkage of giant egg-lecithin vesicles." *Biophysical Journal* **34**(1): 95–109.

Bruckner, R. J., S. S. Mansy, A. Ricardo, L. Mahadevan and J. W. Szostak (2009). "Flip-Flop-induced relaxation of bending energy: Implications for membrane remodeling." *Biophysical Journal* **97**(12): 3113–3122.

Budin, I. and J. W. Szostak (2011). "Physical effects underlying the transition from primitive to modern cell membranes." *Proceedings of the National Academy of Sciences of the United States of America* **108**(13): 5249–5254.

Canham, P. B. (1970). "Minimum energy of bending as a possible explanation of biconcave shape of human red blood cell." *Journal of Theoretical Biology* **26**(1): 61–81.

Castellan, G. W. (1964). *Physical Chemistry.* Boston, MA: Addison-Wesley Publishing Company.

Chandler, D. (2005). "Interfaces and the driving force of hydrophobic assembly." *Nature* **437**(7059): 640–647.

Chen, D. and M. M. Santore (2014). "Large effect of membrane tension on the fluid–solid phase transitions of two-component phosphatidylcholine vesicles." *Proceedings of the National Academy of Sciences* **111**(1): 179–184.

Chen, I. A., R. W. Roberts and J. W. Szostak (2004). "The emergence of competition between model protocells." *Science* **305**(5689): 1474–1476.

Cheng, H.-T., Megha and E. London (2009). "Preparation and properties of asymmetric vesicles that mimic cell membranes: Effect upon lipid raft formation and transmembrane helix orientation." *Journal of Biological Chemistry* **284**(10): 6079–6092.

Chiantia, S., P. Schwille, A. S. Klymchenko and E. London (2011). "Asymmetric GUVs prepared by M beta CD-mediated lipid exchange: An FCS study." *Biophysical Journal* **100**(1): L01-L03.

Csonka, L. N. (1989). "physiological and genetic responses of bacteria to osmotic-stress." *Microbiological Reviews* **53**(1): 121–147.

Dai, J. W., M. P. Sheetz, X. D. Wan and C. E. Morris (1998). "Membrane tension in swelling and shrinking molluscan neurons." *Journal of Neuroscience* **18**(17): 6681–6692.

Deamer, D. W. and J. Bramhall (1986). "Permeability of lipid bilayers to water and ionic solutes." *Chemistry and Physics of Lipids* **40**(2–4): 167–188.

DeClue, M. S., P.-A. Monnard, J. A. Bailey, S. E. Maurer, G. E. Collis, H.-J. Ziock, S. Rasmussen and J. M. Boncella (2009). "Nucleobase mediated, photocatalytic vesicle formation from an ester precursor." *Journal of the American Chemical Society* **131**(3): 931–933.

Deuling, H. J. and W. Helfrich (1976). "Curvature elasticity of fluid membranes - catalog of vesicle shapes." *Journal De Physique* **37**(11): 1335–1345.

Dimitrov, D. S. and M. I. Angelova (1988). "Lipid swelling and liposome formation mediated by electric-fields." *Bioelectrochemistry and Bioenergetics* **19**(2): 323–336.

Discher, B. M., Y. Y. Won, D. S. Ege, J. C. M. Lee, F. S. Bates, D. E. Discher and D. A. Hammer (1999). "Polymersomes: Tough vesicles made from diblock copolymers." *Science* **284**(5417): 1143–1146.

Dix, J. A. and A. S. Verkman (2008). Crowding effects on diffusion in solutions and cells. *Annual Review of Biophysics.* **37**: 247–263.

Dobereiner, H. G., J. Kas, D. Noppl, I. Sprenger and E. Sackmann (1993). "Budding and fission of vesicles." *Biophysical Journal* **65**(4): 1396–1403.

Doeven, M. K., J. H. A. Folgering, V. Krasnikov, E. R. Geertsma, G. van den Bogaart and B. Poolman (2005). "Distribution, lateral mobility and function of membrane proteins incorporated into giant unilamellar vesicles." *Biophysical Journal* **88**(2): 1134–1142.

Dominak, L. M. and C. D. Keating (2007). "Polymer encapsulation within giant lipid vesicles." *Langmuir* **23**(13): 7148–7154.

Dominak, L. M. and C. D. Keating (2008). "Macromolecular crowding improves polymer encapsulation within giant lipid vesicles." *Langmuir* **24**(23): 13565–13571.

Eastman, S. J., M. J. Hope and P. R. Cullis (1991). "Transbilayer transport of phosphatidic acid in response to transmembrane pH gradients." *Biochemistry* **30**(7): 1740–1745.

Ellis, R. J. (2001). "Macromolecular crowding: Obvious but underappreciated." *Trends in Biochemical Sciences* **26**(10): 597–604.

Ellis, R. J. and A. P. Minton (2003). "Cell biology: Join the crowd." *Nature* **425**(6953): 27–28.

Ertel, A., A. G. Marangoni, J. Marsh, F. R. Hallett and J. M. Wood (1993). "Mechanical properties of vesicles. I. Coordinated analysis of osmotic swelling and lysis." *Biophysical Journal* **64**(2): 426–434.

Estes, D. J. and M. Mayer (2005). "Giant liposomes in physiological buffer using electroformation in a flow chamber." *Biochimica et Biophysica Acta (BBA)-Biomembranes* **1712**(2): 152–160.

Everett, J., A. Zlotnick, J. Tennyson and P. W. Holloway (1986). "Fluorescence quenching of cytochrome b5 in vesicles with an asymmetric transbilayer distribution of brominated phosphatidylcholine." *Journal of Biological Chemistry* **261**(15): 6725–6729.

Eytan, G. D. (1982). "Use of liposomes for reconstitution of biological functions." *Biochimica et Biophysica Acta (BBA) - Reviews on Biomembranes* **694**(2): 185–202.

Farge, E. and P. F. Devaux (1992). "Shape changes of giant liposomes induced by an asymmetric transmembrane distribution of phospholipids." *Biophysical Journal* **61**(2): 347–357.

Fettiplace, R. and D. A. Haydon (1980). "Water permeability of lipid-membranES." *Physiological Reviews* **60**(2): 510–550.

Fischer, A., A. Franco and T. Oberholzer (2002). "Giant vesicles as microreactors for enzymatic mRNA synthesis." *Chembiochem* **3**(5): 409–417.

Girard, P., J. Pécréaux, G. Lenoir, P. Falson, J.-L. Rigaud and P. Bassereau (2004). "A new method for the reconstitution of membrane proteins into giant unilamellar vesicles." *Biophysical Journal* **87**(1): 419–429.

Givli, S., H. Giang and K. Bhattacharya (2012). "Stability of multicomponent biological membranes." *SIAM Journal on Applied Mathematics* **72**(2): 489–511.

Gregoriadis, G. (1995). "Engineering liposomes for drug delivery: Progress and problems." *Trends in Biotechnology* **13**(12): 527–537.

Gummadi, S., S. Hrafnsdóttir, J. Walent, W. Watkins and A. Menon (2003). Reconstitution and assay of biogenic membrane-derived phospholipid flippase activity in proteoliposomes. *Membrane Protein Protocols: Expression, Purification, and Characterization*, B. S. Selinsky. Humana Press, 271–279.

Haleva, E. and H. Diamant (2008). "Critical swelling of particle-encapsulating vesicles." *Physical Review Letters* **101**(7): 078104.

Hallett, F. R., J. Marsh, B. G. Nickel and J. M. Wood (1993). "Mechanical-properties of vesicles.2. A model for osmotic swelling and lysis." *Biophysical Journal* **64**(2): 435–442.

Hamada, T., Y. Kishimoto, T. Nagasaki and M. Takagi (2011). "Lateral phase separation in tense membranes." *Soft Matter* **7**(19): 9061–9068.

Hamill, O. P. and B. Martinac (2001). "Molecular basis of mechanotransduction in living cells." *Physiological Reviews* **81**(2): 685–740.

Hanczyc, M. M., S. M. Fujikawa and J. W. Szostak (2003). "Experimental models of primitive cellular compartments: Encapsulation, growth, and division." *Science* **302**(5645): 618–622.

Hardy, M. D., J. Yang, J. Selimkhanov, C. M. Cole, L. S. Tsimring and N. K. Devaraj (2015). "Self-reproducing catalyst drives repeated phospholipid synthesis and membrane growth." *Proceedings of the National Academy of Sciences* **112**(27): 8187–8192.

Haswell, E. S. and P. E. Verslues (2015). "The ongoing search for the molecular basis of plant osmosensing." *Journal of General Physiology* **145**(5): 389–394.

He, X., M. Aizenberg, O. Kuksenok, L. D. Zarzar, A. Shastri, A. C. Balazs and J. Aizenberg (2012). "Synthetic homeostatic materials with chemo-mechano-chemical self-regulation." *Nature* **487**(7406): 214–218.

Heimburg, T. (2007). Membrane structure. *Thermal Biophysics of Membranes*, Wiley-VCH Verlag GmbH & Co. KGaA: 15–27.

Helenius, A., M. Sarvas and K. Simons (1981). "Asymmetric and symmetric membrane reconstitution by detergent elimination." *European Journal of Biochemistry* **116**(1): 27–35.

Helfrich, W. (1973). "Elastic properties of lipid bilayers - theory and possible experiments." *Zeitschrift Fur Naturforschung C-a Journal of Biosciences* C **28**(11-1): 693–703.

Herold, C., C. Leduc, R. Stock, S. Diez and P. Schwille (2012). "Long-range transport of giant vesicles along microtubule networks." *ChemPhysChem* **13**(4): 1001–1006.

Hoekstra, F. A., E. A. Golovina and J. Buitink (2001). "Mechanisms of plant desiccation tolerance." *Trends in Plant Science* **6**(9): 431–438.

Hope, M. J., M. B. Bally, G. Webb and P. R. Cullis (1985). "Production of large unilamellar vesicles by a rapid extrusion procedure - characterization of size distribution, trapped volume and ability to maintain a membrane-potential." *Biochimica Et Biophysica Acta* **812**(1): 55–65.

Hope, M. J., T. E. Redelmeier, K. F. Wong, W. Rodrigueza and P. R. Cullis (1989). "Phospholipid asymmetry in large unilamellar vesicles induced by transmembrane pH gradients." *Biochemistry* **28**(10): 4181–4187.

Hotani, H. (1984). "Transformation pathways of liposomes." *Journal of Molecular Biology* **178**(1): 113–120.

Iglic, A. (1997). "A possible mechanism determining the stability of spiculated red blood cells." *Journal of Biomechanics* **30**(1): 35–40.

Julicher, F. and R. Lipowsky (1993). "Domain-induced budding of vesicles." *Physical Review Letters* **70**(19): 2964–2967.

Julicher, F. and R. Lipowsky (1996). "Shape transformations of vesicles with intramembrane domains." *Physical Review E* **53**(3): 2670–2683.

Kang, Y. J., H. S. Wostein and S. Majd (2013). "A simple and versatile method for the formation of arrays of giant vesicles with controlled size and composition." *Advanced Materials* **25**(47): 6834–6838.

Karatekin, E., O. Sandre, H. Guitouni, N. Borghi, P.-H. Puech and F. Brochard-Wyart (2003). "Cascades of transient pores in giant vesicles: Line tension and transport." *Biophysical Journal* **84**(3): 1734–1749.

Kas, J. and E. Sackmann (1991). "Shape transitions and shape stability of giant phospholipid-vesicles in pure water induced by area-to-volume changes." *Biophysical Journal* **60**(4): 825–844.

Keating, C. D. (2012). "Aqueous phase separation as a possible route to compartmentalization of biological molecules." *Accounts of Chemical Research* **45**(12): 2114–2124.

Keighron, J. D. and C. D. Keating (2011). Towards a minimal cytoplasm. *The Minimal Cell*. Netherlands: Springer, 3–30.

Kempf, B. and E. Bremer (1998). "Uptake and synthesis of compatible solutes as microbial stress responses to high-osmolality environments." *Archives of Microbiology* **170**(5): 319–330.

Koch, A. L. (1984). "Shrinkage of growing escherichia-coli-cells by osmotic challengE." *Journal of Bacteriology* **159**(3): 919–924.

Koslov, M. M. and V. S. Markin (1984). "A theory of osmotic lysis of lipid vesicles." *Journal of Theoretical Biology* **109**(1): 17–39.

Koster, G., M. VanDuijn, B. Hofs and M. Dogterom (2003). "Membrane tube formation from giant vesicles by dynamic association of motor proteins." *Proceedings of the National Academy of Sciences* **100**(26): 15583–15588.

Kramer, R. (2009). "Osmosensing and osmosignaling in *Corynebacterium glutamicum*." *Amino Acids* **37**(3): 487–497.

Kung, C. (2005). "A possible unifying principle for mechanosensation." *Nature* **436**(7051): 647–654.

Kung, C., B. Martinac and S. Sukharev (2010). Mechanosensitive channels in microbes. *Annual Review of Microbiology* S. Gottesman and C. S. Harwood. **64**: 313–329.

Langer, R. (1998). "Drug delivery and targeting." *Nature* **392**(6679): 5–10.

Leneveu, D. M., R. P. Rand and V. A. Parsegian (1976). "Measurement of forces between lecithin bilayerS." *Nature* **259**(5544): 601–603.

Lim, H. W. G., M. Wortis and R. Mukhopadhyay (2002). "Stomatocyte-discocyte-echinocyte sequence of the human red blood cell: Evidence for the bilayer-couple hypothesis from membrane mechanics." *Proceedings of the National Academy of Sciences of the United States of America* **99**(26): 16766–16769.

Limozin, L., M. Bärmann and E. Sackmann (2003). "On the organization of self-assembled actin networks in giant vesicles." *The European Physical Journal E* **10**(4): 319–330.

Lipowsky, R. (1991). "The conformation of membranes." *Nature* **349**(6309): 475–481.

Lipowsky, R. (1992). "Budding of membranes induced by intramembrane domains." *Journal De Physique Ii* **2**(10): 1825–1840.

Luisi, P. L., T. P. de Souza and P. Stano (2008). "Vesicle behavior: In search of explanations." *Journal of Physical Chemistry B* **112**(46): 14655–14664.

Luisi, P. L., F. Ferri and P. Stano (2006). "Approaches to semi-synthetic minimal cells: A review." *Naturwissenschaften* **93**(1): 1–13.

Mai, Y. and A. Eisenberg (2012). "Self-assembly of block copolymers." *Chemical Society Reviews* **41**(18): 5969–5985.

Mathai, J. C., S. Tristram-Nagle, J. F. Nagle and M. L. Zeidel (2008). "Structural determinants of water permeability through the lipid membrane." *Journal of General Physiology* **131**(1): 69–76.

Menger, F. M. and M. I. Angelova (1998). "Giant vesicles: Imitating the cytological processes of cell membranes." *Accounts of Chemical Research* **31**(12): 789–797.

Miao, L., U. Seifert, M. Wortis and H. G. Dobereiner (1994). "Budding transitions of fluid-bilayer vesicles - the effect of area-difference elasticity." *Physical Review E* **49**(6): 5389–5407.

Mika, J. T. and B. Poolman (2011). "Macromolecule diffusion and confinement in prokaryotic cells." *Current Opinion in Biotechnology* **22**(1): 117–126.

Minton, A. P. (2007). "The effective hard particle model provides a simple, robust, and broadly applicable description of nonideal behavior in concentrated solutions of bovine serum albumin and other nonassociating proteins." *J Pharm Sci* **96**(12): 3466–3469.

Montes, L. R., A. Alonso, F. M. Goñi and L. A. Bagatolli (2007). "Giant unilamellar vesicles electroformed from native membranes and organic lipid mixtures under physiological conditions." *Biophysical Journal* **93**(10): 3548–3554.

Morales-Penningston, N. F., J. Wu, E. R. Farkas, S. L. Goh, T. M. Konyakhina, J. Y. Zheng, W. W. Webb and G. W. Feigenson (2010). "GUV preparation and imaging: Minimizing artifacts." *Biochimica Et Biophysica Acta-Biomembranes* **1798**(7): 1324–1332.

Mui, B. L., P. R. Cullis, E. A. Evans and T. D. Madden (1993). "Osmotic properties of large unilamellar vesicles prepared by extrusion." *Biophysical Journal* **64**(2): 443–453.

Mui, B. L. S., H. G. Dobereiner, T. D. Madden and P. R. Cullis (1995). "Influence of transbilayer area asymmetry on the morphology of large unilamellar vesicles." *Biophysical Journal* **69**(3): 930–941.

Mukhopadhyay, R., G. Lim and M. Wortis (2002). "Echinocyte shapes: Bending, stretching, and shear determine spicule shape and spacing." *Biophysical Journal* **82**(4): 1756–1772.

Needham, D. and R. S. Nunn (1990). "Elastic-deformation and failure of lipid bilayer-membranes containing cholesterol." *Biophysical Journal* **58**(4): 997–1009.

Oberholzer, T., M. Albrizio and P. L. Luisi (1995). "Polymerase chain reaction in liposomes." *Chemistry & Biology* **2**(10): 677–682.

Oberholzer, T., K. H. Nierhaus and P. L. Luisi (1999). "Protein Expression in Liposomes." *Biochemical and Biophysical Research Communications* **261**(2): 238–241.

Ogłęcka, K., P. Rangamani, B. Liedberg, R. S. Kraut and A. N. Parikh (2014). "Oscillatory phase separation in giant lipid vesicles induced by transmembrane osmotic differentials." *eLife* **3**: e03695.

Perozo, E., A. Kloda, D. M. Cortes and B. Martinac (2002). "Physical principles underlying the transduction of bilayer deformation forces during mechanosensitive channel gating." *Nature Structural Biology* **9**(9): 696–703.

Peterlin, P. and V. Arrigler (2008). "Electroformation in a flow chamber with solution exchange as a means of preparation of flaccid giant vesicles." *Colloids and Surfaces B: Biointerfaces* **64**(1): 77–87.

Peterlin, P., V. Arrigler, E. Haleva and H. Diamant (2012). "Law of corresponding states for osmotic swelling of vesicles." *Soft Matter* **8**(7): 2185–2193.

Poolman, B., J. J. Spitzer and J. A. Wood (2004). "Bacterial osmosensing: Roles of membrane structure and electrostatics in lipid-protein and protein-protein interactions." *Biochimica Et Biophysica Acta-Biomembranes* **1666**(1–2): 88–104.

Popescu, D. and A. G. Popescu (2008). "The working of a pulsatory liposome." *Journal of Theoretical Biology* **254**(3): 515–519.

Portet, T., S. E. Gordon and S. L. Keller (2012). "Increasing membrane tension decreases miscibility temperatures; an experimental demonstration via micropipette aspiration." *Biophysical Journal* **103**(8): L35-L37.

Pott, T., H. Bouvrais and P. Meleard (2008). "Giant unilamellar vesicle formation under physiologically relevant conditions." *Chem Phys Lipids* **154**(2): 115–119.

Rand, R. P. (2004). "Probing the role of water in protein conformation and function." *Philosophical Transactions of the Royal Society of London Series B-Biological Sciences* **359**(1448): 1277–1284.

Rand, R. P., V. A. Parsegian and D. C. Rau (2000). "Intracellular osmotic action." *Cellular and Molecular Life Sciences* **57**(7): 1018–1032.

Reeves, J. P. and R. M. Dowben (1969). "Formation and properties of thin-walled phospholipid vesicles." *Journal of Cellular Physiology* **73**(1): 49–60.

Reid, C. and R. P. Rand (1997). "Probing protein hydration and conformational states in solution." *Biophysical Journal* **72**(3): 1022–1030.

Richmond, D. L., E. M. Schmid, S. Martens, J. C. Stachowiak, N. Liska and D. A. Fletcher (2011). "Forming giant vesicles with controlled membrane composition, asymmetry, and contents." *Proceedings of the National Academy of Sciences* **108**(23): 9431–9436.

Rigaud, J.-L., B. Pitard and D. Levy (1995). "Reconstitution of membrane proteins into liposomes: Application to energy-transducing membrane proteins." *Biochimica et Biophysica Acta (BBA) - Bioenergetics* **1231**(3): 223–246.

Roux, A., G. Cappello, J. Cartaud, J. Prost, B. Goud and P. Bassereau (2002). "A minimal system allowing tubulation with molecular motors pulling on giant liposomes." *Proceedings of the National Academy of Sciences* **99**(8): 5394–5399.

Sachs, F. and M. V. Sivaselvan (2015). "Cell volume control in three dimensions: Water movement without solute movement." *Journal of General Physiology* **145**(5): 373–380.

Safinya, C. R. and K. K. Ewert (2012). "Materials chemistry liposomes derived from molecular vases." *Nature* **489**(7416): 372–374.

Sajbidor, J. (1997). "Effect of some environmental factors on the content and composition of microbial membrane lipids." *Critical Reviews in Biotechnology* **17**(2): 87–103.

Sandre, O., L. Moreaux and F. Brochard-Wyart (1999). "Dynamics of transient pores in stretched vesicles." *Proceedings of the National Academy of Sciences* **96**(19): 10591–10596.

Scherfeld, D., N. Kahya and P. Schwille (2003). "Lipid dynamics and domain formation in model membranes composed of ternary mixtures of unsaturated and saturated phosphatidylcholines and cholesterol." *Biophysical Journal* **85**(6): 3758–3768.

Schmid, E. M., D. L. Richmond and D. A. Fletcher (2015). "Reconstitution of proteins on electroformed giant unilamellar vesicles." *Methods in cell biology* **128**: 319–338.

Seifert, U. (1997). "Configurations of fluid membranes and vesicles." *Advances in Physics* **46**(1): 13–137.

Seifert, U., K. Berndl and R. Lipowsky (1991). "Shape transformations of vesicles - phase-diagram for spontaneous-curvature and bilayer-coupling models." *Physical Review A* **44**(2): 1182–1202.

Sheetz, M. P. and S. J. Singer (1974). "Biological-membranes as bilayer couples - molecular mechanism of drug-erythrocyte interactions." *Proceedings of the National Academy of Sciences of the United States of America* **71**(11): 4457–4461.

Shimanouchi, T., H. Umakoshi and R. Kuboi (2009). "Kinetic study on giant vesicle formation with electroformation method." *Langmuir* **25**(9): 4835–4840.

Spitzer, J. J. and B. Poolman (2005). "Electrochemical structure of the crowded cytoplasm." *Trends Biochem Sci* **30**(10): 536–541.

Stewart, M. P., J. Helenius, Y. Toyoda, S. P. Ramanathan, D. J. Muller and A. A. Hyman (2011). "Hydrostatic pressure and the actomyosin cortex drive mitotic cell rounding." *Nature* **469**(7329): 226–230.

Stroka, K. M., H. Y. Jiang, S. H. Chen, Z. Q. Tong, D. Wirtz, S. X. Sun and K. Konstantopoulos (2014). "Water permeation drives tumor cell migration in confined microenvironments." *Cell* **157**(3): 611–623.

Sukharev, S. I., B. Martinac, V. Y. Arshavsky and C. Kung (1993). "2 types of mechanosensitive channels in the *Escherichia coli* cell-envelope - solubilization and functional reconstitution." *Biophysical Journal* **65**(1): 177–183.

Svetina, S., M. Brumen and B. Zeks (1985). "Lipid bilayer elasticity and the bilayer couple interpretation of red-cell shape transformations and lysis." *Studia Biophysica* **110**(1–3): 177–184.

Svetina, S. and B. Zeks (1989). "Membrane bending energy and shape determination of phospholipid-vesicles and red blood-cells." *European Biophysics Journal with Biophysics Letters* **17**(2): 101–111.

Svetina, S. and B. Zeks (2002). "Shape behavior of lipid vesicles as the basis of some cellular processes." *Anatomical Record* **268**(3): 215–225.

Szostak, J. W., D. P. Bartel and P. L. Luisi (2001). "Synthesizing life." *Nature* **409**(6818): 387–390.

Tao, F. and P. Yang (2015). "Ca-mediated electroformation of cell-sized lipid vesicles." *Scientific Reports* **5**: 9839.

Taylor, P., C. Xu, P. D. I. Fletcher and V. N. Paunov (2003). "A novel technique for preparation of monodisperse giant liposomes." *Chemical Communications* (14): 1732–1733.

Tsai, F.-C., B. Stuhrmann and G. H. Koenderink (2011). "Encapsulation of active cytoskeletal protein networks in cell-sized liposomes." *Langmuir* **27**(16): 10061–10071.

Uline, Mark J., M. Schick and I. Szleifer (2012). "Phase behavior of lipid bilayers under tension." *Biophysical Journal* **102**(3): 517–522.

Veatch, S. L. and S. L. Keller (2003). "Separation of liquid phases in giant vesicles of ternary mixtures of phospholipids and cholesterol." *Biophysical Journal* **85**(5): 3074–3083.

Veatch, S. L. and S. L. Keller (2005). "Miscibility phase diagrams of giant vesicles containing sphingomyelin." *Physical Review Letters* **94**(14): 148101.

Viallat, A., J. Dalous and M. Abkarian (2004). "Giant lipid vesicles filled with a gel: Shape instability induced by osmotic shrinkage." *Biophysical Journal* **86**(4): 2179–2187.

Walde, P., K. Cosentino, H. Engel and P. Stano (2010). "Giant vesicles: Preparations and applications." *Chembiochem* **11**(7): 848–865.

Walde, P. and S. Ichikawa (2001). "Enzymes inside lipid vesicles: Preparation, reactivity and applications." *Biomolecular Engineering* **18**(4): 143–177.

Weinberger, A., F.-C. Tsai, G. H. Koenderink, T. F. Schmidt, R. Itri, W. Meier, T. Schmatko, A. Schröder and C. Marques (2013). "Gel-assisted formation of giant unilamellar vesicles." *Biophysical Journal* **105**(1): 154–164.

Wolff, J., S. Komura and D. Andelman (2015). "Budding of domains in mixed bilayer membranes." *Physical Review E* **91**(1): 012708.

Wood, J. (2010). *Osmotic Stress. In Bacterial Stress Responses, Second Edition*. Washington, DC: ASM Press, 133–156.

Wood, J. M. (1999). "Osmosensing by bacteria: Signals and membrane-based sensors." *Microbiology and Molecular Biology Reviews* **63**(1): 230–262.

Wood, J. M. (2011). "Bacterial osmoregulation: A paradigm for the study of cellular homeostasis." *Annual Review of Microbiology* **65**: 215–238.

Wood, J. M. (2015). "Bacterial responses to osmotic challenges." *Journal of General Physiology* **145**(5): 381–388.

Yamashita, Y., M. Oka, T. Tanaka and M. Yamazaki (2002). "A new method for the preparation of giant liposomes in high salt concentrations and growth of protein microcrystals in them." *Biochimica Et Biophysica Acta-Biomembranes* **1561**(2): 129–134.

Yanagisawa, M., M. Imai and T. Taniguchi (2008). "Shape deformation of ternary vesicles coupled with phase separation." *Physical Review Letters* **100**(14): 148102.

Yancey, P. H., M. E. Clark, S. C. Hand, R. D. Bowlus and G. N. Somero (1982). "Living with water-stress - evolution of osmolyte systeMS." *Science* **217**(4566): 1214–1222.

Zhang, Y. M. and C. O. Rock (2008). "Membrane lipid homeostasis in bacteria." *Nature Reviews Microbiology* **6**(3): 222–233.

Zhou, H. X., G. Rivas and A. P. Minton (2008). "Macromolecular crowding and confinement: Biochemical, biophysical, and potential physiological consequences." *Annu Rev Biophys* **37**: 375–397.

Zhu, T. F. and J. W. Szostak (2009). "Coupled growth and division of model protocell membranes." *Journal of the American Chemical Society* **131**(15): 5705–5713.

Zimmerberg, J. and V. A. Parsegian (1986). "Polymer inaccessible volume changes during opening and closing of a voltage-dependent ionic channel." *Nature* **323**(6083): 36–39.

Zimmerman, S. B. and A. P. Minton (1993). "Macromolecular crowding - biochemical, biophysical, and physiological consequences." *Annual Review of Biophysics and Biomolecular Structure* **22**: 27–65.

11 Cationic Liposomes as Spatial Organizers of Nucleic Acids in One, Two, and Three Dimensions: Liquid Crystal Phases with Applications in Delivery and Bionanotechnology

Cyrus R. Safinya, Kai K. Ewert, and Youli Li
University of California, Santa Barbara, California

Joachim O. Rädler
Ludwig-Maximilians-Universität, Munich, Germany

CONTENTS

11.1 Introduction 195
11.2 Structures of Cationic Liposomes Complexed with Long DNA 197
    11.2.1 The Role of the Spontaneous Curvature of the Membrane, Resulting from Lipid Shape, in Controlling the Spatial Organization of DNA in Cationic Liposome–DNA Complexes 199
    11.2.2 Transfection Efficiency of Lamellar and Non-lamellar Cationic Liposome–DNA Complexes 200
11.3 Cationic Liposomes Complexed with Short Nucleic Acids 201
    11.3.1 The Gyroid Cubic Lipid Phase Containing Short RNA Molecules 201
    11.3.2 Two-Dimensional Packing of Short DNA with Nonpairing Overhangs in Cationic Liposome–DNA Complexes 201
    11.3.3 A Three-Dimensional Columnar Phase of Stacked Blunt Short DNA Induced by Coherent Membrane Undulations in Fluid Membrane/DNA Complexes 202
11.4 Concluding Remarks and Future Directions 204
Acknowledgements 206
References 206

11.1 INTRODUCTION

Liposomes consist of closed self-assemblies of lipid bilayers, including spherical membranes (Figure 11.1a i). They were discovered in the early 1960s by A. D. Bangham and R. W. Horne in electron microscopy images, which revealed the structural similarities between liposomes (also known as vesicles) and the plasma membranes of 5–7 nm thickness that surround biological cells (Bangham and Horne 1964). Their observations were consistent with the hypothesis that lipids are a major constituent of biological membranes (Alberts et al. 2015, Gelbart et al. 2012, Lipowsky and Sackmann 1995, Pollard et al. 2017). Liposomes provide a natural environment for elucidation of the functions of membrane-associated and integral membrane proteins (Rigaud et al. 1988, Shen et al. 1993). In addition to the interest in liposomes from a biological perspective, liposomes and multilamellar vesicles have been employed in fundamental studies of the statistical mechanical properties of model fluctuating membranes (Aronovitz and Lubensky 1988, Betz and Sykes 2012, Khurana 1989, König et al. 1992, Lei et al. 1995, Nelson and Peliti 1987, Rädler et al. 1995). This has included studies for the purpose of understanding the structural nature of phases and phase transitions in membranes (Nallet et al. 1993, Porte 1992, Roux and Safinya 1988, Safinya et al. 1986, 1989, Safinya 1989, 2006). Shape fluctuations in liposomes suspended in aqueous solution under equilibrium conditions have also received extensive attention in biophysical studies (Chiruvolu et al. 1994, Käs and Sackmann 1991, Zidovska et al. 2009c, d) (for a comprehensive review, see Chapter 7 by A. Zidovska).

Liposomes are promising and versatile synthetic carriers (vectors) of drugs in therapeutic applications due to their

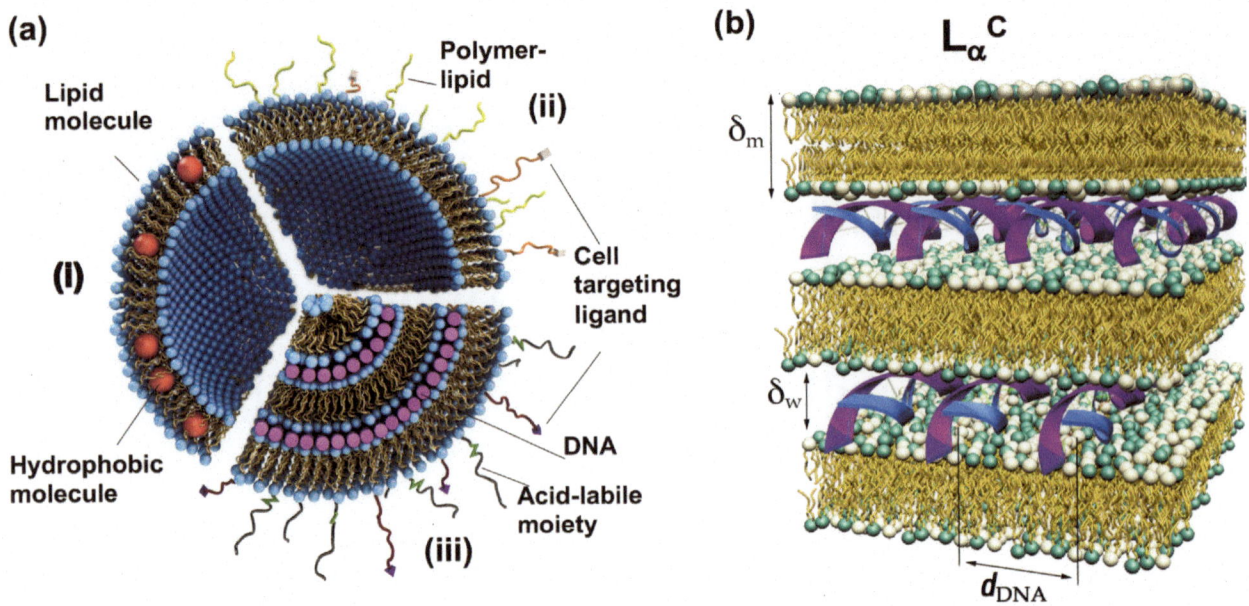

FIGURE 11.1 (a) The evolution of liposomes from their discovery by A. D. Bangham and R. W. Horne in the 1960s. (i) A unilamellar liposome consisting of a self-assembly of amphiphilic lipid molecules. The liposome can trap hydrophobic molecules (red spheres) within its hydrophobic bilayer and hydrophilic molecules in its aqueous interior. (ii) A "stealth" liposome, where the lipid bilayer contains a small percentage of polymer–lipids to enable the surface-modified liposome to avoid immune cells. Such liposomes may also incorporate cell-targeting ligand groups (e.g., peptides) attached to the distal end of the polymer–lipid (shown as white rectangular blocks). (iii) A cationic liposome–DNA complex consisting of an onion-like multilamellar structure with DNA (purple rods) sandwiched between cationic membranes. In addition to cell-targeting ligands attached to polymer–lipids, surface functionalization may also include polymer–lipids with acid-labile, hydrolysable groups for shedding of the polymer in late endosomes upon uptake of complexes by cells. (b) Mixing of DNA and cationic liposomes (CLs) results in the spontaneous formation of CL–DNA complexes with equilibrium self-assembled structures. The schematic shows the local nanometre-scale structure of the lamellar (L_α^C) phase of CL–DNA complexes, derived from synchrotron X-ray diffraction, with alternating lipid bilayers and DNA monolayers. ((a) is adapted and modified from Safinya and Ewert (2012), with permission from the Nature Publishing Group, and adapted from Safinya et al. (2014) by permission of The Royal Society of Chemistry (RSC) on behalf of the Centre National de la Recherche Scientifique (CNRS) and the RSC. (b) Reprinted and modified from Rädler et al. 1997 with permission from AAAS.)

ability to sequester a wide range of molecules. This includes hydrophobic drugs within the lipid bilayer region and hydrophilic nucleic acids (e.g., plasmid DNA for gene delivery) trapped inside their aqueous interior (Figure 11.1a i) (Allen and Cullis 2013, Lasic 1997, Sercombe et al. 2015, Sofias et al. 2017). In most delivery applications *in vivo*, poly(ethylene) glycol–lipids (PEG-lipids) are incorporated in the lipid bilayer in the brush conformation to coat the outer surface of the liposome (Figure 11.1a ii). The PEG sheath provides a repulsive barrier thought to prevent the attachment of blood protein markers that precipitate the rapid removal of liposomes by the mononuclear phagocytic immune system (Lasic and Martin 1995). Thus, PEGylation vastly increases the lifetime for liposome circulation *in vivo*. The development of PEGylated liposomes has also led to the development of so-called targeted delivery where a ligand directed towards a specific cell receptor is attached to the distal end of a fraction of the PEG chains (Figure 11.1a ii).

After the initial success in delivery applications with PEGylated liposomes, a novel type of delivery vector for nucleic acids was introduced by P. Felgner and coworkers (1987). In contrast to earlier liposomes, which were charge-neutral or negative, the vectors introduced by Felgner et al. consisted of cationic liposomes (CLs, consisting of mixtures of cationic and neutral lipids) electrostatically complexed with long strands of anionic DNA encoding for specific therapeutic genes. Felgner et al. hypothesized that CLs would simultaneously bind the long anionic DNA chain and the anionic sulphated groups on surface proteoglycans of mammalian cells. The landmark paper of Felgner et al. (1987) has, in large part, led to the more than 110 ongoing gene therapy clinical trials worldwide that use CL vectors to address single gene diseases such as cystic fibrosis and multigene diseases including cancer (Ginn et al. 2018). Simultaneously, a very large number of laboratories worldwide are engaged in research with CL vectors with the aim of improving efficacy both at the *in vitro* (cell) level and *in vivo* (Bielke and Erbacher 2010, Ewert et al. 2004, 2005, 2010, 2016, Foldvari et al. 2016, Safinya and Ewert 2012, Safinya et al. 2014, Yin et al. 2014). In addition to plasmid DNA gene delivery, short-interfering RNAs (siRNAs; 19-27 bps, with $2nt$ 3′-overhangs) complexed with

CLs are studied in a variety of settings for optimization of sequence-specific gene silencing (Bouxsein et al. 2007, Gindy et al. 2012, Lares et al. 2010, Leal et al. 2010, 2011, Ozcan et al. 2015, Rupaimoole and Slack 2017).

Engineered viral vectors are far more efficient than nonviral vectors *in vivo* (Ginn et al. 2018); however, they are also far less safe due to the possibility of immune reactions (Thomas et al. 2003) or insertional mutagenesis, with vectors that integrate within the host genome disrupting genes or regulatory sequences and causing cancer (Hacein-Bey-Abina et al. 2008). The relative safety of lipid vectors results from the lack of viral genes to cause disease and the lack of proteins, which renders the nonviral vectors relatively nonimmunogenic. Significantly, CL vectors can carry very large pieces of DNA including large human genes and regulatory sequences exceeding 100 kbps (Harrington et al. 1997). This is because CL–DNA complexation is a self-assembling process (Koltover et al. 1999, Rädler et al. 1997, 1998, Safinya 2001). In contrast, the nanometre-sized capsid particles of viruses limit the maximum size of therapeutic DNA in engineered viral vectors, and to date, viral vectors have been limited to delivering cDNA molecules where the long intron sections of the gene are absent.

As we describe in this chapter, the mechanisms of transfection via CL–DNA complexes remain poorly understood even after more than three decades of worldwide research. Future work, which combines custom lipid synthesis and state-of-the-art techniques enabling elucidation of interactions of CL–nucleic acid (CL–NA) complexes with cellular membranes, in particular along the endocytic pathway leading to cytosol delivery, is required to optimize transfection efficiency.

Aside from applications in molecular delivery, CL–NA complexes are of interest to the fields of nucleic acid nanomaterials and bionanotechnology. This arises from the ability to use lipids with distinct molecular shape to control the collective spatial organization of DNA and RNA molecules. As we describe in this chapter, synchrotron X-ray diffraction has revealed the distinct nanometre-scale structures of CL–NA complexes. The first arrangement reported was for CL–DNA complexes that spontaneously form a liquid crystal lamellar phase (labelled L_α^C), where DNA is sandwiched between fluid cationic membranes with weak cross-bilayer correlations (Koltover et al. 1999, Rädler et al. 1997, 1998, Salditt et al. 1997, 1998). In complexes where the membrane is in the ordered (gel) phase with chain-frozen lipids, DNA rods may form a columnar phase with long-range position and orientation order within the multilayer (Artzner et al. 1998, Koynova and MacDonald 2004, McManus et al. 2004). For certain noncylindrical-shaped lipids, CL–DNA complexes exhibit the inverted hexagonal phase (labelled H_{II}^C), with DNA encapsulated within inverse lipid tubules (Koltover et al. 1998), or the H_I^C phase, where DNA rods are adsorbed on hexagonally arranged cylindrical micelles and form a continuous substructure with honeycomb symmetry (Ewert et al. 2006b). Remarkably, DNA molecules are confined in one dimension (1D), 2D, and 3D in the liquid crystal H_{II}^C, L_α^C, and H_I^C phases, respectively.

This chapter further describes how short nucleic acids (short DNA (sDNA) and siRNA), when complexed with CLs, may form entirely new phases of matter. CLs complexed with sDNA with nonpairing overhangs may result in finite-length stacks of sDNA adsorbed on cationic membranes (Bouxsein et al. 2011). For siRNA with two nonpairing overhangs, bicontinuous cubic phases may be constructed. These include the gyroid ($Q_{II}^{G,NA}$) phase, with siRNA molecules confined to curved surfaces with negative Gaussian curvature (Kang et al. 2016, Kang and Leal 2016, Kim et al. 2018, Leal et al. 2010, 2011, 2013). With blunt sDNA confined in fluid membranes, a new type of sDNA organization has been discovered that consists of a 3D columnar phase of stacked sDNA (Bouxsein et al. 2019). Unlike 3D columnar phases of long rod-like DNA in ordered multilayer membrane systems (Artzner et al. 1998), this phase is stabilized by macroscopic, coherent 3D undulations of the fluid multilayer membranes.

11.2 STRUCTURES OF CATIONIC LIPOSOMES COMPLEXED WITH LONG DNA

The self-assembled structures of CL–DNA complexes were first solved by the quantitative methods of X-ray diffraction. Using high-resolution synchrotron X-ray scattering, it was found that mixtures of CLs and linear λ-phage DNA (≈ 48 kbps) spontaneously precipitate a topological transition which breaks the closed spherical membrane structure of vesicles as a result of the electrostatic interactions between CLs and DNA. This leads to a higher-ordered multilamellar structure with DNA sandwiched between fluid cationic bilayer membranes (Figure 11.1b, labelled L_α^C) (Rädler et al. 1997). In these initial experiments, the cationic liposomes consisted of mixtures of cationic dioleoyl-trimethylammonium propane (DOTAP) and neutral (zwitterionic) dioleoyl-phosphatidylcholine (DOPC). Polarized optical microscopy revealed that the CL–DNA complexes were birefringent with textures characteristic of smectic liquid crystal phases. Macroscopic, unaligned samples contain both onion-like structures (Figure 11.1a iii) and nearly flat, layered regions that include bilayer disclinations. Furthermore, synchrotron studies (Rädler et al. 1997) showed that highly dilute solutions of CL–DNA complexes—typical of concentrations used in cryogenic TEM (Majzoub et al. 2015b)—had the same structure as more concentrated solutions of complexes.

The structural studies also revealed that linear DNA sandwiched between membranes was packed in distinct regimes, depending on the overall charge of the CL–DNA complex. In experiments that varied the lipid-to-DNA weight ratio, the DNA–DNA interaxial spacing (d_{DNA}) was found to abruptly decrease from $d_{DNA} \approx 44$ Å for positive complexes to $d_{DNA} \approx 37$ Å for negative complexes. In these distinctly different packing regimes, charged complexes coexist with excess giant liposomes in the positive regime and with excess DNA in the negative regime (Koltover et al. 1999, Rädler et al. 1997). This observation of charge reversal of complexes is consistent with theoretical models of CL–DNA complexes (Bruinsma 1998, Harries et al. 1998).

The driving force for higher-order self-assembly is the release of counterions upon complexation. This includes positive and negative ions released into solution that were originally bound to DNA, due to Manning condensation (Manning 1969), and within the Guoy–Chapman layer near cationic membranes of liposomes (Israelachvili 2011), respectively. Thus, complexation results in the cationic membranes (nearly) neutralizing the phosphate groups on the DNA (Rädler et al. 1997, Wagner et al. 2000). Concordantly with the concept of neutralization, molecular dynamics simulation exhibits close proximity of the cationic lipid head groups with the phosphate groups of the DNA backbone and likewise proximity of choline groups of the zwitterionic PC lipid with DNA phosphates (Bandyopadhyay et al. 1999). Taken together, the structure and phase behaviour of CL–DNA complexes has turned out to be remarkably complex and different from the loosely organized "bead-on-string model" hypothesized by Felgner et al. in their seminal paper, picturing the DNA strand as decorated with distinctly attached cationic liposomes (Felgner et al. 1987).

In a second series of synchrotron studies, it was found that the symmetry of CL–DNA liquid crystal phases may be altered by the choice of the shape of the lipid, which determines the spontaneous curvature of the membrane. In these experiments, the neutral zwitterionic lipid DOPC was replaced with dioleoyl-phosphatidylethanolamine (DOPE), a neutral lipid with a much smaller head group. The diffractions peaks in small-angle X-ray scattering (SAXS) scans of the CL–DNA complexes were found to index precisely on a 2D hexagonal lattice with unit cell spacing $a \approx 67.4$ Å (Koltover et al. 1998). This structure is closely related to the inverted hexagonal (H_{II}) phase of pure DOPE in excess water (Gruner 1989, Seddon 1990), but now with the water space inside the inverted lipid micelles filled by DNA (Figure 11.2a, labelled H_{II}^C). Assuming an average lipid monolayer thickness of 20 Å, the diameter of the micellar void in the H_{II}^C phase is close to 28 Å, sufficient for a DNA molecule with approximately two hydration shells. Thus, in the H_{II}^C phase, the DNA molecules are surrounded by a lipid monolayer with the DNA/lipid inverted cylindrical micelles arranged on a hexagonal lattice (Figure 11.2a) (Koltover et al. 1998, May and Ben-Shaul 1997).

After the discovery of the H_{II}^C phase, there was renewed interest to develop other types of nonlamellar phases of CL–DNA complexes but now based on lipids with head groups larger than the lipids that gave rise to the lamellar phases, i.e., DOPC and DOTAP. Aside from the interest in discovering lipid–DNA complexes with new arrangements, the motivation for these studies was also, in part, the finding that nonlamellar complexes are highly efficient transfection vectors for DNA delivery to cells *in vitro* (Ahmad et al. 2005, Lin et al. 2003, Zidovska et al. 2009a, b). This led to the design and synthesis of a novel multivalent cationic lipid, labelled MVLBG2, with an extremely large head group size with charge +16 e (Ewert et al. 2006a, b). The space-filling molecular model of MVLBG2 with its large branched head group is shown in Figure 11.2b together with DOTAP as a comparison. Synchrotron X-ray studies showed that MVLBG2/DOPC–DNA complexes at ≈25 mol% MVLBG2 consisted of highly charged MVLBG2/DOPC cylindrical

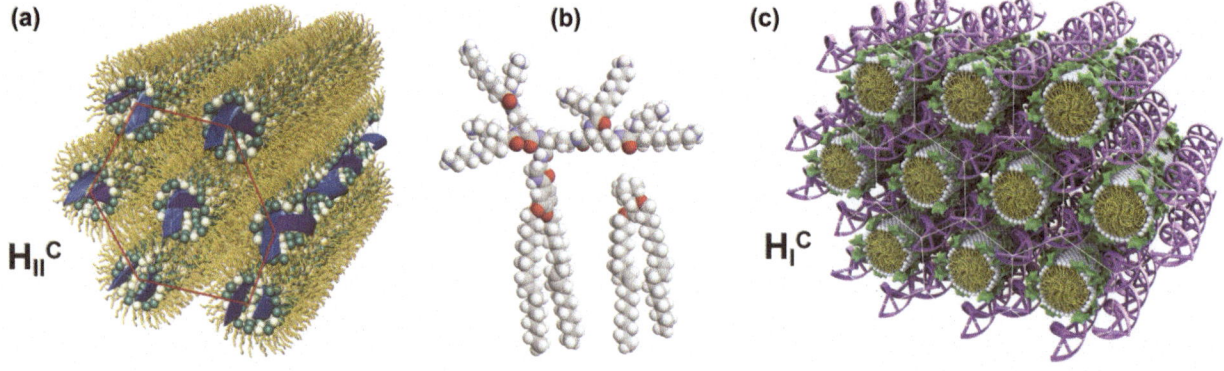

FIGURE 11.2 Mixing DNA and cationic liposomes (CLs) results in the spontaneous formation of CL–DNA complexes with equilibrium self-assembled structures. The schematics show the local structure of the interior of CL–DNA complexes on the nanometre scale as derived from synchrotron X-ray diffraction. (a) The inverted hexagonal (H_{II}^C) phase of CL–DNA complexes, composed of DNA inserted within inverse lipid tubules that are arranged on a hexagonal lattice. (b) Molecular models of the hexadecavalent cationic lipid MVLBG2 (charge of the dendritic head group: +16 e) and univalent lipid DOTAP (+1 e head group charge). (c) The hexagonal (H_I^C) phase of MVLBG2/DOPC–DNA complexes, where the large lipid head group of the multivalent lipid MVLBG2 leads to the formation of rod-like lipid micelles arranged on a hexagonal lattice, with DNA inserted within the interstices in honeycomb symmetry. ((a) Reprinted from Koltover et al. 1998 with permission from AAAS. (b) and (c) reprinted with permission from Ewert et al. 2006b. Copyright 2006 by the American Chemical Society.)

micelles coated with DNA and arranged on a 2D hexagonal lattice with an interaxial unit cell spacing of 81.5 Å (Figure 11.2c, labelled H_I^C) (Ewert et al. 2006b). The extremely large head group of MVLBG2 stabilizes the cylindrical lipid micelles. The overall structure that arises from the analysis of the scattering data is consistent with hydrated DNA rods, with a diameter ≈ 25 Å, forming a honeycomb lattice in the interstices of the lipid micelle arrangement. These dimensions are consistent with an estimated diameter of the rod-shaped micelles of around 41 Å, close to the thickness of a bilayer formed by the lipids. In addition to DNA, the interstitial space contains the head groups, water, and counterions. Further structural studies after the discovery of the H_I^C phase showed that MVLs with large dendritic head groups (Ewert et al. 2006a) may also form distorted H_I^C phases (i.e., where the peaks index to a distorted 2D hexagonal lattice) (Zidovska et al. 2009b).

Taken together, synchrotron X-ray scattering data show that DNA resides in a continuous 3D substructure in the H_I^C phase (Figure 11.2c), whereas DNA occupies 1D and 2D space in the H_{II}^C (Figure 11.2a) and L_α^C (Figure 11.1b) phases, respectively. As we describe in the next section, the spontaneous curvature of the confining membrane, which is dependent on the shapes of lipid molecules within the membrane, is the main predictor of the spatial organization of DNA with distinct dimensionality.

11.2.1 The Role of the Spontaneous Curvature of the Membrane, Resulting from Lipid Shape, in Controlling the Spatial Organization of DNA in Cationic Liposome–DNA Complexes

To qualitatively understand the relationship between lipid shape and the symmetry of self-assembled CL–DNA structures, we turn to W. Helfrich's model of membrane curvature elasticity, which is predictive of biological membrane shapes (Helfrich 1973). The elastic energy per unit area of a lipid membrane in terms of the membrane curvatures $C_1 = 1/R_1$ and $C_2 = 1/R_2$, the spontaneous curvature C_0, and the membrane bending (κ) and Gaussian (κ_G) elastic moduli is $F/A = 0.5 \kappa (C - C_0)^2 + \kappa_G C_1 C_2$. Here, R_1 and R_2 are the membrane radii, $C = C_1 + C_2$ is the mean curvature, and $C_1 C_2$ is the Gaussian curvature. The "shape" of a lipid molecule determines the spontaneous (i.e., preferred) curvature of the membrane $C_0 = 1/R_0$ (with R_0 the spontaneous radius of curvature) (Figure 11.3a) (Helfrich 1973, Israelachvili et al. 1977, Israelachvili 2011, Safran 1994). The elastic moduli κ and κ_G are, in principle, derivable from the interactions between neighbouring lipids in the membrane (Szleifer et al. 1988).

The first term in F/A is the cost of bending the membrane away from the preferred curvature C_0. To minimize the elastic cost, lipids with a cylindrical shape, i.e., with head

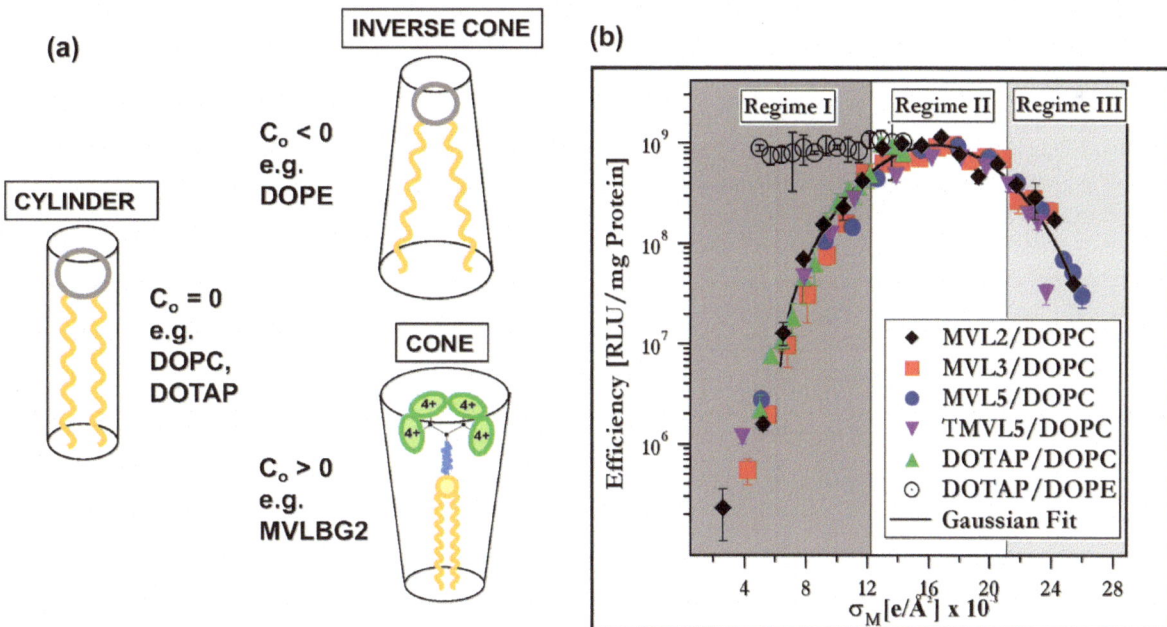

FIGURE 11.3 (a) Three common shapes of lipid molecules with spontaneous curvature $C_0 < 0$ (e.g., DOPE), $C_0 = 0$ (e.g., DOPC), and $C_0 > 0$ (e.g., MVLBG2 shown in Figure 11.2b). (b) Transfection efficiency (TE) plotted versus the membrane charge density σ_M (defined in the text). The data points are for DNA complexes prepared with the multivalent lipids MVL2 (black diamond, valence $Z = 2$), MVL3 (red square, valence $Z = 3$), MVL5 (blue circle, valence $Z = 5$), TMVL5 (purple inverted triangle, valence $Z = 5$), and univalent DOTAP (green triangle, valence $Z = 1$). All data were taken at a cationic lipid/DNA charge ratio equal to 2.8. The data show that TE of the lamellar (L_α^C) complexes follows a universal, bell-shaped curve as a function of σ_M (the solid line is a Gaussian fit to the data). Data for DOTAP/DOPE–DNA complexes in the H_{II}^C phase (open circles) deviate from the universal curve. This implies that the inverted hexagonal phase has a different transfection mechanism. The figure shows three regimes of transfection efficiency as discussed in the text. ((b) Reprinted with permission from Ahmad et al. 2005. Copyright 2005, John Wiley & Sons.)

group area equal to the tail area (Figure 11.3a, $C_0=0$), tend to assemble into lamellar bilayer structures. Thus, since DOTAP and DOPC have cylindrical shapes, complexation of DOTAP/DOPC membranes with DNA will give rise to the L_α^C phase (Figure 11.1b). Similarly, lipids with an inverse cone shape, where the head group area is smaller than the tail area (Figure 11.3a, $C_0<0$), will favour assembly into inverse cylindrical micelles forming the inverse hexagonal H_{II} phases at sufficiently high lipid concentrations (Gruner 1989, Israelachvili et al. 1977, Israelachvili 2011, Seddon 1990). The H_{II}^C phase of CL–DNA complexes described above (Figure 11.2a) consisted of uniform mixtures of DOPE ($C_0<0$) and DOTAP where the DOPE content was large enough, i.e., of order 50 mol% and larger, to favour inverse cylindrical micelles consistent with the Helfrich model. Cone-shaped lipids, with a head group area larger than their tail area (Figure 11.3a, $C_0>0$), normally assemble into cylindrical micellar structures consistent with the H_I^C phase of CL–DNA complexes (Figure 11.2c) that contained the multivalent lipid MVLBG2 (+16 e charge) with a very large head group (Figure 11.2b).

The second term in Helfrich's model describes membranes with either a negative Gaussian modulus, $\kappa_G<0$, which tend to form spherical shapes with positive Gaussian curvature $C_1C_2>0$, or a positive Gaussian modulus, $\kappa_G>0$, which favour saddle-splay-shaped surfaces with $C_1C_2<0$. The latter include the surfaces of bicontinuous cubic phases (Conn et al. 2008, Lindblom et al. 1979). For a comprehensive discussion of cubic and inverse hexagonal phases, we refer the reader to Chapter 3 by C. Conn and J. Seddon.

It is interesting to note that when the membrane bending modulus κ is much larger than k_BT, which is the situation for membranes described in this chapter, Helfrich's statistical mechanical model (Helfrich 1973) is expected to agree with the molecular model of J. Israelachvili (Israelachvili et al. 1977, Israelachvili 2011). Israelachvili's model connects the shape of a lipid molecule to the overall shape of the self-assembled lipids (i.e., the spontaneous curvature). On the other hand, when $\kappa \approx k_BT$, bicontinuous isotropic structures, with large thermal fluctuations, may be stabilized, including microemulsion (Andelman et al. 1987) and sponge phases (Cates et al. 1988), which are not predicted within the Israelachvili model.

Several groups have developed quantitative theoretical models to explain the experimental findings of thermodynamically stable phases of CL–DNA complexes with distinct structures and symmetries described above. This includes a numerical model of the structure and thermodynamics of CL–DNA complexes incorporating the electrostatic interactions between charged membranes and DNA (Harries et al. 1998) and an analytical model in the limit of low cationic lipid concentration (Bruinsma 1998), which explained the physics of overcharging of CL–DNA complexes observed in the experimental studies (Rädler et al. 1997). According to this model, in the regime where CL–DNA complexes coexist with excess DNA, the free energy is lowered if a fraction of DNA chains enter the complexes. This leads to negative overcharging of complexes and release of counterions bound to linear DNA within the 3D volume of the complex. Similarly, the free energy is lowered in the regime where complexes coexist with excess liposomes when some cationic lipids (and partially associated counterions) from liposomes are incorporated in the complex, rendering it positive.

11.2.2 Transfection Efficiency of Lamellar and Non-lamellar Cationic Liposome–DNA Complexes

Cationic liposomes are important carriers of nucleic acids for delivery applications in gene therapeutics (Foldvari et al. 2016, Ginn et al. 2018; see also Section 11.4). Transfection efficiency (TE) studies, measuring the ability of complexes to transfer DNA into cells and achieve gene expression, show that CL–DNA complexes interact with cells in a manner that depends on their underlying liquid crystalline structures (Ahmad et al. 2005, Ewert et al. 2006b, Lin et al. 2003, Zidovska et al. 2009a, b). To establish the relationship between the chemical and physical properties of CL–DNA complexes and TE in cells, a series of multivalent lipids (MVLs) were synthesized with head group charge between +2 e and +5 e (Ewert et al. 2002, 2004, 2006a). X-ray diffraction shows that the MVL/DOPC–DNA complexes form the lamellar L_α^C phase (Ahmad et al. 2005). Figure 11.3b displays TE of the MVLs over nearly four decades of dynamic range versus the membrane charge density of complexes (σ_M; the average charge per unit area of membrane). Remarkably, TE for MVLs with charges ranging from +2 e (MVL2) to +5 e (MVL5, TMVL5) falls onto a universal, bell-shaped curve. TE of lamellar L_α^C phase DOTAP/DOPC–DNA complexes, i.e., using the univalent cationic lipid DOTAP, also falls onto the universal curve (but only covers the initial part because of the limited charge density that can be achieved).

This bell-shaped curve for lamellar complexes reveals three distinct regimes of interactions between complexes and cells (Figure 11.3b) (Ahmad et al. 2005). TE increases with increasing σ_M in regime I at low σ_M and shows saturated behaviour (regime II) with further increases in σ_M. Notably, at even higher σ_M (regime III), TE reverses behaviour and decreases with further increases in σ_M. A simple hypothesis (Ahmad et al. 2005) proposes that complexes with low σ_M remain trapped in the endosome in regime I. At the opposite end at high σ_M (regime III), which is readily accessible with MVLs, complexes escape by overcoming a kinetic barrier to fusion with the endosomal membrane. Nevertheless, a reduced level of efficiency is found, which may result from the inability of DNA to dissociate from the highly charged membranes in the cytosol. Regime II at intermediate σ_M reflects the best compromise between the opposing demands on σ_M for endosomal escape and dissociation of the complex in the cytosol.

For nonlamellar complexes, one observes high transfection efficiency with no dependence on σ_M. This includes DOTAP/DOPE–DNA complexes exhibiting the H_{II}^C phase (open circles, Figure 11.3b) and MVLBG2/DOPC–DNA complexes of the H_I^C phase (Ewert et al. 2006b). Thus, while the molecular mechanisms of CL–DNA complexes navigating intracellular pathways remain to be clarified, it is evident that the interactions between CL–DNA complexes and cells are distinctly different for lamellar and nonlamellar complexes.

11.3 CATIONIC LIPOSOMES COMPLEXED WITH SHORT NUCLEIC ACIDS

11.3.1 The Gyroid Cubic Lipid Phase Containing Short RNA Molecules

To explore the possibilities of other novel lipid–nucleic acid structures, Leal et al. studied the phase diagrams of cationic DOTAP and the neutral, cubic phase-forming lipid 1-monooleoyl-glycerol (GMO; a single-tail, neutral lipid) in water with and without nucleic acid with synchrotron X-ray scattering (Leal et al. 2010, 2011, 2013). Their study revealed a bicontinuous double-gyroid cubic phase structure for complexes in mixtures of DOTAP/GMO CLs and short nucleic acid. The nucleic acid, confined to the water channels, consisted of siRNA, and the gyroid phase was thus labelled $Q_{II}^{G,siRNA}$ (Figure 11.4a). The study by Leal et al. was motivated by the desire to develop bicontinuous cubic phase complexes as vectors for delivery of siRNA into cells for gene silencing applications. Typical siRNAs have a length of 19–25 base pairs with two 3′-nucleotide overhangs. Upon complexation with cationic liposomes, CL–siRNA complexes induce the RNA interference pathway, leading to posttranscriptional, sequence-dependent gene silencing (Bouxsein et al. 2007, Caplen et al. 2001, Elbashir et al. 2001). In addition to its widespread applications in functional genomics, siRNA technology promises to revolutionize biotechnology and therapeutics (Hannon and Rossi 2004, Hoy 2018, Karagiannis and El-Osta 2005, Sioud 2004).

The reason cubic phase-forming lipids were used is that, as mentioned above, the membranes favour saddle-splay-shaped surfaces with negative Gaussian curvature ($C_1C_2 < 0$). Biological membrane pores are also comprised of surfaces with negative Gaussian curvature. Thus, one may expect that cubic phase complexes should promote membrane fusion with cellular membranes, because such a process leads naturally to the formation of pores with the desired negative Gaussian curvature (Porte 1992, Siegel 1999, Yang and Huang 2002). The cubic phase complexes in the Leal studies, which exist at high molar fraction of GMO ($0.75 \leq \Phi_{GMO} \leq 0.95$), showed efficient sequence-specific gene silencing (Figure 11.4b, open squares) and low cytotoxicity (Leal et al. 2010). In contrast, silencing by lamellar L_α^{siRNA} complexes is low in the regime of low membrane charge density (Figure 11.4b, open circles), where L_α^{siRNA} complexes remain trapped in endosomes (see Section 11.2.2 and regime I in Figure 11.3b). This finding by Leal et al. is consistent with the above-mentioned hypothesis that cubic phase-forming lipids drive fusion of the membranes of the complex and the endosome by favouring pore formation. Pores, in turn, allow for efficient cytoplasmic delivery of siRNA and subsequent gene silencing. (For a comprehensive review of peptide- and protein-induced membrane pores, see Chapter 8 by M. W. Lee, N. W. Schmidt, and G. C. L. Wong.)

11.3.2 Two-Dimensional Packing of Short DNA with Nonpairing Overhangs in Cationic Liposome–DNA Complexes

Parallel to studies of CLs complexed with short siRNA, the self-assembling properties of short DNA (sDNA) molecules, with nonpairing overhangs, were studied in synchrotron scattering experiments of CL–sDNA complexes with sDNA sandwiched between cationic membranes (Bouxsein et al. 2011). Here, DNA end-to-end hydrophobic stacking attractions (Nakata et al. 2007, Zanchetta et al. 2008) compete with repulsive interactions mediated through thymine (T) overhangs of varying lengths (10T, 5T, and 2T). For 48 and 24 bp, sDNA was orientationally ordered and organized into a 2D nematic LC phase (Chaikin and Lubensky 1995, de Gennes and Prost 1993). This is consistent with the prediction of a well-known model of nematic LC ordering by L. Onsager, where anisotropic rods, with length L and width D, exhibit an entropic transition from a disordered liquid to the oriented nematic phase when $L/D \geq 4$ (Onsager 1949).

Remarkably, for very short 11 bp DNA ($L/D \approx 1.9$) with 2T overhangs, where Onsager theory predicts that LC ordering should not occur because $L/D < 4$, end-to-end attractions between sDNA rods lead to a novel 2D columnar nematic phase. This is shown schematically in Figure 11.4: shortening of the sDNA overhangs from 10T to 2T leads to a transition from an isotropic phase with random orientation (Figure 11.4c displays a layer inside a CL–sDNA-10T complex), to a novel 2D nematic phase where the building blocks consist of 1D stacks of ~four 11 bp DNA-2T rods (Figure 11.4d).

These findings may have important implications in the general area of DNA-based nanotechnology. To date, all DNA-directed assembly or crystallization of colloidal particles into larger-scale materials, including metallic or nonmetallic nanoparticles (NPs) with optical and optoelectronic properties, has been based on recognition through H-bonding of complementary base pairs (Elghanian et al. 1997, Maye et al. 2006, Mirkin et al. 1996, Nykypanchuk

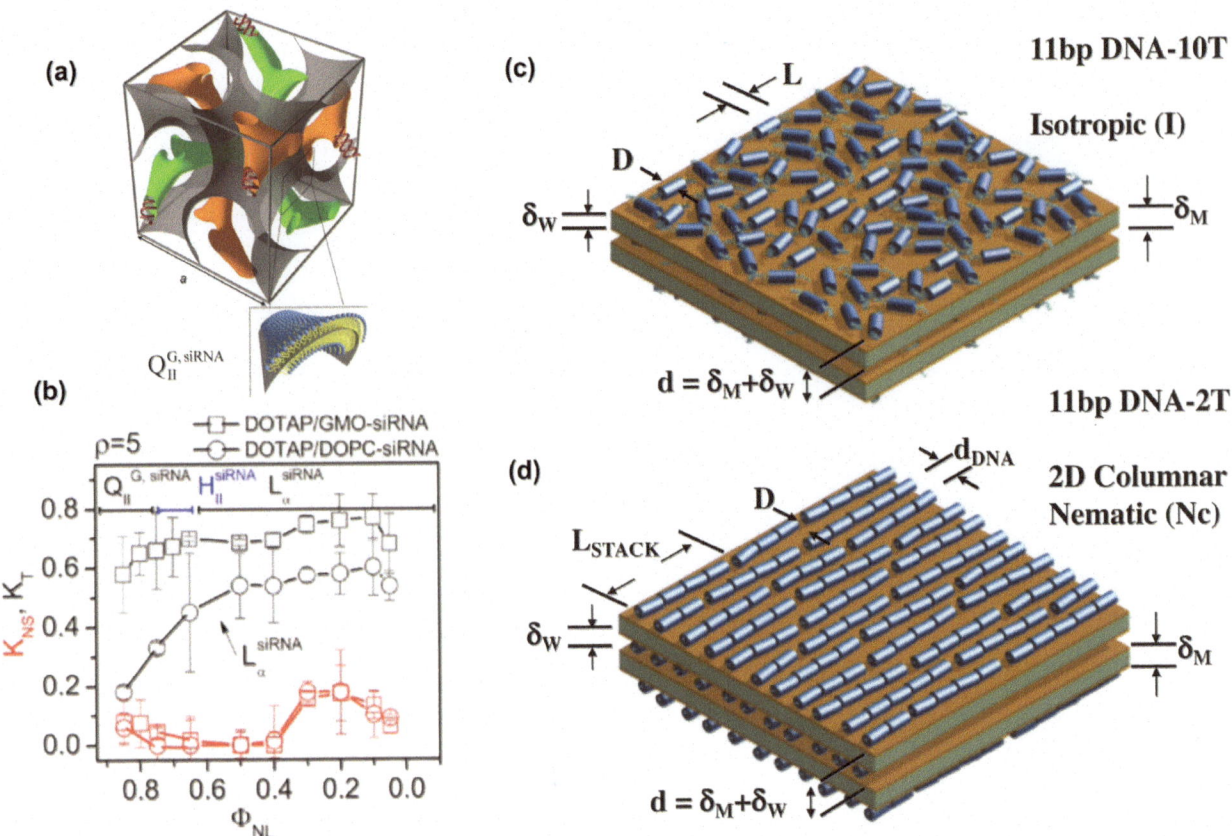

FIGURE 11.4 (a) Schematic showing incorporation of short-interfering RNA (siRNA) within the two (green and orange) water channels of a double gyroid lipid cubic phase (labelled $Q_{II}^{G,siRNA}$) with space group Ia3d deduced from synchrotron X-ray scattering. A lipid bilayer surface separates the two intertwined but independent water channels. For clarity in the cartoon, the bilayer, with negative Gaussian curvature $C_1C_2<0$, is represented by a grey surface that corresponds to a thin layer in the centre of the membrane (see enlarged inset). The gyroid phase for DOTAP/GMO–siRNA complexes is stable for GMO mole fractions (Φ_{GMO}) between 0.75 and 0.975. (The nonionic lipid glycerol monooleate (GMO, 1-monooleoyl-glycerol) is a cubic phase-forming lipid.) (b) Total (specific and nonspecific) gene knockdown (K_T, black lines and symbols) and nonspecific gene knockdown (K_{NS}, red lines and symbols) for DOTAP/GMO–siRNA (squares) and DOTAP/DOPC–siRNA complexes (circles) versus mole fraction of neutral lipid (Φ_{NL}). Remarkably, at low cationic lipid content with $\Phi_{GMO} \geq 0.75$, DOTAP/GMO–siRNA complexes in the gyroid $Q_{II}^{G,siRNA}$ cubic phase show significantly higher sequence-specific gene silencing over complexes in the lamellar L_α^{siRNA} phase. (c,d) Drawings of the distinct packing phases of short DNA (sDNA) rods (with nonsticky overhangs) in cationic liposome–sDNA (CL–sDNA) complexes. The complexes are multilamellar assemblies of alternating cationic membranes (thickness δ_M) and water layers (thickness δ_W) containing sDNA molecules. The unit cell dimension is $d = \delta_W + \delta_M$, and d_{DNA} is the average interhelical spacing. The isotropic phase, lacking position and orientation order (c), is stable for 11 bp DNA-10T (11 bp DNA core with 10-T nonsticky overhangs). Remarkably, a novel 2D columnar nematic phase is observed for 11 bp DNA-2T rods (d, with short 2T nonsticky overhangs). The onset of DNA end-to-end interactions for 11 bp DNA with the short 2T nonsticky overhangs leads to the formation of a distribution of 1D stacks of sDNA rods, which become the building blocks of the nematic phase. The 1D stacks have an average size of four rods, giving an effective length to width ratio $L/D \approx 7.3$ consistent with Onsager's theory of nematic liquid crystal phase formation. ((a) and (b) Reprinted with permission from Leal et al. 2010. Copyright 2010 by the American Chemical Society. (c) and (d) reprinted with permission from Bouxsein et al. 2011. Copyright 2011 by the American Chemical Society.)

et al. 2008, Seeman 2003). In these systems, NP building blocks are bound together through complementary sticky ends tethered to apposing NP surfaces, or via noncomplementary projecting domains bridged by an intermediate oligonucleotide with sticky ends. In contrast, the assembly described above offers a new direction in DNA-based directed assembly, where macromolecular building blocks, containing short base pairs chemically linked to their surfaces, are driven to assemble due to end-to-end stacking interactions.

11.3.3 A Three-Dimensional Columnar Phase of Stacked Blunt Short DNA Induced by Coherent Membrane Undulations in Fluid Membrane/DNA Complexes

The formation of a highly organized lipid–nucleic acid multilayer phase was recently reported in fluid membrane CL–nucleic acid complexes when the nucleic acid consisted of blunt sDNA (Bouxsein et al. 2019). In this system, strong end-to-end interactions of blunt sDNA (11, 24, and

48 bps) (Nakata et al. 2007, Zanchetta et al. 2008) stabilize stacks of sDNA, which are stable over a wide temperature range between room temperature and ≈70°C. Schematics of stacked blunt sDNA columns sandwiched between fluid lipid bilayers are depicted in the top parts of Figure 11.5a and b. In this new phase, the sDNA stacks form a 3D columnar phase with the columns indexing on a rectangular lattice (Figure 11.5b, labelled $R_\alpha^{sDNA,3D}$). The $R_\alpha^{sDNA,3D}$ phase is found only at relatively high mole fractions of neutral lipid Φ_{NL}, for $0.5 < \Phi_{NL} \leq 0.75$ with 11 bp and 24 bp sDNA, and $0.52 < \Phi_{NL} \leq 0.75$ with 48 bp sDNA. For these relatively high values of Φ_{NL}, the wall-to-wall spacing between sDNA columns is of the order of the diameter of the columns (Figure 11.5b). At lower mole fractions of Φ_{NL} (≈ 0.5 and less), with column wall-to-wall separation much less than the diameter of the columns, the system transitions to a 2D columnar phase (Figure 11.5a) similar to the phase in CL–DNA complexes with long λ-DNA (48,502 bps) (Figure 11.1b).

The finding of a 3D columnar phase with long-range orientation and position order in a fluid multilayer membrane system is unexpected because the electrostatic interactions between sDNA columns across layers are screened by freely diffusing cationic lipids near the phosphate groups of sDNA. 3D columnar phases of long DNA in CL–DNA complexes have previously been reported in systems where the lipid bilayer is in the ordered gel phase with suppressed thermal undulations (Artzner et al. 1998, Koynova and MacDonald 2004, McManus et al. 2004). Although cationic lipids exhibit limited mobility (slowed diffusion) in the gel phase because of chain ordering, they are most likely locally segregated beneath the DNA (i.e., forming strips microphase separated from neutral lipids), which would result in rigid undulations due to the difference in head group size between the cationic and neutral lipids. Spatial correlations between the rigid undulations from layer to layer would lead to the 3D columnar phase.

The robustness of the $R_\alpha^{sDNA,3D}$ columnar phase—as measured by the strength of the diffraction peaks resulting from the structure (Bouxsein et al. 2019)—increases as the membrane bending rigidity (κ) is reduced by the introduction of long-chain alcohol cosurfactants in the fluid bilayer (Safinya et al. 1989, Szleifer et al. 1988). This is a counterintuitive result since thermal undulations normally destroy long-range order (Chaikin and Lubensky 1995, de Gennes and Prost 1993). Taken together the experimental findings are in line with the predictions of a model by Schiessel and Aranda-Espinoza where undulations are the source of the existence of the 3D columnar phase (Schiessel 1998, Schiessel and Aranda-Espinoza 2001). In their model, membrane undulations (due to partial wrapping around sDNA columns), phase-lock from layer to layer, with coherent "crystal-like" undulations precipitating the onset of long-range position and orientation order of sDNA columns within the 3D multilayer phase (Figure 11.5b). Further support for the model is the finding that the new phase is enhanced with *decreasing* κ and stable only at *large* d_{sDNA}.

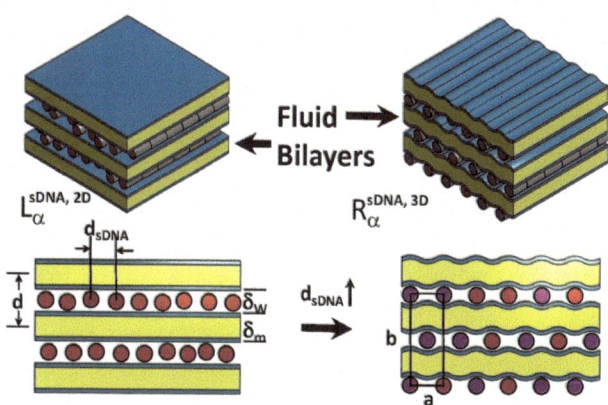

FIGURE 11.5 Drawings of fluid membrane–nucleic acid complexes with the nucleic acid consisting of stacked blunt sDNA, which spontaneously form distinct multilamellar phases depending on the mole fraction of neutral lipid (Φ_{NL}). The top and bottom parts show two views of the multilayers. The multilamellar assemblies consist of alternating fluid cationic membranes (thickness δ_M) and water layers (thickness δ_W) containing blunt sDNA molecules that stack to form columns (short cylinders, see top drawings). The unit cell dimension is $d = \delta_W + \delta_M$, and d_{sDNA} is the average interhelical spacing between the columns. (a) The multilamellar $L_\alpha^{sDNA,2D}$ phase is stable for Φ_{NL} less than ≈ 0.5. In the $L_\alpha^{sDNA,2D}$ phase, the 2D columns, with finite-sized domains, have no position or orientation correlations from layer to layer. This phase is similar to the previously observed 2D smectic-like organization of long DNA in CL–DNA complexes (see Figure 11.1b). (b) The multilamellar $R_\alpha^{sDNA,3D}$ phase, which forms after Φ_{NL} is increased above a certain threshold Φ_{NL} (Φ_{NL} between 0.5 and 0.75 for 11 bp and 24 bp blunt sDNA, and between 0.52 and 0.75 for 48 bp blunt sDNA). In this 3D columnar phase (with a centred rectangular lattice), the rods of stacked blunt sDNA exhibit position and orientation correlations across the bilayers, resulting from phase locking of undulations of fluid membranes. (Reprinted with permission from Bouxsein et al. 2019. Copyright 2019 by the American Chemical Society.)

These conditions stabilize the phase by reducing the elastic cost of partially wrapping the membrane around the sDNA columns per unit area ($\propto \kappa h^2 / d_{sDNA}^4$, with h^2 equal to the square of the sum of the amplitudes of the inner and outer monolayer undulations) to less than the resulting gain in electrostatic attractions. Partial wrapping is also enhanced by the preferential accumulation of screening cationic lipids underneath the DNA rods that occurs because the head group of DOTAP is smaller than that of DOPC. There is evidence that lateral segregation occurs in supported membranes of DOTAP/DOPC in the presence of single DNA molecules as shown by fluorescence imaging of labelled DOTAP (Kahl et al. 2009).

We note that for confined DNA between flat bilayer membranes, i.e., without coupling via membrane undulations, equilibrium theories predict a rather unique "sliding columnar phase", where rigid DNA rods have long-range orientational but no positional correlations across the bilayers (Golubović and Golubović 1998, Golubović et al.

2000, O'Hern and Lubensky 1998). However, in these flat membrane models, the addition of sufficiently strong DNA–DNA electrostatic interactions across membranes can drive the system into a 3D columnar phase.

The $R_\alpha^{sDNA,3D}$ phase may be thought of as a 3D crystal of membrane undulations where the repeating unit cell consists of a curved membrane ribbon (coupled to a column of sDNA) with the ribbon width equal to the DNA–DNA interaxial distance and the ribbon length set by the membrane size. In this description, it is analogous to the bicontinuous cubic phase with long-range position and orientation order of curved membranes and the $P_{\beta'}$ rippled phase of multilayer membranes (Sirota et al. 1988, Wack and Webb 1989, Zasadzinski et al. 1988). Note, however, that the lipid chains in the $R_\alpha^{sDNA,3D}$ phase are in the fluid (α) state despite long-range DNA order. In cationic lipid mixtures (DMTAP/DMPC) composed of lipids with saturated C_{14} chains, there exists a distinct thermotropic phase transition from the gel-phase $R_\beta^{DNA,3D}$ to the fluid $L_\alpha^{DNA,2D}$ phase (Artzner et al. 1998, Koynova and MacDonald 2004, McManus et al. 2004). In these mixtures, the first-order chain-melting phase transition temperature depends on the cationic lipid mole fraction but is only weakly dependent on the DNA content (Zantl et al. 1999).

11.4 CONCLUDING REMARKS AND FUTURE DIRECTIONS

Cationic liposomes (CLs) are promising vectors in therapeutic applications owing to their ability to sequester hydrophobic and hydrophilic drugs and nucleic acids. Engineered viral vectors, while far more efficient at nucleic acid (NA) delivery *in vivo*, are limited to delivery of relatively small therapeutic NAs, i.e., typically cDNA lacking introns and regulatory sequences, due to the size constraints imposed by viral capsids. As we described in this chapter, CLs form distinct liquid crystalline self-assemblies, including lamellar L_α^C, inverse hexagonal H_{II}^C, and hexagonal H_I^C nanostructures, when complexed with DNA. Thus, through self-assembly, CLs may complex with extremely large pieces of NA, including full-length human genes (exons and introns) and regulatory sequences exceeding 100 kbps. Nevertheless, to improve efficacy of nonviral CL vectors for *in vivo* applications, key intracellular barriers have to be overcome. The most significant intracellular barrier is endosomal escape into the cytosol after CL–drug or CL–NA nanoparticles (NPs) have been internalized by cells.

Important insights into mechanisms of endosomal escape may be derived from real-time imaging studies of spatial and temporal distribution of NPs inside cells. Measurements of distributions of NPs within early endosomes (EEs) in real time have traditionally proven difficult due to the subresolution size and short lifetime of EEs. A novel way of overcoming this hurdle is by spatiotemporal tracking of CL–NA NPs via colocalization of NPs with members of the Rab-GTPase protein family, including suitable Rab mutants (Majzoub et al. 2015a, 2016a–c). Rab-GTPases are markers of the membranes of organelles and organizers of membrane trafficking in cells (Stenmark and Olkkonen 2001, Stenmark 2009, Wandinger-Ness and Zerial 2014, Zerial and McBride 2001). A GTP-hydrolysis-deficient Rab5 mutant, Rab5-Q79L, which extends the size and lifetime of EEs and yields giant early endosomes (GEEs), allows one to resolve and localize individual NPs found within the GEE lumen (Majzoub et al. 2015a, Stenmark et al. 1994).

Figure 11.6a and c (with expanded views in b and d) shows the presence of numerous GEEs labelled with mutant Rab5-Q79L-GFP inside PC3 (prostate cancer) cells transfected with CL–DNA NPs (Majzoub et al. 2015a). The NPs in Figure 11.6a (and b) contain a polymer coat (from incorporation of PEG–lipids, see Figure 11.1a ii), while those in Figure 11.6c (and d) have a peptide–polymer coat (via RGD–PEG–lipid, see Figure 11.1a iii) with the RGD motif targeting cell surface integrins and leading to cell internalization. Aside from GEEs, much smaller EEs are also visible (see, e.g., dashed arrow in Figure 11.6d). GEEs tend to show nonuniform GFP fluorescence in their perimeter, indicative of membrane sections rich in Rab5-Q79L-GFP (red arrows, Figure 11.6b and d). This may result from events where EEs have fused with GEEs. Figure 11.6b also shows evidence of medium-sized GEEs that contain NPs undergoing fusion with a large GEE (yellow arrow and (ii)). The CL uptake observed in the large GEE (Figure 11.6b, iv) is expected because CLs coexist with NPs for overall positively charged NPs (Rädler et al. 1997).

The two types of NPs, namely, with and without RGD tagging, show distinct interactions with cell components. In particular, NPs lacking RGD show far fewer colocalization events with GEEs relative to NPs displaying RGD on their surface (compare Figure 11.6b i–iii to Figure 11.6d v–viii). Remarkably, the availability of GEEs in cells expressing Rab5-Q79L-GFP allows one to spatially resolve individual RGD-tagged NPs within GEEs (Figure 11.6d v–viii). The images show that individual RGD-tagged NPs may adhere to GEE membranes in the case of NPs that have high membrane charge density (Figure 11.6d v and white arrows).

Taken together, the micrographs in Figure 11.6 depict NPs inside giant endosomes and also interacting with the inner membrane wall of GEEs. The lipid composition of GEEs results from fusion of EEs where individually the heterogeneous EE lipid composition is derived from the plasma membranes. Thus, this type of imaging study should lead to important insights into the nature of interactions between NPs and encapsulating endosomes and allow, through visualization, an assessment of whether the NPs are efficient at endosomal escape. We expect that similar studies in the future will unravel the mechanisms by which NPs escape endosomes and navigate intracellular pathways, which is a crucial property of NPs designed for delivery of therapeutic molecules.

Clinical applications of lipid NP vectors are currently ongoing, and we expect significant progress in cancer and

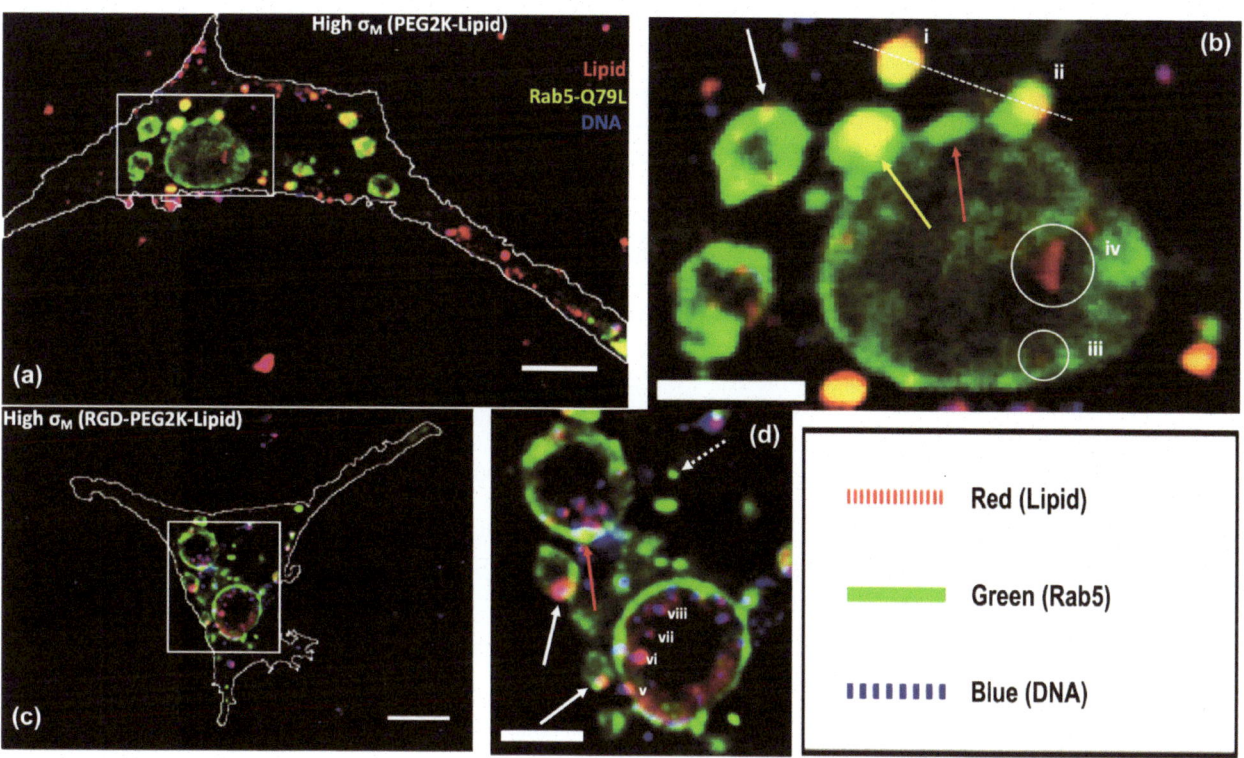

FIGURE 11.6 Colocalization of cationic liposome–DNA nanoparticles (CL–DNA NPs) with early endosomes (EEs) and giant early endosomes (GEEs). Rab5-Q79L, a nonhydrolysing mutant form of Rab5-GTP, gives rise to large (of order 5 μm) endosomes with spatially resolvable NPs. In the images, red, green, and blue colours indicate TRITC (Lipid), GFP (Endosomes), and Cy5 (DNA) fluorescence, respectively. All cells were fixed after 1 hour of incubation at 4°C followed by 1 hour of incubation at 37°C. Almost all intracellular CL–DNA NPs are found within Rab5-Q79L-GFP-labelled endosomes. (a, c) Fluorescent micrographs and cell boundaries (white outlines) of cells with (a) PEGylated NPs at high membrane charge density (σ_M) with NP lipid composition MVL5/DOPC/PEG2000-lipid at a molar ratio of 50/40/10. (c) RGD-tagged high-σ_M NPs (MVL5/DOPC/RGD-PEG2000-lipid at a molar ratio 50/45/5). (b, d) are high magnification views of the boxed regions in (a) and (c). (b) shows NPs (i, ii, iii) and liposomes (iv) found in EEs (i, ii) and GEEs (iii). (d) shows evidence of individually resolvable NPs (v, vi, vii, viii) inside the lumen of a GEE, including a NP interacting with the inner membrane of the GEE (v). The yellow arrow in (b) points to an EE containing a NP that is fusing with a GEE. Red arrows in (b, d) show GFP-rich regions of the GEE membrane. Dashed arrow in (d) shows smaller EEs, similar to what is observed with wild-type Rab5. Solid white arrows in (d) point to clear examples of NPs adhering to the GEE membrane. Scale bars in (a, c) and (b, d) are 10 and 5 μm, respectively. (Reprinted from Majzoub et al. 2015a with permission from Elsevier.)

gene therapeutics in the near future. The first FDA approval of an siRNA therapeutic based on an ionizable cationic lipid nanoparticle formulation (patisiran/ONPATTRO™, Alnylam Pharmaceuticals) occurred in August 2018 (Hoy 2018, Kulkarni et al. 2018). The clinical application of this formulation, which contains siRNA, is in silencing of the expression of the mutated transthyretin (TTR) gene in liver, which when left unchecked leads to build-up of amyloid protein in peripheral nerves. (Mutations in the TTR gene lead to hereditary TTR-mediated amyloidosis (hATTR) in adults.) Moreover, mRNA-based therapies have recently emerged with great potential in both therapy and vaccination (Pardi et al. 2018). Cationic and ionizable lipid formulations, originally optimized for siRNA delivery, have proven equally effective for mRNA delivery (Yin et al. 2014). Most notably, in December 2020, the mRNA vaccines developed by Pfizer/BioNTech (Mulligan et al. 2020, Polack et al. 2020) and Moderna (Baden et al. 2020, Jackson et al. 2020) became the first vaccines approved against COVID-19.

Building on decades of work on both the lipid vector and the mRNA payload (Nance and Meier 2021), these CL-based vaccines realized the promise of rapid construction and deployment to provide a path out of the pandemic caused by SARS-CoV-2. Both vaccines are complexes of mRNA and a lipid mixture similar to that of patisiran (Corbett et al. 2020), albeit employing ionizable cationic lipids with branched rather than unsaturated tails. Single-stranded mRNA, like DNA and short double-stranded siRNA, forms lamellar CL–NA complexes when complexed with DOPC/DOTAP (Ziller et al. 2018). However, the interactions in CL–mRNA complexes, compared with the CL–DNA complexes discussed in this chapter, are more complex in several aspects. Firstly, the electrostatic interactions of mRNA with cationic lipid membranes are biased by hydrophobic interaction of the exposed nucleic acid bases, and secondly, mRNA is more flexible and prone to form secondary structures (Michanek et al. 2010). The exact structural features of mRNA in cationic lipid moieties are an unresolved issue.

Nevertheless, the various distinct properties of bulk CL–NA phases described in this chapter are expected to help our understanding of the governing principles of self-assembly in lipid nanoparticles. We expect that many new lipid vectors will be approved for clinical applications in the near future for treatment of a broad range of genetic diseases.

In parallel to the interest in CL–NA liquid crystal complexes for therapeutics, the systems are also of interest in fundamental studies of biomolecular materials systems under confinement. As we described, long and short DNA and RNA may be electrostatically coupled to thin lipid layers to occupy 1D, 2D, and 3D space in the H_{II}^C, L_α^C, and H_I^C phases, respectively. Statistical mechanical theories of condensed matter systems show that increased thermal fluctuations at lower dimensions fundamentally alter ordering of molecular assemblies (Aronovitz and Lubensky 1988, Chaikin and Lubensky 1995, de Gennes and Prost 1993, Khurana 1989, Nelson and Peliti 1987). Thus, CL–DNA phases with distinct symmetries are suitable models for studies of the structure and dynamics of confined polyelectrolytes as the dimension of the confining volume decreases from three to two and one.

ACKNOWLEDGEMENTS

The support by the National Institutes of Health under Awards R01GM130769 and R01GM59288 (CRS and KKE, confocal microscopy study of CL–DNA NPs and their pathways), the US Department of Energy (DOE), Office of Basic Energy Sciences, Division of Materials Sciences and Engineering, under Award DE-FG02-06ER46314 (CRS, YL, KKE, self- and directed assembly in biomimetic charged biomolecular materials systems studied by synchrotron X-ray scattering and cryogenic TEM), and the US National Science Foundation (NSF) under Award DMR-1807327 (CRS, membrane phase behaviour) is gratefully acknowledged. The support by the Excellence Cluster Nanosystems Initiative Munich (NIM) and the German Federal Ministry of Education, Research and Technology (BMBF) under the cooperative project 05K2018-2017-06716 Medisoft is gratefully acknowledged (JOR). Synchrotron X-ray scattering studies were carried out at the Stanford Synchrotron Radiation Lightsource, a Directorate of SLAC National Accelerator Laboratory and an Office of Science User Facility operated for the U.S. DOE Office of Science by Stanford University.

REFERENCES

Ahmad, A., H. M. Evans, K. Ewert, C. X. George, C. E. Samuel, and C. R. Safinya. 2005. New multivalent cationic lipids reveal bell curve for transfection efficiency versus membrane charge density: Lipid-DNA complexes for gene delivery. *J. Gene. Med.* 7:739–48.

Alberts, B., A. D. Johnson, J. Lewis et al. 2015. *Molecular Biology of the Cell*. Garland Science, New York.

Allen, T. M. and P. R. Cullis. 2013. Liposomal drug delivery systems: From concept to clinical applications. *Adv. Drug Delivery Rev.* 65:36–48.

Andelman, D., M. E. Cates, D. Roux, and S. A. Safran. 1987. Structure and phase equilibria of microemulsions. *J. Chem. Phys.* 87:7229–41.

Aronovitz, J. A. and T. C. Lubensky. 1988. Fluctuations of Solid Membranes. *Phys. Rev. Lett.* 60:2634–7.

Artzner, F., R. Zantl, G. Rapp, and J. O. Rädler. 1998. Observation of a rectangular columnar phase in condensed lamellar cationic lipid-DNA complexes. *Phys. Rev. Lett.* 81:5015–8.

Baden, L. R., H. M. El Sahly, B. Essink et al. 2020. Efficacy and Safety of the mRNA-1273 Sars-Cov-2 Vaccine. *N. Engl. J. Med.* 384:403–16.

Bandyopadhyay, S., M. Tarek, and M. L. Klein. 1999. Molecular dynamics study of a lipid–DNA complex. *J. Phys. Chem. B* 103:10075–80.

Bangham, A. D. and R. W. Horne. 1964. Negative staining of phospholipids and their structural modification by surface-active agents as observed in the electron microscope. *J. Mol. Biol.* 8:660–8.

Betz, T. and C. Sykes. 2012. Time resolved membrane fluctuation spectroscopy. *Soft Matter* 8:5317–26.

Bielke, W. and C. Erbacher, editors. 2010. *Nucleic Acid Transfection*. Springer, Berlin.

Bouxsein, N. F., C. l. Leal, C. S. McAllister et al. 2011. Two-dimensional packing of short DNA with nonpairing overhangs in cationic liposome–DNA complexes: From onsager nematics to columnar nematics with finite-length columns. *J. Am. Chem. Soc.* 133: 7585–95.

Bouxsein, N. F., C. Leal, C. S. McAllister et al. 2019. 3D columnar phase of stacked short DNA organized by coherent membrane undulations. *Langmuir* 35:11891–901.

Bouxsein, N. F., C. S. McAllister, K. K. Ewert, C. E. Samuel, and C. R. Safinya. 2007. Structure and gene silencing activities of monovalent and pentavalent cationic lipid vectors complexed with siRNA. *Biochemistry* 46:4785–92.

Bruinsma, R. 1998. Electrostatics of DNA-cationic lipid complexes: Isoelectric instability. *Eur. Phys. J. B* 4:75–88.

Caplen, N. J., S. Parrish, F. Imani, A. Fire, and R. A. Morgan. 2001. Specific inhibition of gene expression by small double-stranded RNAs in invertebrate and vertebrate systems. *Proc. Natl. Acad. Sci. U. S. A.* 98:9742–7.

Cates, M. E., D. Roux, D. Andelman, S. T. Milner, and S. A. Safran. 1988. Random surface model for the L3-phase of dilute surfactant solutions. *Europhys. Lett.* 5:733–9.

Chaikin, P. M. and T. C. Lubensky. 1995. *Principles of Condensed Matter Physics*. Cambridge University Press, Cambridge.

Chiruvolu, S., H. E. Warriner, E. Naranjo et al. 1994. A phase of liposomes with entangled tubular vesicles. *Science* 266:1222–5.

Conn, C. E., O. Ces, A. M. Squires et al. 2008. A pressure-jump time-resolved x-ray diffraction study of cubic-cubic transition kinetics in monoolein. *Langmuir* 24:2331–40.

Corbett, K. S., D. Edwards, S. R. Leist et al. 2020. Sars-Cov-2 mRNA Vaccine Development Enabled by Prototype Pathogen Preparedness. bioRxiv:2020.06.11.145920. DOI: 10.1101/2020.06.11.145920.

de Gennes, P. G. and J. Prost. 1993. *The Physics of Liquid Crystals*. Clarendon Press, Oxford.

Elbashir, S. M., J. Harborth, W. Lendeckel, A. Yalcin, K. Weber, and T. Tuschl. 2001. Duplexes of 21-nucleotide RNAs mediate RNA interference in cultured mammalian cells. *Nature* 411:494–8.

Elghanian, R., J. J. Storhoff, R. C. Mucic, R. L. Letsinger, and C. A. Mirkin. 1997. Selective colorimetric detection of polynucleotides based on the distance-dependent optical properties of gold nanoparticles. *Science* 277:1078–81.

Ewert, K., A. Ahmad, H. Evans, and C. Safinya. 2005. Cationic lipid-DNA complexes for non-viral gene therapy: Relating supramolecular structures to cellular pathways. *Expert Opin. Biol. Ther.* 5:33–53.

Ewert, K., A. Ahmad, H. M. Evans, H.-W. Schmidt, and C. R. Safinya. 2002. Efficient synthesis and cell-transfection properties of a new multivalent cationic lipid for nonviral gene delivery. *J. Med. Chem.* 45:5023–9.

Ewert, K., N. L. Slack, A. Ahmad et al. 2004. Cationic lipid-DNA complexes for gene therapy: Understanding the relationship between complex structure and gene delivery pathways at the molecular level. *Curr. Med. Chem.* 11:133–49.

Ewert, K. K., H. M. Evans, N. F. Bouxsein, and C. R. Safinya. 2006a. Dendritic cationic lipids with highly charged headgroups for efficient gene delivery. *Bioconjugate Chem.* 17:877–88.

Ewert, K. K., H. M. Evans, A. Zidovska, N. F. Bouxsein, A. Ahmad, and C. R. Safinya. 2006b. A columnar phase of dendritic lipid-based cationic liposome-DNA complexes for gene delivery: Hexagonally ordered cylindrical micelles embedded in a DNA honeycomb lattice. *J. Am. Chem. Soc.* 128:3998–4006.

Ewert, K. K., V. R. Kotamraju, R. N. Majzoub et al. 2016. Synthesis of linear and cyclic peptide–PEG–lipids for stabilization and targeting of cationic liposome–DNA complexes. *Bioorg. Med. Chem. Lett.* 26:1618–23.

Ewert, K. K., A. Zidovska, A. Ahmad et al. 2010. Cationic liposome–nucleic acid complexes for gene delivery and silencing: Pathways and mechanisms for plasmid DNA and siRNA. *Top. Curr. Chem.* 296:191–226.

Felgner, P. L., T. R. Gadek, M. Holm et al. 1987. Lipofection: A highly efficient, lipid-mediated DNA-transfection procedure. *Proc. Natl. Acad. Sci. U. S. A.* 84:7413–7.

Foldvari, M., D. W. Chen, N. Nafissi, D. Calderon, L. Narsineni, and A. Rafiee. 2016. Non-viral gene therapy: Gains and challenges of non-invasive administration methods. *J. Controlled Release* 240:165–90.

Gelbart, W. M., A. Ben-Shaul, and D. Roux. 2012. *Micelles, Membranes, Microemulsions, and Monolayers*. Springer, New York.

Gindy, M. E., A. M. Leone, and J. J. Cunningham. 2012. Challenges in the pharmaceutical development of lipid-based short interfering ribonucleic acid therapeutics. *Expert Opin. Drug Delivery* 9:171–82.

Ginn, S. L., A. K. Amaya, I. E. Alexander, M. Edelstein, and M. R. Abedi. 2018. Gene therapy clinical trials worldwide to 2017: An update. *J. Gene Med.* 20:e3015.

Golubović, L. and M. Golubović. 1998. Fluctuations of quasi-two-dimensional smectics intercalated between membranes in multilamellar phases of DNA-cationic lipid complexes. *Phys. Rev. Lett.* 80:4341–4.

Golubović, L., T. C. Lubensky, and C. S. O'Hern. 2000. Structural properties of the sliding columnar phase in layered liquid crystalline systems. *Phys. Rev. E* 62:1069–94.

Gruner, S. M. 1989. Stability of lyotropic phases with curved interfaces. *J. Phys. Chem.* 93:7562–70.

Hacein-Bey-Abina, S., A. Garrigue, G. P. Wang et al. 2008. Insertional oncogenesis in 4 patients after retrovirus-mediated gene therapy of SCID-X1. *J. Clin. Invest.* 118:3132–42.

Hannon, G. J. and J. J. Rossi. 2004. Unlocking the potential of the human genome with RNA interference. *Nature* 431:371–8.

Harries, D., S. May, W. M. Gelbart, and A. Ben-Shaul. 1998. Structure, stability, and thermodynamics of lamellar DNA-lipid complexes. *Biophys. J.* 75:159–73.

Harrington, J. J., G. Van Bokkelen, R. W. Mays, K. Gustashaw, and H. F. Willard. 1997. Formation of de novo centromeres and construction of first-generation human artificial microchromosomes. *Nat. Genet.* 15:345–55.

Helfrich, W. Z. 1973. Elastic properties of lipid bilayers: Theory and possible experiments. *Z. Naturforsch., C: J. Biosci.* 28:693–703.

Hoy, S. M. 2018. Patisiran: First global approval. *Drugs* 78:1625–31.

Israelachvili, J. N. 2011. *Intermolecular and Surface Forces*. Elsevier, Amsterdam.

Israelachvili, J. N., D. J. Mitchell, and B. W. Ninham. 1977. Theory of self-assembly of lipid bilayers and vesicles. *Biochim. Biophys. Acta, Biomembr.* 470:185–201.

Jackson, L. A., E. J. Anderson, N. G. Rouphael et al. 2020. An mRNA Vaccine against Sars-Cov-2 — Preliminary Report. *N. Engl. J. Med.* 383:1920–31.

Kahl, V., M. Hennig, B. Maier, and J. O. Rädler. 2009. Conformational dynamics of DNA-electrophoresis on cationic membranes. *Electrophoresis* 30:1276–81.

Kang, M., H. Kim, and C. Leal. 2016. Self-organization of nucleic acids in lipid constructs. *Curr. Opin. Colloid Interface Sci.* 26:58–65.

Kang, M. and C. Leal. 2016. Soft nanostructured films for actuated surface-based siRNA delivery. *Adv. Funct. Mater.* 26:5610–20.

Karagiannis, T. C. and A. El-Osta. 2005. RNA interference and potential therapeutic applications of short interfering RNAs. *Cancer Gene Ther.* 12:787–95.

Käs, J. and E. Sackmann. 1991. Shape transitions and shape stability of giant phospholipid vesicles in pure water induced by area-to-volume changes. *Biophys. J.* 60:825–44.

Khurana, A. 1989. Fluid membranes repel one another; solid membranes may not crumple. *Physics Today* 42:17–21.

Kim, H., J. Sung, Y. Chang, A. Alfeche, and C. Leal. 2018. Microfluidics synthesis of gene silencing cubosomes. *ACS Nano* 12:9196–205.

Koltover, I., T. Salditt, J. O. Rädler, and C. R. Safinya. 1998. An inverted hexagonal phase of cationic liposome-DNA complexes related to DNA release and delivery. *Science* 281:78–81.

Koltover, I., T. Salditt, and C. R. Safinya. 1999. Phase diagram, stability, and overcharging of lamellar cationic lipid- DNA self-assembled complexes. *Biophys. J.* 77:915–24.

König, S., W. Pfeiffer, T. Bayerl, D. Richter, and E. Sackmann. 1992. Molecular dynamics of lipid bilayers studied by incoherent quasi-elastic neutron scattering. *J. Phys. II* 2:1589–615.

Koynova, R. and R. C. MacDonald. 2004. Columnar DNA superlattices in lamellar o-ethylphosphatidylcholine lipoplexes: Mechanism of the gel-liquid crystalline lipid phase transition. *Nano Lett.* 4:1475–9.

Kulkarni, J. A., P. R. Cullis, and R. van der Meel. 2018. Lipid nanoparticles enabling gene therapies: From concepts to clinical utility. *Nucleic Acid Therapeutics* 28:146–57.

Lares, M. R., J. J. Rossi, and D. L. Ouellet. 2010. RNAi and small interfering RNAs in human disease therapeutic applications. *Trends Biotechnol.* 28:570–9.

Lasic, D. D., editor. 1997. *Liposomes in Gene Delivery*. CRC Press, Boca Raton.

Lasic, D. D. and F. Martin, editors. 1995. *Stealth Liposomes*. CRC Press, Boca Raton.

Leal, C., N. F. Bouxsein, K. K. Ewert, and C. R. Safinya. 2010. Highly efficient gene silencing activity of siRNA embedded in a nanostructured gyroid cubic lipid matrix. *J. Am. Chem. Soc.* 132:16841–7.

Leal, C., K. K. Ewert, N. F. Bouxsein, R. S. Shirazi, Y. Li, and C. R. Safinya. 2013. Stacking of short DNA induces the gyroid cubic-to-inverted hexagonal phase transition in lipid-DNA complexes. *Soft Matter* 9:795–804.

Leal, C., K. K. Ewert, R. S. Shirazi, N. F. Bouxsein, and C. R. Safinya. 2011. Nanogyroids incorporating multivalent lipids: Enhanced membrane charge density and pore forming ability for gene silencing. *Langmuir* 27:7691–7.

Lei, N., C. R. Safinya, and R. F. Bruinsma. 1995. Discrete harmonic model for stacked membranes: Theory and experiment. *J. Phys. II* 5:1155–63.

Lin, A. J., N. L. Slack, A. Ahmad, C. X. George, C. E. Samuel, and C. R. Safinya. 2003. Three-dimensional imaging of lipid gene-carriers: Membrane charge density controls universal transfection behavior in lamellar cationic liposome-DNA complexes. *Biophys. J.* 84:3307–16.

Lindblom, G., K. Larsson, L. Johansson, K. Fontell, and S. Forsen. 1979. The cubic phase of monoglyceride-water systems. Arguments for a structure based upon lamellar bilayer units. *J. Am. Chem. Soc.* 101:5465–70.

Lipowsky, R. and E. Sackmann, editors. 1995. *Structure and Dynamics of Membranes*. Elsevier, Amsterdam.

Majzoub, R. N., C.-L. Chan, K. K. Ewert, B. F. B. Silva, K. S. Liang, and C. R. Safinya. 2015a. Fluorescence microscopy colocalization of lipid–nucleic acid nanoparticles with wildtype and mutant Rab5-GFP: A platform for investigating early endosomal events. *Biochim. Biophys. Acta, Biomembr.* 1848:1308–18.

Majzoub, R. N., K. K. Ewert, E. L. Jacovetty et al. 2015b. Patterned threadlike micelles and DNA-tethered nanoparticles: A structural study of PEGylated cationic liposome–DNA assemblies. *Langmuir* 31:7073–83.

Majzoub, R. N., K. K. Ewert, and C. R. Safinya. 2016a. Quantitative intracellular localization of cationic lipid-nucleic acid nanoparticles with fluorescence microscopy. In: *Methods in Molecular Biology, Non-Viral Gene Delivery Vectors: Methods and Protocols*. G. Candiani, editor. Springer/Humana Press, New York, pp. 77–108.

Majzoub, R. N., K. K. Ewert, and C. R. Safinya. 2016b. Cationic liposome–nucleic acid nanoparticle assemblies with applications in gene delivery and gene silencing. *Philos. Trans. R. Soc., A* 374.

Majzoub, R. N., E. Wonder, K. K. Ewert, V. R. Kotamraju, T. Teesalu, and C. R. Safinya. 2016c. Rab11 and lysotracker markers reveal correlation between endosomal pathways and transfection efficiency of surface-functionalized cationic liposome–DNA nanoparticles. *J. Phys. Chem. B* 120:6439–53.

Manning, G. S. 1969. Limiting laws and counterion condensation in polyelectrolyte solutions I. *Colligative Properties. J. Chem. Phys.* 51:924–33.

May, S. and A. Ben-Shaul. 1997. DNA-lipid complexes: Stability of honeycomb-like and spaghetti-like structures. *Biophys. J.* 73:2427–40.

Maye, M. M., D. Nykypanchuk, D. van der Lelie, and O. Gang. 2006. A simple method for kinetic control of DNA-induced nanoparticle assembly. *J. Am. Chem. Soc.* 128:14020–1.

McManus, J. J., J. O. Rädler, and K. A. Dawson. 2004. Observation of a rectangular columnar phase in a DNA–calcium–Zwitterionic lipid complex. *J. Am. Chem. Soc.* 126:15966–7.

Michanek, A., N. Kristen, F. Höök, T. Nylander, and E. Sparr. 2010. RNA and DNA interactions with zwitterionic and charged lipid membranes — A DSC and QCM-D study. *Biochim. Biophys. Acta, Biomembr.* 1798:829–38.

Mirkin, C. A., R. L. Letsinger, R. C. Mucic, and J. J. Storhoff. 1996. A DNA-based method for rationally assembling nanoparticles into macroscopic materials. *Nature* 382:607–9.

Mulligan, M. J., K. E. Lyke, N. Kitchin et al. 2020. Phase I/II Study of Covid-19 RNA Vaccine Bnt162b1 in Adults. *Nature* 586:589–93.

Nakata, M., G. Zanchetta, B. D. Chapman et al. 2007. End-to-end stacking and liquid crystal condensation of 6- to 20-base pair DNA duplexes. *Science* 318:1276–9.

Nallet, F., R. Laversanne, and D. Roux. 1993. Modelling X-ray or neutron scattering spectra of lyotropic lamellar phases: Interplay between form and structure factors. *J. Phys. II* 3:487–502.

Nance, K. D. and J. L. Meier. 2021. Modifications in an Emergency: The Role of N1-Methylpseudouridine in COVID-19 Vaccines. *ACS Cent. Sci.* 7:748–56.

Nelson, D. R. and L. Peliti. 1987. Fluctuations in membranes with crystalline and hexatic order. *J. Phys. (Paris)* 48:1085–92.

Nykypanchuk, D., M. M. Maye, D. van der Lelie, and O. Gang. 2008. DNA-guided crystallization of colloidal nanoparticles. *Nature* 451:549–52.

O'Hern, C. S. and T. C. Lubensky. 1998. Sliding columnar phase of DNA-lipid complexes. *Phys. Rev. Lett.* 80:4345–8.

Onsager, L. 1949. The effects of shape on the interaction of colloidal particles. *Ann. N. Y. Acad. Sci.* 51:627–59.

Ozcan, G., B. Ozpolat, R. L. Coleman, A. K. Sood, and G. Lopez-Berestein. 2015. Preclinical and clinical development of siRNA-based therapeutics. *Adv. Drug Delivery Rev.* 87:108–19.

Pardi, N., M. J. Hogan, F. W. Porter, and D. Weissman. 2018. mRNA vaccines — a new era in vaccinology. *Nat. Rev. Drug Discovery* 17:261–79.

Polack, F. P., S. J. Thomas, N. Kitchin et al. 2020. Safety and Efficacy of the Bnt162b2 mRNA Covid-19 Vaccine. *N. Engl. J. Med.* 383:2603–15.

Pollard, T. D., W. C. Earnshaw, J. Lippincott-Schwartz, and G. Johnson. 2017. *Cell Biology*. Elsevier, New York.

Porte, G. 1992. Lamellar phases and disordered phases of fluid bilayer membranes. *J. Phys.: Condens. Matter* 4:8649–70.

Rädler, J. O., T. J. Feder, H. H. Strey, and E. Sackmann. 1995. Fluctuation analysis of tension-controlled undulation forces between giant vesicles and solid substrates. *Phys. Rev. E* 51:4526–36.

Rädler, J. O., I. Koltover, A. Jamieson, T. Salditt, and C. R. Safinya. 1998. Structure and interfacial aspects of self-assembled cationic lipid-DNA gene carrier complexes. *Langmuir* 14:4272–83.

Rädler, J. O., I. Koltover, T. Salditt, and C. R. Safinya. 1997. Structure of DNA-cationic liposome complexes: DNA intercalation in multilamellar membranes in distinct interhelical packing regimes. *Science* 275:810–4.

Rigaud, J. L., M. T. Paternostre, and A. Bluzat. 1988. Mechanisms of membrane protein insertion into liposomes during reconstitution procedures involving the use of detergents. 2. Incorporation of the light-driven proton pump bacteriorhodopsin. *Biochemistry* 27:2677–88.

Roux, D. and C. R. Safinya. 1988. A synchrotron X-ray study of competing undulation and electrostatic interlayer interactions in fluid multimembrane lyotropic phases. *J. Phys. (Paris)* 49:307–18.

Rupaimoole, R. and F. J. Slack. 2017. MicroRNA therapeutics: Towards a new era for the management of cancer and other diseases. *Nat. Rev. Drug Discovery* 16:203–22.

Safinya, C. R. 1989. Rigid and fluctuating surfaces: A series of synchrotron X-ray scattering studies of interacting stacked membranes. In: *Phase Transitions in Soft Condensed Matter.* T. Riste and D. Sherrington, editors. Springer, Boston, pp. 249–70.

Safinya, C. R. 2001. Structures of lipid-DNA complexes: Supramolecular assembly and gene delivery. *Curr. Opin. Struct. Biol.* 11:440–8.

Safinya, C. R. 2006. Biophysics and biomolecular materials. In: *The New Physics: For the Twenty-First Century.* G. Fraser, editor. Cambridge University Press, Cambridge and New York.

Safinya, C. R., D. Roux, G. S. Smith et al. 1986. Steric interactions in a model multimembrane system: A synchrotron X-ray study. *Phys. Rev. Lett.* 57:2718–21.

Safinya, C. R. and K. K. Ewert. 2012. Materials chemistry: Liposomes derived from molecular vases. *Nature* 489:372–4.

Safinya, C. R., K. K. Ewert, R. N. Majzoub, and C. Leal. 2014. Cationic liposome-nucleic acid complexes for gene delivery and gene silencing. *New J. Chem.* 38:5164–72.

Safinya, C. R., E. B. Sirota, D. Roux, and G. S. Smith. 1989. Universality in interacting membranes: The effect of cosurfactants on the interfacial rigidity. *Phys. Rev. Lett.* 62:1134–7.

Safran, S. A. 1994. *Statistical Thermodynamics of Surfaces, Interfaces, and Membranes.* Addison-Wesley, Reading.

Salditt, T., I. Koltover, J. O. Rädler, and C. R. Safinya. 1997. Two-dimensional smectic ordering of linear DNA chains in self-assembled DNA-cationic liposome mixtures. *Phys. Rev. Lett.* 79:2582–5.

Salditt, T., I. Koltover, J. O. Rädler, and C. R. Safinya. 1998. Self-assembled DNA--cationic-lipid complexes: Two-dimensional smectic ordering, correlations, and interactions. *Phys. Rev. E* 58:889–904.

Schiessel, H. 1998. Bending of charged flexible membranes due to the presence of macroions. *Eur. Phys. J. B* 6:373–80.

Schiessel, H. and H. Aranda-Espinoza. 2001. Electrostatically induced undulations of lamellar DNA-lipid complexes. *Eur. Phys. J. E* 5:499–506.

Seddon, J. M. 1990. Structure of the inverted hexagonal (H_{II}) phase, and non-lamellar phase transitions of lipids. *Biochim. Biophys. Acta* 1031:1–69.

Seeman, N. C. 2003. DNA in a material world. *Nature* 421:427–31.

Sercombe, L., T. Veerati, F. Moheimani, S. Y. Wu, A. K. Sood, and S. Hua. 2015. Advances and challenges of liposome assisted drug delivery. *Front. Pharmacol.* 6:286.

Shen, Y., C. R. Safinya, K. S. Liang, A. F. Ruppert, and K. J. Rothschild. 1993. Stabilization of the membrane protein bacteriorhodopsin to 140°C in two-dimensional films. *Nature* 366:48–50.

Siegel, D. P. 1999. The modified stalk mechanism of lamellar/inverted phase transitions and its implications for membrane fusion. *Biophys. J.* 76:291–313.

Sioud, M. 2004. Therapeutic siRNAs. *Trends Pharmacol. Sci.* 25:22–8.

Sirota, E. B., G. S. Smith, C. R. Safinya, R. J. Plano, and N. A. Clark. 1988. X-ray scattering studies of aligned, stacked surfactant membranes. *Science* 242:1406–9.

Sofias, A. M., M. Dunne, G. Storm, and C. Allen. 2017. The battle of "nano" paclitaxel. *Adv. Drug Delivery Rev.* 122:20–30.

Stenmark, H. 2009. Rab GTPases as coordinators of vesicle traffic. *Nat. Rev. Mol. Cell Biol.* 10:513–25.

Stenmark, H., R. G. Parton, O. Steele-Mortimer, A. Lütcke, J. Gruenberg, and M. Zerial. 1994. Inhibition of rab5 GTPase activity stimulates membrane fusion in endocytosis. *EMBO J.* 13:1287–96.

Stenmark, H. and V. M. Olkkonen. 2001. The Rab GTPase family. *Genome Biol.* 2:reviews3007.1-reviews.7.

Szleifer, I., D. Kramer, A. Ben-Shaul, D. Roux, and W. M. Gelbart. 1988. Curvature elasticity of pure and mixed surfactant films. *Phys. Rev. Lett.* 60:1966–9.

Thomas, C. E., A. Ehrhardt, and M. A. Kay. 2003. Progress and problems with the use of viral vectors for gene therapy. *Nat. Rev. Genet.* 4:346–58.

Wack, D. C. and W. W. Webb. 1989. Synchrotron X-ray study of the modulated lamellar phase $P_{\beta'}$ in the lecithin-water system. *Phys. Rev. A* 40:2712–30.

Wagner, K., D. Harries, S. May, V. Kahl, J. O. Rädler, and A. Ben-Shaul. 2000. Direct evidence for counterion release upon cationic lipid–DNA condensation. *Langmuir* 16:303–6.

Wandinger-Ness, A. and M. Zerial. 2014. Rab proteins and the compartmentalization of the endosomal system. *Cold Spring Harbor Perspect. Biol.* 6:a022616.

Yang, L. and H. W. Huang. 2002. Observation of a membrane fusion intermediate structure. *Science* 297:1877–9.

Yin, H., R. L. Kanasty, A. A. Eltoukhy, A. J. Vegas, J. R. Dorkin, and D. G. Anderson. 2014. Non-viral vectors for gene-based therapy. *Nat. Rev. Genet.* 15:541–55.

Zanchetta, G., T. Bellini, M. Nakata, and N. A. Clark. 2008. Physical polymerization and liquid crystallization of RNA oligomers. *J. Am. Chem. Soc.* 130:12864–5.

Zantl, R., L. Baicu, F. Artzner, I. Sprenger, G. Rapp, and J. O. Rädler. 1999. Thermotropic phase behavior of cationic lipid–DNA complexes compared to binary lipid mixtures. *J. Phys. Chem. B* 103:10300–10.

Zasadzinski, J., J. Schneir, J. Gurley, V. Elings, and P. Hansma. 1988. Scanning tunneling microscopy of freeze-fracture replicas of biomembranes. *Science* 239:1013–5.

Zerial, M. and H. McBride. 2001. Rab proteins as membrane organizers. *Nat. Rev. Mol. Cell Biol.* 2:107–17.

Zidovska, A., H. M. Evans, A. Ahmad, K. K. Ewert, and C. R. Safinya. 2009a. The role of cholesterol and structurally related molecules in enhancing transfection of cationic liposome–DNA complexes. *J. Phys. Chem. B* 113:5208–16.

Zidovska, A., H. M. Evans, K. K. Ewert et al. 2009b. Liquid crystalline phases of dendritic lipid-DNA self-assemblies: Lamellar, hexagonal, and DNA bundles. *J. Phys. Chem. B* 113:3694–703.

Zidovska, A., K. K. Ewert, J. Quispe, B. Carragher, C. S. Potter, and C. R. Safinya. 2009c. The effect of salt and pH on block liposomes studied by cryogenic transmission electron microscopy. *Biochim. Biophys. Acta, Biomembr.* 1788:1869–76.

Zidovska, A., K. K. Ewert, J. Quispe, B. Carragher, C. S. Potter, and C. R. Safinya. 2009d. Block liposomes from curvature-stabilizing lipids: Connected nanotubes, -rods, or -spheres. *Langmuir* 25:2979–85.

Ziller, A., S. S. Nogueira, E. Hühn et al. 2018. Incorporation of mRNA in lamellar lipid matrices for parenteral administration. *Mol. Pharmaceutics* 15:642–51.

12 Lipids in DNA, RNA, and Peptide Delivery for *In Vivo* Therapeutic Applications

Tyler Goodwin and Leaf Huang
University of North Carolina, Chapel Hill, North Carolina

CONTENTS

12.1 Introduction 211
12.2 Rational Design of Lipid Vectors to Overcome Extracellular and Intracellular Challenges 212
12.3 Current Lipid Vectors for Delivery of DNA, RNA, and Peptide 214
12.4 Applications for Lipid Vectors in the Field of DNA, RNA, and Peptide Therapeutics..........................218
12.5 Past and Present Clinical Trials..219
12.6 Conclusion 220
Acknowledgement..220
References..220

12.1 INTRODUCTION

The field of drug delivery has experienced significant expansion and growth over the past several decades. Traditional drug discovery was seen in conjunction with drug development which consisted of synthesizing small molecule (MW < 500 g mol^{-1}) drugs with a desired pharmacokinetic/pharmacodynamics (PK/PD) profile. It was found to be more advantageous and convenient to develop small molecule drugs that demonstrated optimal pharmacokinetics, biodistribution, and minimal toxicity. These small molecule drugs are easy to manipulate through organic chemistry techniques allowing pharmaceutical companies to produce libraries of these drugs with improved PK/PD profiles. Due to this convenient method for developing drugs, libraries of small molecules were developed with outstanding results in treating numerous ailments with no need for a drug delivery vector. This field drove what is now known as "large pharma". These were great accomplishments by the drug development community. However, over the past decades, these libraries of small molecules have shown limited to no effect in many serious diseases due to lack of specific/cell-selective delivery. Furthermore, many of these diseases have evolved to overcome drug treatments through positive selection and evolution. Therefore, life-threatening diseases are pushing the drug development industry to produce much more toxic/potent compounds that come with adverse off-target side effects. This challenge has brought the need to understand and research what drives the establishment and progression of these diseases. Through understanding the unique genetic drivers of these diseases we can better assess whether they can be selectively targeted through novel therapeutic approaches. Understanding each disease and its genetic phenotype results in promising unique/disease-specific targets that may help develop potential treatments. The distinctive characteristic factors of each disease have driven the macromolecule drug development field to establish peptides and/or nucleotides that hold specific activity towards a unique phenotype. These macromolecule therapies show high potency towards many serious diseases. However, they are limited due to their intrinsic chemical properties. The body contains many enzymes that work quickly to degrade these macromolecules as soon as they are administered. Additionally, the immune system is well established to process these foreign macromolecules before they reach their necessary site of action. These extracellular barriers are just a few that need to be overcome for macromolecule treatments to reach the target cells. These challenges are what is now driving the field of drug delivery. Nucleic acid and peptide drugs need to overcome many barriers in the body that are not possible without a delivery vector.

The field of drug delivery has progressed in developing vectors capable of carrying macromolecules and shielding these macromolecules from the numerous barriers present *in vivo*. The knowledge gained over decades has allowed many researchers to create and improve lipid vectors for the delivery of nucleotides and peptides. Therapeutic nucleotide–lipid formulations have shown great promise as a therapeutic agent against numerous ailments including genetic disorders, chronic and acute diseases, as well as many cancers. An increased understanding of the field has catalysed efficiency to new levels. Delivery of plasmid

DNA (pDNA), messenger RNA (mRNA), oligonucleotide (miRNA and siRNA), and peptides into cells can be well characterized and has yielded promising results in preclinical and clinical trials. Lipid vectors exhibit many desired attributes including cell specificity through the addition of targeting ligands, minimal immunogenicity compared with viral vectors, as well as sufficient cargo release into the cytoplasm of the targeted cell population through endosomal escape mechanisms. Improved safety profile of these lipid vectors over current viral vectors allows many options and techniques to be used to overcome the barriers mentioned below. Current lipid vectors have shown a robust capability to condense and deliver various nucleic acid molecules ranging in size from several nucleotides (siRNA) to several thousand nucleotides (pDNA). However, even with these strides, the field of lipid vectors has many areas that need to be improved, particularly in endosome escape, nucleotide release in cytoplasm, nuclear uptake, nuclear release (pDNA), and expression which are all lagging behind viral vector capabilities. The physiological barriers, strategies to overcome these barriers, as well as promising clinical trials will be discussed in further detail throughout this chapter.

12.2 RATIONAL DESIGN OF LIPID VECTORS TO OVERCOME EXTRACELLULAR AND INTRACELLULAR CHALLENGES

An understanding of the cargo, the necessary site of action, and the cells barriers to prevent delivery of the macromolecule therapy are needed to rationally design an efficient lipid vector delivery system. Therefore, the physiochemical and biological properties of DNA, RNA, and peptide therapies need to be addressed. To simplify this discussion, this chapter will mainly discuss the delivery of RNA (siRNA, miRNA, and mRNA) and DNA. The main route of administration to treat a disease that has disseminated throughout the body is through intravenous injection (i.v.). The administration of RNA or DNA via i.v. allows the immediate introduction of the nucleic acids into the blood circulation. However, the administration of nucleic acids into direct circulation presents many extra and intracellular barriers that can drastically hamper the therapeutic effects. In the extracellular physiological environment, RNA and DNA are radically destabilized through enzymatic degradation by endogenous nucleases and cleared by the reticuloendothelial system (RES). Subsequently, RNA and DNA have an anionic hydrophilic sugar–phosphate backbone in which these physiochemical properties are unfavourable for uptake by cells, primarily due to the anionic charge at the surface of the cell membrane. Immunogenicity is another concern after systemic administration of nucleic acids, in which the immune response can drastically decrease the potency of the delivered gene as well as cause autoimmune toxicities. Virus-derived vectors for gene therapy are efficient in gene delivery and transfer; however, some safety issues that have occurred in the past have limited the use of viral vectors in gene therapy. Therefore, the rational design of nonviral lipid vectors focuses on overcoming and protecting the therapeutic cargo from these extracellular barriers. To design promising vectors for RNA and DNA therapies, we must first understand the therapeutic cargo and optimal sites for delivery. The primary difference between these two cargos is the final desired target location, DNA (nuclear) versus RNA (cytoplasm).

Firstly, we will discuss RNA, in particular, oligo-RNA (siRNA or miRNA). Short-interfering RNA (siRNA) and micro-RNA (miRNA) have gained much interest in the treatment of diseases over the past decade. The discovery of interfering RNAs' ability to knock down a specific protein's level via interfering with RNA translation has considerable clinical implications. RNA interference is an endogenous process in eukaryotic cells in which short-oligo RNA molecules catalyse the degradation of specific, complementary messenger RNA (mRNA) sequences. The mechanism of action entails the complexation of the fragments of siRNA with the RNA-induced gene silencing complex (RISC) (Rand et al., 2004). The degradation of the sense strand (passenger) via the Argonaute 2 (Ago2) protein allows the single antisense (guide) to activate the RISC complex in order to pursue and bind the complementary mRNA to the antisense strand of the siRNA (Ameres et al., 2007). The guide strand is designed based on the thermodynamic properties of the siRNA's 5′ end, in which R2D2 binds to the more stable 5′ end, allowing Dicer to bind to the less stable end. After binding to mRNA, Ago2 mediates the cleavage of the mRNA between nucleotides 10 and 11 from the 5′ end of the antisense strand (Tomari, 2005). This process is catalytic in nature allowing the RISC complex to pursue and degrade additional mRNA. This catalytic process leads to outstanding knockdown of the target gene (Hutvagner, 2002). The ability of siRNA to specifically block the synthesis of any protein responsible for any disease allows for incredible potential in modern medicine. However, the discovery of this potent and site-specific macromolecule has come up short of its potential in clinical settings. The physiochemical and biological barriers that have stunted siRNA's clinical success include the nucleases present *in vivo*, the lack of specificity to target tissues, and the intracellular delivery (endosome escape) to the target cell's cytoplasm where the RISC complex and mRNA are located.

Therefore, strategies have been deployed to overcome these biological barriers. The lipid nanoparticles are one such strategy leading the way in delivering therapeutic siRNA for systemic applications. The goal at present in RNA delivery is to create a lipid vector capable of protecting cargo from extracellular enzyme degradation and RES uptake, while facilitating active targeting, uptake, and ultimately endosomal release into the target cell cytoplasm. This has been accomplished by many research groups working with lipid vectors. One such research group demonstrated high *in vivo* liver (hepatocyte) silencing by incorporating ionizable cationic lipids into a lipid vector–siRNA system. At doses as low as 5.0 μg siRNA kg^{-1} body weight following intravenous injection in mice showed therapeutic levels of silencing (Jayaraman et al., 2012). These systems demonstrate low to no toxic side effects with promising clinical applications. Throughout this chapter, we will discuss the intelligent design of these lipid vectors and how to overcome the physicochemical and biological barriers and actively deliver the desired RNA cargo. We will also discuss promising clinical applications in the delivery of not only therapeutic siRNA cargo but also numerous other nucleotide macromolecules such as mRNA and miRNA that also need cytoplasm delivery.

As mentioned earlier, the main difference between the delivery of RNA and DNA is the final cellular destination needed for the therapeutic cargo to work. Two approaches for gain-of-function therapies are either mRNA or pDNA. The use of mRNA is a popular option that can yield gain-of-function through cytoplasm delivery and has been a successful approach for vaccine development in the case of combatting coronavirus; however, since pDNA does have the challenge of nuclear delivery, we will focus on pDNA delivery rather than mRNA. Once exogenous genes (pDNA) enter the human biological system, they are recognized by the RES as foreign pathogens and cleared quickly from blood circulation. The RES and nucleases drastically reduce the therapy from reaching the target cells. It has been reported that the half-life of naked DNA in the bloodstream is no more than several minutes. Upon injection, DNA is rapidly degraded by enzymes and eliminated from the plasma due to extensive uptake by the liver (Kawabata et al., 1995). Therefore, to overcome this rapid degradation *in vivo* and avoid the RES uptake and enzyme degradation, many chemical modification and delivery methods have been established to protect and stabilize nucleic acids. These modifications ultimately increase the stability and efficacy of the RNA/DNA therapy. Throughout this chapter, we will discuss the strategies used to improve the stability and prolong the circulation of lipid delivery vectors.

One such strategy is the incorporation of helper lipids into the lipid membrane. An example for improving the stability of the lipid membrane is the incorporation of cholesterol. Cholesterol reduces the mobility of the phospholipids and increases the packing density (Semple et al., 1996). Another additive that can aid in prolonging the circulation of the nanoparticle by avoiding RES uptake is through the incorporation of an outer layer coating of polyethylene glycol (PEG). This method is the typical strategy used in the majority of lipid nanoparticle formulations to avoid the immune system and escape RES uptake (Jokerst et al., 2011). PEGylation has been widely used to stabilize lipid nanoparticles and their payloads through physical, chemical, and biological mechanisms. The use of lipid–PEG conjugates, such as DSPE-PEG, allows the lipid portion to insert into the lipid membrane positioning the amphiphilic PEG to remain on the outside as a shield against RES uptake. The density and loading of PEG onto the surface of these lipid nanoparticles play a crucial role in improving the circulation time and avoiding the RES uptake. Typically, it has been found that loading of low PEG densities (usually <5 mol %) will result in PEG assembly into a condensed, short, mushroom conformation. As the PEG is increased, the PEG will reassemble into an extended brush conformation in which the PEG corona yields a densely packed shield around the lipid membrane (Guo and Huang, 2011). It has been shown that increased PEGylation leads to a significant increase in the circulation half-life of liposomes (Li et al., 2010). However, due to the detergent-like property of many lipid–PEG conjugates, the lipid membrane may be unstable with higher loading. Li and Huang discovered that through incorporation of a core-like structure, such as liposome–polycation–pDNA (LPD), nanoparticles could be stable at higher DSPE–PEG loading (Li et al., 2010). The LPD nanoparticle is stabilized by electrostatic interactions between the negatively charged nucleic acid–protamine complex core and positively charged lipid bilayer. This core–surface type of liposome was able to support the bilayer and tolerate a high level of PEG–DSPE (10 mol %) with a relatively dense PEG brush structure on the surface. Most importantly, these liposomes were not taken up by the liver Kupffer cells (Li et al., 2009). Recently, the asymmetrical lipid bilayer core structure of the lipid calcium phosphate (LCP) nanoparticles was studied to determine the effects of PEG density (Liu et al., 2014). It was found that delivery to hepatocytes was dependent on both the concentration of PEG and the surface lipids. It was found that LCP vectors could discriminate uptake from hepatocytes (20 mol % PEG) to Kupffer cells (5 mol % PEG) by decreasing PEG concentration on the particle surface. It was also found that the use of positively charged lipid 1,2-dioleoyl-3-trimethylammonium-propane (DOTAP) exhibited higher accumulation in the hepatocytes than LCP vectors with neutral lipid dioleoylphosphatidylcholine (Liu et al., 2014).

The extracellular barriers are just a few challenges that are being overcome through the use of lipid nanoparticle carriers. The intracellular delivery of RNA (cytoplasm) or DNA (nucleus) has been the most challenging barriers to date. It has been reported that typically >95% of cells in culture internalize nanoparticle vectors; however, only a small fraction, typically <50%, express the transgene (Mark, 2003). Internalization of the gene delivery vectors and release of the cargo at the desired target site are challenges that must be overcome to elicit a desired level of gene expression or silencing. The intracellular barriers that must be overcome include endosomal escape before lysosomal degradation, nucleic acid unpacking from vectors, and release in the cytoplasm (RNA), translocation across the nuclear membrane (for DNA) and release in the nucleus (DNA). The intracellular barriers are clearly shown in Figure 12.1, in which a cartoon illustrates the intelligent design of the previously mentioned LCP nanoparticle to overcome these barriers and deliver DNA into the nucleus of the cell. Delivery and active targeting strategies to reach a desired cell type have predominately been through the

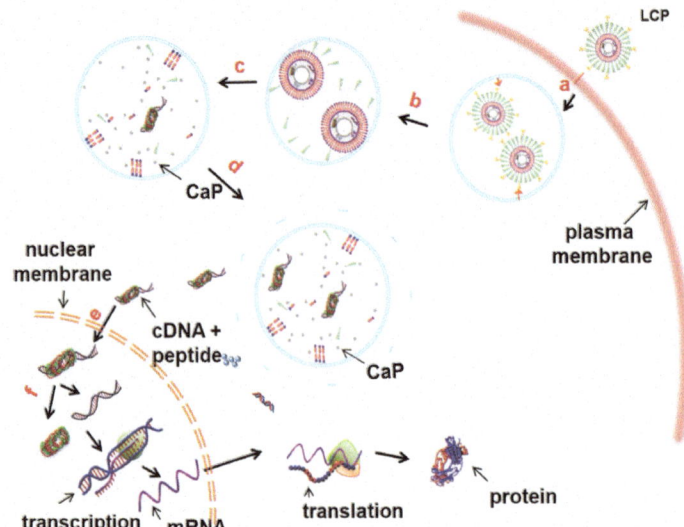

FIGURE 12.1 Proposed mechanism for intracellular delivery of DNA by LCP. Step-wise scheme for nonviral acid-sensitive vector (lipid calcium phosphate, or LCP), in which the vector is internalized through receptor mediated endocytosis, destabilized as endosome's pH decreases and releases the DNA–peptide complex into the cytoplasm. The DNA–peptide complex enters the nucleus through the nuclear pore, and the DNA–peptide complex dissociates and releases free DNA, which is transcribed to mRNA, migrates to the cytoplasm to be translated, and results in desired protein synthesis (Hu et al., 2013). (The original figure was prepared by Bethany DiPrete.)

incorporation of targeting moieties such as galactose, RGD, folate, anisamide, or other ligands that bind to overexpressed receptors on the desired cell surface. Through the incorporation of these targeting ligands, the vector is taken into the cell through receptor-mediated endocytosis. Escape from endosomal compartments is thought to represent a major obstacle. The incorporation of cationic lipids such as DOTAP is believed to form ion pairs with anionic lipids within the endosome membrane. This pairing leads to the disruption of the endosomal membrane via formation of a hexagonal (H_{II}) phase, which allows nucleic acid–cationic lipid complexes to be released into the cells cytoplasm (Hafez et al., 2001). Significant intracellular hurdles beyond endosomal escape include limited nuclear entry (Dean et al., 2005) and inefficient intranuclear release of plasmid for transcription (Hama et al., 2006). These two barriers only pertain to the delivery of pDNA in which nuclear delivery and release are vital to gene expression. Therefore, the development of tunable cationic peptides used in condensing the large DNA molecule for lipid nanoparticle packaging aids in facilitating nuclear import and release. These peptides have been crucial and need to be investigated more thoroughly to improve the ability of DNA to be delivered/released in the nucleus. As shown in Table 12.1, the incorporation of a monocyclic CR8C peptide into LCP nanoparticles resulted in a significant improvement in liver luciferase expression compared with other nanoparticle platforms (Hu et al., 2013). It was found that the affinity of monocyclic CR8C to pDNA is less than that of linear CR8C. Therefore, it is hypothesized that the mc-CR8C increased the amount of pDNA released in the nucleus, resulting in higher expression. However, the

TABLE 12.1
Comparison of Improved Hepatic Luciferase Gene Expression in Various Nonviral Gene Delivery Vectors (Hu et al., 2013)

Nonviral Vector	Dose (mg DNA kg^{-1})	Luc Expression (RLU/mg protein)
Poly(amine-co-ester)	0.5 (i.t.)a	1.5×10^5
Bifunctional dendrimer	2.5 (i.v.)	7.5×10^5
Ethyl-alkylated PEI	2.5 (i.v.)	1.0×10^6
R8-GALA-MEND	2.5 (i.v.)	1.3×10^6
LCP(mc-CR8C)Gal	0.3 (i.v.)	4.6×10^7
Hydrodynamic injection	0.3 (i.v.)	4.8×10^9

[a] Intratumoral tissue injection.

ultimate standard, hydrodynamic injection of DNA for liver expression is still 100-fold higher than that of LCP. Therefore, there is much more room for improving these lipid gene delivery vectors.

12.3 CURRENT LIPID VECTORS FOR DELIVERY OF DNA, RNA, AND PEPTIDE

Over the past several decades, numerous methods have been reported to formulate nucleic acids into lipid nanoparticles. These methods can be divided into three major lipid categories based on the physicochemical properties of the lipid used in the formulation. These categories include cationic, neutral, or ionizable lipids (Table 12.2). The most prevalent

TABLE 12.2
Representative Formulations and Characterizations of Lipid Vectors for Delivery of Nucleic Acids *In Vitro* or *In Vivo*

Preparation Procedure	Lipid Composition	Trapping Efficiency	Particle Size	*In Vivo* PK or *In Vitro* Serum Stability	*In Vitro* or *In Vivo* Delivery Efficiency	Ref.
Bulk mixing	DOTMA/DOPE (Lipofectin™)	N.D.	N.D.	N.D.	ED50 < 30 nm for inhibition of intracellular adhesion molecule (ICAM-1) expression in HUVEC cells	Bennett et al. (1992)
Lipid film hydration	DDAB:PC: Chol (5:16:8)	>90%	Heterogeneous size distribution (200 nm–10 μm)	Two-compartment PK model $t1/2\alpha = 24.5$ min $t1/2\beta = 11.36$ hours	Inhibition of Raf-1 (52%) expression in SQ-20B cells dosed at 10 μM ON; inhibition of Raf-1 in liver (51%), kidney (42%) and variable levels in SQ-20B xenograft (37% 57%) after 5 daily iv injections (6 mg kg^{-1})	Gokhale et al. (1997)
Bulk mixing	DE: DOPE: Chol (2:1:1)	N.D.	<200 nm	Aggregation in the presence of 10% serum	Reducing NF-κB/DNA binding activity by 58% in RAW264.7 macrophages after LPS stimulation dosed at 400 nM	De Rosa et al. (2008)
Bulk mixing of lipidoid NP and NA in 35% ethanol (pH 5.2) followed by buffer exchange/dialysis	98 N12–5:Chol:C16 Ceramide-PEG (42:48:10)	>95%	50 nm	Stable in serum	Decreased miR-122 level in the liver after three consecutive injections of anti-miR122 dosed at 5 mg kg^{-1}	Akinc et al. (2008)
Lyophilization-rehydration	DOPC	>95%	100 nm	Two-compartment PK model $t1/2\alpha = 8.1$ minutes $t1/2\beta = 3.9$ hours	Inhibition of Bcl-2 expression by 44% in lymphoma cells. No *in vivo* knockdown study has been performed	Tormo et al. (1998), Gutierrez Puente (1999)
Minimal volume entrapment	EPC:Chol: DSPE–PEG (2:1:0.1)	20%	200 nm	Stable in the presence of 10% serum at 37°C	N.D.	Stuart et al. (2000)
Reverse-phase evaporation	DSPC:Chol:DOTAP:DSPE–PGE	80%–90%	70–120 nm	T1/2 = 4h	Inhibition of c-myb expression by 70% in neuroblastoma cells dosed at 100 μg ml^{-1} [42]. Reduction of c-myc expression in melanoma xenograft when dosed at 2.5 mg kg^{-1}	Stuart et al. (2000), Pagnan et al. (2000), Pastorino et al. (2003), Brignole et al. (2004)
Reverse-phase evaporation	DOPE:OA: Chol (10:5:2)	10%	170 nm	N.D.	ED50 = 40 nM against Friend retrovirus infection in Dunni cells	Ropert et al. (1992)
Reverse-phase evaporation with detergent dialysis	DOPE:OA: Chol (2:1:2)	10%–12%	N.D.	N.D.	Enhanced uptake of ON delivered by streptavidin–biotin-coupled immunoliposomes by twofold compared with ordinary liposomes	Ma (1996)
Nanoparticles step-wise bulk mixing	DOTAP/Chol/DSPE-PGE/ Protamine	>90%	64 nm	Stable in Serum	Downregulation of surviving expression in H1299 cells by 87% when dosed at 1 uM	Li et al. (2006)
Step-wise bulk mixing	EPC/DC-Chol/DSPE–PEG/ protamine	90%	90 nm	Stable in Serum	Inhibition of Bcl-2 expression in transferrin receptor expressing leukaemia cell lines MV4–11 (41%), K562 (62%) and Raji (50%)	Yang et al. (2009)

DDAB, Dimethyldioctadecylammonium bromide; DE, (2,3-didodecyloxypropyl)(2-hydroxyethyl) dimethylammonium bromide; OA, oleic acid; EPC, egg-PC. Adapted from Wang *et al.* (2015).

of these lipids used in nucleotide delivery is the cationic lipids. These lipids are frequently used for the delivery of nucleotides due to their multiple cationic head groups. The cationic head group allows for endosomal escape through formation/complexation with the endosome membrane via H_{II} phase. Cationic lipids also allow for improved packaging of the anionic cargo. These cationic lipids naturally interact with polyanionic nucleic acids and form what is known as lipoplexes. The ability to incorporate different cationic lipids to improve and optimize endosome escape and nucleic acid packing is beneficial in formulations. A varying number of cationic lipids and hybrids can be incorporated in order to optimize a formulation for a specific treatment. Some of these include monovalent lipids such as N(1-(2,3-dioleyloxy) propyl)-N,N,N-trimethylammonium-chloride (DOTMA) (Felgner et al., 1987) and 1,2-dioleyl-3-trimethylammonium-propane (DOTAP) (Alexander and Akhurst, 1995), multivalent lipids such as dioctadecylamidoglycylspermine (DOGS) (Remy et al., 1995), and cationic lipid derivatives such as 3β-(N-(N',N'-dimethylaminoethane)-carbamoyl) cholesterol (DC-Chol) (Gao and Huang, 1991). The varying hydrophobic chains and degree of saturation of the lipids also provide tunable lipid vectors with different characteristics. Through decreased chain length and varying degrees of unsaturation, the lipid membrane increases in fluidity, allowing for improved ion pair formation and membrane fusion. It has been demonstrated that the myristoyl (C14) chain is optimal for transfection compared with C16 and C18 chains (Felgner et al., 1994). The unsaturated alkyl chains display considerably higher lipid fluidity often leading to a higher transfection efficiency compared with saturated alkyl chain lipids (Yuba et al., 2012).

As mentioned above, the cationic lipid vectors have many advantages in nucleic acid encapsulation and delivery compared to other lipid constructs. However, a major drawback with cationic lipid vectors is the low half-life in circulation due to high RES uptake. The use of PEGylation is one such remedy to help avoid RES uptake. However, an alternative approach to prevent cationic liposome-related issues such as low circulation half-life and cytotoxicity is the use of neutral liposomes as carriers (Stuart et al., 2000). Commonly used neutral lipids are phosphatidylcholine (PC), phosphatidylethanolamine (PE), and cholesterol. Neutral lipids have been demonstrated as effective helper lipids when incorporated in cationic formulations to achieve higher transfection efficiency. Neutral lipids have also been investigated as independent carriers for the delivery of nucleic acids. In contrast to cationic lipids, neutral lipids are void of the positive charges that promote attractive interaction with nucleotide phosphate backbone to efficiently encapsulate into the lipid vector. Therefore, this physiochemical difference makes it more challenging to achieve high loading efficiency. However, this does allow neutral lipid vectors to reduce their interactions with serum proteins (opsonization), which increases the circulation half-life compared with cationic lipids. Another barrier that must be overcome is endosomal escape. It is hypothesized that the endocytic pathway is the primary pathway for the lipid vectors to enter the cells. It was demonstrated that the acidic endosomal environment is a critical factor that causes the leakage of liposomal contents into the cytoplasm (Straubinger et al., 1983). It was found that the higher-molecular-weight, charged molecules escape the endosome at a slower rate. Therefore, a zwitterionic lipid such as DOPE is often included in the neutral liposomal formulation to enhance the transfection efficiency. This is believed to be possible via the acidic endosomal compartment inducing DOPE to transform to an inverted hexagonal H_{II} phase which more readily fuses with the anionic lipid layer and destabilizes the endosome membranes (Koltover et al., 1998).

Neutral lipids as nucleic acid carriers have shown great promise in extending serum clearance time. However, as mentioned above, the delivery efficiency is often subpar compared with that of cationic lipids. Therefore, a more efficient lipid vector is one that consists of tunable characteristics in different physiological environments. There are many different physiological environments that these lipid vectors experience *in vivo*. One such environment feature that can be taken advantage of is pH. Although the pH found in the majority of physiological environments that the lipid vector would experience is around a pH of 7.4, the endosome approaches pH around 5.5 as it progresses towards the lysosome. Therefore, the pKa of the head group, and the lipid's ability to induce hexagonal H_{II} phase structure with membrane lipids are crucial. Designing lipid vectors with a pKa <7.0 can maintain a cationic charge during nucleic acid loading in acid buffer, subsequently losing the cationic charge throughout the lifetime of the physiological circulation and finally gaining the cationic charge again once it is introduced to the acidic environment found in the endosomal compartment. This design would allow for optimal encapsulation, increased half-life in circulation, and endosomal escape through the formation of ion pairs between the endosomal membrane and the carrier lipid. Many researchers are looking to design lipid vectors with these tunable properties. One such group uses a structure–activity relationship (SAR) as the guideline to direct rational lipid design. The design was based on the hypothesis that ionizable cationic lipids disrupt the endosome through ion pair formation with anionic lipids in the endosomal membrane. Therefore, a lipid head group with pKa <7.0 should achieve better encapsulation and endosome escape at acidic pH, while avoiding opsonization and RES uptake by maintaining a neutral surface charge at physiological pH (Semple at al., 2010). A series of derivatives were developed based on the ionizable cationic lipid 1,2-dilinoleyloxy-3-dimethylaminopropane (DLinDMA) which is a highly effective lipid at delivering siRNA in rodents and nonhuman primates (Zimmermann et al., 2006; Akinc et al., 2009). The derivatives were synthesized by varying the head group of DLinDMA, the hydrocarbon chains, and the linker. Through SAR experiments, it was determined that a linoleyl lipid containing two double bonds in the hydrocarbon chain was the most efficient. Subsequently, it was determined that alkoxy-containing lipids showed higher activity than ester-containing lipids in terms of chemical and enzymatic stability. This is probably due to the fast hydrolysis rate *in vivo* of the ester-containing lipids. Through investigating the size, acid dissociation constant, and number of ionizable groups, the dimethylamino groups in the DLinDMA showed increased activity compared with piperazino, morpholino, trimethylamine, and bis-dimethylamino groups. In subsequent studies, Jayaraman et al. maintained the unsaturated dilinoleyl chain and modified the head groups to investigate the structure–activity correlation (Jayaraman et al., 2012). The DLin-MC3-DMA lipid with an amino head group pKa of 6.44 yielded the highest delivery. This pKa allowed the lipid nanoparticles to display minimal charges during blood circulation (pH 7.4) and became cationic in the acidic endosome (pH 5.5). These lipid nanoparticles demonstrated rapid elimination rates in circulation and tissue while maintaining high delivery efficiency.

Several researchers use an empirical approach to develop and study optimal characteristics necessary for the delivery of nucleotide therapies. One such group synthesized lipids based on one-step conjugation addition of alkyl-acrylates or alkyl-acrylamides to primary or secondary amines (Akinc et al., 2008). Through their analysis, they concluded that the optimal lipids for transfection maintain an amide linkage, at least two alkyl tails, an 8–12 carbon chain tail, and one tail that is not substituted by amine reactants and contains one secondary amine. A similar one-step synthesis approach was adopted by Love et al. This group established an epoxide-derived lipidoid library using a rapid ring opening reaction between amine substrates and epoxide (Love et al., 2010). The resulting compounds are amine-containing alcohols with nonpolar hydrocarbon tails. They produced a potency two magnitudes higher than the top compound screened by Akinc et al. in regards to hepatocyte-targeted siRNA delivery efficiency. These formulated nanoparticles were able to knock down Factor VII expression by 50% in the liver at a dose of 10.0 $\mu g\ kg^{-1}$ via tail vein injection. The knockdown duration lasted up to 20 days when dosed at 0.1 mg kg^{-1}. This, C12–200, nanoparticle was shown to avoid the endosomal pathway and enter the cells via micropinocytosis. Through micropinocytosis, this nanoparticle system avoided the lysosomal degradation of its cargo (Love et al., 2010). Furthermore, another group established a compound library by reacting amino acids or lysine-based dipeptides with aldehydes, acrylates, or epoxides (Dong et al., 2014). The lead compound, cKK-E12, was found to silence the Factor VII in mouse livers with an ED50 of approximately 2.0 $\mu g\ kg^{-1}$, a fivefold increase compared with C12–200. Moreover, this compound showed a 500-fold increase in specificity for gene silencing in hepatocytes compared with endothelial cells or immune cells (Dong et al., 2014). The development of these three libraries has demonstrated improved efficacy and specificity for the siRNA delivery system through taking an empirical approach. This strategy of applying simple synthesis to develop and screen large amounts of lipids can be used to develop novel materials with high specificity and potency for the delivery of all nucleotide and peptide therapies.

The final and most recent lipid vector nanotechnology to be discussed can be classified as a hybrid between liposomes and other nanoformulation technologies. These novel formulations take advantage of outer-lipid properties to help in evading RES, enzymatic degradation, and target selectivity. Additionally, incorporating core nanoformulations with properties optimized for release and trafficking of the cargo to the desired cellular location. These multicomponent strategies are referred to as composite nanoparticles. Lipid composite nanoparticles are some of the most investigated formulations due to the versatility of the lipids used. The formulation of nucleic acids into a core-membrane-structured lipid nanoparticle has shown much promise in recent years. These nanoparticle formulations are achieved through a multiple-step, self-assembly procedure. Firstly, the anionic charge of nucleic acids is complexed with the cationic polypeptides (protamine) with excess anionic charge remaining to form a negatively charged polyplex. The anionic charge of this polyplex allows for a cationic lipid surface coating. The lipid nanoparticles were then further stabilized by postinsertion of DSPE-PEG. The core structure supports the lipid coating of the cationic liposome, allowing up to 20 mol % of the DSPE-PEG to be inserted into the membrane. Although the PEGylated LPD showed compromised cellular uptake compared with non-PEGylated cationic LPD, the presence of a targeting ligand could increase the active uptake efficiency and selectivity of the system (Li et al., 2006).

Another composite nanoparticle discussed earlier and shown in Figure 12.1 takes advantage of an acid-sensitive calcium phosphate core. The strategy used in this formulation consists of condensing pDNA with a cationic peptide, such as monocyclic-CR8C. The peptide used in this technology needs to efficiently condense the DNA to accomplish high encapsulation efficiency, while also allowing release of the pDNA once it has been imported into the nucleus. Premature release before nuclear entry or high binding affinity with low nuclear release will radically reduce expression levels. This condensed peptide/pDNA complex is entrapped in a calcium phosphate core in which the phosphate head group of DOPA is inserted to cover the core through ionic interactions. Cationic, neutral, or ionizable lipids along with helper lipids such as cholesterol and DSPE-PEG are inserted to make an asymmetrical bilayer lipid membrane. These lipid vectors are about 40–60 nm in size containing the loaded pDNA/peptide complex. Hu et al. formulated the LCP nanocore with approximately 40% DOTAP, 20% DSPE-PEG, 40% cholesterol (molar ratio), as well as galactose as a hepatocyte targeting ligand covalently bound to 10% of the PEG. This group achieved approximately 50% pDNA encapsulation efficiency and yielded 4.6×10^7 RLU/mg protein, luciferase expression in the liver, 24 hours post administration of 0.3 mg pDNA/kg via tail vein injection into mice (Hu et al., 2013). This is one of the most promising nonviral vectors for liver transfection (Table 12.1). The use of linear CR8C was also investigated for its ability to condense, traffic, and release pDNA into the nucleus. The linear CR8C was found to have a higher binding affinity to pDNA. Therefore, slower pDNA release in the nucleus ultimately resulted in lower pDNA expression levels compared with the mc-CR8C peptide (Hu et al., 2013). Through the research accomplished over the past few decades, lipid vectors such as these mentioned above have improved drastically in their design and ability to deliver macromolecules to the desired cell type. Recently, numerous studies in Dr. Huang's lab have shown *in vivo* mouse proof of concept gene and peptide delivery for the treatment of several liver and pancreatic cancers via nanocore-based lipid vectors, LCP and LPD (Goodwin et al., 2016–2017, Miao et al., 2017). These lipid vectors may still have a long way before they can be compared with the viral vectors in expression efficiency, but that does not mean these lipid vectors cannot play a crucial role in the treatment of serious diseases in the near future.

12.4 APPLICATIONS FOR LIPID VECTORS IN THE FIELD OF DNA, RNA, AND PEPTIDE THERAPEUTICS

Since 2015, over 2400 gene therapy clinical trials have been conducted and/or are in progress. This number is substantially higher if RNAi therapy clinical trials are also included. The number of clinical trials is still increasing due to the promising opportunity to obtain gain-of-function in genetic diseases or to knockdown factors which are causing severe disorders such as cancers, viral infection (HBV), macular degeneration, and many others (Kanasty et al., 2013). The majority of cancer research has put focus on understanding which defective or missing genes increase an individual's risk of cancer. Over 60% of the gene therapy clinical trials conducted have been in the field of cancer (Giacca, 2010). Cancer nucleotide therapy has three main strategies to improve cancer treatments. Firstly, gene therapy can directly influence specific defective genes that drive cancer at the molecular level. Secondly, gene therapy can prevent cancer proliferation by training the immune system to identify and target these cancer cells. Lastly, the delivery of RNAi allows disruption/knockdown of chemotherapy resistant pathways increasing cancer sensitivity to current chemotherapy treatments. Lipid vector-based nucleotide therapies can directly eliminate the cancer cells, improve the efficacy of the immune systems to recognize and destroy cancer cells, or knock down resistant pathways allowing for increased chemotherapy sensitivity.

One of the driving characteristics of solid tumours that allows these nanoparticles to preferentially accumulate in tumours is the phenomena known as enhanced permeability and retention (EPR). This irregular morphology in the tumour is a result of the tumours rapid growth, in which blood and lymphatic vessels are formed with non-tight junctions causing leaky vessels and low interstitial pressure inside the tumour. These factors allow increased nanoparticle accumulation and retention. This EPR effect leads to higher accumulation of nanoparticles at the tumour site and subsequently decreased off-target effects (Maeda et al., 2012, 2013, Matsumura and Maeda, 1986). However, the severity and scale of EPR in tumours varies drastically between tumour types. This variation and inconsistency has hindered the development of effective drugs and delivery systems (Prabhakar et al., 2013). Vascular permeability is the key factor involved in the EPR effect in cancer. It is well accepted that vascular endothelial growth factor (VEGF) enhances the vascular permeability of tumour vessels. A recent study detailed the delivery of VEGF siRNA and gemcitabine monophosphate (GMP) via targeted LCP nanoparticle formulation in which approximately 40% of tumour cells became apoptotic, drastically reduced tumour cell proliferation, and significantly decreased the tumour microvessel density (Zhang et al., 2013). Recently, ALN-VSP, a human trial of RNAi therapy targeting VEGF and kinesin spindle protein in cancer patients, was performed using lipid vectors as a delivery system (Tabernero et al., 2013). Through tumour and liver biopsies, the researchers found traces of drug in the tumour, siRNA-mediated mRNA cleavage in the liver, downregulation of the targeted gene, and antitumor activity. To our knowledge, this was the first human trial to use lipid vectors to deliver siRNA as a cancer therapy, and as such presented a proof-of-concept that RNAi therapeutics/lipid vectors can show significant efficacy and is a safe option in humans. We will discuss the ALN-VSP formulation in more detail in Section 12.5.

Diseases residing in the liver are another application that lipid vector delivery systems are impacting including fatty liver, viral liver infections, as well as liver cancers. Currently, many liver diseases have no effective therapy and call for drastic measures of liver transplantation. These are obvious diseases in which lipid vectors can improve. Current transplantation is hindered by donor shortage, surgical risks, long-term immunosuppression, and cost. Therefore, safer and more efficient therapies are necessary. Through the development and studies of lipid vectors, it is well known that the liver is the predominate organ in which these carriers naturally accumulate. Therefore, lipid vectors are able to deliver nucleotide-based therapies more specifically to the liver with minimal toxicity and immunogenicity. However, it has been found that many lipid vectors are taken up in the Kupffer cells of the liver, which subsequently lowers expression due to the high degradation rates of the therapy in these cells. However, taking advantage of special membrane receptors located on liver cells (hepatocytes), lipid vectors can be modified with PEGylation and targeting moieties (galactose) in order to deliver the targeted therapy preferentially to the hepatocytes of the liver. Several attempts have shown potential success in liver disease. Several groups have demonstrated the ability to target certain cell or tissue populations through incorporation of targeting moieties. These groups were able to target such cell populations expressing high levels of collagen type VI receptor (Du et al., 2007), mannose-6-phosphate receptor (Adrian et al., 2007), and galactose receptor (Mandal et al., 2007). One research group attempted to treat liver fibrosis/cirrhosis through delivering gp46 siRNA, the rat homolog of human heat shock protein 47, to hepatic stellate cells via a vitamin A-coupled liposome. In their study, five treatments with the collagen-specific liposomes drastically decreased and almost completely resolved the liver fibrosis and prolonged the survival of rats (Sato et al., 2008).

12.5 PAST AND PRESENT CLINICAL TRIALS

The progression in gene therapy since Friedmann et al. first proposed the use of genes for human genetic disease in 1972 has been at times incredible but has still fallen short of its once believed potential (Friedmann and Roblin 1972). In the past decades, many researchers have incorporated the use of lipid vectors in pursuit of delivering nucleic acid agents to target various types of diseases. Many of these projects have reached late stages of clinical trials. Recently, Allovectin 7, a locally administered lipid–gene formulation consisting of DMRIE-DOPE and pDNA, failed to meet its efficacy end points in a phase III clinical trial for treatment of advanced metastatic melanoma. Nonetheless, this proved that lipid vectors can be formulated to deliver pDNA while maintaining a promising safety profile. Therefore, unique and creative lipid formulations continue to be developed clinically, including DOTAP–cholesterol, GL67A–DOPE–DMPE–polyethylene glycol (PEG), and GAP–DMORIE–DPyPE. One such company, Alnylam Pharmaceutical, is developing lipid vectors carrying nucleic acids via enhanced stabilization via GalNAc-conjugated siRNA delivery technology as well as other combinations of ionizable cationic lipids. Three products from Alnylam Pharmaceuticals are currently in clinical trials. ALN-VSP, mentioned above, is categorized as a small nucleic acid lipid particle (SNALP), in which siRNA against vascular endothelial growth factor VEGF and kinesin spindle protein (KSP) is delivered to treat liver cancer (Tabernero et al., 2013). ALNTTRsc is targeted to TTR for treatment of transthyretin-mediated amyloidosis, and ALN-PCS02 is targeted to proprotein convertase subtilisin/kexin type 9 (PCSK9) to lower cholesterol for treatment of hypercholesterolemia. ALN-VSP's lipid membrane consists of a mixture of ionizable cationic lipid DLinDMA, a fusogenic lipid DSPC, cholesterol, and a low molar concentration of PEG-C-DMA. This formulation in which siRNA is encapsulated through destabilization of the lipids in 40% ethanol and incubation with the siRNA, yielding a nanoparticle under 100 nm in diameter, and has completed phase I trials with a promising safety profile. The second formulation mentioned, ALN-TTR02, known as Patisiran, recently released clinical data that resulted in the treatment/lipid vector achieving sustained serum TTR protein knockdown of 96% with a mean TTR knockdown of about 80%. This ALN-TTR02 formulation, using MC3 ionizable lipid, has recently passed the phase III clinical trial for the treatment of transthyretin (TTR) amyloidosis. Studies in nonhuman primates, resulted in a 75% decrease in serum TTR levels 7 days after a single i.v. dose of 0.1 mg siRNA kg^{-1} body weight (Coelho et al., 2013). Animals receiving a single 0.3 mg kg^{-1} dose of siRNA yielded over a 70% suppression in serum TTR levels for 28 days, post administration. Similar results were found in human studies in which an average of 50% knockdown of TTR levels by day 3 was observed after a single 0.3 mg kg^{-1} dose of siRNA and over a 50% reduction in protein levels 28 days post administration was found (Coelho et al., 2013). Furthermore, the formulation was well tolerated, with no liver and kidney toxicity. The final formulation, ALN-PCS, is designed to specifically target proprotein convertase subtilisin/kexin type 9 (PCSK9) produced by liver. This formulation was studied for the treatment of hypercholesterolemia. Knockdown of PCSK9 in nonhuman primates by the siRNA formulation resulted in lower serum LDL cholesterol levels and increased expression of LDLR in the liver (Frank-Kamenetsky et al., 2008). A recent study on 32 participants in a phase I trial demonstrated that a single dose of ALN-PCS (0.4 mg siRNA kg^{-1} body weight) led to a 70% decrease in serum PCSK9 levels 3 days post administration (Fitzgerald et al., 2014). Furthermore, this treatment resulted in a 40% reduction in serum LDL cholesterol levels over a month post administration. ALN-PCS was well tolerated with similar safety profiles as mentioned for the other formulations.

Recent advances in lipid-based formulation technologies have led to numerous siRNA lipid formulations in various stages of clinical trials. The SNALP technology, patented by Tekmira Pharmaceuticals which has become Arbutus Therapeutics, and sublicensed to Alnylam to develop the formulations mentioned above, is one of the most widely used lipid-based nucleic acid delivery approaches for systemic administration in clinical trials. Lipid vesicles encapsulating nucleic acids are formed instantaneously by mixing lipids dissolved in ethanol with an aqueous solution of nucleic acids in a controlled, stepwise manner. Using this method, SNALP encapsulates nucleic acids with high efficiency (95%) in uniform LNPs, which are effective in delivering gene therapeutics. However, following failure of a few of Tekmira's lead product candidates, such as TKMPLK1, targets polo-like kinase 1 (PLK1), a protein involved in tumour cell proliferation and TKM-Ebola, a formulation for the treatment of infection from the Zaire strain of Ebola virus (ZEBOV), Tekmira was forced to close and change into Arbutus Therapeutics. Although Tekmira's clinical trials were not as successful as Alynlam's trial, there are many unique and creative lipid formulations being used to improve and further the field of nucleic acid delivery. These current formulations show that within the near future, we will have unique and creative lipid vectors being used to deliver siRNA, mRNA, or pDNA, not only in clinical trials, but reaching the market to truly help treat patients with these severe diseases.

12.6 CONCLUSION

In the past decades, substantial advances have been achieved in lipid vector formulations for nucleic-acid-based therapy, including the development of promising ionizable lipids, composite nanocore-lipid vectors, and improved potency and stability of nucleic acids. Furthermore, the advances in genomics have allowed researchers to understand the genetic makeup of many diseases and provided a range of new targets for genetic medicine. The ability for many different nucleic acid cargos ranging from siRNA to pDNA to be incorporated into lipid vectors opens the doors for numerous therapeutic applications. However, these vectors will need further investigation to truly understand the physicochemical relationship these vectors have in vivo, as well as to understand the potential and limitations these vectors have in targeting certain tissues and organs. The recent development of nanoparticle with unique and creative formulations allows in vivo delivery of nucleotide therapies to achieve high efficiency, low toxicity, and promising efficacy. Recent progress in the clinical development of these nucleic acid formulations has generated considerable excitement, especially in the field of RNAi delivery. However, even with the advances and promising results, the ability to deliver and obtain gain-of-function through mRNA or DNA expression is still needing creative thoughts and ideas. The delivery and release of these pDNA therapies into the nucleus bring about many challenges. Unique and intelligently designed condensing peptides are one such area that needs to be investigated further. The therapeutic potential of these technologies is so close to touching the lives of so many patients in need. The progression and impact of the lipid vector field will soon have a resounding impact on the lives of so many.

ACKNOWLEDGEMENT

The original work in this lab has been supported by NIH grants CA149363, CA149387, CA151652, and DK100664.

REFERENCES

Adrian, J. E., Poelstra, K., Scherphof, G. L., Meijer, D. K., van Loenen-Weemaes, A. M., Reker-Smit, C., et al. (2007). Effects of a new bioactive lipid-based drug carrier on cultured hepatic stellate cells and liver fibrosis in bile duct-ligated rats. *J. Pharmacol Experiment. Ther.* 321(2), 536–543. DOI: 10.1124/jpet.106.117945.

Akinc, A., Goldberg, M., Qin, J., Dorkin, J. R., Gamba-Vitalo, C., Maier, M., et al. (2009). Development of lipidoid-siRNA formulations for systemic delivery to the liver. *Mol. Ther.* 17(5), 872–879. DOI: 10.1038/mt.2009.36.

Akinc, A., Zumbuehl, A., Goldberg, M., Leshchiner, E. S., Busini, V., Hossain, N., et al. (2008). A combinatorial library of lipid-like materials for delivery of RNAi therapeutics. *Nat. Biotechnol.* 26(5), 561–569. DOI: 10.1038/Nbt1402.

Alexander, M. Y., Akhurst, R. J. (1995) Liposome-mediated gene transfer and expression via the skin, *Hum. Mol. Genet.* 4, 2279–2285.

Ameres, S. L. Martinez, J., Schroeder, R. (2007). Molecular basis for target RNA recognition and cleavage by human RISC. *Cell*, 130, 101–112.

Bennett, C. F., Chiang, M. Y., Chan, H, Shoemaker, J. E., Mirabelli, C. K. (1992) Cationic lipids enhance cellular uptake and activity of phosphorothioate antisense oligonucleotides, *Mol. Pharmacol.* 41, 1023–1033.

Brignole, C., Pastorino, F., Marimpietri, D., Pagnan, G., Pistorio, A., Allen, T. M., et al. (2004). Immune cell-mediated antitumor activities of GD2-targeted liposomal c- myb antisense oligonucleotides containing CpG motifs, *J. Natl. Cancer Inst.* 96, 1171–1180.

Coelho, A., Adams, D., Silva, L., Lozeron, P., Hawkins, M., Mant, T. et al. (2013). Safety and efficacy of RNAi therapy for transthyretin amyloidosis. *New England J. Med.* 369, 819–829.

Dean, D. A., Strong, D. D., Zimmer, W. E. (2005). Nuclear entry of nonviral vectors. *Gene Ther.* 12(11), 881–890. DOI: 10.1038/Sj.Gt.3302534.

De Rosa, G., De Stefano, D., Laguardia, V., Arpicco, S., Simeon, V., Carnuccio, R., et al. Novel cationic liposome formulation for the delivery of an oligonucleotide decoy to NF-kappaB into activated macrophages, *Eur. J. Pharm. Biopharm.: Off. J. Arbeitsgemeinschaft Pharm. Verfahrenstechnik eV* 70 (2008) 7–18.

Dong, Y., Love, K. T., Dorkin, J. R., Sirirungruang, S., Zhang, Y., Chen, D., et al. (2014). Lipopeptide nanoparticles for potent and selective siRNA delivery in rodents and nonhuman primates. *Proc. Nat. Acad. Sci. USA*, 111(11), 3955–3960. DOI: 10.1073/pnas.1322937111.

Du, S. L., Pan, H., Lu, W. Y., Wang, J., Wu, J., Wang, J. Y. (2007). Cyclic Arg-Gly- Asp peptide-labeled liposomes for targeting drug therapy of hepatic fibrosis in rats. *J. Pharmacol. Experiment. Ther.*, 322(2), 560–568. DOI: 10.1124/jpet.107.122481.

Felgner, J. H., Kumar, R., Sridhar, C. N. Wheeler, C. J., Tsai, Y. J., Border, R. et al. (1994). Enhanced gene delivery and mechanism studies with a novel series of cationic lipid formulations, *J. Biol. Chem.* 269, 2550–2561.

Felgner, P. L., Gadek, T. R., Holm, M., Roman, R., Chan, H. W., Wenz, M. et al. (1987) Lipofection: A highly efficient, lipid-mediated DNA-transfection procedure, *Proc. Natl. Acad. Sci. U. S. A.* 84, 7413–7417.

Fitzgerald, K., Frank-Kamenetsky, M., Shulga-Morskaya, S., Liebow, A., Bettencourt, B. R., Sutherland, J. E., et al. (2014). Effect of an RNA interference drug on the synthesis of proprotein convertase subtilisin/kexin type 9 (PCSK9) and the concentration of serum LDL cholesterol in healthy volunteers: A randomised, single-blind, placebo-controlled, phase 1 trial. *Lancet*, 383(9911), 60–68. DOI: 10.1016/S0140-6736(13)61914-5.

Frank-Kamenetsky, G., Grefhorst, A., Anderson, R., Racie, T.S., Bramlage, A., Akinc, A. et al. (2008) Therapeutic RNAi targeting PCSK9 acutely lowers plasma cholesterol in rodents and LDL cholesterol in nonhuman primates. *Proc. Nat. Acad. Sci. USA*, 105, 11915–11920.

Friedmann, T., Roblin, R. (1972). Gene therapy for human genetic disease? *Science* 175(4025), 949–955.

Gao, X. Huang, L. (1991). A novel cationic liposome reagent for efficient transfection of mammalian cells, *Biochem. Biophys. Res. Commun.* 179, 280–285.

Giacca, M. (2010). *Gene Therapy*. Springer, Italia.

Gokhale, P. C., Soldatenkov, V., Wang, F. H., Rahman, A., Dritschilo, A., Kasid, U. (1997). Antisense raf oligodeoxyribonucleotide is protected by liposomal encapsulation and inhibits Raf-1 protein expression *in vitro* and *in vivo*: Implication for gene therapy of radioresistant cancer, *Gene Ther.* 4, 1289–1299.

Goodwin, T. J., Huang, L. (2017, May 2). Investigation of phosphorylated adjuvants co-encapsulated with a model cancer peptide antigen for the treatment of colorectal cancer and liver metastasis. *Vaccine 35*(19), 2550–2557. DOI: 10.1016/j.vaccine.2017.03.067. Epub2017 Apr 3.

Goodwin, T. J., Shen, L., Hu, M., Li, J., Feng, R., Dorosheva, O., Liu, R., Huang, L. (2017, October). Liver specific gene immunotherapies resolve immune suppressive ectopic lymphoid structures of liver metastasis and prolong survival. *Biomaterials*, *141*, 260–271 DOI: 10.1016/j.biomaterials.2017.07.007.Epub 2017 Jul6.

Goodwin, T. J., Zhou, Y., Musetti, S. N., Liu, R., Huang, L. (2016, November 9). Local and transient gene expression primes the liver to resist cancer metastasis. *Sci. Trans. Med.* 8(364), 364ra153.

Guo, S., Huang, L. (2011). Nanoparticles escaping RES and endosome: Challenges for siRNA delivery for Cancer therapy. *J. Nanomater.* 12. DOI: 10.1155/2011/742895.

Gutierrez-Puente, Y., Tari, A. M., Stephens, C., Rosenblum, M., Guerra, R. T., Lopez-Berestein, G. (1999). Safety, pharmacokinetics, and tissue distribution of liposomal P-ethoxy antisense oligonucleotides targeted to Bcl-2, *J. Pharmacol. Exp. Ther.* 291, 865–869.

Hafez, I. M., Maurer, N., Cullis, P. R. (2001). On the mechanism whereby cationic lipids promote intracellular delivery of polynucleic acids. *Gene Ther.* 8(15), 1188–1196. DOI: 10.1038/sj.gt.3301506.

Hama, S., Akita, H., Ito, R., Mizuguchi, H., Hayakawa, T., Harashima, H. (2006). Quantitative comparison of intracellular trafficking and nuclear transcription between adenoviral and lipoplex systems. *Mol. Ther.* 13(4), 786–794. DOI: 10.1016/J.Ymthe.2005.10.007.

Hu, Y., Haynes, M. T., Wang, Y., Liu, F., Huang, L. (2013). A highly efficient synthetic vector: Nonhydrodynamic delivery of DNA to hepatocyte nuclei *in vivo*. *ACS Nano*, 7(6), 5376–5384. DOI: 10.1021/nn4012384.

Hutvágner, Z. (2002). A microRNA in a multiple-turnover RNAi enzyme complex. *Science*, 297, 2056–2060.

Jayaraman, M., Ansell, S. M., Mui, B. L., Tam, Y. K., Chen, J., Du, X., et al. (2012). Maximizing the potency of siRNA lipid nanoparticles for hepatic gene silencing *in vivo*. *Angewandte Chemie Inter. Ed.* 51(34), 8529–8533. DOI: 10.1002/anie.201203263.

Jokerst, J. V., Lobovkina, T., Zare, R. N., Gambhir, S. S. (2011). Nanoparticle PEGylation for imaging and therapy. *Nanomedicine (London)*, 6(4), 715–728. DOI: 10.2217/nnm.11.19.

Kanasty, R. Dorkin, J. R., Vegas, A., Anderson, D. (2013). Delivery materials for siRNA therapeutics. *Nat. Mater.* 12, 967–977.

Kawabata, K., Takakura, Y., Hashida, M. (1995). The fate of plasmid DNA after intravenous injection in mice: Involvement of scavenger receptors in its hepatic uptake. *Pharmac. Res.*, 12(6), 825–830.

Koltover, I. T. Salditt, J. O. Radler, C. R. Safinya. (1998). An inverted hexagonal phase of cationic liposome–DNA complexes related to DNA release and delivery, *Science 281*, 78–81.

Li, S. D., Huang, L. (2006). Targeted delivery of antisense oligodeoxynucleotide and small interference RNA into lung cancer cells, *Mol. Pharm. 3*, 579–588.

Li, S. D., Huang, L. (2009). Nanoparticles evading the reticuloendothelial system: Role of the supported bilayer. *Biochimica et Biophysica Acta*, *1788*(10), 2259–2266. DOI: 10.1016/j.bbamem.2009.06.022.

Li, S. D., Huang, L. (2010). Stealth nanoparticles: High density but sheddable PEG is a key for tumor targeting. *J. Control. Release*, 145(3), 178–181. DOI: 10.1016/j.jconrel.2010.03.016.

Liu, Y., Hu, Y., Huang, L. (2014). Influence of polyethylene glycol density and surface lipid on pharmacokinetics and biodistribution of lipid-calcium-phosphate nanoparticles. *Biomaterials*, *35*(9), 3027–3034. DOI: 10.1016/j.biomaterials.2013.12.022.

Love, K. T., Mahon, K. P., Levins, C. G., Whitehead, K. A., Querbes, W., Dorkin, J. R., et al. (2010). Lipid-like materials for low-dose, *in vivo* gene silencing. *Proc. Nat. Acad. Sci. USA*, 107(5), 1864–1869. DOI: 10.1073/pnas.0910603106.

Ma, D. D., Wei, A. Q., (1996). Enhanced delivery of synthetic oligonucleotides to human leukaemic cells by liposomes and immunoliposomes. *Leuk. Res.* 20, 925–930.

Maeda, H. (2012). Macromolecular therapeutics in cancer treatment: The EPR effect and beyond. *J. Control. Release*, 164(2), 138–144. DOI: 10.1016/j.jconrel.2012.04.038.

Maeda, H., Nakamura, H., Fang, J. (2013). The EPR effect for macromolecular drug delivery to solid tumors: Improvement of tumor uptake, lowering of systemic toxicity, and distinct tumor imaging *in vivo*. *Adv. Drug Delivery Rev.* 65(1), 71–79. DOI: 10.1016/j.addr.2012.10.002.

Mandal, A. K., Das, S., Basu, M. K., Chakrabarti, R. N., Das, N. (2007). Hepatoprotective activity of liposomal flavonoid against arsenite-induced liver fibrosis. *J. Pharmacol. Experiment. Ther.* 320(3), 994– 1001. DOI: 10.1124/jpet.106.114215.

Mark, H. F. (2003). Encyclopedia of polymer science and technology. 3rd ed. (Book), *Choice: Current Reviews for Academic Libraries*, *41*(2), 316.

Matsumura, Y., Maeda, H. (1986). A new concept for macromolecular therapeutics in cancer chemotherapy: Mechanism of tumoritropic accumulation of proteins and the antitumor agent smancs. *Cancer Res.* 46(12 Pt 1), 6387–6392.

Miao, L., Li, J., Liu, Q., Feng, R., Das, M., Lin, C. M., Goodwin, T. J., Dorosheva, O., Liu, R., Huang, L. (2017, September 26). Transient and local expression of chemokine and immune checkpoint traps to treat pancreatic cancer. *ACS Nano*, 11(9), 8690–8706. DOI:10.1021/acsnano.7b01786. Epub 2017 Aug 28.

Pagnan, G., Stuart, D. D., Pastorino, F., Raffaghello, L., Montaldo, P. G., Allen, T. M., et al. (2000). Delivery of c-myb antisense oligodeoxynucleotides to human neuroblastoma cells via disialoganglioside GD(2)-targeted immunoliposomes: Antitumor effects, *J. Natl. Cancer Inst.* 92, 253–261.

Pastorino, F., Brignole, C., Marimpietri, D., Pagnan, G., Morando, A., Ribatti, D., et al. (2003). Targeted liposomal c-myc antisense oligodeoxynucleotides induce apoptosis and inhibit tumor growth and metastases in human melanoma models, *Clin. Cancer Res.: Off. J. Am. Assoc. Cancer Res.* 9, 4595–4605.

Prabhakar, U., Maeda, H., Jain, R. K., Sevick-Muraca, E. M., Zamboni, W., Farokhzad, O. C., et al. (2013). Challenges and key considerations of the enhanced permeability and retention effect for nanomedicine drug delivery in oncology. *Cancer Res.* 73(8), 2412–2417. DOI: 10.1158/0008-5472.CAN-12-4561.

Rand, T. A., Ginalski, K., Grishin, N. V. Wang, X. (2004). Biochemical identification of Argonaute 2 as the sole protein required for RNA-induced silencing complex activity. *Proc. Nat. Acad. Sci. USA*, *101*, 14385–14389.

Remy, J. S., C. Sirlin, P. Vierling, J. P. Behr. (1995). Gene transfer with a series of lipophilic DNA-binding molecules, *Bioconjug. Chem. 5*, 647–654.

Ropert, C., Lavignon, M., Dubernet, C., Couvreur, P., Malvy, C., (1992). Oligonucleotides encapsulated in pH sensitive liposomes are efficient toward friend retrovirus, *Biochem. Biophys. Res. Commun. 183*, 879–885.

Sato, Y., Murase, K., Kato, J., Kobune, M., Sato, T., Kawano, Y., et al. (2008). Resolution of liver cirrhosis using vitamin A-coupled liposomes to deliver siRNA against a collagen specific chaperone. *Nat. Biotechnol. 26*(4), 431–442. DOI: 10.1038/nbt1396.

Semple, S. C., Akinc, A., Chen, J., Sandhu, A. P., Mui, B. L., Cho, C. K., et al. (2010). Rational design of cationic lipids for siRNA delivery. *Nat. Biotechnol., 28*(2), 172–176. DOI: 10.1038/nbt.1602.

Semple, S. C., Chonn, A., Cullis, P. R. (1996). Influence of cholesterol on the association of plasma proteins with liposomes. *Biochemistry, 35*(8), 2521–2525. DOI: 10.1021/bi950414i.

Straubinger, R. M., K. Hong, D. S. Friend, D. Papahadjopoulos. (1983). Endocytosis of liposomes and intracellular fate of encapsulated molecules: Encounter with a low pH compartment after internalization in coated vesicles, *Cell 32*, 1069–1079.

Stuart, D. D., T. M. Allen. (2000). A new liposomal formulation for antisense oligodeoxynucleotides with small size, high incorporation efficiency and good stability, *Biochim. Biophys. Acta* 1463 (2000) 219–229.

Tabernero, J., Shapiro, G. I., LoRusso, P. M., Cervantes, A., Schwartz, G. K., Weiss, G. J., et al. (2013). First-in-humans trial of an RNA interference therapeutic targeting VEGF and KSP in cancer patients with liver involvement. *Cancer Discov. 3*(4), 406–417. DOI: 10.1158/2159-8290.CD-12-0429.

Tomari, Z. (2005). Perspective: Machines for RNAi. *Genes Develop. 19*, 517–529.

Tormo, M., Tari, A. M., McDonnell, T. J., Cabanillas, F., Garcia-Conde, J., Lopez-Berestein, G. (1998). Apoptotic induction in transformed follicular lymphoma cells by Bcl-2 down-regulation, *Leuk. Lymphoma 30*, 367–379.

Wang, Y., Miao, L., Satterlee, A., Huang, L. (2015). Delivery of Oligonucleotides with Lipid Nanoparticles. *Adv. Drug Delivery Rev. 87*, 68–80.

Yang, X., Koh, C. G., Liu, S., Pan, X., Santhanam, R., Yu, B., et al. (2009). Transferrin receptor targeted lipid nanoparticles for delivery of an antisense oligodeoxyribonucleotide against Bcl-2, *Mol. Pharm.* 6 (2009) 221–230.

Yuba, E., Y. Nakajima, K. Tsukamoto, S. Iwashita, C. Kojima, A. Harada, et al. (2012). Effect of unsaturated alkyl chains on transfection activity of poly(amidoamine) dendron bearing lipids. *J. Control. Release: Off. J. Control. Release Soc. 160*, 552–560.

Zhang, Y., Schwerbrock, N. M., Rogers, A. B., Kim, W. Y., Huang, L. (2013). Co- delivery of VEGF siRNA and gemcitabine monophosphate in a single nanoparticle formulation for effective treatment of NSCLC. *Mol. Ther. 21*(8), 1559–1569. DOI: 10.1038/mt.2013.120.

Zimmermann, T. S, A. C. Lee, A. Akinc, B. Bramlage, D. Bumcrot, M. N. Fedoruk, et al. (2006). RNAi-mediated gene silencing in non-human primates, *Nature 441*, 111– 114.

13 Electrostatics of Lipid Membranes Interacting with Oppositely Charged Macromolecules

Guilherme Volpe Bossa
North Dakota State University, Fargo, North Dakota

Klemen Bohinc
University of Ljubljana, Ljubljana, Slovenia

Sylvio May
North Dakota State University, Fargo, North Dakota

CONTENTS

13.1 Introduction 223
13.2 Reminder of Continuum Mean-Field Electrostatics 224
13.3 Electrostatic Protein Adsorption onto a Lipid Membrane 227
    13.3.1 Role of Lipid Mobility 227
    13.3.2 Protein-Induced Membrane Phase Separation 229
    13.3.3 Role of Charge Regulation 230
13.4 Lipid–DNA Complexes 231
    13.4.1 Cationic Lipid–DNA Complexes 231
    13.4.2 Zwitterionic Lipids and Divalent Cations 233
    13.4.3 Adsorption of DNA onto a Zwitterionic Lipid Monolayer 234
13.5 Macroion-Induced Membrane Bending 235
13.6 Amphipathic Peptide–Membrane Interaction 237
Acknowledgements 238
Bibliography 238

13.1 INTRODUCTION

Because a substantial fraction of all naturally occurring lipids carry an excess negative charge, it will not come as a surprise that electrostatic interactions of biomembranes with oppositely charged proteins and peptides play an important role for numerous biological processes (Vance and Vance 2008). However, the electrostatic properties of anionic lipids are quite complex and result not only from the lipid's net charge (Langner and Kubica 1999). For example, the most abundant anionic phospholipid in the plasma membrane is phospatidylserine (PS). The excess charge of PS originates from a negatively charged carboxyl group in the outer head group region. Beyond that, the head group of PS carries a phosphate-amino group ($P^-–N^+$) dipole, which adds a large dipole moment of about 20 Debye, resulting from two opposite elementary charges separated, roughly, by 0.4 nm. The head group is able to orient water molecules in a manner that the resulting dipole potential due to the oriented water molecules is opposite and, in fact, overcompensates that of the $P^-–N^+$ dipole. As a result, the electrostatic potential inside a lipid membrane is positive and not, as both PS's excess charge and the $P^-–N^+$ dipole would suggest, negative (Brockman 1994). It is also important to realize that lipids are asymmetrically distributed in the plasma membrane. Despite being the most abundant anionic lipid, there is very little PS in the plasma membrane's outer leaflet. An exception are apoptotic cells which signal phagocytes to recognize and remove the cell by exposing PS on their outer surface.

Besides naturally occurring lipids, numerous artificial lipids have been synthesized for biotechnological applications and drug delivery systems: electrostatic properties are, here too, among the key factors of lipid design. For example, a multitude of cationic lipids have been synthesized as condensing agent of negatively charged DNA. The resulting self-assembled cationic lipid–DNA complexes are readily taken up by living cells, thus offering a lipid-based vector to facilitate gene therapy. Of course, electrostatic attraction can be mediated equally well by other cationic molecules, and indeed, multivalent metal cations, polymers, dendrimers, polyamines, peptides, and nanoparticles have also been used as condensing agents for DNA (Elouahabi and Ruysschaert 2005).

The interaction of lipid membranes with oppositely charged macromolecules is influenced not only by electrostatics but also by various other factors, including the hydrophobic effect and hydrogen bond formation, polarization and charge transfer, protonation and dissociation equilibria, and van der Waals interactions. For example, phosphatidic acid (PA) is the structurally most simple glycerophospholipid. PA's head group is a phosphomonoester that can adjust its protonation state close to physiological pH, switching between one or two negative excess charges and thus allowing PA to regulate its electrostatic interactions with globular proteins. PA is not very abundant in the plasma membrane (a few mole percent at most). Yet, its signalling function as a second messenger requires the binding to specific target proteins. Electrostatic interactions alone would be insufficient to mediate binding of PA in favour of other much more prevalent anionic lipids (such as PS). There is evidence that part of the pH-dependent binding specificity to globular proteins arises from a combination of electrostatic interactions and hydrogen bond formation known as the electrostatic–hydrogen bond switch (Kooijman and Burger 2009). Hence, here and in many other instances, electrostatics acts in conjunction with other specific or unspecific interactions.

The electrostatic interaction of lipids with macroions in an aqueous solution depends also on the chemical nature of the mobile (typically salt) ions that are contained in the solvent. This *ion specificity* has been discovered long ago (Hofmeister 1888) and is still not completely understood. Size variations between different salt ions impose steric interactions, but this alone does not explain ion specificity. What needs to be considered in addition is that mobile ions differ in the ability to order water molecules in their immediate vicinity. This ability depends on the ordering ion's size and charge distribution: kosmotropes are typically highly charged and small; they order water. In contrast, chaotropes are often larger in size; they disorder water. At sufficiently high ion densities (typically beyond 0.1 M), ordered water molecules surrounding each ion (arranged in hydration shells) start interacting with each other. This has experimentally observable consequences. For example, water-soluble proteins adjust their maximal solubility in water as function of the type and amount of added salt. Typically, kosmotropic ions decrease ("salting out") and chaotropic ions increase ("salting in") protein solubility.

As pointed out in the preceding two paragraphs, non-electrostatic interactions often modify pure electrostatic aspects of interacting macroions. Modelling aims at understanding these modifications. In the first place, however, we wish to understand bare electrostatics and its implications for the interaction between lipid membranes and oppositely charged macroions. As we shall demonstrate in this chapter, electrostatic effects alone can lead to a variety of interesting and complex phenomena, and they are often sufficient to rationalize experimental findings. Perhaps the most notable aspect of electrostatics in aqueous solutions is the fact that water always contains mobile charge carriers (microions) – in most cases salt ions. The microions tend to form a diffuse layer around a given macroion so as to most efficiently screen the macroion charges, without sacrificing too much of their translational entropy. When two macroions interact with each other, the mobile charges (microions) are able to readjust their locations. This readjustment is crucial and, in fact, is the main task of modelling the electrostatic interaction between two macroions. For example, if two oppositely charged macroions are brought into close contact, their charges can neutralize each other without the need to form a diffuse counterion cloud any more. Consequently, the previously immobilized counterions of each macroion can be released into the aqueous solution. In fact, the corresponding increase in translational entropy of the involved counterions constitutes the main driving force for oppositely charged macroions to attract each other (Harries, May, and Ben-Shaul 2013). If one of the macroions is a charged lipid membrane in its fluid state, the problem may become even more intricate because fluid membranes are soft and have numerous degrees of freedom. In particular, charged lipids are able to migrate within each of the two membrane leaflets, rendering the location of the charged lipids an additional degree of freedom. Moreover, membranes can undergo bending deformations, form pores, and adjust their local thickness – all in response to interacting with a macroion.

Subject of the present chapter is the modelling of electrostatic interactions between lipid membranes and oppositely charged macroions. We focus almost exclusively on the most simple approach – the classical continuum Poisson–Boltzmann theory. (Approaches beyond Poisson–Boltzmann theory as well as computer simulations are covered in different chapters of this book.) Section 13.2 briefly recapitulates Poisson–Boltzmann theory and discusses electrostatic interactions of macroions across a lipid bilayer. Sections 13.3 and 13.4 elaborate on membrane–protein and membrane–DNA interactions, followed by brief accounts of macroion-induced membrane bending (Section 13.5) and interactions of lipid membranes with amphipathic peptides (Section 13.6).

13.2 REMINDER OF CONTINUUM MEAN-FIELD ELECTROSTATICS

Most charged biomacromolecules, including DNA, proteins, and lipid membranes, share the same structural motif, where a core of low dielectric constant with ε_L roughly between 2 and 5 is separated from the aqueous solution, which has a large dielectric constant of $\varepsilon_W \approx 80$. Electric charges reside in biomacromolecules predominantly at the dielectric interface; in case of DNA and most proteins, they are firmly attached to the interface implying lateral

Charged Macromolecules on Membranes

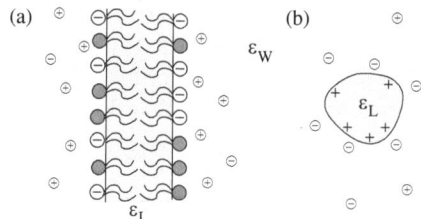

FIGURE 13.1 A diffuse ion cloud forms in the vicinity of macroions immersed in an aqueous solution. In the illustration, two macroions are displayed, a negatively charged lipid membrane (a) and a positively charged protein (b). Both the membrane and protein enclose a region of low dielectric constant ε_L (much smaller than the dielectric constant of water $\varepsilon_W = 80$), with surface charges at the dielectric interface. These charges can be attached to the surface at fixed positions or be laterally mobile. The diffuse ion cloud consists of small monovalent co- and counterions that are present with bulk concentration n_0.

immobility and thus a fixed local surface charge density σ. In contrast, charged lipids in biomembranes can migrate in the lateral direction, leading to a constant electrochemical potential.

Electrostatic properties of bare lipid membranes and their interactions with each other are discussed in the preceding chapter of this book, which also introduces the mean-field-level Poisson–Boltzmann approach and applies it to the planar geometry. This chapter focuses on the same modelling approach, yet applied to lipid membranes that interact electrostatically with other macromolecules. We briefly recall the Poisson–Boltzmann equation and its corresponding free energy. To this end, we consider the generic situation illustrated in Figure 13.1, where a charged membrane (a) interacts with an oppositely charged protein (b). Both macroions (membrane and protein) are embedded in an aqueous solution containing monovalent salt of bulk concentration n_0. Let n_+ and n_- be the local concentrations of mobile cations and anions, respectively, in the vicinity of the macroions. The Poisson–Boltzmann approach assumes these concentrations are a function of the average local electrostatic potential Φ and obey a Boltzmann distribution $n_\pm = n_0 \exp(\mp e\Phi / k_B T)$, where e denotes the elementary charge, k_B the Boltzmann constant, and T the absolute temperature. The concentrations $n_\pm$ determine the local volume charge density $\rho = e(n_+ - n_-)$ which upon insertion into the Poisson equation $\varepsilon_0 \varepsilon_W \nabla^2 \Phi = -\rho$ (where ε_0 is the vacuum permittivity) gives rise to the Poisson–Boltzmann equation

$$\nabla^2 \Phi = \frac{2en_0}{\varepsilon_W \varepsilon_0} \sinh\left(\frac{e\Phi}{k_B T}\right). \quad (13.1)$$

Equation (13.1) is a nonlinear partial differential equation for the electrostatic potential Φ that must be solved with respect to appropriate boundary conditions reflecting the fixed charge densities or, for mobile surface charges, fixed electrochemical potentials on the macroions. For example, if the surface charge density σ on one of the macroions is fixed, then the derivative of the potential Φ in the normal direction n at the macroion surface A (pointing into the aqueous solution) reads $(\partial \Phi / \partial n)_A = -\sigma / (\varepsilon_W \varepsilon_0)$. We point out that this simple boundary condition makes use of the large change in dielectric constant, $\varepsilon_W \gg \varepsilon_L$, across the macroion–electrolyte interface.

The Boltzmann distributions for $n_\pm$ are based on a mean-field assumption, valid only if ion–ion correlations can be neglected. They can be derived via the minimization of a mean-field electrostatic free energy $F_{el} = U_{el} - TS$ with its electrostatic (U_{el}) and entropic ($-TS$) contributions. The minimization approach is useful because, first, it provides access to the thermodynamic free energy and, second, it can be generalized to more complex systems, where the involved ions (macroions or mobile salt ion) exhibit additional nonelectrostatic interactions or internal degrees of freedom. An appropriate electrostatic mean-field free energy expression, referred to as charging free energy, for the system shown in Figure 13.1 is

$$F_{el} = \frac{\varepsilon_W \varepsilon_0}{2} \int_V dv (\nabla \Phi)^2 + k_B T \int_V dv$$
$$\times \left[n_+ \ln \frac{n_+}{n_0} - n_+ + n_- \ln \frac{n_-}{n_0} - n_- + 2n_0 \right], \quad (13.2)$$

where both volume integrals run over the aqueous solution. F_{el} consists of the sum of the energy stored in the electrostatic field $-\nabla \Phi = (-\partial \Phi/\partial x, -\partial \Phi/\partial y, -\partial \Phi/\partial z)$ (the first integral in Eq. 13.2) and the free energy of two ideal gases kept at constant chemical potential (the second integral in Eq. 13.2), one for the cations and the other for the anions. Note that $F_{el}[n_+, n_-, \Phi(n_+, n_-)]$ in Eq. (13.2) depends on the yet unknown local ion concentrations $n_\pm$ and on the electrostatic potential Φ, which itself is also a function of $n_\pm$ through Poisson's equation. Hence, using Poisson's equation, we can perform the variation of F_{el} with respect to n_+ and n_-, yielding

$$\delta F_{el} = \int_V dv \left[\delta n_+ \left(k_B T \ln \frac{n_+}{n_0} + e\Phi \right) \right.$$
$$\left. + \delta n_- \left(k_B T \ln \frac{n_-}{n_0} - e\Phi \right) \right] + \int_A da \, \Phi \delta \sigma. \quad (13.3)$$

Thermal equilibrium demands the vanishing of δF_{el} and thus entails the Boltzmann distributions $n_\pm = n_0 \exp(\mp e\Phi / k_B T)$. Note that the final integral in Eq. (13.3) extends over the surfaces of all involved macroions. Whenever σ for a given macroion is fixed, $\delta\sigma = 0$, and the surface integral vanishes. If σ is not fixed, there must be additional contributions to the free energy F_{el} associated with the energy of altering the surface charge density. These additional contributions enter the variation of the free energy and give rise to modified boundary conditions for the Poisson–Boltzmann equation. For example, for unrestricted mobility of the surface charges, the electrostatic surface potential adopts a constant value and the relevant free energy reads $\tilde{F}_{el} = F_{el} - \int_A da\,\Phi\sigma$.

The two cases, fixed surface charge density and fixed surface potential, constitute two thermodynamic limits. The free energy values of all other possible scenarios – surface charges with restricted mobility – are located between $\tilde{F}_{el}$ (the lower limit of unrestricted surface charge mobility) and F_{el} (the upper limit of completely immobile surface charges). Specific examples for intermediate cases will be discussed in Section 13.4.

The following three remarks prepare the subsequent sections of the present chapter. First, Eq. (13.1) suggests to introduce the dimensionless electrostatic potential $\Psi = e\Phi/k_BT$ in terms of which the Poisson–Boltzmann equation reads $l_D^2 \nabla^2 \Psi = \sinh\Psi$, with the Debye screening length $l_D = \sqrt{k_BT\varepsilon_W\varepsilon_0/(2e^2n_0)}$. At room temperature and given molarity M (in moles per liter where 1 M corresponds to a concentration of 0.6/nm³), the Debye screening length is $l_D = 0.3\,\text{nm}/\sqrt{M}$. Hence, under physiological conditions (at 100 mM salt), $l_D \approx 1$ nm. It is also common to define the Bjerrum length $l_B = e^2/(4\pi\varepsilon\varepsilon_0 k_BT)$, at which two elementary charges in a medium of dielectric constant ε have an interaction energy of k_BT. Note that the dielectric constant of water $\varepsilon_W = 80$ implies $l_B = 0.7$ nm at room temperature. Second, for sufficiently small potentials $|\Psi| \ll 1$ (where $\Psi = 1$ corresponds to a potential of $\Phi = 25$ mV at room temperature), we can linearize the Poisson–Boltzmann model. The result is referred to as Debye–Hückel theory; it is based on the linear Poisson–Boltzmann equation $l_D^2 \nabla^2 \Psi = \Psi$ as well as the corresponding free energy $F_{el} = (1/2) \int_A da\,\sigma\Phi$ for fixed σ and $\tilde{F}_{el} = -(1/2) \int_A da\,\sigma\Phi$ for fixed Φ. The two free energy expressions follow from the corresponding nonlinear relations $F_{el} = \int_A da \int_0^\sigma d\tilde{\sigma}\Phi(\tilde{\sigma})$ and $\tilde{F}_{el} = -\int_A da \int_0^\Phi d\tilde{\Phi}\sigma(\tilde{\Phi})$ by assuming a linear relationship between σ and Φ. Third, the Poisson–Boltzmann model works reasonably well if the mobile salt ions are monovalent. For higher valencies, ion–ion correlations (which are neglected on the mean-field level) become important. Nevertheless, generalization of the Poisson–Boltzmann model to di- or higher valent salt ions is straightforward; the corresponding Poisson–Boltzmann equation is $2I l_D^2 \nabla^2 \Psi = -\sum_i z_i n_i^0 \exp(-z_i\Psi)$ with the ionic strength $I = (1/2)\sum_i z_i^2 n_i^0$ and Debye screening length $l_D = \sqrt{k_BT\varepsilon_W\varepsilon_0/(2e^2 I)}$. Here the index i runs through all ionic species, each with valence z_i and bulk concentration n_i^0. Electroneutrality in the bulk of the aqueous solution requires $\sum_i z_i n_i^0 = 0$.

Examples for applying the Poisson–Boltzmann equation to planar membranes are presented in the preceding chapter. Here, we add one single example that illustrates the degree of electrostatic interactions across a lipid membrane: How does a lipid membrane inserted in between two oppositely charged macroions affect their attractive interaction? Consider two planar macroions of sufficiently large lateral area A with fixed surface charge densities σ and $-\sigma$, separated by a distance $d+l$, with a lipid membrane of thickness l centred in between as shown schematically in Figure 13.2. The membrane is represented by a slab of dielectric constant ε_L that is impenetrable to salt ions and carries the same charge density σ_m on each interface. It is most instructive to focus on the Debye–Hückel limit, where a straightforward calculation of the charging free energy per unit area reveals

$$\frac{F_{el}}{Ak_BT} = \frac{4\pi l_B l_D/e^2}{1 + 2H \coth\left(\dfrac{d}{2l_D}\right)} \times$$

$$\left\{(\sigma^2 + \sigma_m^2)\left[2H + \coth\left(\dfrac{d}{2l_D}\right)\right] + \frac{2H\sigma_m^2}{\sinh\left(\dfrac{d}{2l_D}\right)}\right\}, \quad (13.4)$$

expressed in terms of the coupling parameter $H = (\varepsilon_L l_D)/(\varepsilon_W l)$. Obviously, in the absence of the inserted

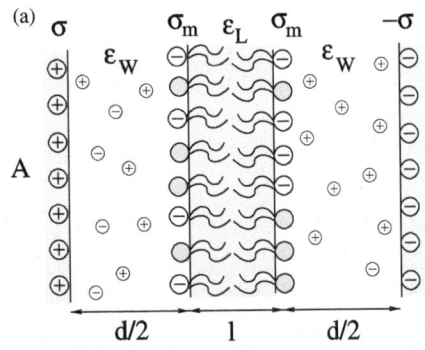
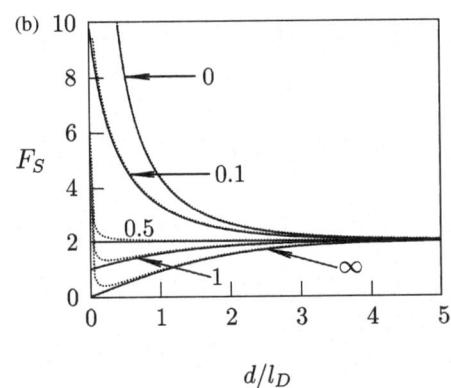

FIGURE 13.2 (a) Schematic illustration of two oppositely charged macroion surfaces with surface charge densities σ and $-\sigma$, sandwiching an electrolyte and a lipid membrane (a slab of width l and dielectric constant ε_L with surface charge density σ_m on both interfaces). (b) Corresponding scaled free energy $F_S = F_{el}e^2/(4\pi l_B l_D A k_B T \sigma^2)$ according to Eq. (13.4) as function of the scaled distance d/l_D. Solid lines: $\sigma_m = 0$; dotted lines: $\sigma_m/\sigma = 0.32$. The corresponding value of H is indicated for each pair of curves.

membrane (that is for $H \to \infty$ and $\sigma_m = 0$), we obtain the attractive free energy $F_{el}/(Ak_BT) = 4\pi l_B l_D (\sigma/e)^2 \tanh[d/(2l_D)]$. For $\sigma_m = 0$ and $H = 1/2$, we obtain $F_{el}/(Ak_BT) = 4\pi l_B l_D (\sigma/e)^2$, which is independent of d and corresponds to the charging free energy of two isolated planar surfaces with surface charge densities σ. That is, an uncharged dielectric slab with $H = 1/2$ cancels the interaction of the two macroions out. In fact, $H > 1/2$ is required to retain an attractive interaction between the two oppositely charged macroions. Any nonvanishing σ_m increases the required value for H even further.

In the limit $H = 0$, the dielectric slab is completely impermeable to the electric field, rendering the free energy $F_{el}/(Ak_BT) = 4\pi l_B l_D \left[(\sigma/e)^2 + (\sigma_m/e)^2 \right] \coth[d/(2l_D)]$ repulsive. See Figure 13.2 for a plot of the free energy according to Eq. (13.4) as function of the macroion separation.

Lipid membranes have a dielectric constant of about $\varepsilon_L \approx 4$ and a hydrophobic thickness $l \approx 4$ nm. Hence, $H = 1/2$ requires a Debye screening length $l_D = 40$ nm, corresponding to a 50 μM salt solution. Salt concentrations below about 50 μM are thus a prerequisite for electrostatic macroion attraction to propagate through a lipid bilayer. At physiological conditions $H \approx 1/80$, and electrostatic interactions at the two leaflets of a lipid bilayer are effectively decoupled. This decoupling, in fact, allows electrostatic modelling of macroion membrane interactions to focus on a single membrane leaflet.

13.3 ELECTROSTATIC PROTEIN ADSORPTION ONTO A LIPID MEMBRANE

Protein adsorption onto a biomembrane is a routine event for many fundamental cellular processes such as signalling, vesicle trafficking, membrane fusion, apoptosis, and viral entry. In some cases, such as viral entry or membrane fusion, proteins interact with the host membrane to induce membrane bending or a structural membrane destabilization. In other cases, such as signalling, adsorption reduces the dimensionality of the space the protein diffuses through, thus increasing the effective concentration and facilitating the recognition of membrane-bound substrates or binding partners (McLaughlin and Aderem 1995).

The adsorption is driven in most cases by a mix of electrostatic and hydrophobic interactions and regulated by a specific interaction mechanism. For example, annexins are a structurally diverse family of proteins that appear to be involved in a range of processes including membrane organization, protein sorting, scaffolding, and the transport and formation of vesicles. Annexins bind preferentially to negatively charged phospholipids, yet in a distinct calcium-dependent manner where calcium located at specific binding sites at the membrane-facing side of the protein coordinates with the phosphate groups of membrane lipids (Gerke and Moss 1997). There are additional electrostatic and hydrophobic interactions that modulate the binding strength of annexins with lipid membranes. Another example is the myristoylated alanine-rich C kinase substrate (MARCKS), a natively unfolded protein whose 25-residue basic effector domain provides a binding site for lipid membranes (Arbuzova, Murray, and McLaughlin 1998). Beyond the mere ionic interaction between the 13 positively charged residues of the effector domain with acidic phospholipids, MARCKS also inserts its myristate chain into the host membrane. The binding constant of MARCKS is regulated via phosphorylation by protein kinase C, which introduces three additional phosphates into the effector domain. Finally, an example for a very specific interaction mechanism is encountered in the influenza virus. The virus surface contains the protein hemagglutinin (HA) that binds to glycolipids on the surface of animal cells. The binding occurs specifically with sialic acid, a group of monosaccharides with a backbone consisting of nine carbon atoms that are widely found in glycolipids on the surfaces of animal cells.

Biological membranes contain a large number of different lipids, of which a substantial fraction is anionic (Langner and Kubica 1999). For example, 10%–30% of the lipids in the inner leaflet of the plasma membrane are anionic. Phosphatidylserine is the most abundant anionic lipid, but others such as phosphatidylinositol and phosphatidic acid are physiologically important as indicated by their tight metabolic control. The outer leaflet of the plasma membrane is composed mostly of zwitterionic lipids, especially phosphatidylcholine (PC) and sphingolipids. These lipids together with a protective layer of glycocalyx tend to render the outer surface inert with regard to interactions with molecules outside the cell. An important role is also played by cholesterol, which is present in the plasma membrane at large molar fractions (20%–30%) but unknown distribution between the inner and outer leaflets.

13.3.1 Role of Lipid Mobility

Consider the adsorption of a basic protein onto a mixed lipid membrane that contains a certain fraction of anionic lipids. For any fixed orientation and distance between the protein and membrane, one can calculate the charging free energy F_{el} of the protein–membrane system through solving the Poisson–Boltzmann equation within the aqueous phase. This conceptional approach has been implemented employing detailed atomistic models of the lipid bilayer and protein (Honig and Nicholls 1995), in all cases with fixed charge distributions of all involved macroions. Examples include the myristoylated N-terminus of the tyrosine kinase Src (Murray et al. 1998) or simple basic peptides (BenTal et al. 1997). Another approach is more fundamental and focuses on the continuum version of the Poisson–Boltzmann model, using highly approximate representations of the involved macroions as simple geometric bodies with continuous charge distributions. Its origin goes back to the calculation of interactions between colloids, such as between a sphere and a plane (Warszynski and Adamczyk 1997).

Let us discuss the interaction of a single sphere (of radius R) with an extended planar surface in more detail. A sphere of low dielectric constant and a uniform positive charge density

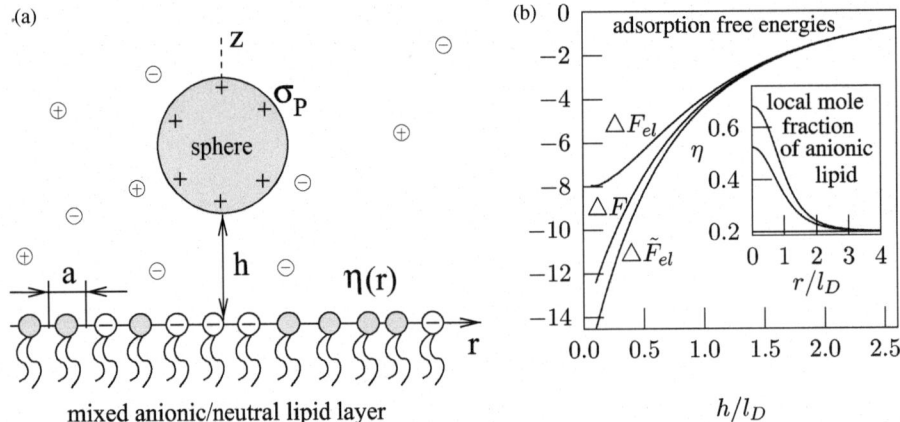

FIGURE 13.3 (a) Schematic illustration of a spherical particle at distance h away from a planar lipid layer, consisting of an anionic/neutral lipid mixture with local mole fraction η and average mole fraction ϕ of the anionic lipid. The spherical particle carries a surface charge density σ_P, and each lipid has a cross-sectional area a. (b) Adsorption free energies (in units of k_BT) as function of the scaled distance h/l_D for $R = 1$ nm, $\phi = 0.2$, $a = 0.65$ nm^2, $l_D = 1$ nm, and $\sigma_P = 0.7e/a$. The three curves refer to uniform membrane surface charge density (ΔF_{el}), ideal lipid demixing (ΔF), and fixed surface potential ($\Delta \tilde{F}_{el}$). The inset displays the corresponding local mole fractions $\eta(r)$ as function of the distance r away from the z-axis. The largest variation in η (upper curve in the inset) corresponds to the case of fixed surface potential, whereas for fixed and uniform surface charge density (bottom curve in the inset), $\eta = 0.2$ is constant. (Adopted with permission May, Harries, and Ben-Shaul 2000a.)

σ_P on its surface is a convenient generic representation of a globular protein. Similarly, we model the membrane as a flat surface, containing a binary anionic/neutral lipid mixture with mole fraction ϕ of the (monovalently charged) anionic lipid (see Figure 13.3 for a schematic illustration). We are interested in calculating the mean-field electrostatic interaction free energy as function of the sphere-to-plane distance h. To this end, we solve the Poisson–Boltzmann equation $l_D^2 \nabla^2 \psi = \sinh \psi$ within the aqueous space and calculate the corresponding charging free energy.

An important question addresses the distribution of the charged lipids in the membrane. In a gel-phase membrane, the lipids would be unable to move laterally, implying a uniform surface charge density $\sigma_L = -\phi \, e/a$, where a denotes the cross-sectional area per lipid (assumed to be the same for both lipid species). In this case, the Poisson–Boltzmann equation would have to be solved within the aqueous phase, imposing the boundary conditions of fixing both σ_P and σ_L. The corresponding charging free energy is then F_{el} according to Eq. (13.2). A different scenario results if the charged lipids are mobile, being able to adjust their local mole fraction η on the membrane. Completely unrestricted lipid mobility would imply a constant membrane surface potential Φ_0 and thus a free energy $\tilde{F}_{el} = F_{el} - \Phi_0 \int_A da\sigma_L$, where the surface integration runs over the lipid layer. However, any demixing of the membrane lipids entails at least an entropy penalty, associated with changing the local composition η away from the average ϕ. In the most simple case, the mixing properties of the lipids are ideal, implying to add the ideal demixing free energy contribution.

$$F_{mix} = \frac{k_B T}{a} \int_A da \left[\eta \ln \frac{\eta}{\phi} + (1-\eta) \ln \frac{1-\eta}{1-\phi} + \lambda(\eta - \phi) \right] \quad (13.5)$$

to the total free energy $F = F_{el} + F_{mix}$, where the integral runs over the lipid layer and F_{el} is given in Eq. (13.2). Note that the Lagrange multiplier λ ensures conservation $(1/a)\int_A da\, \eta$ of the total number of charged lipids. If the chemical potential of the lipids is fixed (that is, when the surface density of proteins on the membrane is small), then λ becomes identical to the dimensionless membrane surface potential $\Psi_0 = -2\,\text{arsinh}(2\pi l_B l_D \phi/a)$. Minimization of the total free energy proceeds analogously to Eq. (13.3). Note, however, that the term $\int_A da\Phi\delta\sigma$ in Eq. (13.3) only vanishes for the sphere (which carries fixed surface charge density σ_P) but not for the lipid layer, where the surface charge density $\sigma_L = -\eta e/a$ is able to adjust subject to the demixing free energy penalty. In fact, the minimization of F with respect to η yields for the local mole fraction of anionic lipids the distribution

$$\eta = \frac{1}{1 + \dfrac{1-\phi}{\phi} e^{(\lambda - e\Phi/k_BT)}}, \quad (13.6)$$

which is similar to a Fermi-Dirac distribution and enters the boundary condition $(\partial \Phi/\partial z)_A = -\sigma_L/(\varepsilon_w \varepsilon_0) = \eta e/(\varepsilon_w \varepsilon_0 a)$ for solving the Poisson–Boltzmann equation. Recall from the discussion after Eq. (13.1) that $(\partial \Phi/\partial z)_A$ denotes the derivative of the potential Φ in the normal direction of the lipid layer. Obviously, if the dimensionless potential $\Psi = e\Phi/k_BT$ is very negative, anionic lipids deplete ($\eta \to 0$), whereas for large positive Ψ, they accumulate up to saturation ($\eta \to 1$). Demixing is suppressed for $\Psi = \lambda$, implying $\eta = \phi$.

An example for the calculation of the electrostatic adsorption free energy $\Delta F(h) = F(h) - F(h \to \infty)$ (and analogous definitions for $\Delta F_{el}(h)$ and $\Delta \tilde{F}_{el}(h)$) is displayed in the right diagram of Figure 13.3. Here, the membrane contains 20% anionic lipids ($\phi=0.2$), and the cross-sectional area per lipid is $a=0.65\,\text{nm}^2$. The adsorbing spherical particle has a radius of $R=1\,\text{nm}$ and a surface charge density $\sigma_P=0.7\,e/a$. The three adsorption free energies (ΔF_{el} for fixed $\sigma_L=-\phi\,e/a$, ΔF for ideal lipid demixing, and $\Delta \tilde{F}_{el}$ for fixed surface potential $\Psi_0=-2\,\text{arsinh}(2\pi l_B l_D \phi/a)$) deviate significantly – by several $k_B T$ for small distances h – due to the different degrees of lipid demixing. The differences between the three cases are also visible in the inset of the right diagram of Figure 13.3, which shows the local mole fraction η as function of the distance r away from the axis of cylindrical symmetry. The three curves refer to fixing Ψ_0 (top), ideal lipid demixing (middle), and fixing a uniform σ_L (bottom). The model also predicts that doubling the mole fraction of the anionic lipid (that is, 40% instead of 20%) largely suppresses demixing. These notions are in good agreement with experimental results probing the adsorption of cytochrome c onto phosphatidylcholine/phosphatidylglycerol (PC/PG) model membranes containing 10%–40% of the anionic lipid PG (Heimburg, Angerstein, and Marsh 1999), where lipid demixing was observed only for small but not for large mole fractions of PG.

The example in Figure 13.3 refers to monovalently charged lipids but can be generalized to include anionic lipids of higher valence, besides neutral and monovalently charged ones (Haleva, Ben-Tal, and Diamant 2004). This is of relevance for the polyvalent lipid phosphatidylinositol 4,5-bisphosphate (PIP$_2$), which is present at very small concentrations (about 1% of all lipids) in the inner leaflet of the plasma membrane. PIP$_2$ competes with PS for interactions with basic proteins. Both the Poisson–Boltzmann theory (Wang et al. 2004, Haleva, Ben-Tal, and Diamant 2004) and Monte Carlo simulations (Tzlil, Murray, and Ben-Shaul 2008) predict a dramatic increase of the local PIP$_2$ concentration in the vicinity of the protein, effectively replacing and releasing all PS that would otherwise engage in a weaker interaction with the protein. These theoretical results agree qualitatively with experimental findings employing MARCKS and basic peptides (Gambhir et al. 2004).

13.3.2 Protein-Induced Membrane Phase Separation

Even in the absence of protein, lipid membranes are able to form domains or undergo phase separation. For example, model membranes containing certain lipid mixtures involving cholesterol, an unsaturated and a saturated lipid, exhibit macroscopic phase separation (Simons and Vaz 2004) into a liquid-ordered (l_o) and a liquid-disordered (l_d) phase. The separation is driven by the more favourable interaction of cholesterol with saturated lipid chains as compared with unsaturated ones (Elson et al. 2010) and has been described, for example, in terms of cholesterol engaging in the formation of binary complexes with the saturated lipid (Radhakrishnan and McConnell 2005). Studying phase separation in model membranes is motivated by our continuing lack of understanding domain formation in biomembranes, where – according to the membrane raft hypothesis – small and highly dynamic lipid domains function as signalling platforms (Lingwood and Simons 2010). A multitude of studies indicate that domain formation (Epand 2008) and even phase separation (Hammond et al. 2005, Lingwood et al. 2008) in lipid membrane can be controlled by adsorbed proteins. For example, the peripheral protein α-synuclein has been reported to induce macroscopic phase separation on supported PC/PG membranes (Pandey et al. 2009). Similarly, annexin A2 induces the formation of large domains in supported bilayers composed of PC/PG (Menke, Gerke, and Steinem 2005).

Monte Carlo simulations (Reynwar and Deserno 2008) and phenomenological models (Loew, Hinderliter, and May 2009, Fan, Sammalkorpi, and Haataja 2010) have been employed to elucidate the mechanisms behind protein-induced phase separation or domain formation of mixed membranes. We briefly recapitulate a method (Mbamala, Ben-Shaul, and May 2005) to include nonideal mixing of the membrane into the framework of mean-field electrostatic modelling. The system is identical as that illustrated in Figure 13.3, yet with the difference that the total free energy $F=F_{el}+F_{mix}+F_{non\text{-}id}$ includes the nonideal contribution:

$$\frac{F_{non\text{-}id}}{k_B T} = -\frac{\chi}{a}\int_A da(\eta-\phi)^2 + \frac{\chi}{6}\int_A da(\nabla\eta)^2, \quad (13.7)$$

in addition to the electrostatic free energy F_{el} (see Eq. 13.2) and ideal mixing free energy of the binary membrane F_{mix} (see Eq. 13.5). Here again, η is the local mole fraction of the charged lipids, which are mobile but subject to nonelectrostatic next-neighbour lipid–lipid interactions described by the nonideality parameter χ. The same nonideality parameter appears in the Bragg–Williams approximation of a lattice gas (Safran 1994). Positive χ corresponds to an effective attraction between lipids of the same type, thus favouring domain formation. For $\chi>2$, the nonelectrostatic lipid–lipid interactions alone would be predicted to induce phase separation. However, additional electrostatic repulsion between the anionic lipids increases that value to $\chi>3.7$ (Gelbart and Bruinsma 1997). The second integral in Eq. (13.7) accounts for the energy penalty of compositional gradients within the membrane. This term, which also originates in the presence of nonelectrostatic lipid–lipid interactions, entails a line tension at the interface between the membrane patch of the protein adsorption site and the bulk membrane. Note that with the additional energy contribution in Eq. (13.7), minimization of the total free energy F yields besides the Poisson–Boltzmann equation also a boundary condition at the membrane:

$$a\frac{\chi}{3}\nabla^2\eta = \ln\frac{\eta(1-\eta)}{\phi(1-\eta)} - 2\chi(\eta-\phi) + \lambda - \Psi, \quad (13.8)$$

which replaces Eq. (13.6) when nonideal mixing is taken into account. Here, both the Poisson–Boltzmann equation and its boundary condition at the membrane surface are differential equations that must be solved self-consistently. What the approach stipulated by Eqs. (13.7) and (13.8) reveals is that a binary membrane can indeed be triggered to undergo phase separation by purely electrostatically driven protein adsorption, somewhat unexpectedly even if χ is smaller than 2. In fact, it is the reduction of the total line tension energy that serves as the driving force for individual protein-decorated membrane patches to coalesce and form an individual phase (Mbamala, Ben-Shaul, and May 2005).

13.3.3 Role of Charge Regulation

Migration of charged lipids is not the only mechanism by which a lipid membrane can regulate its local charge density. Adjusting the charge of a lipid head group by protonation/deprotonation or other dissociation equilibria constitutes another biologically and technologically important lipid degree of freedom. Implications of this charge regulation mechanism for the interaction between lipid membranes and peripheral proteins are only beginning to emerge. For example, upon glucose starvation, yeast downregulates its metabolism. A key step in this process, the release of the transcription factor Opi1 from the endoplasmic reticulum, is triggered by a decrease in intracellular pH. The lipid phosphatidic acid (PA) appears to act as a pH sensor and is involved in regulating the affinity of Opi1 (Young et al. 2010). As already mentioned in the introduction, PA carries either one or two negative charges close to physiological pH. Hence, a more acidic environment tends to protonate PA's phosphomonoester head group, $PA^{2-}+H^+ \to PA^-$, implying weaker electrostatic interaction with cationic residues of adsorbed proteins and thus a larger probability to release the protein from the membrane into the solution. Modelling studies suggest that this purely electrostatic mechanism is not sensitive enough to allow protein adsorption to be regulated by a pH change of about two units, but the conjunction of charge regulation with the PA-specific electrostatic hydrogen bond switch increases the sensitivity (Loew, Kooijman, and May 2013).

Charge regulation has been discussed in the preceding chapter for a bare membrane; here, we suffice with a brief discussion relevant to a membrane interacting with an oppositely charged macroion. We focus on the protonation/deprotonation equilibrium of an acidic lipid (AL) of the type $AL^- + H^+ \leftrightarrow AL$. Assume a bare binary lipid layer contains acidic lipids with mole fraction ϕ_0 and neutral lipids with mole fraction $1-\phi_0$. When the lipid layer has a dimensionless surface potential Ψ_0, a fraction η_0 of the acidic lipids carry a negative charge, and the remaining fraction $1-\eta_0$ remains uncharged. The charge density of the bare lipid layer is then $\sigma_0 = -\phi_0\eta_0 e/a$, where e denotes the elementary charge and a is the cross-sectional area per lipid (assumed to be the same for both lipid types and independent of the protonation state of the acidic lipid). Note also that within the Poisson–Boltzmann framework, the surface potential of the lipid layer is $\Psi_0 = -2ar\sinh(2\pi l_B l_D \phi_0 \eta_0/a)$. When a positively charged protein adsorbs onto the surface of the acidic lipid layer, it will render the local membrane potential Ψ less negative than Ψ_0 (that is, $\Psi-\Psi_0 > 0$). That induces acidic lipids to migrate into the protein adsorption region and to become more deprotonated. Equivalently expressed, the fraction ϕ of acidic lipids within the protein adsorption region and the fraction η of the acidic lipids that carry a charge become larger than the corresponding values for the bare membrane, ϕ_0 and η_0, respectively. To quantify how charge regulation and lipid migration act together, we consider the free energy of a lipid layer.

$$\frac{F_{\text{ch-reg}}}{k_B T} = \frac{1}{a}\int_{A_p} da \left\{ \phi \ln\frac{\phi}{\phi_0} + (1-\phi)\ln\frac{1-\phi}{1-\phi_0} \right.$$
$$\left. +\phi\left[\eta\ln\frac{\eta}{\eta_0} + (1-\eta)\ln\frac{1-\eta}{1-\eta_0} - \eta(\Psi-\Psi_0) \right] \right\}, \quad (13.9)$$

where the integration runs over the protein adsorption region A_p (elsewhere on the lipid layer $\eta=\eta_0$ and $\phi=\phi_0$). The surface potential on the lipid layer Ψ may vary within the adsorption region but, as pointed out above, will generally be less negative than Ψ_0 if the absorbing protein is basic. Minimization of the free energy $F_{\text{ch-reg}}$ with respect to ϕ and η yields

$$\eta = \frac{1}{1+\frac{1-\eta_0}{\eta_0}e^{-(\Psi-\Psi_0)}}, \quad \phi = \frac{1}{1+\frac{1-\phi_0}{\phi_0}e^{-\ln\left[1+\eta_0\left(e^{\Psi-\Psi_0}-1\right)\right]}}. \quad (13.10)$$

Using the expressions for η and ϕ in Eq. (13.10) gives rise to a surface charge density of the lipid layer within the protein adsorption zone of

$$\sigma = -\phi\eta\frac{e}{a} = -\frac{1}{1+\frac{1-\phi_0\eta_0}{\phi_0\eta_0}e^{-(\Psi-\Psi_0)}}\frac{e}{a}, \quad (13.11)$$

which is the same for a binary lipid layer with bulk mole fractions $\phi_0\eta_0$ and $1-\phi_0\eta_0$ of negatively charged and neutral lipids, respectively. Hence, a lipid layer with fraction ϕ_0 of acidic lipids of which a fraction η_0 either carries a fixed charge or is charged due to a charge regulation process adopts the same elevated surface charge density at the adsorption region of an identical basic protein. This statement, however, neglects any cooperative effects due to nonelectrostatic interactions.

Charge regulation is important not only for the target membrane but also for the adsorbing protein (Lund and Jönsson 2005, Hartvig et al. 2011). A large fraction (typically about one-third) of the water-exposed amino acids in globular proteins are charged, some of which have pK values close to physiological pH. For example, the imidazole side chain of histidine has an apparent pK of about 6.0. Histidine-rich proteins such as cofilin or hisactophilin (both are actin-binding proteins) will thus adjust their charging state as function of pH. Hisactophilin is known to bind a negatively charged membrane at pH 6.5, whereas at pH 7.5, binding is largely suppressed due to the larger degree of histidine deprotonation (Lund, Åkesson, and Jönsson 2005). We point out that charge regulation of a membrane-adsorbing protein is affected by the anionic lipids through the electrostatic potential they produce at the protein surface. In the most complex case, both membrane and protein are able to adjust their charges through charge regulation, which adds to the ability of the acidic lipids to migrate towards the protein adsorption site. All these effects (together with amino acid side chain flexibility and dipole moments as well as salt content in the aqueous solution) affect the (dimensionless) membrane surface potential Ψ that we have used in Eqs. (13.9–13.11).

13.4 LIPID–DNA COMPLEXES

The usage of DNA as a drug to treat diseases is known as gene therapy. Different variants of employing DNA range from simply replacing a gene to encoding for a functional protein. The main challenge of expressing DNA inside the cell nucleus is to transfer the DNA into the cell. Naked DNA is unlikely to cross the plasma membrane without facilitation such as electroporation (Herweijer and Wolff 2003). Carrier-based approaches include the use of viruses as transport vehicle. Viruses are highly efficient in depositing genetic material into their host cells but also induce an immuno-reaction. An increasingly viable alternative is offered by nonviral methods which employ a DNA-condensing agent – often a cationic lipid or polymer – to form a self-assembled carrier. The transfection efficiency of nonviral vectors is generally lower than that of their viral counterparts, but their lack of immunogenicity, low cost of large-scale production, and susceptibility to technological advance offer advantages that have sparked substantial research efforts.

13.4.1 Cationic Lipid–DNA Complexes

Virtually all naturally occurring charged lipids are anionic. In fact, cationic lipids and their complexes with oppositely charged macroions pose a significant toxicity concern (Yew and Scheule 2005). Still, cationic lipids are being used as condensing agent for DNA, and the resulting cationic lipid–DNA complexes – also referred to as *lipoplexes* (Felgner et al. 1987) – are among the most promising nonviral delivery vehicles of DNA into cells for gene therapy applications.

When mixed together outside the cell under conditions of low ionic strength, the strong electrostatic interaction between DNA and cationic lipids leads to the formation of condensed, often micron-sized, structures that will be internalized by cells through endocytosis (Rejman et al. 2004). To reduce cytotoxicity and to control the morphology and transfection efficiency of the resulting complexes, an additional zwitterionic lipid – a so-called helper lipid (Hirsch-Lerner et al. 2005) – is usually added. Two principal choices of the helper lipid are PC and phosphatidylethanolamine (PE). These two lipids differ in their tendency to induce curvature in a lipid layer. The effective cross-sectional sizes of head group and chain region are comparable for PC; the lipid thus tends to self-assemble into planar structures, namely the planar bilayer (the L_α phase, displayed in Figure 13.4 and marked by the letter B). A similar tendency is also observed for most monovalent cationic lipids such as trimethylammonium propane. In contrast, PE, which lacks the methylation of the amine in PC, has a smaller head group and thus favours to form curved lipid layers such as the inverse hexagonal phase (the H_{II} phase, marked by the letter I in Figure 13.4). The two major lipoplex structures mimic the lipid morphologies of their bare lipid counterparts. One – formed when PC is used as helper lipid – is the sandwich structure (Rädler et al. 1997, Lasic et al. 1997), which consists of a lamellar membrane stack with intercalated DNA molecules (the L_α^C phase, marked by the letter S in Figure 13.4). The other – formed when PE is used as helper lipid – is the honeycomb structure (Koltover et al. 1998), in which the DNA fills the hydrophilic tubes of an inverse hexagonal lipid phase (the H_{II}^C phase, marked by the letter H in Figure 13.4). Among other morphological structures that have been observed are micellar (Gershon et al.

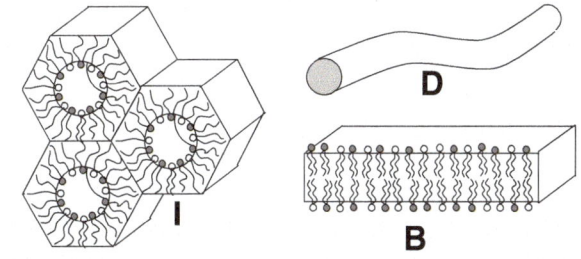

FIGURE 13.4 Schematic illustrations of two cationic lipid–DNA complexes, honeycomb ("H") and sandwich ("S"), together with their corresponding uncomplexed lipid phases: inverse-hexagonal phase ("I") and planar bilayer ("B"). Finally, "D" denotes naked DNA. The shaded circular regions correspond to DNA cross sections. The lipid layers are mixed, consisting of cationic and uncharged helper lipids. (Adopted with permission May, Harries, and Ben-Shaul 2000b.)

1993) and cubic (Koynova and MacDonald 2003) lipid–DNA complex, as well as isolated DNA wrapped by a highly curved single lipid bilayer (Sternberg, Sorgi, and Huang 1994). Additional structures are predicted by computer simulations (Farago and Gronbech-Jensen 2009) but have not yet been observed experimentally.

Consider adding DNA to a binary lipid mixture, consisting of PC as the helper lipid and a cationic species. The thermodynamic properties of this system can be characterized in terms of two parameters, the cationic lipid-to-DNA charge ratio ρ and the mole fraction ϕ of the cationic lipid. Because both the cationic and helper lipid favour the planar aggregation geometry, the only observed lipoplex is the sandwich structure (Rädler et al 1997). The most stable sandwich is adopted at isoelectric conditions, $\rho=1$, where the charge of the cationic lipids matches that of the DNA, and virtually all excess counterions can be released (Wagner et al. 2000). If ρ deviates somewhat from isoelectricity, the system still forms a single sandwich phase. This is possible because the sandwich structure is able to accommodate variable amounts of lipid by adjusting the distance d between neighbouring DNA molecules in each gallery. As is illustrated in the left diagram of Figure 13.5, at certain ratios $\rho_1 < 1$ and $\rho_2 > 1$, the single-phase behaviour changes into the coexistence of two phases. As ρ decreases below 1, neighbouring DNA strands further approach each other, and at $\rho=\rho_1$, the ensuing electrostatic DNA–DNA repulsion prevents additional DNA from entering into the complex. At this point, a new phase of uncomplexed DNA forms. Similarly, if ρ increases beyond 1, the DNA–DNA distance grows and thus increases the electrostatic repulsion between the cationic lipid bilayers in between neighbouring DNA strands. Here again, the system splits into two phases, the sandwich and an excess lipid bilayer. The phase evolution with the corresponding DNA-to-DNA distance d between neighbouring DNA molecules is illustrated in Figure 13.5a. The same phase sequence is predicted for different mole fractions ϕ of the cationic lipid. The dependencies $\rho_1(\phi)$ and $\rho_2(\phi)$ thus translate into a ϕ vs. ρ phase diagram, which is displayed in Figure 13.5b.

Modelling the phase diagram in Figure 13.5b is based on a numerical minimization of the total free energy, which contains contributions from the sandwich structure, uncomplexed bilayer, and DNA. Specifically, the free energy of the uncomplexed bilayer of lateral area A is given by the charging free energy of a planar lipid layer $F_{el} = Ak_BT/(\pi l_B l_D) \int_0^{2\pi l_B l_D (\sigma/e)} dp \, \mathrm{arsinh}(p)$. That of the DNA is computed using the numerical solution of the Poisson–Boltzmann equation in cylindrical geometry, with the DNA being represented by a straight and uniformly charged rod. Finally, for the sandwich structure, the Poisson–Boltzmann equation is solved within a unit cell, shown as the shaded rectangular regions in Figure 13.5a, thereby accounting for the mobility of the cationic lipid within the membrane plane analogously to Eq. (13.6). The corresponding free energy of the sandwich (the sum of electrostatic free energy according to Eq. (13.2) and mixing contribution as in Eq. (13.5) is then obtained as function of average lipid composition in the sandwich and DNA-to-DNA distance d.

Similar calculations of phase diagrams have been carried out for two other systems (May, Harries, and Ben-Shaul 2000b), one where the membrane is very soft (see Figure 13.6a), which can be achieved by adding a cosurfactant such as hexanol (Safinya et al. 1989), and another one with a helper lipid that prefers to form the inverse-hexagonal phase (see Figure 13.6b). Both cases give rise to nonbilayer structures, either the inverse-hexagonal phase or the honeycomb complex. Generally, nonbilayer phases involve curved lipid layers, which can be described by the elastic curvature free energy (Helfrich 1973)

$$F_c = \int_A da \left[\frac{\kappa}{2}(c_1 + c_1 - c_0)^2 + \bar{\kappa} c_1 c_2 \right], \quad (13.12)$$

where c_1 and c_2 are the local principal curvatures at a given point of the membrane surface A; κ and $\bar{\kappa}$ are the bending stiffness and Gaussian modulus, respectively; and c_0 is the spontaneous curvature. (Note that stability with respect to

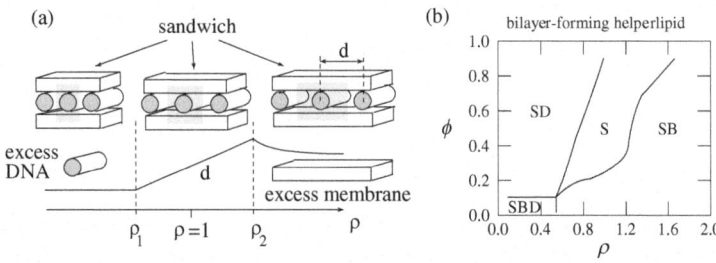

FIGURE 13.5 (a) Schematic illustration of the phase evolution involving the sandwich structure, coexisting either with excess DNA (for a lipid-to-DNA charge ratio $\rho<\rho_1$) or with excess membrane (for $\rho>\rho_2$). The point $\rho=1$ denotes the isoelectric point; d denotes the DNA-to-DNA distance as indicated. The shaded rectangular regions mark cross sections of the unit cell in which the Poisson–Boltzmann equation is solved. (b) Calculated phase diagram for a cationic lipid–DNA mixture with a bilayer-forming helper lipid (such as PC), shown as function of the cationic lipid-to-DNA charge ratio ρ and mole fraction ϕ of the cationic lipid. The notation is "S" for sandwich, "B" for uncomplexed bilayer, and "D" for naked DNA (see Figure 13.4). "SBD" refers to three-phase coexistence between "S", "B", and "D". (Adopted with permission May, Harries, and Ben-Shaul 2000b.)

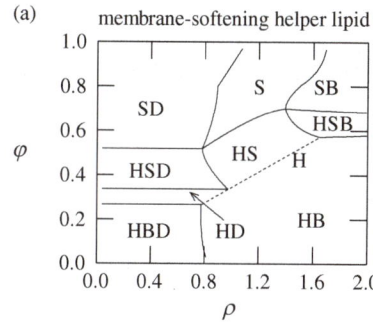 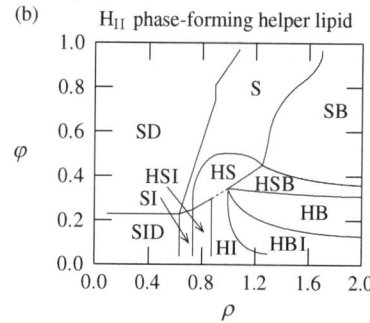

FIGURE 13.6 Calculated phase diagrams for a cationic lipid–DNA mixture with a membrane-softening helper lipid (such as added cosurfactant) (a), and an inverse-hexagonal (H_{II}) phase-forming helper lipid (b), shown as function of the cationic lipid- to-DNA charge ratio ρ and mole fraction ϕ of the cationic lipid. The notation is "S" for sandwich, "H" for honeycomb, "B" for uncomplexed bilayer, "I" for the bare inverse-hexagonal phase, and "D" for naked DNA (see Figure 13.4). The broken line refers to a single phase consisting of "H". "HSB" is three-phase coexistence between "H", "S", and "B", etc. (Adopted with permission May, Harries, and Ben-Shaul 2000b.)

curvature deformations demands $-2\bar{\kappa} < \bar{\kappa} < 0$.) A mixed membrane consisting of lipids with different propensity to form curved structures will have a composition-dependent spontaneous curvature $c_0 = c_0(\phi)$. This includes the case of a binary mixture consisting of a bilayer-forming cationic lipid and a helper lipid that tends to form the H_{II} phase.

The nature of the helper lipid has a significant influence on the predicted phase behaviour: nonbilayer phases form especially for small ϕ^m Figure 13.6. Unlike the sandwich structure, the honeycomb has no internal degrees of freedom other than its lipid composition; the single honeycomb phase is thus represented by a single line in the phase diagram (the broken line in both diagrams of Figure 13.6). Note that complete association of all DNA with all lipids takes place in a region of the phase diagram close to $\rho = 1$ and at sufficiently large ϕ. Excerpt for $\rho = 1$, the complexes are "overcharged", with an excess of either DNA or cationic lipid charges. The complexity of the phase diagrams in Figure 13.6 results from a subtle interplay between electrostatic and elastic interactions: cationic lipids drive complex formation with DNA but favour a planar aggregation geometry that does not optimally match the cylindrical shape of the DNA. Conversely, the helper lipid is electrostatically silent but able to "help" the membrane adopt a DNA-matching curvature. Various predictions of the phase diagrams in Figures 13.5 and 13.6 have been verified experimentally (Simberg et al. 2001).

13.4.2 Zwitterionic Lipids and Divalent Cations

Cationic lipids are potent and yet also toxic DNA condensing agents. A promising nontoxic alternative are zwitterionic lipids in conjunction with divalent metal cations such as Ca^{2+} or Mg^{2+}; these biocompatible lipids possess a large P^-–N^+ head group dipole that, when interacting with a divalent metal cation, is rendered effectively cationic (McManus, Rädler, and Dawson 2003). Indeed, in salt solutions containing calcium, zwitterionic lipids have not only been observed to engage in complex formation, but these complexes are able to transfect DNA (Bruni et al. 2011).

Compared to cationic lipids, it is much less obvious what drives the association between DNA and zwitterionic lipids, and what role the divalent cation plays in this process. In fact, recent modelling suggests that the association is not a process driven by counterion release but by a redistribution of calcium from DNA to the zwitterionic lipids (Mengistu, Bohinc, and May 2009). Let us briefly discuss an electrostatic model for zwitterionic lipids and their interaction with DNA in the presence of divalent cations.

A zwitterionic P^-–N^+ head group such as in PC can, approximately, be modelled as a set of two charges of opposite sign that are separated by a fixed distance $l \approx 0.5$ nm. The negative charge represents the head group's phosphate group (P^-) and the positive charge the amino group (N^+). The former is attached to the hydrocarbon–water interface ($x = 0$ if x points normal into the electrolyte), whereas the latter can freely rotate on the surface of a hemisphere in the aqueous solution. Hence, while the negative charges appear as an effective surface charge density at $x = 0$, the positive head group charge is, on average, smeared out and resides with a certain probability distribution $P(x)$ within the region $0 < x < l$. To find the probability $P(x)$, we add an additional conformational entropy contribution for all lipids contained within the lipid layer's lateral area A:

$$F_{\text{conf}} = \frac{k_B T}{a} \int_A da \frac{1}{l} \int_0^l dx P(x) \ln P(x) \quad (13.13)$$

to the total free energy $F = F_{\text{el}} + F_{\text{conf}}$ (with F_{el} according to Eq. 13.2). Minimization of F yields a Boltzmann distribution for the probability $P(x) = e^{-\Psi(x)}/q$ where $q = (1/l) \int_0^l dx e^{-\Psi(x)}$ ensures proper normalization of $P(x)$. The Boltzmann factor then appears in the modified Poisson–Boltzmann equation as an additional contribution $\sim e^{-\Psi(x)}$ within the head group region $0 < x < l$ (Mbamala, Fahr, and May 2006). This simple zwitterionic head group model has been used to study the divalent cation-mediated interaction of DNA with a zwitterionic lipid layer (Mengistu, Bohinc, and May 2009). Here, the modified Poisson–Boltzmann equation reads

$$\frac{\nabla^2 \Psi}{8\pi l_B} = n_0 \sinh\Psi + m_0\left(e^{\Psi} - e^{-2\Psi}\right) - \frac{e^{-\Psi}}{2laq}, \qquad (13.14)$$

where n_0 is the concentration of monovalent salt, m_0 is the concentration of the added divalent cation, $a = 0.65$ nm^2 is the cross-sectional area per zwitterionic lipid, and q is the partition sum as defined above. Note that the final term on the right-hand side of Eq. (13.14) is present only within the head group region but vanishes outside. Note also that the presence of DNA (modelled as a straight rod with surface charge density $\sigma^{\text{DNA}} = -0.6$ e/a that is parallel to the lipid layer) affects different lipids differently, dependent on their distance to the DNA. This renders the Poisson–Boltzmann equation a partial differential equation. Calculating the difference of the free energy F for the adsorbed and desorbed DNA yields the adsorption free energy ΔF, displayed in Figure 13.7 as function of the bulk concentration m_0 of added divalent cations. Different curves correspond to different bulk concentrations n_0 of monovalent salt as indicated. In the absence of divalent cations (see the inset in the left diagram of Fig. 13.7) adsorption is generally unfavorable; $\Delta F > 0$ unless a large concentration of monovalent salt is present. Adding divalent cations renders ΔF negative, and the gain in adsorption free energy is larger if less monovalent salt is present. The adsorption mechanism that the Poisson–Boltzmann model suggests is illustrated in Figure 13.7. Desorbed DNA immobilizes divalent counterions, one Ca^{2+} ion for each pair of phosphate groups on the DNA. Because the area density of phosphate groups on the lipid layer ($1/a$) is larger than that on the DNA ($0.6/a$), it would be more favourable for the calcium ions to associate with the lipids instead of DNA. The mechanism to achieve this is to adsorb DNA onto the zwitterionic lipid layer, thereby allowing Ca^{2+} to relocate from the DNA to the lipid phosphate groups. The phosphate groups of the DNA then become associated with the (positively charged) choline groups of the lipid. Hence, this mechanism is based entirely on an exchange of calcium ions and the charges of choline moieties between the phosphate groups of the DNA and lipids, and no release of counterions takes place.

13.4.3 Adsorption of DNA onto a Zwitterionic Lipid Monolayer

As discussed in the preceding subsection, the adsorption of DNA onto a zwitterionic lipid layer can be triggered by adding divalent cations. Langmuir monolayers at the air–water interface are a suitable system to study DNA adsorption. Gromelski and Brezesinski (2006) have investigated the influence of DNA and added divalent cations (Ca^{2+} and Mg^{2+}) on the pressure–area isotherm of a monolayer consisting of the zwitterionic lipid dimyristoylphosphatidyl-ethanolamine (DMPE). At the air–water interface, a DMPE monolayer undergoes a second-order transition from a tilted (for low applied lateral pressure) to an untilted (for high applied lateral pressure) liquid-ordered state of their lipid

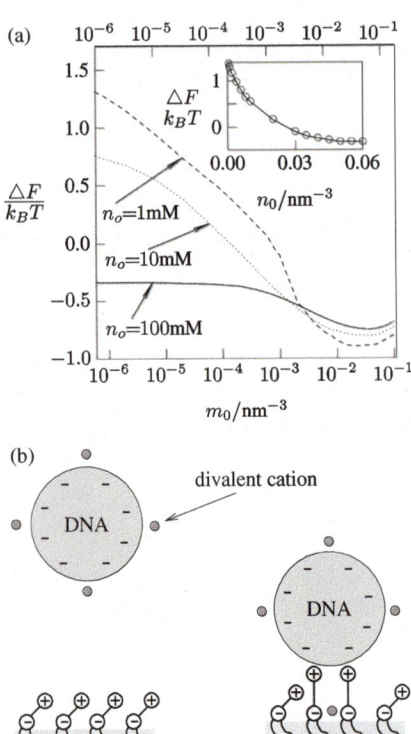

FIGURE 13.7 (a) The free energy ΔF of adsorbing DNA (per 1 nm unit length) onto a planar zwitterionic lipid layer as function of divalent cation bulk concentration m$_0$. Different curves correspond to different bulk concentration n$_0$ of monovalent salt as indicated. The inset shows ΔF as a function of n$_0$ in the absence of divalent cations (m$_0$ = 0). The DNA radius is 1 nm and its charge density $\sigma^{\text{DNA}} = -0.6$ e/a, where $a = 0.65$ nm^2 is the cross-sectional area per zwitterionic lipid. (b) Schematic illustration of the physical mechanism that underlies the adsorption of DNA onto a zwitterionic lipid layer. (Reprinted with permission Mengistu, Bohinc, and May 2009, Copyright 2009 American Chemical Society.)

hydrocarbon chains. Gromelski and Brezesinski demonstrated that when divalent cations are present, DNA adsorbs onto the zwitterionic lipid monolayer and decreases the tilt transition pressure. In the absence of both DNA and divalent cations, they observed the second-order phase transition at a lateral pressure of about 32 mN m^{-1}. If the DNA and divalent cations were added, then the transition from the tilted to the untilted ordered phase was shifted to a smaller lateral pressure of about 28 mN m^{-1}. The lowering of the tilt transition pressure indicates that the DNA condenses the lipid monolayer laterally (see Figure 13.8). Neither DNA nor divalent cations alone are able to individually cause a significant shift of the transition pressure. The shift in the tilt transition pressure is a nonspecific effect: it was observed that the shift was very similar irrespective of whether Ca^{2+} or Mg^{2+} was used. On the other hand, the interaxial distance between the neighbouring DNA strands is ion specific: Ca^{2+} decreases the interaxial distance more than Mg^{2+} due to its smaller hydrodynamic radius.

The transition from the tilted to the untilted ordered phase of lipid hydrocarbon chains upon increasing the

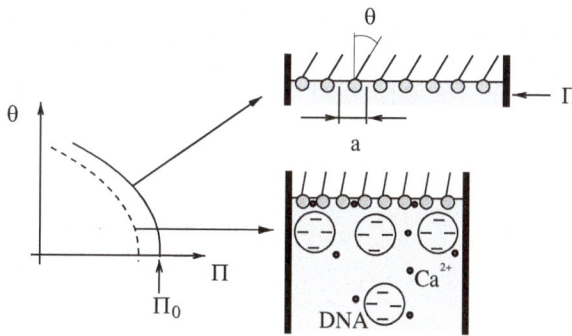

FIGURE 13.8 Schematic illustration of the second-order phase transition that occurs for increasing lateral pressure Π from a tilted (tilt angle $\theta>0$) to an untilted ($\theta=0$) ordered phase. The solid line in the pressure vs. tilt angle diagram corresponds to the absence of DNA. The critical lateral pressure in the absence of DNA is denoted by Π_0. The addition of DNA and divalent cations shifts the transition pressure from the tilted to the untilted ordered phase to a smaller lateral pressure (dashed line).

applied lateral pressure has also been modelled using a combination of Poisson–Boltzmann and Landau theory (Bohinc, Brezesinski, and May 2012). In the absence of DNA, the tilt transition for a zwitterionic monolayer can be described based on a Landau model with two order parameters, the cross-sectional area per lipid a and the tilt angle θ. The Landau-type mean-field free energy per lipid can then be written as

$$f_L(a,\theta) = \frac{B}{2}\left(\frac{a}{a_0}-1\right)^2 + \Delta\Pi a - \alpha\theta^2\left(\frac{a}{a_0}-1\right) + \frac{c}{2}\theta^4,$$
(13.15)

where the first term describes the area compressibility of the monolayer at a certain reference area a_0 per lipid; B is the corresponding compression modulus. The second term represents the pressure contribution beyond the reference pressure Π_0, where $\Delta\Pi = \Pi - \Pi_0$ denotes the pressure difference between the actual applied lateral pressure Π and the reference pressure Π_0. The last two terms represent a fourth-order expansion in terms of the tilt angle θ: the third term accounts for the area-dependent tilt elasticity with a coupling constant α (with $0<\alpha<\sqrt{Bc}$) between cross-sectional area and tilt angle, and the fourth term contains a constant $c>0$ and thus stabilizes the tilt. In thermal equilibrium, the free energy $f_L(a,\theta)$ adopts a minimum with respect to the cross-sectional area per lipid a and tilt angle θ. Minimization yields the tilt angle

$$\theta = \begin{cases} 0, & \text{for } \Delta\Pi > 0 \\ \pm\sqrt{\dfrac{-\Delta\Pi a_0 \alpha}{Bc-\alpha^2}}, & \text{for } \Delta\Pi > 0 \end{cases}$$
(13.16)

and the cross-sectional area per lipid

$$a = \begin{cases} a_0 - \dfrac{a_0^2}{B}\Delta\Pi & \text{for } \Delta\Pi > 0 \\ a_0 - \dfrac{a_0^2}{B-\alpha^2/c}\Delta\Pi & \text{for } \Delta\Pi < 0. \end{cases}$$
(13.17)

In the absence of DNA, the untilted phase corresponds to $\Pi>\Pi_0$, whereas the tilted phase is adopted for $\Pi<\Pi_0$. Hence, the reference values a_0 and Π_0 refer to the transition point in the absence of DNA. In the presence of DNA, the total free energy $f(a,\theta) = f_L(a,\theta) + f_{el}[d(a),a] - f_{el}^{(0)}(a)$ accounts for the divalent cation-mediated adsorption of DNA. Here, f_{el} (measured per lipid) is the electrostatic free energy of the monolayer with adsorbed DNA, and $f_{el}^{(0)}$ subtracts the electrostatic free energy of a bare monolayer. The distance $d(a)$ between adsorbed neighbouring DNA strands is determined by the thermal equilibrium between monolayer-adsorbed and free DNA in the aqueous solution (both must have the same chemical potential). Calculation of $f_{el}[d(a), a]$ for any given $d(a)$ proceeds along the lines of the preceding subsection, which has focused on the divalent cation-mediated adsorption of isolated DNA on a zwitterionic lipid layer of fixed lateral area. Numerical minimization of $f(a, \theta)$ predicts a relation between the tilt angle θ and $\Delta\Pi$, leading to a shift of the transition pressure. These calculations show that the adsorbed DNA induces a 10% reduction of the electrostatic contribution to the lateral pressure exerted by the monolayer. This indeed corresponds to a DNA-induced monolayer condensation and a corresponding shift of the tilt transition pressure to smaller values.

13.5 MACROION-INDUCED MEMBRANE BENDING

Lipid membranes are soft, fluid-like materials, able to adjust their shape in response to external forces. Membranes are engaged in bending deformations in a variety of biological processes including fusion and fission, endo- and exocytosis, viral entry into or budding of newly formed virions out of their host cells. Membrane bending is often facilitated by specialized proteins; an example is the banana-shaped BAR (Bin/Amphiphysin/Rvs) domain that is able to tubulate lipid vesicles (Farsad et al. 2001). Living cells can take up drug delivery vehicles such as nanoparticles or lipoplexes through specific receptor-mediated or unspecific endocytosis (Rejman et al. 2004). Yet, even for model systems – vesicles or liposomes – have endocytosis-like uptake processes of charged or uncharged nanoparticles been observed (Le Bihan et al. 2009, Tahara et al. 2012). It is no surprise that the physics underlying the internalization of particles into lipid vesicles has been studied in a number of theoretical investigations. In the following, we briefly discuss how membrane bending affects electrostatic membrane–macroion interactions.

Even in the absence of a macroion does membrane bending incur a free energy change. Specifically, when a bare charged lipid layer is subject to a bending deformation, there will be an electrostatic contribution to the curvature elastic moduli as defined in Eq. (13.12). Note that c_1 and c_2 are measured with respect to the so-called neutral surface at which bending does not affect the lateral area (Kozlov, Leikin, and Markin 1989). If the location of the charges coincides with the position of the neutral surface, the electrostatic contributions to the bending stiffness and Gaussian modulus are (Lekkerkerker 1989)

$$\kappa_{el} = \frac{k_B T}{2\pi} \frac{l_D}{l_B} \frac{(q-1)(q+2)}{q(q+1)},$$

$$\bar{\kappa}_{el} = -\frac{2k_B T}{\pi} \frac{l_D}{l_B} \int_{2/(1+q)}^{1} \frac{\ln z}{z-1} dz, \quad (13.18)$$

respectively, and the spontaneous curvature is $c_0^{el} = k_B T \ln[(q+1)/2]/(\kappa_{el}\pi l_B)$. Here, $p = 2\pi l_B l_D |\sigma|/e$ and $q = \sqrt{p^2 + 1}$. Figure 13.9 shows κ_{el} and $\bar{\kappa}_{el}$ as function of the mole fraction ϕ of charged lipids in a binary charged/uncharged lipid layer. Note that the electrostatic contribution to the bending stiffness is positive and typically smaller than (or on the order of) $k_B T$. However, it can be significantly larger if the neutral surface does not coincide with the location of the charges (Winterhalter and Helfrich 1988, May 1996). Note also the negative sign of $\bar{\kappa}_{el}$, thus disfavouring the formation of saddle curvatures ($c_1 = -c_2$). This is a consequence of the reduced space available to the diffuse counterion cloud for such a deformation.

When a uniformly charged macroion such as a protein or DNA adsorbs onto an initially flat, oppositely charged lipid bilayer, the macroion tends to curve the membrane so as to locally match its shape. Membrane bending is favourable from an electrostatic point of view but also entails a (nonelectrostatic) curvature energy penalty (see Eq. 13.12). If the former outweighs the latter, the macroion can become completely wrapped by the membrane (Harries, Ben-Shaul, and Szleifer 2004). This is the case, for example, when DNA interacts with a soft cationic membrane or one with a macroion-matching spontaneous curvature (see the phase diagrams in Figure 13.5). We can illustrate the interplay between electrostatic attraction and membrane curvature deformation for a cationic lipid bilayer that wraps around a DNA molecule (Sternberg, Sorgi, and Huang 1994) using a simple capacitor model, consisting of two concentric cylinders. The inner cylinder (of radius R^{DNA}) represents the DNA; it has a charge density $\sigma^{DNA} = e/(2\pi R^{DNA} l_{DNA})$, where $l_{DNA} = 0.17$ nm is the average distance between phosphate groups along double-stranded DNA. The outer cylinder (of radius R) describes a lipid layer with principal curvatures $c_1 = 1/R$ and $c_2 = 0$, corresponding bending stiffness κ, and vanishing spontaneous curvature $c_0 = 0$. Electroneutrality demands a charge density $\sigma = \sigma^{DNA} R^{DNA}/R$ on the lipid layer. The free energy (per unit length L of the DNA) is the sum of electrostatic capacitor energy and bending energy:

$$\frac{F}{L} = k_B T \frac{l_B}{l_{DNA}^2} \ln\left(\frac{R}{R^{DNA}}\right) + 2\pi R \frac{\kappa}{2}\left(\frac{1}{R}\right)^2. \quad (13.19)$$

Minimization yields $R = \pi l_{DNA}^2/(k_B T l_B)$. For $\kappa = 10\ k_B T$, we obtain $R = 1.3$ nm, barely larger than the radius of the DNA, $R^{DNA} = 1.0$ nm.

From a physical point of view, the engulfment of a spherical colloid by a lipid vesicle is of special interest as a simple model for endocytosis. For this process, phenomenological models consider the interplay between colloid–membrane adhesion and membrane bending energies. For increasing adhesion energy, the engulfment proceeds discontinuously from an unwrapped or partially wrapped to a fully wrapped state (Deserno 2004). Extended models have included the shape (Dasgupta, Auth, and Gompper 2013) and softness (Yi, Shi, and Gao 2011) of the colloid as well as the dynamics of the wrapping process (Gao, Shi, and Freund 2005). Electrostatic interactions as the driving force for colloid binding and subsequent wrapping have been studied in a number of simulations (Fošnarič et al. 2009, Li and Gu 2010). Figure 13.7 shows the wrapping of a colloid by a charged vesicle that contains a small number of oppositely charged lipids. The vesicle is modelled as a

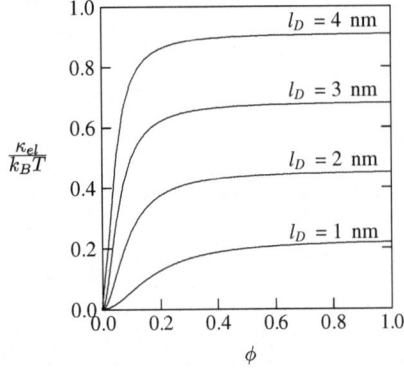
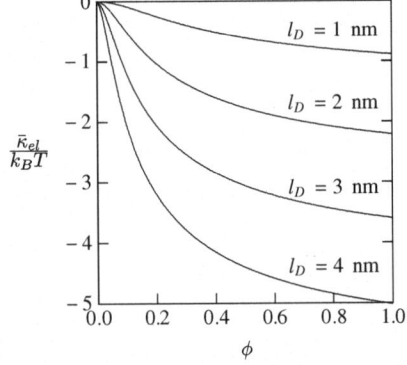

FIGURE 13.9 The electrostatic contributions to the bending stiffness κ_{el} and Gaussian modulus $\bar{\kappa}_{el}$ of a binary charged/uncharged lipid layer according to Eq. (13.18), plotted as function of the mole fraction ϕ of charged (anionic or cationic) lipids. Note that ϕ determines the surface charge density $|\sigma| = \phi e/a$ of the lipid layer, where $a = 0.65$ nm^2 denotes the cross-sectional area per lipid. Different curves correspond to different values of the Debye screening length l_D as indicated.

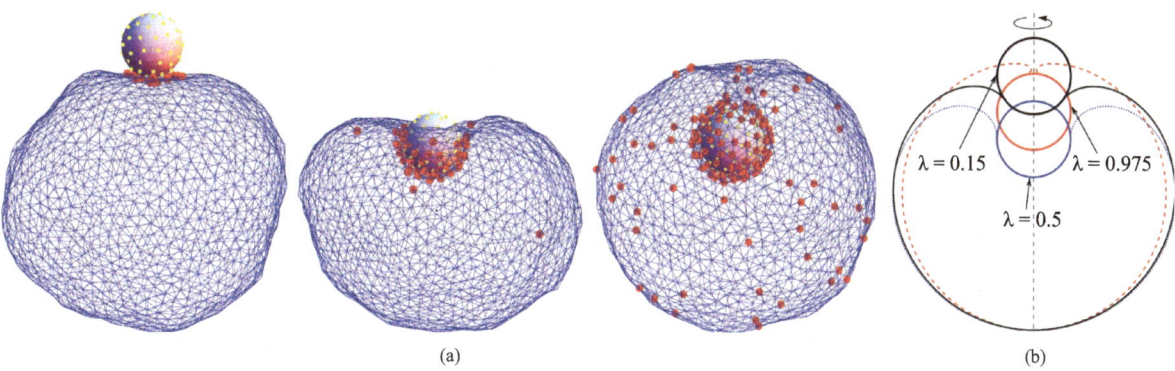

FIGURE 13.10 The wrapping of a spherical colloid by a vesicle modelled on the basis of Monte Carlo simulations. Shown are three snapshots (a) of representative colloid–vesicle configurations obtained from simulations for different numbers of charged vertices on the membrane. The shape of the vesicle is represented by a triangulated surface; mobile charges on the vesicle and fixed charges on the colloid are indicated by dots. (b) shows cross-sections of colloid–vesicle complexes obtained by minimizing the bending energy of the vesicle; λ denotes the degree of wrapping. (Reproduced with permission Fošnarič et al. 2009.)

triangulated closed surface whose fluidity is ensured by a bond-flip mechanism. Charges reside only on a small number of vertices, yet they are mobile and can thus migrate towards the oppositely charged colloid. The three different colloid–vesicle complexes in Figure 13.10 contain increasing numbers of charged lipids, inducing a sequence from an unwrapped state to partial wrapping and full engulfment. The right diagram in Figure 13.10 shows different degrees of wrapping according to a phenomenological model that is based on minimizing the vesicle shape for various electrostatic colloid adsorption strengths (Fošnarič et al. 2009).

13.6 AMPHIPATHIC PEPTIDE–MEMBRANE INTERACTION

Antimicrobial peptides are part of the innate immune system; they constitute a diverse group of cationic peptides with antimicrobial activity, often executed through destabilization and lysis of the target membrane (Brogden 2005). Many of these peptides tend to be unstructured in solution, adopting an amphipathic, alpha-helical structure upon interaction with a lipid bilayer. Examples include melittin, magainin, protegrins, alamethicin, and cecropin. Here, hydrophobic residues on one side of the helix face polar amino acids on the opposite side, with some of the polar residues being cationic (typically lysine or arginine). Their positive excess charge helps amphipathic peptides to target bacterial membranes that typically carry more anionic lipids (such as phosphatidyl-glycerol and cardiolipin) than mammalian cells (with their mostly zwitterionic lipids such as PC and sphingolipids in the plasma membrane's outer leaflet).

There is considerable evidence that the initial adsorption of amphipathic peptides is mainly due to electrostatic interactions, including the increase in adsorption when the membrane or peptide contains more anionic lipids or cationic residues, respectively, or when the ionic strength of the solution is lowered (Seelig 2004). Recall that the electrostatic adsorption is driven by the entropy gain associated with the release of counterions; the migration of anionic lipids into the adsorption site of the peptide, membrane bending deformations, and a range of nonelectrostatic interactions (especially hydrophobic interactions) further facilitate the adsorption process. For example, upon binding to the membrane, the peptide's hydrophobic face allows them to partially penetrate into the hydrocarbon core of the target membrane (Dathe et al. 1996) and to either translocate through the bilayer or initiate membrane destabilization through either a so-called "carpet" mechanism (Shai 1999) or membrane pore formation (Bechinger 1997). The carpet mechanism is believed to result in membrane solubilization, where a high amount of peptides initially cover the membrane surface in a carpet-like manner and above a certain density start to form mixed lipid–peptide micelles. Membrane solubilization is well known for short-chain surfactants such as hexanol and is driven there by the more favourable packing of mixed short and long hydrocarbon chains in micellar geometries.

Two pore types have been suggested: the "toroidal-pore" and the "barrel-stave" pore; both types are illustrated in Figure 13.11. The toroidal-pore forms predominantly for highly charged peptides such as magainin, where electrostatic repulsion between neighbouring peptides widens the pore radius and induces lipids to intercalate. The head groups of the intercalating anionic lipids benefit from favourable electrostatic interactions with the peptides. At the same time, the intercalating lipids experience an unfavourable chain packing geometry, similar to that in a membrane rim. Electrostatic modelling suggests that the optimal peptide-to-peptide distance tends to allow for the intercalation of one single lipid head group (Zemel, Fattal, and Ben-Shaul 2003) and decreases if added salt increases electrostatic screening. Weakly charged peptides such as alamethicin form barrel-stave pores. Here, the role of the peptides is to mainly reduce the line tension of the pore rim; electrostatics plays no particular role in this process. The number of peptides in a pore can vary as evidenced by the observation of multiple conductance levels (Boheim 1974).

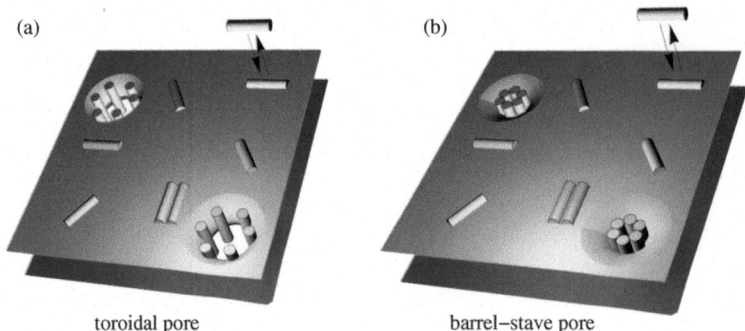

FIGURE 13.11 Schematic illustration of pore formation in lipid bilayers induced by amphipathic peptides: toroidal (a) and barrel-stave (b). Helical peptides are represented by cylinders. Some monomeric peptides and a peptide dimer are also shown.

We point out that Molecular Dynamics simulations suggest that pores possess a high degree of disorder, exhibiting structural irregularities (Sengupta et al. 2008) and variable peptide orientations, and often evade the simple scheme of barrel-stave or toroidal (Leontiadou, Mark, and Marrink 2006).

The formation of a pore involves the self-assembly of membrane-adsorbed peptides and a subsequent reorientation of the helix' long axis from parallel to perpendicular with respect to the membrane surface. The peptides then span through the entire membrane, directing their hydrophobic side chains towards the hydrocarbon core of the hosting lipid bilayer. Pore formation through peptide self-assembly is a cooperative process that starts at a critical concentration similar to the CMC at the onset of micelle formation (Israelachvili, Mitchell, and Ninham 1977). As there, the critical concentration reflects a favourable pore formation enthalpy, which must be on the order of several $k_B T$ for each of the involved peptides. Where does this energy gain come from? An isolated adsorbed peptide protrudes partially into the hydrocarbon core of the host membrane; this entails an energy penalty due to a perturbation in lipid chain packing. Detailed chain packing calculations suggest that peptides can avoid this penalty by associating into a pore, where the chain perturbation, measured per peptide, is lower by about 4–5 $k_B T$ (Zemel, Ben-Shaul, and May 2005).

A class of peptides that has attracted considerable interest are cell-penetrating peptides (Richard et al. 2003). These peptides, which are generally rich in cationic amino acids such as lysine or arginine, are able to translocate through cell and model membranes. An example is the TAT peptide from the HIV transactivator protein TAT, which has six arginine and two lysine residues, was the first discovered cell-penetrating peptide, and remains one of the best studied. No consensus has been reached regarding the uptake mechanism of TAT (and other cell-penetrating) peptides. Evidence exists not only for endocytotic pathways (specific or unspecific) but also for direct translocation mechanisms. It has been pointed out (Schmidt et al. 2010) that the translocation activity of cell-penetrating peptides is likely related to their ability to produce negative Gaussian curvature (that is a negative sign of the product $c_1 c_2$ in Eq. 13.12).

Identification of the physical mechanism how lysine or arginine promotes negative Gaussian curvature is a work in progress.

ACKNOWLEDGEMENTS

S.M. thanks Avinoam Ben-Shaul, Daniel Harries, David Andelman, and Gerard Wong for numerous insightful discussions. G.V.B. acknowledges a doctoral scholarship from CAPES Foundation/Brazil Ministry of Education (Grant No. 9466/13-4).

BIBLIOGRAPHY

Arbuzova, A., D. Murray, and S. McLaughlin. 1998. MARCKS, membranes, and calmodulin: Kinetics of their interaction. *Biochim. Biophys. Acta, Biomembr.* 1376(3): 369–379.

Bechinger, B. 1997. Structure and functions of channel-forming peptides: Magainins, cecropins, melittin and alamethicin. *J. Membr. Biol.* 156(3): 197–211.

BenTal, N., B. Honig, C. Miller, and S. McLaughlin. 1997. Electrostatic binding of proteins to membranes. Theoretical predictions and experimental results with charybdotoxin and phospholipid vesicles. *Biophys. J.* 73(4): 1717–1727.

Boheim, G. 1974. Statistical analysis of alamethicin channels in black lipid membranes. *J. Membr. Biol.* 19(1): 277–303.

Bohinc, K., G. Brezesinski, and S. May. 2012. Modeling the influence of adsorbed DNA on the lateral pressure and tilt transition of a zwitterionic lipid monolayer. *Phys. Chem. Chem. Phys.* 14(30): 10613–10621.

Brockman, H. 1994. Dipole potential of lipid-membranes. *Chem. Phys. Lipids* 73(1–2): 57–79.

Brogden, K. A. 2005. Antimicrobial peptides: Pore formers or metabolic inhibitors in bacteria? *Nat. Rev. Microbiol.* 3(3): 238–250.

Bruni, P., O. Francescangeli, M. Marini, G. Mobbili, M. Pisani, and A. Smorlesi. 2011. Can neutral liposomes be considered as genetic material carriers for human gene therapy? *Mini-Reviews in Organic Chem.* 8(1): 38–48.

Dasgupta, S., T. Auth, and G. Gompper. 2013. Wrapping of ellipsoidal nano-particles by fluid membranes. *Soft Matter* 9(22): 5473–5482.

Dathe, M., M. Schümann, T. Wieprecht, A. Winkler, M. Beyermann, E. Krause, K. Matsuzaki, O. Murase, and M. Bienert. 1996. Peptide helicity and membrane surface charge modulate the balance of electrostatic and hydrophobic interactions with lipid bilayers and biological membranes. *Biochemistry* 35(38): 12612–12622.

Deserno, M. 2004. Elastic deformation of a fluid membrane upon colloid binding. *Phys. Rev. E* 69(3): 031903.

Elouahabi, A., and J. M. Ruysschaert. 2005. Formation and intracellular trafficking of lipoplexes and polyplexes. *Mol. Therapy* 11(3): 336–347.

Elson, E. L., E. Fried, J. E Dolbow, and G. M. Genin. 2010. Phase separation in biological membranes: Integration of theory and experiment. *Annu. Rev. Biophys.* 39: 207.

Epand, R. M. 2008. Proteins and cholesterol-rich domains. *Biochim. Biophys. Acta, Biomembr.* 1778(7–8): 1576–1582.

Fan, J., M. Sammalkorpi, and M. Haataja. 2010. Formation and regulation of lipid microdomains in cell membranes: Theory, modeling, and speculation. *FEBS Lett.* 584(9): 1678–1684.

Farago, O., and N. Gronbech-Jensen. 2009. Simulation of self-assembly of cationic lipids and DNA into structured complexes. *J. Am. Chem. Soc.* 131(8): 2875–2881.

Farsad, K., N. Ringstad, K. Takei, S. R. Floyd, K. Rose, and P. D. Camilli. 2001. Generation of high curvature membranes mediated by direct endophilin bilayer interactions. *J. Cell Biol.* 155(2): 193–200.

Felgner, P. L., T. R. Gadek, M. Holm, R. Roman, H. W. Chan, M. Wenz, J. P. Northrop, G. M. Ringold, and M. Danielsen. 1987. Lipofection: A highly efficient, lipid-mediated DNA-transfection procedure. *Proc. Natl. Acad. Sci. U. S. A.* 84: 7413–7417.

Fošnarič, M., A. Iglič, D. M. Kroll, and S. May. 2009. Monte Carlo simulations of complex formation between a mixed fluid vesicle and a charged colloid. *J. Chem. Phys.* 131(10): 105103.

Gambhir, A., G. Hangyas-Mihalyne, I. Zaitseva, D. S. Cafiso, J. Y. Wang, D. Murray, S. N. Pentyala, S. O. Smith, and S. McLaughlin. 2004. Electrostatic sequestration of PIP2 on phospholipid membranes by basic/aromatic regions of proteins. *Biophys. J.* 86(4): 2188–2207.

Gao, H. J., W. D. Shi, and L. B. Freund. 2005. Mechanics of receptor-mediated endocytosis. *Proc. Natl. Acad. Sci. U. S. A.* 102(27): 9469–9474.

Gelbart, W. M., and R. Bruinsma. 1997. Compositional-mechanical instability of interacting mixed lipid membranes. *Phys. Rev. E* 55(1): 831–835.

Gerke, V., and S. E. Moss. 1997. Annexins and membrane dynamics. *Biochim. Biophys. Acta, Mol. Cell Res.* 1357(2): 129–154.

Gershon, H., R. Ghirlando, S. B. Guttman, and A. Minsky. 1993. Mode of formation and structural features of DNA-cationic liposome complexes used for transfection. *Biochemistry* 31: 7110–7119.

Gromelski, S., and G. Brezesinski. 2006. DNA condensation and interaction with Zwitterionic phospholipids mediated by divalent cations. *Langmuir* 22(14): 6293–6301.

Haleva, E., N. Ben-Tal, and H. Diamant. 2004. Increased concentration of polyvalent phospholipids in the adsorption domain of a charged protein. *Biophys. J.* 86(4): 2165–2178.

Hammond, A. T., F. A. Heberle, T. Baumgart, D. Holowka, B. Baird, and G. W. Feigenson. 2005. Crosslinking a lipid raft component triggers liquid ordered-liquid disordered phase separation in model plasma membranes. *Proc. Natl. Acad. Sci. U. S. A.* 102(18): 6320–6325.

Harries, D., A. Ben-Shaul, and I. Szleifer. 2004. Enveloping of charged proteins by lipid bilayers. *J. Phys. Chem. B* 108(4): 1491–1496.

Harries, D., S. May, and A. Ben-Shaul. 2013. Counterion release in membrane–biopolymer interactions. *Soft Matter* 9(39): 9268–9284.

Hartvig, R. A., M. van de Weert, J. Ostergaard, L. Jorgensen, and H. Jensen. 2011. Protein adsorption at charged surfaces: The role of electrostatic interactions and interfacial charge regulation. *Langmuir* 27(6): 2634–2643.

Heimburg, T., B. Angerstein, and D. Marsh. 1999. Binding of peripheral proteins to mixed lipid membranes: Effect of lipid demixing upon binding. *Biophys. J.* 76: 2575–2586.

Helfrich, W. 1973. Elastic properties of lipid bilayers: Theory and possible experiments. *Z. Naturforsch. C* 28:693–703.

Herweijer, H., and J. A. Wolff. 2003. Progress and prospects: Naked DNA gene transfer and therapy. *Gene Ther.* 10(6): 453–458.

Hirsch-Lerner, D., M. Zhang, H. Eliyahu, M. E. Ferrari, C. J. Wheeler, and Y. Barenholz. 2005. Effect of "helper lipid" on lipoplex electrostatics. *Biochim. Biophys. Acta, Biomembr.* 1714(2): 71–84.

Hofmeister, F. 1888. Zur Lehre von der Wirkung der Salze. *Arch. Exptl. Pathol. Pharmakol.* 24: 247–260.

Honig, B., and A. Nicholls. 1995. Classical electrostatics in biology and chemistry. *Science* 268(5214): 1144–1149.

Israelachvili, J. N., J. Mitchell, and B. W. Ninham. 1977. Theory of self-assembly of lipid bilayers and vesicles. *Biophys. Biochim. Acta.* 470: 185–201.

Koltover, I., T. Salditt, J. O. Rädler, and C. R. Safinya. 1998. An inverted hexagonal phase of cationic liposome-DNA complexes related to DNA release and delivery. *Science* 281:78–81.

Kooijman, E. E., and K. N. J. Burger. 2009. Biophysics and function of phosphatidic acid: A molecular perspective. *Biochim. Biophys. Acta, Mol. Cell Biol. Lipids* 1791(9): 881–888.

Koynova, R., and R. C. MacDonald. 2003. Mixtures of cationic lipid O-ethylphosphatidylcholine with membrane lipids and DNA: Phase diagrams. *Biophys. J.* 85: 2449–2465.

Kozlov, M. M., S. L. Leikin, and V. S. Markin. 1989. Elastic properties of interfaces - elasticity moduli and spontaneous geometric characteristics. *J. Chem. Society-Faraday Transactions II* 85:277–292.

Langner, M., and K. Kubica. 1999. The electrostatics of lipid surfaces. *Chem. Phys. Lipids* 101(1): 3–35.

Lasic, D. D., H. Strey, M. C. A. Stuart, R. Podgornik, and P. M. Frederik. 1997. The structure of DNA-liposome complexes. *J. Am. Chem. Soc.* 119: 832–833.

Le Bihan, O., P. Bonnafous, L. Marak, T. Bickel, S. Trépout, S. Mornet, F. De Haas, H. Talbot, J.-C. Taveau, and O. Lambert. 2009. Cryo-electron tomography of nanoparticle transmigration into liposome. *J. Struct. Biol.* 168(3): 419–425.

Lekkerkerker, H. N. W. 1989. Contribution of the electric double layer to the curvature elasticity of charged amphiphilic monolayers. *Physica A.* 159: 319–328.

Leontiadou, H., A. E. Mark, and S. J. Marrink. 2006. Antimicrobial peptides in action. *J. Am. Chem. Soc.* 128(37): 12156–12161.

Li, Yang, and Ning Gu. 2010. Thermodynamics of charged nanoparticle adsorption on charge-neutral membranes: A simulation study. *J. Phys. Chem. B* 114(8): 2749–2754.

Lingwood, D., J. Ries, P. Schwille, and K. Simons. 2008. Plasma membranes are poised for activation of raft phase coalescence at physiological temperature. *Proc. Natl. Acad. Sci. U. S. A.* 105(29): 10005–10010.

Lingwood, D., and K. Simons. 2010. Lipid rafts as a membrane-organizing principle. *Science* 327(5961): 46–50.

Loew, S., A. Hinderliter, and S. May. 2009. Stability of protein-decorated mixed lipid membranes: The interplay of lipid-lipid, lipid-protein, and protein-protein interactions. *J. Chem. Phys.* 130(4): 045102.

Loew, S., E. E Kooijman, and S. May. 2013. Increased pH-sensitivity of protein binding to lipid membranes through the electrostatic-hydrogen bond switch. *Chem. Phys. Lipids* 169:9–18.

Lund, M., T. Å. Kesson, and B. Jonsson. 2005. Enhanced protein adsorption due to charge regulation. *Langmuir* 21(18): 8385–8388.

Lund, M., and B. Jönsson. 2005. On the charge regulation of proteins. *Biochemistry* 44(15): 5722–5727.

May, S. 1996. Curvature elasticity and thermodynamic stability of electrically charged membranes. *J. Chem. Phys.* 105: 8314–8322.

May, S., D. Harries, and A. Ben-Shaul. 2000a. Lipid demixing and protein-protein interactions in the adsorption of charged proteins on mixed membranes. *Biophys. J.* 79(4): 1747–1760.

May, S., D. Harries, and A. Ben-Shaul. 2000b. The phase behavior of cationic lipid-DNA complexes. *Biophys. J.* 78(4): 1681–1697.

Mbamala, E. C., A. Ben-Shaul, and S. May. 2005. Domain formation induced by the adsorption of charged proteins on mixed lipid membranes. *Biophys. J.* 88(3): 1702–1714.

Mbamala, E. C., A. Fahr, and S. May. 2006. Electrostatic model for mixed cationic-zwitterionic lipid bilayers. *Langmuir* 22(11): 5129–5136.

McLaughlin, S., and A. Aderem. 1995. The myristoyl-electrostatic switch - a modulator of reversible protein-membrane interactions. *Trends Biochem. Sci.* 20(7): 272–276.

McManus, J. J., J. O. Rädler, and K. A. Dawson. 2003. Does calcium turn a Zwitterionic lipid cationic? *J. Phys. Chem. B* 107(36): 9869–9875.

Mengistu, D. H., K. Bohinc, and S. May. 2009. Binding of DNA to Zwitterionic lipid layers mediated by divalent cations. *J. Phys. Chem. B* 113(36): 12277–12282.

Menke, M., V. Gerke, and C. Steinem. 2005. Phosphatidylserine membrane domain clustering induced by annexin A2/S100A10 heterotetramer. *Biochemistry* 44(46): 15296–15303.

Murray, D., L. H. Matsumoto, C. A. Buser, J. Tsang, C. T. Sigal, N. Ben-Tal, B. Honig, M. D. Resh, and S. McLaughlin. 1998. Electrostatics and the membrane association of Src: Theory and experiment. *Biochem.* 37(8): 2145–2159.

Pandey, A. P., F. Haque, J. C. Rochet, and J. S. Hovis. 2009. Clustering of alpha-synuclein on supported lipid bilayers: Role of anionic lipid, protein, and divalent ion concentration. *Biophys. J.* 96(2): 540–551.

Radhakrishnan, A., and H. McConnell. 2005. Condensed complexes in vesicles containing cholesterol and phospholipids. *Proc. Natl. Acad. Sci. U. S. A.* 102(36): 12662–12666.

Rädler, J. O., I. Koltover, T. Salditt, and C. R. Safinya. 1997. Structure of DNA-cationic liposome complexes: DNA intercalation in multilamellar membranes in distinct interhelical packing regimes. *Science* 275:810–814.

Rejman, J., V. Oberle, I. S. Zuhorn, and D. Hoekstra. 2004. Size-dependent internalization of particles via the pathways of clathrin- and caveolae-mediated endocytosis. *Biochem. J.* 377: 159–169.

Reynwar, B. J., and M. Deserno. 2008. Membrane composition-mediated protein-protein interactions. *Biointerphases* 3(2): FA117–FA124.

Richard, J. P., K. Melikov, E. Vives, C. Ramos, B. Verbeure, M. J. Gait, L. V Chernomordik, and B. Lebleu. 2003. Cell-penetrating peptides a reevaluation of the mechanism of cellular uptake. *J. Biol. Chem.* 278(1): 585–590.

Safinya, C. R., E. B. Sirota, D. Roux, and G. S. Smith. 1989. Universality in interacting membranes: The effect of cosurfactants on the interfacial rigidity. *Phys. Rev. Lett.* 62: 1134–1137.

Safran, S. A. 1994. *Statistical Thermodynamics of Surfaces, Interfaces, and Membranes.* Addison-Wesley Reading, MA.

Schmidt, N., A. Mishra, G. H. Lai, and G. C. L. Wong. 2010. Arginine-rich cell-penetrating peptides. *FEBS Lett.* 584(9): 1806–1813.

Seelig, J. 2004. Thermodynamics of lipid–peptide interactions. *Biochim. Biophys. Acta, Biomembr.* 1666(1): 40–50.

Sengupta, D., H. Leontiadou, A. E. Mark, and S.-J. Marrink. 2008. Toroidal pores formed by antimicrobial peptides show significant disorder. *Biochim. Biophys. Acta, Biomembr.* 1778(10): 2308–2317.

Shai, Y. 1999. Mechanism of the binding, insertion and destabilization of phospholipid bilayer membranes by α-helical antimicrobial and cell non-selective membrane-lytic peptides. *Biochim. Biophys. Acta, Biomembr.* 1462(1): 55–70.

Simberg, D., D. Danino, Y. Talmon, A. Minsky, M. E. Ferrari, C. J. Wheeler, and Y. Barenholz. 2001. Phase behavior, DNA ordering, and size instability of cationic lipoplexes - Relevance to optimal transfection activity. *J. Biol. Chem.* 276(50): 47453–47459.

Simons, K., and W. L. C. Vaz. 2004. Model systems, lipid rafts, and cell membranes. *Ann. Rev. Biophys. Biomol. Struct.* 33: 269–295.

Sternberg, B., F. L. Sorgi, and L. Huang. 1994. New structures in complex-formation between DNA and cationic liposomes visualized by freeze-fracture electron-microscopy. *FEBS Lett.* 356: 361–366.

Tahara, K., S. Tadokoro, Y. Kawashima, and N. Hirashima. 2012. Endocytosis-like uptake of surface-modified drug nanocarriers into giant unilamellar vesicles. *Langmuir* 28(18): 7114–7118.

Tzlil, S., D. Murray, and A. Ben-Shaul. 2008. The "Electrostatic-Switch" mechanism: Monte Carlo study of MARCKS-membrane interaction. *Biophys. J.* 95(4): 1745–1757.

Vance, J. E, and D. E. Vance. 2008. *Biochemistry of Lipids, Lipoproteins and Membranes.* Fifth edition. Elsevier, Amsterdam.

Wagner, K., D. Harries, S. May, V. Kahl, J. O. Rädler, and A. Ben-Shaul. 2000. Direct evidence for counterion release upon cationic lipid-DNA condensation. *Langmuir* 16(2): 303–306.

Wang, J. Y., A. Gambhir, S. McLaughlin, and D. Murray. 2004. A computational model for the electrostatic sequestration of PI(4,5)P-2 by membrane-adsorbed basic peptides. *Biophys. J.* 86(4): 1969–1986.

Warszynski, P., and Z. Adamczyk. 1997. Calculations of double-layer electrostatic interactions for the sphere/plane geometry. *J. Colloid Interface Sci.* 187(2): 283–295.

Winterhalter, M., and W. Helfrich. 1988. Effect of surface charge on the curvature elasticity of membranes. *J. Phys. Chem.* 92(24): 6865–6867.

Yew, N. S., and R. K. Scheule. 2005. Toxicity of cationic lipid-DNA complexes. *Nonviral Vectors Gene Ther.* 53: 189–214.

Yi, Xin, Xinghua Shi, and Huajian Gao. 2011. Cellular uptake of elastic nanoparticles. *Phys. Rev. Lett.* 107(9): 098101.

Young, B. P., J. J. H. Shin, R. Orij, J. T. Chao, S. C. Li, X. L. Guan, A. Khong, E. Jan, M. R. Wenk, W. A. Prinz, G. J. Smits, and C. J. R. Loewen. 2010. Phosphatidic acid is a pH biosensor that links membrane biogenesis to metabolism. *Science* 329(5995): 1085–1088.

Zemel, A., A. Ben-Shaul, and S. May. 2005. Perturbation of a lipid membrane by amphipathic peptides and its role in pore formation. *Eur. Biophys. J.* 34(3): 230–242.

Zemel, A., D. R. Fattal, and A.-n. Ben-Shaul. 2003. Energetics and self-assembly of amphipathic peptide pores in lipid membranes. *Biophys. J.* 84(4): 2242–2255.

14 Lipid-Based Bioanalytical Sensors

Marta Bally and Hudson Pace
Chalmers University of Technology, Göteborg, Sweden
Umeå University, Umeå, Sweden

Fredrik Höök
Chalmers University of Technology, Göteborg, Sweden

CONTENTS

14.1 Introduction 241
14.2 Immobilization of Lipid Assemblies at a Sensor Surface 243
    14.2.1 Planar Lipid Membranes 243
        14.2.1.1 Formation of Planar Lipid Bilayers 243
        14.2.1.2 Patterns of Planar Lipid Membranes 244
    14.2.2 Surface-Tethered Lipid Vesicles 245
14.3 Surface-Based Biosensing Techniques 246
    14.3.1 Label-Mediated Detection of Biomolecular Recognition 246
        14.3.1.1 Optical Sensing 246
        14.3.1.2 Electrochemical Sensing 246
    14.3.2 Label-Free Biosensors 246
        14.3.2.1 Optical Biosensors 246
        14.3.2.2 Acoustic Biosensors 247
    14.3.3 Probing Transport through the Membrane 247
        14.3.3.1 Electrochemical Sensing 247
        14.3.3.2 Optical Sensing of Membrane Transport 248
14.4 Membrane–Protein-Based Biosensors 248
    14.4.1 Integration of Functional Membrane Proteins in the Sensing Platform 248
    14.4.2 Biosensing Approaches to Detect Interactions between Membrane Proteins and Ligands 250
    14.4.3 Probing Membrane–Protein-Mediated Transport across the Membrane 251
        14.4.3.1 Biosensors to Study Ion Channels 252
        14.4.3.2 Engineering Sensing Platforms Based on Natural Pores 252
14.5 Probing Lipid Membrane-Mediated Interactions 254
    14.5.1 Probing Glycolipid-Mediated Interactions 254
    14.5.2 Probing Enzymatic or Peptide-Mediated Membrane Disruption Events 256
    14.5.3 Single-Vesicle Assays to Probe Membrane-Curvature–Specific Proteins 256
14.6 Liposomes as Nanoscale Labels for Signal Amplification 256
14.7 Lipid-Based Nanoreactors 259
14.8 Exploiting Force-Driven Membrane Component Manipulation for Biosensing 260
14.9 Concluding Remarks 261
References 261

14.1 INTRODUCTION

Over the past decades, sensors for the detection and characterization of specific biomolecular interactions have seen their application field grow considerably. Generally speaking, a biosensor is a bioanalytical tool combining a biological or biologically inspired recognition element with a transducing element capable of converting a biomolecular binding event into a detectable signal, typically of optical, electrical, magnetic, or gravimetric nature. By providing information on biomolecular interactions (*e.g.*, between a receptor and a ligand, an antigen and an antibody, complementary oligonucleotides, the action of an enzyme on a substrate, etc.), biosensors have the capability of supplying fundamental understanding to the immensely complex biochemical interaction patterns orchestrating life. They therefore play an increasingly important role in fundamental biological as well as medical and pharmaceutical research. Some biosensors also allow for the detection and identification of target molecules in complex biological samples. This has stimulated the development of portable point-of-care and field-oriented devices for medical diagnostics but also food and environmental monitoring (Cooper 2002, 2006; Grieshaber et al. 2008; Kirsch et al. 2013; Rodriguez-Mozaz et al. 2006; Sadana and Sadana 2010).

The lipid bilayer is one of the most remarkable structures in living organisms. Through evolutionary design, this 5–10 nm thin, two-dimensional fluid has evolved to become the physical boundary of all living cells, acting as a compartmentalizing hydrophobic barrier preventing solutes from freely diffusing in and out of cells and their organelles. With the help of specialized membrane proteins responsible for selective transport of ions and signals across this barrier, the compartmentalization provided by the cell membrane makes it possible for cells to maintain energy production, biosynthesis, and communication. To grasp the stunning nature of cellular membranes, it is instructive to consider that the outer membrane of the simplest cell contains millions of lipid molecules, each with a footprint of around 1 nm^2, and that about half its mass is made up of different membrane-associated proteins, each with a footprint of around 20 nm^2. Furthermore, since cellular membranes contain up to several hundred different types of lipids and proteins, the combinatorial possibilities of static and dynamic protein–protein, lipid–lipid, and protein–lipid associations are essentially limitless. In view of this complexity and of the essential role of the cell membrane in mediating the interaction of a cell with the extracellular space, it is not surprising that biomolecular components associated with the cell membrane have emerged as the central drug target class. However, applications of bioanalytical sensors and screening technologies for studies of the cell membrane and its components often fail to adequately match the needs, as reflected by our significantly lower understanding of the nature and function membrane-residing proteins than of water-soluble proteins. Consequently, well-defined lipid structures formed by spontaneous self-assembly of their amphiphilic molecular building blocks have attracted much attention for a variety of biotechnical and bioanalytical applications. Of the different three-dimensional lipid assemblies formed in aqueous solution, lamellar phases and in particular lipid bilayers are, due to their structural similarity with native cellular membranes, most commonly used in biosensing applications. Bilayers and multilayers thereof can take a variety of shapes: they can be flat and supported onto sensor substrates, forming so-called supported lipid bilayers (Figure 14.1a), they can form nanodiscs, *i.e.*, circular areas of lipid bilayers stabilized by a membrane scaffolding protein (Figure 14.1c), or adopt tubular or spherical configurations (Figure 14.1b). In the latter case, the lipids form so-called vesicles with diameters that can range from a few tens of nanometres to tens of micrometres.

The increased interest in lipid assemblies for biosensor applications can be understood from a number of reasons. As already indicated, lipids, and in particular phospholipids, are the major building blocks of cellular membranes. Phospholipid assemblies can therefore provide a native-like environment which is essential in preventing denaturation of fragile membrane-residing proteins. Phospholipid bilayers have also been widely used for the preparation of experimental model membrane systems that make it possible to mimic the physicochemical environment of the cell membrane and to reproduce, *in vitro*, its two-dimensional fluidity (Chan et al. 2007). These minimal cell membrane mimics have therefore made it possible to study the collective behaviour of the membrane's basic components from which in-depth insights have been gained regarding, for example, phase transitions and the effect of lipid and protein mobility or clustering on biomolecular interaction processes (Castellana et al. 2006; Chan et al. 2007; Jackman et al. 2012a; Tresset 2009; Yu and Groves 2010).

Beyond their role as cell membrane mimics and carriers for membrane proteins, lipid self-assemblies, and in particular liposomes (artificially prepared lipid vesicles), have also shown promise as nanoscale components in bioanalytical assays. This is in part motivated by the fact that phosphatidylcholine bilayers exhibit excellent nonfouling properties (Glasmastar et al. 2002), a characteristic essential to the development of highly specific and accurate sensing platforms. Selected functionalities can be further added to the membrane surface by self-insertion of hydrophobic domains or through the incorporation of a small amount of functional lipids allowing for the immobilization of biomolecules or dyes on the membrane surface. Liposomes also provide a confined aqueous cavity making it possible to encapsulate a variety of molecules in their interior. Based on the latter features, liposomes have been shown to function as promising candidates to amplify biosensor signals (Bally et al. 2009) and to generate miniaturized reactors for single-molecule measurements of bio(chemical) reactions (Christensen et al. 2007, 2010).

There are numerous reviews on the use of surface-based sensors to study lipid assemblies (Cooper 2004), both on single lipid vesicles as small-scale chemical reactors and sensor elements (Christensen et al. 2007) and on planar lipid bilayers for studies of transport processes (Demarche et al. 2011; Knoll et al. 2008; Koper 2007; Sinner et al. 2008; Tiefenauer and Demarche 2012). We do not intend in this chapter to give a complete description of this vast field but instead focus on providing an overview on the creation, application, and potential of lipid assemblies in the development of surface-based affinity sensors. Consequently, we focus on those sensors that make it possible to probe the formation of a biomolecular complex between a suspended target and a sensor-bound capture molecule. Such sensors often require the immobilization of the lipid assembly at the sensing interface. We therefore

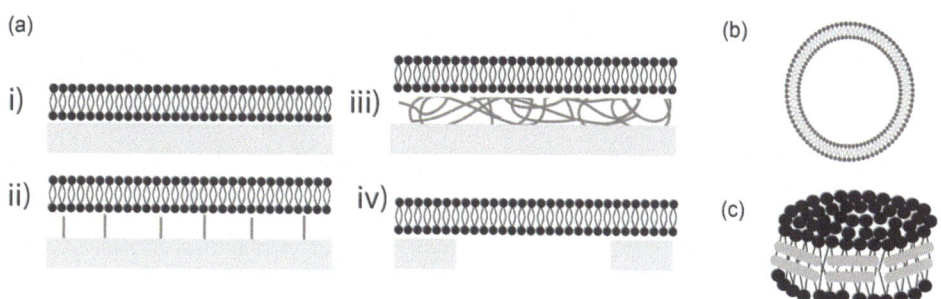

FIGURE 14.1 Lipid assemblies. Schematic representation of (a) planar lipid bilayers: (i) supported lipid bilayer, (ii) tethered lipid bilayer, (iii) bilayer on a polymer cushion, and (iv) free spanning bilayer over a nanopore; (b) unilamellar vesicle; and (c) lipid nanodisc.

first provide an overview of the most promising surface-tethering approaches (Section 14.2), before presenting the major biosensing techniques used in conjunction with lipid-based sensors (Section 14.3). We then illustrate their application in membrane protein sensing (Section 14.4) but also in probing lipid–protein interactions (Section 14.5). In the last part of this chapter, we also illustrate the potential of lipid assemblies as analytical components of the sensing device through their use as labels for signal amplification (Section 14.6) or as nanoreactors for single-molecule studies (Section 14.7). Finally, we discuss how external forces can be used to manipulate constituents of lipid membranes both to create arrays of their components and for direct biosensing applications (Section 14.8).

14.2 IMMOBILIZATION OF LIPID ASSEMBLIES AT A SENSOR SURFACE

The development of lipid-based biosensing interfaces relies on the availability of protocols for the immobilization of lipid assemblies at the sensor surface. Ideally, the immobilization procedure should be simple and versatile, while simultaneously ensuring a good stability of the lipid assemblies. In addition to this, surface functionalization strategies offering the possibility of generating patterns of heterogeneous composition are essential to the development of lipid-based microarrays for high-throughput screening of biomolecular interactions (Bally et al. 2010). In this section, we provide an overview of the most common methods available for the sensor immobilization of planar lipid membranes and liposomes with special focus on the methods allowing for the patterning of such lipid assemblies.

14.2.1 Planar Lipid Membranes

14.2.1.1 Formation of Planar Lipid Bilayers

For decades, planar lipid bilayers have played an important role in lipid membrane research and have greatly contributed to the development of lipid-based sensors. Black lipid membranes suspended over micron-sized apertures have been used since the 1960s to study transport processes across membranes (Suzuki and Takeuchi 2008; Zagnoni 2012). While their potential in providing fundamental understanding on the function of membrane proteins, ion channels in particular, is undeniable, the presence of organic solvent residues originating from solvent-based deposition can be problematic for the analysis of sensitive proteins. Moreover, the lack of stability of these lipid assemblies often limits their implementation as biosensors.

The stability of bilayers can be increased upon immobilization onto a solid support, forming supported lipid bilayers (Figure 14.1a). This can be achieved without leaving residual organic solvents by either the Langmuir–Blodgett/Schaefer technique or by substrate-induced vesicle fusion (Nielsen 2009; Zagnoni 2012). An additional advantage of supported lipid mono- or bilayers is the emerging possibility of investigating the assemblies with a large number of surface-sensitive techniques, the most common being (i) ellipsometry, surface plasmon resonance, and waveguide spectroscopy, to gain information about bound mass, film thickness, and optical density (see Section 14.3.2.1); (ii) quartz crystal microbalance, which adds information about both bound mass and viscoelastic film properties (Cho et al. 2010; Janshoff et al. 2000; Rodahl et al. 1996) (see Section 14.3.2.2); (iii) scanning probe techniques, which provide information about the topographical and structural properties with nm resolution (Dufrene et al. 2000; Goksu et al. 2009); and (iv) electrical impedance techniques; which are helpful in characterizing the electrical properties of supported lipid bilayers and make it possible to perform ion translocation measurements (Demarche et al. 2011; Grieshaber et al. 2008) (see Section 14.3.3.1). While these methods operate in so-called label-free mode, valuable information can be gained by using a small fraction of fluorescently labelled lipids or membrane-associated proteins in combination with fluorescence microscopy methods (e.g., total internal reflection fluorescence microscopy [TIRF] and fluorescence recovery after photobleaching [FRAP]). The latter techniques can provide information on lipid bilayer formation kinetics (Johnson et al. 2002), biomolecular binding reactions (Bally et al. 2011a; Gunnarsson et al. 2008, 2015; Kunze et al. 2013), and spatial diffusivity of species within the membrane (Axelrod et al. 1976; Jonsson et al. 2008; Meyvis et al. 1999).

Thanks to this vast arsenal of surface sensitive tools, membranes deposited directly onto solid supports have a unique advantage in the study of both peripheral membrane proteins and lipid–protein interactions, as illustrated in Section 14.5. However, their main drawback is that the lipid layer is only separated by a thin (~1–2 nm) water layer from the solid support (Bayerl et al. 1990; Johnson et al. 1991). This close contact between the lipid layer and solid support makes it difficult to incorporate bulky transmembrane proteins without denaturation or inhibition of their lateral mobility. To overcome this issue and ensure fluidity of the membrane's components, a number of approaches have been proposed. One strategy relies on forming the bilayer on a highly hydrated polymer layer, often referred to as polymer cushion, mimicking the extra cellular matrix (Figure 14.1a ii) (Goennenwein et al. 2003; Jeon et al. 2008; Kang et al. 2007; Kugler et al. 2002; Sackmann and Tanaka 2000; Sugihara et al. 2010; Tanaka and Sackmann 2005). Polyelectrolyte layers, carbohydrates, and peptides have all been proposed for this purpose. In the case of a surface-tethered bilayer, the bilayer is instead stabilized by surface anchors, e.g., thiolipids or thiopeptides bound to either a gold substrate or via silane coupling to an oxide substrate (Figure 14.1a iii). The spacing can be increased using lipids with a macromolecular hydrophilic head group spacer of predefined length (e.g., based on polyethylene glycol or oligopeptides) (Andersson et al. 2007; Lahiri etal. 2000; Naumann et al. 2002; Stora et al. 1999; Terrettaz et al. 2009, 2003). The advantage of the latter approaches resides in the improved sealing properties, an essential feature to study transport across ion channels. In some cases, air stability of both polymer-cushioned (Albertorio et al. 2005) and surface-tethered (Deng et al. 2008) bilayers have been reported), opening up possibilities of long-term storage which is essential for commercial applications. The formation, characterization, and application of polymer-cushioned or surface-tethered membranes have been reviewed in a number of articles over the past two decades, to which the reader is referred for further details (Castellana et al. 2006; Chan et al. 2007; Danelon et al. 2008; Jackman et al. 2012b; Knoll et al. 2008; Koper 2007; Sinner and Knoll 2001; Tanaka and Sackmann 2005; Tiefenauer and Demarche 2012).

For some applications, the lack of accessibility to the lower leaflet remains a major drawback of solid-supported membranes. This can be a requirement when studying transport across the membrane especially when biomolecular interactions that occur in response to a transport event take place on both sides. For example, in the case of G-protein-coupled receptors (GPCRs), which is a major target in drug discovery, the receptor needs to be accessible both from the extracellular side, where it becomes activated upon ligand binding, and from the intracellular side, where it interacts with the G-protein complex. Additionally, when probing charge translocation through ion channels, ion accumulation between bilayer and substrate should be avoided as it can hamper the current flow through the channel. To accommodate the need for stability and accessibility, progress has been made towards the preparation of supported lipid bilayers onto a variety of nanoporous substrates (Figure 14.1a iv), as further detailed in a number of reviews (Demarche et al. 2011; Janshoff et al. 2006; Reimhult and Kumar 2008; Tiefenauer and Demarche 2012; Tiefenauer and Studer 2008). As an alternative, stabilization of black lipid membranes has been achieved after filling micropores with hydrogels or polymers (Costello et al. 1999; Jeon et al. 2008; Kang et al. 2007; Tiefenauer and Demarche 2012). The formation of a stable and highly insulating (i.e., defect free) bilayer has long remained a major issue in the development of electrochemical sensors, especially with respect to meeting the need to probe with single ion channel sensitivity. However, supported lipid bilayers with gigaohm resistance have been reported (Andersson et al. 2007; Keizer et al. 2007; Schmitt et al. 2006; Studer et al. 2009; Sugihara et al. 2010). This allows for nonensemble averaged (i.e., single ion channel) measurements of ion activity which are essential in drug development and the development of stochastic sensors as further detailed in Section 14.4.3.2.

14.2.1.2 Patterns of Planar Lipid Membranes

The need of platforms allowing for high-throughput screening of membrane-related biomolecular interactions has called for the development of patterning methods for supported lipid bilayers with the aim of producing lipid-based microarrays and lab-on-a-chip devices (Bally et al. 2010; Castellana et al. 2006; Mazur et al. 2017). Towards this goal, a variety of membrane patterning methods have been proposed over the past decades. These include the formation of membrane corrals by polymer lift-off (Orth et al. 2003) or using chemical or topographical grids acting as diffusion barriers (Groves et al. 1997, 1998; Kam and Boxer, 2000, 2001), backfilling of supported lipid membranes after the creation of membrane holes by photochemistry (Morigaki et al. 2001; Yee et al. 2004), nanoshaving with an AFM tip (Shi et al. 2008), as well as membrane patterning by lipid blotting, stamping, or microcontact printing (Hovis and Boxer 2000, 2001; Majd and Mayer 2005; Tanaka et al. 2004). While these approaches allow for the creation of well-defined membrane spots, multiplexing remains limited and practically cumbersome. Printing of lipid assemblies using microcapillary tubes (Cremer et al. 1999) or by noncontact robotic spotters (Binkert et al. 2009; Kaufmann et al. 2011; Yamazaki et al. 2005) represents a more versatile alternative to the above-mentioned techniques as they allow for the creation of membrane patch arrays of heterogeneous composition, required for multiplexing. Arrays of lipid bilayers membranes deposited in microwells (Binkert et al. 2009), onto prepatterned surfaces (Cremer et al. 1999; Yamada et al. 2013; Yamazaki et al. 2005), or onto polymeric tether layers (Deng et al. 2008; Lahiri et al. 2000) have been prepared using such techniques. More recently, Kaufmann et al. demonstrated noncontact printing of vesicle solutions through a thin water film, an approach which allows for a simple and rapid production of heterogeneous arrays of supported lipid bilayer patches, without prepatterning or prefunctionalization of the surface (Kaufmann et al. 2011). Most interestingly, the use of trehalose in the vesicle spotting buffer offers the possibility of creating vesicle that can be vitrified for storage in dehydrated state. These droplets will upon rehydration and rinsing with buffer lead to the creation of SLB patches of good quality (Hinman et al. 2015; Wilkop et al. 2014). Stepping up in the complexity of SLB structures, Ainla et al. further presented a microfluidic toolbox to draw two-dimensional supported lipid membrane networks. Their multifunctional pipette approach allows for the creation of complex patterns of supported lipid bilayers and therefore offers unprecedented possibilities to study membrane-related processes (Ainla et al. 2013).

14.2.2 Surface-Tethered Lipid Vesicles

As an alternative to the above-mentioned planar membranes, intact lipid vesicles have also been immobilized onto solid supports in a variety of biosensing assays. An ideal vesicle immobilization platform should allow for the specific capture of the vesicles while preventing nonspecific adsorption of both the vesicles and the biomolecules used within the assay. For specific applications, in particular in conjunction with curvature sensing, it is also important that vesicle deformation is suppressed. While vesicle monolayers are the most common, in some cases vesicle multilayers were formed at the sensing interface with the aim of increasing the sensor's loading capacity (Burgel et al. 2010; Graneli et al. 2007).

A number of surface passivation approaches have been proposed; they include backfilling with proteins such as BSA, the use of phosphatidylcholine supported lipid bilayers, or passivation via surface-grafted poly(ethylene) glycol (PEG) chains. Most of the conventional coupling chemistries, including the biotin–streptavidin linkage (Bolinger et al. 2004, 2008; Boukobza et al. 2001; Cisse et al. 2007; Kalyankar et al. 2006; Pick et al. 2005; Rossetti et al. 2005; Stamou et al. 2003; Tabaei et al. 2012, 2013), antibody-antigen linkages (Silin et al. 2006), and covalent coupling of chemically derivatized liposomes (Ma et al. 2012; Schonherr et al. 2004), have been used to immobilize vesicles. As an alternative, self-insertion of hydrophobic moieties bound to the chip can also be used (Cooper et al. 2000). Oligonucleotide tags have also been proposed as particularly promising vesicle anchors. Tagging of vesicles with oligonucleotide strands can be readily achieved either by self-insertion of cholesterol–DNA conjugates into the membrane (Bailey et al. 2009; Pfeiffer and Hook 2004; Stadler et al. 2004) or by covalent coupling of end-derivatized oligonucleotide strands onto functional vesicles (Chaize et al. 2006; Stadler et al. 2006; Yoshina-Ishii et al. 2005). This approach further allows for the creation of heterogeneous vesicle arrays by oligonucleotide-directed assembly of the tagged vesicles onto conventional oligonucleotide arrays (Chaize et al. 2006; Dusseiller et al. 2005; Stadler et al. 2006; Svedhem et al. 2003). This is illustrated, for example, in a proof-of-concept study which shows how DNA-labelled vesicle populations can be directed at predefined spots on a DNA microarray with little cross-reactivity leading to the creation of a functional vesicle array (see Figure 14.2a).

Provided that they are sufficiently separated (*i.e.*, the distance between the two vesicles should be larger than the limit of resolution of the microscope, preferably ≥1 μm), individual fluorescent vesicles can be readily visualized by fluorescence microscopy. This has led to the

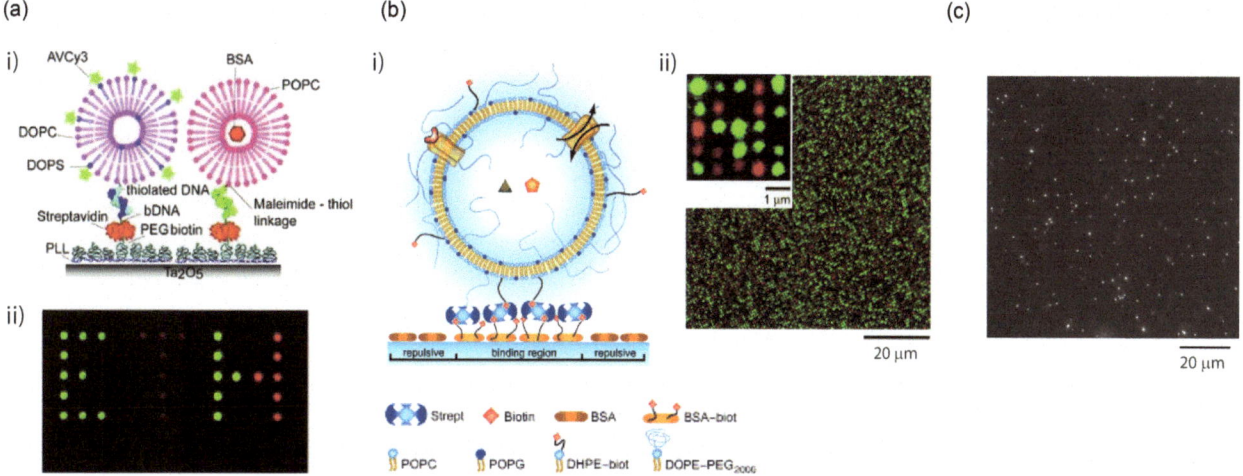

FIGURE 14.2 Lipid–vesicle arrays. (a) Oligonucleotide-directed immobilization of DNA-tagged vesicles onto an oligonucleotide array. (i) Liposomes carrying complementary oligonucleotides and either annexin-Cy3 on their surface or BSAAlexaFluor647 in their interior are hybridized to complementary strands on the array surface. (ii) Microscopy image of an array after incubation with both liposome populations. E: DNA1 complementary to the oligonucleotide on the annexin-CY3 liposomes. T: DNA2 complementary to the oligonucleotide on the BSAAlexaFluor 647 liposomes. H: labelled streptavidin spot. Underline: streptavidin controls for nonspecific binding. (Reproduced with permission from Stadler et al. 2006.) (b) Ordered array of single vesicles. (i) Biotinylated vesicles are immobilized on BSA–biotin patches. (ii) Fluorescence images of individual arrayed liposomes. (Reproduced with permission from Stamou et al. 2003.) (c) Random vesicle array: Fluorescence-microscopy-based visualization of individual vesicles.

development of a number of single-vesicle assays for the detection of biomolecular binding reactions occurring at the vesicle surface or within its aqueous inner cavity, as further detailed in Sections 14.5.3 and 14.7. Such assays have also made it possible to shed light on heterogeneities occurring within vesicle populations, for example, in the case of reconstituted proteoliposomes (Mathiasen et al. 2014). Single-vesicle biochips can be obtained either by random adsorption (Figure 14.2c) or, if maximum vesicle density is desired, using conventional patterning techniques (Christensen et al. 2010, 2013). In the latter case, small equidistant vesicle-adhesive patches embedded into a nonfouling background are first generated using, for example, photolithography or microcontact printing approaches (Figure 14.2b) (Christensen et al. 2007; Michel et al. 2002; Stamou et al. 2003). Alternatively, the vesicles can be trapped in micro- or nanowells (Dahlin et al. 2008; Kalyankar et al. 2006; Wittenberg et al. 2011).

14.3 SURFACE-BASED BIOSENSING TECHNIQUES

A number of surface-based signal transduction principles have been successfully utilized for read-out in lipid-based biosensors. Traditionally, the detection of a specific analyte in a complex sample (for example, in environmental monitoring or medical diagnostics) has often relied on the use of optical or electrochemical labels for signal generation (see Section 14.3.1); however, methods allowing for the direct "label-free" characterization of binding events using optical or acoustic sensing principles are emerging and show great promise (see Section 14.3.2). Beyond the characterization of binding events, the possibility of probing transport processes across the membrane has proven particularly attractive as many membrane proteins are channels which modulate transport of ions and larger molecules across membranes. The methods for characterization of transport processes are briefly reviewed in Section 14.3.3.

14.3.1 LABEL-MEDIATED DETECTION OF BIOMOLECULAR RECOGNITION

14.3.1.1 Optical Sensing

Surface-based biosensors relying on the transduction of optical signals are today the most widespread. Many of those require either direct labelling of the molecule of interest or the use of a reporter molecule, which binds to the target after it has been captured to the sensor surface, in a so-called sandwich assay configuration. Optical transduction relies on the generation of colorimetric, fluorescent, or chemiluminescent signals where the signal originates either directly from dye labels or from the activity of enzymes linked to the reporter molecule. The detection of fluorescence is typically performed using either microscopes combined with spectral filters, fluorescence-based microarray scanners or spectrophotometers. The detection of a colorimetric signal can be performed by naked eye or using conventional scanners.

14.3.1.2 Electrochemical Sensing

With few exceptions, electrochemical sensing of biomolecular interactions relies on labelling of a reporter molecule with an appropriate electroactive label or a redox-active enzyme. Different approaches for signal transduction are available; for example, a measurement can be performed by recording a current resulting from the oxidation/reduction of the electroactive species (amperometric measurement). Alternatively, an accumulation of charge at the working electrode as compared with the reference electrode is measured (potentiometric measurement). In a conductometric experiment, the conductance (*i.e.*, the ability of a medium to conduct an electrical current between the two electrodes) is monitored. Liposomes are being used more frequently in conjunction with amplification of an electrochemical signal as detailed in Section 14.6. In this case, the electroactive marker is released upon liposome lysis after target-mediated vesicle immobilization. Read-out is carried out either by voltammetry, *i.e.*, by measuring the change of current when the potential is changed, or by potentiometric detection (Grieshaber et al. 2008).

14.3.2 LABEL-FREE BIOSENSORS

Devices allowing for the detection of biomolecular binding events without the need of external markers have an undeniable potential in the study of ligand–receptor interactions. Label-free biosensors make it possible to monitor in real time the binding reaction between a receptor at a sensor surface and a ligand in solution, yielding information on the binding affinity and binding kinetics of the interaction under investigation. In such a biosensor, the physical or chemical changes induced upon a binding event at the sensing interface are directly transduced into a detectable signal.

14.3.2.1 Optical Biosensors

Label-free optical biosensors often take advantage of an evanescent field generated at the interface between the sensor surface and the analyte solution to probe changes in the refractive index in close proximity of the sensor. This change in refractive index can then be related to the amount of analyte bound. Furthermore, analysis of real-time binding data yields information on affinity and interaction kinetics (Cooper 2004). A variety of configurations have been proposed to generate an evanescent field at the biosensing interface. In the case of surface plasmon resonance (SPR), probably the most common optical biosensor, the evanescent field is generated upon excitation of surface plasmons in a thin metal coating (typically gold) via reflection of a light beam at the gold surface. A change in refractive index at the sensor surface leads to a change in the conditions required to excite surface plasmons – typically the illumination angle. The change in resonant angle can then be related to the change in interfacial refractive index (Besenicar et al. 2006; Mozsolits and Aguilar 2002; Patching 2014). Other optical biosensors take advantage of a waveguide and of total internal reflection of light to generate the evanescent field at the interface between a high and low refractive medium (Cooper 2002; Fan et al. 2008).

14.3.2.2 Acoustic Biosensors

Acoustic sensors, such as the ones based on a piezoelastic quartz crystal excited to its resonance frequency, are a popular alternative to the above-mentioned optical techniques.

The quartz crystal microbalance measures a decrease in the resonance frequency of an oscillating quartz crystal upon mass deposition at the sensing interface. A change in resonance frequency is therefore directly related to the film mass, including the water hydrodynamically coupled to the oscillation. In the context of biosensing, it can therefore be used to monitor adlayer formation and biomolecule binding events in real time. In addition to this, monitoring of the energy dissipation, *i.e.*, the damping of the crystal's oscillation, provides insights into the viscoelastic and structural properties of interfacial films (Janshoff et al. 2000; Rodahl et al. 1996).

The popularity of the quartz crystal microbalance with dissipation (QCM-D) technique in the context of lipid-based biosensors is widely related to its unique capability of distinguishing the signal generated by a supported lipid bilayer from the one of a layer of nonruptured adsorbed vesicles. Indeed, nonruptured vesicles will exhibit a highly dissipative behaviour (*i.e.*, induce high damping) due to their viscoelastic nature, while a supported lipid bilayer (SLB) is a rigid film with a change in the dissipation energy near zero. These differences make it possible to monitor in real time the kinetics of the vesicle rupture process during the formation of a SLB as further detailed in Figure 14.3.

14.3.3 Probing Transport through the Membrane

14.3.3.1 Electrochemical Sensing

Measurement of transmembrane current flow is the method of choice for the detection of membrane–protein-controlled transport of charged species (ions, small molecules, macromolecules …) across membranes. This is because the lipid bilayer itself can be viewed as a resistive (dielectric) layer, *i.e.*, impermeable to the transport of charged particles, with selective transport of specific molecules ensured by highly specialized membrane protein channels. In lipid-based biosensors, currents across the membrane are typically measured using the voltage-clamp method (Demarche et al. 2011; Janshoff et al. 2006; Zagnoni 2012). Thanks to the highly insulating nature of the lipid bilayer, single ion-channel current measurements can be performed with this technique as been demonstrated (Schmitt et al. 2006). This requires, however, the production of high-quality, defect-free bilayers with good sealing properties to reach resistances in the gigaohm range.

An alternative method to voltage clamping is electrical impedance spectroscopy (EIS). This technique probes the

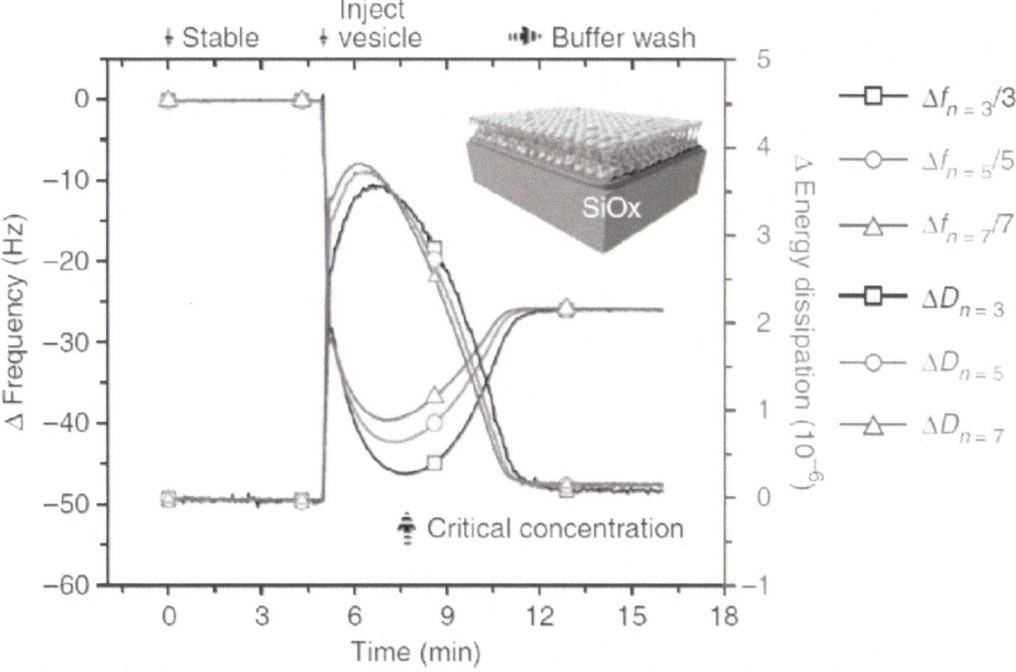

FIGURE 14.3 Formation of a supported lipid bilayer by vesicle rupture probed with the quartz-crystal microbalance with dissipation technique (QCM-D). Adsorption of intact vesicles leads to a decrease in frequency and an increase in energy dissipation. At a critical concentration, the vesicles fuse into a more rigid supported lipid bilayer, resulting in a decrease in energy dissipation and increase in frequency, due to release of water from the vesicles. (Reproduced with permission from Cho et al. 2010).

capacitance and resistance/conductivity of the membrane by applying a sinusoidal voltage at different frequencies and recording the current response. In such electrochemical studies, a conductive surface supporting a tethered bilayer acts as a working electrode in connection with separate counter and reference electrodes. The capacitance of the bilayer is directly related to its quality, allowing EIS to provide information complementary to that obtained using optical or acoustic methods (Grieshaber et al. 2008). EIS has been used in conjunction with both SPR (Heyse et al. 1998; Knoll et al. 2008; Stora et al. 1999) and QCM (Briand et al. 2010) measurements to correlate binding events with changes in the membrane integrity and channel activity.

14.3.3.2 Optical Sensing of Membrane Transport

The optical characterization of transport processes represents a valuable alternative to the above-mentioned electrochemical approaches and is essential in the characterization of both transport of uncharged molecules and slow transportation processes (Peters 2003). When confined aqueous volumes delimited by a membrane are used (*i.e.*, vesicles or bilayer-coated microwells), the fluorescence-based detection of transport processes becomes feasible. For instance, both pH- (Kuyper et al. 2006) and Ca^{2+}- (Hemmler et al. 2005) sensitive dyes have been used to probe transport processes across membranes. In addition, transportation of sugar alcohols across the membrane of surface-immobilized vesicle via aquaglyceroporins has been probed by sensing the change in refractive index inside the vesicles using SPR (Branden et al. 2010).

14.4 MEMBRANE–PROTEIN-BASED BIOSENSORS

Membrane-associated proteins can be categorized into two dominating classes: integral membrane proteins (transmembrane proteins) and peripheral membrane proteins (membrane-bound proteins). Combinations of these two classes are often found in signalling complexes, and both play important roles within the physiology of the cell. However, much more is known about peripheral membrane proteins due to their greater water solubility and relatively weak interactions with the membrane surface. Transmembrane proteins are much more difficult to study due to their degree of hydrophobicity, which requires disruption of the membrane for protein isolation (traditionally through the use of detergents, nonpolar solvents, and often denaturing salts) (Josic et al. 2007). Transmembrane proteins have been reported to be encoded by as many as 7500 genes, *i.e.*, about 30% of our genome, based on genomic sequencing predictions (Fagerberg et al. 2010). Their vital roles in mediating interactions between cells and their environment, as well as in regulating transfer processes across the different intracellular compartments, have made this protein class of great interest to the pharmaceutical industry. Indeed, membrane proteins are currently one of the most common drug targets and remain the principal targets in contemporary drug discovery (Josic et al. 2007; Overington et al. 2006; Patching 2014; Rask-Andersen et al. 2011).

The critical role of membrane proteins in cellular function and their potential as drug targets motivates the development of sensors for fundamental studies of their function and for the screening of drug candidates. Integration of transmembrane proteins into sensor devices has, however, proven highly challenging due to the intrinsic fragility of these molecules. Additionally, recent studies have shown that the activity of many transmembrane proteins can be modulated by the lipid composition of their membrane environment (Denning et al. 2013; Koshy et al. 2015; Scott et al. 2012). In this section, we provide an overview over lipid-based biosensors employed to study membrane proteins. We start with addressing the issues of incorporation of the proteins into a suitable lipid environment (Section 14.4.1) before presenting some examples of biosensors that have been used to probe membrane-protein–mediated binding processes (Section 14.4.2). Finally, we focus on ion channels and transporters and briefly describe recent progress towards the development of assays to probe membrane–protein-mediated transport events (Section 14.4.3).

14.4.1 INTEGRATION OF FUNCTIONAL MEMBRANE PROTEINS IN THE SENSING PLATFORM

The immobilization of functional integral membrane proteins at the sensing interface remains a major bottleneck in the development of membrane–protein-based biosensors. A variety of protocols have been developed for the reconstitution of integral membrane proteins into liposomes leading to the formation of so-called proteoliposomes (Seddon et al. 2004). Even though incorporation of selected transmembrane proteins has been demonstrated, membrane protein reconstitution remains a relatively cumbersome procedure which needs to be optimized for each protein individually (Ollivon et al. 2000; Prive 2007). An alternative is the direct isolation of cell membrane extracts by either chemical (Costello et al. 2013; Liu et al. 2017; Pick et al. 2005) or physical (Gunnarsson et al. 2011; Pace et al. 2015; Rao et al. 2002; Silin et al. 2006) means, leading to the creation of so-called native membrane vesicles (NMVs). With this approach, the target protein is not "purified" away from its native lipid and protein neighbours and then reconstituted into a lipid assembly; instead the membrane containing the target protein is isolated and purified producing vesicles of significant complexity and with a great similarity to their precursor cellular membranes in terms of their native-like lipid, carbohydrate, and protein composition.

Both proteoliposomes and NMVs can be directly immobilized at sensing interfaces, as described in Section 14.2.2. Such approaches are well suited to study protein–ligand interaction, provided that the amount of membrane protein is sufficient to generate a detectable signal upon ligand binding. However, since the membrane is only physically accessible from outside of the vesicles, the possibility to study membrane protein transport or signalling reactions is limited. This calls for the development of methods facilitating the deposition of protein-containing membranes onto solid supports or even porous substrates which ensure liquid-handling access from both sides (Tiefenauer and Demarche 2012). This represents yet another challenging task as vesicles of complex composition containing bulky membrane proteins do not fuse readily onto solid supports. To overcome this issue, a number of procedures have been proposed as reviewed, for example, in Demarche et al. (2011), Hardy et al. (2013), Hirano-Iwata et al. (2008), Nielsen (2009), and Tiefenauer and Demarche (2012). They include surface capturing of detergent-stabilized proteins followed by backfilling with lipids (Giess et al. 2004; Karlsson and Lofas 2002; Stenlund et al. 2003), or fusion of proteoliposomes into preformed supported lipid bilayers (Simonsson et al. 2010; Studer et al. 2011; Woodbury and Miller 1990). The first approach potentially yields a high protein density on the sensor surface but only works if the protein is stable enough. The latter approach is gentler, but the number of proteins incorporated in the bilayer is typically low. It can, however, be increased after addition of fusogenic compounds to the proteoliposome and the lipid membrane (e.g., DNA zippers (Simonsson et al. 2010) and specific ion-channel-forming proteins (Studer et al. 2011; Woodbury and Miller 1990)). Direct formation of native-like SLBs by fusion of NMVs is possible but depends on the vesicle composition and requires optimization of the experimental parameters. Vesicle rupture can be facilitated by either adding fusion-prone phosphatidylcholine liposomes to surface-bound native vesicles (Costello et al. 2013; Dodd et al. 2008), through the use of osmotic stress generated either by salt gradients in the case of a suspended membrane (Woodbury and Miller 1990) or by addition of soluble PEG to surface-immobilized proteoliposomes (Elie-Caille et al. 2005; Malinin et al. 2002). Recently, the use of a polyelectrolyte cushion to drive the direct rupturing of plasma membrane vesicles to form a polymer-supported lipid bilayer which was made of 100% native membrane material was also reported (Liu et al. 2018). We have recently proposed two versatile methods to create native-like SLBs, which contain a mixture of native membrane components and synthetic lipids to form a SLB with a diluted native-like lipid, carbohydrate, and protein composition. The first approach, illustrated in Figure 14.4a, is based on hydrodynamically driving a pure-lipid bilayer edge against a layer of nonruptured adsorbed NMVs in a microfluidic channel, allowing for the creation of supported native membranes with low dilution of the native constituents (Jonsson et al. 2011; Simonsson et al. 2011). Alternatively, hybrid vesicles can be created prior to the formation of the native-like SLB. In this case, hybrid vesicles generated through the sonication-facilitated merger of PEGylated phosphatidylcholine vesicles with mechanically isolated NMVs undergo

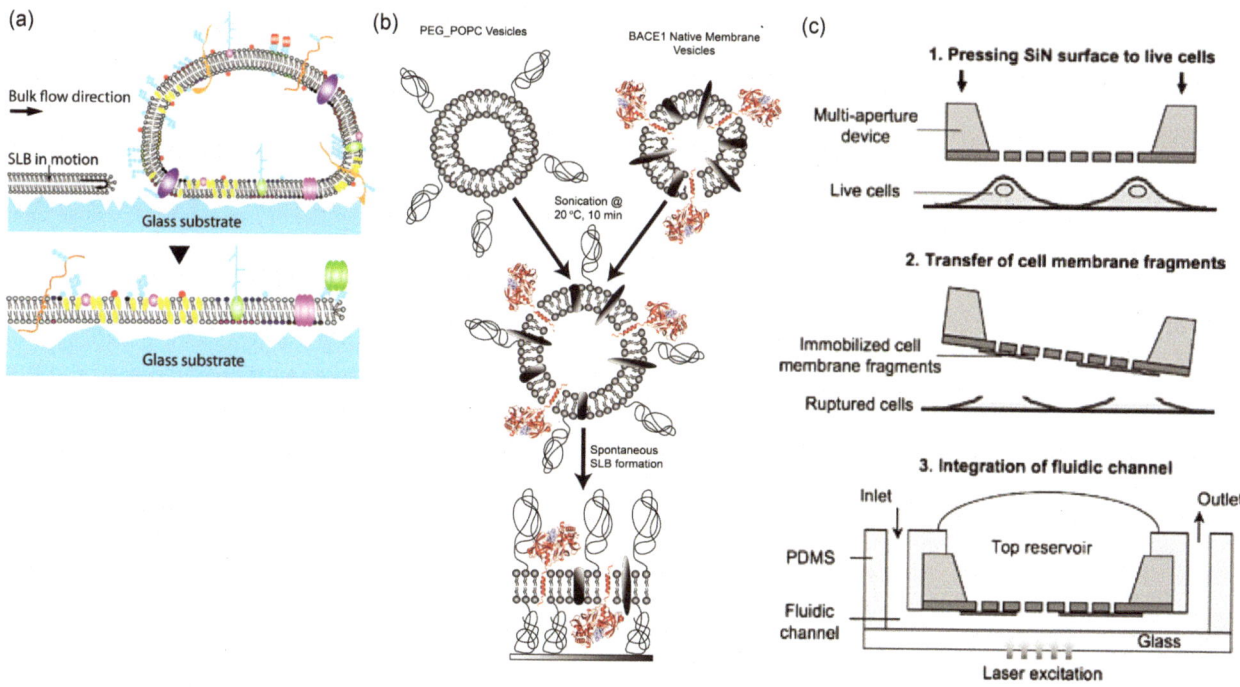

FIGURE 14.4 Formation of native supported lipid membranes. (a) Schematics of the bilayer-edge-driven fusion approach. A liquid bulk flow is used to generate a hydrodynamic force that drives forward the edge of a supported lipid bilayer in a microfluidic channel. Upon contact, the native vesicle ruptures and incorporates into the bilayer. (Reproduced with permission from Simonsson et al. 2011.) (b) schematic of the merger of PEG_POPC vesicles with BACE1-containing native membrane vesicles to produce hybrid vesicles, which subsequently can undergo spontaneous SLB formation to produce polymer-cushioned SLBs. (c) Formation of native membranes supported onto a porous substrate by cell membrane ripping. A porous substrate coated with poly-L-Lysine is pressed against the cell and retracted resulting in the creation of suspended cell membrane sheets. (Reproduced with permission from Danelon et al. 2006.)

spontaneous substrate-induced vesicle rupture and create native-like polymer-supported lipid bilayers (Figure 14.4b). (Pace et al. 2015; Peerboom 2018). These SLBs have been shown to maintain both transmembrane protein mobility and enzymatic functionality (Pace et al. 2015). While not maintaining the high degree of membrane component orientation described in previous examples, the facile production of these hybrid vesicles and their amenability with liquid-handling systems make this approach an interesting alternative in future biosensor applications. Additionally, the high degree of transmembrane protein mobility reported indicates that external forces (e.g., hydrodynamic or electrophoretic) could be applied to these native-like polymer-supported bilayers for the accumulation and/or separation of membrane components within the lipid bilayer environment (see Section 14.8).

In an alternative approach, Vogel et al. have demonstrated that surface-supported membrane sheets can be transferred to poly-L-lysine-coated slides by ripping off the apical side of a whole adherent cell. With this approach, the intracellular leaflet is exposed for further analysis (Perez et al. 2006). Furthermore, accessibility to both leaflets of the membrane was demonstrated after the creation of membrane sheets on a nanoporous substrate (Figure 14.4c) (Danelon et al. 2006).

14.4.2 Biosensing Approaches to Detect Interactions between Membrane Proteins and Ligands

With the increased availability of methods facilitating the functional immobilization of membrane proteins, sensors allowing for the characterization of the recognition process between a ligand and a membrane protein have become popular tools. They are particularly valuable in the context of drug discovery where compounds binding to ligand-gated ion channels, GPCRs, or cytokine receptors need to be identified and their interaction with the membrane-bound receptor carefully characterized (Cooper 2002).

The use of SPR in the context of studying the interaction between membrane proteins and either their natural ligands or drug candidates has been recently reviewed by S. Patching (2014). As detailed in this review, a variety of assay configurations have been dedicated to the characterization of both GPCR and non-GPCR membrane protein systems (Patching 2014). For example, in one of the pioneering works on SPR applications for membrane protein sensing, Vogel et al. have illustrated how the method can be used for structural and functional investigation of the light-activated GPCR rhodopsin. After tethering the detergent-stabilized GPCR to the surface via a biotin–streptavidin linkage and subsequent bilayer reconstitution, the authors investigated, in real-time, the light-induced receptor activation, G-protein desorption, and receptor deactivation (Bieri et al. 1999). In the context of diagnostic applications, Vega et al. report an SPR-based biosensing device for rheumatoid arthritis, allowing for the nanomolar detection of the chemokine CXCL12 from patient urine samples. Its receptor, CXCR4, was immobilized though covalent coupling of lentiviral particles obtained from cells expressing the receptor, to a self-assembled monolayer (Figure 14.5a) (Vega et al. 2013).

The label-free detection of small-molecule binding events, as in, e.g., Vega et al. (2013) to a surface can be challenging due to both the low molecular weight of the analyte and to difficulties associated with the production of a dense receptor layer at the sensor surface, a requirement to generate detectable binding signals. As membrane–protein-containing lipid assemblies are relatively bulky and will therefore produce a greater mass change, reversed assay configurations have been proposed. In this case, detergent solubilized proteins (Harding et al. 2006; Liao et al. 2013) or membrane–protein-containing nanodiscs (Gluck et al. 2011; Ritchie et al. 2011) are added to the ligand-functionalized sensor surface. Along the lines of the above-mentioned assay configuration, we have recently presented a fluorescence-based method to probe the interactions between a surface-bound ligand and a membrane-bound receptor embedded in NMVs. Here, we visualize individual fluorescent vesicles using total internal fluorescence (TIRF) microscopy by simply labelling the lipid membrane in which the membrane protein resides. By avoiding the labelling of the protein directly, the risk of affecting the binding kinetics of the interaction partners is diminished. We further take advantage of the surface-confined illumination generated by the evanescent field of this set-up, to illuminate the ligand-bound fluorescent vesicles while discriminating them from the ones in solution (Figure 14.5b). Our assay, termed equilibrium fluctuation analysis, makes it therefore possible to probe, under stagnant and equilibrium conditions, the kinetics of a ligand–receptor interaction and to determine the affinity constant. Because each vesicle can be visualized independently, the assay exhibits a single membrane protein sensitivity which can make it possible to probe ligand–receptor interactions without the need for membrane protein enrichment in the vesicles by overexpression or reconstitution (Gunnarsson et al. 2011; Wahlsten et al. 2015). Recently, we have also demonstrated how this single-membrane protein sensitivity makes it possible to employ a so-called inhibition-in-solution assays (ISAs) to determine the affinity of high-affinity drug candidates ($K_d < nM$), which cannot be characterized using conventional label-free surface-based sensors, such as SPR, due to sensitivity limitations (Gunnarsson et al. 2015).

Fluorescence remains also the major read-out strategy for microarray applications where the possibility of acquiring binding data in a high-throughput manner prevails over the need to follow binding kinetics in real time (Bally et al. 2006). Together with the growing popularity of microarray assays, efforts towards the development of high-throughput tools to study membrane protein targeting for drug discovery purposes are on the rise. The direct printing of GPCRs-containing cell membrane fractions on solid

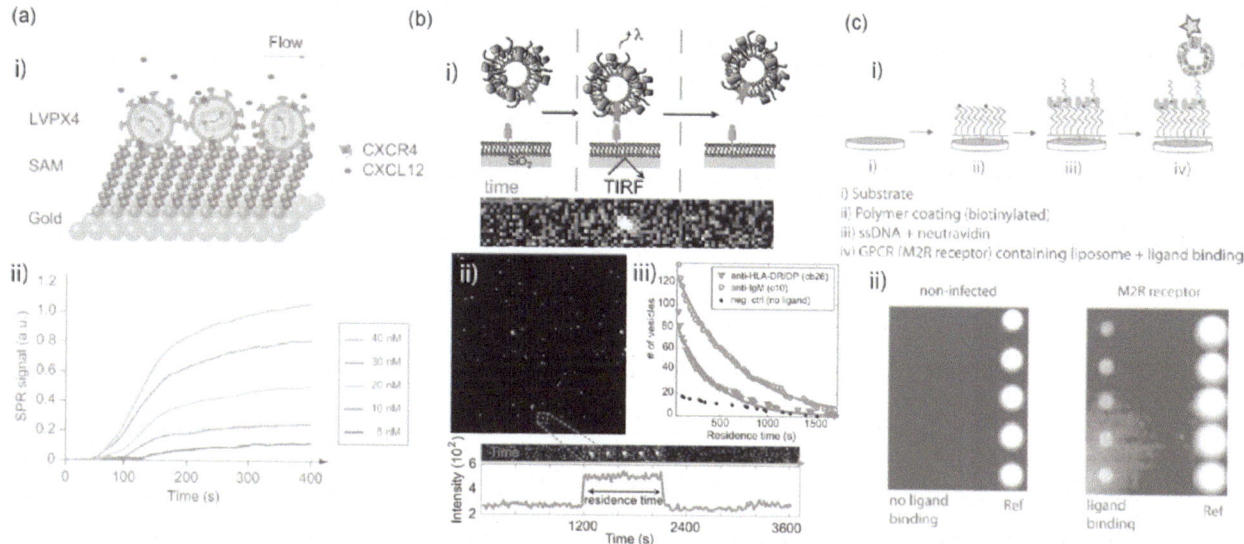

FIGURE 14.5 Biosensors to probe the ligand–membrane protein interactions. (a) Biosensor for the detection of chemokines from the urine of patient samples. (i) Lentiviral particles (LPX4) from a cell overexpressing the receptor CXCR4 are immobilized via covalent coupling to a self-assembled monolayer (SAM) on gold. (ii) SPR response at different chemokine ligand (CXCL12) concentrations. (Reprinted with permission from Vega et al. 2013.) (b) Equilibrium fluctuation analysis of the interaction between a native vesicle and a surface-bound ligand. (i) Total internal reflection fluorescence microscopy (TIRF) allows for the visualization of single fluorescently labelled vesicles bound to the surface while discriminating them form the background. This allows for the monitoring of binding and unbinding events under equilibrium conditions. (ii) Analysis of time lapse movies yields the vesicle residence time as illustrated by the kymograph. (iii) Analysis of the residence is used to probe the dissociation behaviour and to determine k_{off} after fitting the data as shown for two ligands (triangles and empty circle) and compared with a negative control (no ligand, black dot). (Reprinted with permission from Gunnarsson et al. 2011.) (c) Liposome microarray to probe binding of fluorescent ligands. (i) The array is obtained by spotting complexes of streptavidin and biotinylated oligonucleotides onto a biotinylated polymer and further immobilization of native vesicles carrying a complementary oligonucleotide. (ii) Fluorescent ligand binding to a liposome array composed of native vesicles from cells (left) not expressing the muscarinic receptor (M_2R) and (right) overexpressing it. (Adapted with permission from Bailey et al. 2009.)

supports was first reported by Lahiri et al. (Fang et al. 2002; Hong et al. 2006). The ligand binding properties (Fang et al. 2002) and functional activation (Hong et al. 2006) of the GPCR were preserved as demonstrated by fluorescence imaging. We have recently demonstrated that GCPR arrays can also be obtained after functional immobilization of native vesicles via oligonucleotide tethers while preserving the receptor activity (Bailey et al. 2009). This approach is particularly gentle to the fragile transmembrane proteins as native vesicle immobilization can be carried out in solution without any harsh spotting conditions and drying steps (Figure 14.5c).

14.4.3 Probing Membrane–Protein-Mediated Transport across the Membrane

There is a large diversity to the family of transmembrane proteins concerned with the transport of molecules and ions across cellular membranes. Beyond the membrane-based ligand–protein binding assays, such as the ones presented in Section 14.4.2, reliable assays for the characterization of transport processes across the membrane are also needed. Depending on their mechanism of action, membrane channels can be classified into ion channels, carriers, and transporters. The first two types of proteins are passive channels driven by concentration gradients; no direct energy input is therefore required for transport. In the case of the active "transporters", transport occurs against a gradient and coincides with energy consumption. Ion-channels transport ions and can open and close upon an electrical stimulus (voltage gated channel), a mechanical stimulus (mechanosensitive channel) or upon binding of a ligand (ligand-gated channel). Carrier-mediated transport, on the other hand, involves recognition of the molecule to be transported by the membrane protein followed by translocation to the other side of the membrane and release (Demarche et al. 2011). Generally speaking, ion channels exhibit rather fast translocation (up to 10^8 ions s^{-1}), while carriers and transporter are much slower (10–10^4 molecules s^{-1}).

In this section, we start with briefly providing examples of biosensing platforms allowing for the characterization of ligand binding events while simultaneously probing channel-mediated transport across membranes (Section 14.4.3.1). We then present the potential of both natural and artificial pores as components of versatile electrochemical sensors for the detection of biomolecular interactions (Section 14.4.3.2). For more detailed reviews of the techniques and biosensors allowing for the detection of transport across membranes, the reader is referred to one of the numerous reviews on the topic (Bayley and Cremer 2001; de la Escosura-Muniz et al. 2012; Demarche et al. 2011; Guidelli and Becucci 2011; Han et al. 2007;

Hirano-Iwata et al. 2008; Jackman et al. 2012b; Janshoff et al. 2006; Knoll et al. 2008; Koper 2007; Naumann et al. 2011; Peters 2003; Steller et al. 2012; Suzuki and Takeuchi 2008; Zagnoni 2012).

14.4.3.1 Biosensors to Study Ion Channels

The development of ion channel screening methods and the associated possibility of probing ligand-induced changes in ion transport processes have several motivations. First, ligand–receptor interactions play a key role in the function of ligand-gated ion channels. These channels undergo a conformational change upon binding of a ligand, which forces the channel to switch from a closed to an open state. Ligands are typically small molecules such as a neurotransmitter or a hormone. Second, channel dysfunction is a major cause of disease (*e.g.*, epilepsy, diabetes, hypertension, and other cardiac conditions) (Cooper 2004). The strategy of many therapeutic agents targeting ion channels is to either interfere with the ligand-gating process or through a channel-blocking action. However, nondesired interactions of drug compounds with these proteins can be a common cause of adverse side effects of the pharmaceutical in question and are therefore a topic of study. A well-known example is the cardiac-arrhythmia-causing interaction of drug candidates with the hERG potassium channel, which has forced drug companies to terminate drug development projects at late stages (Netzer et al. 2003). This further highlights the need of high-throughput screening devices to probe channel activity.

A number of biosensors for the electrochemical detection of ion currents in conjunction with ligand binding have been presented. Stora et al., for example, showed how a combination between optical and electrochemical monitoring can be used to provide information on ligand binding in conjunction with ion channel monitoring (Stora et al. 1999). In their study, bilayers containing the porin OmpF were tethered at the surface of a gold sensor, and SPR was used to monitor ligand binding while simultaneous EIS measurements provided information on the induced changes of conductivity through the channel. The same protein has also been studied by voltage clamping using nanoporous alumina with pores in the 60 nm range (Figure 14.6a). The membrane protein retained both its activity and its normal functioning mode, characterized by trimeric conductance states. As visible in Figure 14.6b, the antibiotic ampicillin was shown to inhibit ion flow, further illustrating the potential of such an assay platform as a drug screening device (Schmitt et al. 2006). Using immobilized attoliter lipid vesicles, a fluorescence-based assay was developed by Pick et al. (2005) who probed Ca^{2+} fluxes from native vesicles carrying a ligand-gated ionotropic receptor (5-HT_3) to monitor ligand binding and receptor activity, leading to calcium influx into the vesicle.

14.4.3.2 Engineering Sensing Platforms Based on Natural Pores

With the aim of taking advantages of nanoscopic membrane pores to detect and analyse a broad range of compounds

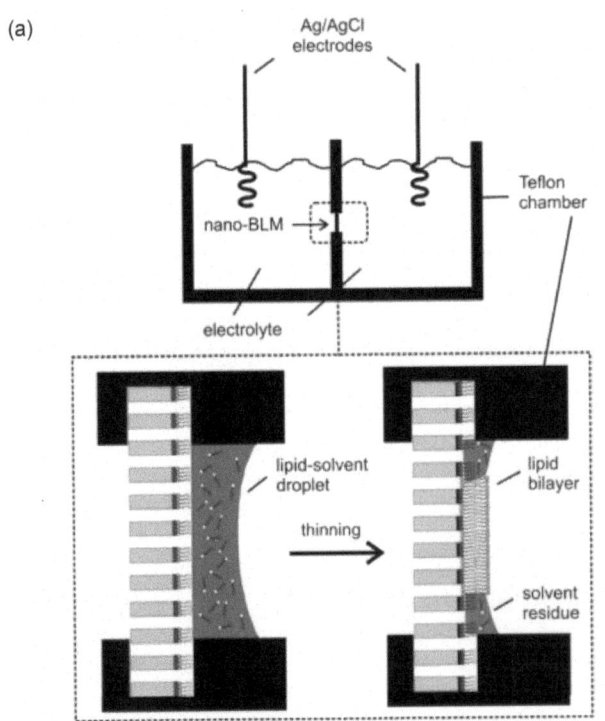
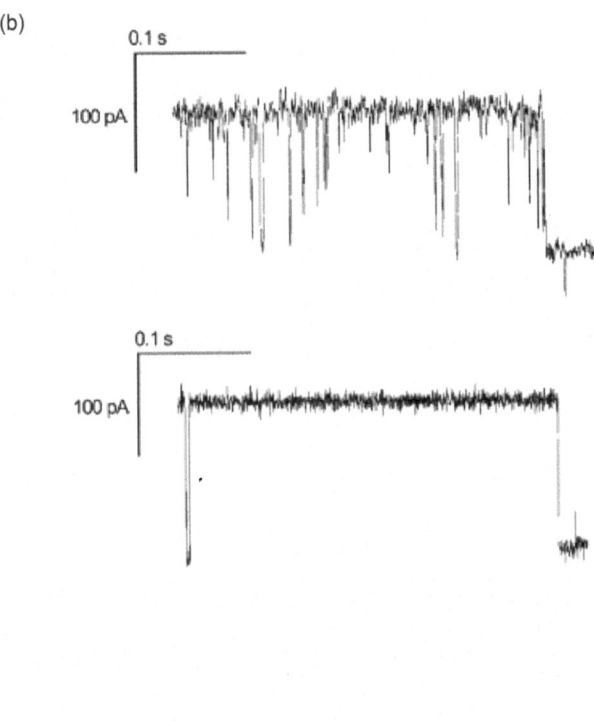

FIGURE 14.6 Voltage-clamp-based monitoring of ion-channel activity. (a) Experimental set-up and solvent-based deposition of a lipid bilayer on a nanoporous substrate. (b) Single OmpF channel activity monitored across a bilayer supported on a nanoporous substrate in presence of ampicillin which blocks the channel (top) or in absence of the antibiotic (bottom). (Reprinted with permission from Schmitt et al. 2006.)

Lipid-Based Bioanalytical Sensors

(*e.g.*, small molecules, proteins, and DNA), attempts have been made to use relatively simple ion channels (*e.g.*, the self-inserting peptide gramicidin or staphylococcal α-haemolysin) as sensory elements in electrochemical biosensors. Here, the general idea is to chemically modify natural or artificial ion channels in such a way that binding of the analyte results in a change in pore conductivity. In such an assay, the molecular recognition element (*i.e.*, the engineered ion channel) becomes responsible both for the specific detection of the target and for signal amplification, based on inducing large changes in the total ion flux by a single molecule. Using direct electronic read-out allows for the sensitive detection of any ligand modulating the ion channel activity (Bayley and Cremer 2001; de la Escosura-Muniz et al. 2012). Along these lines, Cornell et al. made a pioneering contribution when developing a generic sensing strategy termed "ion channel switch" based on the incorporation of dimeric gramicidin A (gA) into a tethered supported lipid bilayer (Cornell et al. 1997). As illustrated in Figure 14.7a, the assay is configured in such a way that ligand binding to a receptor molecule induces a more likely separation between the two mobile gA dimers resulting in a decrease in conductance through the membrane. In brief, receptor molecules are coupled both to the gA in the outer leaflet and directly to immobile (surface tethered) lipids in the membrane. In the absence of a ligand, the gA monomers in the two leaflets transiently assemble into a conducting dimer. However, a ligand-binding-induced cross-linking of both receptors leads to a decrease in mobility of the gA in the upper leaflet and hence to an overall decrease in membrane conductance that can be monitored using, *e.g.*, impedance spectroscopy. This sensor design has, for example, been used for the detection of influenza from respiratory samples with sensitivities comparable with ELISA (Oh et al. 2008).

The group of H. Vogel proposed a different strategy, based on the design of synthetic ligand-gated ion channels (SLICs). These synthetic channels are composed of a pore-forming domain build-up from melittin segments and a ligand-binding peptide domain for specific recognition (Figure 14.7b) (Terrettaz and Vogel 2005). In a more recent study, Terrettaz et al. showed that this approach,

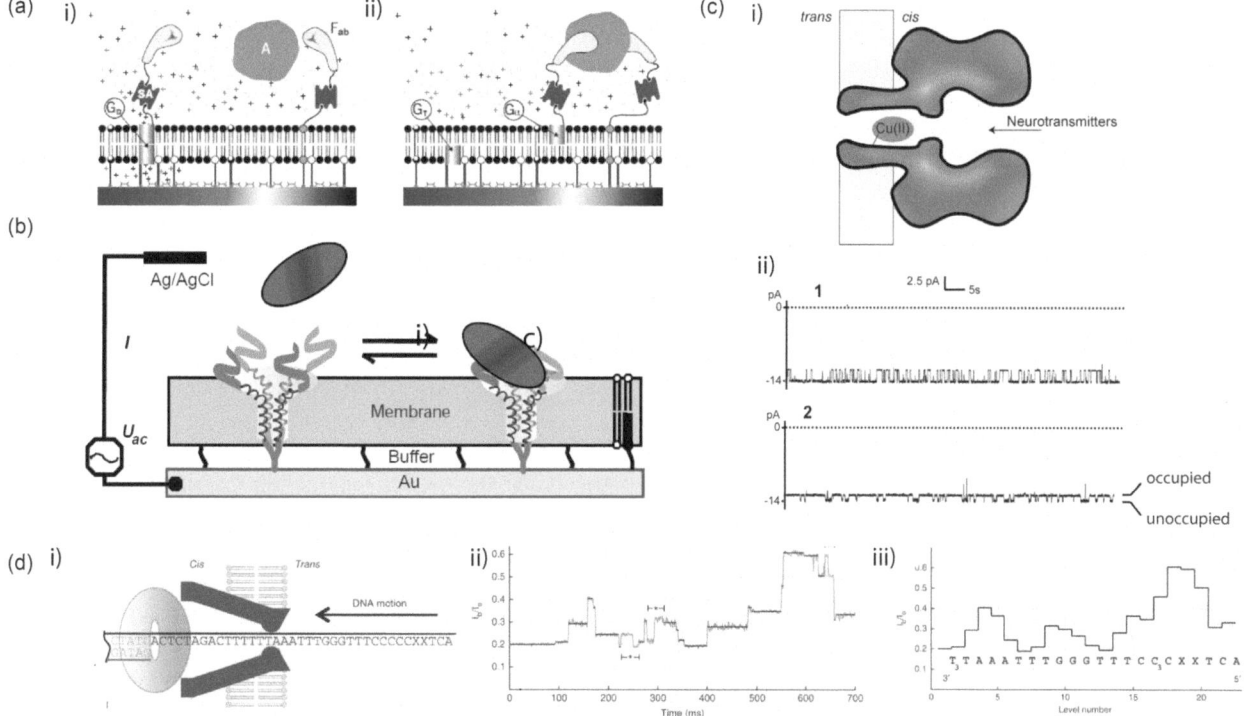

FIGURE 14.7 Sensing platforms based on engineered membrane proteins. (a) Assay set-up of the ion channel switch, based on the dimeric gramicidin A (gA). The mobile dimer in the outer leaflet and immobile bilayer lipids are modified with capture probes. In absence of target, transient gA dimers are formed, leading to an increase in conductance. Target recognition prevents dimer formation, leading to a decrease in conductivity across the membrane. (Reproduced with permission from Cornell et al. 1997.) (b) Schematic representation of the synthetic ligand-gated ion channels (SLICs) approach. These synthetic channels are composed of a pore-forming domain and a ligand-binding peptide domain for specific recognition of the target molecule. Target binding leads to changes in the conductivity across the membrane. (Adapted from Terrettaz et al. 2009.) (c) Stochastic sensor based on an engineered α-haemolysin pore. Transient binding of neurotransmitters to the copper-modified pore leads to a change in conductivity as shown in (ii). Different neurotransmitters can be distinguished by differences in current block amplitude and dwell time. (Reproduced with permission from Boersma et al. 2012.) (d) Nanopore-based DNA sequencing. (i) The mutated protein nanopore MspA is used. phi29 DNA polymerase controls the rate of DNA translocation. (ii) Typical trance of the read-out and (iii) extracted mean current levels. (Reproduced with permission from Manrao et al. 2012.)

combined with impedance measurements, is well suited for the sensitive detection of malaria antibodies, label-free, in a whole blood sample. This highlights the potential of this type of sensing strategy as an interesting candidate for the development of simple point-of-care devices (Terrettaz et al. 2009).

The possibility of observing, in real time, the activity of individual ion channels (see Section 14.3.2) has also led to the development of so-called stochastic sensors enabling detection of molecules with single-molecule sensitivity. Stochastic sensors are often based on the use α-haemolysin genetically modified to recognize a target ligand of interest. In this case, an ionic current through the pore-containing membrane is generated when a potential is applied across the membrane. Reversible binding of the analyte results in stochastic changes in ion current, which can be monitored on an individual basis (Figure 14.7c ii). Analysis of the frequency of these individual events can be related back to the analyte concentration in solution. Furthermore, the duration of each interaction event is directly related to the affinity, while the amplitude can be associated to the physicochemical properties of the molecule under investigation. Affinity and amplitude can therefore be used as specific signatures for the ligands under investigation (Bayley and Cremer 2001; Jackman et al. 2012b). This type of α-haemolysin-based stochastic sensors has been proposed for the detection of a variety of small molecules (Gu et al. 1999; Guan et al. 2005), DNA (Howorka et al. 2001), and proteins (Rotem et al. 2012; Schmitt et al. 2006). For example, using an aptamer-modified α-haemolysin, Rotem et al. reported on a sensor for the detection of thrombin with nM sensitivity (Rotem et al. 2012). In another study, a copper-modified α-haemolysin nanopore was used to detect copper(II) chelating neurotransmitters (Figure 14.7c) (Boersma et al. 2012). Using the same pore, different families of neurotransmitters could be discriminated based on their specific kinetic signature, illustrating the potential of the sensor to detect, in real time, neurotransmitter secretion. Another attractive development of nanopore-based sensors lies in the field of oligonucleotide sequencing (Branton et al. 2008; Venkatesan and Bashir 2011). The possibility of translocating single-stranded DNA or RNA molecules through pores by applying an electric potential has generated a lot of excitement and paved the way to rapid and cheap DNA sequencing, based on the read-out of nucleotide-specific changes in ion current as the oligonucleotide sequence passes through the channel (Bayley 2015). Successful nanopore-based DNA sequencing using a protein nanopore has recently been demonstrated (Figure 14.7d) (Laszlo et al. 2014; Manrao et al. 2012).

While the potential of stochastic biosensors based on engineered membrane proteins is being more and more recognized for DNA sequencing, effort has been put into the development of robust and portable devices for stochastic single-ion-channel sensing of other biomolecular interactions (Kang et al. 2007; Shim and Gu 2007; White et al. 2007). Single-protein nanopore chips for the detection of triphosphate (IP_3) were, for example, produced by encapsulating the lipid membrane in an agarose gel (Kang et al. 2007; Shim and Gu 2007). The device was shown to remain functional over weeks. Alternative devices with similar stabilities are based on the use of a glass-nanopore-supported lipid membrane (White et al. 2007).

14.5 PROBING LIPID MEMBRANE-MEDIATED INTERACTIONS

While much effort has been put into the study of membrane proteins with affinity-based sensors (as shown in Section 14.4), lipid-membrane-mediated interactions – in particular protein–membrane interactions – have been by far less studied. Nevertheless, it is becoming increasingly evident that the function of the lipids reaches far beyond their role as a mechanical support for the cell's barrier. Cell surface phospholipids and glycolipids (GLs) have indeed been shown to play a key role in mediating a number of biological processes by interacting with a variety of biomolecules.

The cell membrane is a highly fluid and heterogeneous environment in which lipids, carbohydrates, and proteins diffuse freely but also cluster into so-called microdomains or rafts. A number of studies have highlighted the fact that the lipid orientation, arrangement, and mobility in this fluid bilayer environment greatly affect the binding affinity of other biomolecules (Evans et al. 1999). Indeed, many lipid-mediated interactions, in particular those involving glycolipids, are known to be associated with the establishment of multiple weak ligand–receptor bonds with the interacting biomolecule, leading to a significant increase in avidity (accumulated increase in affinity caused by multiple binding sites) (Mammen et al. 1998). This calls for the development of sensing strategies in which lipids or GLs are presented in a native-like and membrane-mimetic architecture.

In this section, we have chosen three relevant cases involving the direct interaction between biomolecules and the lipid membrane, some of which rely on avidity. We start by focussing on bimolecular interactions mediated by cell surface GLs (Section 14.5.1). We then discuss membrane disruption by membrane-active enzymes or peptides (Section 14.5.2) before briefly presenting approaches to investigate membrane-curvature-sensing proteins (Section 14.5.3).

14.5.1 PROBING GLYCOLIPID-MEDIATED INTERACTIONS

GL-mediated interactions play a key role in a number of biological processes. GL–protein interactions are involved in, *e.g.*, immune recognition or cell–cell signalling, while GL–GL interactions have been shown to play a key role in, *e.g.*, embryonic development and cancer metastasis (Bucior et al. 2004; Hakomori 2000). A number of pathogens, including toxins and viruses, have also been shown to take advantage of the cell surface GLs to initiate their interaction with a cell; GLs are also often targets of self-antibodies in autoimmune diseases. This makes these glycans interesting candidates in the design of biosensors for diagnostics and environmental monitoring applications (Edwards et al. 2006; Galban-Horcajo et al. 2014; Horlacher and Seeberger 2008). Additionally, GL-containing membranes have been used to study the multivalent nature of carbohydrate-mediated interactions (Jayaraman et al. 2013).

SPR (Borch et al. 2008) or QCM-D (Bally et al. 2012; Nasir et al. 2014; Parveen et al. 2017; Parveen et al. 2019; Rydell et al. 2009) have been used to probe the binding of viruses) and toxins (Borch et al. 2008; Terrettaz et al. 1993) to GL-containing supported lipid bilayers (Bally et al. 2012; Rydell et al. 2009) or sensor-immobilized nanodiscs (Borch et al. 2008), in an attempt to gain fundamental understanding on the interplay between the physicochemical characteristics of the membrane and its ability to bind proteins. This includes investigations focussed on the multivalent nature of the interaction. Our recent study, for example, highlighted the critical role of fine-tuned galactosylceramide clustering in promoting a strong and detectable attachment of norovirus virus-like particles (VLPs) to a GL-containing model membrane (Bally et al. 2012). In another study, we have further shown how the equilibrium fluctuation analysis (see Section 14.4.2) can be used to probe and quantify the affinity between a GL-containing membrane and virus particles. In our assay, norovirus VLPs were bound on the bottom of a glass microwell while binding and release events of single fluorescent GL-containing liposomes to individual VLPs were imaged under equilibrium conditions using TIRF microscopy. The data were analysed to yield binding energies (Figure 14.8) (Bally et al. 2011a). Kunze et al. also used the equilibrium fluctuation approach to probe extremely weak GL–GL interactions ($K_D \sim$ mM), further highlighting the capacity of the single-molecule sensitivity of the technique to unravel weak and therefore poorly studied membrane interactions (Kunze et al. 2013).

A number of carbohydrate microarrays, including arrays of glycolipids, have been developed with the hope of contributing rapid developments in glycomics (Galban-Horcajo et al. 2014; Horlacher and Seeberger 2008; Liu et al. 2009; Oyelaran and Gildersleeve 2009; Shin et al. 2005; Wu et al. 2009). With few exceptions, these arrays are based on direct conjugation of the glycans onto a solid support (Arigi et al. 2012; Fukui et al. 2002; Shin et al. 2005). They do therefore not take into account the influence of the structural properties of the cell membrane on the recognition process. To provide a membrane-mimetic presentation of the glycans, a few groups have put effort into reproducing a GL-presenting lipid bilayer environment in a microarray

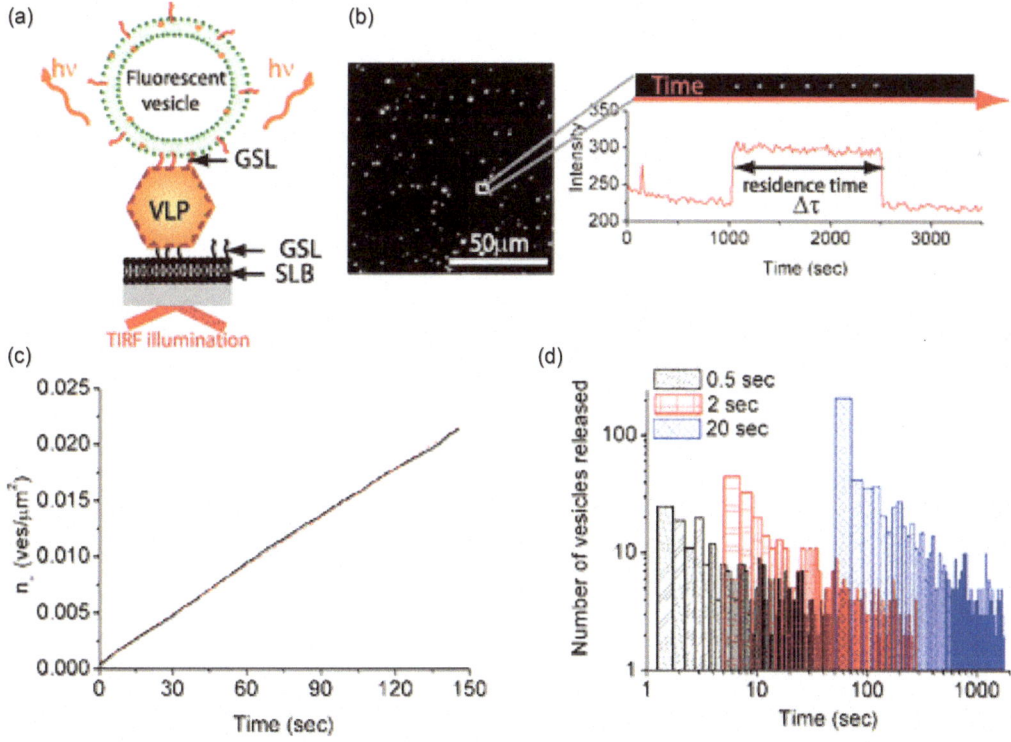

FIGURE 14.8 Equilibrium fluctuation analysis of the interaction between sensor-bound virus-like particles (VLPs) and glycolipid (GSL)-containing membranes. (a) Schematic representation of the assay. VLPs immobilized on a supported lipid bilayer containing GSLs interact with fluorescently labelled vesicles containing the GSL of interest. TIRF-based illumination is used to track surface-bound vesicles. (b) A typical TIRF image of surface-bound vesicles together with a kymograph and the intensity profile of an image area containing a single vesicle. (c) Amount of newly arrived vesicles as a function of time yielding information on k_{on}. (d) Vesicle residence times for different time-lapse series yielding information on the dissociation behaviour. (Reproduced with permission from Bally et al. 2011a.)

format. Ma et al., for example, produced a vesicle array to study gangliosides (Ma et al. 2012), while others demonstrated the creation of GL-containing membrane patches with preserved fluidity (Fang et al. 2003; Yamazaki et al. 2005).

14.5.2 Probing Enzymatic or Peptide-Mediated Membrane Disruption Events

All affinity-based sensing approaches presented so far rely on the detection of highly specific interactions between membrane-bound biomolecules and soluble counterparts. In this section, we focus on less specific recognition events mediated by hydrophobic and electrostatic interactions with the lipid components. In particular, we consider those biomolecules known to disrupt membranes. These include membrane-degrading enzymes such as phospholipases, toxins exhibiting a phospholipase activity, and membrane-disrupting peptides. Phospholipases are a class of interface-activated enzymes that hydrolyse phospholipids in lipid membranes. They play a key role in phospholipid homeostasis, signal transduction, and inflammatory responses and are important disease markers in pancreatic, coronary artery, and heart conditions, as well as in a number of neurodegenerative diseases (Dennis et al. 2011). Lytic peptides are an important class of antibiotics compounds and part of our innate immune system.

In the past two decades, surface-based sensing tools such as SPR (Papo and Shai 2003; Uesugi et al. 2007) and QCM-D (Jackman et al. 2010; Justesen et al. 2004; Mechler et al. 2007) combined with supported lipid bilayers or tethered vesicles have been used to study peptide and protein binding together with bilayer disruption. Furthermore, a variety of lipid-based sensors for the detection of phospholipases and their activity have been described, targeting diagnostic applications. Phospholipid monolayers formed on hydrophobic liquid crystals surfaces (Brake et al. 2003) have, for example, been used to detect enzymatic degradation of the lipid monolayer assembly by phospholipases (Hartono et al. 2008) and phospholipase-like toxins (Hartono et al. 2009) as well as to screen for their inhibitors. Signal transduction is here based on the loss of orientation of the liquid crystals due to the disruption of the lipid monolayer which is monitored using cross-polarized light. This relatively simple set-up allows for label-free monitoring of enzymatic membrane degradation with picomolar sensitivity. Using a surface-based assay, we have recently presented a single-vesicle assay for the detection and characterization of single-enzyme kinetics based on the visualization of individual vesicles with TIRF microscopy. Thanks to the detection of the digestion of individual vesicles carrying fluorescent lipids without ensemble averaging, the assay was used to characterize single-enzyme kinetics of phospholipase A2 (PLA2) in cerebrospinal fluid from Alzheimer's patients. Besides allowing for the subpicomolar detection of PLA2 (Tabaei et al. 2013), the kinetic analysis of single enzymatic events allows for a decoupling of the contributions of enzymatic activity and enzyme concentration to the overall response. It also makes it possible to determine the Michaelis–Menten parameters without varying the substrate concentration (Rabe et al. 2015), which is challenging for interfacial enzymes like PLA2 since the substrate concentration cannot be easily varied. A similar experimental set-up using vesicles carrying a dye within the membrane and another dye in their aqueous interior was used to probe the membrane curvature-dependent pore formation and membrane rupture induced by an antiviral peptide, as shown in Figure 14.9 (Tabaei et al. 2012).

14.5.3 Single-Vesicle Assays to Probe Membrane-Curvature–Specific Proteins

The visualization of single vesicles immobilized on a sensor surface further opens the door to the development of methodologies to probe, with high-throughput, lipid–protein interactions as function of curvature, as suggested at the end of the previous section. Towards this end, Stamou et al. have demonstrated that analysis of the fluorescence intensity emitted by a membrane-bound dye can be related to the size of the immobilized vesicle (Kunding et al. 2008). Through this calibration, vesicle arrays with vesicle-size heterogeneities can be used to probe the curvature dependence of a number of biological processes (Christensen et al. 2010), although the analytical expressions utilized should rather be those presented in reference (Tabaei et al. 2013). In this context, single-vesicle arrays were used to probe the curvature-dependent binding of various biomolecules, including BAR-domain-containing proteins, to elucidate mechanisms underlying this process (Hatzakis et al. 2009). Alternatively, we have shown that a nanofluidic device can be used to visualize vesicles in solution and quantify their fluorescence intensity, a feature that was exploited to probe the curvature-dependent binding of cell-penetrating peptides to fluorescent liposomes (Friedrich et al. 2017).

14.6 LIPOSOMES AS NANOSCALE LABELS FOR SIGNAL AMPLIFICATION

The lipid-based biosensors presented in Sections 14.4 and 14.5 rely on the use of lipid assemblies for the functional immobilization of a biomolecular receptor at a sensor surface. In Section 14.6, we instead focus on the use of liposomes as nanoscale labels for signal generation in biosensing assays. The growing interest in this application of liposomes stems from the versatility of these lipid assemblies, *i.e.*, from the flexibility with which their physicochemical properties and their biomolecular content can be tailored. Liposomes can accommodate a large number of dyes or molecules within their aqueous interior, in the hydrophobic core of the bilayer lamella, or coupled to the head group of its lipid constituents, making them promising candidates as signal-amplification reagents. For signal generation, the use of dyes (Bally et al. 2011b; Connelly et al.

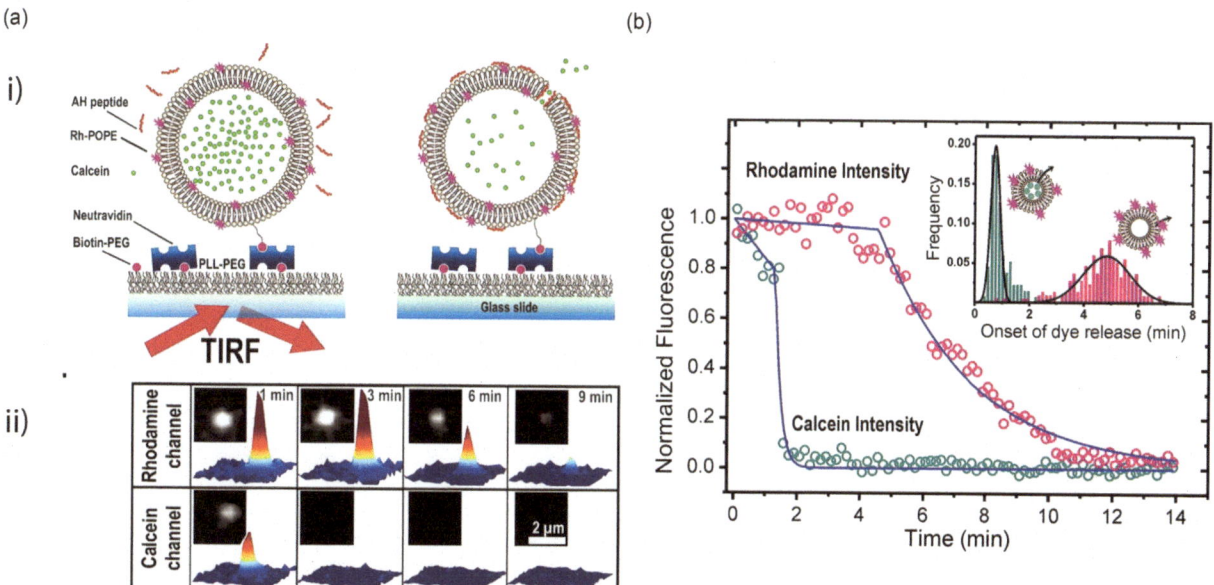

FIGURE 14.9 Single-vesicle array to study a membrane-disrupting peptide. (a) Schematic representation of the assay. (i) Calcein-loaded vesicles containing rhodamine–POPE in their membrane are immobilized via biotin–streptavidin. (ii) Membrane disruption and content release are monitored by measuring the change in red or green fluorescence of individual vesicles. (b) Fluorescence intensity time trace for an individual vesicle together with the evaluation of the onset time for membrane disruption and content release (insert). (Reproduced with permission from Tabaei et al. 2012.)

2012; Edwards et al. 2007b, 2008, 2010, 2013; Gunnarsson et al. 2011; Ho et al. 2007; Kwakye and Baeumner 2003; Zaytseva et al. 2005b), enzymes (Genc et al. 2011), electrochemically active species (Liu and Lin 2007; Viswanathan et al. 2012; Wang et al. 2011; Zhong et al. 2011), and oligonucleotides (Edwards et al. 2007a; Ou et al. 2009) has been reported. The surface of the liposome can further be easily functionalized with a variety of biomolecular recognition elements which include glycolipid (Bally et al. 2013; Edwards et al. 2007b; Ho et al. 2007), membrane protein receptors (as detailed in Section 14.4), oligonucleotides (Edwards et al. 2013; Gunnarsson et al. 2008; Kwakye and Baeumner 2003; Zaytseva et al. 2005b), antibodies (Connelly et al. 2012; Edwards et al. 2007a, 2013), aptamers (Edwards et al. 2010), or small molecule targets (Ho et al. 2002). In view of the excellent nonfouling properties of bilayer surfaces (Glasmastar et al. 2002), liposomes also exhibit low nonspecific binding, an essential feature in ensuring good assay specificity. Many liposome labels are also compatible with conventional biosensor detectors; colorimetric, chemiluminescent, fluorescence-based, or electrochemical signal transduction is commonly reported in conjunction with liposomes as detailed in recent reviews (Bally et al. 2009; Edwards et al. 2006, 2012; Liu and Boyd 2013; Mazur et al. 2017). Depending on the assay format, detection is performed either directly or after liposome lysis and concurrent release of an encapsulated dye.

The most widespread applications of lipid vesicle labels are probably related to signal amplification in portable and cheap point-of-care devices for on-site diagnostics or environmental monitoring. In this context, a variety of semiquantitative liposome-based lateral flow assays, also termed strip-based assays, have been developed and are available as commercial products (Gerber et al. 1990). In a lateral flow assay, such as the one shown in Figure 14.10a, the reagents migrate from the sample loading patch to an analyte capture zone by capillary action in nitrocellulose or polyethersulfone membranes. During migration, analyte and vesicle-labelled reporter molecules are allowed to react. After capture of the so-formed complexes at the analyte zone carrying anti-analyte antibodies, the colorimetric signal is detected either visually, using a scanner or by measuring the absorbance with a hand-held reflectometer. Electrochemiluminescent signal generation (i.e., the generation of a luminescent signal during an electrochemical reaction) has also been also reported (Yoon et al. 2003). Such sensors were developed for the detection of food- and waterborne pathogens (Baeumner et al. 2003; Park and Durst 2000), toxins (Ahn-Yoon et al. 2004; Ho et al. 2002), spores (Baeumner et al. 2004b), and allergens (Wen et al. 2005). The limit of detection is typically in the nanomolar range, as shown for the detection of oligonucleotides (Baeumner et al. 2004a).

For the development of portable devices, flow-through assays based on flow-injection analysis or on microfluidics represent an alternative to the membrane-stripe sensors. Here, the reagents are passed in a continuous buffer through a microcapillary (Ho et al. 2007) or microfluidic channel (Bunyakul et al. 2009; Connelly et al. 2012; Kwakye and Baeumner 2003; Nugen et al. 2009; Zaytseva et al. 2005a) functionalized with capture molecules (antibodies); readout is performed after lysis of the liposome and subsequent

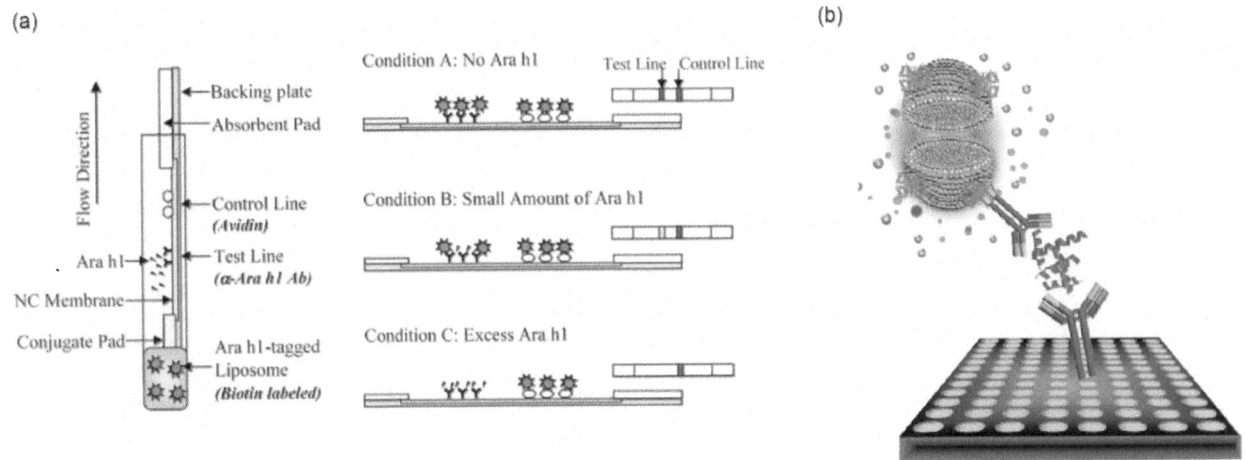

FIGURE 14.10 Bioanalytical assays based on liposome labels. (a) Lateral flow assay for the detection of a peanut allergen (Ara h1) in a competitive binding assay. The reagents migrate along the nitrocellulose strip by capillary action, and binding of the allergen competes with binding of biotinylated vesicles carrying the antigen. (Reproduced with permission from Wen et al. 2005.) (b) Signal amplification in a microtiter plate assay after lysis of dye-containing liposomes. (Reproduced with permission from Edwards et al. 2013.)

transport of its fluorescent or electroactive cargo to the site of detection (Bunyakul et al. 2009). Microfluidics-based sensors make it possible to perform bioanalytical assays with low sample consumption and at low cost, while displaying increased sensitivity with decreased assay times. Furthermore, they can be integrated into micro total analysis systems (Connelly et al. 2012; Nugen et al. 2009). Beaumner et al. have presented a microfluidic biosensor based on the use of a magnetic particles with assay sensitivities in the low picomolar range for the detection of both cholera toxin (Bunyakul et al. 2009) and oligonucleotides (Zaytseva et al. 2005b). In their assay, the biomolecular complex is first formed in solution at the surface of magnetic nanoparticles. The beads are then immobilized and concentrated at the channel surface using a magnet, before a fluorescent dye (Bunyakul et al. 2009; Connelly et al. 2012; Kwakye and Baeumner 2003; Zaytseva et al. 2005a) or electroactive compound (Bunyakul et al. 2009; Goral et al. 2006; Nugen et al. 2009) is released from the liposomes by lysis and detected. Electrochemical-based detection has several advantages over the fluorescence-based approaches, including ease-of-fabrication, low price, compatibility with well-established microfabrication technology, and integration capability of the detectors for a compact instrument design. A number of sensors taking advantage of liposomes for the encapsulation of electroactive compounds (Viswanathan et al. 2012), including redox enzymes (Genc et al. 2011) have been proposed. Viswanthan et al., for example, presented a gold-nanoelectrodes-based disposable sensor for the detection of carcinoembryonic antigen in saliva and serum, or of Mucin-16, an ovarian cancer marker in human serum samples (Viswanathan et al. 2012). In both cases, a sandwich assay using liposomes containing ferrocene carboxylic acid as the reporter was performed, and the signal was detected by voltammetry after the vesicles were lysed. The performance was comparable with a standard ELISA test (Viswanathan et al. 2012). Electrochemical assay multiplexing was achieved by Zhong et al. (2011) who demonstrated this possibility by using liposomes containing ascorbic acid and uric acid.

Liposomal labels have also proven useful in a number of multiplexed bioanalytical assays. They have, for example, been presented as an alternative to enzyme-linked immunosorbent assays in liposome-based microtiter plate experiments (Figure 14.10b) as early as in the 1990s. Their sensitivity is often better than standard ELISA assays (Edwards et al. 2013; Rongen et al. 1997); they are potentially cheaper and require shorter assay times (Edwards et al. 2007b). Liposomes have also been proposed as labels to amplify the signal in fluorescence-based microarrays (Bally et al. 2011b; Hwang et al. 2007; Wiese 2003). In our study on the use of fluorescent liposomes in microarray experiments (Bally et al. 2011b), we exemplify how a signal evaluation based on the possibility of counting vesicles one-by-one can lead to an increase in signal-to-noise values and thus an improvement in assay sensitivity as compared with an evaluation based on measuring the average intensity of a spot.

Another method that takes advantage of the visualization of individual liposomal labels is the equilibrium fluctuation analysis technique presented in Sections 14.4.2 and 14.5.1, in the context of ligand–membrane (Gunnarsson et al. 2011), virus–glycolipid (Bally et al. 2011a), and glycolipid–glycolipid (Kunze et al. 2013) interactions. This concept was used for the detection of oligonucleotide strands in the low fM regime. Analysis of the kinetic behaviour further allows for the accurate discrimination of single-nucleotide polymorphisms (Gunnarsson et al. 2008). Using the experimental set-up illustrated in Figure 14.8a, we further showed that such an assay can be used for the detection of norovirus-like particles with femtomolar sensitivity (Bally et al. 2013). In the latter study, we

Lipid-Based Bioanalytical Sensors

demonstrated that careful analysis of the dynamic behaviour of specific and nonspecific binding events could be used to increase the accuracy and the sensitivity of the assay by further suppressing the nonspecific background signal. This method is suggested to be generally applicable, provided that the events to be discriminated display sufficiently different dissociation kinetics.

14.7 LIPID-BASED NANOREACTORS

Lipid assemblies, in particular lipid vesicles and lipid nanotubes, are also generating increased interest for use in nanofluidics and as nanoreactors; they allow for the handling of minute amounts of materials, for the study of biochemical reactions in confined environments or for single-molecule studies. Their intrinsic inert character makes them particularly well suited to gently handle biomolecules. The inner cavity of a vesicle can be used to maintain individual (or a few) biomolecules in a well-defined location. This makes it possible to maintain the molecule's rotational freedom, which is greatly impeded when it is directly immobilized on a surface, and to avoid surface-induced denaturation.

Large unilamellar vesicles with an inner aqueous compartment in the zeptoliter range can be viewed as ultraminiaturized reactors to be interrogated with surface-sensitive techniques after immobilization. The visualization of individual liposomes allows for high-throughput studies of processes occurring within the individual vesicles (Christensen et al. 2010, 2013). Surface-tethered vesicles have been used in the context of single-molecule studies (Boukobza et al. 2001), to investigate, for example, heterogeneities in the protein-folding behaviour (Okumus et al. 2004; Rhoades et al. 2003). Lipid containers have also been found attractive in the context of reproducing basic cellular functions towards the creation of protocells (Stano et al. 2011). In the context of probing biomolecular interactions in a confined lipid environment, Ha et al. probed ribozyme folding, as well as transient interactions between proteins and DNA by Förster resonance energy transfer (Cisse et al. 2007; Okumus et al. 2009). Here, the vesicle was rendered porous by addition of the pore-forming toxin α hemolysin, and the pores were small enough to retain both protein and ssDNA inside the vesicles while allowing for exchange of buffer and small molecules, such as ATP, between the vesicle's interior and exterior (Figure 14.11a). The encapsulation of the biomolecules makes it possible to study interactions on a single-molecule level or at high local reagent concentrations (up to micomolar), an important asset when probing weak and transient recognition events such as the relatively weak protein–protein interactions between a copper–chaperone and the Wilson disease protein (Benitez et al. 2008, 2010). Towards the use of surface-bound vesicles as individual nanoreactors, Christensen et al. showed that subattoliter volumes can be mixed by fusion of vesicles of opposite charge making it possible to perform in parallel a multitude of miniaturized reactions on a random vesicle array (Figure 14.11b) (Christensen et al. 2012).

Efforts have also been put into engineering more complex lipid-based nanofluidic systems. Boliger et al. (2004, 2008) presented a reactor set-up based on multiple small unilamellar vesicles encapsulated into a larger one. Taking advantage of the fact that leakage out of the vesicles mainly occurs at the phase transition temperature, the authors demonstrate that temperature can be used to sequentially trigger content release of small vesicles of different composition. Orwar et al. used micropipettes and micromanipulators to draw complex nanofluidic networks consisting of giant vesicles connected by lipid nanotubes (Jesorka et al. 2011; Karlsson et al. 2004, 2006). The possibility of selectively injecting different components in the vesicles and of controlling the flow of reagents between the compartments makes it possible to use such devices

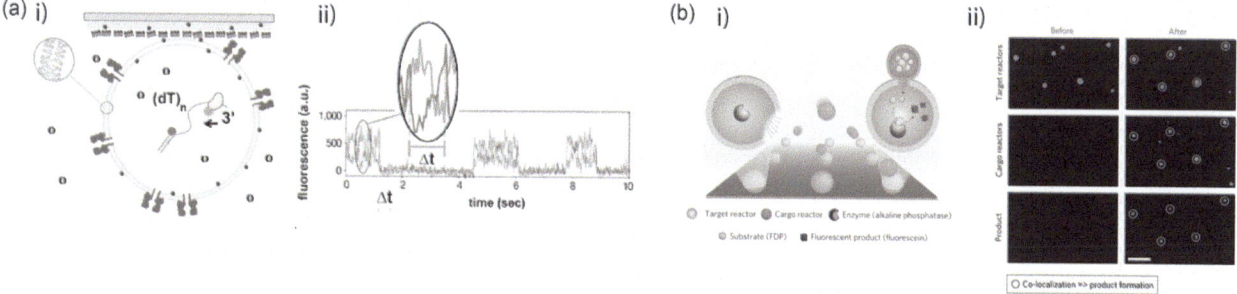

FIGURE 14.11 Lipid assemblies as nanocontainers. (a) Nanocontainer for single-molecule studies (i) Assay set-up for the visualization of the DNA translocation activity of Rep Helicase. The surface-bound vesicles are rendered porous by the presence of a-hemolysin pores which allow for transport of small reagents such as ions and ATP across the membrane. (ii) FRET-based visualization of the interaction between enzyme and DNA carrying a donor and receptor dye respectively. (Reproduced with permission from Okumus et al. 2009.) (b) Mixing of subattoliter volumes using vesicles. (i) Schematic representation of an experiment: liposomes containing alkaline phosphatase are immobilized on a surface prior addition of liposomes of opposite charge containing a profluorescent substrate. (ii) Fusion of target and cargo reactor is visualized by FRET since both vesicle types carry acceptor- or donor-labelled lipids (top and middle). Content mixing and fluorescent product formation is visualized for each individual vesicle (bottom). (Reproduced with permission from Christensen et al. 2012.)

to perform complex chemical and biochemical reactions (Karlsson et al. 2006).

14.8 EXPLOITING FORCE-DRIVEN MEMBRANE COMPONENT MANIPULATION FOR BIOSENSING

One of the hallmarks of the cellular membrane is its dynamic fluid nature, *i.e.*, the ability of its lipid, protein, and carbohydrate components to diffuse laterally for the association and dissociation of complexes vital to cellular physiology via signal transduction, adhesion, etc. In order to better understand the motion and organization of membrane components within their native environment, researchers began investigating methods of manipulating, or moving, the components through the use of an external force. In the 1970s, Poo et al. demonstrated that fluorescently labelled concanavalin A receptors on the surface of an embryonic muscle cell membrane could be redistributed by an external electric field (Poo and Robinson 1977). These results conjured up ideas on how such phenomena could be exploited for the accumulation of low-abundance membrane receptors to increase their local concentration and hence the signal-to-noise ratio when probing ligand–receptor interactions. While subjecting intact cells to an electric field provided the inspiration, it was the emergence of SLBs (Tamm and Mcconnell 1985) as functional model systems which allowed for the detailed investigations of how external forces can manipulate membrane components. Knowledge gained from such fundamental studies was later exploited for biotechnological applications (e.g., transport, separations, biosensing, etc.).

Early studies of SLB electrophoresis demonstrated the accumulation of charged lipids against a barrier (Groves et al. 1995), the ability to separate lipids of opposite charge (Stelzle et al. 1992), and eventually the ability to separate lipids with the same polarity (Daniel et al. 2007). These technological advancements open the possibility to create arrays from mixtures of membrane components. Indeed, both the separation of two cell receptors (gangliosides) into separate bands (Pace et al. 2013) and the ability to focus (i.e. concentrate and separate) a mixture of membrane-bound proteins into bands (Figure 14.12a) (Liu et al. 2011) have been shown. Both demonstrate how such technology could potentially be exploited to create arrays of receptors from complex membrane component mixtures. Such receptor-array-based biosensors could be valuable in studying how different pathogens or pharmaceuticals interact with specific receptors.

Electric fields are not the only external forces that can be utilized for the manipulation and separation of SLB components. The use of surface acoustic waves (SAWs) technology to create differences in SLB density has been demonstrated to focus (*i.e.*, concentrate and separate) membrane-bound proteins (Figure 14.12b) (Neumann et al. 2010) and lipid-anchored single-stranded DNA (Hennig et al. 2011). Hydrodynamic forces have also been employed for the separation of membrane-bound proteins through differences in the size and number of interactions with lipid-anchored ligands (Figure 14.12c) (Jonsson et al. 2011).

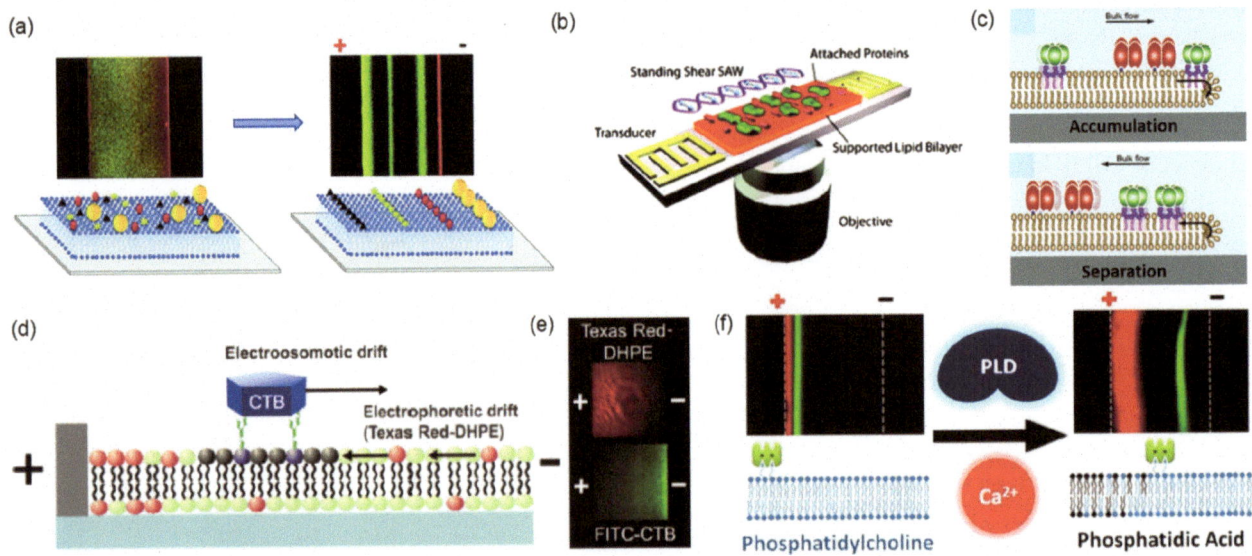

FIGURE 14.12 Examples of SLB manipulation and how they have been exploited to make biosensors. The use of (a) an electric field, (b) a standing acoustic wave, or (c) hydrodynamic forces to focus (*i.e.*, accumulate and separate) lipids and membrane-bound proteins into arrays of discrete bands from compositionally homogeneous SLBs. (Reproduced with permission from Jonsson et al. 2011; Liu et al. 2011; Neumann et al. 2010 (d) The use of an electric field to create a Texas Red-DHPE gradient which is modulated by the binding of a protein (CTB) to ganglioside lipids within the SLB. (Reproduced with permission from Lee and Nam 2012.) (e) Fluorescent gradients of the lipid Texas Red-DHPE and FITC-cholera toxin B (CTB) during a typical experiment. (Reproduced with permission from Lee and Nam 2012.) (f) The use of an electric field to focus bands of Texas Red-DHPE and FITC-Streptavidin before and after the addition of phospholipase D (PLD) and calcium ions. (Reproduced with permission from Liu et al. 2014.)

While these technologies show promise in the formation of receptor arrays and thus hold value in biosensor platform production, there are also a few reports in which these SLB manipulation strategies have already been directly exploited as a biosensor.

To date, the clearest examples of utilizing SLB manipulation directly as a biosensor have been the works by Lee et al. (Lee and Nam 2012; Liu et al. 2014), both of which exploited the use of an electric field for making a gradient against a barrier. Lee and coworkers showed that the binding of cholera toxin B-subunit to its lipid receptor, GM1, resulted in the deformation of a fluorescent lipid gradient which could be used to probe indirectly the amount of the protein that had bound to the surface (Figure 14.12d and e) (Lee and Nam 2012). In a slightly more elegant use of the technology, Liu and coworkers monitored the production of phosphatic acid (PA) from phosphatidylcholine lipids within SLBs by reactions with phospholipase D (PLD) (Liu et al. 2014) (Figure 14.12f). The production of PA altered the composition of the SLB which resulted in the modulation of the position of a band of fluorescently labelled, lipid-anchored protein. The position of this fluorescent protein band within the SLB was used to probe the production of PA and thus monitor an enzymatic reaction in a "label-free" manner.

While the use of concentration gradients generated by SLB manipulation is the only direct use of these technologies in biosensing to date, the potential these technologies hold for the creation of complex receptor arrays must be stressed. Particularly their combination with the previously discussed creation of cell-derived SLBs (see Section 14.4.1) is expected to potentially open a whole new platform for pathogen– and pharmaceutical–receptor discovery.

14.9 CONCLUDING REMARKS

In this chapter, we have provided an overview on the diversity of applications that make use in one way or another of lipid assemblies in the context of biosensor and bioanalytical assay technology. Their potential in the field is undeniable, and accordingly, much effort has been put into taking advantage of their unique properties. Versatile methodologies for the immobilization of lipid assemblies at sensing interfaces are becoming available, and much progress is being made towards the functional incorporation of integral membrane proteins therein. Accordingly, a number of biosensors probing ligand binding to both lipids and membrane proteins, as well as transport across membrane channels, have been presented. In addition, researchers have taken advantage of liposomes in designing more robust bioanalytical assays towards easy-to-use point-of-care devices, as well as in the context of single-molecule studies in confined volumes.

The innovation and diversity of the approaches adopted towards the creation of lipid-based sensors has been illustrated through the various examples presented herein. Yet, a number of bottlenecks remain to be resolved before many of these technologies become commonly available in biotech, biomedical, and pharmaceutical environments. The stability of lipid assemblies remains often a major issue. This is particularly true for sensing applications relying on free-standing lipid bilayers. In addition to this, the efficient and functional incorporations of membrane proteins into artificial lipid assemblies remain a difficult and cumbersome task. Protocols often need to be optimized for each type of protein individually, and the end user will often have to compromise between, for example, controlled protein orientation, high protein density, or transmembrane protein mobility. Finally, implementation of lipid assemblies in drug discovery, incontestably one of the most promising large-scale applications of lipid-based biosensors, further requires the development of reliable, automatable, and simple protocols for the production of high-throughput systems for screening applications. Nevertheless, as highlighted in this chapter, the field is progressing rapidly, and the concerted effort of physicists, chemists, biochemists, and engineers in academic, clinical, and pharmaceutical settings has generated a wealth of novel methodologies bringing us every day closer to these aims.

REFERENCES

Ahn-Yoon, S., T. R. DeCory, and R. A. Durst. 2004. Ganglioside-liposome immunoassay for the detection of botulinum toxin. *Anal Bioanal Chem* 378:68–75.

Ainla, A., I. Gozen, B. Hakonen, and A. Jesorka. 2013. Lab on a Biomembrane: Rapid prototyping and manipulation of 2D fluidic lipid bilayer circuits (vol 3, 2743, 2013). *Sci Rep-Uk* 3: 1–7.

Albertorio, F., A. J. Diaz, T. L. Yang et al. 2005. Fluid and air-stable lipopolymer membranes for biosensor applications. *Langmuir* 21:7476–7482.

Andersson, M., H. M. Keizer, C. Y. Zhu, D. Fine, A. Dodabalapur, and R. S. Duran. 2007. Detection of single ion channel activity on a chip using tethered bilayer membranes. *Langmuir* 23:2924–2927.

Arigi, E., O. Blixt, K. Buschard, H. Clausen, and S. B. Levery. 2012. Design of a covalently bonded glycosphingolipid microarray. *Glycoconjugate J* 29:1–12.

Axelrod, D., D. E. Koppel, J. Schlessinger, E. Elson, and W. W. Webb. 1976. Mobility measurement by analysis of fluorescence photobleaching recovery kinetics. *Biophys J* 16:1055–1069.

Baeumner, A. J., R. N. Cohen, V. Miksic, and J. H. Min. 2003. RNA biosensor for the rapid detection of viable *Escherichia coli* in drinking water. *Biosensors & bioelectronics* 18:405–413.

Baeumner, A. J., C. Jones, C. Y. Wong, and A. Price. 2004a. A generic sandwich-type biosensor with nanomolar detection limits. *Anal Bioanal Chem* 378:1587–1593.

Baeumner, A. J., B. Leonard, J. McElwee, and R. A. Montagna. 2004b. A rapid biosensor for viable B-anthracis spores. *Anal Bioanal Chem* 380:15–23.

Bailey, K., M. Bally, W. Leifert, J. Voros, and T. McMurchie. 2009. G-protein coupled receptor array technologies: Site directed immobilisation of liposomes containing the H-1-histamine or M-2-muscarinic receptors. *Proteomics* 9:2052–2063.

Bally, M., K. Bailey, K. Sugihara, D. Grieshaber, J. Voros, and B. Stadler. 2010. Liposome and lipid bilayer arrays towards biosensing applications. *Small* 6:2481–2497.

Bally, M., M. Graule, F. Parra, G. Larson, and F. Hook. 2013. A virus biosensor with single virus-particle sensitivity based on fluorescent vesicle labels and equilibrium fluctuation analysis. *Biointerphases* 8:4

Bally, M., A. Gunnarsson, L. Svensson, G. Larson, V. P. Zhdanov, and F. Hook. 2011a. Interaction of single viruslike particles with vesicles containing glycosphingolipids. *Phys Rev Lett* 107:188103.

Bally, M., M. Halter, J. Voros, and H. M. Grandin. 2006. Optical microarray biosensing techniques. *Surf Interface Anal* 38:1442–1458.

Bally, M., G. E. Rydell, R. Zahn et al. 2012. Norovirus GII.4 virus-like particles recognize galactosylceramides in domains of planar supported lipid bilayers. *Angew Chem Int Edit* 51:12020–12024.

Bally, M., S. Syed, A. Binkert, E. Kauffmann, M. Ehrat, and J. Voros. 2011b. Fluorescent vesicles for signal amplification in reverse phase protein microarray assays. *Analytical biochemistry* 416:145–151.

Bally, M., and J. Voros. 2009. Nanoscale labels: Nanoparticles and liposomes in the development of high-performance biosensors. *Nanomedicine-Uk* 4:447–467.

Bayerl, T. M., and M. Bloom. 1990. Physical-properties of single phospholipid-bilayers adsorbed to micro glass-beads - a new vesicular model system studied by H-2-nuclear magnetic-resonance. *Biophys J* 58:357–362.

Bayley, H. 2015. Nanopore sequencing: From imagination to reality. *Clin Chem* 61:25–31.

Bayley, H., and P. S. Cremer. 2001. Stochastic sensors inspired by biology. *Nature* 413:226–230.

Benitez, J. J., A. M. Keller, and P. Chen. 2010. Nanovesicle trapping for studying weak protein interactions by single-molecule fret. *Meth Enzymol, Vol 472: Single Mol Tools, Pt A: Fluorescence Based Approach* 472:41–60.

Benitez, J. J., A. M. Keller, P. Ochieng et al. 2008. Probing transient copper Chaperone-Wilson disease protein interactions at the single-molecule level with nanovesicle trapping. *J Am Chem Soc* 130:2446–2447.

Besenicar, M., P. Macek, J. H. Lakey, and G. Anderluh. 2006. Surface plasmon resonance in protein-membrane interactions. *Chem Phys Lipids* 141:169–178.

Bieri, C., O. P. Ernst, S. Heyse, K. P. Hofmann, and H. Vogel. 1999. Micropatterned immobilization of a G protein-coupled receptor and direct detection of G protein activation. *Nat Biotechnol* 17:1105–1108.

Binkert, A., P. Studer, and J. Voros. 2009. A microwell array platform for picoliter membrane protein assays. *Small* 5:1070–1077.

Boersma, A. J., K. L. Brain, and H. Bayley. 2012. Real-time stochastic detection of multiple neurotransmitters with a protein nanopore. *Acs Nano* 6:5304–5308.

Bolinger, P. Y., D. Stamou, and H. Vogel. 2004. Integrated nanoreactor systems: Triggering the release and mixing of compounds inside single vesicles. *J Am Chem Soc* 126:8594–8595.

Bolinger, P. Y., D. Stamou, and H. Vogel. 2008. An integrated self-assembled nanofluidic system for controlled biological chemistries. *Angew Chem Int Edit* 47:5544–5549.

Borch, J., F. Torta, S. G. Sligar, and P. Roepstorff. 2008. Nanodiscs for immobilization of lipid bilayers and membrane receptors: Kinetic analysis of cholera toxin binding to a glycolipid receptor. *Anal Chem* 80:6245–6252.

Boukobza, E., A. Sonnenfeld, and G. Haran. 2001. Immobilization in surface-tethered lipid vesicles as a new tool for single biomolecule spectroscopy. *J Phys Chem B* 105:12165–12170.

Brake, J. M., M. K. Daschner, Y. Y. Luk, and N. L. Abbott. 2003. Biomolecular interactions at phospholipid-decorated surfaces of liquid crystals. *Science* 302:2094–2097.

Branden, M., S. R. Tabaei, G. Fischer, R. Neutze, and F. Hook. 2010. Refractive-index-based screening of membrane-protein-mediated transfer across biological membranes. *Biophys J* 99:124–133.

Branton, D., D. W. Deamer, A. Marziali et al. 2008. The potential and challenges of nanopore sequencing. *Nat Biotechnol* 26:1146–1153.

Briand, E., M. Zach, S. Svedhem, B. Kasemo, and S. Petronis. 2010. Combined QCM-D and EIS study of supported lipid bilayer formation and interaction with pore-forming peptides. *Analyst* 135:343–350.

Bucior, I., and M. M. Burger. 2004. Carbohydrate-carbohydrate interactions in cell recognition. *Curr Opin Struc Biol* 14:631–637.

Bunyakul, N., K. A. Edwards, C. Promptmas, and A. J. Baeumner. 2009. Cholera toxin subunit B detection in microfluidic devices. *Anal Bioanal Chem* 393:177–186.

Burgel, S. C., O. Guillaume-Gentil, L. M. Zheng, J. Voros, and M. Bally. 2010. Zirconium Ion mediated formation of liposome multilayers. *Langmuir* 26:10995–11002.

Castellana, E. T., and P. S. Cremer. 2006. Solid supported lipid bilayers: From biophysical studies to sensor design. *Surf Sci Rep* 61:429–444.

Chaize, B., M. Nguyen, T. Ruysschaert et al. 2006. Microstructured liposome array. *Bioconjugate Chem* 17:245–247.

Chan, Y. H. M., and S. G. Boxer. 2007. Model membrane systems and their applications. *Curr Opin Chem Biol* 11:581–587.

Cho, N. J., C. W. Frank, B. Kasemo, and F. Hook. 2010. Quartz crystal microbalance with dissipation monitoring of supported lipid bilayers on various substrates. *Nat Protoc* 5:1096–1106.

Christensen, A. L., C. Lohr, S. M. Christensen, and D. Stamou. 2013. Single vesicle biochips for ultra-miniaturized nanoscale fluidics and single molecule bioscience. *Lab Chip* 13:3613–3625.

Christensen, S. M., P. Y. Bolinger, N. S. Hatzakis, M. W. Mortensen, and D. Stamou. 2012. Mixing subattolitre volumes in a quantitative and highly parallel manner with soft matter nanofluidics. *Nat Nanotechnol* 7:51–55.

Christensen, S. M., and D. Stamou. 2007. Surface-based lipid vesicle reactor systems: Fabrication and applications. *Soft Matter* 3:828–836.

Christensen, S. M., and D. G. Stamou. 2010. Sensing-applications of surface-based single vesicle arrays. *Sensors-Basel* 10:11352–11368.

Cisse, I., B. Okumus, C. Joo, and T. J. Ha. 2007. Fueling protein-DNA interactions inside porous nanocontainers. *P Natl Acad Sci USA* 104:12646–12650.

Connelly, J. T., S. Kondapalli, M. Skoupi, J. S. L. Parker, B. J. Kirby, and A. J. Baeumner. 2012. Micro-total analysis system for virus detection: Microfluidic pre-concentration coupled to liposome-based detection. *Anal Bioanal Chem* 402:315–323.

Cooper, M. A. 2002. Optical biosensors in drug discovery. *Nat Rev Drug Discov* 1:515–528.

Cooper, M. A. 2004. Advances in membrane receptor screening and analysis. *J Mol Recognit* 17:286–315.

Cooper, M. A. 2006. Optical biosensors: Where next and how soon? *Drug Discov Today* 11:1061–1067.

Cooper, M. A., A. Hansson, S. Lofas, and D. H. Williams. 2000. A vesicle capture sensor chip for kinetic analysis of interactions with membrane-bound receptors. *Analyt Biochem* 277:196–205.

Cornell, B. A., V. L. B. BraachMaksvytis, L. G. King et al. 1997. A biosensor that uses ion-channel switches. *Nature* 387:580–583.

Costello, D. A., C. Y. Hsia, J. K. Millet, T. Porri, and S. Daniel. 2013. Membrane fusion-competent virus-like proteoliposomes and proteinaceous supported bilayers made directly from cell plasma membranes. *Langmuir* 29:6409–6419.

Costello, R. F., I. R. Peterson, J. Heptinstall, and D. J. Walton. 1999. Improved gel-protected bilayers. *Biosens Bioelectron* 14:265–271.

Cremer, P. S., and T. L. Yang. 1999. Creating spatially addressed arrays of planar supported fluid phospholipid membranes. *J Am Chem Soc* 121:8130–8131.

Dahlin, A. B., M. P. Jonsson, and F. Hook. 2008. Specific self-assembly of single lipid vesicles in nanoplasmonic apertures in gold. *Adv Mater* 20: 1436–+.

Danelon, C., J. B. Perez, C. Santschi, J. Brugger, and H. Vogel. 2006. Cell membranes suspended across nanoaperture arrays. *Langmuir* 22:22–25.

Danelon, C., S. Terrettaz, O. Guenat, M. Koudelka, and H. Vogel. 2008. Probing the function of ionotropic and G protein-coupled receptors in surface-confined membranes. *Methods* 46:104–115.

Daniel, S., A. J. Diaz, K. M. Martinez, B. J. Bench, F. Albertorio, and P. S. Cremer. 2007. Separation of membrane-bound compounds by solid-supported bilayer electrophoresis. *J Am Chem Soc* 129: 8072–8073.

de la Escosura-Muniz, A., and A. Merkoci. 2012. Nanochannels Preparation and Application in Biosensing. *Acs Nano* 6:7556–7583.

Demarche, S., K. Sugihara, T. Zambelli, L. Tiefenauer, and J. Voros. 2011. Techniques for recording reconstituted ion channels. *Analyst* 136:1077–1089.

Deng, Y., Y. Wang, B. Holtz et al. 2008. Fluidic and air-stable supported lipid bilayer and cell-mimicking microarrays. *J Am Chem Soc* 130:6267–6271.

Denning, E. J., and O. Beckstein. 2013. Influence of lipids on protein-mediated transmembrane transport. *Chem Phys Lipids* 169:57–71.

Dennis, E. A., J. Cao, Y.-H. Hsu, V. Magrioti, and G. Kokotos. 2011. Phospholipase A2 enzymes: Physical structure, biological function, disease implication, chemical inhibition, and therapeutic intervention. *Chemical Reviews* 111:6130–6185.

Dodd, C. E., B. R. G. Johnson, L. J. C. Jeuken, T. D. H. Bugg, R. J. Bushby, and S. D. Evans. 2008. Native E. coli inner membrane incorporation in solid-supported lipid bilayer membranes. *Biointerphases* 3:Fa59-Fa67.

Dufrene, Y. F., and G. U. Lee. 2000. Advances in the characterization of supported lipid films with the atomic force microscope. *Bba-Biomembranes* 1509:14–41.

Dusseiller, M. R., B. Niederberger, B. Stadler, D. Falconnet, M. Textor, and J. Voros. 2005. A novel crossed microfluidic device for the precise positioning of proteins and vesicles. *Lab Chip* 5:1387–1392.

Edwards, K. A., and A. J. Baeumner. 2006. Liposomes in analyses. *Talanta* 68:1421–1431.

Edwards, K. A., and A. J. Baeumner. 2007a. DNA-oligonucleotide encapsulating liposomes as a secondary signal amplification means. *Anal Chem* 79:1806–1815.

Edwards, K. A., O. R. Bolduc, and A. J. Baeumner. 2012. Miniaturized bioanalytical systems: Enhanced performance through liposomes. *Curr Opin Chem Biol* 16:444–452.

Edwards, K. A., K. L. Curtis, J. L. Sailor, and A. J. Baeumner. 2008. Universal liposomes: Preparation and usage for the detection of mRNA. *Anal Bioanal Chem* 391:1689–1702.

Edwards, K. A., and J. C. March. 2007b. GM(1)-functionalized liposomes in a microtiter plate assay for cholera toxin in Vibrio cholerae culture samples. *Anal Biochem* 368:39–48.

Edwards, K. A., K. J. Meyers, B. Leonard, and A. J. Baeumner. 2013. Superior performance of liposomes over enzymatic amplification in a high-throughput assay for myoglobin in human serum. *Anal Bioanal Chem* 405:4017–4026.

Edwards, K. A., Y. Wang, and A. J. Baeumner. 2010. Aptamer sandwich assays: Human alpha-thrombin detection using liposome enhancement. *Anal Bioanal Chem* 398:2645–2654.

Elie-Caille, C., O. Fliniaux, J. Pantigny, J. C. Maziere, and C. Bourdillon. 2005. Self-assembly of solid-supported membranes using a triggered fusion of phospholipid-enriched proteoliposomes prepared from the inner mitochondrial membrane. *Langmuir* 21:4661–4668.

Evans, S. V., and C. R. MacKenzie. 1999. Characterization of protein-glycolipid recognition at the membrane bilayer. *J Mol Recognit* 12:155–168.

Fagerberg, L., K. Jonasson, G. von Heijne, M. Uhlen, and L. Berglund. 2010. Prediction of the human membrane proteome. *Proteomics* 10:1141–1149.

Fan, X. D., I. M. White, S. I. Shopova, H. Y. Zhu, J. D. Suter, and Y. Z. Sun. 2008. Sensitive optical biosensors for unlabeled targets: A review. *Anal Chim Acta* 620:8–26.

Fang, Y., A. G. Frutos, and J. Lahiri. 2002. Membrane protein microarrays. *J Am Chem Soc* 124:2394–2395.

Fang, Y., A. G. Frutos, and J. Lahiri. 2003. Ganglioside microarrays for toxin detection. *Langmuir* 19:1500–1505.

Friedrich, R., S. Block, M. Alizadehheidari et al. 2017. A nano flow cytometer for single lipid vesicle analysis. *Lab Chip* 17:830–841.

Fukui, S., T. Feizi, C. Galustian, A. M. Lawson, and W. G. Chai. 2002. Oligosaccharide microarrays for high-throughput detection and specificity assignments of carbohydrate-protein interactions. *Nat Biotechnol* 20:1011–1017.

Galban-Horcajo, F., S. Halstead, R. McGonigal, and H. Willison. 2014. The application of glycosphingolipid arrays to autoantibody detection in neuroimmunological disorders. *Curr Opin Chem Biol* 18:78–86.

Genc, R., D. Murphy, A. Fragoso, M. Ortiz, and C. K. O'Sullivan. 2011. Signal-enhancing thermosensitive liposomes for highly sensitive immunosensor development. *Anal Chem* 83:563–570.

Gerber, M. A., M. F. Randolph, and K. K. Demeo. 1990. Liposome immunoassay for rapid identification of group-a streptococci directly from throat swabs. *J Clin Microbiol* 28:1463–1464.

Giess, F., M. G. Friedrich, J. Heberle, R. L. Naumann, and W. Knoll. 2004. The protein-tethered lipid bilayer: A novel mimic of the biological membrane. *Biophys J* 87:3213–3220.

Glasmastar, K., C. Larsson, F. Hook, and B. Kasemo. 2002. Protein adsorption on supported phospholipid bilayers. *J Colloid Interf Sci* 246:40–47.

Gluck, J. M., B. W. Koenig, and D. Willbold. 2011. Nanodiscs allow the use of integral membrane proteins as analytes in surface plasmon resonance studies. *Analy Biochem* 408:46–52.

Goennenwein, S., M. Tanaka, B. Hu, L. Moroder, and E. Sackmann. 2003. Functional incorporation of integrins into solid supported membranes on ultrathin films of cellulose: Impact on adhesion. *Biophys J* 85:646–655.

Goksu, E. I., J. M. Vanegas, C. D. Blanchette, W. C. Lin, and M. L. Longo. 2009. AFM for structure and dynamics of biomembranes. *Bba-Biomembranes* 1788:254–266.

Goral, V. N., N. V. Zaytseva, and A. J. Baeumner. 2006. Electrochemical microfluidic biosensor for the detection of nucleic acid sequences. *Lab Chip* 6:414–421.

Graneli, A., J. J. Benkoski, and F. Hook. 2007. Characterization of a proton pumping transmembrane protein incorporated into a supported three-dimensional matrix of proteoliposomes. *Analytical biochemistry* 367:87–94.

Grieshaber, D., R. MacKenzie, J. Voros, and E. Reimhult. 2008. Electrochemical biosensors - Sensor principles and architectures. *Sensors-Basel* 8:1400–1458.

Groves, J. T., and S. G. Boxer. 1995. Electric field-induced concentration gradients in planar supported bilayers. *Biophys J* 69:1972–1975.

Groves, J. T., N. Ulman, and S. G. Boxer. 1997. Micropatterning fluid lipid bilayers on solid supports. *Science* 275:651–653.

Groves, J. T., N. Ulman, P. S. Cremer, and S. G. Boxer. 1998. Substrate-membrane interactions: Mechanisms for imposing patterns on a fluid bilayer membrane. *Langmuir* 14:3347–3350.

Gu, L. Q., O. Braha, S. Conlan, S. Cheley, and H. Bayley. 1999. Stochastic sensing of organic analytes by a poreforming protein containing a molecular adapter. *Nature* 398:686–690.

Guan, X. Y., L. Q. Gu, S. Cheley, O. Braha, and H. Bayley. 2005. Stochastic sensing of TNT with a genetically engineered pore. *Chembiochem* 6:1875–1881.

Guidelli, R., and L. Becucci. 2011. Ion transport across biomembranes and model membranes. *J Solid State Electr* 15:1459–1470.

Gunnarsson, A., L. Dexlin, P. Wallin et al. 2011. Kinetics of ligand binding to membrane receptors from equilibrium fluctuation analysis of single binding events. *J Am Chem Soc* 133:14852–14855.

Gunnarsson, A., P. Jonsson, R. Marie, J. O. Tegenfeldt, and F. Hook. 2008. Single-molecule detection and mismatch discrimination of unlabeled DNA targets. *Nano Lett* 8:183–188.

Gunnarsson, A., Q. Snijder, J. Hicks, J. Gunnarsson, F. Hook, and S. Geschwinder. 2015. Drug discovery at the single molecule level: Inhibition-in-solution assay of membranereconstituted beta-secretase using single-molecule imaging. *Anal Chem* 87:4100–4103.

Hakomori, S. 2000. Traveling for the glycosphingolipid path. *Glycoconjugate J* 17:627–647.

Han, X. J., A. Studer, H. Sehr et al. 2007. Nanopore arrays for stable and functional free-standing lipid bilayers. *Adv Mater* 19: 4466–+.

Harding, P. J., T. C. Hadingham, J. M. McDonnell, and A. Watts. 2006. Direct analysis of a GPCR-agonist interaction by surface plasmon resonance. *Eur Biophys J Biophy* 35:709–712.

Hardy, G. J., R. Nayak, and S. Zauscher. 2013. Model cell membranes: Techniques to form complex biomimetic supported lipid bilayers via vesicle fusion. *Curr Opin Colloid In* 18:448–458.

Hartono, D., X. Y. Bi, K. L. Yang, and L. Y. L. Yung. 2008. An air-supported liquid crystal system for real-time and labelfree characterization of phospholipases and their inhibitors. *Adv Funct Mater* 18:2938–2945.

Hartono, D., S. L. Lai, K. L. Yang, and L. Y. L. Yung. 2009. A liquid crystal-based sensor for real-time and label-free identification of phospholipase-like toxins and their inhibitors. *Biosensors & bioelectronics* 24:2289–2293.

Hatzakis, N. S., V. K. Bhatia, J. Larsen et al. 2009. How curved membranes recruit amphipathic helices and protein anchoring motifs. *Nat Chem Biol* 5:835–841.

Hemmler, R., G. Bose, R. Wagner, and R. Peters. 2005. Nanopore unitary permeability measured by electrochemical and optical single transporter recording. *Biophys J* 88:4000–4007.

Hennig, M., M. Wolff, J. Neumann, A. Wixforth, M. F. Schneider, and J. O. Radler. 2011. DNA concentration modulation on supported lipid bilayers switched by surface acoustic waves. *Langmuir* 27:14721–14725.

Heyse, S., T. Stora, E. Schmid, J. H. Lakey, and H. Vogel. 1998. Emerging techniques for investigating molecular interactions at lipid membranes. *Bba-Rev Biomembranes* 1376:319–338.

Hinman, S. S., C. J. Ruiz, G. Drakakaki, T. E. Wilkop, and Q. Cheng. 2015. On-demand formation of supported lipid membrane arrays by trehalose-assisted vesicle delivery for SPR imaging. *Acs Appl Mater Inter* 7:17122–17130.

Hirano-Iwata, A., M. Niwano, and M. Sugawara. 2008. The design of molecular sensing interfaces with lipid-bilayer assemblies. *Trac-Trend Anal Chem* 27:512–520.

Ho, J. A. A., and R. D. Wauchope. 2002. A strip liposome immunoassay for aflatoxin B-1. *Anal Chem* 74:1493–1496.

Ho, J. A. A., L. C. Wu, M. R. Huang, Y. J. Lin, A. J. Baeumner, and R. A. Durst. 2007. Application of gangliosidesensitized liposomes in a flow injection immunoanalytical system for the determination of cholera toxin. *Anal Chem* 79:246–250.

Hong, Y. L., B. L. Webb, S. Pai et al. 2006. G-protein-coupled receptor Microarrays for multiplexed compound screening. *J Biomol Screen* 11:435–438.

Horlacher, T., and P. H. Seeberger. 2008. Carbohydrate arrays as tools for research and diagnostics. *Chem Soc Rev* 37:1414–1422.

Hovis, J. S., and S. G. Boxer. 2000. Patterning barriers to lateral diffusion in supported lipid bilayer membranes by blotting and stamping. *Langmuir* 16:894–897.

Hovis, J. S., and S. G. Boxer. 2001. Patterning and composition arrays of supported lipid bilayers by microcontact printing. *Langmuir* 17:3400–3405.

Howorka, S., S. Cheley, and H. Bayley. 2001. Sequence-specific detection of individual DNA strands using engineered nanopores. *Nat Biotechnol* 19:636–639.

Hwang, S. Y., Y. Kumada, G. H. Seong, J. Choo, S. Katoh, and E. K. Lee. 2007. Characteristics of a liposome immunoassay on a poly(methyl methacrylate) surface. *Anal Bioanal Chem* 389:2251–2257.

Jackman, J. A., and N. J. Cho. 2012a. Model membrane platforms for biomedicine: Case study on antiviral drug development. *Biointerphases* 7:18.

Jackman, J. A., N. J. Cho, R. S. Duran, and C. W. Frank. 2010. Interfacial binding dynamics of bee venom phospholipase A(2) investigated by dynamic light scattering and quartz crystal microbalance. *Langmuir* 26:4103–4112.

Jackman, J. A., W. Knoll, and N. J. Cho. 2012b. Biotechnology applications of tethered lipid bilayer membranes. *Materials* 5:2637–2657.

Janshoff, A., H. J. Galla, and C. Steinem. 2000. Piezoelectric mass-sensing devices as biosensors - An alternative to optical biosensors? *Angew Chem Int Edit* 39:4004–4032.

Janshoff, A., and C. Steinem. 2006. Transport across artificial membranes - an analytical perspective. *Anal Bioanal Chem* 385:433–451.

Jayaraman, N., K. Maiti, and K. Naresh. 2013. Multivalent glycoliposomes and micelles to study carbohydrate-protein and carbohydrate-carbohydrate interactions. *Chem Soc Rev* 42:4640–4656.

Jeon, T. J., N. Malmstadt, J. L. Poulos, and J. J. Schmidt. 2008. Black lipid membranes stabilized through substrate conjugation to a hydrogel. *Biointerphases* 3:Fa96-Fa100.

Jesorka, A., N. Stepanyants, H. J. Zhang, B. Ortmen, B. Hakonen, and O. Orwar. 2011. Generation of phospholipid vesicle-nanotube networks and transport of molecules therein. *Nat Protoc* 6:791–805.

Johnson, J. M., T. Ha, S. Chu, and S. G. Boxer. 2002. Early steps of supported bilayer formation probed by single vesicle fluorescence assays. *Biophys J* 83:3371–3379.

Johnson, S. J., T. M. Bayerl, D. C. Mcdermott et al. 1991. Structure of an adsorbed dimyristoylphosphatidylcholine bilayer measured with specular reflection of neutrons. *Biophys J* 59:289–294.

Josic, D., and J. G. Clifton. 2007. Mammalian plasma membrane proteomics. *Proteomics* 7:3010–3029.

Justesen, P. H., T. Kristensen, T. Ebdrup, and D. Otzen. 2004. Investigating porcine pancreatic phospholipase A(2) action on vesicles and supported planar bilayers using a quartz crystal microbalance with dissipation. *J Colloid Interf Sci* 279:399–409.

Kalyankar, N. D., M. K. Sharma, S. V. Vaidya et al. 2006. Arraying of intact liposomes into chemically functionalized microwells. *Langmuir* 22:5403–5411.

Kam, L., and S. G. Boxer. 2000. Formation of supported lipid bilayer composition arrays by controlled mixing and surface capture. *J Am Chem Soc* 122:12901–12902.

Kam, L., and S. G. Boxer. 2001. Cell adhesion to protein-micropatterned-supported lipid bilayer membranes. *J Biomed Mater Res* 55:487–495.

Kang, X. F., S. Cheley, A. C. Rice-Ficht, and H. Bayley. 2007. A storable encapsulated bilayer chip containing a single protein nanopore. *J Am Chem Soc* 129:4701–4705.

Karlsson, M., M. Davidson, R. Karlsson et al. 2004. Biomimetic nanoscale reactors and networks. *Annu Rev Phys Chem* 55:613–649.

Karlsson, O. P., and S. Lofas. 2002. Flow-mediated on-surface reconstitution of G-protein coupled receptors for applications in surface plasmon resonance biosensors. *Analytical biochemistry* 300:132–138.

Karlsson, R., A. Karlsson, A. Ewing et al. 2006. Chemical analysis in nanoscale surfactant networks. *Anal Chem* 78:5960–5968.

Kaufmann, S., J. Sobek, M. Textor, and E. Reimhult. 2011. Supported lipid bilayer microarrays created by non-contact printing. *Lab Chip* 11:2403–2410.

Keizer, H. M., B. R. Dorvel, M. Andersson et al. 2007. Functional ion channels in tethered bilayer membranes - Implications for biosensors. *Chembiochem* 8:1246–1250.

Kirsch, J., C. Siltanen, Q. Zhou, A. Revzin, and A. Simonian. 2013. Biosensor technology: Recent advances in threat agent detection and medicine. *Chem Soc Rev* 42:8733–8768.

Knoll, W., I. Koper, R. Naumann, and E. K. Sinner. 2008. Tethered bimolecular lipid membranes - A novel model membrane platform. *Electrochim Acta* 53:6680–6689.

Koper, I. 2007. Insulating tethered bilayer lipid membranes to study membrane proteins. *Mol Biosyst* 3:651–657.

Koshy, C., and C. Ziegler. 2015. Structural insights into functional lipid-protein interactions in secondary transporters. *Bba-Gen Subjects* 1850:476–487.

Kugler, R., and W. Knoll. 2002. Polyelectrolyte-supported lipid membranes. *Bioelectrochemistry* 56:175–178.

Kunding, A. H., M. W. Mortensen, S. M. Christensen, and D. Stamou. 2008. A fluorescence-based technique to construct size distributions from single-object measurements: Application to the extrusion of lipid vesicles. *Biophys J* 95:1176–1188.

Kunze, A., M. Bally, F. Hook, and G. Larson. 2013. Equilibrium-fluctuation-analysis of single liposome binding events reveals how cholesterol and Ca2+ modulate glycosphingolipid trans-interactions. *Sci Rep-Uk* 3.

Kuyper, C. L., J. S. Kuo, S. A. Mutch, and D. T. Chiu. 2006. Proton permeation into single vesicles occurs via a sequential two-step mechanism and is heterogeneous. *J Am Chem Soc* 128:3233–3240.

Kwakye, S., and A. Baeumner. 2003. A microfluidic biosensor based on nucleic acid sequence recognition. *Anal Bioanal Chem* 376:1062–1068.

Lahiri, J., P. Kalal, A. G. Frutos, S. T. Jonas, and R. Schaeffler. 2000. Method for fabricating supported bilayer lipid membranes on gold. *Langmuir* 16:7805–7810.

Laszlo, A. H., I. M. Derrington, B. C. Ross et al. 2014. Decoding long nanopore sequencing reads of natural DNA. *Nat Biotechnol* 32:829–833.

Lee, Y. K., and J. M. Nam. 2012. Electrofluidic lipid membrane biosensor. *Small* 8:832–837.

Liao, W. S., H. H. Cao, S. Cheunkar et al. 2013. Small-molecule arrays for sorting g-protein-coupled receptors. *J Phys Chem C* 117:22362–22368.

Liu, C. M., D. Huang, T. L. Yang, and P. S. Cremer. 2014. Monitoring phosphatidic acid formation in intact phosphatidylcholine bilayers upon phospholipase D catalysis. *Anal Chem* 86:1753–1759.

Liu, C. M., C. F. Monson, T. L. Yang, H. Pace, and P. S. Cremer. 2011. Protein separation by electrophoretic-electroosmotic focusing on supported lipid bilayers. *Anal Chem* 83:7876–7880.

Liu, G. D., and Y. H. Lin. 2007. Nanomaterial labels in electrochemical immunosensors and immunoassays. *Talanta* 74:308–317.

Liu, H. Y., W. L. Chen, C. K. Ober, and S. Daniel. 2018. Biologically complex planar cell plasma membranes supported on polyelectrolyte cushions enhance transmembrane protein mobility and retain native orientation. *Langmuir* 34:1061–1072.

Liu, H. Y., H. Grant, H. L. Hsu et al. 2017. Supported planar mammalian membranes as models of *in vivo* cell surface architectures. *Acs Appl Mater Inter* 9:35526–35538.

Liu, Q. T., and B. J. Boyd. 2013. Liposomes in biosensors. *Analyst* 138:391–409.

Liu, Y., A. S. Palma, and T. Feizi. 2009. Carbohydrate microarrays: Key developments in glycobiology. *Biol Chem* 390:647–656.

Ma, Y., I. Sobkiv, V. Gruzdys, H. L. Zhang, and X. L. Sun. 2012. Liposomal glyco-microarray for studying glycolipid-protein interactions. *Anal Bioanal Chem* 404:51–58.

Majd, S., and M. Mayer. 2005. Hydrogel stamping of arrays of supported lipid bilayers with various lipid compositions for the screening of drug-membrane and protein-membrane interactions. *Angew Chem Int Edit* 44:6697–6700.

Malinin, V. S., P. Frederik, and B. R. Lentz. 2002. Osmotic and curvature stress affect PEG-induced fusion of lipid vesicles but not mixing of their lipids. *Biophys J* 82:2090–2100.

Mammen, M., S. K. Choi, and G. M. Whitesides. 1998. Polyvalent interactions in biological systems: Implications for design and use of multivalent ligands and inhibitors. *Angew Chem Int Edit* 37:2755–2794.

Manrao, E. A., I. M. Derrington, A. H. Laszlo et al. 2012. Reading DNA at single-nucleotide resolution with a mutant MspA nanopore and phi29 DNA polymerase. *Nat Biotechnol* 30:349–U174.

Mathiasen, S., S. M. Christensen, J. J. Fung et al. 2014. Nanoscale high-content analysis using compositional heterogeneities of single proteoliposomes. *Nature methods* 11:931–934.

Mazur, F., M. Bally, B. Stadler, and R. Chandrawati. 2017. Liposomes and lipid bilayers in biosensors. *Adv Colloid Interface Sci* 249:88–99.

Mechler, A., S. Praporski, K. Atmuri, M. Boland, F. Separovic, and L. L. Martin. 2007. Specific and selective peptide-membrane interactions revealed using quartz crystal microbalance. *Biophys J* 93:3907–3916.

Meyvis, T. K. L., S. C. De Smedt, P. Van Oostveldt, and J. Demeester. 1999. Fluorescence recovery after photobleaching: A versatile tool for mobility and interaction measurements in pharmaceutical research. *Pharm Res-Dordr* 16:1153–1162.

Michel, R., I. Reviakine, D. Sutherland et al. 2002. A novel approach to produce biologically relevant chemical patterns at the nanometer scale: Selective molecular assembly patterning combined with colloidal lithography. *Langmuir* 18:8580–8586.

Morigaki, K., T. Baumgart, A. Offenhausser, and W. Knoll. 2001. Patterning solid-supported lipid bilayer membranes by lithographic polymerization of a diacetylene lipid. *Angew Chem Int Edit* 40:172–174.

Mozsolits, H., and M. I. Aguilar. 2002. Surface plasmon resonance spectroscopy: An emerging tool for the study of peptide-membrane interactions. *Biopolymers* 66:3–18.

Nasir, W., J. Nilsson, S. Olofsson, M. Bally, and G. E. Rydell. 2014. Parvovirus B19 VLP recognizes globoside in supported lipid bilayers. *Virology* 456–457:364–369.

Naumann, R., T. Baumgart, P. Graber, A. Jonczyk, A. Offenhausser, and W. Knoll. 2002. Proton transport through a peptide-tethered bilayer lipid membrane by the H+-ATP synthase from chloroplasts measured by impedance spectroscopy. *Biosensors & bioelectronics* 17:25–34.

Naumann, R. L. C., C. Nowak, and W. Knoll. 2011. Proteins in biomimetic membranes: Promises and facts. *Soft Matter* 7:9535–9548.

Netzer, R., U. Bischoff, and A. Ebneth. 2003. HTS techniques to investigate the potential effects of compounds on cardiac ion channels at early-stages of drug discovery. *Curr Opin Drug Di De* 6:462–469.

Neumann, J., M. Hennig, A. Wixforth, S. Manus, J. O. Radler, and M. F. Schneider. 2010. Transport, Separation, and Accumulation of Proteins on Supported Lipid Bilayers. *Nano Lett* 10:2903–2908.

Nielsen, C. H. 2009. Biomimetic membranes for sensor and separation applications. *Anal Bioanal Chem* 395:697–718.

Nugen, S. R., P. J. Asiello, J. T. Connelly, and A. J. Baeumner. 2009. PMMA biosensor for nucleic acids with integrated mixer and electrochemical detection. *Biosensors & bioelectronics* 24:2428–2433.

Oh, S. Y., B. Cornell, D. Smith, G. Higgins, C. J. Buffell, and T. W. Kok. 2008. Rapid detection of influenza A virus in clinical samples using an ion channel switch biosensor. *Biosensors & bioelectronics* 23:1161–1165.

Okumus, B., S. Arslan, S. M. Fengler, S. Myong, and T. Ha. 2009. Single molecule nanocontainers made porous using a bacterial toxin. *J Am Chem Soc* 131:14844–14849.

Okumus, B., T. J. Wilson, D. M. J. Lilley, and T. Ha. 2004. Vesicle encapsulation studies reveal that single molecule ribozyme heterogeneities are intrinsic. *Biophys J* 87:2798–2806.

Ollivon, M., S. Lesieur, C. Grabielle-Madelmont, and M. Paternostre. 2000. Vesicle reconstitution from lipid-detergent mixed micelles. *Biochim Biophys Acta* 1508:34–50.

Orth, R. N., J. Kameoka, W. R. Zipfel et al. 2003. Creating biological membranes on the micron scale: Forming patterned lipid bilayers using a polymer lift-off technique. *Biophys J* 85:3066–3073.

Ou, L. J., S. J. Liu, X. Chu, G. L. Shen, and R. Q. Yu. 2009. DNA encapsulating liposome based rolling circle amplification immunoassay as a versatile platform for ultrasensitive detection of protein. *Anal Chem* 81:9664–9673.

Overington, J. P., B. Al-Lazikani, and A. L. Hopkins. 2006. Opinion - How many drug targets are there? *Nat Rev Drug Discov* 5:993–996.

Oyelaran, O., and J. C. Gildersleeve. 2009. Glycan arrays: Recent advances and future challenges. *Curr Opin Chem Biol* 13:406–413.

Pace, H., L. S. Nystrom, A. Gunnarsson et al. 2015. Preserved transmembrane protein mobility in polymer-supported lipid bilayers derived from cell membranes. *Anal Chem* 87:9194–9203.

Pace, H. P., J. K. Hannestad, A. Armonious et al. 2018. Structure and Composition of Native Membrane Derived Polymer-Supported Lipid Bilayers. *Anal Chem* 90:13065–13072.

Pace, H. P., S. D. Sherrod, C. F. Monson, D. H. Russell, and P. S. Cremer. 2013. Coupling supported lipid bilayer electrophoresis with matrix-assisted laser desorption/ionization-mass spectrometry imaging. *Anal Chem* 85:6047–6052.

Papo, N., and Y. Shai. 2003. Exploring peptide membrane interaction using surface plasmon resonance: Differentiation between pore formation versus membrane disruption by lytic peptides. *Biochemistry-Us* 42:458–466.

Park, S., and R. A. Durst. 2000. Immunoliposome sandwich assay for the detection of Escherichia coli O157: H7. *Analytical biochemistry* 280:151–158.

Parveen, N., S. Block, V. P. Zhdanov, G. E. Rydell, and F. Hook. 2017. Detachment of Membrane Bound Virions by Competitive Ligand Binding Induced Receptor Depletion. *Langmuir* 33:4049–4056.

Parveen, N., G. E. Rydell, G. Larson et al. 2019. Competition for Membrane Receptors: Norovirus Detachment via Lectin Attachment. *J Am Chem Soc* 141:16303–16311.

Patching, S. G. 2014. Surface plasmon resonance spectroscopy for characterisation of membrane protein-ligand interactions and its potential for drug discovery. *Bba-Biomembranes* 1838:43–55.

Peerboom, N., E. Schmidt, E. Trybala et al. 2018. Cell Membrane Derived Platform To Study Virus Binding Kinetics and Diffusion with Single Particle Sensitivity. *ACS Infect Dis* 4:944–953.

Perez, J. B., K. L. Martinez, J. M. Segura, and H. Vogel. 2006. Supported cell-membrane sheets for functional fluorescence imaging of membrane proteins. *Adv Funct Mater* 16:306–312.

Peters, R. 2003. Optical single transporter recording: Transport kinetics in microarrays of membrane patches. *Annu Rev Bioph Biom* 32:47–67.

Pfeiffer, I., and F. Hook. 2004. Bivalent cholesterol-based coupling of oligonucletides to lipid membrane assemblies. *J Am Chem Soc* 126:10224–10225.

Pick, H., E. L. Schmid, A. P. Tairi, E. Ilegems, R. Hovius, and H. Vogel. 2005. Investigating cellular signaling reactions in single attoliter vesicles. *J Am Chem Soc* 127:2908–2912.

Poo, M. M., and K. R. Robinson. 1977. Electrophoresis of Concanavalin-a Receptors Along Embryonic Muscle-Cell Membrane. *Nature* 265:602–605.

Prive, G. G. 2007. Detergents for the stabilization and crystallization of membrane proteins. *Methods* 41:388–397.

Rabe, M., S. R. Tabaei, H. Zetterberg, V. P. Zhdanov, and F. Hook. 2015. Hydrolysis of a lipid membrane by single enzyme molecules: Accurate determination of kinetic parameters. *Angew Chem Int Ed Engl* 54:1022–1026.

Rao, N. M., V. Silin, K. D. Ridge, J. T. Woodward, and A. L. Plant. 2002. Cell membrane hybrid bilayers containing the G-protein-coupled receptor CCR5. *Analytical biochemistry* 307:117–130.

Rask-Andersen, M., M. S. Almen, and H. B. Schioth. 2011. Trends in the exploitation of novel drug targets. *Nat Rev Drug Discov* 10:579–590.

Reimhult, E., and K. Kumar. 2008. Membrane biosensor platforms using nano- and microporous supports. *Trends Biotechnol* 26:82–89.

Rhoades, E., E. Gussakovsky, and G. Haran. 2003. Watching proteins fold one molecule at a time (vol 100, pg 3197, 2003). *P Natl Acad Sci USA* 100:7418–7418.

Ritchie, T. K., H. Kwon, and W. M. Atkins. 2011. Conformational analysis of human ATP-binding cassette transporter ABCB1 in lipid nanodiscs and inhibition by the antibodies MRK16 and UIC2. *J Biol Chem* 286:39489–39496.

Rodahl, M., and B. Kasemo. 1996. A simple setup to simultaneously measure the resonant frequency and the absolute dissipation factor of a quartz crystal microbalance. *Rev Sci Instrum* 67:3238–3241.

Rodriguez-Mozaz, S., M. J. L. de Alda, and D. Barcelo. 2006. Biosensors as useful tools for environmental analysis and monitoring. *Anal Bioanal Chem* 386:1025–1041.

Rongen, H. A. H., A. Bult, and W. P. vanBennekom. 1997. Liposomes and immunoassays. *J Immunol Methods* 204:105–133.

Rossetti, F. F., M. Bally, R. Michel, M. Textor, and I. Reviakine. 2005. Interactions between titanium dioxide and phosphatidyl serine-containing liposomes: Formation and patterning of supported phospholipid bilayers on the surface of a medically relevant material. *Langmuir* 21:6443–6450.

Rotem, D., L. Jayasinghe, M. Salichou, and H. Bayley. 2012. Protein detection by nanopores equipped with aptamers. *J Am Chem Soc* 134:2781–2787.

Rydell, G. E., A. B. Dahlin, F. Hook, and G. Larson. 2009. QCM-D studies of human norovirus VLPs binding to glycosphingolipids in supported lipid bilayers reveal strain-specific characteristics. *Glycobiology* 19:1176–1184.

Sackmann, E., and M. Tanaka. 2000. Supported membranes on soft polymer cushions: Fabrication, characterization and applications. *Trends Biotechnol* 18:58–64.

Sadana, A., and N. Sadana. 2010. Handbook of biosensors and biosensor kinetics. Elsevier.

Schmitt, E. K., M. Vrouenraets, and C. Steinem. 2006. Channel activity of OmpF monitored in nano-BLMs. *Biophys J* 91:2163–2171.

Schonherr, H., D. I. Rozkiewicz, and G. J. Vancso. 2004. Atomic force microscopy assisted immobilization of lipid vesicles. *Langmuir* 20:7308–7312.

Scott, J. L., C. A. Musselman, E. Adu-Gyamfi, T. G. Kutateladze, and R. V. Stahelin. 2012. Emerging methodologies to investigate lipid-protein interactions. *Integr Biol-Uk* 4:247–258.

Seddon, A. M., P. Curnow, and P. J. Booth. 2004. Membrane proteins, lipids and detergents: Not just a soap opera. *Bba-Biomembranes* 1666:105–117.

Shi, J. J., J. X. Chen, and P. S. Cremer. 2008. Sub-100nm Patterning of supported bilayers by nanoshaving lithography. *J Am Chem Soc* 130: 2718–+.

Shim, J. W., and L. Q. Gu. 2007. Stochastic sensing on a modular chip containing a single-ion channel. *Anal Chem* 79:2207–2213.

Shin, I., S. Park, and M. R. Lee. 2005. Carbohydrate microarrays: An advanced technology for functional studies of glycans. *Chem-Eur J* 11:2894–2901.

Silin, V. I., E. A. Karlik, K. D. Ridge, and D. J. Vanderah. 2006. Development of surface-based assays for transmembrane proteins: Selective immobilization of functional CCR5, a G protein-coupled receptor. *Analytical biochemistry* 349:247–253.

Simonsson, L., A. Gunnarsson, P. Wallin, P. Jonsson, and F. Hook. 2011. Continuous lipid bilayers derived from cell membranes for spatial molecular manipulation. *J Am Chem Soc* 133:14027–14032.

Simonsson, L., P. Jonsson, G. Stengel, and F. Hook. 2010. Site-specific DNA-controlled fusion of single lipid vesicles to supported lipid bilayers. *Chemphyschem* 11:1011–1017.

Sinner, E.-K., and W. Knoll. 2001. Functional tethered membranes. *Curr Opin Chem Biol* 5:705–711.

Sinner, E.-K., I. Köper, and W. Knoll. 2008. Preface. *Biointerphases* 3:FA1-FA2.

Stadler, B., M. Bally, D. Grieshaber, J. Voros, A. Brisson, and H. M. Grandin. 2006. Creation of a functional heterogeneous vesicle array via DNA controlled surface sorting onto a spotted microarray. *Biointerphases* 1:142–145.

Stadler, B., D. Falconnet, I. Pfeiffer, F. Hook, and J. Voros. 2004. Micropatterning of DNA-tagged vesicles. *Langmuir* 20:11348–11354.

Stamou, D., C. Duschl, E. Delamarche, and H. Vogel. 2003. Self-assembled microarrays of attoliter molecular vessels. *Angew Chem Int Edit* 42:5580–5583.

Stano, P., P. Carrara, Y. Kuruma, T. P. de Souza, and P. L. Luisi. 2011. Compartmentalized reactions as a case of soft-matter biotechnology: Synthesis of proteins and nucleic acids inside lipid vesicles. *J Mater Chem* 21:18887–18902.

Steller, L., M. Kreir, and R. Salzer. 2012. Natural and artificial ion channels for biosensing platforms. *Anal Bioanal Chem* 402:209–230.

Stelzle, M., R. Miehlich, and E. Sackmann. 1992. 2-Dimensional microelectrophoresis in supported lipid bilayers. *Biophys J* 63:1346–1354.

Stenlund, P., G. J. Babcock, J. Sodroski, and D. G. Myszka. 2003. Capture and reconstitution of G protein-coupled receptors on a biosensor surface. *Analytical biochemistry* 316:243–250.

Stora, T., J. H. Lakey, and H. Vogel. 1999. Ion-channel gating in transmembrane receptor proteins: Functional activity in tethered lipid membranes. *Angew Chem Int Edit* 38:389–392.

Studer, A., S. Demarche, D. Langenegger, and L. Tiefenauer. 2011. Integration and recording of a reconstituted voltage-gated sodium channel in planar lipid bilayers. *Biosensors & bioelectronics* 26:1924–1928.

Studer, A., X. J. Han, F. K. Winkler, and L. X. Tiefenauer. 2009. Formation of individual protein channels in lipid bilayers suspended in nanopores. *Colloid Surface B* 73:325–331.

Sugihara, K., J. Voros, and T. Zambelli. 2010. A gigaseal obtained with a self-assembled long-lifetime lipid bilayer on a single polyelectrolyte multilayer-filled nanopore. *Acs Nano* 4:5047–5054.

Suzuki, H., and S. Takeuchi. 2008. Microtechnologies for membrane protein studies. *Anal Bioanal Chem* 391:2695–2702.

Svedhem, S., I. Pfeiffer, C. Larsson, C. Wingren, C. Borrebaeck, and F. Hook. 2003. Patterns of DNA-labeled and scFv-antibody-carrying lipid vesicles directed by material-specific immobilization of DNA and supported lipid bilayer formation on an Au/SiO$_2$ template. *Chembiochem* 4:339–343.

Tabaei, S. R., M. Rabe, H. Zetterberg, V. P. Zhdanov, and F. Hook. 2013. Single lipid vesicle assay for characterizing single-enzyme kinetics of phospholipid hydrolysis in a complex biological fluid. *J Am Chem Soc* 135:14151–14158.

Tabaei, S. R., M. Rabe, V. P. Zhdanov, N. J. Cho, and F. Hook. 2012. Single vesicle analysis reveals nanoscale membrane curvature selective pore formation in lipid membranes by an antiviral alpha-helical peptide. *Nano Lett* 12:5719–5725.

Tamm, L. K., and H. M. Mcconnell. 1985. Supported phospholipid-bilayers. *Biophys J* 47:105–113.

Tanaka, M., and E. Sackmann. 2005. Polymer-supported membranes as models of the cell surface. *Nature* 437:656–663.

Tanaka, M., A. P. Wong, F. Rehfeldt, M. Tutus, and S. Kaufmann. 2004. Selective deposition of native cell membranes on biocompatible micropatterns. *J Am Chem Soc* 126:3257–3260.

Terrettaz, S., S. Follonier, S. Makohliso, and H. Vogel. 2009. A synthetic membrane protein in tethered lipid bilayers for immunosensing in whole blood. *J Struct Biol* 168:177–182.

Terrettaz, S., M. Mayer, and H. Vogel. 2003. Highly electrically insulating tethered lipid bilayers for probing the function of ion channel proteins. *Langmuir* 19:5567–5569.

Terrettaz, S., T. Stora, C. Duschl, and H. Vogel. 1993. Protein-binding to supported lipid-membranes - investigation of the cholera-toxin ganglioside interaction by simultaneous impedance spectroscopy and surface-plasmon resonance. *Langmuir* 9:1361–1369.

Terrettaz, S., and H. Vogel. 2005. Investigating the function of ion channels in tethered lipid membranes by impedance spectroscopy. *Mrs Bull* 30:207–210.

Tiefenauer, L., and S. Demarche. 2012. Challenges in the development of functional assays of membrane proteins. *Materials* 5:2205–2242.

Tiefenauer, L. X., and A. Studer. 2008. Nano for bio: Nanopore arrays for stable and functional lipid bilayer membranes (Mini Review). *Biointerphases* 3:Fa74-Fa79.

Tresset, G. 2009. The multiple faces of self-assembled lipidic systems. *PMC biophysics* 2:3.

Uesugi, Y., J. Arima, M. Iwabuchi, and T. Hatanaka. 2007. Sensor of phospholipids in Streptomyces phospholipase D. *Febs J* 274:2672–2681.

Vega, B., A. Calle, A. Sanchez et al. 2013. Real-time detection of the chemokine CXCL12 in urine samples by surface plasmon resonance. *Talanta* 109:209–215.

Venkatesan, B. M., and R. Bashir. 2011. Nanopore sensors for nucleic acid analysis. *Nat Nanotechnol* 6:615–624.

Viswanathan, S., C. Rani, and C. Delerue-Matos. 2012. Ultrasensitive detection of ovarian cancer marker using immunoliposomes and gold nanoelectrodes. *Anal Chim Acta* 726:79–84.

Wahlsten, O., A. Gunnarsson, L. S. Nystrom, H. Pace, S. Geschwindner, and F. Hook. 2015. Equilibrium-fluctuation analysis for interaction studies between natural ligands and single G protein-coupled receptors in native lipid vesicles. *Langmuir* 31:10774–10780.

Wang, H. Y., D. Y. Sun, Z. A. Tan, W. Gong, and L. Wang. 2011. Electrochemiluminescence immunosensor for alpha-fetoprotein using Ru(bpy)(3)(2+)-encapsulated liposome as labels. *Colloid Surface B* 84:515–519.

Wen, H. W., W. Borejsza-Wysocki, T. DeCory, and R. Durst. 2005. Development of a competitive liposome-based lateral flow assay for the rapid detection of the allergenic peanut protein Ara h1. *Anal Bioanal Chem* 382:1217–1226.

White, R. J., E. N. Ervin, T. Yang et al. 2007. Single ion-channel recordings using glass nanopore membranes. *J Am Chem Soc* 129:11766–11775.

Wiese, R. 2003. Analysis of several fluorescent detector molecules for protein microarray use. *Luminescence* 18:25–30.

Wilkop, T. E., J. Sanborn, A. E. Oliver, J. M. Hanson, and A. N. Parikh. 2014. On-demand self-assembly of supported membranes using sacrificial, anhydrobiotic sugar coats. *J Am Chem Soc* 136:60–63.

Wittenberg, N. J., H. Im, T. W. Johnson et al. 2011. Facile assembly of micro- and nanoarrays for sensing with natural cell membranes. *Acs Nano* 5:7555–7564.

Woodbury, D. J., and C. Miller. 1990. Nystatin-induced liposome fusion - a versatile approach to ion channel reconstitution into planar bilayers. *Biophys J* 58:833–839.

Wu, C. Y., P. H. Liang, and C. H. Wong. 2009. New development of glycan arrays. *Org Biomol Chem* 7:2247–2254.

Yamada, M., H. Imaishi, and K. Morigaki. 2013. Microarrays of phospholipid bilayers generated by inkjet printing. *Langmuir* 29:6404–6408.

Yamazaki, V., O. Sirenko, R. J. Schafer et al. 2005. Cell membrane array fabrication and assay technology. *Bmc Biotechnol* 5.

Yee, C. K., M. L. Amweg, and A. N. Parikh. 2004. Direct photochemical patterning and refunctionalization of supported phospholipid bilayers. *J Am Chem Soc* 126:13962–13972.

Yoon, C. H., J. H. Cho, H. I. Oh et al. 2003. Development of a membrane strip immunosensor utilizing ruthenium as an electro-chemilumine scent signal generator. *Biosensors & bioelectronics* 19:289–296.

Yoshina-Ishii, C., G. P. Miller, M. L. Kraft, E. T. Kool, and S. G. Boxer. 2005. General method for modification of liposomes for encoded assembly on supported bilayers. *J Am Chem Soc* 127:1356–1357.

Yu, C. H., and J. T. Groves. 2010. Engineering supported membranes for cell biology. *Med Biol Eng Comput* 48:955–963.

Zagnoni, M. 2012. Miniaturised technologies for the development of artificial lipid bilayer systems. *Lab Chip* 12:1026–1039.

Zaytseva, N. V., V. N. Goral, R. A. Montagna, and A. J. Baeumner. 2005a. Development of a microfluidic biosensor module for pathogen detection. *Lab Chip* 5:805–811.

Zaytseva, N. V., R. A. Montagna, and A. J. Baeumner. 2005b. Microfluidic biosensor for the serotype-specific detection of Dengue virus RNA. *Anal Chem* 77:7520–7527.

Zhong, Z. Y., N. Peng, Y. Qing et al. 2011. An electrochemical immunosensor for simultaneous multiplexed detection of neuron-specific enolase and pro-gastrin-releasing peptide using liposomes as enhancer. *Electrochim Acta* 56:5624–5629.

15 Lipids in Dermal Applications: Cosmetics and Pharmaceutics

Jérôme Bibette
Universite Paris Sciences et Lettres (PSL), Paris, France

Abdou Rachid Thiam
Universite Paris Sciences et Lettres (PSL), Paris, France
Sorbonne Universite, Universite de Paris, Paris, France

CONTENTS

15.1 Introduction 271
15.2 Skin Morphology and Properties 272
   15.2.1 Basic Description of Skin Structure 272
   15.2.2 Skin Sensitivity to Lipid Changes 273
   15.2.3 Dermal Uptake 274
15.3 Formulation of Lipid Nanocarriers 276
   15.3.1 Liposomes and Derivatives 276
      15.3.1.1 Release Mechanisms 276
      15.3.1.2 Liposomes Formulation and Lipid Choice 277
   15.3.2 Emulsions in Dermal Applications 278
      15.3.2.1 Surfactants 279
      15.3.2.2 Nanoemulsions and Microemulsions 279
      15.3.2.3 Lipid Ingredients 280
      15.3.2.4 Destabilization 280
   15.3.3 Lipid Nanoparticles 281
      15.3.3.1 Solid Lipid Nanoparticles and Nanostructured Lipid Carriers 281
      15.3.3.2 Application of Lipid Nanoparticle Products and Release 283
15.4 Conclusion 283
Acronyms 283
References 284

15.1 INTRODUCTION

Cosmetic and pharmaceutical products are used to meet aesthetical and therapeutic needs. The products are constantly required to provide efficient, specific, and safe delivery of cargos or drugs to the target body sites. For this aim, most cargo and drug molecules are carried by excipients to prevent their denaturation and favour optimal administration (Egbaria and Weiner, 1990; Kwon and Okano, 1996; Schreier and Bouwstra, 1994; Uekama et al., 1998). Presently, most excipients, called carriers or vehicles, are colloid-based particles (Cevc, 2004; Gallardo et al., 2000; Kwon and Okano, 1996; Nielloud, 2000; Uekama et al., 1998) acutely formulated to improve shelf life and delivery of the therapeutic molecules or cargos. When applied, the amounts of vehicles and/or cargo molecules that reach the target are generally extremely low, as the major part of the molecules is dispersed in the body. The risk of having drugs accumulating in nontargeted body parts can lead to undesired effects (Allen and Cullis, 2004). Over the years, the goal for cosmetic and pharmaceutical fields has evolved to design smart and safe vehicles (Nielloud, 2000). Their strategy has led to the formulation of carriers with tunable physical and biological properties to improve uptake of therapeutic molecules in tissues such as of ocular, cutaneous, subcutaneous, or intestinal organs (Gupta et al., 2002; Kawashima et al., 1989; Lostritto and Silvestri, 1987). Particularly, since the skin is by far the largest and more accessible body organ, it is appealing to formulate carriers that can directly use the skin as the portal for local or systemic delivery by topical medication (Bharadwaj et al., 2011; Cevc, 2004; Costes et al., 2009; Schreier and Bouwstra, 1994). For cosmetics, the skin is in most cases the essential target. For dermal and transdermal delivery, efficient delivery of cargos to/through the skin can be reached by the formulation of nanocarriers exhibiting physical chemistry and biophysical features that are in harmony with skin properties (Gallardo et al., 2000; Marti-Mestres and Nielloud, 2000; Nielloud, 2000). Selective delivery can be achieved by the formulation of bioactive carriers, which bear ligands, or signalling motifs, that can be selectively recognized by cells (Allen, 2002; Schiffelers et al., 2004).

In addition to the industrial processing limitations, the main restriction regarding the design of carriers and the choice of cargos remains the toxicity of the compounds (Allen and Cullis, 2004; Marti-Mestres and Nielloud, 2000). Components used to produce carriers are chosen following specific guidelines, e.g., fixed by national pharmacopoeia charts or the international cosmetic ingredient dictionary, which establish an updated list of materials providing good benefit with minimized risk (Marti-Mestres and Nielloud, 2000). Following these guidelines, various carrier types, all colloid-based, were developed (Egbaria and Weiner, 1990; Elsayed et al., 2007; Honeywell-Nguyen et al., 2006; Kogan and Garti, 2006; Kwon and Okano, 1996; Nielloud, 2000; Osborne et al., 1991; Schreier and Bouwstra, 1994; Sinico and Fadda, 2009), but few were suitable for the market. Presently the main carriers include nano- and microemulsions, vesicles, lipid nanoparticles (LNPs), emulgels, and liquid-core capsules (Ashara et al., 2014; Bodmeier et al., 1997; Cevc, 2004; Egbaria and Weiner, 1990; Elsayed et al., 2007; Eslamian and Shekarriz, 2009; Fabiilli et al., 2013; Gercek et al., 2007; Gou et al., 2008, 2010; Gupta et al., 2002, 2006; Gursoy and Benita, 2004; Hamidi et al., 2008, 2011; Himes et al., 2010; Honeywell-Nguyen et al., 2006; Jeong et al., 2012; Kawashima et al., 1989; Kogan and Garti, 2006; Kwon and Okano, 1996; Lee and Kim, 2006; Li et al., 2009; Lostritto and Silvestri, 1987; Mehnert and Mäder, 2001; Morishige et al., 2005; Müller et al., 2000, 2002; Osborne et al., 1991; Patravale and Ambarkhane, 2003; Sadurni et al., 2005; Schreier and Bouwstra, 1994; Settel, 1966; Sinico and Fadda, 2009; Smith et al., 2000; Thaci et al., 2002; Trotta et al., 2005; van der Ven et al., 2001; Vasiljevic et al., 2006; Weiner et al., 1973; Wissing and Müller, 2003). These formulations allow the generation of nanocarriers which have distinct advantages and shortcomings, depending on the target tissue, starting from the incorporation of cargos in carriers to the release (Müller et al., 2000). Most of the carriers are lipid based (Alvarez and Rodriguez, 2000; Cevc et al., 1996; Hippalgaonkar et al., 2010). The advantage of lipids is that they are one of the major components of the body and are not recognized thereby as foreign molecules, as they will be stored or degraded (Brune et al., 1992; Feingold, 2009; Grice and Segre, 2011).

This chapter is dedicated to present the strategies used for formulating lipid-based dermal products in the pharmaceutical and cosmetic fields. We will first introduce vesicles and emulsions, as soft carriers, before solid and semisolid nanocarriers, i.e., lipid nanoparticles, for which the formulation of an emulsion is an intermediate step. To understand how the carriers propagate though the skin, it is first necessary to have a brief description of the morphological and physical aspects of the skin, made of complex lipid layers.

15.2 SKIN MORPHOLOGY AND PROPERTIES

15.2.1 Basic Description of Skin Structure

The skin is formed by a superposition of three layers: the outer layer is called epidermis, beneath which lays the dermis covering the innermost subcutaneous layer also called hypodermis. A simplified sketch of skin layers is shown in Figure 15.1, and numerous documentations are available with detailed descriptions (Bouwstra et al., 2003; Darlenski et al., 2011; Elias, 2007; Rawlings and Harding, 2004).

The epidermis consists of several layers whose total thickness can vary between 50 μm and 2 mm depending on the body part. The outermost layer of the epidermis is the stratum corneum (SC) (Candi et al., 2005), which is made of flat compacted dead cells, corneocytes, and plays a crucial barrier function in regulating molecular transport through the skin (Bouwstra and Gooris, 2010; Christophers and Kligman, 1964; Elias, 2007; Imokawa et al., 1989; Lampe et al., 1983; Zettersten et al., 1997). The SC prevents transdermal water loss (abnormal body water loss) and protects the underneath and vulnerable skin layers from the external environment. Corneocytes are embedded in a high lipophilic region enriched with long-chain ceramides (more than six different ceramides), sphingolipids, cholesterol, and saturated free fatty acids (Coderch et al., 2003; Elias, 2007; Imokawa et al., 1989, 1991; Jungersted et al., 2010a, b; Meckfessel and Brandt, 2014; Mojumdar et al., 2014; Torin Huzil et al., 2011), which mixture form relatively rigid structures and weakly permeable barriers (Mouritsen, 2005), called the lipid matrix. Soft body parts, e.g., face or abdomen, have more triglycerides and fatty acids (Table 15.1) than rough skins, e.g., plantar, more enriched with cholesteryl esters and ceramides (Lampe et al., 1983; Pappas, 2008).

The dermis, a more fluid layer than the epidermis, embeds a large network of blood and lymph vessels and sebaceous or oil glands (Haake et al., 2001). Hair follicles and nerves are also localized in the dermis (Figure 15.1). The dermis floods the epidermis with nutrients and offers a possible route for drugs and cargos delivery through blood vessels or hair follicles.

The hypodermis is mainly composed of fat cells such as adipose tissues, which are neutral lipid or fat storing cells. The hypodermis is also important for preventing transdermal water loss, thanks to its greasy nature, and a decrease of body temperature. Likewise, it contains a lot of blood vessels, nerves, as in the dermis, and also collagens fibres necessary for providing a consistent skin structure with the two upper layers (Hendriks et al., 2006).

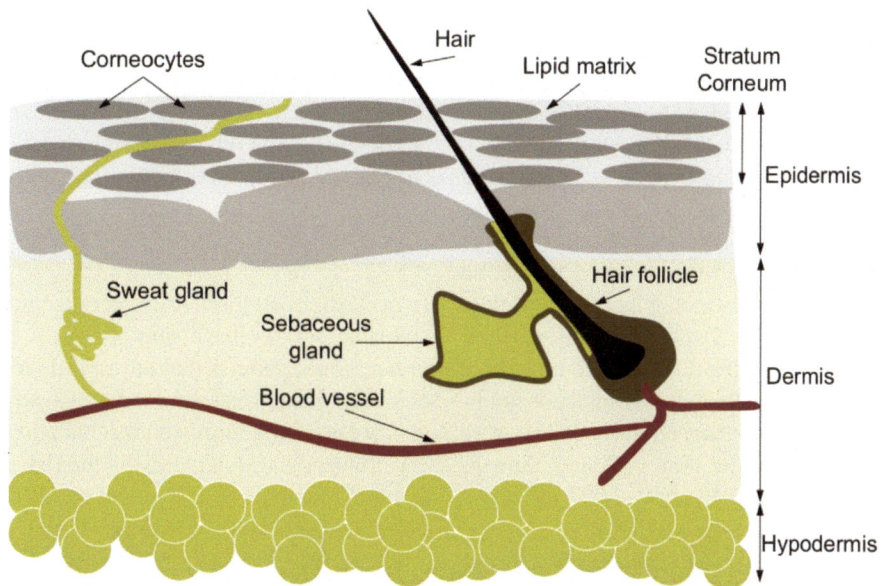

FIGURE 15.1 Minimal schematic of skin layers. The upper skin layer is the epidermis covering the dermis beneath which is the hypodermis. The epidermis itself is also formed by many layers, and its upper protective one is the stratum corneum. Hair follicles and sweat glands are mainly in the dermis. Blood vessels are found in the dermis and the hypodermis.

TABLE 15.1

Main Skin Lipid Components and Relative Fraction (Bouwstra et al., 2003; Darlenski et al., 2011; Elias, 2007; Rawlings and Harding, 2004)

Lipids in Soft Skins	Fraction (%)
Triglycerides	20
Sphingolipids	20
Ceramides	15
Free sterols	15
Free fatty acids	15
Wax esters	5
Alkanes	3
Squalene	5
Glycosphingolipids	2

15.2.2 Skin Sensitivity to Lipid Changes

Skin pathologies may originate from damages of one of the three sublayers, resulting, for example, in recurrent epidermis itching or deeper skin disorders affecting the dermis or hypodermis (Abe, 1978; Abramovits and Elias, 2014; Ahcan et al., 2003; Bouwstra et al., 2009; Carneiro et al., 2011; Chan et al., 2011; Choi et al., 2007; Choi and Maibach, 2005; Duque et al., 2005; Elias and Schmuth, 2009; Elias and Wakefield, 2010; Findley and Grice, 2014; Greenberg et al., 1989; Hiruma et al., 2002; Jungersted et al., 2010a; Proksch et al., 2003, 2006, 2007, 2008; Williams and Elias, 1980; Yoshida et al., 1991). Cosmetic and pharmaceutical dermal products can be used to exclusively target skin layers or more severe causes by having systemic effects (Cevc, 2004; Schreier and Bouwstra, 1994; Williams and Elias, 1980, 1981). Systemic treatments are indeed possible by targeting the skin portal which has the unique attribute of being in close contact with almost all organs (e.g., lungs, muscles, nerves, vessels) (Cevc et al., 1996).

Skin health is particularly dependent on lipid levels and types throughout its constitutive layers (Bouwstra et al., 2003; Bouwstra and Ponec, 2006; Choi and Maibach, 2005; Coderch et al., 2003; Imokawa et al., 1989; Jungersted et al., 2010a; Man et al., 1993, 1995; MaoQiang et al., 1996; Meckfessel and Brandt, 2014; Miyazaki et al., 2005; Mojumdar et al., 2014; Mouritsen, 2005; Pappas, 2008; Picardo et al., 1991; Proksch et al., 2003; Vasireddy et al., 2007; Wang et al., 1997; Zettersten et al., 1997). Itchy, irritating, and cracking skins are caused by lipid deficiency, while greasy skins reflect an excess of fluid lipids. These extreme conditions appear for derelict skins and represent unfavourable conditions for makeups, as they latter would unevenly spread or smear; greasy skins are additionally prone to develop acnes (Abramovits and Gonzalez-Serva, 2000; Grice and Segre, 2011; Zouboulis, 2004). The right distribution of lipids across the skin layers will improve the skin softening, lubrication, and waterproof capacity, and overall provide a good body comfort.

There are surface lipids (surfactants, e.g., phospholipids), forming biological membranes such as corneocytes, and bulk lipids that are natural fat lipids present in the skin SC and in the whole body (Abe, 1978; Abramovits and Gonzalez-Serva, 2000; Clarys and Barel, 1995; Feingold, 2009; Pappas, 2008; Picardo et al., 1991; Yosipovitch et al., 2007). Sebaceous glands in the dermis layer produce fat lipids, generally lipid waxes, squalene oil, triglycerides, and sterol esters, forming the so-called sebum (Abramovits and Gonzalez-Serva, 2000; Zouboulis, 2004). Excess sebum and abnormal lipid alteration will cause modification of skin layers chemistry and physical properties (Honeywell-Nguyen et al., 2006; Maghraby et al., 2008; Maziere, 1997). Such ramifications can, for example, arise from elevated activity of follicular bacterial lipases, which hydrolyse triglycerides in the dermis and generate a surplus of free fatty acids; this reaction will lead to undesired metabolic signatures such as acne lesions (Abramovits and Gonzalez-Serva, 2000; Findley and Grice, 2014; Grice, 2014; Grice and Segre, 2011; Kong et al., 2009). Deregulation of whole body lipid metabolism, e.g., due to improper food consumption (Miyazaki et al., 2005), or enhanced lipolytic activities occurring in tissues in close to the skin, will lead likewise to acne lesions (Feingold, 2009; Li et al., 2007; Maziere, 1997; Miyazaki et al., 2005; Stone et al., 2004).

The presence of specific lipids in dermal products can enable to locally maintain adequate lipid composition (by supplying or removing lipids) (Bouwstra et al., 2003; Carneiro et al., 2011; Cevc, 2004; Elsayed et al., 2007; Pardeike et al., 2009; Schreier and Bouwstra, 1994). Such lipids are generally recovered from plants, animals, or genetically engineered yeast cells that produce specific lipids. Products can be also applied to target sebaceous glands for modulating their activity; the therapeutic compounds assist sebaceous glands to produce the lipids needed or inhibit the activity of follicular bacteria (Abramovits and Gonzalez-Serva, 2000). Products developed for such aims are found in many topical products (applicable to any visible body part) (Elsayed et al., 2007; Osborne et al., 1991; Smith et al., 2000; Weiner et al., 1973). They are generally in colloidal form such as gels, emulsions, lotions, ointments, and creams, all containing carriers encapsulating therapeutic cargos targeting for the sebaceous glands (Ashara et al., 2014; Atrux-Tallau et al., 2013; Carneiro et al., 2011; Cevc, 2004; Kogan and Garti, 2006; Osborne et al., 1991; Panwar et al., 2011; Pardeike et al., 2009; Weiner et al., 1973).

A classic example of skin products are moisturizers (Flynn et al., 2001; Guo and Ehrlich, 2014; Kraft and Lynde, 2005; Loden, 2003a, b; Simion et al., 2005; Spencer, 1988; Subramanyan and Johnson, 2005; Wallisch, 1975), which are applied for adequate skin lubrication or hydration. They are used for administering therapeutic agents or preventing transdermal water loss by the application of creamy films to the SC. Moisturizers can be oil or water based depending on the target skin (Abramovits and Gonzalez-Serva, 2000). For aging or dry skins, which are often associated to eczemas, oil moisturizers containing antioxidants and oils made of long-chain fatty acids, e.g., petroleum jellies or ceramides, are used (Darmstadt et al., 2002; Guo and Ehrlich, 2014; Loden, 2003a; Rawlings et al., 2004; Spencer, 1988; Subramanyan and Johnson, 2005). For greasy skins containing excess sebum, water-based moisturizers, mainly containing detergents, surfactants, are applied for removing excess sebum, without impairing the normal lubrication state of deeper skin layers (Abramovits and Gonzalez-Serva, 2000; Black et al., 1998). In this situation, therapeutic compounds such as retinoids, especially isotretinoin, or highly polyunsaturated fatty acids, or metabolites of vitamin A, are delivered to the skin with moisturizers to limit the production of sebum with the ultimate goal of silencing acnes (Abramovits and Gonzalez-Serva, 2000).

15.2.3 Dermal Uptake

In oral administration, drug carriers are required to be bioavailable and not have side effects. A major limitation in oral administration is that only a little fraction of excipients and cargos will reach the target. Consequently, oral products are usually administered at high dose to increase delivery efficiency, which increases the risk of side effects such as high blood spikes (Hafeez et al., 2013). The use of skin portal for local and systemic treatments is hence privileged whenever possible, which supports the prospection of efficient topical treatment approaches, as in the transdermal delivery field (Bharadwaj et al., 2011; Hafeez et al., 2013; Naik et al., 2000; Prausnitz and Langer, 2008; Prausnitz et al., 2004). Transdermal delivery has at least two advantages: it minimizes the chances that the applied product undergoes undesired metabolic pathways; it offers the possibility of a continuous local delivery, which enables avoiding internal irritations, and a cessation of administration at any time point.

There are two main transdermal delivery (TDD) methods: transappendageal and transepidermal pathways (Maghraby et al., 2008; Prausnitz and Langer, 2008; Prausnitz et al., 2004; Scheuplein and Blank, 1971). Transappendageal is the permeation of dermal products through sweat glands, hair follicles, and sebaceous glands. The effective surface available for delivery by this mean is much lower than by the transepidermal delivery, which uses all skin surfaces. Transepidermal delivery (representative of the TDD approach) is the delivery of vehicles loads by crossing skin strata. The main limiting step in TDD is the crossing of the SC (Hafeez et al., 2013; Prausnitz et al., 2004), which is why many skin delivery strategies are essentially based on permeabilizing first the SC (Carneiro et al., 2011; Cross et al., 2003; El Maghraby et al., 2005; Li et al., 2007; Man et al., 1993; Prausnitz et al., 2004; Rodriguez et al., 2009; Schurer and Elias, 1991). The permeation is achieved by reversibly disrupting the SC and fluidifying the lipid matrix by surfactant-based solvents such as short acyl-chain alcohols, polypropylene glycol, oleic or linoleic acid, lecithin, isopropyl myristate urea, and menthol (Ashara et al., 2014; Elias and Menon, 1991; Goodman and Barry, 1986; Potts and Guy, 1992; Schaefer and Hensby, 1990), with avoiding to deeply injure the dermis or hypodermis. The materials used to make the carriers can serve in the meantime as permeabilizing agents. For example, DMPC or DOPE phospholipids contained in nanocarriers can mix with the skin lipid matrix to loosen their structure and facilitate transport and delivery (Bouwstra and Honeywell-Nguyen, 2002; Cevc, 1996, 2004; Elsayed et al., 2007; Foldvari et al.,

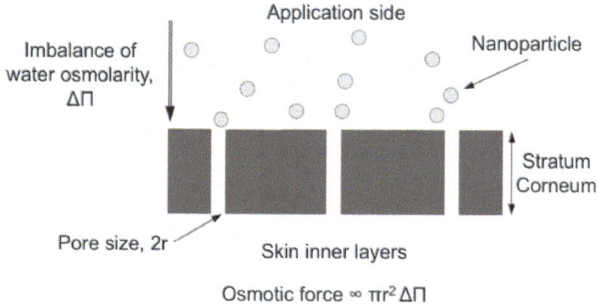

FIGURE 15.2 Illustration of nanoparticles uptake through the stratum corneum. An inward water flux is generated due to water molecules experiencing an osmolarity mismatch, $\Delta\Pi$. The flux generates a driving force leading particles to cross the SC. This principle called permeation is used for favouring skin occlusion for example. Permeation is described in Box 15.1.

2010; Kirjavainen et al., 1999; Lasch et al., 1995; Maghraby et al., 2008; Schreier and Bouwstra, 1994; Zellmer et al., 1995) (see Figure 15.2 and Box 15.1 for more details on the mechanism of permeation). The SC permeabilization strategy method has improved the efficiency of delivery of compounds and enabled even the delivery big macromolecular compounds. Few examples of this permeabilization strategy exist, and an example is the dermal patches (e.g., nicotine patches); they solubilize the sebum and soften the duct (El Maghraby et al., 2005; Maghraby et al., 2008) to locally deliver cargos through blood vessels (e.g., nicotine) or follicular regions of the dermis to limit bacterial colonies.

BOX 15.1

Permeation is the mechanism by which two apposed compartments separated by a permeable barrier equilibrate their chemical potentials by transferring their solution or solute; it is used to favour transfer and uptake of carriers or water inside the skin by crossing the SC and to favour skin occlusion, for example (Balaz, 2009; Schneider et al., 2009). The direction of solute or solution transfer is from the compartment of high to the one of low potential. The material transfer directionality is termed an osmotic flux and is analogous to the generation of and electrical current from high to low potential electrodes in electronics. Following this analogy, the osmotic flux (Hafeez et al., 2013; Sparr et al., 2013) J represents the electrical current and can be obtained from an analogous Joule law:

$$J = P\Delta\Pi$$

P represents the permeability of the permeants, solutes or solutions, to the barrier. P is dependent on the barrier thickness, which will be mainly the SC for the skin, but also on the molecules' activity in the barrier. It is analogous to the electrical conductance.

$\Delta\Pi$ is the concentration gradient between the compartments reflecting the potential difference. Dermal products can be designed to modulate the permeation by tuning the molecules affinity with the SC (Bouwstra and Honeywell-Nguyen, 2002; Bouwstra et al., 2003; Cevc, 2004; Zellmer et al., 1995). Permeation occurs because the water molecules inside and outside the skin have different activities or potential. Aqueous-based product can induce a water flux towards deeper skin layer and meantime generate a dragging force $\pi r^2 \Delta\Pi$, r being skin pore radius, favouring nanocarriers to cross the SC (Otto et al., 2009).

Another mechanism used to modulate the SC barrier function is the occlusion mechanism. Application of particular topical products or patches causes skin occlusion, or obstruction, resulting in enhanced skin hydration (Cevc and Blume, 1992; Jenning et al., 2000; Pardeike et al., 2009; Zhai and Maibach, 2002). This happens by overlaying the skin with a film of particular fat lipids, polymers, or petrolatum oil products (Honeywell-Nguyen et al., 2003; Zhai and Maibach, 2002). In TDD, increasing skin hydration favours the penetration of the cargos contained in occluding particles through the SC. Since skin hydration increases the humidity of the hydrophobic SC lipid matrix, the physical chemistry of the cargos and carriers, such as molecular weight or partition coefficient, needs to be acutely chosen for their efficient permeation.

Skin occlusion also prevents transdermal water loss, which occurs by a misbalance of water chemical potential in skin tissues compared with the surrounding environment (Cevc and Blume, 1992; Zhai and Maibach, 2002). Dermal products such as moisturizers can be formulated to reverse the order of water chemical potentials after their application to the SC (Darlenski et al., 2011), by changing water osmolarity with pH, nanocarriers concentration, etc. An inward water flux will be generated to hydrate the skin; the flux will in the meantime constitute a driving force for the nanocarriers to cross the SC (Cevc and Gebauer, 2003), Figure 15.2 and Box 15.1. For this process, the particle size is an important parameter for crossing skin pores. The SC or skin average pore diameters are, respectively, around 200 and 50 nm (Cevc, 2004; Tang et al., 2001; Verma et al., 2003; Wang et al., 1998). Formulated carriers size is generally beneath ~500 nm to 1 μm; crossing all skin layers is achieved by making carriers with a size lower than 250 nm (Kohli and Alpar, 2004; Müller et al., 2000; Verma et al., 2003). In fact, carriers can be bigger than skin pores because either the pores can be expanded by the carriers, or the latter are ultradeformable (Maghraby et al., 2008; Paul et al., 1995). By using deformable carriers of size bigger than 1 μm, a resistance to skin penetration will build up; external forces such as inward water fluxes will be necessary for their efficient travelling. For particles of size below $\ll 1$ μm, a Fick law will describe the uptake rate (Box 15.1).

In short, increasing skin permeability, by fluidifying its lipid matrix and thereby enlarging skin pore sizes, will favour efficient uptake; likewise, formulating small carriers, e.g., with a molecular weight beneath 500 Da (Bos and Meinardi, 2000), will facilitate uptake.

Finally, carriers have to release the therapeutic compounds at their target. Having an accurate release mechanism represents a big challenge and a key parameter taken into account during the design of carriers. To release their content, the carriers can directly fuse with cell membranes (see Figure 15.4b) or simply disintegrate once they reach the skin; the cargos can reach deeper skin layers by water permeation (see Box 15.1) or diffusion through skin strata (Carneiro et al., 2011; Cevc, 2004; Honeywell-Nguyen et al., 2003; Kirjavainen et al., 1999; Maghraby et al., 2008; Melero et al., 2014; Pardeike et al., 2009; Sinico and Fadda, 2009) (see Figure 15.4a). The release mechanism depends on the material the carriers are made up with, in respect of their capacity to favour membrane fusion, for example. The release can also be specific by functionalizing the surface of the nanocarriers, for example, by ligands recognized by endocytic proteins such as opsonins (see Figure 15.4c). Subsequent cellular internalization of the nanocarriers will lead to release of the therapeutic cargos by the degradation of the carriers.

15.3 FORMULATION OF LIPID NANOCARRIERS

Skin pore sizes are below 200 nm (Cevc, 2004; Tang et al., 2001; Verma et al., 2003; Wang et al., 1998). Efficient uptake and delivery to the skin was observed for particle size beneath ~300 nm, which represents an ideal cut-off size for nanocarriers formulation (Schneider et al., 2009), (Kohli and Alpar, 2004; Müller et al., 2000; Verma et al., 2003). From the formulation to storage and utilization, dermal products must be stable and not experience size variations. This limitation, in addition to the solubility and delivery requirements of therapeutic cargos, and the necessity of using biodegradable lipids of low toxicity, makes the formulation and functionalization of nanocarriers challenging.

Most dermal products are made of liposomes, emulsion droplets as soft nanocarriers, and solid or semisolid nanocarriers, which are made from emulsions (Atrux-Tallau et al., 2013; Cevc, 1996; Elsayed et al., 2007; Kogan and Garti, 2006; Müller et al., 2000; Osborne et al., 1991; Pardeike et al., 2009; Sinico and Fadda, 2009). Understanding the basis of emulsion stability and properties is critic for the formulation of nanocarriers. We present here the chronological order of development of lipid-based nanocarriers used in cosmetic and pharmaceutical products; this choice might be unable to better appreciate how formulation issues were progressively overcome to enable the formulation of stable and efficient dermal products.

15.3.1 LIPOSOMES AND DERIVATIVES

Liposomes were the first generation of nanocarriers developed for the delivery of poorly soluble therapeutics (Alvarez and Rodriguez, 2000; Cevc et al., 1996; Egbaria and Weiner, 1990; Elsayed et al., 2007; Sinico and Fadda, 2009). They are small lipid vesicles (50 to <1000 nm), with inner and outer aqueous phases separated by a surfactant lipid bilayer having the same morphology as cell membranes (Figure 15.3). Liposomes can be formed by the dehydration of phospholipids/surfactants followed by rehydration and shearing steps (e.g. sonication or extrusion) (Alvarez and Rodriguez, 2000; Sinico and Fadda, 2009). There also exist multilamellar liposomes, which are liposomes whose membrane is a sandwich of phospholipid bilayers. This second class of liposomes is also used in drug delivery (El Maghraby et al., 2005; Maghraby et al., 2008). In the following, we will focus on the first class of liposomes.

The structure of liposomes offers the encapsulation of hydrophilic and hydrophobic cargos in the same carrier moreover, respectively, inside the liposomes and within their membranes. These assets of liposomes were introduced for TDD in the early 1980s after the uptake of specific drugs was noticed to happen exclusively at the epidermis and dermis of white rabbit skins (Elsayed et al., 2007; Mezei and Gulasekharam, 1980; Schreier and Bouwstra, 1994). Identification of these delivery properties led to the formulation of the famous Doxil liposomes, which are liposomes coated with a cushion of polyethylene glycol (PEG) and containing the anticancer drug doxorubicin used against ovarian cancer (Allen, 1997; Cʹeh et al., 1997; Lasic, 1996).

15.3.1.1 Release Mechanisms

The delivery mechanisms of active compounds to/through the skin by liposomes are diverse and not completely clear. Different behaviours exist but, of general agreement, liposomes in most cases remain arrested in the SC. Only a little fraction of an applied liposome solution actually

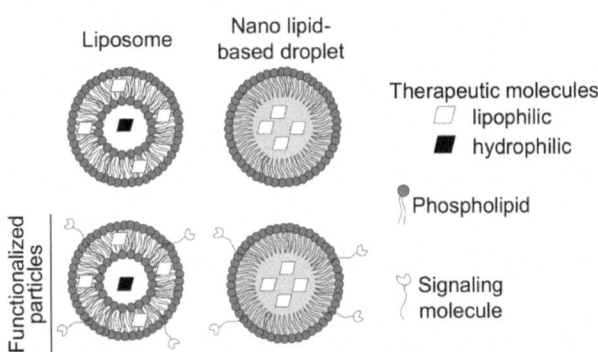

FIGURE 15.3 Design of drug or active nanocarriers. The content of the carriers can be either aqueous (left), to encapsulate hydrophilic compounds, which is the case of liposomes, or lipophilic (right), to encapsulate hydrophobic compounds, which is the case for lipid nanoparticles. The carriers can also be functionalized with ligands for specific cellular recognition.

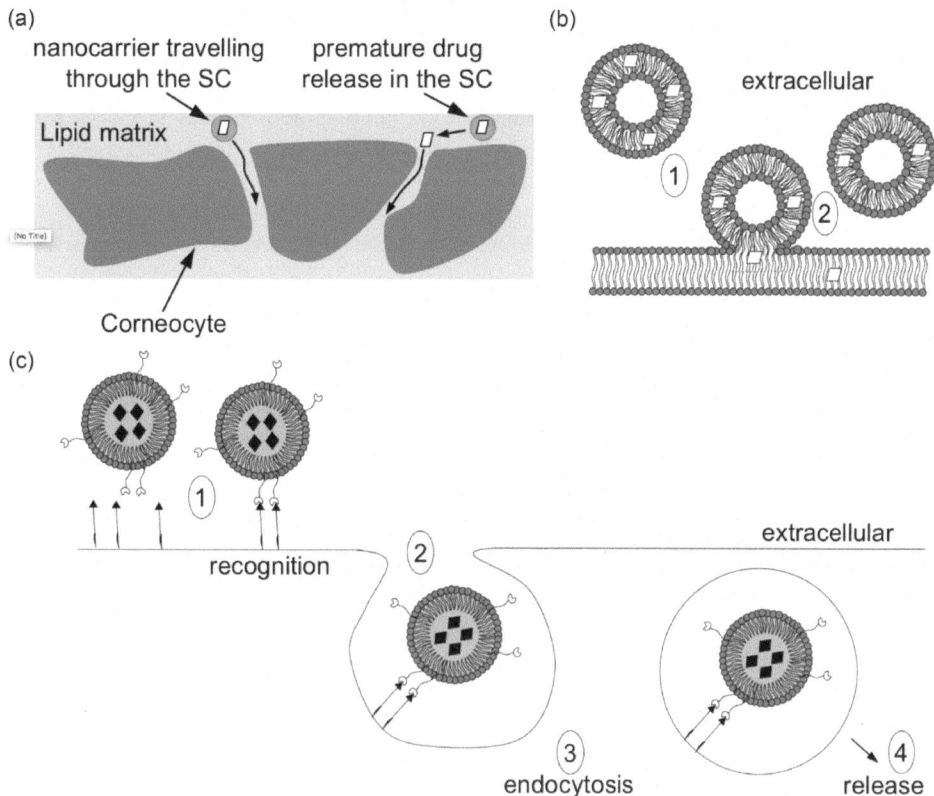

FIGURE 15.4 Uptake steps and delivery mechanisms of the carriers. (a) Nanocarriers reach skin innermost layers through pores in the SC, by permeation for instance. (b) Delivery can occur by nonspecific contact of carriers with cell membranes, provoking a subsequent active content release. Example of a liposomes carrier is shown. (c) Ligand-functionalized nanocarriers are specifically recognized at the cell surface and internalized by endocytosis. The carriers and content are released this way to the cell.

reaches hair follicles, mostly when liposomes are small enough (100 nm) (Barry, 2002; Hoffman, 1998; Verma et al., 2003). Most part of the active compounds is likely released in the SC before freely diffusing in the lipid matrix to reach deeper skin layers (Zellmer et al., 1995). This release mechanism is explained by the bursting of liposomes at the SC (Bouwstra and Honeywell-Nguyen, 2002; Honeywell-Nguyen et al., 2006; Kirjavainen et al., 1999; Maghraby et al., 2008; Schreier and Bouwstra, 1994; Zellmer et al., 1995); bursting would be triggered by the contact of the polar head group of the liposome phospholipids with the SC lipid matrix. Alternatively, a dose of therapeutics contained in the liposome bilayers could be released upon transient contacts of the liposomes with the outer membrane leaflet of the corneocytes or deeper skin cells, as sketched in Figure 15.4b and described as hemifusion (Lasic, 1998). This mechanism would indeed provide a slow and sustained release rate throughout the skin strata without liposomes effectively penetrating the SC; this delivery profile is used for optimal and prolonged deep skin release of anaesthesia such as lidocaine or prilocaine.

The two release mechanisms may both occur and offer advantageous delivery profiles for local treatments and minimized side effects (Bouwstra and Honeywell-Nguyen, 2002; El Maghraby et al., 2005; Elsayed et al., 2007; Lasch et al., 1995; Sinico and Fadda, 2009). Both mechanisms reveal the importance of well choosing the lipids in providing liposomes which are stable enough but able to hemifuse or burst upon contact with the SC (Lasic, 1998; Mouritsen, 2011). The lipid choice will define to which extent each mechanism happens. For example, the bursting of liposomes is proposed to occur with DMPC phospholipid liposomes, as the phospholipids tend to quickly mix with the SC lipid matrix after application (Kirjavainen et al., 1996, 1999); DMPC mixing with the SC has additionally loosened the lipid matrix, which is a favourable condition for the permeation of cargos (Rodriguez et al., 2010). DPPC liposomes instead have much less effect and exhibited slower release profile (Maghraby et al., 2008). These two situations exemplify at a glance how lipid composition controls rate and depth (Müller et al., 2002) of cargo uptake, i.e., by favouring liposomes hemifusion or bursting (Yokomizo and Sagitani, 1996a, b).

15.3.1.2 Liposomes Formulation and Lipid Choice

DPPC was first used for making liposome carriers because they can form gel phases, which stabilize the liposomes but which are disadvantageous for cargos incorporation and release (Mezei and Gulasekharam, 1980). Cholesterol was added to the liposomes to fluidify them; it also increases the bilayer thickness, which is beneficial for increasing the load of lipophilic compounds (Kirjavainen et al., 1996; Maghraby et al., 2008). The ratio between DPPC and cholesterol was adjusted to optimize incorporation level and release. The choice of a respective ratio of 2/1 increased by fivefold the release of triamcinolone acetonide drug in the epidermis and dermis of rabbit skins as compared with pure DPPC liposomes (Mezei and Gulasekharam, 1980). The release yield was optimized with the next generation of liposomes termed niosomes, i.e., liposomes which bilayer is made of nonionic surfactants, with principally a mono-acyl chain and cholesterol. Niosomes offered better incorporation of compounds and skin uptake (Cevc, 1996; Schreier and Bouwstra, 1994; Tabbakhian et al., 2006).

In general, the presence of cholesterol increases cargos incorporation level but mostly plays a role in the release efficacy. Cholesterol has an intrinsic negative curvature meaning that its hydrophilic head area is smaller than its hydrophobic tail (Figure 15.5) (Mouritsen, 2011). This shape of cholesterol is favourable for liposome destabilization; cholesterol mediates hemifusion (Karatekin et al., 2003; Thiam et al., 2013) and will favour bursting in the lipid matrix at very high amounts (Maghraby et al., 2008). This property of cholesterol in controlling release can be probably obtained with other surfactants and phospholipids having similar negative curvature (Figure 15.5) (Mouritsen, 2011). Having too much of these lipid types will, however, lead to unstable and nonstorable liposomal products.

In short, lipid surfactants favouring stability will retard release and vice versa. A trade-off has to be considered for the lipid choice of liposomes. By tuning the lipids ratio (e.g., negatively vs null or positively curved lipids), liposomes can be optimized in respect of stability, skin type, cargo type, and delivery efficiency.

The following liposome generations were better tailored and functionalized: long-lived liposomes were obtained with the combination of nonsaturated phospholipids, e.g., DSPC, with cholesterol and lipid surfactants, mimicking skin/SC lipids for biocompatibility, and lipid-grafted PEGs (Allen and Cullis, 2013; Klibanov et al., 1990; Torchilin, 2005); functionalization was performed by attaching specific peptides or antibodies recognized by cells (Figure 15.3). This class of liposomes encompasses stealth liposomes which membranes are rigid due to the nature of their lipids and functionalization (Cˇeh et al., 1997), and penetrate the skin and deliver cargos at very slow rates (Cˇeh et al., 1997). In contrast, deformable liposomes, generally bigger than 100 nm, have better skin uptake and are able to completely cross the skin (El Maghraby et al., 2001a, b, 2006; Honeywell-Nguyen et al., 2003); they are used for systemic treatments when applied at nonocclusive regimes (e.g., low liposome concentration) (Cevc and Blume, 1992). Permeation is the motor of the liposomes transport and is also aided in this context by setting a water gradient across the skin (Figure 15.2 and Box 15.1 for details). The contribution of liposome deformability to the delivery kinetics and depth of cargos is ensured by an acute compromise between liposome mechanical stability and delivery property (Lasic, 1998). A balance is obtained by mixing phospholipids with other surfactants increasing liposomes elasticity to improve release depth. A good trade-off is obtained with DOPC liposomes made with sodium cholate, Span 80 or Tween 80, as nonbiologically occurring surfactants, or with oleic acid as a biological surfactant. The resulting ultradeformable liposomes were termed transfersomes, representing now an important segment of TDD (Cevc, 2004; Pierre and Costa, 2011).

Another strategy for optimal formulation and release is the formulation of highly stable liposomes, important for storage, and content release is triggered by enzymes that alter the liposome lipid composition and stability. For example, liposomes can be formed, for example, with lipids such as PC and LPC, and skin lipases, as follicle bacteria lipases, can cleave them into free fatty acids. These fatty acids favour hemifusion and bursting and solubilization of the SC lipid matrix (Arouri et al., 2013; Arouri and Mouritsen, 2013; Lasic, 1998), which promotes release.

These various strategies mirror the extensive utilization and improvement of liposomes as nanocarriers. However, it remains a main commercially important limitation, which is their poor loading capacity.

15.3.2 Emulsions in Dermal Applications

An emulsions is colloidal solution of drops of one liquid phase (dispersed phase) immersed into another immiscible liquid medium (continuous phase). Emulsions such as nanoemulsions and microemulsions, owing to their droplets size ≤ 200 nm (nanodroplets), are very attractive for the encapsulation and delivery of pharmaceutical and cosmetic ingredients (Al-Bawab and Friberg, 2006; Atrux-Tallau et al., 2013; Kogan and Garti, 2006; Osborne et al., 1991). They offer cheap formulation means providing a high loading capacity of compounds, as opposed to liposomes, with a variety of techniques. The nanodroplets can be also

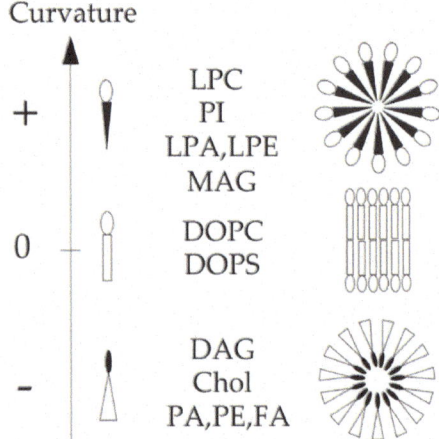

FIGURE 15.5 Examples of biologically occurring surface lipids and their favoured curvature. For phospholipids, they are considered here to have 18 carbons in their acyl chains.

functionalized for enhanced biological response (Atrux-Tallau et al., 2013).

15.3.2.1 Surfactants

Surfactants are amphiphilic molecules containing hydrophilic and lipophilic parts, e.g., phospholipids. They can be ionic, nonionic, or polymeric. To meet their dual affinity, surfactants absorb at oil and water interface and minimize energy. They form a protective steric or electrostatic monolayer around droplets. Emulsion droplets are for most not thermodynamically stable, as the liquid phases tend to spontaneously separate (Bibette et al., 2002). Ageing of emulsions leads to their destabilization that occurs by coalescence, which is the fusion of drops following a pore nucleation between them, or Ostwald ripening, which is a molecular destabilization process by the diffusion through the continuous phase of molecules or micelles from smaller to bigger droplets (Bibette et al., 2002; Kabalnov et al., 1996; Kabalnov and Shchukin, 1992; Roger et al., 2015). Emulsions can, however, acquire kinetic stability with the presence of surfactants.

At low concentrations, surfactants are free in solvents, be it aqueous or oil phases depending on the surfactant solubility preference. The solubility phase can be predicted by considering surfactants hydrophilic and lipophilic balance (HLB) introduced by Griffin in early fifties (Bibette et al., 2002; Kabalnov and Wennerström, 1996). The HLB compares the importance of the volume occupied by the hydrophilic head and lipophilic tail of surfactants. It is expressed as values comprised between 0 and 20. HLBs < 8 are mostly oil-soluble surfactants favouring the formation water droplets in oil (W/O). HLBs over 12 correspond to water-soluble surfactants forming oil/water (O/W) emulsions (Bibette et al., 2002; Leal-Calderon et al., 2007). Certain surfactants, e.g., PEGs, have temperature-sensitive HLBs, because variation of the size of the surfactant polar head by solvation is tuned by heat (Shinoda and Saito, 1968; Shlnoda and Kunleda, 1983). Temperature-switchable surfactants' HLB can be used to switch between W/O and O/W emulsions. The transition temperature is called phase inversion temperature (PIT) (Bibette et al., 2002; Shinoda and Saito, 1969). This principle is used to generate nanodroplets in nanoemulsions.

At increased concentrations, surfactants form thermodynamically stable structures in the continuous liquid phase. The concentration at which these structures appear is called the critical micellar concentration (CMC). Surfactants first form micelles, small spheres of roughly 10 nm diameter having a core immiscible with the continuous phase (Bibette et al., 2002; Leal-Calderon et al., 2007; Shlnoda and Kunleda, 1983). They can form bicontinuous structures at very high concentrations. Water-soluble surfactants, i.e., surfactants of high HLBs, will form direct micelles that mask the hydrophobic tails of the surfactants from water molecules; oil-soluble surfactants will form reverse micelles (Bibette et al., 2002). Direct micelles can be swollen by the disperse oil phase, to form thermodynamically stable emulsions termed microemulsion (nanodroplets size ≥10–20 nm) (Langevin, 1988; McClements, 2012b). The stability of this system needs a significant decrease of the surface tension – the energy cost per area for generating the O/W interface – and the bending modulus of droplets surface energy cost for curving the O/W interface (De Gennes and Taupin, 1982; Kabalnov and Wennerström, 1996). Reaching these mechanical properties is facilitated by the presence of cosurfactants which are generally smaller than surfactants and cover the gaps between surfactants at the drops surface (Kozlov and Helfrich, 1992; van Buuren et al., 1996). Cosurfactants decrease surface tension by providing a better interface coverage and increasing the interface deformability (De Gennes and Taupin, 1982; Kozlov and Helfrich, 1992; Langevin, 1988; van Buuren et al., 1996). Efficient cosurfactants are nearly soluble in both aqueous and oil phases.

These various properties surfactants need to be taken into account for controlling droplets stability and deformability, which are critical parameters for the encapsulation of therapeutic molecules into nanodroplets and their release.

15.3.2.2 Nanoemulsions and Microemulsions

Nanoemulsions and microemulsions look optically translucent (droplets size <200 nm) but are fundamentally different (McClements, 2012b). Nanoemulsions are thermodynamically unstable emulsions, while microemulsions are stable. The free energy of a nanoemulsions formation is higher than that of the separated liquid phases (Mason et al., 2006), whereas it is lower or null for microemulsions (McClements, 2012b). The thermodynamic stability of microemulsions can be understood by considering that the entropy configuration term of the free energy, $\Delta G = \gamma \Delta A - T\Delta S$, prevails over the interfacial free energy, $T\Delta S > \gamma \Delta A$ (T, temperature; S, entropy; γ, surface tension; A, area), making the free energy negative. A problem with this argument is that at the scale of microemulsion droplet, curvature becomes so important that definition of surface tension does not hold, as tension is a 'macroscopic' measurement of interfaces for which curvature does not account. Nonetheless, although not fundamentally true, the above energy argument gives insights on the stability of microemulsions.

Both nano and microemulsions are O/W mixtures plus surfactants and potentially cosurfactants, but microemulsions will require much more of the latter to minimize surface tension, and $\gamma \Delta A$, so as to keep the predominance of $T\Delta S$ (Bibette et al., 2002; De Gennes and Taupin, 1982; Langevin, 1988; McClements, 2012b); ultralow and even transiently negative values of surface tensions and bending modulus can be reached by having excess surfactant and cosurfactant in microemulsions (Kozlov and Helfrich, 1992).

In practice, starting with the same type of material, the formation of nanoemulsions will require an important energy input provided, for example, by stirred-media mills, parallelized microfluidic devices, sonication, or high-pressure homogenization (McClements, 2011), while the fabrication of microemulsions will be almost spontaneous (Anton et al., 2008; McClements, 2012b; Sainsbury et al., 2014); microemulsions might simply need a slight stirring or heating processes due to kinetic energy barriers such as slow surfactant diffusion kinetics (McClements, 2012b). This ease for making microemulsions is the main asset used in self-microemulsifying drug delivery systems (SMEDDS) for optimizing the encapsulation of therapeutic ingredients in lipid-based O/W emulsions (Lawrence and Rees, 2000, 2012). For example, mildly stirring isotropic mixtures of drugs, triglycerides, surfactants, and cosurfactants with concomitant addition of water allow forming O/W microemulsions with nanodroplets encapsulating the drugs (Mahapatra and Murthy, 2014).

15.3.2.3 Lipid Ingredients

A variety of oil combinations, surfactants, and cosurfactants are used for the formulation of dermal products. Their acute mixture, based on their physical chemistry match with cargos, is important as it dictates further distribution of the cargos within the skin layers. Lipids used for the formulation for nano- and microemulsions in dermal products are medium- and long-chain triglycerides of various saturation levels (Lawrence and Rees, 2000, 2012). A list of oils used in typical derma products is presented in Table 15.2. Medium-chain triglycerides were initially preferred oils because of their high fluidity and solubilization of lipophilic therapeutic compounds (Balaz, 2009; Kogan and Garti, 2006; McCarron et al., 2008). Triglyceride-derived metabolites, such as free fatty acids, diglycerides and monoglycerides, became afterwards attractive (Constantinides, 1995; Constantinides and Scalart, 1997; Prajapati et al., 2012). These molecules can serve as oil phases and behave in the meantime as surfactants and/or cosurfactants helping for the formation and stabilization of nanodroplets (Kogan and Garti, 2006; Lawrence and Rees, 2012; Prajapati et al., 2012; van Buuren et al., 1996). In addition, these triglyceride-derived metabolites are biologically occurring and hence can be degraded in the body by enzymes such as apolipoproteins lipases (Buszello and Müller, 2000). For surfactants, synthetic ones are avoided because of possible adverse effects, and biodegradable surfactants or biosurfactants are privileged. Biosurfactants are generally made from plants or microorganisms (Banat et al., 2000): glycolipids, mainly mannosylerythritol, rhamnolipids, and sophorolipids are examples of extensively used biosurfactants (Lourith and Kanlayavattanakul, 2009; Varvaresou and Iakovou, 2015) in cosmetics, due to their softening properties through increasing intercellular interactions in the skin. Other restrictions exist for surfactant choice in dermal products. Anionic or cationic surfactants are, for example, irritant and cytotoxic for the skin, compared with nonionic or amphoteric surfactants (Marti-Mestres and Nielloud, 2000). Examples of used nonionic and amphoteric surfactants, and cosurfactants, are listed in Table 15.2. They include polyol esters, polyoxyl esters, polyoxamers, alkanolamides (Marti-Mestres and Nielloud, 2000), and phospholipids used as nonionic biological surfactants. For pharmaceutics, cationic surfactants can be added in some cases because they have antiseptic, antibacterial, and antifungal properties (Vieira and Carmona-Ribeiro, 2006) and improve the cleansing of lesions. Examples of cationic surfactants are cetrimides, stearylamines, DC-cholesterol, or DOTMA (Marti-Mestres and Nielloud, 2000). Finally, as for stealth vesicles, PEGs represent the most biocompatible surfactant used to stabilize and functionalize nanodroplets (Cevc, 2004). PEGs are anchored to lipids on the nanodroplets and form brushes that can be conjugated to antigens for recognition (Figure 15.3). The properties of PEGs enable to modulate droplets surface protection and functionalization by playing with the polymer surface density and brush length (Delmas et al., 2012), which are important features for modulating the delivery of the nanodroplet content.

15.3.2.4 Destabilization

Microemulsions are in principle stable and should not display size changes during shelf storage. In reality, if droplets are larger than a certain size, attuned to the curvature of the surfactants, they tend to coalesce and lead to formation of lamellar structures and multiple emulsions which are

---

TABLE 15.2

Example Oils, Surfactants, and Cosurfactants Used in Cosmetic and Pharmaceutical Products (Kogan and Garti, 2006; Lawrence and Rees, 2012; Marti-Mestres and Nielloud, 2000)

Triglycerides and derivatives	Capric triglyceride, castor oil, squalene, oleic acid, medium-chain triglyceride, coconut oil, corn oil mono-, di-, triglyceride, medium-chain triglyceride, palm seed oil, mixture of mono- and diglycerides of caprylic/capric acid, medium-chain mono- and diglycerides, corn oil, olive oil, sesame oil, hydrogenated soybean oil, hydrogenated vegetable oils, soybean oil, peanut oil, beeswax
Surfactants and cosurfactants	Polyola esters (Polysorbate-20–40–60–65–80–85, sorbitan monooleate-monostearate-trioleate-monopalmitate, mono-diglyceride), Tween 80, polyoxyethylated oleic glycerides, polyoxyethylated glycerides, polyglyceryl-3 methylglucose distearate, soybean lecithin, sodium cholate, poloxamer 188, polyoxyl-40-hydrogenated castor oil, PEG-8–40–50 stearate, polyvinyl alcohol, sophorolipids, rhamnolipids, mannosyl natural surfactants: lysophospholipides, phosphatidylcholine (PC), phosphatidylethanolamine (PE), phosphatidylinositol (PI), phosphatidic acid (PA), sphingomyelin (SM)

thermodynamically stable (McClements, 2012a, b). Other factors may impede the stability of microemulsions over time such as contamination of the surfactants or their degradation by oxidation, dilution, or temperature variations. In short, all common steps undergone for large-scale production of microemulsions droplets for products development can ruin stability. In the case of nanoemulsions, droplets are only kinetically stabilized and will destabilize during long-time storage (Bibette et al., 2002; McClements, 2012b).

Emulsions can destabilize by coalescence or ripening (Bibette et al., 2002). Coalescence is the fusion of droplet. Ostwald ripening is the disappearance of small droplets that transfer material to bigger droplets. The transfer is a molecular or micellar diffusion process through the continuous phase. The direction of the transfer is set by the difference between the Laplace pressures of the droplets and goes from high-pressure droplets to low-pressure ones. The Laplace pressure ΔP is the counterbalancing pressure built up in droplets, as they are subjected to an isotropic compression due to surface tension γ ($\Delta P = 2\gamma/r$, r being droplets radius). Coalescence and ripening mechanisms lead to phase separation of emulsions.

For long-term storage, coalescence can be preceded by segregation phenomena such as droplets flocculation or settling, which occur because of a density mismatch $\Delta \rho$ between continuous and disperse phases. The speed of segregation is approximately given by $v = \dfrac{2\pi}{9} \dfrac{\Delta \rho g}{\eta} r^2$ (Stokes' law; η, viscosity of the continuous phase; g, gravity constant; r, droplet radius). For nanoemulsions and microemulsions, such phenomena are annihilated for products with reasonable storage time because of $v \propto r^2$ and r being extremely small <200 nm. Coalescence is significantly hindered because the interaction between nanodroplets is below $k_B T$, the thermal energy, meaning that droplets contact under Brownian motion is too brief for them to fuse (Delmas et al., 2012).

The smaller the nanodroplets, the higher their Laplace pressures. The main mechanism destabilizing nanoemulsions is hence ripening. Even a slight mismatch of droplet sizes in emulsions will result in accelerated ripening over storage. Molecules of smaller drops, of higher Laplace pressure, solely or hidden into micelles in presence of excess surfactants, will diffuse through the continuous phase to enlarge bigger ones by an accelerating rate. The rate of the resulting phase separation will depend on the solubility (s_0) of the droplet molecules, e.g., the oil, in the continuous phase, e.g., water, their molar mass (M), diffusion coefficient (D), density (d), and surface tension γ. The ripening rate follows $\omega = \dfrac{dr^3}{dt} = \dfrac{8 D \gamma M s_0}{9 d^2 RT}$ (R, gas constant; T, temperature; r, characteristic radius of the emulsion droplets at a given ripening time). The signature of ripening is thus the linear increase of the cube of the radius over time.

Ripening can be prevented by (1) diminishing surface tension, by working with surfactants and cosurfactants, to get closer to microemulsion formulation conditions, or (2) for O/W emulsions, choosing low aqueous-soluble oils such as long-chain triglycerides (Anton et al., 2008), (3) choosing high molar mass oils such as triglycerides, and (4) having a rigid surfactant monolayer at the nanodroplets interface, as it is the first barrier to cross for diffusing into the continuous phase (Delmas et al., 2012). For oil droplets, an efficient blockade of ripening is possible by encapsulating species of lower solubility in the water phase than the oil molecules (Delmas et al., 2012); such species could be simply the therapeutic compounds to deliver. Indeed, in this situation, the oil phase is first transferred to big droplets by ripening; consequently, the chemical potential of the encapsulated species in the small and shrinking droplets will be dropping down, while the potential of the species in the big droplets will be increasing. The encapsulated species thus sense a chemical potential misbalance that is spontaneously corrected by a backward diffusion of oil molecules from the big to small droplets. Hence, the system reaches an equilibrium state for which ripening is blocked to satisfy the chemical potential balance for both oil molecules and encapsulated species (Delmas et al., 2012).

Emulsions offered good alternatives for the design of functionalized dermal products. Their formulation is still subtle for optimizing the incorporation of cargos along with nanodroplets functionalization. Long-term storage is also still challenging for keeping integrity of the cargos. Most cargos can escape from nanodroplets over time because the protective nanodroplets monolayer can be transiently or steadily lost. Tunable solid or semisolid lipid nanoparticles were hence developed to overcome these limitations with emulsions (Müller et al., 2000; Puglia and Bonina, 2012). This generation of nanocarriers became rapidly attractive for controlled release.

15.3.3 Lipid Nanoparticles

For improved shelf life and drug storage, as compared with emulsions or liposomes, a strategy was adopted in early nineties to develop hydrophobic solid core nanocarriers, referred as lipid nanoparticles (LNPs) (Müller et al., 2000; Pardeike et al., 2009), which do not undergo coalescence or ripening. Solid lipid nanoparticles (SLNs) were the first generation of LNPs, later improved to nanostructured lipid carriers (NLCs) (Müller et al., 2002; Pardeike et al., 2009; Wissing and Müller, 2003), which offered better encapsulation and delivery properties. As for many other novel body products, LNP formulations were immediately adopted in cosmetics before being implemented in the pharmaceutical field for safety reasons.

15.3.3.1 Solid Lipid Nanoparticles and Nanostructured Lipid Carriers

SLNs are nano-sized particles with a core made of a lipid blend that crystallizes at room temperature. Such lipids are generally long saturated chains of triglycerides, waxes, very long fatty acid chains, and partial glycerides. Table 15.3 lists example lipids used for making solid nanoparticles. There

TABLE 15.3
Examples of Crystal-Forming Lipids Used in Dermal Products (Pietkiewicz and Sznitowska, 2004)

Lipids with low crystallizing temperature	Animal waxes (e.g., beewax, Chinese wax) PEG-8 beeswax, vegetable waxes (e.g., soy, jojoba, castor), carnauba wax, palm oil, petroleum jellies, paraffin wax, palmitates, stearylamines, stearic acids, glyceryl monostearate, glyceryl cocoate, capric triglyceride oleic acid, castor oils, squalene, capric triglyceride

are various fabrication processes of SLNs (Charcosset et al., 2005; Gasco, 1997; Heurtault et al., 2002; Hu et al., 2002; Liedtke et al., 2000; Mehnert and Mäder, 2001; Müller et al., 2002; Pardeike et al., 2009; Priano et al., 2007; Puglia et al., 2008; Schubert and Müller-Goymann, 2003; Sjöström and Bergenståhl, 1992; Trotta et al., 2003; Wissing and Müller, 2003), but the most popular is the hot pressure homogenization (Gasco, 1997; Liedtke et al., 2000; Mehnert and Mäder, 2001; Muller et al., 2007; Müller et al., 2002; Pardeike et al., 2009; Priano et al., 2007; Wissing and Müller, 2003). In the presence of surfactants and cosurfactants, lipids are heated over their fluid transition temperature within the aqueous phase by a high-pressure homogenizer to generate O/W nanodroplets; cargos contained in the lipid phase will be encapsulated in the droplets. Immediate following cooling of the emulsion will lead to the formation of SLNs (Müller et al., 2000, 2002; Pardeike et al., 2009; Wissing and Müller, 2003). Three types of SLNs can be generated with this method (Mehnert and Mäder, 2001): active-enriched shell, active-enriched core, and homogeneous blend (Delmas et al., 2012; Figure 15.6). High lipophilic active compounds encapsulation generally result in homogeneous SLNs (Delmas et al., 2012); during the cooling step, if the cargos precipitate before the crystallization of the lipids, active-enriched cores will appear, while the opposite case will lead to active-enriched shell. The proportion of three SLNs that form depends on the nature of the active compound, lipids, and surfactants (Pardeike et al., 2009). For example, some surfactants can particularly modulate the precipitation rate of the therapeutic actives, e.g., by hiding them into micelles and preventing their clustering (Müller et al., 2000; Muller et al., 2007).

SLNs offered good nanoparticle stability and industrial cheap fabrication processes but show three main shortcomings that required improvement for market survival (Müller et al., 2000, 2002). First, SLNs can carry only a low fraction of cargos. Second, most compounds get expelled from the nanoparticles during long-term storage (Figure 15.6). Indeed, the lipids tend over time to form a high degree ordered crystal (Delmas et al., 2012; McClements, 2012a; Müller et al., 2000) to thereupon exclude intrusions such as cargos. Third, SLNs can represent at maximum 30% of the aqueous phase, as the nanoparticles will otherwise aggregate (Müller et al., 2000). These issues do not occur with emulsions and are overcome with the formulation of semisolid particles.

Nanostructured lipid carriers (NLCs) are developed as a perfect compromise between emulsions and SLNs (Delmas et al., 2012; Muller et al., 2007). NLCs consist on nanoparticles with a semisolid core made of a mixture of solid and liquid phases of lipids in typical ratio of 70/30–99/1

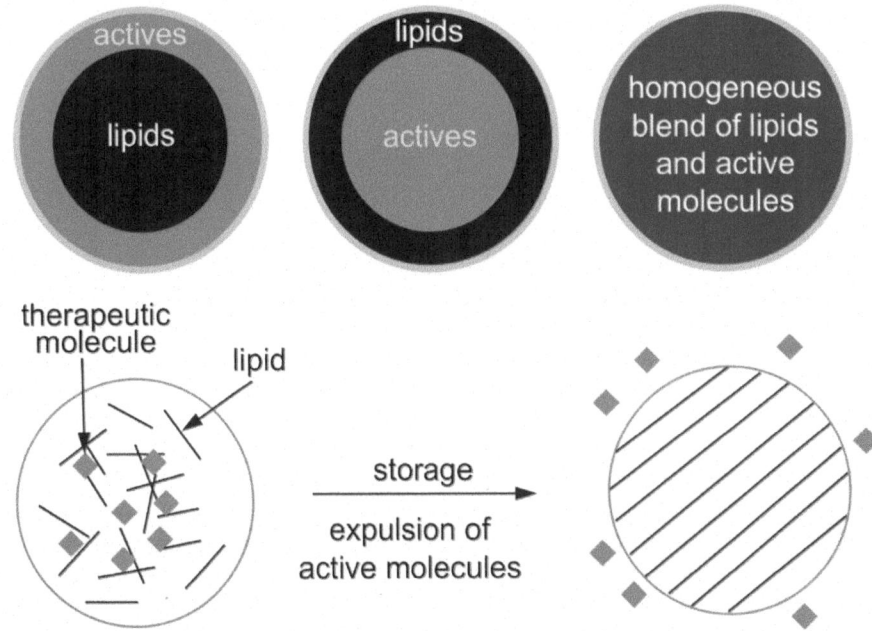

FIGURE 15.6 Different structures of solid lipid nanoparticle carriers. Upper panel, actives and lipids can demix to core or shell actives, or form a homogeneous mixture by choosing well-tuned thermodynamic mixtures. Bottom panel, shelf storage leads to actives expulsion.

depending on the lipid type. The presence of the liquid lipid phase prevents the perfect crystallization of the core of the particles by introducing defects. Examples of liquid lipids are medium-chain triglycerides, diglycerides, and fatty acids (Müller et al., 2002). Three NLC types can be generated with the same protocol of SLN formation: imperfect crystals, amorphous crystals, and oil in wax particles (Figure 15.7). The defects in these NLCs can be kinetically amorphous stable pockets (Müller et al., 2002) with no specific localization or droplets trapped in the crystal (Figure 15.7). NLCs offer a maximum encapsulation of active compounds, expulsion of these is noticeably reduced, and shelf storage is improved (Mehnert and Mäder, 2001). Finally, NLCs can be added up to 50% in the aqueous phase and, compared with SLNs, offer a perfect compromise between long-term stability and the encapsulation of cargos (Muller et al., 2007; Pardeike et al., 2009).

15.3.3.2 Application of Lipid Nanoparticle Products and Release

LNPs offer practical large-scale production and cheap root materials for dermal applications. They have very good occluding properties (Wissing and Müller, 2003), as compared with O/W nanoemulsions or microemulsions that are softer materials (Delmas et al., 2012; Mehnert and Mäder, 2001), and therefore provide better skin hydration. Owing to their solid or semisolid opaque cores, LPNs can also prevent the degradation or oxidation of encapsulated light-sensitive cargos and vitamins (Dingler et al., 1998; Jenning et al., 2000; Puglia et al., 2008; Puglia and Bonina, 2012; Teeranachaideekul et al., 2007). In clinical interventions, LNPs are successfully used to treat follicle bacteria, as they show good uptake to hair follicles (Lademann et al., 2007). They are also used for treating dermal pathologies such as skin mycosis, eczemas, and psoriasis vulgaris disease (Liu et al., 2007; Queille-Roussel et al., 2001). This success of LNPs is owed in part to their high loading capacity, especially with LNCs, and their stability and controlled release.

For cosmetic products such as fragrances, UV blocker particles, or insect repellents, a slowly and continuously release of cargos is required (Pardeike et al., 2009). In pharmaceutical skin products, the same release profile is also often required to prevent the fast propagation of therapeutics into blood vessels, which provokes blood peaks. LNPs also fulfil such needs. The release rate of cargos of LNPs occurs in two steps. NLCs have a fast release speed that slows down (Muller et al., 2007), while SLNs start with a slow release that accelerates after few hours (Müller et al., 2000). These release profiles are complementary and different from nanoemulsions, for example, which provide a constant and rapid release rate (Dingler et al., 1998; Pardeike et al., 2009; Teeranachaideekul et al., 2007). Together, SLN and NLCs provide a better controlled release of cargos through the skin (Pardeike et al., 2009), and tunable release profiles are best obtained with NLCs having amorphous cores (Delmas et al., 2012). Release kinetics is controlled by the crystallization level of the particles, which is controlled by surfactants.

15.4 CONCLUSION

Lipids are the main body components regulating many physical appearances such as skin softness and waterproof. Understanding skin properties and its lipid constituents is important for designing optimal cosmetic and pharmaceutical skin products. Pioneer methods have been developed based on soft matter to design appropriate products. Many challenges are still to overcome essentially regarding controlled release. A promising area is the design of metabolic compounds based on specific lipids whose uptake will endow to the skin its good health and appearance and prevent the development of skin pathologies.

ACRONYMS

γ:	surface tension
Chol:	cholesterol
CMC:	critical micellar concentration
DAG:	diacylglycerol
DMPC:	dimyristoyl–glycerol–phosphocholine
DOPE:	dioleoyl–glycero–phosphoethanolamine
DPPC:	dipalmiloyldiphosphatidylcholine
DOTMA:	dioleyloxypropyl trimethylammonium
FA:	free fatty acids
HLB:	hydrophilic lipophilic balance
k_B:	Boltzmann constant
LNPs:	lipid nanoparticles
LPA:	lysophosphatidic acid
LPC:	lysophosphatidylcholine
LPE:	lysophosphatidylethanolamine
MAG:	monoacylglycerol
NLC:	nanostructured lipid carriers
O/W:	oil-in-water emulsion
PC:	phosphatidylcholine
PEG:	polyethyleneglycol
PI:	phophoinositol
PS:	phophatidylserine

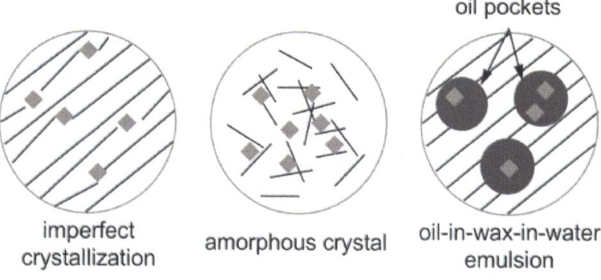

FIGURE 15.7 Defects-increased actives encapsulation in nanostructured lipid carriers. The crystallization of lipids can be imperfect and leaves room for actives encapsulation. An amorphous crystal can be formed leading to kinetically stable defects where actives are trapped. Oil pockets containing the actives can be trapped in the lipid crystal.

PA:	phosphatidic acid
PE:	phosphatidylethanolamine
PIT:	phase inversion temperature
S:	entropy
SC:	stratum corneum
SLN:	solid lipid nanoparticles
SMEDDS:	self-microemulsifying drug delivery systems
T:	temperature
TDD:	transdermal delivery
W/O:	water-in-oil emulsion

REFERENCES

Abe, T. (1978). Studies on skin surface-barrier functions - transepidermal water-loss and skin surface-lipids during childhood. *Chemical & Pharmaceutical Bulletin 26*, 1659–1665.

Abramovits, W., and Elias, P. (2014). Skin barrier repair in atopic dermatitis therapy. *Atopic Dermatitis and Eczematous Disorders*, 159–177.

Abramovits, W., and Gonzalez-Serva, A. (2000). Sebum, cosmetics, and skin care. *Dermatologic Clinics 18*, 617–620.

Ahcan, U., El-Tonsy, M.H., El-Din, H., Al-Sawy, A.E., and Ralca, S. (2003). Treatment of various vascular malformations of the dermal and hypodermal layers of the skin with a multiple wavelength aesthetic laser system. *Lasers in Surgery and Medicine*, 57–57.

Al-Bawab, A., and Friberg, S.E. (2006). Some pertinent factors in skin care emulsion. *Advances in Colloid and Interface Science 123*, 313–322.

Allen, T.M. (1997). Liposomes. *Drugs 54*, 8–14.

Allen, T.M. (2002). Ligand-targeted therapeutics in anticancer therapy. *Nature Reviews Cancer 2*, 750–763.

Allen, T.M., and Cullis, P.R. (2004). Drug delivery systems: Entering the mainstream. *Science 303*, 1818–1822.

Allen, T.M., and Cullis, P.R. (2013). Liposomal drug delivery systems: from concept to clinical applications. *Advanced Drug Delivery Reviews 65*, 36–48.

Alvarez, A.M.R., and Rodriguez, M.L.G. (2000). Lipids in pharmaceutical and cosmetic preparations. *Grasas Aceites 51*, 74–96.

Anton, N., Benoit, J.-P., and Saulnier, P. (2008). Design and production of nanoparticles formulated from nano-emulsion templates—a review. *Journal of Controlled Release 128*, 185–199.

Arouri, A., Hansen, A.H., Rasmussen, T.E., and Mouritsen, O.G. (2013). Lipases, liposomes and lipid-prodrugs. *Current Opinion in Colloid & Interface Science 18*, 419–431.

Arouri, A., and Mouritsen, O.G. (2013). Membrane-perturbing effect of fatty acids and lysolipids. *Progress in Lipid Research 52*, 130–140.

Ashara, K.C., Paun, J.S., Soniwala, M.M., Chavada, J.R., and Mori, N.M. (2014). Micro-emulsion based emulgel: A novel topical drug delivery system. *Asian Pacific Journal of Tropical Disease 4*, S27–S32.

Atrux-Tallau, N., Delmas, T., Han, S.H., Kim, J.W., and Bibette, J. (2013). Skin cell targeting with self-assembled ligand addressed nanoemulsion droplets. *International Journal of Cosmetic Science 35*, 310–318.

Balaz, S. (2009). Modeling kinetics of subcellular disposition of chemicals. *Chemical Reviews 109*, 1793–1899.

Banat, I.M., Makkar, R.S., and Cameotra, S.S. (2000). Potential commercial applications of microbial surfactants. *Applied Microbiology and Biotechnology 53*, 495–508.

Barry, B.W. (2002). Drug delivery routes in skin: A novel approach. *Advanced Drug Delivery Reviews 54*, S31–S40.

Bharadwaj, S., Garg, V.K., Sharma, P., Bansal, M., and Kumar, N. (2011). Recent advancement in transdermal drug delivery system. *International Journal of Pharma Professional's Research 2*, 6–9.

Bibette, J., Leal-Calderon, F., Schmitt, V., and Poulin, P. (2002). Emulsion science - Basic principles. An overview - Introduction. *Springer Tracts in Modern Physics 181*, 1–4.

Black, D., Diridollou, S., Lagarde, J.-M., and Gall, Y. (1998). Skin care products for normal, dry and greasy skin. *Textbook of Cosmetic Dermatology 2*, 125–150.

Bodmeier, R., Chen, H., Davidson, R.G., and Hardee, G.E. (1997). Microencapsulation of antimicrobial ceftiofur drugs. *Pharmaceutical Development and Technology 2*, 323–334.

Bos, J.D., and Meinardi, M.M.H.M. (2000). The 500 Dalton rule for the skin penetration of chemical compounds and drugs. *Experimental Dermatology 9*, 165–169.

Bouwstra, J.A., and Gooris, G.S. (2010). The lipid organization in human stratum corneum and model systems. *Open Dermatology Journal 4*, 10–13.

Bouwstra, J.A., Gooris, G.S., Groen, D., and Ponec, M. (2009). The skin barrier in healthy and diseased state. *Chemistry and Physics of Lipids 160*, S16–S16.

Bouwstra, J.A., and Honeywell-Nguyen, P.L. (2002). Skin structure and mode of action of vesicles. *Advanced Drug Delivery Reviews 54*, S41–S55.

Bouwstra, J.A., Honeywell-Nguyen, P.L., Gooris, G.S., and Ponec, M. (2003). Structure of the skin barrier and its modulation by vesicular formulations. *Progress in Lipid Research 42*, 1–36.

Bouwstra, J.A., and Ponec, M. (2006). The skin barrier in healthy and diseased state. *Bba-Biomembranes 1758*, 2080–2095.

Brune, K.A., Gotz, F., and Winkelmann, G. (1992). Degradation of lipids by bacterial lipases. *Microbial Degradation of Natural Products*, 142, pp. 243–266.

Buszello, K., and Müller, B.W. (2000). Emulsions as drug delivery systems. *Pharmaceutical Emulsions and Suspensions*, 191–228.

Candi, E., Schmidt, R., and Melino, G. (2005). The cornified envelope: A model of cell death in the skin. *Nature Reviews Molecular Cell Biology 6*, 328–340.

Carneiro, R., Salgado, A., Raposo, S., Marto, J., Simoes, S., Urbano, M., and Ribeiro, H.M. (2011). Topical emulsions containing ceramides: Effects on the skin barrier function and anti-inflammatory properties. *European Journal of Lipid Science and Technology 113*, 961–966.

Čeh, B., Winterhalter, M., Frederik, P.M., Vallner, J.J., and Lasic, D.D. (1997). Stealth® liposomes: From theory to product. *Advanced Drug Delivery Reviews 24*, 165–177.

Cevc, G. (1996). Transfersomes, liposomes and other lipid suspensions on the skin: Permeation enhancement, vesicle penetration, and transdermal drug delivery. *Critical Reviews in Therapeutic Drug Carrier Systems 13*, 257–388.

Cevc, G. (2004). Lipid vesicles and other colloids as drug carriers on the skin. *Advanced Drug Delivery Reviews 56*, 675–711.

Cevc, G., and Blume, G. (1992). Lipid vesicles penetrate into intact skin owing to the transdermal osmotic gradients and hydration force. *Biochimica et Biophysica Acta 1104*, 226–232.

Cevc, G., Blume, G., Schatzlein, A., Gebauer, D., and Paul, A. (1996). The skin: A pathway for systemic treatment with patches and lipid-based agent carriers. *Advanced Drug Delivery Reviews 18*, 349–378.

Cevc, G., and Gebauer, D. (2003). Hydration-driven transport of deformable lipid vesicles through fine pores and the skin barrier. *Biophysical Journal 84*, 1010–1024.

Chan, A., Holleran, W.M., Ferguson, T., Crumrine, D., Goker-Alpan, O., Schiffmann, R., Tayebi, N., Ginns, E.I., Elias, P.M., and Sidransky, E. (2011). Skin ultrastructural findings in type 2 Gaucher disease: Diagnostic implications. *Molecular Genetics and Metabolism 104*, 631–636.

Charcosset, C., El-Harati, A., and Fessi, H. (2005). Preparation of solid lipid nanoparticles using a membrane contactor. *Journal of Controlled Release 108*, 112–120.

Choi, E.H., Man, M.Q., Xu, P., Xin, S.J., Liu, Z.L., Crumrine, D.A., Jiang, Y.J., Fluhr, J.W., Feingold, K.R., Elias, P.M., et al. (2007). Stratum corneum acidification is impaired in moderately aged human and murine skin. *Journal of Investigative Dermatology 127*, 2847–2856.

Choi, M.J., and Maibach, H.I. (2005). Role of ceramides in barrier function of healthy and diseased skin. *American Journal of Clinical Dermatology 6*, 215–223.

Christophers, E., and Kligman, A.M. (1964). Visualization of the cell layers of the stratum corneum. *Journal of Investigative Dermatology 42*, 407–409.

Clarys, P., and Barel, A. (1995). Quantitative evaluation of skin surface lipids. *Clinics in Dermatology 13*, 307–321.

Coderch, L., Lopez, O., de la Maza, A., and Parra, J.L. (2003). Ceramides and skin function. *American Journal of Clinical Dermatology 4*, 107–129.

Constantinides, P.P. (1995). Lipid microemulsions for improving drug dissolution and oral absorption - physical and biopharmaceutical aspects. *Pharmaceutical Research 12*, 1561–1572.

Constantinides, P.P., and Scalart, J.P. (1997). Formulation and physical characterization of water-in-oil microemulsions containing long- versus medium-chain glycerides. *International Journal of Pharmaceutics 158*, 57–68.

Costes, B., Raj, V.S., Michel, B., Fournier, G., Thirion, M., Gillet, L., Mast, J., Lieffrig, F., Bremont, M., and Vanderplasschen, A. (2009). The major portal of entry of koi herpesvirus in cyprinus carpio is the skin. *Journal of Virology 83*, 2819–2830.

Cross, S.E., Magnusson, B.M., Winckle, G., Anissimov, Y., and Roberts, M.S. (2003). Determination of the effect of lipophilicity on the *in vitro* permeability and tissue reservoir characteristics of topically applied solutes in human skin layers. *Journal of Investigative Dermatology 120*, 759–764.

Darlenski, R., Kazandjieva, J., and Tsankov, N. (2011). Skin barrier function: Morphological basis and regulatory mechanisms. *Journal of Clinical Medicine 4*, 36–45.

Darmstadt, G.L., Mao-Qiang, M., Chi, E., Saha, S.K., Ziboh, V.A., Black, R.E., Santosham, M., and Elias, P.M. (2002). Impact of topical oils on the skin barrier: possible implications for neonatal health in developing countries. *Acta Paediatrica 91*, 546–554.

De Gennes, P.G., and Taupin, C. (1982). Microemulsions and the flexibility of oil/water interfaces. *The Journal of Physical Chemistry 86*, 2294–2304.

Delmas, T., Atrux-Tallau, N., Goutayer, M., Han, S.H., Kim, J.W., and Bibette, J. (2012). Nanoemulsions: Preparation, stability and application in biosciences. *Nanomaterials in Drug Delivery, Imaging, and Tissue Engineering*, 1–58. doi : 10.1002/9781118644591.ch1

Dingler, A., Hildebrand, G., Niehus, H., and Müller, R. (1998). Cosmetic anti-aging formulation based on vitamin E-loaded solid lipid nanoparticles. Paper Presented at: *Proceedings of International Symposium on Control Release Bioact Mater.*

Duque, M.I., Yosipovitch, G., Freedman, B.I., and Elias, P.M. (2005). Skin barrier function in ESRF patients and relationship to itch. *Journal of Investigative Dermatology 124*, A69–A69.

Egbaria, K., and Weiner, N. (1990). Liposomes as a topical drug delivery system. *Advanced Drug Delivery Reviews 5*, 287–300.

El Maghraby, G.M., Williams, A.C., and Barry, B.W. (2001a). Skin delivery of 5-fluorouracil from ultradeformable and standard liposomes *in-vitro*. *Journal of Pharmacy and Pharmacology 53*, 1069–1077.

El Maghraby, G.M., Williams, A.C., and Barry, B.W. (2001b). Skin hydration and possible shunt route penetration in controlled estradiol delivery from ultradeformable and standard liposomes. *Journal of Pharmacy and Pharmacology 53*, 1311–1322.

El Maghraby, G.M., Williams, A.C., and Barry, B.W. (2006). Can drug-bearing liposomes penetrate intact skin? *Journal of Pharmacy and Pharmacology 58*, 415–429.

El Maghraby, G.M.M., Campbell, M., and Finnin, B.C. (2005). Mechanisms of action of novel skin penetration enhancers: Phospholipid versus skin lipid liposomes. *International Journal of Pharmaceutics 305*, 90–104.

Elias, P.M. (2007). The skin barrier as an innate immune element. *Seminars in Immunopathology 29*, 3–14.

Elias, P.M., and Menon, G.K. (1991). Structural and lipid biochemical correlates of the epidermal permeability barrier. *Advances in Lipid Research 24*, 1–26.

Elias, P.M., and Schmuth, M. (2009). Abnormal skin barrier in the etiopathogenesis of atopic dermatitis. *Current Opinion in Allergy and Clinical Immunology 9*, 437–446.

Elias, P.M., and Wakefield, J.S. (2010). Skin barrier function. *Nutrition for healthy skin*, 35–48.

Elsayed, M.M.A., Abdallah, O.Y., Naggar, V.F., and Khalafallah, N.M. (2007). Lipid vesicles for skin delivery of drugs: Reviewing three decades of research. *International Journal of Pharmaceutics 332*, 1–16.

Eslamian, M., and Shekarriz, M. (2009). Recent advances in nanoparticle preparation by spray and micro-emulsion methods. *Recent Patents on Nanotechnology 3*, 99–115.

Fabiilli, M.L., Piert, M.R., Koeppe, R.A., Sherman, P.S., Quesada, C.A., and Kripfgans, O.D. (2013). Assessment of the biodistribution of an [(18) F]FDG-loaded perfluorocarbon double emulsion using dynamic micro-PET in rats. *Contrast Media & Molecular Imaging 8*, 366–374.

Feingold, K.R. (2009). The outer frontier: The importance of lipid metabolism in the skin. *Journal of Lipid Research 50*, S417–S422.

Findley, K., and Grice, E.A. (2014). The skin microbiome: A focus on pathogens and their association with skin disease. *PLoS Pathogens 10*, e1004436.

Flynn, T.C., Petros, J., Clark, R.E., and Viehman, G.E. (2001). Dry skin and moisturizers. *Clinics in Dermatology 19*, 387–392.

Foldvari, M., Badea, I., Wettig, S., Baboolal, D., Kumar, P., Creagh, A.L., and Haynes, C.A. (2010). Topical delivery of interferon alpha by biphasic vesicles: Evidence for a novel nanopathway across the stratum corneum. *Molecular Pharmaceutics* 7, 751–762.

Gallardo, V., Ruiz, M., and Delgado, A. (2000). Pharmaceutical suspensions and their applications. *Pharmaceutical Emulsions and Suspensions* 105, 409–464.

Gasco, M. (1997). Solid lipid nanospheres from warm micro-emulsions: Improvements in SLN production for more efficient drug delivery. *Pharmaceutical Technology Europe* 9, 52–58.

Gercek, A., Yildirim, O., Konya, D., Bozkurt, S., Ozgen, S., Kilic, T., Sav, A., and Pamir, N. (2007). Effects of parenteral fish-oil emulsion (Omegaven) on cutaneous wound healing in rats treated with dexamethasone. *JPEN Journal of Parenteral and Enteral Nutrition* 31, 161–166.

Goodman, M., and Barry, B. (1986). Action of skin permeation enhancers azone, oleic acid and decylmethyl sulphoxide: permeation and DSC studies. *Journal of Pharmacy and Pharmacology* 38, 71P–71P.

Gou, M.L., Gong, C.Y., Zhang, J., Wang, X.H., Wang, X.H., Gu, Y.C., Guo, G., Chen, L.J., Luo, F., Zhao, X., et al. (2010). Polymeric matrix for drug delivery: Honokiol-loaded PCL-PEG-PCL nanoparticles in PEG-PCL-PEG thermosensitive hydrogel. *Journal of Biomedical Materials Research Part A* 93A, 219–226.

Gou, M.L., Li, X.Y., Dai, M., Gong, C.Y., Wang, X.H., Xie, Y., Deng, H.X., Chen, L.J., Zhao, X., Qian, Z.Y., et al. (2008). A novel injectable local hydrophobic drug delivery system: Biodegradable nanoparticles in thermo-sensitive hydrogel. *International Journal of Pharmaceutics* 359, 228–233.

Greenberg, E.R., Baron, J.A., Stevens, M.M., Stukel, T.A., Mandel, J.S., Spencer, S.K., Elias, P.M., Lowe, N., Nierenberg, D.N., Bayrd, G., et al. (1989). The skin-cancer prevention study - design of a clinical-trial of beta-carotene among persons at high-risk for nonmelanoma skin-cancer. *Controlled Clinical Trials* 10, 153–166.

Grice, E.A. (2014). The skin microbiome: potential for novel diagnostic and therapeutic approaches to cutaneous disease. *Seminars in Cutaneous Medicine and Surgery* 33, 98–103.

Grice, E.A., and Segre, J.A. (2011). The skin microbiome. *Nature Reviews Microbiology* 9, 244–253.

Guo, A., and Ehrlich, A. (2014). Dry skin: There's More to consider than moisturizers. *Journal of the American Geriatrics Society* 62, S21–S22.

Gupta, P., Vermani, K., and Garg, S. (2002). Hydrogels: From controlled release to pH-responsive drug delivery. *Drug Discovery Today* 7, 569–579.

Gupta, S., Sanyal, S.K., Datta, S., and Moulik, S.P. (2006). Preparation of prospective plant oil derived micro-emulsion vehicles for drug delivery. *Indian Journal of Biochemistry & Biophysics* 43, 254–257.

Gursoy, R.N., and Benita, S. (2004). Self-emulsifying drug delivery systems (SEDDS) for improved oral delivery of lipophilic drugs. *Biomedicine & Pharmacotherapy = Biomedecine & Pharmacotherapie* 58, 173–182.

Haake, A., Scott, G.A., and Holbrook, K.A. (2001). Structure and function of the skin: overview of the epidermis and dermis. *The Biology of the Skin* 2001, 19–45.

Hafeez, A., Jain, U., Singh, J., Maurya, A., and Rana, L. (2013). Recent advances in transdermal drug delivery system (TDDS): An overview. *Journal of Scientific and Innovative Research* 2, 695–709.

Hamidi, M., Azadi, A., and Rafiei, P. (2008). Hydrogel nanoparticles in drug delivery. *Advanced Drug Delivery Reviews* 60, 1638–1649.

Hamidi, M., Rafiei, P., Azadi, A., and Mohammadi-Samani, S. (2011). Encapsulation of valproate-loaded hydrogel nanoparticles in intact human erythrocytes: A novel nanocell composite for drug delivery. *Journal of Pharmaceutical Sciences* 100, 1702–1711.

Hendriks, F.M., Brokken, D., Oomens, C.W.J., Bader, D.L., and Baaijens, F.P.T. (2006). The relative contributions of different skin layers to the mechanical behavior of human skin in vivo using suction experiments. *Medical Engineering & Physics* 28, 259–266.

Heurtault, B., Saulnier, P., Pech, B., Proust, J.-E., and Benoit, J.-P. (2002). A novel phase inversion-based process for the preparation of lipid nanocarriers. *Pharmaceutical Research* 19, 875–880.

Himes, R., Lee, S., McMenigall, K., and Russell-Jones, G.J. (2010). Reduction in inflammation in the footpad of carrageenan treated mice following the topical administration of anti-TNF molecules formulated in a micro-emulsion. *Journal of Controlled Release : Official Journal of the Controlled Release Society* 145, 210–213.

Hippalgaonkar, K., Majumdar, S., and Kansara, V. (2010). Injectable lipid emulsions—advancements, opportunities and challenges. *AAPS PharmSciTech* 11, 1526–1540.

Hiruma, T., Yabe, H., Sato, Y., Sutoh, T., and Kaneko, S. (2002). Differential effects of the hiba odor on CNV and MMN. *Biological Psychology* 61, 321–331.

Hoffman, R.M. (1998). Topical liposome targeting of dyes, melanins, genes, and proteins selectively to hair follicles. *Journal of Drug Targeting* 5, 67–74.

Honeywell-Nguyen, P.L., Groenink, H.W.W., and Bouwstra, J.A. (2006). Elastic vesicles as a tool for dermal and transdermal delivery. *Journal of Liposome Research* 16, 273–280.

Honeywell-Nguyen, P.L., Groenink, H.W.W., de Graaff, A.M., and Bouwstra, J.A. (2003). The in vivo transport of elastic vesicles into human skin: Effects of occlusion, volume and duration of application. *Journal of Controlled Release* 90, 243–255.

Hu, F., Yuan, H., Zhang, H., and Fang, M. (2002). Preparation of solid lipid nanoparticles with clobetasol propionate by a novel solvent diffusion method in aqueous system and physicochemical characterization. *International Journal of Pharmaceutics* 239, 121–128.

Imokawa, G., Abe, A., Jin, K., Higaki, Y., Kawashima, M., and Hidano, A. (1991). Decreased level of ceramides in stratum corneum of atopic dermatitis: An etiologic factor in atopic dry skin. *Journal of Investigative Dermatology* 96: 523–526.

Imokawa, G., Akasaki, S., Minematsu, Y., and Kawai, M. (1989). Importance of intercellular lipids in water-retention properties of the stratum-corneum - induction and recovery study of surfactant dry skin. *Archives of Dermatological Research* 281, 45–51.

Jenning, V., Gysler, A., Schafer-Korting, M., and Gohla, S.H. (2000). Vitamin A loaded solid lipid nanoparticles for topical use: Occlusive properties and drug targeting to the upper skin. *European Journal of Pharmaceutics and Biopharmaceutics* 49, 211–218.

Jeong, C.W., Ju, J., Lee, D.W., Lee, S.H., and Yoon, M.H. (2012). Lipid-emulsion propofol less attenuates the regulation of body temperature than micro-emulsion propofol or sevoflurane in the elderly. *Yonsei Medical Journal* 53, 198–203.

Jungersted, J.M., Hellgren, L.I., Hogh, J.K., Drachmann, T., Jemec, G.B.E., and Agner, T. (2010a). Ceramides and barrier function in healthy skin. *Acta dermato-venereologica* 90, 350–353.

Jungersted, J.M., Scheer, H., Mempel, M., Baurecht, H., Cifuentes, L., Hogh, J.K., Hellgren, L.I., Jemec, G.B.E., Agner, T., and Weidinger, S. (2010b). Stratum corneum lipids, skin barrier function and filaggrin mutations in patients with atopic eczema. *Allergy 65*, 911–918.

Kabalnov, A., Tarara, T., Arlauskas, R., and Weers, J. (1996). Phospholipids as emulsion stabilizers: 2. Phase behavior versus emulsion stability. *Journal of Colloid and Interface Science 184*, 227–235.

Kabalnov, A., and Wennerström, H. (1996). Macroemulsion stability: The oriented wedge theory revisited. *Langmuir: The ACS Journal of Surfaces and Colloids 12*, 76–292.

Kabalnov, A.S., and Shchukin, E.D. (1992). Ostwald ripening theory: Applications to fluorocarbon emulsion stability. *Advances in Colloid and Interface Science 38*, 69–97.

Karatekin, E., Sandre, O., Guitouni, H., Borghi, N., Puech, P.H., and Brochard-Wyart, F. (2003). Cascades of transient pores in giant vesicles: Line tension and transport. *Biophysical Journal 84*, 1734–1749.

Kawashima, Y., Niwa, T., Handa, T., Takeuchi, H., Iwamoto, T., and Itoh, Y. (1989). Preparation of prolonged-release spherical micro-matrix of ibuprofen with acrylic polymer by the emulsion-solvent diffusion method for improving bioavailability. *Chemical & Pharmaceutical Bulletin 37*, 425–429.

Kirjavainen, M., Urtti, A., Jaaskelainen, I., Suhonen, T.M., Paronen, P., ValjakkaKoskela, R., Kiesvaara, J., and Monkkonen, J. (1996). Interaction of liposomes with human skin in vitro - The influence of lipid composition and structure. *Biochimica et Biophysica Acta 1304*, 179–189.

Kirjavainen, M., Urtti, A., Valjakka-Koskela, R., Kiesvaara, J., and Monkkonen, J. (1999). Liposome-skin interactions and their effects on the skin permeation of drugs. *European Journal of Pharmaceutical Sciences 7*, 279–286.

Klibanov, A.L., Maruyama, K., Torchilin, V.P., and Huang, L. (1990). Amphipathic polyethyleneglycols effectively prolong the circulation time of liposomes. *FEBS Letters 268*, 235–237.

Kogan, A., and Garti, N. (2006). Microemulsions as transdermal drug delivery vehicles. *Advances in Colloid and Interface Science 123*, 369–385.

Kohli, A.K., and Alpar, H.O. (2004). Potential use of nanoparticles for transcutaneous vaccine delivery: effect of particle size and charge. *International Journal of Pharmaceutics 275*, 13–17.

Kong, H.H., Grice, E.A., Conlan, S.P., Deming, C., Young, A.C., Bouffard, G.G., Blakesley, R.W., Green, E.D., Murray, P.R., Mijares, L., et al. (2009). Diversity profile of the human skin microbiome. *Journal of Investigative Dermatology 129*, S43–S43.

Kozlov, M.M., and Helfrich, W. (1992). Effects of a cosurfactant on the stretching and bending elasticities of a surfactant monolayer. *Langmuir: The ACS Journal of Surfaces and Colloids 8*, 2792–2797.

Kraft, J., and Lynde, C. (2005). Moisturizers: What they are and a practical approach to product selection. *Skin Therapy Letter 10*, 1–8.

Kwon, G.S., and Okano, T. (1996). Polymeric micelles as new drug carriers. *Advanced Drug Delivery Reviews 21*, 107–116.

Lademann, J., Richter, H., Teichmann, A., Otberg, N., Blume-Peytavi, U., Luengo, J., Weiß, B., Schaefer, U.F., Lehr, C.-M., and Wepf, R. (2007). Nanoparticles–an efficient carrier for drug delivery into the hair follicles. *European Journal of Pharmaceutics and Biopharmaceutics 66*, 159–164.

Lampe, M.A., Burlingame, A., Whitney, J., Williams, M.L., Brown, B.E., Roitman, E., and Elias, P.M. (1983). Human stratum corneum lipids: characterization and regional variations. *Journal of Lipid Research 24*, 120–130.

Langevin, D. (1988). Microemulsions. *Accounts of Chemical Research 21*, 255–260.

Lasch, J., Zellmer, S., Pfeil, W., and Schubert, R. (1995). Interaction of liposomes with the human skin lipid barrier: hSCLLs as model system - DSC of intact stratum corneum and in situ CLSM of human skin. *Journal of Liposome Research 5*, 99–108.

Lasic, D.D. (1996). Doxorubicin in sterically stabilized liposomes. *Nature 380*, 561–562.

Lasic, D.D. (1998). Novel applications of liposomes. *Trends in Biotechnology 16*, 307–321.

Lawrence, M.J., and Rees, G.D. (2000). Microemulsion-based media as novel drug delivery systems. *Advanced Drug Delivery Reviews 45*, 89–121.

Lawrence, M.J., and Rees, G.D. (2012). Microemulsion-based media as novel drug delivery systems. *Advanced Drug Delivery Reviews 64*, 175–193.

Leal-Calderon, F., Schmitt, V., and Bibette, J. (2007). *Emulsion Science: Basic Principles* (Springer). ISBN 978-0-387-39683-5

Lee, T.R., and Kim, J.H. (2006). COLL 102-Biocompatible hydrogel-coated gold nanoparticles for in vivo drug delivery. *Abstracts of Papers of the American Chemical Society 232*.

Li, W.M., Sandhoff, R., Kono, M., Zerfas, P., Hoffmann, V., Ding, B.C.H., Proia, R.L., and Deng, C.X. (2007). Depletion of ceramides with very long chain fatty acids causes defective skin permeability barrier function, and neonatal lethality in ELOVL4 deficient mice. *International Journal of Biological Sciences 3*, 120–128.

Li, X.Y., Zheng, X.L., Wei, X.W., Guo, G., Gou, M.L., Gong, C.Y., Wang, X.H., Dai, M., Chen, L.J., Wei, Y.Q., et al. (2009). A novel composite drug delivery system: Honokiol nanoparticles in thermosensitive hydrogel based on Chitosan. *Journal of Nanoscience and Nanotechnology 9*, 4586–4592.

Liedtke, S., Wissing, S., Müller, R., and Mäder, K. (2000). Influence of high pressure homogenisation equipment on nanodispersions characteristics. *International Journal of Pharmaceutics 196*, 183–185.

Liu, J., Hu, W., Chen, H., Ni, Q., Xu, H., and Yang, X. (2007). Isotretinoin-loaded solid lipid nanoparticles with skin targeting for topical delivery. *International Journal of Pharmaceutics 328*, 191–195.

Loden, M. (2003a). Role of topical emollients and moisturizers in the treatment of dry skin barrier disorders. *American Journal of Clinical Dermatology 4*, 771–788.

Loden, M. (2003b). The skin barrier and use of moisturizers in atopic dermatitis. *Clinics in Dermatology 21*, 145–157.

Lostritto, R.T., and Silvestri, S.L. (1987). Temperature and cosurfactant effects on lidocaine release from submicron oil in water emulsions. *Journal of Parenteral Science and Technology: A Publication of the Parenteral Drug Association 41*, 220–224.

Lourith, N., and Kanlayavattanakul, M. (2009). Natural surfactants used in cosmetics: Glycolipids. *International Journal of Cosmetic Science 31*, 255–261.

Maghraby, G.M., Barry, B.W., and Williams, A.C. (2008). Liposomes and skin: From drug delivery to model membranes. *European Journal of Pharmaceutical Sciences 34*, 203–222.

Mahapatra, A.K., and Murthy, P.N. (2014). Self-emulsifying drug delivery systems (SEDDS): An update from formulation development to therapeutic strategies. *Self 6*, 546–568.

Man, M.Q., Feingold, K.R., and Elias, P.M. (1993). Exogenous lipids influence permeability barrier recovery in acetone-treated murine skin. *Archives of Dermatology 129*, 728–738.

Man, M.Q., Feingold, K.R., Jain, M., and Elias, P.M. (1995). Extracellular processing of phospholipids is required for permeability barrier homeostasis in murine skin. *Journal of Investigative Dermatology 104*, 562–562.

MaoQiang, M., Feingold, K.R., Wang, F.S., Thornfeldt, C.R., and Elias, P.M. (1996). A natural lipid mixture improves barrier function and hydration in human and murine skin. *Journal of Cosmetic Science 47*, 157–166.

Marti-Mestres, G., and Nielloud, F. (2000). Main surfactants used in the pharmaceutical field. Pharmaceutical Emulsions and Suspensions. *Pharmaceutical Emulsions and Suspensions 105*, 1–18.

Mason, T.G., Wilking, J.N., Meleson, K., Chang, C.B., and Graves, S.M. (2006). Nanoemulsions: Formation, structure, and physical properties. *Journal of Physics: Condensed Matter 18*, R635–R666.

Maziere, J.C. (1997). Layers of the epidermis and metabolism of lipids in relation to the hydric barrier function of the skin. *OCL: Oléagineux Corps Gras Lipides 4*, 258–265.

McCarron, P.A., Donnelly, R.F., and Al-Kassas, R. (2008). Comparison of a novel spray congealing procedure with emulsion-based methods for the micro-encapsulation of water-soluble drugs in low melting point triglycerides. *Journal of Microencapsulation 25*, 365–378.

McClements, D.J. (2011). Edible nanoemulsions: Fabrication, properties, and functional performance. *The Royal Society of Chemistry 7*, 2297–2316.

McClements, D.J. (2012a). Crystals and crystallization in oil-in-water emulsions: Implications for emulsion-based delivery systems. *Advances in Colloid and Interface Science 174*, 1–30.

McClements, D.J. (2012b). Nanoemulsions versus microemulsions: terminology, differences, and similarities. *The Royal Society of Chemistry 8*, 1719–1729.

Meckfessel, M.H., and Brandt, S. (2014). The structure, function, and importance of ceramides in skin and their use as therapeutic agents in skin-care products. *Journal of the American Academy of Dermatology 71*, 177–184.

Mehnert, W., and Mäder, K. (2001). Solid lipid nanoparticles: Production, characterization and applications. *Advanced Drug Delivery Reviews 47*, 165–196.

Melero, A., Ourique, A.F., Guterres, S.S., Pohlrnann, A.R., Lehr, C.M., Beck, R.C.R., and Schaefer, U. (2014). Nanoencapsulation in lipid-core nanocapsules controls mometasone furoate skin permeability rate and its penetration to the deeper skin layers. *Skin Pharmacology and Physiology 27*, 217–228.

Mezei, M., and Gulasekharam, V. (1980). Liposomes-a selective drug delivery system for the topical route of administration I. Lotion Dosage Form. *Life Sciences 26*, 1473–1477.

Miyazaki, M., Dobrzyn, A., Elias, P.M., and Ntambi, J.M. (2005). Stearoyl-CoA desaturase-2 gene expression is required for lipid synthesis during early skin and liver development. *Proceedings of the National Academy of Sciences of the United States of America 102*, 12501–12506.

Mojumdar, E.H., Helder, R.W.J., Gooris, G.S., and Bouwstra, J.A. (2014). Monounsaturated fatty acids reduce the barrier of stratum corneum lipid membranes by enhancing the formation of a hexagonal lateral packing. *Langmuir: The ACS Journal of Surfaces and Colloids 30*, 6534–6543.

Morishige, I., Toorisaka, E., Hirata, M., Ohtake, T., and Hano, T. (2005). Preparation of various metal-silicate micro-balloons using W/O/W emulsion. *Journal of Microencapsulation 22*, 291–301.

Mouritsen, O.G. (2005). Powerful and strange lipids at work. *Front Collect*, 197–207. ISBN 978-3-319-22613-2 I, DOI: 10.1007/978-3-319-22614-9

Mouritsen, O.G. (2011). Lipids, curvature, and nano-medicine. *European Journal of Lipid Science and Technology 113*, 1174–1187.

Müller, R.H., MaÈder, K., and Gohla, S. (2000). Solid lipid nanoparticles (SLN) for controlled drug delivery–a review of the state of the art. *European Journal of Pharmaceutics and Biopharmaceutics 50*, 161–177.

Müller, R.H., Petersen, R.D., Hornmoss, A., and Pardeike, J. (2007). Nanostructured lipid carriers (NLC) in cosmetic dermal products. *Advanced Drug Delivery Reviews 59*, 522–530.

Müller, R.H., Radtke, M., and Wissing, S.A. (2002). Solid lipid nanoparticles (SLN) and nanostructured lipid carriers (NLC) in cosmetic and dermatological preparations. *Advanced Drug Delivery Reviews 54*, S131–S155.

Naik, A., Kalia, Y.N., and Guy, R.H. (2000). Transdermal drug delivery: Overcoming the skin's barrier function. *Pharmaceutical Science & Technology Today 3*, 318–326.

Nielloud, F. (2000). *Pharmaceutical Emulsions and Suspensions: Revised and Expanded* (CRC Press: Boca Raton, FL).

Osborne, D.W., Ward, A.J.I., and Oneill, K.J. (1991). Microemulsions as topical drug delivery vehicles – in vitro transdermal studies of a model hydrophilic drug. *Journal of Pharmacy and Pharmacology 43*, 451–454.

Otto, A., Du Plessis, J., and Wiechers, J. (2009). Formulation effects of topical emulsions on transdermal and dermal delivery. *International Journal of Cosmetic Science 31*, 1–19.

Panwar, A., Upadhyay, N., Bairagi, M., Gujar, S., Darwhekar, G., Jain, D., Yadav, A., Sharma, A., Jain, D.K., and Singh, S. (2011). Emulgel: A review. *Asian Journal of Pharmacy and Life Science* ISSN *2231*, 4423.

Pappas, A. (2009). Epidermal surface lipids. *Dermato-endocrinology 1*(2), 72–76.

Pardeike, J., Hommoss, A., and Muller, R.H. (2009). Lipid nanoparticles (SLN, NLC) in cosmetic and pharmaceutical dermal products. *International Journal of Pharmaceutics 366*, 170–184.

Patravale, V.B., and Ambarkhane, A.V. (2003). Study of solid lipid nanoparticles with respect to particle size distribution and drug loading. *Die Pharmazie 58*, 392–395.

Paul, A., Cevc, G., and Bachhawat, B.K. (1995). Transdermal immunization with large proteins by and means of ultradeformable drug carriers. *European Journal of Immunology 25*, 3521–3524.

Picardo, M., Zompetta, C., Deluca, C., Cirone, M., Faggioni, A., Nazzaroporro, M., Passi, S., and Prota, G. (1991). Role of skin surface-lipids in UV-induced epidermal-cell changes. *Archives of Dermatological Research 283*, 191–197.

Pierre, M.B.R., and Costa, I.d.S.M. (2011). Liposomal systems as drug delivery vehicles for dermal and transdermal applications. *Archives of Dermatological Research 303*, 607–621.

Pietkiewicz, J., and Sznitowska, M. (2004). The choice of lipids and surfactants for injectable extravenous microspheres. *Die Pharmazie-An International Journal of Pharmaceutical Sciences 59*, 325–326.

Potts, R.O., and Guy, R.H. (1992). Predicting skin permeability. *Pharmaceutical Research 9*, 663–669.

Prajapati, H.N., Dalrymple, D.M., and Serajuddin, A.T.M. (2012). A comparative evaluation of mono-, di- and triglyceride of medium chain fatty acids by lipid/surfactant/water phase diagram, solubility determination and dispersion testing for application in pharmaceutical dosage form development. *Pharmaceutical Research 29*, 285–305.

Prausnitz, M.R., and Langer, R. (2008). Transdermal drug delivery. *Nature Biotechnology 26*, 1261–1268.

Prausnitz, M.R., Mitragotri, S., and Langer, R. (2004). Current status and future potential of transdermal drug delivery. *Nature Reviews Drug Discovery 3*, 115–124.

Priano, L., Esposti, D., Esposti, R., Castagna, G., De Medici, C., Fraschini, F., Gasco, M.R., and Mauro, A. (2007). Solid lipid nanoparticles incorporating melatonin as new model for sustained oral and transdermal delivery systems. *Journal of Nanoscience and Nanotechnology 7*, 3596–3601.

Proksch, E., Brandner, J.M., and Jensen, J.M. (2008). The skin: An indispensable barrier. *Experimental Dermatology 17*, 1063–1072.

Proksch, E., Folster-Holst, R., and Jensen, J.M. (2006). Skin barrier function, epidermal proliferation and differentiation in eczema. *Journal of Dermatological Science 43*, 159–169.

Proksch, E., Jensen, J.M., and Elias, P.M. (2003). Skin lipids and epidermal differentiation in atopic dermatitis. *Clinics in Dermatology 21*, 134–144.

Proksch, E., Jensen, J.M., Elias, P., and Folster-Holst, R. (2007). Substantial improvement of the skin barrier in atopic dermatitis after treatment with pimecrofimus but not with betamethasone. *Journal of the American Academy of Dermatology 56*, Ab3-Ab3.

Puglia, C., Blasi, P., Rizza, L., Schoubben, A., Bonina, F., Rossi, C., and Ricci, M. (2008). Lipid nanoparticles for prolonged topical delivery: An *in vitro* and *in vivo* investigation. *International Journal of Pharmaceutics 357*, 295–304.

Puglia, C., and Bonina, F. (2012). Lipid nanoparticles as novel delivery systems for cosmetics and dermal pharmaceuticals. *Expert Opinion on Drug Delivery 9*, 429–441.

Queille-Roussel, C., Paul, C., Duteil, L., Lefebvre, M.C., Rapatz, G., Zagula, M., and Ortonne, J.P. (2001). The new topical ascomycin derivative SDZ ASM 981 does not induce skin atrophy when applied to normal skin for 4 weeks: A randomized, double-blind controlled study. *British Journal of Dermatology 144*, 507–513.

Rawlings, A., Canestrari, D.A., and Dobkowski, B. (2004). Moisturizer technology versus clinical performance. *Dermatologic Therapy 17*, 49–56.

Rawlings, A., and Harding, C. (2004). Moisturization and skin barrier function. *Dermatologic Therapy 17*, 43–48.

Rodriguez, G., Barbosa-Barros, L., Rubio, L., Cocera, M., Diez, A., Estelrich, J., Pons, R., Caelles, J., De la Maza, A., and Lopez, O. (2009). Conformational changes in stratum corneum lipids by effect of bicellar systems. *Langmuir: The ACS Journal of Surfaces and Colloids 25*, 10595–10603.

Rodriguez, G., Rubio, L., Cocera, M., Estelrich, J., Pons, R., de la Maza, A., and Lopez, O. (2010). Application of bicellar systems on skin: Diffusion and molecular organization effects. *Langmuir: The ACS Journal of Surfaces and Colloids 26*, 10578–10584.

Roger, K., Olsson, U., Schweins, R., and Cabane, B. (2015). Emulsion ripening through molecular exchange at droplet contacts. *Angewandte Chemie International Edition 54*, 1452–1455.

Sadurni, N., Solans, C., Azemar, N., and Garcia-Celma, M.J. (2005). Studies on the formation of O/W nano-emulsions, by low-energy emulsification methods, suitable for pharmaceutical applications. *European Journal of Pharmaceutical Sciences : Official Journal of the European Federation for Pharmaceutical Sciences 26*, 438–445.

Sainsbury, F., Zeng, B., and Middelberg, A.P. (2014). Towards designer nanoemulsions for precision delivery of therapeutics. *Current Opinion in Chemical Engineering 4*, 11–17.

Schaefer, H., and Hensby, C. (1990). Skin permeability and models of percutaneous-absorption. *Nato Advanced Science Institutes Series, Series A, Life Sciences 181*, 77–83.

Scheuplein, R.J., and Blank, I.H. (1971). Permeability of the skin. *Physiological Reviews 51*, 702–747.

Schiffelers, R.M., Ansari, A., Xu, J., Zhou, Q., Tang, Q.Q., Storm, G., Molema, G., Lu, P.Y., Scaria, P.V., and Woodle, M.C. (2004). Cancer siRNA therapy by tumor selective delivery with ligand-targeted sterically stabilized nanoparticle. *Nucleic Acids Research 32*.

Schneider, M., Stracke, F., Hansen, S., and Schaefer, U.F. (2009). Nanoparticles and their interactions with the dermal barrier. *Dermato-Endocrinology 1*, 197–206.

Schreier, H., and Bouwstra, J. (1994). Liposomes and niosomes as topical drug carriers - dermal and transdermal drug-delivery. *Journal of Controlled Release 30*, 1–15.

Schubert, M., and Müller-Goymann, C. (2003). Solvent injection as a new approach for manufacturing lipid nanoparticles–evaluation of the method and process parameters. *European Journal of Pharmaceutics and Biopharmaceutics 55*, 125–131.

Schurer, N.Y., and Elias, P.M. (1991). The biochemistry and function of stratum-corneum lipids. *Advances in Lipid Research 24*, 27–56.

Settel, E. (1966). Further observations on the cerumenolytic efficacy of a new micro-emulsion: a review of 287 instances. *Current Therapeutic Research, Clinical and Experimental 8*, 528–530.

Shlnoda, K., and Kunleda, H. (1983). Phase properties of emulsions: PIT and HLB. *Encyclopedia of Emulsion Technology: Basic Theory 1*, 337.

Shinoda, K., and Saito, H. (1968). The effect of temperature on the phase equilibria and the types of dispersions of the ternary system composed of water, cyclohexane, and nonionic surfactant. *Journal of Colloid and Interface Science 26*, 70–74.

Shinoda, K., and Saito, H. (1969). The stability of O/W type emulsions as functions of temperature and the HLB of emulsifiers: The emulsification by PIT-method. *Journal of Colloid and Interface Science 30*, 258–263.

Simion, F.A., Abrutyn, E.S., and Draelos, Z.D. (2005). Ability of moisturizers to reduce dry skin and irritation and to prevent their return. *Journal of Cosmetic Science 56*, 427–444.

Sinico, C., and Fadda, A.M. (2009). Vesicular carriers for dermal drug delivery. *Expert Opinion on Drug Delivery 6*, 813–825.

Sjöström, B., and Bergenståhl, B. (1992). Preparation of submicron drug particles in lecithin-stabilized o/w emulsions I. Model studies of the precipitation of cholesteryl acetate. *International Journal of Pharmaceutics 88*, 53–62.

Smith, E.W., Maibach, H.I., and Surber, C. (2000). Use of emulsions as topical drug delivery systems. *Pharmaceutical Emulsions and Suspensions*, 259–270.

Sparr, E., Millecamps, D., Isoir, M., Burnier, V., Larsson, Å., and Cabane, B. (2013). Controlling the hydration of the skin though the application of occluding barrier creams. *Journal of The Royal Society Interface 10*, 20120788.

Spencer, T.S. (1988). Dry skin and skin moisturizers. *Clinics in Dermatology 6*, 24–28.

Stone, S.J., Myers, H.M., Watkins, S.M., Brown, B.E., Feingold, K.R., Elias, P.M., and Farese, R.V. (2004). Lipopenia and skin barrier abnormalities in DGAT2-deficient mice. *Journal of Biological Chemistry 279*, 11767–11776.

Subramanyan, K., and Johnson, A. (2005). Advanced moisturizers for dry sensitive skin. *Journal of the American Academy of Dermatology 52*, P66–P66.

Tabbakhian, M., Tavakoli, N., Jaafari, M.R., and Daneshamouz, S. (2006). Enhancement. of follicular delivery of finasteride by liposomes and niosomes -1. *In vitro* permeation and *in vivo* deposition studies using hamster flank and ear models. *International Journal of Pharmaceutics 323*, 1–10.

Tang, H., Mitragotri, S., Blankschtein, D., and Langer, R. (2001). Theoretical description of transdermal transport of hydrophilic permeants: Application to low-frequency sonophoresis. *Journal of Pharmaceutical Sciences 90*, 545–568.

Teeranachaideekul, V., Müller, R.H., and Junyaprasert, V.B. (2007). Encapsulation of ascorbyl palmitate in nanostructured lipid carriers (NLC)—effects of formulation parameters on physicochemical stability. *International Journal of Pharmaceutics 340*, 198–206.

Thaci, D., Brautigam, M., Kaufmann, R., Weidinger, G., Paul, C., and Christophers, E. (2002). Body-weight-independent dosing of cyclosporine micro-emulsion and three times weekly maintenance regimen in severe psoriasis. A Randomised Study. *Dermatology 205*, 383–388.

Thiam, A.R., Farese Jr, R.V., and Walther, T.C. (2013). The biophysics and cell biology of lipid droplets. *Nature Reviews Molecular Cell Biology 14*(12), 775–786.

Torchilin, V.P. (2005). Recent advances with liposomes as pharmaceutical carriers. *Nature Reviews Drug Discovery 4*, 145–160.

Torin Huzil, J., Sivaloganathan, S., Kohandel, M., and Foldvari, M. (2011). Drug delivery through the skin: molecular simulations of barrier lipids to design more effective noninvasive dermal and transdermal delivery systems for small molecules, biologics, and cosmetics. Wiley *Interdisciplinary Reviews: Nanomedicine and Nanobiotechnology 3*, 449–462.

Trotta, M., Cavalli, R., Carlotti, M.E., Battaglia, L., and Debernardi, F. (2005). Solid lipid micro-particles carrying insulin formed by solvent-in-water emulsion-diffusion technique. *International Journal of Pharmaceutics 288*, 281–288.

Trotta, M., Debernardi, F., and Caputo, O. (2003). Preparation of solid lipid nanoparticles by a solvent emulsification–diffusion technique. *International Journal of Pharmaceutics 257*, 153–160.

Uekama, K., Hirayama, F., and Irie, T. (1998). Cyclodextrin drug carrier systems. *Chemical Reviews 98*, 2045–2076.

van Buuren, A.R., Tieleman, D.P., de Vlieg, J., and Berendsen, H.J.C. (1996). Cosurfactants lower surface tension of the diglyceride/water interface: A molecular dynamics study. *Langmuir: The ACS Journal of Surfaces and Colloids 12*, 2570–2579.

van der Ven, C., Gruppen, H., de Bont, D.B., and Voragen, A.G. (2001). Emulsion properties of casein and whey protein hydrolysates and the relation with other hydrolysate characteristics. *Journal of Agricultural and Food Chemistry 49*, 5005–5012.

Varvaresou, A., and Iakovou, K. (2015). Biosurfactants in cosmetics and biopharmaceuticals. *Letters in Applied Microbiology 61*, 214–223.

Vasiljevic, D., Parojcic, J., Primorac, M., and Vuleta, G. (2006). An investigation into the characteristics and drug release properties of multiple W/O/W emulsion systems containing low concentration of lipophilic polymeric emulsifier. *International Journal of Pharmaceutics 309*, 171–177.

Vasireddy, V., Uchida, Y., Salem, N., Kim, S.Y., Mandal, M.N.A., Reddy, G.B., Bodepudi, R., Alderson, N.L., Brown, J.C., Hama, H., et al. (2007). Loss of functional ELOVL4 depletes very long-chain fatty acids ($>=$ C28) and the unique omega-O-acylceramides in skin leading to neonatal death. *Human Molecular Genetics 16*, 471–482.

Verma, D., Verma, S., Blume, G., and Fahr, A. (2003). Particle size of liposomes influences dermal delivery of substances into skin. *International Journal of Pharmaceutics 258*, 141–151.

Vieira, D.B., and Carmona-Ribeiro, A.M. (2006). Cationic lipids and surfactants as antifungal agents: Mode of action. *Journal of Antimicrobial Chemotherapy 58*, 760–767.

Wallisch, R. (1975). Cosmetic moisturizers... causes and remedies of dry skin conditions. *Soap and Chemical Specialities 51*, 32–35.

Wang, F.S., Man, M.Q., and Elias, P.M. (1997). A lipid mixture improves skin hydration in ichthyosis vulgaris. *International Journal of Dermatology 36*, 876–877.

Wang, Y.C., Thio, Y.S., and Doyle, F.M. (1998). Formation of semi-permeable polyamide skin layers on the surface of supported liquid membranes. *Journal of Membrane Science 147*, 109–116.

Weiner, E.M., Beiser, S., Giudice, R., Kanat, I.O., Kaplan, E., Kauth, B., and Stone, D. (1973). Treating the dry skin syndrome. Regional experiences with a micro-emulsion lotion. *Journal of the American Podiatry Association 63*, 571–581.

Williams, M.L., and Elias, P.M. (1980). Epidermal pathogenesis of skin fragility in patients receiving systemic retinoids. *Journal of Clinical Research 28*, A137–A137.

Williams, M.L., and Elias, P.M. (1981). Nature of skin fragility in patients receiving retinoids for systemic effect. *Archives of Dermatology 117*, 611–619.

Wissing, S.A., and Müller, R.H. (2003). Cosmetic applications for solid lipid nanoparticles (SLN). *International Journal of Pharmaceutics 254*, 65–68.

Yokomizo, Y., and Sagitani, H. (1996a). The effects of phospholipids on the percutaneous penetration of indomethacin through the dorsal skin of guinea pig *in vitro*.2. The effects of the hydrophobic group in phospholipids and a comparison with general enhancers. *Journal of Controlled Release 42*, 37–46.

Yokomizo, Y., and Sagitani, H. (1996b). Effects of phospholipids on the percutaneous penetration of indomethacin through the dorsal skin of guinea pigs *in vitro*. *Journal of Controlled Release 38*, 267–274.

Yoshida, R., Sakai, K., Okano, T., Sakurai, Y., Bae, Y.H., and Kim, S.W. (1991). Surface-modulated skin layers of thermal responsive hydrogels as on off switches.1. Drug release. *Journal of Biomaterials Science Polymer Edition 3*, 155–162.

Yosipovitch, G., Duque, M.I., Patel, T.S., Ishiuji, Y., Guzman-Sanchez, D.A., Dawn, A.G., Freedman, B.I., Chan, Y.H., Crumrine, D., and Elias, P.M. (2007). Skin barrier structure and function and their relationship to pruritus in end-stage renal disease. *Nephrology Dialysis Transplantation 22*, 3268–3272.

Zellmer, S., Pfeil, W., and Lasch, J. (1995). Interaction of phosphatidylcholine liposomes with the human stratum-corneum. *Bba-Biomembranes 1237*, 176–182.

Zettersten, E.M., Ghadially, R., Feingold, K.R., Crumrine, D., and Elias, P.M. (1997). Optimal ratios of topical stratum corneum lipids improve barrier recovery in chronologically aged skin. *Journal of the American Academy of Dermatology 37*, 403–408.

Zhai, H., and Maibach, H.I. (2002). Occlusion vs. skin barrier function. *Skin Research and Technology 8*, 1–6.

Zouboulis, C.C. (2004). Acne and sebaceous gland function. *Clinics in Dermatology 22*, 360–366.

16 Supported Lipid Bilayers

Theo Lohmüller, Bert Nickel, and Joachim O. Rädler
Ludwig-Maximilians-Universität, Munich, Germany

CONTENTS

16.1 Introduction 293
16.2 Formation of Supported Lipid Bilayer via Vesicle Fusion and Spreading 293
16.3 Lipid and Protein Mobility in Supported Lipid Bilayers 295
16.4 Adsorption of Macromolecules to Charged Supported Lipid Bilayers 296
16.5 Manipulation of Supported Lipid Bilayers by External Fields and Surface-Directed Forces 298
16.6 Polymer-Supported Lipid Bilayers 300
16.7 Supported Lipid Bilayers as Platform to Present Molecules to Living Cells 301
Acknowledgements 302
References 303

16.1 INTRODUCTION

A lipid bilayer is the core element of the plasma membrane, which separates a cell's interior from the exterior space. The plasma membrane is composed of a mixture of mostly phospholipids, sphingolipids, and cholesterol and serves as a fluid matrix for functional proteins that control transport, cell signalling, and motility. For many practical purposes in research, deposition of lipid bilayers on planar solid surfaces is a viable approach to study membrane-related processes with greater precision than possible on living cells. This is because the membrane is held in place on the substrate, yet its natural lateral fluidity is preserved. Hence, so-called (solid) supported lipid bilayers (SLBs) provide the base to reconstitute planar cell surface mimicking model systems on microscopy slides or sensitive solid-state devices. In early work in the group of McConnel, lipid monolayers were transferred from the air–water interface onto alkylated solid substrates using the Langmuir–Schäfer method to create planar lipid model interfaces for the study of immune recognition (Hafeman et al. 1981). Sequential transfer of two lipid monolayers onto planar glass or quartz surfaces results in lipid bilayers with a molecular diffusivity close to the one known for multilayered lipid systems (McConnell et al. 1986, Tamm and McConnell 1985). At the same time, it was discovered that vesicles interacting with suitable solid surfaces undergo rupture and fusion into a coherent layer with identical properties as the transferred lipid bilayer (McConnell et al. 1986, Kalb 1992). Both approaches are consistent with the idea of SLBs as a single continuous lipid bilayer on a solid support. SLBs established themselves as biomimetic interfaces (phantom cells) to study cell recognition and adhesion (Sackmann 1996). As a fluid, planar membrane system, they enabled incorporation of membrane-bound and membrane-spanning proteins. Membrane-based sensors are made of SLBs in combination with surface-sensitive devices enabling quantitative measurements of protein adsorption, ion channels currents, or vesicle–membrane interaction. Chapter 14 by Höök and coauthors in this book is devoted to sensors. In this chapter, we focus on SLBs as functional interfaces for spatiotemporal control of membrane-anchored molecules with the aim to purposefully design interaction with living cells. We describe how SLBs form by vesicle fusion and surface spreading (Section 16.2) and how fluidity and integrity of SLBs are examined (Section 16.3). We then show that macromolecules, adsorbed or anchored to SLBs, exhibit in-plane mobility and close-to-ideal two-dimensional scaling behaviour (Section 16.4). SLBs furthermore enable controlled rearrangement of molecules at surfaces by means of external electrical, mechanical, or hydrodynamic forces as shown in Section 16.5. In order to incorporate transmembrane proteins, polymer-supported membranes are introduced, which are SLBs lifted from the solid surface by a grafted polymer layer (Section 16.6). We finally report recent work using SLBs as a platform to study protein self-organization by reaction–diffusion and rationally designed self-assembly by means of DNA nanostructures (Section 16.7). We summarize how SLBs provide exciting opportunities to interrogate and address living cells by exploiting the fact that proteins can be positioned, structured, and the SLB matrix potentially made switchable.

16.2 FORMATION OF SUPPORTED LIPID BILAYER VIA VESICLE FUSION AND SPREADING

The most practical and commonly used method to create SLBs is via fusion of small unilamellar vesicles (SUVs) on hydrophilic solid surfaces. Lipid vesicles are prepared from lipid films or solvent extraction and are subsequently extensively sonicated to yield SUVs. Sonication induces breakdown of large and multilamellar vesicles into SUVs with a minimum size (as small as approximately 50 nm) that are under mechanical tension. Incubating a solid substrate with SUVs leads to an adherent layer of SUVs on the surface. In some cases, SUVs may also rupture (Kalb 1992, Nollert-Jähnig et al. 1995). It appears that phospholipids are particularly convenient to form continuous lipid bilayer (SLBs) on glass and quartz surfaces or oxidized silicon wafers under favourable buffer conditions that include few µM amounts of Ca^{2+} and ionic strength in the physiological range (~100 mM, pH 7) (Cremer and Boxer 1999). Atomic force microscopy (AFM) imaging confirmed that SLBs are formed by immediate fusion of vesicles at highly attractive surfaces or occurred successively after vesicle adsorption and collective fusion of adherent vesicles when a critical surface coverage was reached (Richter 2003). The kinetics of mass uptake at the surface can be monitored by quartz balance techniques. It is of order 100–300 seconds after incubation with vesicle solutions of 100–700 mg L^{-1} concentration (Reimhult et al. 2006). Attractive membrane substrate interaction is favourable for SLB formation. Electrostatic repulsion of, e.g., weakly negatively charged lipid membranes at glass surfaces does not prevent adhesion since counteracting van der Waals' and hydration forces contribute. However, the tendency for rupture may be reduced due to electrostatic repulsion. Rupture of vesicles is governed by the mechanical rupture strength of the lipid bilayer in the first place, but likely to be largely influenced by surface heterogeneities. Finally, spreading and complete wetting of a lipid bilayer on a solid surface is required to form an extended, defect-free SLB (Cremer and Boxer 1999). On hydrophilic surfaces, when a thin water layer separates the solid surface from the bottom lipid monolayer and frictional coupling is low, bilayer spreads as a continuous bilayer sheet with a micellar edge at the leading rim (Nissen et al. 1999). The spreading phenomenon itself is directly observable when liquid crystalline lipid patches are deposited on a bare glass surface. After addition of buffer, a single bilayer is found to spread at the periphery of the lipid source, and the leading edge velocity, $v(t)$, can be measured by fluorescence microscopy (Figure 16.1a). The edge velocity decreases with time following a square root power law (Figure 16.1b).

$$v(t) = \beta^{1/2} \cdot t^{-1/2} \qquad (16.1)$$

where $\beta = S/b_S$ denotes the spreading coefficient which is determined by the surface free energy gain of the bilayer,

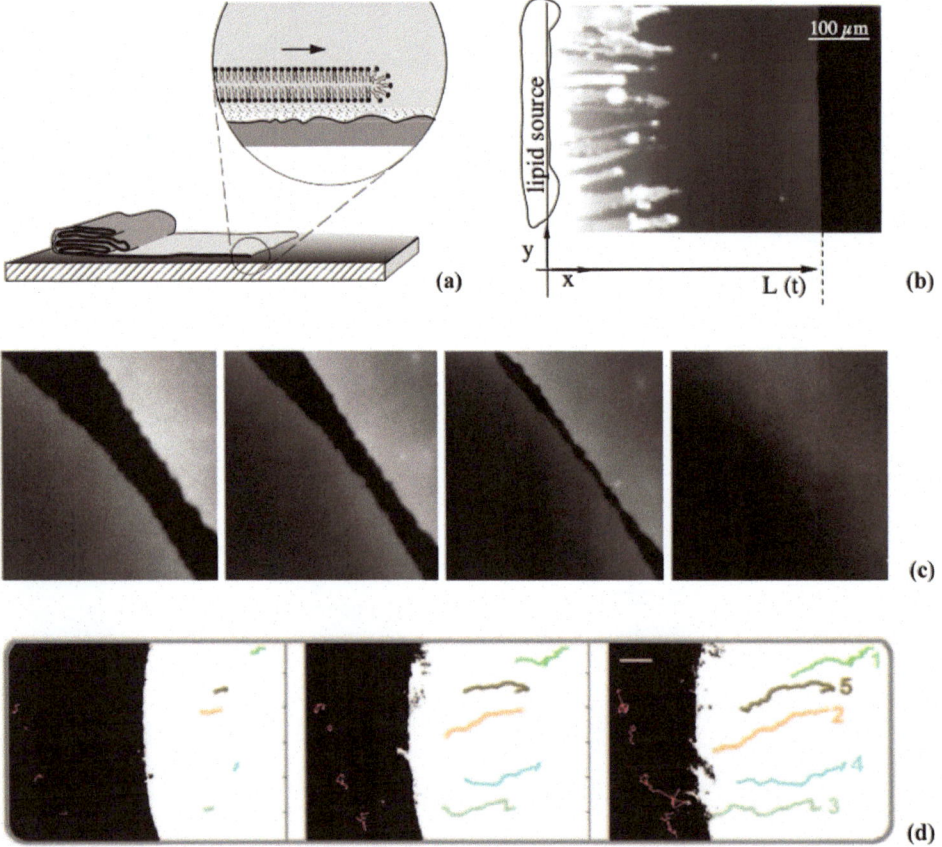

FIGURE 16.1 Lipid bilayer spreading. (a) Schematic drawing of a lipid bilayer spreading on a glass surface from a liquid crystalline source of lipid. (b) A corresponding fluorescence micrograph of the situation shown in (a). (c) A series of fluorescence images showing the closure of a gap by lipid bilayer spreading. (d) Tracks of beads attached to the upper monolayer indicate a rolling over motion of the bilayer. (Reprinted in part from Nissen et al. 1999 and Sanii et al. 2009 with permission. Copyright 1999, EDP Sciences and Copyright 2009, Wiley, respectively.)

$S\left[J/m^2\right]$ and the bilayer friction coefficient, b_S $[Pa \cdot s/m]$. For example, the spreading coefficient for DMPC bilayer on glass is $26 \pm 3\,\mu m^2 s^{-1}$. That means that a 1 µm gap would be closed in 10 ms, but lipid spreading over a 1 cm distance requires 10 days. The spreading coefficient increases with temperature and depends on ionic strength and membrane charge (Nissen et al. 1999). Note that two distinct modes of lipid bilayer spreading exist that are governed by independent friction coefficients. If the bilayer slides as a solid plate across the surface, dominating friction arises from shear viscosity of the water gap separating the bilayer from the surface. Monolayer–monolayer friction dominates if the bilayer roles over the surface, i.e., by sliding the upper (distal) lipid monolayer on top of the lower (proximal) lipid monolayer that is facing the substrate. Experiments using beads as marker for the upper monolayer flow show that both processes are superimposed (Figure 16.1d) (Sanii et al. 2009). The reason for this phenomenon is that frictional coupling of the two monolayers is of the same order of magnitude as the friction between the bottom monolayer and the substrate as described in the following paragraph.

In order to coat hydrophobic surfaces, e.g., silanized glass or silicon, with a lipid monolayer, solvent-controlled precipitation is a viable alternative to vesicle fusion. Here, the hydrophobic surface is exposed to a solution of lipids in ethanol (Miller et al. 1990). The solution is subsequently diluted with an equal volume of an aqueous solution and rinsed after some waiting time. Hohner et al. showed that the same procedure also forms bilayers on hydrophilic glass surfaces (Hohner et al. 2010). In the study, the lipid phase in solvent was measured using small-angle X-ray scattering and the lipid surface coating via attenuated total reflection–Fourier transform infrared, both as a function of ethanol/water weight ratio. The phase behaviour reveals the following typical order of bulk structures for lipid in alcohol with increasing water content: inverse micelles, monomers, micelles, and vesicles. At the solid surface, SLBs form at the micelle-to-vesicle transition. The advantage of the solvent-controlled SLB preparation is that hydrophilic and hydrophobic surfaces are coated equally well. On glass, the mobility of lipids is equally high for SLBs prepared by solvent exchange and by vesicle fusion (Hohner et al. 2010). Even on hydrophobic plastics, lipid bilayers with high lipid mobility can form (Hochrein et al. 2006a). However, a potential caveat of the solvent exchange approach is that alcohol residuals could remain in the SLB.

16.3 LIPID AND PROTEIN MOBILITY IN SUPPORTED LIPID BILAYERS

An important property of SLBs is the fact that the bilayer preserves lipid mobility despite the close proximity of the substrate's surface. Also proteins embedded in or attached to the membrane remain mobile. The order of magnitude of lipid diffusion in SLBs is in the range of free membranes and hence indicates the presence of a continuously extended lipid bilayer over the area of the experiment. Mobility in SLBs has been mostly measured by means of fluorescence recovery after photobleaching (Axelrod et al. 1976). When fluorescently labelled lipids are added to the supported bilayer, the time evolution of fluorescence in a laser spot after a bleaching pulse is determined by in-plane lipid diffusion yielding the diffusion coefficient (Tamm and McConnell 1985). Alternatively, the fluorescence profile inside the view of a field stop can be analysed as a continuous photobleaching and diffusion process using digital image processing (Dietrich et al. 1997) (see also Figure 16.2a). The advantage that SLBs are planar and remain in focus enables sensitive time-resolved microscopy measurements. In fact, SLBs

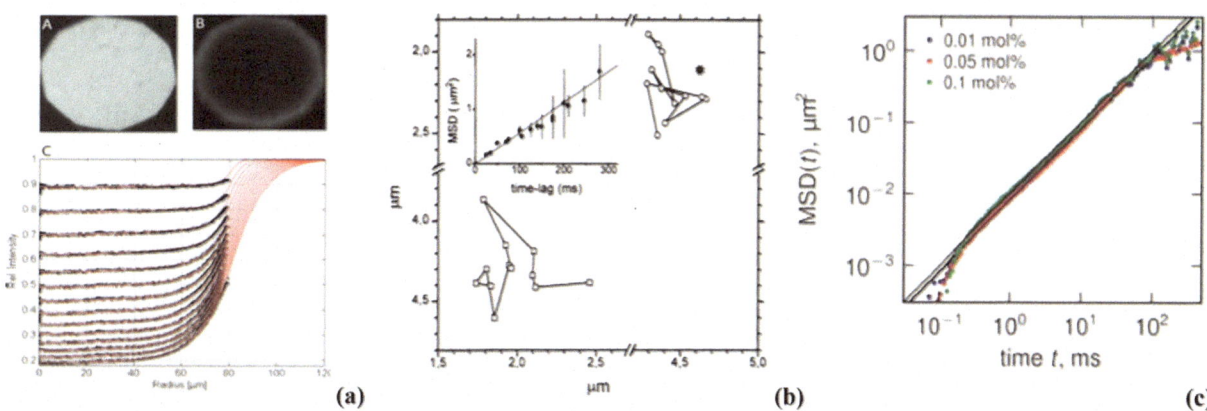

FIGURE 16.2 Diffusion measurements. (a) Continuous photobleaching of a supported lipid membrane (SLB) prepared by solvent exchange. (b) Trace of single lipid tracked by single chromophore fluorescence microscopy and (c) mean square displacement (MSD) of streptavidin bound to biotin lipids in SLBs directly converted from the time autocorrelation function measured by FCS. (Reprinted in part from Hohner et al. 2010, Schmidt et al. 1996 and Horton et al. 2010 with permission. Copyright 2010, Springer, Copyright 1996, Natl Acad Sci, Copyright 2009, Elsevier, respectively.)

were the first systems where tracking of single lipid molecules was achieved using either gold nanoparticles (Lee et al. 1991) or covalently bound fluorophores (Schmidt et al. 1996) as a label. In both cases, the self-diffusion coefficient D is derived using the mean-square displacement relation, $x^2 = 4Dt$, effective for two-dimensional Brownian motion of single particles. The same relation underlies fluorescence correlation spectroscopy, which determines the lipid self-diffusion constant from intensity fluctuations in a confocal observation volume (Korlach et al. 1999, Horton et al. 2010). The measured self-diffusion constant provides insight into the membrane viscosity and size of the inclusion as diffusion is linked to molecular drag via the Einstein relation:

$$D = \frac{k_B T}{\lambda} \quad (16.2)$$

The drag coefficient of proteins in a membrane was theoretically described by Saffmann and Dellbrück in a hydrodynamic continuum model that treated the molecular inclusions as cylindrical particles with radius r_p in a viscous slab of bulk viscosity, η_M, and finite thickness d_M surrounded by two infinite half-spaces of water with viscosity, η_W. An approximate analytical expression for the diffusion coefficient is (Saffmann and Dellbrück 1975)

$$D_{\text{Saffmann}} = \frac{k_B T}{4\pi \eta_M d_M} \cdot \left(\ln\left(\frac{\eta_M d_M}{\eta_W r_p} \right) - 0.5772 \right) \quad (16.3)$$

The diffusion coefficient is dominated by the viscosity and thickness of the membrane and has only a weak dependence on the size of the embedded molecule. Indeed the bulk viscosity of lipid membranes is about 0.1 Pa·s and hence two orders of magnitude large than for water (1 mPa·s at 20°C). The hydrodynamic screening length, which characterizes the penetration of the membrane flow into the adjacent water is $\delta = d_M \eta_M / 2\eta_W$, which is of order 2 μm. The Saffmann–Dellbrück formula is frequently used to estimate the hydrodynamic size of proteins based on measured diffusion constants. Note, however, that the Saffmann formula is only valid in the particular size range of particles that are larger than the membrane thickness, but smaller than the screening length, $d_M < r_p < \delta$. For peptides and transmembrane proteins that are smaller than the membrane thickness, the measured diffusion coefficients exhibit Stokes-like, $D \sim k_B T / \eta_M r_p$, size dependence (Gambin 2006). In the other extreme, large phase-separated lipid domains, $r_p \gg \delta$, enter a regime, where the size of the disk is so large that the drag is dominated by the viscosity of the surrounding water $D \sim k_B T / \eta_W r_p$ (Klingler 1993, Petrov and Schwille 2008).

The diffusion coefficient of SLBs is affected by the presence of the substrate, which is separated by a water film considerably smaller than the hydrodynamic length scale of particles in a free membrane. The effect of proximity in substrate-coupled membranes has been studied by Evans (Merkel et al. 1989). In the limit of close proximity, i.e., thin lubricating water film of thickness, d_w, between the SLB and the solid the diffusion coefficient is approximated by $D \approx k_B T \cdot d_w / \eta_w r_p^2$ (Sackmann 1996). In this limit, the diffusion in both monolayers is strongly coupled. If, however, the proximal monolayer of the SLB is strongly coupled to the surface, there is still finite lipid mobility in the distal monolayer assuming that the lipids chains are in a liquid state. In this case, monolayer–monolayer friction determines diffusion in the distal monolayer. The frictional coupling coefficient, b_s, between the proximal and distal monolayers in SLBs has been measured in a seminal work by Merkel, Evans, and Sackmann (Merkel et al. 1989) and is of order $b_S = 10^7 - 10^8 \text{Pa} \cdot \text{s m}^{-1}$ in the case of a DMPC or DOPC monolayer coupled to a silane-covered surface. It is noteworthy that in some studies, SLBs exhibit only mild reduction of diffusion constant in presence of a dense membrane-bound protein layer. Such SLB–protein layer systems are an interesting playground to study anomalous diffusion and crowding of anchored proteins as function of coverage. For example, in the study by Horton et al., the surface coverage of the protein avidin on the lipid bilayer was controlled by varying the concentration of biotinylated lipid anchors (Horton et al. 2010).

In the group of Steven Boxer, various micropatterning approaches were developed in order to corral-supported membranes by means of artificial diffusion barriers (Hovis and Boxer 2000, Boxer 2000). Using PDMS moulds, material can be removed from SLBs or protein blotted by microcontact printing prior to the formation of SLBs. The surfaces with geometric barriers provide two-dimensional container, where SLBs can be assembled with arbitrary concentration and composition. The laterally structured lipid corals build the basis for membrane assays on-chip for diagnostic or separation purposes (Boxer 2000). To this end, it is important to have profound knowledge about the mechanism by which SLBs spontaneously form on solid surfaces.

16.4 ADSORPTION OF MACROMOLECULES TO CHARGED SUPPORTED LIPID BILAYERS

Next, we discuss the conformational dynamics of macromolecules adsorbed to SLBs. DNA is a linear macromolecule that forms a semiflexible double helix with phosphate groups homogenously spaced along the backbone. DNA molecules are attracted by cationic lipid membranes and adsorb flat onto the membrane without penetrating into the hydrophobic part of the bilayer (see also Chapters 11 and 13 of this book). Since lipids in SLBs are mobile, the DNA molecules adsorbed onto the fluid lipid interface are also able to diffuse laterally in two dimensions. Maier et al. analysed the contour of sparsely distributed λ-phage DNA adsorbed to cationic SLBs (Maier and Rädler 1999). As shown in

Figure 16.3a, fluorescently labelled DNA molecules were bound to glass-supported cationic lipid membranes, SLBs, containing 10% cationic lipid at salt concentrations lower than 50 mM. As DNA molecules coupled to SLBs, they exhibit remarkably slow conformational dynamics, which can be imaged in time-lapse movies using fluorescence microscopy. It turns out that on average the observed DNA contours reflect the statistical properties of polymers in two dimensions. Intercalating dyes (TOTO) label DNA molecules homogeneously, and hence, fluorescence images are resolution-limited representations of the actual molecular conformation. The contours are examined in terms of the second moment of the real-space fluorescence distribution of a single molecule (Figure 16.3c). The radius of gyration of the fluorescence distribution shows power law scaling $R^2 \sim N^{2\nu}$ as a function of size, i.e., the number of base pairs N of DNA fragments. The exponent $\nu = 0.79 \pm 0.04$ is in accordance with the theoretically predicted value $\nu = 3/4$ for self-avoiding walks in two dimensions (Figure 16.3d). The snapshots of actual shapes of DNA molecules allow to assess the so-called average asphericity, Δ_2, of polymer conformations. The rotationally invariant quantity Δ_2 is defined as the ratio of the long and short axes of the gyration tensor of polymers (Figure 16.3c). Experiment showed $\Delta_2 = 0.61$ for DNA on cationic SLBs in good agreement with simulations of polymers with excluding volume interactions in two dimensions (Maier and Rädler 2001). The centre-of-mass diffusion showed dynamical Rouse-like scaling, $D_m \sim N^{-1}$, and the rotational relaxation times, $\tau_r \sim N^{\mu}$ with $\mu = 2.6 \pm 0.4$. The Rouse scaling indicates that friction is proportional to the polymer contour length as would be expected in the limit of strong hydrodynamic coupling to the substrate. Zhang and Granik showed that indeed lipid diffusion of charged lipids is slaved to oppositely charged macromolecules adsorbed to SLBs (Zhang and Granik 2005). In the case of DNA absorbed to cationic membranes, the local segregation of cationic lipid underneath the DNA strands can be visualized using fluorescently labelled cationic lipid (Kahl et al. 2009).

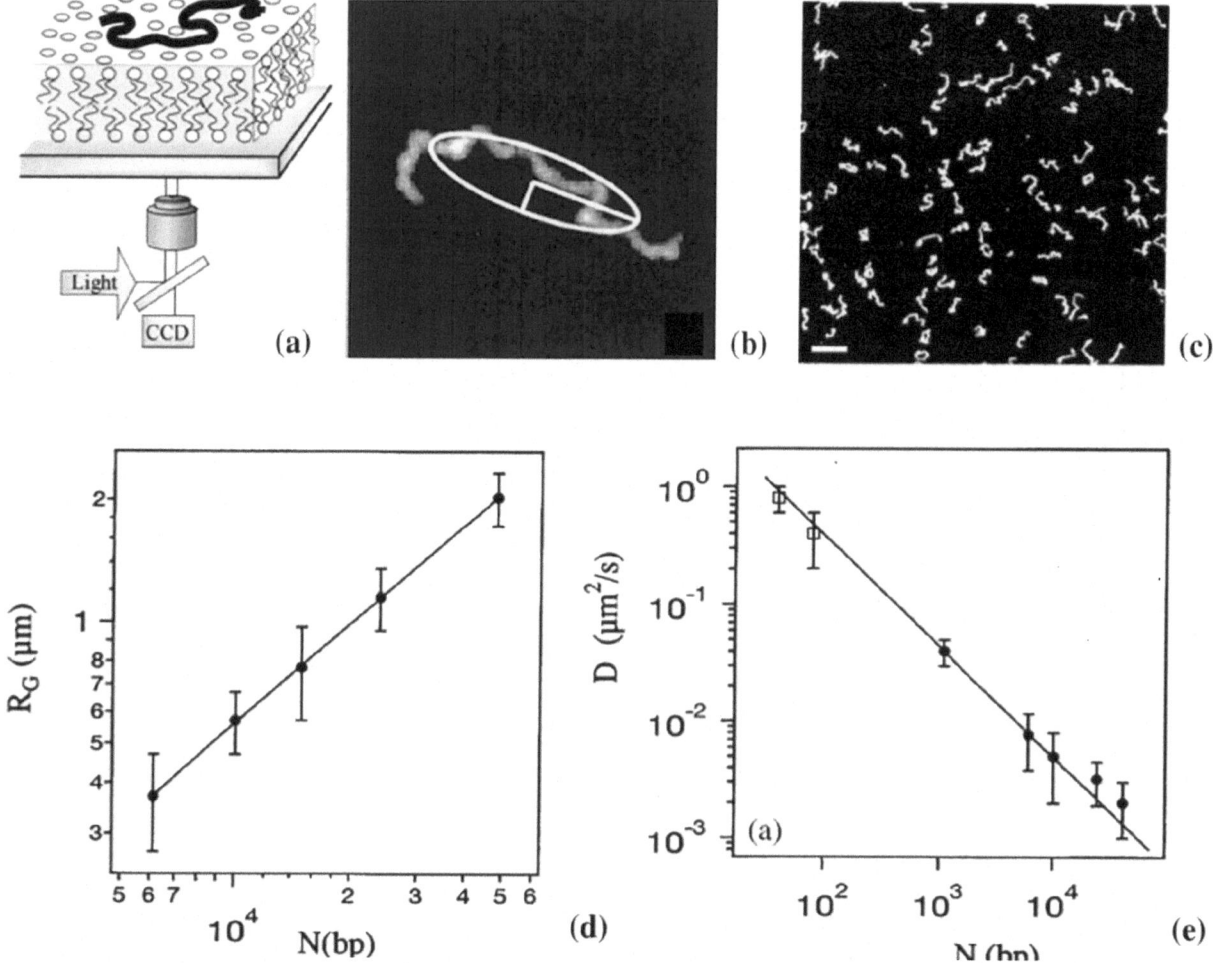

FIGURE 16.3 (a) Schematic drawing of DNA adsorbed onto a cationic supported lipid bilayer (SLB) on glass. (b) Fluorescence micrograph of a single λ-phage DNA molecule. White lines indicate long and short axis of the gyration tensor. (c) Ensemble of λ-phage DNA molecules. (d) Scaling behaviour of the radius of gyration as a function of the number of base pairs. The straight line exhibits slope 0.78, (e) DNA centre of mass self-diffusion coefficient versus number of base pairs. (Reprinted in part from Maier and Rädler 1999, 2000, 2001 with permission. Copyright 1999, The American Physical Society. Copyright 2000, 2001, American Chemical Society.

A different condition exists when cationic SLBs are fully saturated by adsorbed DNA molecules. Figure 16.4 shows an AFM image of DNA tightly packed on a cationic lipid membrane in the gel phase (Clausen-Schaumann and Gaub 1999). The fingerprint texture reveals close packing of DNA with a characteristic DNA–DNA strand separation (Fang and Yang 1997), which is consistent with close DNA packing in multilayered lipoplexes (see also Chapter 11 of this book). Interestingly at microscopic length scale, the conformation of large DNA molecules like lambda phage DNA exhibits a compact segregated conformation due to excluded volume interaction with neighbouring DNA chains. The conformation of a single chain in this 2D polymer melt is measurable, when a small percentage of labelled DNA molecules are adsorbed together with many unlabelled molecules as shown in Figure 16.4b. The chain conformation is compact and considerably smaller than the radius of gyration of isolated chains. While in three-dimensional polymer melt chains exhibit essentially ideal behaviour, theory indeed predicts that in two dimensions, chains in a polymer melt are segregated and only slightly swollen (De Gennes 1979). Thus, DNA bound to fluid membranes demonstrates some fundamental aspects of polymer behaviour in lower dimensions. An obvious thought along this line is to confine DNA to one dimension. This can be achieved by depositing cationic SLBs on rippled solid surfaces. Due to geometry a charged, corrugated membrane represents electrostatic line traps that adsorb anionic macromolecules. Regions with negative surface curvature are more attractive than regions with positive curvature. As a result, DNA can be seen to align at the inner edge of rectangular microgroves (Hochrein et al. 2006b). The detailed electrostatics of macroion adsorption to charged membranes is described in detail in Chapter 13 of this book. In short,

mobile cationic lipids act like counterions in the Poisson–Boltzmann Equation with the constraint that lipids are confined to the two-dimensional plane of the membrane. In case of SLBs on corrugated substrates, the geometric constrains are more intricate but feasible for numerical simulations. There is one more aspect that deserves attention. Since lipids exhibit a finite flip-flop kinetics between both monolayers, the distribution of charged lipids in a SLB is not a priori symmetric due to the presence of the underlying substrate. There is indeed evidence that cationic lipids on negatively charged glass surfaces are asymmetrically distributed (Kasbauer et al. 1999). Likewise, anionic lipids in SLBs on glass surfaces exhibit a prevalence of the anionic lipid in the distal monolayer as shown by neutron reflectometry (Stanglmaier et al. 2012). Consequently, when SLBs are prepared by vesicle fusion, the molar lipid ratio of charged and neutral lipids in the vesicle stock solution is not necessarily identical to the molar ratio in the distal monolayer of the SLBs formed on the substrate.

16.5 MANIPULATION OF SUPPORTED LIPID BILAYERS BY EXTERNAL FIELDS AND SURFACE-DIRECTED FORCES

A continuing challenge in SLBs technology is to devise strategies to sort, guide, and organize the lateral distribution of lipids and membrane-bound molecules within the bilayer. Several strategies have been reported to accurately and effectively manipulate SLBs by electric fields, temperature, surface topology, surface acoustic waves (SAWs), and light.

The lateral diffusion of charged lipids in a bilayer can be biased by an external electrical field. Like charges repel each other, and phospholipids bearing an equal charge

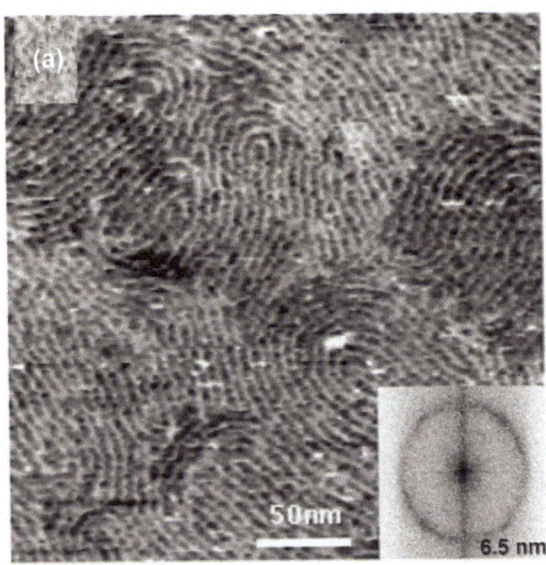
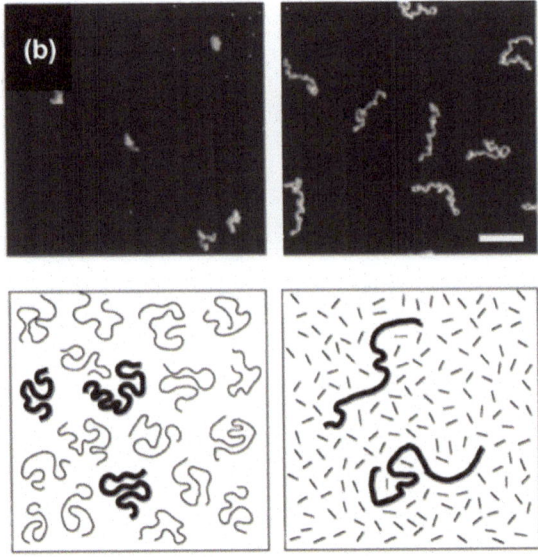

FIGURE 16.4 (a) AFM image of DNA adsorbed to a cationic supported lipid bilayer (SLB) on glass. (b) Fluorescence micrograph of fluorescently labelled λ-phage DNA molecules on a SLB fully saturated by unlabelled DNA (left: full length DNA, right: DNA fragments) (Reprinted in part from Clausen-Schaumann and Gaub 1999 and Maier and Rädler 1999 with permission. Copyright 1999A, American Chemical Society. Copyright 1999B, The American Physical Society.) AFM, atomic force microscopy.

thus favour a homogeneous distribution within an SLB. In presence of an electric field, however, the charged lipids are subject to a directional drift. The first demonstration of lipid redistribution by electrophoresis in cell membranes was reported Poo et al. (Poo and Robinson 1977). Eventually, it was shown that the electrical manipulation of charged lipids works in SLBs as well. In their seminal paper, Stelzle et al. demonstrated 2D microelectrophoresis for the first time, by comparing drift velocities and diffusion coefficients of charged and uncharged lipids in SLBs (Stelzle et al. 1992). Electrophoresis also allows to control lipid demixing and the formation of concentration gradients of charged lipid molecules within a bilayer membrane (Groves and Boxer 1995, Groves et al. 1998), and in a study by Kahl et al., microelectrophoresis of DNA on cationic SLBs was demonstrated (Kahl et al. 2009). Surprisingly, DNA molecules undergoing electrophoresis on a fluid SLB exhibit reptation-like motion, and the electrophoretic mobility follows a $1/N$ size dependence in analogy to the biased reptation model in gels.

A further development in SLB-based electrophoresis was the combination of electric fields and diffusion barriers to segregate lipid mixtures and membrane components by their diffusive properties and charge (van Oudenaarden and Boxer 1999). In the realization of a "geometrical Brownian ratchet", the directional movement of charged phospholipids along an applied electric field was influenced in perpendicular direction by an array of asymmetric titanium oxide membrane barriers. By adding cholesterol in order to slow down lipid mobility and to reduce lipid diffusive mixing, Daniel et al. demonstrated that using electrophoresis on a SLB separation and purification of dye-conjugated lipids is possible, and they achieved even separation of dye isomers (Daniel et al. 2007). In this case, no further introduction of membrane barriers was required to enable the separation process.

Temperature is another means of controlling lipid diffusion and phase separation in lipid membranes. SLBs made from binary mixtures of lipids with a high and a low melting temperature display a homogeneous liquid state at elevated temperatures. Cooling down, however, results in phase separation and microdomain formation (Szmodis et al. 2010). Local temperature gradients may be applied by laser heating of noble metal nanoparticles (Urban et al. 2009). Light absorbed by gold or silver nanorods and spheres is almost instantaneously converted into heat. Applied to SLBs, plasmonic heating may thus be used to control localized gel–fluid phase transitions and temperature-controlled partitioning of membrane components into the different phases and melting regions (Ma et al. 2012).

The physical properties of SLBs are furthermore sensitive to the topology of the underlying support. Lipids prefer assembling into different structures depending on their shape. The packing shape (or geometrical packing parameter) of phospholipid molecules is defined by $v/a_0 l_c$ (where v is the hydrocarbon volume, a_0 is the head group area, and l_c is the chain length) (Israelachvili et al. 1977). For $v/a_0 l_c \sim 1$, the lipids are nearly cylindrically shaped, which promotes the formation of a planar bilayer membrane. In comparison, lipid molecules with a larger head group than tail area (e.g. phosphatidyl choline) display a conical packing shape ($v/a_0 l_c < 1$). This shape supports the formation of more flexible bilayers with a low bending rigidity. The tendency of displaying membrane curvature is generally suppressed if the lipids are equally distributed in both bilayer leaflets and SLBs with a uniform lipid composition will typically not display phase separation on a flat support. On a substrate displaying a microcurvature pattern, however, the separation of lipid molecules into domains according to their aggregation and bending properties can be observed (Parthasarathy et al. 2006, Parthasarathy and Groves 2007). This separation can also be induced gradually, as shown by Sanii et al. (2008), who studied lipid reorganization in SLBs by introducing surface wrinkles on a deformable polydimethylsiloxane (PDMS) substrate. Controlling the stretching conditions and the elasticity of a flexible support should render lipid redistribution and domain formation in an SLB even more tunable and reversible. Yet, controlling the deformation of a polymer support with high accuracy and on fast timescales is rather challenging from an experimental point of view.

A more dynamical and fully reversible approach for generating lateral patterns in SLBs by mechanical manipulation are surface acoustic waves (SAWs) (Wixforth et al. 2004). SAWs are generated by introducing a periodic deformation of a piezoelectric substrate with an alternating voltage (Figure 16.5a). The propagation of surface waves in opposite direction from both ends of the substrate results in the formation of standing waves, which can modulate lateral lipid demixing (Hennig et al. 2009); the separation, transport, and accumulation of proteins (Neumann et al. 2010); and the accumulation of membrane bound DNA (Henning 2011) (Figure 16.5b). Importantly, the integrity of the SLB is not harmed by the SAWs, and the modulation itself is fully reversible (Figure 16.5c–d). An entirely new way of membrane protein transport in SLBs has been recently developed in the Schwille group using an *in vitro* reconstitution assay of the *Escherichia coli* MinCDE system (Zitat Loose et al. 2008). Ramm et al. showed that ATP-driven MinDE oscillations lead to travelling concentration waves that induce a spatially controlled net transport of other, functionally unrelated, membrane-bound proteins (Ramm et al. 2018).

Finally, a reversible control of bilayer properties can be achieved by introducing a photo-controlled molecular switch into the bilayer membrane. The wider applicability of this approach is somewhat restricted by the fact that some kind of photosensitive lipid or membrane compound it required. However, the advantage of using light is that it can be applied with almost unmatched spatiotemporal precision. The most prominent example of a molecular photoswitch is arguably azobenzene, a functional group composed of two phenyl rings linked by an N=N double bond. Azobenzene undergoes a reversible *trans*-to-*cis* photoisomerization upon illumination with UV and blue light, which affects

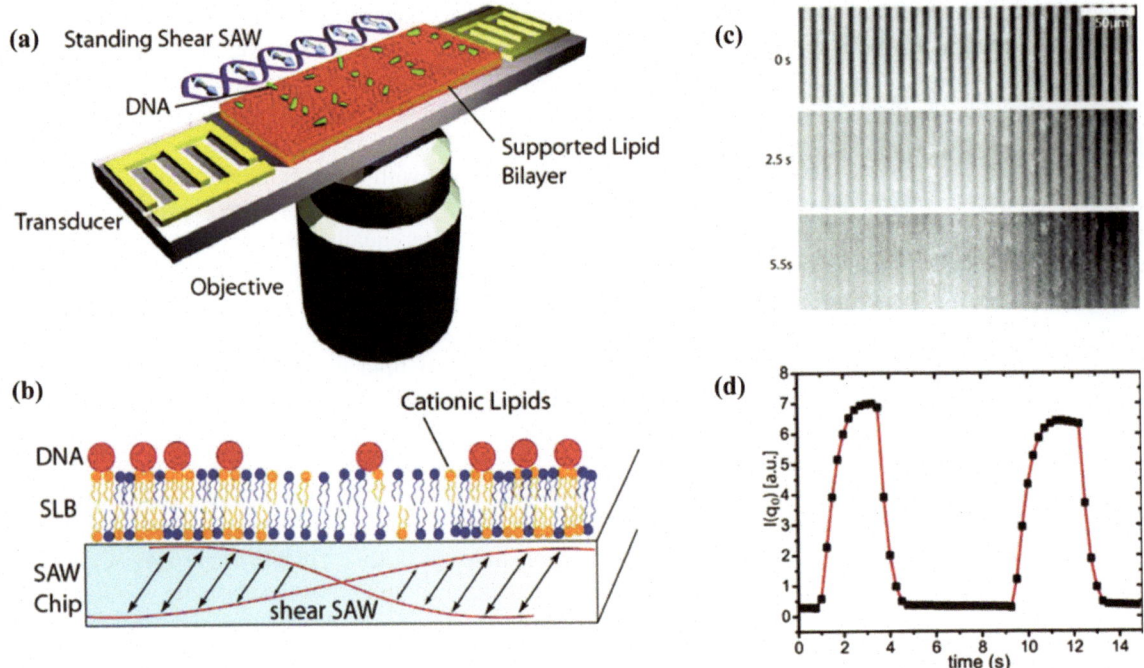

FIGURE 16.5 (a) Schematic of the experimental set-up. A supported bilayer membrane is formed on top of a piezoelectric chip. Interdigital transducers on both ends of the substrate excite a standing surface acoustic wave (SAW). The membrane is investigated from the bottom by epifluorescence microscopy. (b) The standing SAW induces a lateral demixing and of charged lipids and membrane-bound DNA. This is observable by a striped pattern of accumulated and depleted fluorescent DNA. (c) The striped DNA pattern disappears after switching the SAWs off, illustrating the fluidic character of the supported lipid bilayer (SLB). (d) Pattern formation dynamics when the SAW is switched on and off. (Reprinted in part from Hennig et al. 2011 with permission. Copyright 2011, American Chemical Society.)

the lipid shape and the bilayer properties. Synthetic photoswitchable lipid molecules (dubbed photolipids) with an azobenzene group in one or both of the lipid tails have been successfully switched in lipid vesicles by light, e.g., in order to study lipid–lipid interactions (Song et al. 1997, Pernpeintner et al. 2017, Urban et al. 2018). Also SLBs can be prepared entirely from photolipid molecules (Urban et al. 2020). Photocontrol of bilayer fluidity on demand and without interference with the overall experimental conditions (e.g., temperature, salinity, pH) has been demonstrated. The diffusion coefficient can be switched reversibly and also gradually by illuminating the sample with UV or blue light at different intensities. Furthermore, the *trans*-to-*cis* photoisomerization of the azobenzene group in the lipid tails leads to an overall reduction of the bilayer thickness by 10%, which represents a fast and straightforward approach to locally introduce membrane topology.

16.6 POLYMER-SUPPORTED LIPID BILAYERS

Perhaps the most attractive characteristic of SLBs is the fact that the lipid bilayer remains fluid even though the bilayer is tightly bound to a flat solid support. As described above, this property is due to the separation of the lower leaflet from the solid support by an ultrathin layer of water of approximately 1 nm. However, the incorporation of integral transmembrane proteins, such as integrin, requires a larger free volume on both sides of the membrane. In order to stabilize a lipid membrane in a defined distance from the solid surface, Ringsdorf and Sackmann developed polymer cushions presenting hydrophilic amino groups to anchor lipids which are mixed in the supported bilayer (Beyer et al. 1996). In these so-called "tethered supported bilayers", the separation distance is considerably increased by copolymer interlayers, which are chemically fixed on the solid support and at the same time anchored to the SLB. The decoupling of the lipid membrane from the substrate was considered essential for the investigation of membrane proteins spanning the lipid bilayer. However, photobleaching measurements revealed that the anchored lipids reduce the lipid diffusion constant considerably and protein diffusion was not yet demonstrated. Tamm and Wagner pushed the concept of tethered bilayers further by identifying polyethyleneglycol (PEG) as ideal anchor polymer since its interaction with proteins is negligible (Wagner and Tamm 2000). Here, the anchored PEG chain also acts as a cushion. Careful adjustment of the lipid anchor density allowed free membrane protein diffusion within such tethered lipid bilayers (Wagner and Tamm 2000). Since then, further suited polymer materials, including cellulose cushions, have been employed to model, e.g., cell receptor adhesion processes (cf. Tanaka 2005 and references therein). Finally, the goal to develop sophisticated interfaces to control, organize, and study membranes and membrane-associated proteins is in

Supported Lipid Bilayers

reach (Tanaka and Sackmann 2005). On ITO, cellulose cushioned membranes exhibited a five times higher electrical resistance than solid-supported bilayers (Gritsch et al. 1998, Wiegand et al. 2002). The superior sealing of cushioned membranes allowed to detect the cation selectivity of the embedded reconstituted gramicidin ion channels by impedance spectroscopy (Hillebrandt et al. 1999).

Modelling of electrical and structural properties requires detailed microscopic structural knowledge about the cushion and the lipid membrane thickness. The accessibility of cushioned membranes to surfaces has been studied by neutron and X-ray reflectometry. AFM measurements allow to obtain a rather quantitative structural and mechanical description. For example, Hertrich et al. have developed a variant of the PEG-tethered lipid membranes, which uses lipo-PEG anchors, see Figure 16.6a. X-ray and neutron reflectometry experiments allowed to extract a PEG cushion thickness of 5.5 nm and a remarkably high 90% water fraction (Hertrich et al. 2014) from the data and analysis shown in Figure 16.6b,c. In turn, an AFM tip can deform the cushioned membrane up to 4 nm prior to rupture (Hertrich et al. 2014), cf. Figure 16.6d. Neutron reflectometry experiments with in vivo cells on solid support indicate a similar hydration in the cleft adjacent to the plasma membrane highlighting the similarity physical properties of the polymer cushion and the extracellular matrix (Böhm et al. 2019).

16.7 SUPPORTED LIPID BILAYERS AS PLATFORM TO PRESENT MOLECULES TO LIVING CELLS

The capability to insert transmembrane proteins, preserved fluidity, and means to manipulate membrane-attached molecules make SLBs an ideal platform for biomimetic interfaces. Communication and signalling between adjacent cells are often tightly connected to the physical nature of the membrane receptors interactions, such as spatial organization, receptor clustering, transient binding kinetics, and mechanical transduction. Such processes can be emulated with the help of a hybrid live-cell-supported membrane junction, in which an SLB replaces one of two cells in contact. SLBs are unmatched in terms of recapitulating the biophysical and biochemical nature of cell–cell interactions due to their lateral fluidity and the possibility of membrane functionalization with

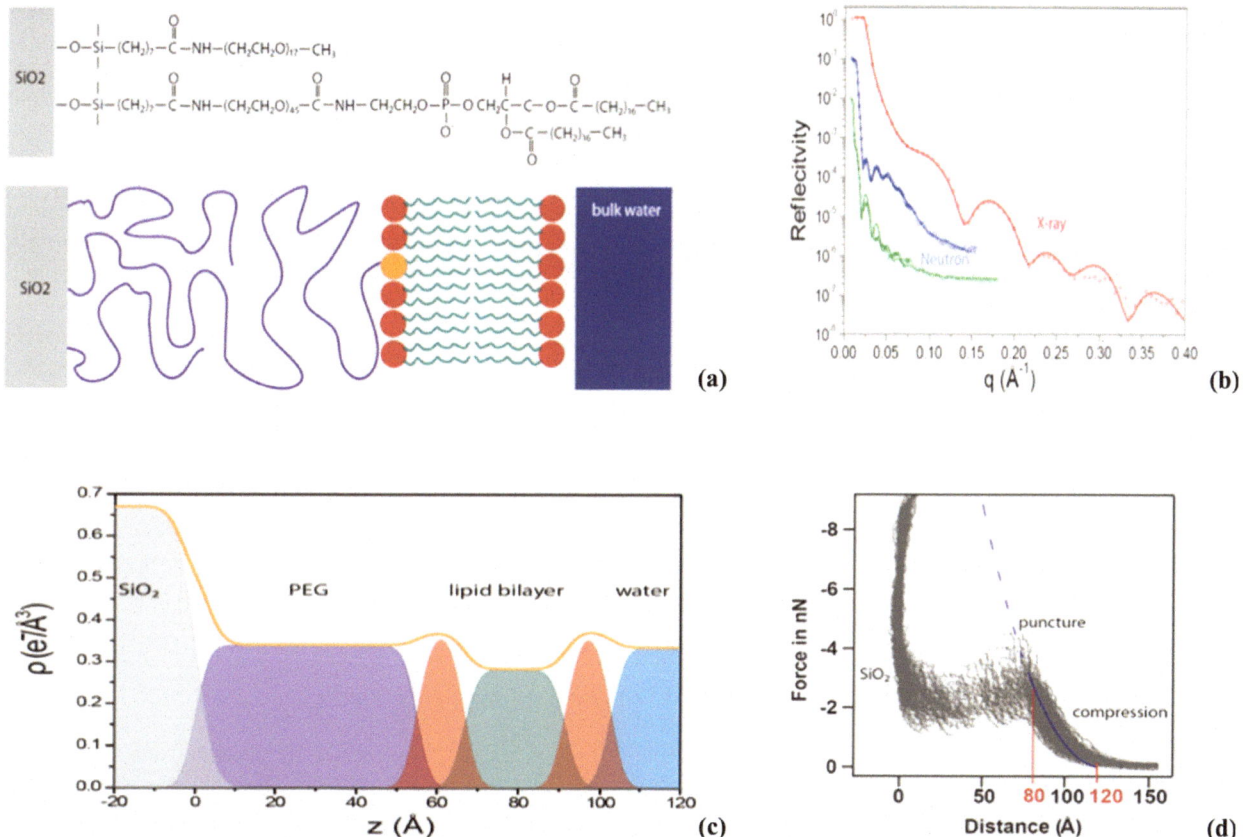

FIGURE 16.6 (a) Schematic of a polymer-supported lipid bilayer. The membrane is tethered by a silane anchor and a PEG–lipid. (b) Normalized neutron and X-ray reflectivity data and fit. Neutron data in D_2O are shown in blue, in D_2O/H_2O mix in green, and X-ray data in red. (c) Electron density profile of the polymer-supported lipid bilayer (a) as obtained from X-ray reflectometry measurements. (d) AFM indentation curves showing compression and puncture of the polymer-supported lipid bilayer. (Reprinted in part from Hertrich et al. 2014 with permission. Copyright 2014, American Chemical Society.) AFM, atomic force microscopy.

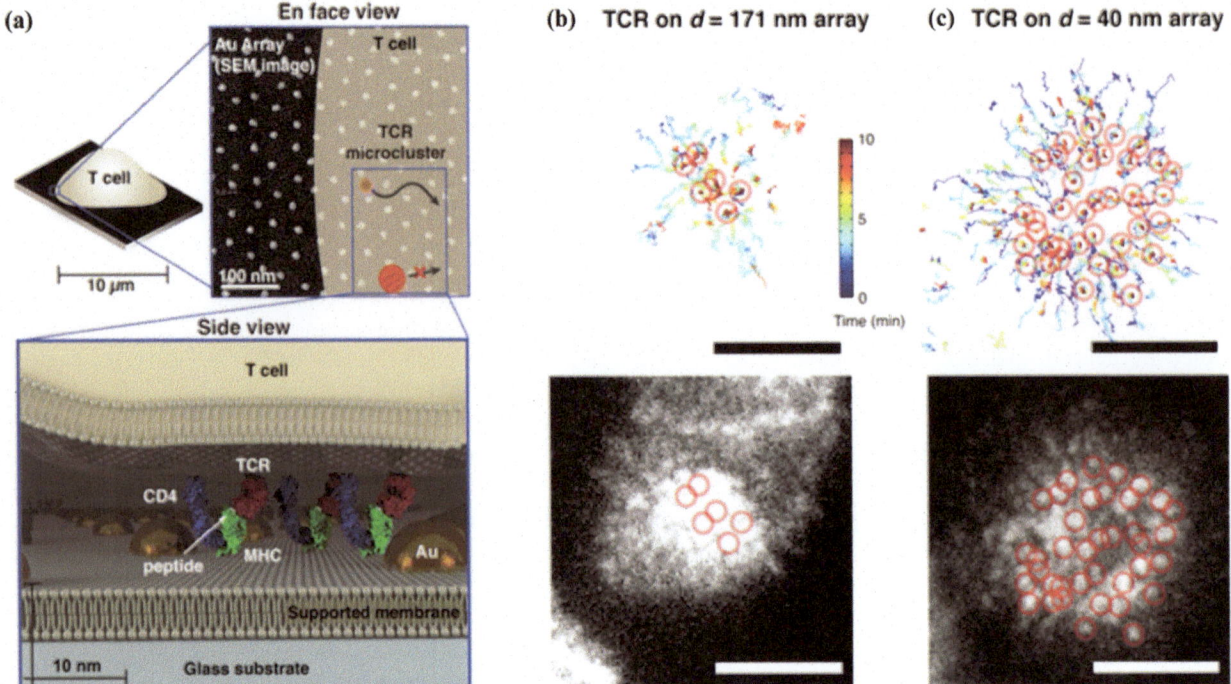

FIGURE 16.7 (a) Schematic of a T cell interacting with a nanopatterned supported lipid bilayer (SLB). Gold nanoparticles arranged in a hexagonal array with an interparticle spacing provide obstacles to the movement of TCR microclusters in the T cell membrane. TCR microclusters with a size larger than the nanoparticle spacing cannot move through the array, which illustrates the concept of cell membrane chromatography. (b,c) Trajectories of the TCR signalling microclusters movement in cells on SLB-nanodot arrays of 171 or 40 nm spacing. Top: Elapsed time (0–10 minutes) of microcluster movement is indicated by colour. Bottom: TIRF images of the TCR microcluster distribution after 10 minutes. (Reprinted in part from Caculitan et al. 2014 with permission. Copyright 2014, American Chemical Society.) TCR, T cell receptor; TIRF, total internal reflection fluorescence.

specific receptor molecules or ligands of interest. In early works, SLBs have been applied to study cell–cell recognition in the immune system (McConnell et al. 1986) and immune cell synapse formation during T cell activation (Grakoui et al. 1999). A particular feature of SLBs is that nanobarriers, such as metal lines (Groves 1997) or nanoparticles (Lohmüller et al. 2011) that are patterned on the support, can be embedded to hybrid live cell–SLB interface to generate fixed barriers and obstacles to the otherwise fluid environment. Such nanopatterned SLBs have been successfully used to probe clustering and spatial reorganization of cell membrane receptors with nanoscale accuracy. A first successful implementation of this idea was presented by Groves et al., who demonstrated spatial control over cell adhesion on patterned SLBs (Groves et al. 2001). Hybrid live-cell-nanopatterned SLBs have been further applied to elucidate of the role of spatial organization in T cell receptor (TCR) signalling (Mossman 2005) and to investigate physical force sensing by EphA2 receptor tyrosine kinase in breast cancer cells (Salaita et al. 2010) (Figure 16.7).

Besides diffusion barriers, immobilized ligands can also be incorporated by embedding a fluid bilayer membrane with arrays of gold nanoparticles. Lohmüller et al. used micellar lithography to prepare fixed anchor points for RGD motifs and Ephrin A1 ligands that were incorporated into the surrounding membrane (Lohmüller et al. 2011, 2013). Cells that were seeded on this platform were thus exposed to a combination of mobile and immobile ligands. Arrays of nanoparticles were further introduced to realize a form of chromatography that can be imposed on the membrane of a living cell (Calculitan 2014). The chromatographic material consists of a hexagonally ordered array of gold nanoparticles, which is fabricated onto the underlying substrate. Individual lipids and proteins in the supported membrane are laterally mobile throughout the array. Movement of larger assemblies, however, is impeded if they exceed the physical dimensions of the array.

ACKNOWLEDGEMENTS

This work was supported by grants from the German Science Foundation (DFG) through the Collaborative Research Center (SFB) 1032 (projects A07 and A08). Further support by the Excellence Cluster Nanosystems Initiative Munich (NIM) and the German Federal Ministry of Education, Research and Technology (BMBF) under the cooperative project 05K2018-2017-06716 Medisoft is gratefully acknowledged (JOR).

REFERENCES

Axelrod, D., D. E. Koppel, J. Schlessinger, E. L. Elson, and W. W. Webb. 1976. Mobility measurements by analysis of fluorescence recovery kinetics. *Biophys. J.* 16:1055–1096.

Beyer, D., G. Elender, W. Knoll, M. Kuhner, S. Maus, H. Ringsdorf, and E. Sackmann. 1996. Influence of anchor lipids on the homogeneity and mobility of lipid bilayers on thin polymer films. *Angew. Chem. Int. Ed.* 35: 1682–1685.

Böhm, P., A. Koutsioubas, J.-F. Moulin, J.O. Rädler, E. Sackmann, and B. Nickel. 2019. Probing the interface structure of adhering cells by contrast variation neutron reflectometry. *Langmuir.* 35: 513–521.

Caculitan, N.G., H. Kai, E.Y. Liu, N. Fay, Y. Yu, T. Lohmüller, G.P. O'Donoghue, and J.T. Groves. 2014. Size-based chromatography of signaling clusters in a living cell membrane. *Nano Lett.* 14: 2293–2298.

Clausen-Schaumann, H., and H.E. Gaub. 1999. DNA adsorption to laterally structured charged lipid membranes. *Langmuir.* 15: 8246–8251.

Cremer, P., and S. Boxer. 1999. Formation and spreading of lipid bilayers on planar glass supports. *J Phys Chem B.* 103: 2554–2559.

Daniel, S., A.J. Diaz, K.M. Martinez, B.J. Bench, F. Albertorio, and P.S. Cremer. 2007. Separation of membrane-bound compounds by solid-supported bilayer electrophoresis. *J. Am. Chem. Soc.* 129: 8072–8073.

Dietrich, C., R. Merkel, and R. Tampe. 1997. Diffusion measurement of fluorescence-labeled amphiphilic molecules with a standard fluorescence microscope *Biophys. J.* 72:1701–1710.

de Gennes, P.G. 1979 *Scaling Concepts in Polymer Physics* (Cornell Univ. Press, Ithaca).

Fang, Y., and J. Yang. 1997. Two-dimensional condensation of DNA molecules on cationic lipid membranes. *J. Phys. Chem. B.* 101: 441–449.

Grakoui, A., S.K. Bromley, C. Sumen, M.M. Davis, A.S. Shaw, P.M. Allen, and M.L. Dustin. 1999. The immunological synapse: A molecular machine controlling T cell activation. *Science.* 285: 221–227.

Gritsch, S., P. Nollert, F. Jahnig, and E. Sackmann. 1998. Impedance spectroscopy of porin and gramicidin pores reconstituted into supported lipid bilayers on indium-tin-oxide electrodes. *Langmuir.* 14: 3118–3125.

Groves, J.T. 1997. Micropatterning fluid lipid bilayers on solid supports. *Science.* 275:651–653.

Groves, J.T., and S.G. Boxer. 1995. Electric field-induced concentration gradients in planar supported bilayers. *Biophys J.* 69: 1972–1975.

Groves, J.T., S.G. Boxer, and H.M. McConnell. 1998. Electric field-induced critical demixing in lipid bilayer membranes. *P Natl Acad Sci USA.* 95: 935–938.

Groves, J.T., L.K. Mahal, and C.R. Bertozzi. 2001. Control of cell adhesion and growth with micropatterned supported lipid membranes. *Langmuir.* 17: 5129–5133.

Hafeman, D.G., V. Von Tscharner, and H.M. McConnell. 1981. Specific antibody-dependent interactions between macrophages and lipid haptens in planar lipid monolayers. *P Natl Acad Sci USA.* 78: 4552–4556.

Hennig, M., J. Neumann, A. Wixforth, J.O. Rädler, and M.F. Schneider. 2009. Dynamic patterns in a supported lipid bilayer driven by standing surface acoustic waves. *Lab Chip.* 9: 3050–3053.

Hennig, M., M. Wolff, J. Neumann, A. Wixforth, M.F. Schneider, and J.O. Rädler. 2011. DNA concentration modulation on supported lipid bilayers switched by surface acoustic waves. *Langmuir.* 27: 14721–14725.

Hertrich, S., F. Stetter, A. Rühm, T. Hugel, and B. Nickel. 2014. Highly hydrated deformable polyethylene glycol-tethered lipid bilayers. *Langmuir.* 30: 9442–9447.

Hillebrandt, H., G. Wiegand, M. Tanaka, and E. Sackmann. 1999. high electric resistance polymer/lipid composite films on indium–tin–oxide electrodes. *Langmuir.* 15: 8451–8459.

Hochrein M.B., J.A. Leierseder, L. Golubovic and J.O. Rädler. 2006b. DNA localization and stretching on periodically microstructured lipid membranes. *Phys. Rev. Lett.* 96.

Hochrein M.B., C. Reich, B. Krause, J.O. Rädler, and B. Nickel. 2006a. Structure and mobility of lipid membranes on a thermoplastic substrate. *Langmuir.* 22: 538–545.

Hohner, A.O., M.P.C. David, and J.O. Rädler. 2010. Controlled solvent-exchange deposition of phospholipid membranes onto solid surfaces. *Biointerphases.* 5: 1–8.

Horton, M.R., F. Höfling, J.O. Rädler, and T. Franosch. 2010. Development of anomalous diffusion among crowding proteins. Soft Matter. 6: 2648.

Hovis JS, and Boxer SG. 2000. Patterning barriers to lateral diffusion in supported lipid bilayer membranes by blotting and stamping. *Langmuir,* 16: 894–897.

Israelachvili, J.N., D.J. Mitchell, and B.W. Ninham. 1977. Theory of self-assembly of lipid bilayers and vesicles. *Biochim Biophys Acta. - Biomembranes.* 470: 185–201.

Kahl, V., M. Hennig, B. Maier, and J.O. Rädler. 2009. Conformational dynamics of DNA-electrophoresis on cationic membranes. Electrophoresis. 30: 1276–1281.

Kalb, E., S. Frey, and L.K. Tamm. 1992. Formation of Supported Planar Bilayers by Fusion of Vesicles to Supported Phospholipid Monolayers. *Biochim Biophys Acta.* 1103: 307–316.

Kasbauer, M., M. Junglas, and T. Bayerl. 1999. Effect of cationic lipids in the formation of asymmetries in supported bilayers. *Biophys J.* 76: 2600–2605.

Klingler, J.F. and McConnell, H.M. 1993. Field-Gradient Electrophoresis of Lipid Domains. *J. Phys. Chem.* 97: 2962–2966.

Korlach, J., P. Schwille, W.W. Webb, and G.W. Feigenson. 1999. Characterization of lipid bilayer phases by confocal microscopy and fluorescence correlation spectroscopy. *P Natl Acad Sci USA.* 96: 8461–8466.

Lee, G.M., A. Ishihara, and K.A. Jacobson. 1991 Direct observation of Brownian motion of lipids in a membrane *Proc. Natl. Acad. Sci. USA.* 88: 6274–6278.

Lohmüller, T., S. Triffo, G.P. O'Donoghue, Q. Xu, M.P. Coyle, and J.T. Groves. 2011. Supported membranes embedded with fixed arrays of gold nanoparticles. *Nano Letters.* 11: 4912–4918.

Lohmüller, T., Q. Xu, and J.T. Groves. 2013. Nanoscale obstacle arrays frustrate transport of EphA2–ephrin-A1 clusters in cancer cell lines. *Nano Lett.* 13:3059–3064.

Loose, M., E. Fischer-Friedrich, J. Ries, K. Kruse, and P. Schwille. 2008. Spatial regulators for bacterial cell division self-organize into surface waves *in vitro. Science.* 320: 789–792.

Ma, H., P.M. Bendix, and L.B. Oddershede. 2012. Large-scale orientation dependent heating from a single irradiated gold nanorod. *Nano Lett.* 12: 3954–3960.

Maier, B., and J.O. Rädler. 1999. Conformation and self-diffusion of single DNA molecules confined to two dimensions. *Phys. Rev. Lett.* 82: 1911–1914.

Maier, B., and J.O. Rädler. 2000. DNA on fluid membranes: A model polymer in two dimensions. *Macromolecules. 33*: 7185–7194.

Maier, B., and J.O. Rädler. 2001. Shape of self-avoiding walks in two dimensions. *Macromolecules. 34*: 5723–5724.

McConnell, H.M., T.H. Watts, R.M. Weis, and A.A. Brian. 1986. Supported planar membranes in studies of cell-cell recognition in the immune-system. *Biochim Biophys Acta. 864*: 95–106.

Merkel, R., E. Sackmann, and E. Evans. 1989. Molecular friction and epitactic coupling between monolayers in supported bilayers. *J. Phys. France. 50*: 1535–1555.

Miller, C., Cuendet, P., Grätzel, M., J. 1990. K+sensitive bilayer supporting electrodes *Electroanal. Chem. 278*, 175–192.

Mossman, K.D. 2005. Altered TCR signaling from geometrically repatterned immunological synapses. *Science. 310*:1191–1193.

Neumann, J., M. Hennig, A. Wixforth, S. Manus, J.O. Rädler, and M.F. Schneider. 2010. Transport, separation, and accumulation of proteins on supported lipid bilayers. *Nano Lett. 10*: 2903–2908.

Nissen, J., S. Gritsch, G. Wiegand, and J.O. Rädler. 1999. Wetting of phospholipid membranes on hydrophilic surfaces - Concepts towards self-healing membranes. *Eur Phys J B. 10*: 335–344.

Nollert, P., H. Kiefer, and F. Jähnig. 1995. Lipid vesicle adsorption versus formation of planar bilayers on solid-surfaces. *Biophys J. 69*: 1447–1455.

Parthasarathy, R., and J.T. Groves. 2007. Curvature and spatial organization in biological membranes. *Soft Matter. 3*: 24–33.

Parthasarathy, R., C. Yu, and J.T. Groves. 2006. Curvature-modulated phase separation in lipid bilayer membranes. *Langmuir. 22*: 5095–5099.

Pernpeintner, C., J.A. Frank, P. Urban, C.R. Roeske, S.D. Pritzl, D. Trauner, and T. Lohmüller. 2017. Light-controlled membrane mechanics and shape transitions of photoswitchable lipid vesicles. *Langmuir. 33*: 4083–4089.

Petrov, E.P., and P. Schwille. 2008. Translational diffusion in lipid membranes beyond the Saffman-Delbruck approximation. *Biophys J. 94*: L41–L43.

Poo, M., and K.R. Robinson. 1977. Electrophoresis of concanavalin a receptors along embryonic muscle cell membrane. *Nature. 265*: 602–605.

Sackmann, E. 1996. Supported membranes: Scientific and practical applications. *Science. 271*: 43–48.

Saffman, P. G., and M. Delbrück. 1975. Brownian motion in biological membranes. *Proc. Natl. Acad. Sci. USA. 72*:3111–3113.

Salaita, K., P.M. Nair, R.S. Petit, R.M. Neve, D. Das, J.W. Gray, and J.T. Groves. 2010. Restriction of receptor movement alters cellular response: Physical force sensing by EphA2. *Science. 327*:1380–1385.

Sanii, B., K. Nguyen, J.O. Rädler, and A.N. Parikh. 2009. Evidence for interleaflet slip during spreading of single lipid bilayers at hydrophilic solids. *ChemPhysChem. 10*: 2787–2790.

Sanii, B., A.M. Smith, R. Butti, A.M. Brozell, and A.N. Parikh. 2008. Bending membranes on demand: Fluid phospholipid bilayers on topographically deformable substrates. *Nano Lett. 8*: 866–871.

Schmidt, T., G.J. Schutz, W. Baumgartner, H.J. Gruber, and H. Schindler. 1996. Imaging of single molecule diffusion. *P Natl Acad Sci USA. 93*: 2926–2929.

Song, X., J. Perlstein, and D.G. Whitten. 1997. Supramolecular aggregates of azobenzene phospholipids and related compounds in bilayer assemblies and other microheterogeneous media: Structure, properties, and photoreactivity. *J. Am. Chem. Soc. 119*: 9144–9159.

Stanglmaier, S., S. Hertrich, K. Fritz, J.F. Moulin, M. Haese-Seiller, J.O. Rädler, and B. Nickel. 2012. Asymmetric distribution of anionic phospholipids in supported lipid bilayers. Langmuir. 28: 10818–10821.

Stelzle, M., R. Miehlich, and E. Sackmann. 1992. Two-dimensional microelectrophoresis in supported lipid bilayers. *Biophysical J. 63*: 1346–1354.

Szmodis, A.W., C.D. Blanchette, M.L. Longo, C.A. Orme, and A.N. Parikh. 2010. Thermally induced phase separation in supported bilayers of glycosphingolipid and phospholipid mixtures. *Biointerphases. 5*: 120–130.

Ramm, B., P. Glock, J. Mücksch, P. Blumhardt, D.A. García-Soriano, M. Heymann, and P. Schwille. 2018. The MinDE system is a generic spatial cue for membrane protein distribution *in vitro*. *Nat Comms*. 9: 411.

Reimhult, E., M. Zäch, F. Höök, and B. Kasemo. 2006. A multitechnique study of liposome adsorption on Au and lipid bilayer formation on SiO_2. *Langmuir. 22*: 3313–3319.

Richter, R., A. Mukhopadhyay, and A. Brisson. 2003. Pathways of lipid vesicle deposition on solid surfaces: A combined QCM-D and AFM study. *Biophys J. 85*: 3035–3047.

Tamm, L.K., and H.M. McConnell. 1985. Supported phospholipid-bilayers. Biophys J. 47: 105–113.

Tanaka, M., and E. Sackmann. 2005. Polymer-supported membranes as models of the cell surface. *Nature. 437*: 656–663.

Urban, A.S., M. Fedoruk, M.R. Horton, J.O. Rädler, F.D. Stefani, and J. Feldmann. 2009. Controlled nanometric phase transitions of phospholipid membranes by plasmonic heating of single gold nanoparticles. *Nano Lett. 9*: 2903–2908.

Urban, P., S.D. Pritzl, D.B. Konrad, J.A. Frank, C. Pernpeintner, C.R. Roeske, D. Trauner, and T. Lohmüller. 2018. Light-controlled lipid interaction and membrane organization in photolipid bilayer vesicles. *Langmuir. 34*: 13368–13374.

Urban, P., S.D. Pritzl, M.F. Ober, C.F. Dirscherl, C. Pernpeintner, D.B. Konrad, J.A. Frank, D. Trauner, B. Nickel, and T. Lohmüller. 2020. A lipid photoswitch controls fluidity in supported bilayer membranes. *Langmuir. 36*: 2629–2634.

van Oudenaarden, A., and S.G. Boxer. 1999. Brownian ratchets: Molecular separations in lipid bilayers supported on patterned arrays. *Science. 285*: 1046–1048.

Wagner, M.L., and L.K. Tamm. 2000. Tethered polymer-supported planar lipid bilayers for reconstitution of integral membrane proteins: Silane-polyethyleneglycol-lipid as a cushion and covalent linker. *Biophys J. 79*: 1400–1414.

Wiegand, G., N. Arribas-Layton, H. Hillebrandt, E. Sackmann, and P. Wagner. 2002. Electrical properties of supported lipid bilayer membranes. *J. Phys. Chem. B. 106*: 4245–4254.

Wixforth, A., C. Strobl, Ch. Gauer, A. Toegl, J. Scriba, and Z. v. Guttenberg. 2004. Acoustic manipulation of small droplets. *Anal Bioanal Chem. 379*: 982–991.

Zhang, L.F., and S. Granick. 2005. Slaved diffusion in phospholipid bilayers. *P Natl Acad Sci USA. 102*: 9118–9121.

17 Artificial Membranes Composed of Synthetic Copolypeptides

Timothy J. Deming
University of California Los Angeles, Los Angeles, California

CONTENTS

17.1 Introduction 305
17.2 Copolypeptide Synthesis 306
    17.2.1 Transition Metal Catalysis 306
    17.2.2 Amine Initiators...307
17.3 Copolypeptide Vesicles 310
    17.3.1 Structural Parameters 310
    17.3.2 Vesicle Assemblies 310
    17.3.3 Functional Vesicles 314
17.4 Conclusions..317
Abbreviations and Symbols 317
References..319

17.1 INTRODUCTION

Biological systems produce a variety of macromolecules that possess the ability to self-assemble into complex, yet highly ordered structures (Branden and Tooze 1991). A subset of these, the proteins, are polypeptide copolymers that derive their properties from precisely controlled sequences and compositions of their constituent amino acid monomers. There has been recent interest in developing synthetic routes for preparation of polypeptides that can self-assemble into precisely defined structures, such as membranes. Polypeptides have many advantages over conventional synthetic polymers since they are able to adopt stable ordered conformations (Voet and Voet 1995). Depending on the amino acid side chain substituents, polypeptides are able to adopt a multitude of conformationally stable regular secondary structures (helices, sheets, turns), tertiary structures (e.g., the β-strand–helix–β-strand unit found in β-barrels), and quaternary assemblies (e.g., collagen microfibrils) (Voet and Voet 1995). The synthesis of polypeptides that can assemble into desirable *de novo* structures is an attractive challenge for polymer chemists.

In recent years, block copolypeptides, which combine different structural and functional peptide elements, have been prepared that are able to form supramolecular assemblies in water, and incorporate some of the functionality found in proteins (ed. Deming 2012). Such polypeptides are well suited for applications where polymer assembly and functional domains need to be at length scales ranging from nanometres to microns. Amphiphilic diblock copolypeptides, when dispersed in water, have been found to be capable of forming peptide based micelles, vesicles, and hydrogels potentially useful in biomedical applications (Deming 2014). The regular secondary structures obtainable within polypeptide segments have been identified as key components that provide opportunities for hierarchical self-assembly unobtainable with conventional block copolymers or small-molecule surfactants.

The most economical and expedient process for synthesis of long polypeptide chains is the polymerization of α-amino acid-N-carboxyanhydrides (NCAs) (Eq. 17.1) (Deming 2013). This method involves the simplest reagents, and high-molecular-weight polymers can be prepared in both good yield and in large quantity with no detectable racemization at the chiral centres. The considerable variety of NCAs that have been synthesized (> 200) allows exceptional diversity in the types of polypeptides that can be prepared (Deming 2013).

$$n \text{ NCA} \xrightarrow{\text{nucleophile or base}} \text{-(NH-CHR-CO)}_n\text{- (polypeptide)} + n\, CO_2 \quad (17.1)$$

Since the late 1940s, NCA polymerizations have been the most common technique used for large-scale preparation of high-molecular-weight polypeptides (Woodward and Schramm 1947). However, these materials have primarily been homopolymers, random copolymers, or graft copolymers that lack the sequence specificity and monodispersity of natural proteins. Attempts to prepare block copolypeptides and hybrid block copolymers using NCAs have traditionally resulted in polymers whose compositions did not match monomer feed compositions and that contained significant homopolymer contaminants (Cardinaux et al. 1977). The main factor limiting the potential of NCA polymerizations has been the presence of side reactions (chain termination and chain transfer) that restrict control over molecular weight, give broad molecular weight distributions, and prohibit formation of well-defined copolymers (Sekiguchi 1981). Progress in elimination of these side reactions over the past 18 years has been a major breakthrough for the polypeptide materials field. This chapter briefly summarizes the developments that have enabled the synthesis of well-defined copolypeptides from controlled and living polymerizations of NCA monomers. Using these methods, designs of block as well as statistical copolypeptides that are able to form vesicle structures in water are described in detail. Both the structural features of the polypeptides, which promote vesicle formation and dictate membrane properties, as well as recent efforts to add functionality to these membranes, are described. This chapter is focused on vesicle-forming copolypeptides, i.e., those with primarily peptide residues in the main chains, and hence excludes the extensive work on hybrid copolymers that contain polypeptide components combined with other synthetic polymer domains.

17.2 COPOLYPEPTIDE SYNTHESIS

17.2.1 Transition Metal Catalysis

NCA polymerizations have been initiated using many different nucleophiles and bases, the most common being primary amines and alkoxide anions (Deming 2013). Primary amines, being more nucleophilic than basic, are good general initiators for polymerization of NCA monomers that provide relatively slow polymerization and are well understood. Tertiary amines, alkoxides, and other initiators that are more basic than nucleophilic have found use since they are in some cases able to prepare polymers of very high molecular weight where primary amine initiators cannot. Strong base initiators generally promote much faster NCA polymerization compared with primary amine initiators, yet fine mechanistic details of these systems are poorly understood. Optimal polymerization conditions have often been determined empirically for each NCA, and thus for many years, there were no universal initiators or conditions that were optimized for all monomers.

A successful strategy for propagation rate enhancement and elimination side reactions in NCA polymerizations has been the use of transition metal complexes as catalysts for addition of NCA monomers to polypeptide chain ends. The use of transition metals to control reactivity has been proven in organic and polymer synthesis as a means to increase reaction selectivity, efficiency, and rate (Collman et al. 1987). Using this approach, a significant advance in the development of a general method for living NCA polymerization was realized in 1997. Highly effective zerovalent nickel and cobalt initiators (i.e., bpyNi(COD) and $(PMe_3)_4Co$) (Deming 1997, 1998, Deming and Curtin 2000) were developed by Deming that allow the living polymerization of many different NCAs into high-molecular-weight polypeptides via an unprecedented activation of the NCAs to generate covalent metal containing propagating species. These propagating species were also found to be highly active for NCA addition and were found to increase polymerization rates more than an order of magnitude compared with amine-initiated polymerizations at 20°C. The metal ions can also be conveniently removed from the polymers by simple precipitation or dialysis of the samples after polymerization.

Mechanistic studies on the initiation process showed that both nickel and cobalt complexes react identically with NCA monomers to form metallacyclic complexes by oxidative addition across the anhydride bonds of NCAs (Deming 1997, 1998, Deming and Curtin 2000). Propagation through the amido-amidate metallacycle was found to occur by initial attack of the nucleophilic amido group on the electrophilic C_5 carbonyl of an NCA monomer (Eq. 17.2). In this manner, the metal is able to migrate along the growing polymer chain, while being held by a robust chelate at the active end. The formation of these chelating metallacyclic intermediates appears to be a general requirement for obtaining living NCA polymerizations using transition metal initiators. These cobalt and nickel complexes are able to produce polypeptides with narrow chain length distributions ($M_w/M_n < 1.1$) and controlled molecular weights ($500 < M_n < 500,000$) (Deming 2013). These polymerizations can be conducted in a variety of solvents (e.g., THF, DMF, EtOAc, dioxane, MeCN, DMAc, nitrobenzene) and over a broad range of temperatures (i.e., 10°C–100°C) with no loss of polymerization control and with dramatic increases in polymerization rate as temperature is increased. By addition of different NCA monomers, the preparation of block copolypeptides of defined sequence and composition is feasible (Deming 2013).

This polymerization system is general and gives controlled polymerization of a wide range of NCA monomers as pure enantiomers (D or L configuration) or as racemic mixtures. In addition to commonly used NCA monomers, such as protected lysine, glutamate, aspartate, and arginine, many hydrophobic amino acid monomers (e.g., leucine, valine, alanine, isoleucine, phenylalanine) as well as other reactive amino acids (e.g., methionine, cysteine, serine, tyrosine, DOPA) and synthetic unnatural amino acids have been successfully polymerized in a controlled manner using cobalt and nickel initiators. There is much current interest in functional and reactive polypeptides, and NCAs bearing more complex functionality have also been polymerized using this methodology (Yu et al. 1999, Hwang and Deming 2001, Schaefer et al. 2006, Kramer and Deming 2010a, 2012a, 2014, Chen et al. 2011, Rhodes and Deming 2013, Yakovlev and Deming 2014, 2015). A key challenge in these recent examples was purification of the highly functional NCAs, which could not be purified by conventional methods (i.e., recrystallization). To solve this problem, Kramer and Deming developed an anhydrous flash column chromatography method for NCA purification that enables one to obtain a wide range of difficult-to-crystallize NCAs in suitable purity for controlled polymerization (Kramer and Deming 2010b) and has made possible the preparation of many new highly functional NCAs.

17.2.2 Amine Initiators

In the past decade, a number of new approaches have been reported to give controlled NCA polymerizations. These approaches share a common theme in that they are all improvements on the use of conventional primary amine polymerization initiators (Eq. 17.3). This approach is attractive since primary amines are readily available and since the initiator does not need to be removed from the reaction after polymerization. In fact, if the polymerization proceeds without any chain breaking reactions, the amine initiator becomes the C-terminal polypeptide end group. In this manner, there is potential to form chain-end-functionalized polypeptides or even hybrid block copolymers if the amine is a macroinitiator. The challenge in this approach is to overcome the numerous side reactions of these systems without the luxury of many adjustable experimental parameters.

In 2004, the group of Hadjichristidis reported the primary amine-initiated polymerization of NCAs under high vacuum conditions (Aliferis et al. 2004). Unlike the vinyl monomers usually polymerized under high vacuum conditions, NCAs cannot be purified by distillation. Consequently, it is unclear if NCAs can be obtained in higher purity by high vacuum recrystallization than by recrystallization under a rigorous inert atmosphere. However, the high vacuum method does allow for better purification of polymerization solvents and the n-hexylamine initiator. It was found that polymerizations of γ-benzyl-L-glutamate NCA, Bn-Glu NCA, and ε-carbobenzyloxy-L-lysine, Z-Lys NCA under high vacuum in DMF solvent displayed all the characteristics of living polymerization systems (Aliferis et al. 2004). Polypeptides could be prepared with control over chain length, chain length distributions were narrow, and block copolypeptides were prepared. This method has been used by Iatrou and coworkers to prepare a number of different block copolypeptides, primarily PBLG segments connected to polymers of lysine, leucine, tryosine, as well as the imino acid proline, and their microphase-separated morphologies have been studied in the bulk state (Mondeshki et al. 2011, Graf et al. 2012).

For the high vacuum method, the authors concluded that the side reactions normally observed in amine-initiated NCA polymerizations are simply a consequence of impurities. Since the main side reactions in NCA polymerizations do not involve reaction with adventitious impurities such as water, but instead reactions with monomer, solvent, or polymer (i.e., termination by reaction of the amine end with an ester side chain, attack of DMF by the amine end, or chain transfer to monomer) (Deming 2013), it is not clear why vacuum conditions are able to inhibit these side reactions. A likely explanation for the polymerization control observed under high vacuum is that CO_2 acts to promote side reactions of growing chains with monomer, polymer, or solvent, and its removal from the reaction medium under vacuum inhibits these reactions and promotes controlled polymerization. A number of early and recent studies support this role of CO_2 as being detrimental to amine-initiated NCA polymerizations, where for some NCAs, it is able to decrease chain propagation rate by reversibly forming a carbamate with the amine end group and may also catalyse side reactions (Thunig et al. 1977, Habraken et al. 2010). Thus, it is reasonable to speculate (*vide infra*) that removal of CO_2 from NCA polymerizations under high vacuum is the dominant factor in enabling controlled chain growth in these systems. Recently, in polymerizations of O-benzyl-L-tyrosine NCA, Bn-Tyr NCA, in DMF, it was determined that although most side reactions are insignificant in the high-vacuum

(17.3)

polymerization, some termination of chains by reaction with DMF solvent does occur (Pickel et al. 2009).

Further insights into amine-initiated NCA polymerizations were also reported in 2004 by the group of Giani and coworkers (Vayaboury et al. 2004). This group studied the polymerization of ε-trifluoroacetyl-L-lysine NCA, TFA-Lys NCA, in DMF using n-hexylamine initiator at different temperatures. Contrary to the high vacuum work, the solvent and initiator were purified using conventional methods, and the polymerizations were conducted under a nitrogen atmosphere on a Schlenk line. After complete consumption of NCA monomer, the crude polymerization mixtures were analysed by GPC and nonaqueous capillary electrophoresis (NACE). A unique feature of this work was the use of NACE to separate and quantify the amount of polymers with different chain ends, which corresponded to living chains (amine end groups) and "dead" chains (carboxylate and formyl end groups from reaction with NCA anions and DMF solvent, respectively, Eqs. 17.4 and 17.5). Not surprisingly, at 20°C, the polymer products consisted of 78% dead chains and only 22% living chains, which illustrates the abundance of side reactions in these polymerizations under conventional conditions.

An intriguing result was found for polymerizations conducted at 0°C where 99% of the chains had living amine chain ends, and only 1% were found to be dead chains. To verify that these were truly living polymerizations, additional NCA monomer was added to these chains at 0°C resulting in increased molecular weight and no increase of the amount of dead chains. While TFA-Lys NCA was the only monomer studied, this work showed that controlled NCA polymerizations can be obtained by lowering temperature. The effect of temperature is not unusual, as similar trends can be found in cationic and anionic vinyl polymerizations (Odian 1991). At elevated temperature, the side reactions have activation barriers similar to chain propagation. When the temperature is lowered, the activation barrier for chain propagation becomes lower than that of the side reactions and chain propagation dominates kinetically. A key limitation of this method is that these polymerizations are very slow at 0°C, often requiring numerous days to obtain polypeptide chains of modest length. An interesting feature of this system is that increased impurity/by-product (i.e., CO_2) levels, as compared with the high vacuum method, did not result in side reactions at low temperature. This result shows that even with CO_2 present, side reactions in amine initiated NCA polymerizations can be made kinetically insignificant at low temperature.

Since these original studies, a number of groups have used and studied low-temperature NCA polymerizations in greater detail. Shao's lab reported the synthesis of block copolypeptides of PBLG with segments of alanine, leucine, and phenylalanine at 0°C and found that greater than 90% of the PBLG chains were active for the second monomer addition using MALDI-MS analysis (Cao et al. 2012). Schouten also reported the controlled polymerization of $tert$-butyl-L-glutamate NCA at 0°C and use of these chains to prepare block copolypeptides with other glutamate ester NCAs (Nguyen et al. 2010). Perhaps the most comprehensive studies of amine-initiated NCA polymerizations at low temperature and/or under vacuum were performed by Heise and coworkers (Habraken et al. 2010). They examined ten different NCA monomers and found that, using MALDI-MS analysis of end groups, most of these, including monomer mixtures for preparation of statistical copolymers, show fewer side reactions at 0°C compared with elevated temperatures. In a follow-up study, they combined low-temperature polymerizations with those run under low pressure in order to identify optimal polymerization conditions (Habraken et al. 2011). Surprisingly, only α-helical favouring monomers (Bn-Glu, alanine, Z-Lys) showed rate accelerations upon reduction in pressure (and consequent CO_2 removal), while nonhelicogenic monomers (β-benzyl-L-aspartate (Bn-Asp), O-benzyl-L-serine, O-benzyl-L-threonine) were not affected by reaction pressure. Thus the use of high vacuum or other methods for CO_2 removal to obtain controlled NCA polymerization seems to be highly monomer dependent. Also the enhancements in polymerization rates seen by removing CO_2 at 20°C were found to be minimal at 0°C, thus indicating there is no advantage for conducting an NCA polymerization under reduced pressure at 0°C. From this study, it was concluded that helicogenic NCA monomers could be polymerized in a controlled manner at 20°C if CO_2 was removed from the reaction mixture, while nonhelicogenic monomers should be polymerized at 0°C for optimal control over polymerization (Habraken et al. 2011). This strategy was validated by preparation of a tetrablock copolypeptide containing glutamate–alanine–lysine–aspartate, i.e., PBLG-A-PZLL-PBLA.

A different innovative approach to controlling amine-initiated NCA polymerizations was reported in 2003 by Schlaad and coworkers (Dimitrov 2003). Their strategy was to avoid formation of NCA anions, which cause significant chain termination after rearranging to isocyanocarboxylates (Deming 2013), through use of primary amine hydrochloride salts as initiators. The reactivity of amine hydrochlorides with NCAs was first explored by the group of Knobler, who found that they can react with NCAs to give single NCA addition products (Knobler et al. 1964, 1969). Use of the hydrochloride salt takes advantage of its diminished reactivity as a nucleophile compared with the parent amine, which effectively halts the reaction after a single NCA insertion by formation of an inert amine hydrochloride in the product. The reactivity of the hydrochloride presumably arises from formation of a small amount of free amine by reversible dissociation of HCl (Eq. 17.6). This equilibrium, which lies heavily towards the dormant amine hydrochloride species, allows for only a very short lifetime of reactive amine species. Consequently, as soon as a free amine reacts with an NCA, the resulting amine end group on the product is immediately protonated and is prevented from further reaction. The acidic conditions also assist elimination of CO_2 from the reactive intermediate and, more importantly, suppress formation of unwanted NCA anions.

To obtain controlled polymerization, and not just single NCA addition reactions, Schlaad's group increased the reaction temperature (40°C–80°C), which was known from Knobler's work to increase the equilibrium concentration of free amine, as well as increase the exchange rate between amine and amine hydrochloride (Knobler et al. 1964, 1969). Using primary amine hydrochloride end-capped polystyrene macroinitiators to polymerize Z-Lys NCA in DMF, Schlaad's group obtained polypeptide hybrid copolymers in 70%–80% yield after 3 days at elevated temperature. Although these polymerizations are slow compared with amine-initiated polymerizations, the resulting polypeptide segments were well defined with very narrow chain length distributions ($M_w/M_n < 1.03$). These distributions were much narrower than those obtained using the free amine macroinitiator, which argues for diminished side reactions during polypeptide synthesis.

The use of amine hydrochloride salts as initiators for controlled NCA polymerizations shows tremendous promise. The concept of fast, reversible deactivation of a reactive species to obtain controlled polymerization is a proven concept in polymer chemistry, and this system can be compared with the persistent radical effect employed in all controlled radical polymerization strategies (Fischer 2001). Like those systems, success of this method requires a carefully controlled matching of the polymer chain propagation rate constant, the amine/amine hydrochloride equilibrium constant, and the forward and reverse exchange rate constants between amine and amine hydrochloride salt. This means it is likely that reaction conditions (e.g., temperature, halide counterion, solvent) will need to be optimized to obtain controlled polymerization for each different NCA monomer, as is the case for most vinyl monomers in controlled radical polymerizations. Within these constraints, it is possible that controlled NCA homopolymerizations utilizing simple amine hydrochloride initiators can be obtained, yet this method may not be advantageous for preparation of block copolypeptides due to the need for monomer-specific optimization.

Another interesting approach to obtain controlled NCA polymerization using silylated amines was reported in 2007 by Cheng. Hexamethyldisilazane (HMDS) was used to initiate polymerizations of either Z-Lys NCA or Bn-Glu NCA in DMF at ambient temperature and was found to give well-defined polypeptides of controlled chain length and low polydispersity in high yield (Lu and Cheng 2007). Addition of a second batch of monomer to completed chains afforded block copolymers. Chain growth in this system does not appear to show any of the common side reactions found in amine-initiated NCA polymerization, which is attributed to the unique properties of the N-trimethylsilyl (TMS) groups. The HMDS is proposed to transfer a TMS group to the NCA followed by addition of the silylamine to the resulting intermediate (Eq. 17.7). This process yields a ring-opened monomer with a TMS-carbamate active end group on the growing chain, similar to processes that occur in group transfer polymerization of vinyl monomers (Webster et al. 1983). This system has an advantage in that it proceeds at much higher rates (ca. 12–24 hours at ambient temperature to obtain DP=100) compared with low temperature or amine hydrochloride initiated polymerizations, yet still is slower than transition-metal-initiated systems (ca. 30–60 minutes at ambient temperature).

17.3 COPOLYPEPTIDE VESICLES

17.3.1 STRUCTURAL PARAMETERS

For assembly into novel supramolecular structures, well-defined copolypeptides are required that have structural domains (i.e., amino acid sequences) whose size and composition can be precisely adjusted. Such materials have historically proven elusive using conventional techniques. There are many early reports on the preparation of block copolypeptides using conventional primary amine initiators (Uralil et al. 1977). Examples include many hydrophilic–hydrophobic and hydrophilic–hydrophobic–hydrophilic di- and triblock copolypeptides (where hydrophilic residues were glutamate and lysine, and hydrophobic residues were leucine (Auer and Doty 1966, Ostroy et al. 1970), valine (Epand and Scheraga 1968), isoleucine (Kubota and Fasman 1975), phenylalanine (Cardinaux et al. 1977), and alanine (Ingwall et al. 1968)) prepared to study conformations of the hydrophobic domain in aqueous solution. More recently, Cameron and coworkers reported the synthesis of novel (α-helix)-b-(β-sheet) block copolypeptides using amine initiation (Gibson and Cameron 2008). These polymers were reported to have polydispersities ranging from 1.47 to 1.60.

The majority of these amine-initiated block copolypeptides were often subjected to only limited characterization (e.g., amino acid compositional analysis) and, as such, their structures, and the presence of homopolymer contaminants, were not conclusively determined. Some copolymers, which had been subjected to chromatography, showed polymodal molecular weight distributions containing substantial high and low molecular weight fractions (Cardinaux et al. 1977). The compositions of these copolymers were found to be different from the initial monomer feed compositions and varied widely for different molecular weight fractions. It appears that most, if not all, block copolypeptides prepared using amine initiators under conventional conditions have structures different than predicted by monomer feed compositions and likely have considerable homopolymer contamination due to the side reactions described above.

Copolypeptides prepared via transition-metal-mediated NCA polymerization are well defined, with the sequence and composition of residues controlled by order and quantity of monomer added to initiating species, respectively. These copolypeptides can be prepared with the same level of control found in anionic and controlled radical polymerizations of vinyl monomers, which greatly expands the potential of polypeptide materials. The unique chemistry of NCAs allows these monomers to be polymerized in any order, which is a challenge in most vinyl copolymerizations, and the robust chain ends allow the preparation of copolypeptides with many block domains (e.g., >2). The robust nature of transition metal initiation was shown by the linear, stepwise synthesis of triblock and pentablock copolypeptides (Eq. 17.8) (Nowak et al. 2006, Li and Deming 2010). The N-TMS amine initiators and amine initiators used under high vacuum and/or low temperature conditions have recently also been used to prepare well-defined copolypeptides (Aliferis et al. 2004, Habraken et al. 2011). The self-assembly of well-defined copolypeptides has also been under extensive investigation in recent years, typically in aqueous media to mimic biological conditions. In the following sections, the assembly of copolypeptides into different types of supramolecular assemblies is described.

17.3.2 VESICLE ASSEMBLIES

Membranes are important materials for many applications, ranging from separations, to devices such as sensors and fuel cells, to encapsulation of sensitive materials, and to biomedical applications such as drug delivery. Vesicles constructed from polymers offer many advantages and opportunities over lipid vesicles for all of these applications (e.g., increased stability, tunable functionality, and permeability) (Rutjes and van Hest 2011). To date, many types of copolypeptide amphiphiles that form stable vesicular assemblies have been developed. The first of these utilized diethylene glycol–modified lysine residues, i.e., K^P vide infra, that impart both nonionic water solubility as well as ordered α-helical conformations to the hydrophilic polypeptide domains (Bellomo et al. 2004). A majority of other materials utilize highly charged polyelectrolyte segments to impart both functionality and flexibility to the membranes. More recently, these copolypeptides have included increasingly complex functionality to assist in cargo loading, vesicle targeting, and vesicle disruption.

In 2004, Deming's lab studied the roles of chain length and block composition on the assembly of uncharged diblock copolypeptide amphiphiles of the general structure: poly(N ε-2-[2-(2-methoxyethoxy)ethoxy]acetyl-L-lysine)-block-poly(L-leucine), or $K^P_x L_y$ (Bellomo et al. 2004) (Figure 17.1). These diblock copolypeptide amphiphiles associate very strongly and possess very low critical aggregation concentrations (i.e., sub μM) in aqueous solution. This property, in most cases, results primarily in the formation of irregular aggregates if the copolymers are simply dispersed in deionized water. A protocol was developed, using organic solvent (THF) and a denaturant (TFA) that allowed annealing of these materials when water is added.

$$(PMe_3)_4Co \xrightarrow{x R^1\text{-NCA}} \xrightarrow{y R^2\text{-NCA}} \xrightarrow{z R^3\text{-NCA}} \text{triblock copolypeptide} \quad (17.8)$$

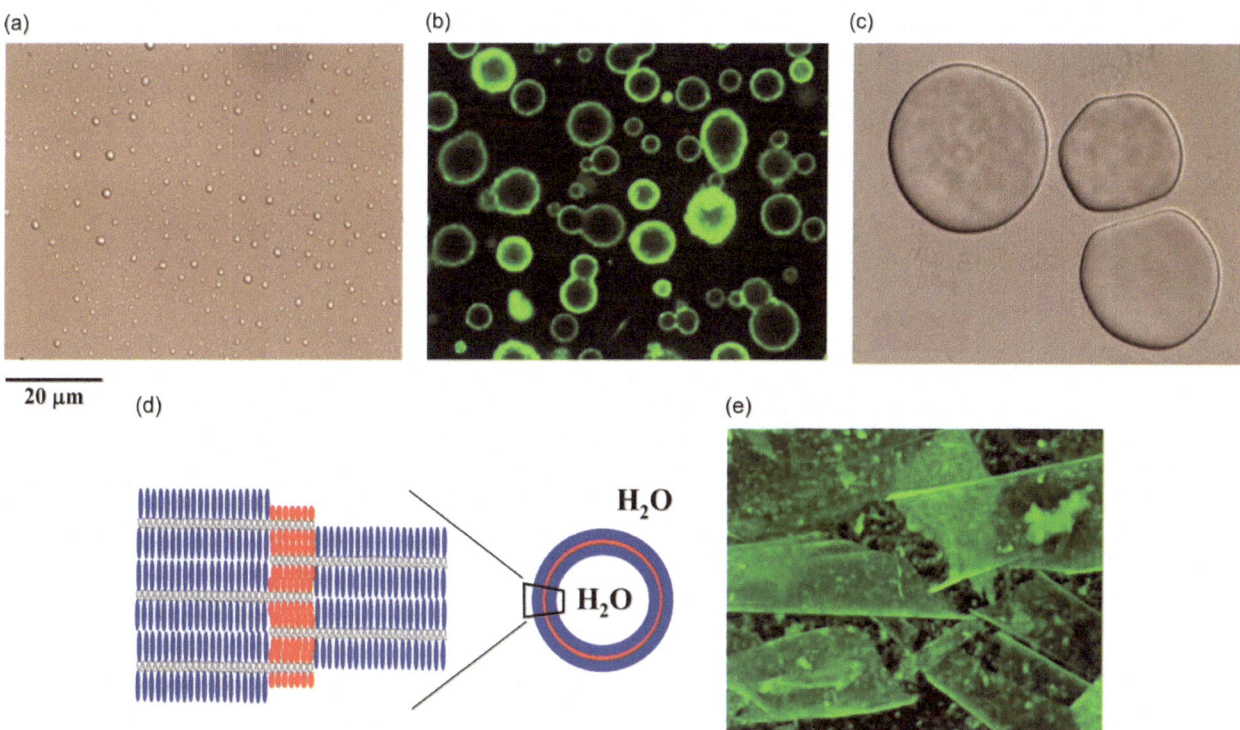

FIGURE 17.1 Optical images of different $K^P_xL_y$ samples for 1.0 wt.% copolymer suspensions in DI water. (a) and (c) = DIC micrographs of $K^P_{60}L_{20}$ and $K^P_{150}L_{40}$ samples, respectively. (b) LSCM image of a $K^P_{100}L_{20}$ suspension visualized with fluorescent dye and a Z-direction slice thickness of 490 nm. (d) Schematic drawing illustrating proposed packing of copolymer chains in vesicle walls. (e) LSCM image of a $K^P_{200}L_{40}$ suspension visualized with fluorescent dye and projection of 40 μm depth composed of 580-nm-thick slices. (Adapted with permission from Bellomo et al. 2004.)

Dialysis of the samples then allows one to obtain regular assemblies in pure water.

Using this procedure, a number of amphiphilic copolymers were studied where the hydrophilic domains were varied from 60 to 200 residues in average length, and the hydrophobic domains were varied from 10 to 75 residues in average length (Bellomo et al. 2004). All block copolypeptides were expected to adopt rod-like conformations due to the strong α-helix forming tendencies of both the leucine and ethylene-glycol-modified lysine residues (Yu et al. 1999). These rod-like conformations provided a low curvature amphiphile interface upon association in water, thus directly tying copolymer conformation to supramolecular structure. Circular dichroism spectroscopy of the copolymers in water confirmed that all samples were fully α-helical. Using differential interference contrast (DIC) optical microscopy, transmission electron microscopy (TEM), laser scanning confocal microscopy (LSCM), and DLS as initial methods to study the assemblies, some trends were identified (Bellomo et al. 2004). When the hydrophobic poly(leucine) domains were less than 20 residues in length, a significant fraction of oblong or irregular micelles (ca. 100 nm diameter) were observed to form by DLS and TEM. When the size of the hydrophilic domain was 100 residues, unilamellar vesicles were observed to form with a size range of approximately 2–15 μm diameter (Figure 17.1). When the hydrophilic domain was increased to 150 residues, the vesicles were much larger in size, approaching 50 μm in diameter. Finally, when the hydrophilic segments were increased to 200 residues long, membrane curvature was hindered such that the major structures formed were extended membrane sheets.

These block copolypeptides, where both hydrophilic and hydrophobic segments were α-helical, gave rise to very stiff membranes, as suggested by the large vesicle diameters and lack of flexibility in the sheets that were formed. Further investigation revealed that these membranes were completely insensitive to osmotic stress, a consequence of their impermeability to water, ions, or other small molecules (Bellomo et al. 2004). They also could not be reduced in size by liposome-type extrusion techniques and could only be made smaller by more aggressive sonication methods. The inability of the uncharged vesicles to pass through small pore diameter filters was likely due to membrane rigidity and virtual absence of chain flexibility. One advantage of these materials for many applications is the media insensitivity of the ethylene glycol coating on the membrane surface. These vesicles were inert towards different ionic media, variations in pH, and the presence of large macromolecules, such as proteins in serum. However, the rigidity of these chains created drawbacks in sample processing, namely the need to use denaturants for vesicle formation, which may be problematic for encapsulation of sensitive materials, and difficulty in preparing nanoscale vesicles due to high membrane rigidity.

In 2005, Lecommandoux's group reported on the self-assembly behaviour of a short, zwitterionic diblock copolypeptide, poly(L-glutamic acid)-b-poly(L-lysine), $E_{15}K_{15}$ (Rodriguez-Hernandez and Lecommandoux 2005). This copolymer has the interesting characteristic that in aqueous solutions near neutral pH (5 < pH < 9), both segments are charged, and the polypeptide is dispersed as soluble chains. However, if pH is lowered to values below pH = 4 or raised above pH = 10, one of the segments is neutralized and the chains self-assemble into small vesicles. By adjustment of pH, vesicles with either anionic (high pH) or cationic (low pH) surfaces could be prepared, hence their description as "schizophrenic" vesicles. It is notable that these chains are soluble in water when both segments are highly charged, considering that the formation of water insoluble polyion complexes between poly(L-lysine) and poly(L-glutamic acid) is well documented (Sela and Katchalski 1959). A key feature of this work is the utilization of short polyelectrolyte segments, which limits such polyion complex formation in dilute solutions.

Also in 2005, Deming's group reported on the assembly of charged diblock copolypeptide amphiphiles, utilizing the structure directing properties of rod-like α-helical segments in only the hydrophobic domains. Specifically, the aqueous self-assembly of a series of poly(L-lysine)-b-poly(L-leucine) block copolypeptides was studied: K_xL_y, where x ranged from 20 to 80, and y ranged from 10 to 30 residues, as well as the poly(L-glutamic acid)-b-poly(L-leucine) block copolypeptide, $E_{60}L_{20}$ (Holowka et al. 2005). In earlier work by Deming's group, it was found that samples with high K to L molar ratios (e.g., $K_{180}L_{20}$) could be dissolved directly into deionized water, yielding transparent hydrogels composed of twisted fibrils (Nowak et al. 2002). Here, it was reasoned that use of shortened charged segments would relax repulsive polyelectrolyte interactions and allow formation of charged polypeptide membranes. Samples were assembled by suspending the polymers in THF/water (1:1), which was followed by dialysis against water to remove all THF. Analysis of these assemblies using DIC optical microscopy revealed the presence of large, sheet-like membranes for $K_{20}L_{20}$, and thin fibrils for $K_{40}L_{20}$. The $K_{60}L_{20}$ sample was most promising, as only large vesicular assemblies were observed by DIC (Holowka et al. 2005).

The $K_{60}L_{20}$ polypeptide vesicles obtained directly from dialysis are polydisperse and range in diameter from ca. 5 μm down to 0.8 μm as determined using DIC and DLS (Figure 17.2). For applications such as drug delivery via

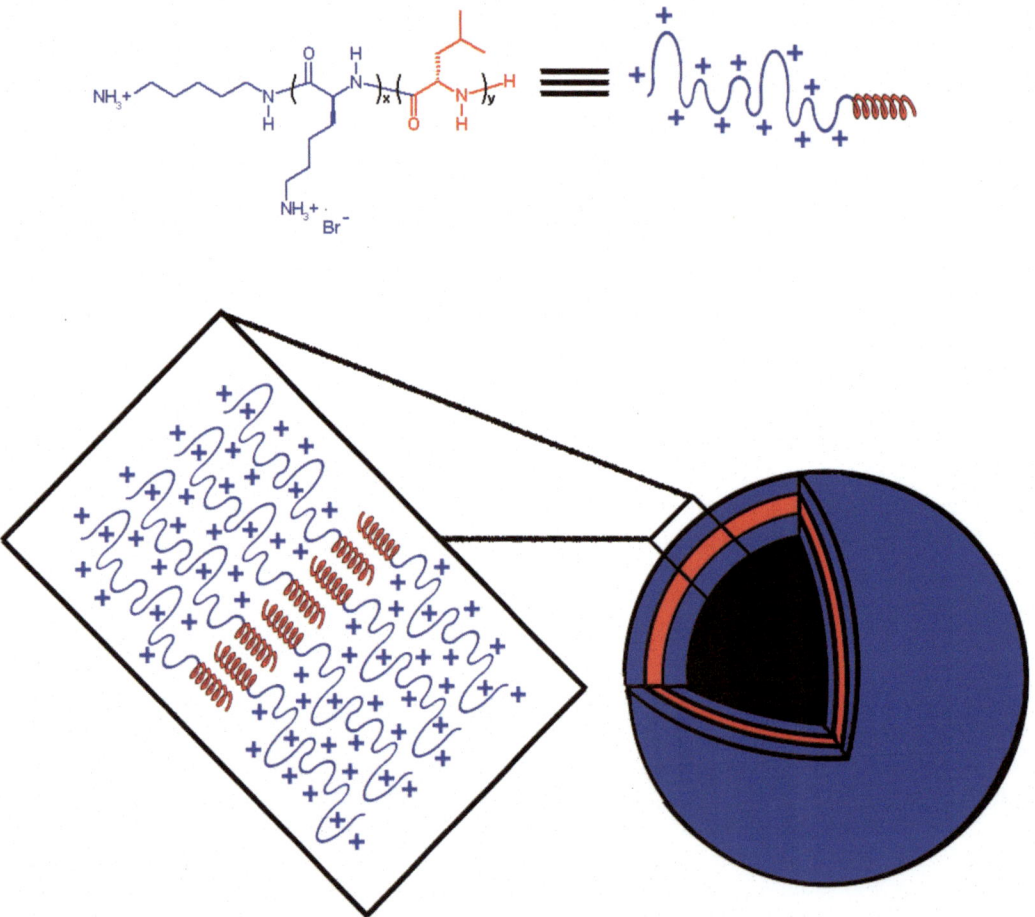

FIGURE 17.2 Schematic drawing showing proposed self-assembly of $K_{60}L_{20}$ into vesicles. (Reprinted with permission from Holowka et al. 2005 copyright 2005 American Chemical Society.)

blood circulation, a vesicle diameter of *ca.* 50–200 nm is desired. It was observed that aqueous suspensions of $K_{60}L_{20}$ vesicles could be extruded through nuclear track-etched polycarbonate (PC) membranes with little loss of polypeptide material. After two passes through a filter, reductions in vesicle diameter to values in close agreement to filter pore size were observed. These results showed that the charged copolypeptide vesicles are readily extruded, allowing good control over vesicle diameter in the tens to hundreds of nanometres range (Figure 17.2). DLS analysis revealed that the extruded vesicles were also less polydisperse than before extrusion and contained no micellar contaminants. The vesicular morphology was confirmed through TEM imaging of the submicron $K_{60}L_{20}$ suspensions. Thus, it appears that the membranes of the $K_{60}L_{20}$ vesicles are more flexible and compliant than those of the purely rod-like uncharged polypeptides described above. Aqueous suspensions of extruded $K_{60}L_{20}$ vesicles were monitored for 6 weeks using DLS and were found to be stable. The vesicles were also found to have high thermal stability. An aqueous suspension of 1 µm vesicles was held at 80°C for 30 minutes, after which no vesicle disruption could be detected (Holowka et al. 2005). Only after heating to 100°C for 30 minutes were the vesicles disrupted, yielding large membrane sheets.

Stability of these highly charged polypeptide vesicles in ionic media is important for use in most applications ranging from personal care products to drug delivery. Although the $K_{60}L_{20}$ vesicles are unstable at high salt concentrations (>1 M), they are stable in 100 mM PBS buffer as well as serum-free DMEM cell culture media (Holowka et al. 2005). Addition of serum, which contains anionic proteins, resulted in vesicle disruption, most likely due to polyion complexation between the serum proteins and the oppositely charged polylysine chains. Accordingly, it was observed that the negatively charged polypeptide vesicles prepared using $E_{60}L_{20}$ are stable in DMEM with 10% foetal bovine serum. These copolypeptides retain much of the stability of the uncharged polypeptide vesicles described earlier but allow straightforward encapsulation and size control due to much simpler processing (Holowka et al. 2005). Another feature of these charged polypeptide vesicles is the potential for facile functionalization of the hydrophilic polypeptide chains at the vesicle surface through either chemical conjugation to amine or carboxylate residues, such as with the cell-targeting protein transferrin (Choe et al. 2013), or by careful choice of charged residues.

Addressing this point, Deming's lab reported the preparation of arginine–leucine (i.e., $R_{60}L_{20}$) vesicles that are able to readily enter cells due to the many guanidinium groups of the arginine segments (Holowka et al. 2007). In this case, the arginine residues played a dual role, where they were both structure directing in vesicle formation, as well as functional for cell binding and entry. Studies on endocytosis and intracellular trafficking of these vesicles revealed that they enter HeLa cells primarily via macropinocytosis (Sun et al. 2011). They were found to then primarily reside in early endosomes, but not in lysosomes, and although some manage to escape into cytoplasm, many are trapped within these compartments. Regardless, another study showed that $R_{60}L_{20}$ vesicles were effective at condensing plasmid DNA and transfecting it into a variety of cell lines, showing the vesicles do have potential for intracellular delivery (Sun et al. 2013). These DNA carriers are advantageous over many other transfection agents due to their low cytotoxicity.

From the pioneering studies on block copolypeptide vesicles described above, design criteria were established for successful vesicle formation: namely use of an α-helical hydrophobic domain connected to a charged hydrophilic domain. Since this original work, many labs have prepared different variants of block copolypeptide vesicles based on this design. In 2007, Hadjichristidis reported lysine–PBLG–lysine (i.e., $K_xPBLG_yK_x$) triblock copolypeptides, where the helical PBLG core favours vesicle formation (Iatrou et al. 2007). In addition, Kimura reported sarcosine-(γ-methyl-L-glutamate) (i.e., Sar_xPMLG_y) block copolypeptides, where the hydrophobic, α-helical PMLG and hydrophilic, disordered Sar segments lead to vesicle formation (Tanisaka et al. 2008). Jing and coworkers prepared vesicle-forming lysine–phenylalanine (K_xF_y) copolypeptides, containing α-helical phenylalanine segments (Sun et al. 2009). These vesicles were also found to be useful in encapsulating haemoglobin and acting as oxygen carriers. Recently, Jan's group also reported vesicle-forming block copolypeptides based on lysine–tyrosine (K_xY_y) sequences (Huang et al, 2012).

Deming's lab also reported the formation of vesicles from dual hydrophilic triblock copolypeptides composed of arginine–glutamate–leucine ($R_xE_yL_z$) or PEGylated lysine–arginine–leucine ($K^P_xR_yL_z$) sequences (Rodriguez et al. 2012). The use of triblock architectures was intended to retain some homoarginine residues for cell uptake but has the majority of the hydrophilic segments be anionic or uncharged to minimize cytotoxicity, all without disrupting vesicle formation. A number of different compositions were prepared, and it was found that, although vesicles exhibiting low cytotoxicity could be formed with a $R_5E_{80}L_{20}$ copolypeptide, the R segments were unable to promote intracellular uptake. With the $K^P_xR_yL_z$ samples, the presence of the "PEGylated" outer blocks was able to diminish cytotoxicity while still allowing the centre R segments to promote cellular uptake (Rodriguez et al. 2012).

Using a different approach towards vesicle formation, Jan and coworkers prepared lysine–glycine (i.e., K_xG_y) copolypeptides, where the polyglycine segment does not adopt a α-helical conformation and has inherent higher flexibility compared with helical segments (Gaspard et al. 2010). Due to the lack of a rigid hydrophobic segment, and due to the hydrophilicity of glycine compared with leucine or phenylalanine, much longer "hydrophobic" segments were needed to drive self-assembly in water and vesicle formation. A $K_{200}G_{50}$ block copolypeptide was found to form vesicles in water using MeOH/H_2O processing and was also mineralized with silica for entrapment of molecules

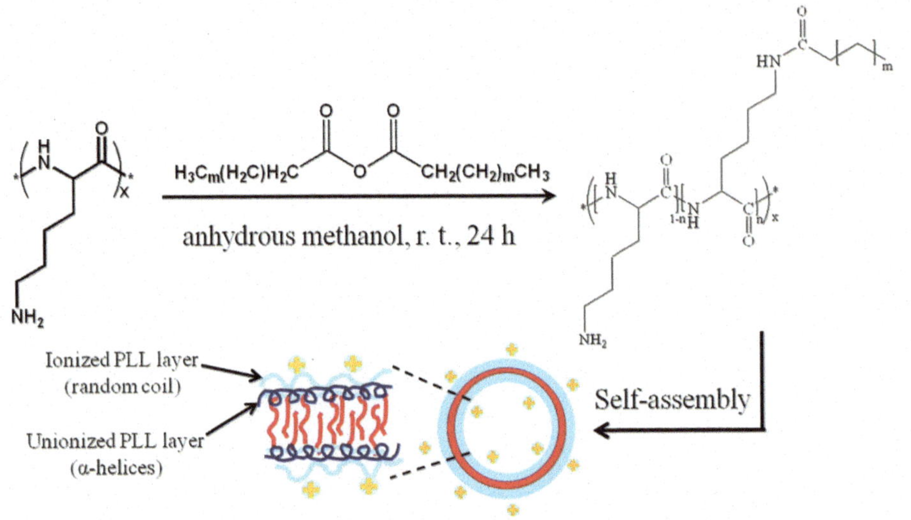

FIGURE 17.3 Schematic illustration of synthesis and self-assembly of alkyl chain-grafted poly(L-lysine) into membranes. (Reprinted with permission from Chen et al. 2015 published by the Royal Society of Chemistry.)

(Lai et al. 2010). In another alternative approach, Lin recently reported lysine-dendronized glutamate copolymers that contained four short glutamate segments connected to each lysine segment (i.e., $K_{40}(E_{15})_4$). These copolymers were found to assemble in water into a variety of structures depending on solution pH (Chen et al. 2013). They found that vesicular structures were formed at pH > 10.

In 2011, Jan's lab reported a new strategy for formation of copolypeptide vesicles that did not rely on the use of amphiphilic block copolypeptides. Instead, they reacted fully hydrophilic polylysine chains (i.e., K_{260} or K_{190}) with substoichiometric quantities of hydrophobic fatty acids to generate amphiphilic statistical copolypeptides containing both lysine and hydrophobically acylated lysine residues (Huang et al. 2011). When modified using hexanoyl groups, vesicles were formed by dissolving the copolypeptides in MeOH followed by dialysis against PBS buffer, and assemblies were found to range in diameter from *ca.* 100 to 350 nm (Figure 17.3). More recent work from this group showed that these vesicles can be functionalized by acylation with sugars (Huang et al. 2013) and that use of longer length fatty acids can lead to increased formation of ordered chain conformations and improved vesicle stabilities (Chen et al. 2014, 2015). These vesicles have been shown to encapsulate myoglobin and doxorubicin and are being developed for drug delivery applications.

17.3.3 FUNCTIONAL VESICLES

Other recent variants of block copolypeptide vesicles have incorporated functionality within one of the segments. In 2010, Deming's lab reported the preparation of lysine–dihydroxyphenylalanine (i.e., $K_{60}DOPA_{20}$)-based vesicles, where the hydrophobic DOPA segments have the added feature of being sensitive towards oxidation (Holowka and Deming 2010). Dihydroxyphenylalanine residues are found naturally in mussel byssus and are important components in the ability of byssal threads to adhere underwater and to cross-link into rigid networks (Waite 1992). In a biomimetic process, $K_{60}DOPA_{20}$ vesicles were oxidized in aqueous media resulting in cross-linking of the vesicle membranes (Eq. 17.9). The resulting membranes were very robust and stable to organic solvents, freeze drying, and osmotic shock. Similar materials, in the form of glutamate–lysine/dihydroxyphenylalanine (i.e., $E_x(K_m/DOPA_n)_y$) copolymers, were reported in 2012 by Qiao (Sulstio 2012), where the hydrophobic domains were statistical copolymers of different ratios (m:n) of lysine and dihydroxyphenylalanine that could be assembled and oxidized to cross-linked vesicles at high pH.

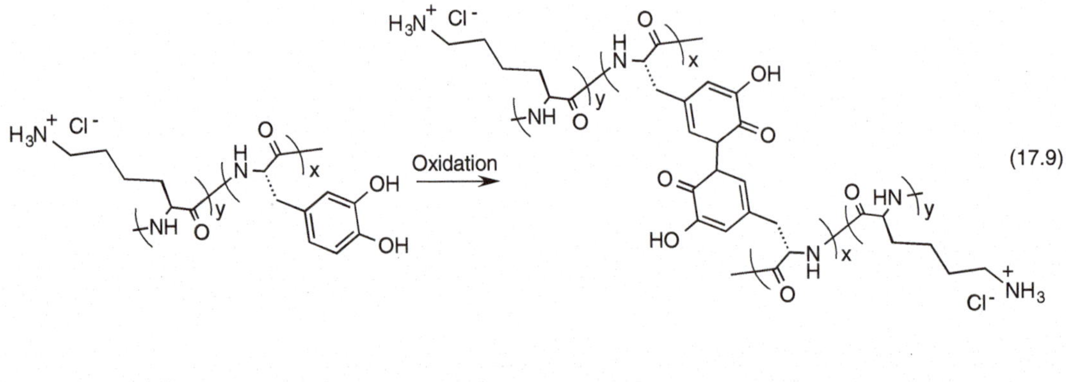

(17.9)

Synthetic Copolypeptide Membranes

$$(PBLG)_x\text{-}[\text{propargyl}]_y \xrightarrow[\text{DMSO}]{\text{galactose-}N_3, \text{Cu(PPh}_3)_3\text{Br}} (PBLG)_x\text{-}[\text{triazole-galactose}]_y \quad (17.10)$$

There is much current interest in synthesis of glycosylated polypeptides, and vesicle-forming amphiphilic copolypeptides that contain sugars in the hydrophilic corona have now also been prepared. In 2012, Lecommandoux and Heise reported the preparation of (Bn-Glu)-propargyl glycine (i.e., $PBLG_{20}PPG_{25}$) diblock copolymers (Huang et al. 2012). The propargyl side chains were then modified by copper-catalysed azide-alkyne cycloaddition with azide-functionalized galactose to give the amphiphilic glycopolypeptide (Eq. 17.10). Since the PPG segment is racemic, it adopts a disordered conformation in glycosylated form. The resulting rod-coil amphiphile was found after DMSO-water processing to assemble into vesicles that were able to bind their complementary lectin. Deming's lab, in 2013, reported a different system prepared from a galactosylated NCA, α,D-galactopyranosyl-L-cysteine (α-gal-C) NCA, and leucine of the composition $(\alpha\text{-gal-C})_{65}L_{20}$, which was able to form vesicles when the side chain thioether functionalities were oxidized to sulfone groups and after THF–water processing (Figure 17.4) (Kramer 2013). The parent glycopolymer, while water soluble, is α-helical, which in the copolymer prohibits formation of small spherical vesicles. The fully oxidized sulfone derivative, i.e., $(\alpha\text{-gal-}C^{O2})_{65}L_{20}$, is more polar, increasing its water solubility, and more importantly gives rise to a disordered conformation yielding a rod-coil amphiphile that assists in vesicle membrane formation.

To find a more economical, flexible, and simple route to functionalize vesicles, Deming's lab has explored the alkylation and oxidation chemistry of poly(L-methionine), M, segments and found these to be efficient and practical methods for creation of a wide array of functionalized polypeptides. Once alkylated or oxidized, hydrophobic M segments become hydrophilic and may be cationic, anionic, zwitterionic, or uncharged, and offer intriguing opportunities for tuning functionality and properties of polypeptide carriers containing M segments (Perlmann and Katchalski 1962, Aiba et al. 1982). Furthermore, some of these modified

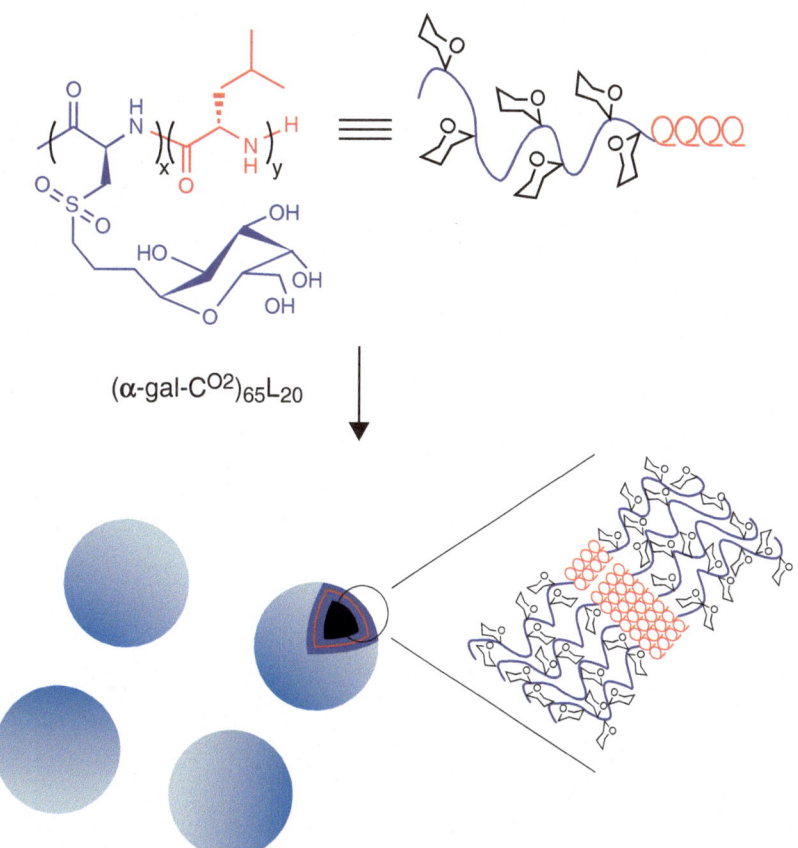

FIGURE 17.4 Schematic showing structures of amphiphilic glycosylated diblock copolypeptides and their assembly into vesicles. (Adapted with permission from Kramer et al. 2013 published by the Royal Society of Chemistry.)

M segments are naturally occurring and possess biological properties that are potentially useful in drug carriers (Toennies 1940, Pitha et al. 1983).

In an effort to create enzyme-responsive polypeptide vesicles, Deming's lab designed amphiphilic copolypeptides containing segments of oxidized methionine residues, which can occur naturally when methionine containing proteins are exposed to reactive oxygen species (ROS). Methionine oxidation in proteins is well known and is believed to help maintain protein activity since these residues act as sacrificial substrates for ROS, preventing irreversible oxidation at critical active site residues such as cysteines (Moskovitz et al. 2001). The possibility to interchange methionine residues between hydrophobic (reduced) and hydrophilic (oxidized) states under biological conditions provided a unique advantage for use of M segments in vesicle assemblies.

Deming designed amphiphilic copolypeptides containing segments of water-soluble methionine sulfoxide, M^O, residues that were prepared by synthesis of a fully hydrophobic precursor diblock copolypeptide, poly(L-methionine)$_{65}$-b-poly(L-leucine$_{0.5}$-stat-L-phenylalanine$_{0.5}$)$_{20}$, $M_{65}(L_{0.5}/F_{0.5})_{20}$, followed by its direct oxidation in water to give the amphiphilic M^O derivative, $M^O_{65}(L_{0.5}/F_{0.5})_{20}$ (Rodriguez et al. 2013). Assembly of $M^O_{65}(L_{0.5}/F_{0.5})_{20}$ in water gave vesicles with average diameters of a few microns that could then be extruded to nanoscale diameters. The M^O segments in the vesicles were found to be substrates for natural intracellular reductase enzymes, which regenerated hydrophobic M segments and resulted in a change in supramolecular morphology that caused vesicle disruption and release of cargos (Figure 17.5).

Initial *in vitro* cell uptake studies revealed that fluorescein-labelled $M^O_{65}(L_{0.5}/F_{0.5})_{20}$ vesicles were not taken up actively by cells, indicating that while the M^O segments provide good biocompatibility, they are inert towards cell membranes. To improve cellular uptake, Deming's lab

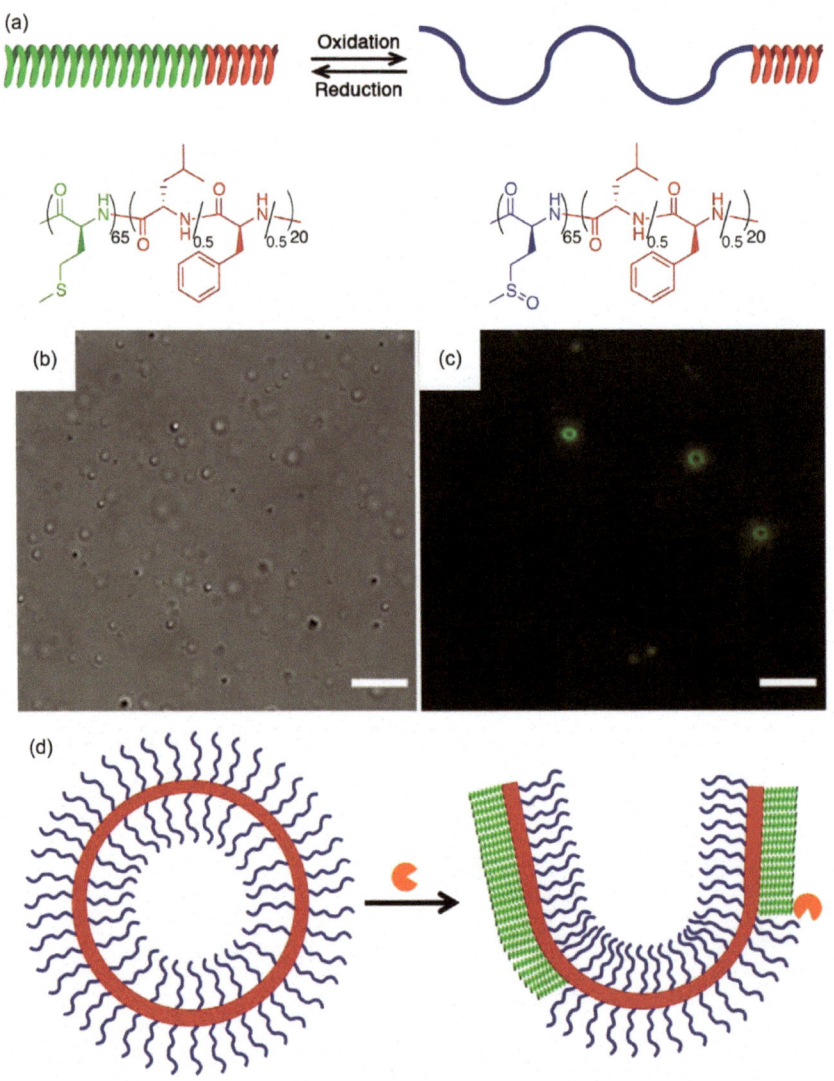

FIGURE 17.5 Schematic showing (a) structure and redox properties, (b) proposed self-assembly of $M^O_{65}(L_{0.5}/F_{0.5})_{20}$ copolypeptides into vesicles, and (c) proposed enzymatic rupture of vesicles. (Adapted with permission from Rodriguez et al. 2013 copyright 2013 American Chemical Society.)

introduced short poly(L-homoarginine), R, segments in the vesicles through the synthesis of the triblock copolypeptides $R_{10}M^O_{55}(L_{0.5}/F_{0.5})_{20}$ and $R_{20}M^O_{55}(L_{0.5}/F_{0.5})_{20}$. Both $R_{10}M^O_{55}(L_{0.5}/F_{0.5})_{20}$ and $R_{20}M^O_{55}(L_{0.5}/F_{0.5})_{20}$ were able to form stable vesicles that could be extruded to diameters less than 200 nm (Rodriguez et al. 2015a). To optimize cell uptake and minimize cytotoxicity, the triblock copolypeptides were mixed with $M^O_{65}(L_{0.5}/F_{0.5})_{20}$ at different ratios to give micron and nanoscale vesicles with tunable R content on vesicle surfaces. Good mixing of the diblock and triblock copolymers within the vesicles was demonstrated by observation of Förster resonance energy transfer (FRET) from fluorescein probes on the diblock chains to tetramethylrhodamine probes on the triblock chains. Increasing the content of either $R_{10}M^O_{55}(L_{0.5}/F_{0.5})_{20}$ or $R_{20}M^O_{55}(L_{0.5}/F_{0.5})_{20}$ in the mixed composition vesicles was found to give improved cell uptake, with the highest degree of uptake found in samples containing 50 mol% triblock or greater.

While $R_{60}L_{20}$ is highly toxic to cells in culture even at low concentrations, pure $R_{10}M^O_{55}(L_{0.5}/F_{0.5})_{20}$ and $R_{20}M^O_{55}(L_{0.5}/F_{0.5})_{20}$ vesicles were found to be much less toxic, confirming that shorter R segments are much better tolerated by cells. Noteworthy are the vesicles composed of 1:1 mixtures of diblock and triblock copolymers, which showed high cell viabilities equivalent to pure $M^O_{65}(L_{0.5}/F_{0.5})_{20}$ and cell only controls. It was found that the nontoxic M^O segments present in the $M^O_{65}(L_{0.5}/F_{0.5})_{20}$ fraction were able to effectively protect cells from the R segments in the triblock chains. Since cell viability was greatest when triblock fractions in the mixed vesicles were 50 mol% or less, and the R_{10} triblocks were less toxic than the R_{20} triblocks, the 1:1 mixture of $R_{10}M^O_{55}(L_{0.5}/F_{0.5})_{20}$ and $M^O_{65}(L_{0.5}/F_{0.5})_{20}$ was selected as an optimized nanovesicle formulation that provides minimal cytotoxicity and good cell uptake (Rodriguez et al. 2015a).

Deming's lab has also recently prepared amphiphilic copolypeptides containing segments of water-soluble methionine sulfonium residues that were derived from a fully hydrophobic precursor diblock copolypeptide, poly(L-methionine)$_{65}$-block-poly(L-leucine$_{0.5}$-stat-L-phenylalanine$_{0.5}$)$_{20}$, $M_{65}(L_{0.5}/F_{0.5})_{20}$, by its direct alkylation in water via reaction with simple alkylating reagents (Rodriguez et al. 2015b). Deming's lab has developed this functionalization reaction as a means to introduce a wide range of functional groups onto polypeptides utilizing the unique reactivity of the thioether group found in methionine. These are chemoselective, broad scope, highly efficient alkylation reactions that can be used in homo- and copolypeptides yielding stable sulfonium derivatives, which mimic methionine sulfoniums found in nature (Kramer and Deming 2012b). Methylation of $M_{65}(L_{0.5}/F_{0.5})_{20}$ gave the cationic methyl-methionine sulfonium derivative, $M^M_{65}(L_{0.5}/F_{0.5})_{20}$, and carboxymethylation gave the zwitterionic carboxymethyl-methionine sulfonium derivative, $M^C_{65}(L_{0.5}/F_{0.5})_{20}$. Assembly of $M^M_{65}(L_{0.5}/F_{0.5})_{20}$ or $M^C_{65}(L_{0.5}/F_{0.5})_{20}$ in water gave rise to vesicles with average diameters of a few microns that could then be extruded to nanoscale diameters (Figure 17.6). While the cationic M^M-based vesicles were found to be cytotoxic, the zwitterionic M^C-based vesicles were found to possess minimal cytotoxicity.

17.4 CONCLUSIONS

Over the past 18 years, vast improvements in NCA polymerizations now allow the synthesis of a variety of copolypeptides of controlled dimensions (molecular weight, sequence, composition, and molecular weight distribution). Many different block copolypeptides have now been prepared and have been used to form self-assembled vesicle structures with a diverse range of promising properties. Some statistical copolypeptides have also been found for vesicles structures at optimized compositions. Early work developed guidelines for formation of vesicular structures, while current work is aimed at increasing the potent functionality and biologically interactive properties of these materials. The ability to easily adjust chain conformation and functionality in copolypeptides, in combination with advanced synthetic methods that enable preparation of complex sequences, has opened up a new, promising field of copolypeptide membranes with a wide range of tunable properties.

ABBREVIATIONS AND SYMBOLS

A:	poly(L-alanine)
Bn-Asp:	β-benzyl-L-aspartate
Bn-Glu:	γ-benzyl-L-glutamate
Bn-Tyr:	O-benzyl-L-tyrosine
bpy:	2,2'-bipyridine
DLS:	dynamic light scattering
DIC:	differential interference contrast
DMAc:	dimethylacetamide
DMEM:	Dulbecco's modified eagle medium
DMF:	dimethylformamide
DMSO:	dimethylsulfoxide
DOPA:	poly(L-dihydroxyphenylalanine)
DP:	degree of polymerization
E:	poly(L-glutamic acid)
EtOAc:	ethyl acetate
F:	poly(L-phenylalanine)
FRET:	Förster resonance energy transfer
G:	poly(glycine)
α-gal-C:	α,D-galactopyranosyl-L-cysteine
α-gal-C^{O2}:	α,D-galactopyranosyl-L-cysteine sulfone
GPC:	gel permeation chromatography
HMDS:	hexamethyldisilazane
K:	poly(L-lysine)
K^P:	poly(ε-2-[2-(2-methoxyethoxy)ethoxy]acetyl-L-lysine)
L:	poly(L-leucine)
LSCM:	laser scanning confocal microscopy
M:	poly(L-methionine)
M^C:	poly(S-carboxymethyl-L-methionine sulfonium)

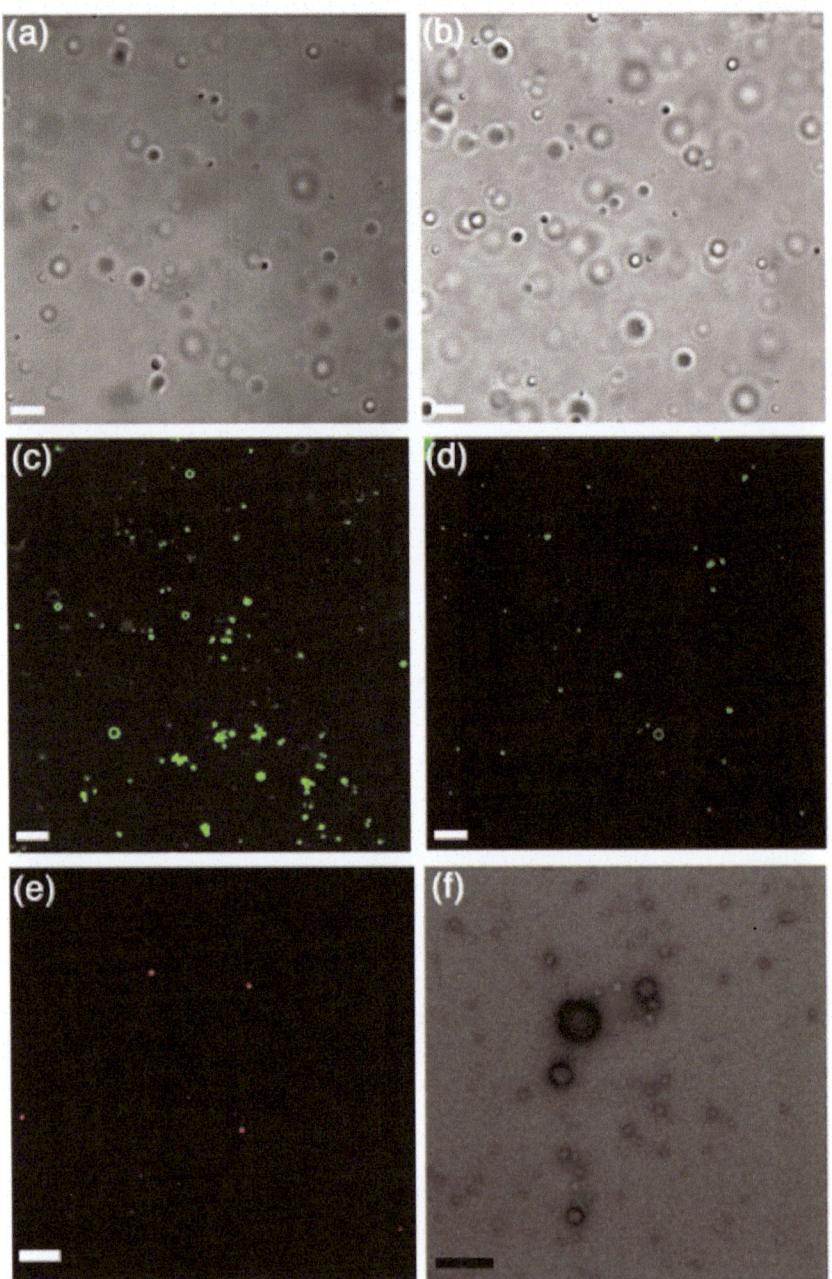

FIGURE 17.6 Images of methionine-sulfonium–containing copolypeptide vesicles. (a) DIC image of a 1% (w/v) $M^M_{65}(L_{0.5}/F_{0.5})_{20}$ aqueous vesicle suspension. (b) DIC image of a 1% (w/v) $M^C_{65}(L_{0.5}/F_{0.5})_{20}$ aqueous vesicle suspension. (c) LSCM image of a fluorescein labelled 1% (w/v) $M^M_{65}(L_{0.5}/F_{0.5})_{20}$ aqueous vesicle suspension. (d) LSCM image of a fluorescein labelled 1% (w/v) $M^C_{65}(L_{0.5}/F_{0.5})_{20}$ aqueous vesicle suspension. (e) LSCM image of a 1% (w/v) $M^M_{65}(L_{0.5}/F_{0.5})_{20}$ aqueous vesicle suspension encapsulating Texas Red dextran. (f) TEM image of negatively stained 0.1% (w/v) $M^M_{65}(L_{0.5}/F_{0.5})_{20}$ aqueous 0.1 μm extruded vesicle suspension. Scale bars: a–e = 5 μm; f = 400 nm. (Reprinted with permission from Rodriguez et al. 2015b.)

M^M:	poly(S-methyl-L-methionine sulfonium)	NACE:	nonaqueous capillary electrophoresis
M^O:	poly(L-methionine sulfoxide)	NCA:	α-amino acid N-carboxyanhydride
M_n:	number average molecular weight	PBLA:	poly(β-benzyl-L-aspartate)
M_w:	weight average molecular weight	PBS:	phosphate-buffered saline
MALDI-MS:	matrix-assisted laser desorption ionization mass spectroscopy	PBLG:	poly(γ-benzyl-L-glutamate)
		PC:	polycarbonate
MeCN:	acetonitrile	PMLG:	poly(γ-methyl-L-glutamate)
MeOH:	methanol	PMe_3:	trimethylphosphine

PPG:	poly(*racemic*-propargylglycine)
PZLL:	poly(ε-carbobenzyloxy-L-lysine)
R:	poly(L-homoarginine)
ROS:	reactive oxygen species
Sar:	poly(sarcosine)
TEM:	transmission electron microscopy
TFA:	trifluoroacetic acid
TFA-Lys:	ε-trifluoroacetyl-L-lysine
THF:	tetrahydrofuran
TMS:	trimethylsilyl
Y:	poly(L-tyrosine)
Z-Lys:	ε-carbobenzyloxy-L-lysine.

REFERENCES

Aiba, S.; Minoura, N. and Fujiwara, Y. 1982. Disordering of helix of oxidized l-methionine containing copolypeptides. *Makromol. Chemie* 183: 1333–42.

Aliferis, T.; Iatrou, H. and Hadjichristidis, N. 2004. Living polypeptides. *Biomacromolecules* 5: 1653–6.

Auer, H. E. and Doty, P. 1966. The synthesis, structure, and optical properties of some copolypeptides containing nonpolar amino acid residues. *Biochemistry* 5: 1708–15.

Bellomo, E.; Wyrsta, M. D.; Pakstis, L.; Pochan, D. J. and Deming, T. J. 2004. Stimuli responsive polypeptide vesicles via conformation specific assembly. *Nat. Mater.* 3: 244–8.

Branden, C,. and Tooze, J. 1991. *Introduction to Protein Structure.* New York: Garland.

Cao, H.; Yao, J.; Shao, Z. 2012. Synthesis of poly (γ-benzyl-L-glutamate) with well-defined terminal structures and its block polypeptides with alanine, leucine and phenylalanine. *Polym. Int.* 61: 774–9.

Cardinaux, F.; Howard, J. C.; Taylor, G. T. and Scheraga, H. A. 1977. Block copolymers of amino-acids. 1. Synthesis and structure of copolymers of L-alanine or L-phenylalanine with D,L-lysine-D7 or D,L-lysine. *Biopolymers* 16: 2005–28.

Chen, B-Y.; Huang, Y-C. and Jan, J-S. 2015. Molecular assembly of alkyl chain-grafted poly(L-lysine) tuned by backbone chain length and grafted alkyl chain. *RSC Adv.* 5: 22783–91.

Chen, B-Y.; Huang, Y-F.; Huang, Y-C.; Wen, T-C. and Jan, J-S. 2014. Alkyl chain-grafted poly(L-lysine) vesicles with tunable molecular assembly and membrane permeability. *ACS Macro Lett.* 3: 220–3.

Chen, C.; Wang, Z. and Li, Z. 2011. Thermoresponsive polypeptides from pegylated poly-L-glutamates. *Biomacromolecules* 12: 2859–63.

Chen, L.; Chen, T.; Fang, W.; et al. 2013. Synthesis and pH-responsive "schizophrenic" aggregation of a linear-dendron-like polyampholyte based on oppositely charged polypeptides. *Biomacromolecules* 14: 4320–30.

Choe, U-J.; Rodriguez, A. R.; Lee, B. S.; Knowles, S. M.; Wu, A. M.; Deming, T. J. and Kamei, D. T. 2013. Endocytosis and intracellular trafficking properties of transferrin-conjugated block copolypeptide vesicles. *Biomacromolecules* 14: 1458–64.

Collman, J. P.; Hegedus, L. S.; Norton, J. R. and Finke, R. G. 1987. *Principles and Applications of Organotransition Metal Chemistry* 2nd Ed., Mill Valley: University Science.

Deming, T. J. 1997. Facile synthesis of block copolypeptides of defined architecture. *Nature* 390: 386–9.

Deming, T. J. 1999. Cobalt and iron initiators for the controlled polymerization of alpha-amino acid-N-carboxyanhydrides. *Macromolecules* 32: 4500–2.

Deming, T. J. (Ed.) 2012. Peptide-based materials. *Top. Curr. Chem.* 310: 1–171.

Deming, T. J. 2013. Synthesis and self-assembly of well defined block copolypeptides via controlled NCA polymerization. *Adv. Polym. Sci.* 262: 1–38.

Deming, T. J. 2014. Preparation and development of block copolypeptide vesicles and hydrogels for biological and medical applications. *WIREs Nanomedicine and Nanobiotechnology* 6: 283–97.

Deming, T. J. and Curtin, S. A. 2000. Chain initiation efficiency in cobalt- and nickel-mediated polypeptide synthesis. *J. Amer. Chem. Soc.* 122: 5710–7.

Dimitrov, I. and Schlaad, H. 2003. Synthesis of nearly monodisperse polystyrene-polypeptide block copolymers via polymerisation of N-carboxyanhydrides. *Chem. Commun.* 2944–5.

Epand, R. E. and Scheraga, H. A. 1968. Conformations of poly-L-valine in solution. *Biopolymers* 6: 1551–71.

Fischer, H. 2001. The persistent radical effect: A principle for selective radical reactions and living radical polymerizations. *Chem. Rev.* 101: 3581–610.

Gaspard J.; Silas, J. A.; Shantz, D. F. and Jan, J-S. 2010. Supramolecular assembly of lysine-b-glycine block copolypeptides at different solution conditions. *Supramolecular Chem.* 22: 178–85.

Gibson, M. I. and Cameron, N. R. 2008. Organogelation of sheet–helix diblock copolypeptides. *Angew. Chem. Int. Ed.* 47: 5160–62.

Graf, R.; Spiess, H. W.; Floudas, G.; Butt, H.-J.; Gkikas, M. and Iatrou, H. 2012. Conformational transitions of poly(l-proline) in copolypeptides with Poly(γ-benzyl-l-glutamate) induced by packing. *Macromolecules* 45: 9326–32.

Habraken, G. J. M.; Peeters, M.; Dietz, C. H. J. T.; Koning, C. E. and Heise, A. 2010. How controlled and versatile is N-carboxyanhydride (NCA) polymerization at 0°C? Effect of temperature on homo-, block- and graft (co)polymerization. *Polym. Chem.* 1: 514–24.

Habraken, G. J. M.; Wilsens, K. H. R. M.; Koning, C. E. and Heise, A. 2011. Optimization of *N*-carboxyanhydride (NCA) polymerization by variation of reaction temperature and pressure. *Polym. Chem.* 2: 1322–30.

Holowka, E. P. and Deming, T. J. 2010. Synthesis and crosslinking of L-DOPA containing polypeptide vesicles. *Macromol. Biosci.* 10: 496–502.

Holowka, E. P.; Pochan, D. J. and Deming, T. J. 2005. Charged polypeptide vesicles with controllable diameter. *J. Amer. Chem. Soc.* 127: 12423–28.

Holowka, E. P.; Sun, V. Z.; Kamei, D. T. and Deming, T. J. 2007. Polyarginine segments in block copolypeptides drive both vesicular assembly and intracellular delivery. *Nat. Mater.* 6: 52–7.

Huang, J.; Bonduelle, C.; Thévenot, J.; Lecommandoux, S. and Heise, A. 2012. Biologically active polymersomes from amphiphilic glycopeptides. *J. Amer. Chem. Soc.* 134: 119–22.

Huang, Y-C.; Arham, M. and Jan, J-S. 2011. Alkyl chain grafted poly(L-lysine): self-assembly and biomedical application as carriers. *Soft Matter* 7: 3975–83.

Huang, Y-C.; Arham, M. and Jan, J-S. 2013. Bioactive vesicles from saccharide- and hexanoyl-modified poly(L-lysine) copolypeptides and evaluation of the cross-linked vesicles as carriers of doxorubicin for controlled drug release. *Eur. Poly. J.* 49: 726–37.

Huang, Y-C.; Yang, Y-S.; Lai, T-Y. and Jan, J-S. 2012. Lysine-block-tyrosine block copolypeptides: Self-assembly, cross-linking, and conjugation of targeted ligand for drug encapsulation. *Polymer* 53: 913–22.

Hwang, J. and Deming, T.J. 2001. Methylated mono- and di(ethyleneglycol)-functionalized β-sheet forming polypeptides. *Biomacromolecules* 2: 17–21.

Iatrou, H.; Frielinghaus, H.; Hanski, S. et al. 2007. Architecturally induced multiresponsive vesicles from well-defined polypeptides: Formation of gene vehicles. *Biomacromolecules* 8: 2173–81.

Ingwall, R. T.; Scheraga, H. A.; Lotan, N.; Berger, A. and Katchalski, E. 1968. Conformational studies of poly-L-alanine in water. *Biopolymers* 6: 331–68.

Knobler, Y.; Bittner, S. and Frankel, M. 1964. Reaction of N-carboxy-alpha-amino-acid anhydrides with hydrochlorides of hydroxylamine O-alkylhydroxylamines+amines syntheses of amino-hydroxamic acids amido-oxy-peptides+alpha-amino-acid amides. *J. Chem. Soc.* 3941–51.

Knobler, Y.; Bittner, S.; Virov, D. and Frankel, M. 1969. Alpha-aminoacyl derivatives of aminobenzoic acid and of amino-oxy-acids by reaction of their hydrochlorides with amino-acid N-carboxyanhydrides. *J. Chem. Soc. C* 1821–4.

Kramer, J. R. and Deming, T. J. 2010a. Glycopolypeptides via living polymerization of glycosylated-L-lysine N-carboxyanhydrides. *J. Amer. Chem. Soc.* 132: 15068–71.

Kramer, J. R. and Deming, T. J. 2010b. General method for purification of β-amino acid-N-carboxyanhydrides using flash chromatography. *Biomacromolecules* 11: 3668–72.

Kramer, J. R. and Deming, T. J. 2012a. Glycopolypeptides with a redox-triggered helix-to-coil transition. *J. Amer. Chem. Soc.* 134:4112–5.

Kramer, J. R. and Deming, T. J. 2012b. Preparation of multifunctional and multireactive polypeptides via methionine alkylation. *Biomacromolecules* 13: 1719–23.

Kramer, J. R. and Deming, T. J. 2014. Multimodal switching of conformation and solubility in homocysteine derived polypeptides. *J. Am. Chem. Soc.* 136: 5547–50.

Kramer, J. R.; Rodriguez, A. R.; Choe, U-J.; Kamei, D. T. and Deming T. J. 2013. Glycopolypeptide conformations in bioactive block copolymer assemblies influence their nanoscale morphology. *Soft Matter* 9: 3389–95.

Kubota, S. and Fasman, G. D. 1975. The β conformation of polypeptides of valine, isoleucine, and threonine in solution and solid-state: Optical and infrared studies. *Biopolymers* 14: 605–31.

Lai J-K.; Chuang T-H.; Jan J-S. and Wang, S-S. 2010. Efficient and stable enzyme immobilization in a block copolypeptide vesicle-templated biomimetic silica support. *Coll. Surf. B: Biointerfaces* 80: 51–8.

Li, Z. and Deming, T. J. 2010. Tunable hydrogel morphology via self-assembly of amphiphilic pentablock copolypeptides. *Soft Matter* 6: 2546–51.

Lu, H. and Cheng, J. 2007. Hexamethyldisilazane-mediated controlled polymerization of alpha-amino acid N-carboxyanhydrides. *J. Amer. Chem. Soc.* 129: 14114–5.

Mondeshki, M.; Spiess, H. W.; Aliferis, T.; Iatrou, H.; Hadjichristidis, N. and Floudas, G. 2011. Hierarchical self-assembly in diblock copolypeptides of poly(γ-benzyl-l-glutamate) with poly poly(l-leucine) and poly(O-benzyl-l-tyrosine). *Eur. Polym. J.* 47: 668–74.

Moskovitz, J.; Bar-Noy, S.; Williams, W. M.; Requena, J.; Berlett, B. S. and Stadtman, E. R. 2001. Methionine sulfoxide reductase (MsrA) is a regulator of antioxidant defense and lifespan in mammals. *Proc. Natl. Acad. Sci. USA* 98: 12920–5.

Nguyen, L-T.; Vorenkamp, E. J.; Daumont, C. J. M.; ten Brinke, G. and Schouten, A. J. 2010. Annealing-induced changes in double-brush langmuir-blodgett films of β-helical diblock copolypeptides. *Polymer* 51:1042–55.

Nowak, A. P.; Breedveld, V.; Pakstis, L.; Ozbas, B.; Pine, D. J.; Pochan, D. J. and Deming, T. J. 2002. Rapidly recovering hydrogel scaffolds from self-assembling diblock copolypeptide amphiphiles. *Nature* 417: 424–8.

Nowak, A. P.; Sato, J.; Breedveld, V. and Deming, T. J. 2006. Hydrogel formation in amphiphilic triblock copolypeptides. *Supramolecular Chemistry* 18: 423–7.

Odian, G. 1991. *Principles of Polymerization 3rd* Ed. New York: Wiley.

Ostroy, S. E.; Lotan, N.; Ingwall, R. T. and Scheraga, H. A. 1970. Effect of side-chain hydrophobic bonding on the stability of homopolyamino acid α-helices: Conformational studies of poly-L-leucine in water. *Biopolymers* 9: 749–64.

Perlmann, G. E. and Katchalski, E. 1962. Conformation of poly-L-methionine and some of its derivatives in solution. *J. Amer. Chem. Soc.* 84: 452–7.

Pickel, D. L.; Politakos, N.; Avgeropoulos, A. and Messman, J. M. 2009. A mechanistic study of alpha-(Amino acid)-N-carboxyanhydride polymerization: Comparing initiation and termination events in high-vacuum and traditional polymerization techniques. *Macromolecules* 42: 7781–8.

Pitha, J.; Szente, L. and Greenberg, J. 1983. Poly-L-methionine sulfoxide: A biologically inert analogue of dimethyl sulfoxide with solubilizing potency. *J. Pharm. Sci.* 72: 665–8.

Rhodes, A. J. and Deming, T. J. 2013. Soluble, clickable polypeptides from azide-containing N-carboxyanhydride monomers. *ACS Macro Lett.* 2: 351–4.

Rodriguez, A. R.; Choe, U-J.; Kamei, D. T. and Deming, T. J. 2012. Fine tuning of vesicle assembly and properties using dual hydrophilic triblock copolypeptides. *Macromol. Biosci.* 12: 805–11.

Rodriguez, A. R.; Choe, U-J.; Kamei, D. T. and Deming, T. J. 2015a. Blending of diblock and triblock copolypeptide amphiphiles yields cell penetrating vesicles with low toxicity. *Macromol. Biosci.* 15: 90–7.

Rodriguez, A. R.; Choe, U-J.; Kamei, D. T. and Deming, T. J. 2015b. Use of methionine alkylation to prepare cationic and zwitterionic block copolypeptide vesicles. *Israel J. Chem.* DOI: 10.1002/ijch.201400116.

Rodriguez, A. R.; Kramer, J. R. and Deming, T. J. 2013. Enzyme-triggered cargo release from methionine sulfoxide containing copolypeptide vesicles. *Biomacromolecules* 14: 3610–4.

Rodriguez-Hernandez, J. and Lecommandoux, S. 2005. Reversible inside-out micellization of pH-responsive and water-soluble vesicles based on polypeptide diblock copolymers. *J. Amer. Chem. Soc.* 127: 2026–7.

Rutjes, F. P. J. T. and van Hest, J. C. M. 2011. Polymeric vesicles in biomedical applications. *Polym. Chem.* 2: 1449–62.

Schaefer, K. E.; Keller, P. and Deming, T. J. 2006. Thermotropic polypeptides bearing side-on mesogens. *Macromolecules* 39: 19–22.

Sekiguchi, H. 1981. Mechanism of N-carboxy-alpha-amino acid anhydride (Nca) polymerization. *Pure Appl. Chem.* 53: 1689–714.

Sela, M. and Katchalski, E. 1959. Biological properties of poly-α-amino acids. *Adv. Protein Chem.* 14: 391–478.

Sun, J.; Huang, Y.; Shi, Q.; Chen, X. and Jing, X. 2009. Oxygen carrier based on hemoglobin/poly(L-lysine)-block-poly(L-phenylalanine) vesicles. *Langmuir* 25: 13726–9.

Sun, V. Z.; Choe, U-J.; Rodriguez, A. R.; Dai, H.; Deming, T. J. and Kamei, D. T. 2013. Transfection of mammalian cells using block copolypeptide vesicles. *Macromol. Biosci.* 13: 539–50.

Sun, V. Z.; Li, Z.; Deming, T. J. and Kamei, D. T. 2011. Intracellular fates of cell penetrating block copolypeptide vesicles. *Biomacromolecules* 12: 10–3.

Tanisaka, H.; Kizaka-Kondoh, S.; Makino, A.; Tanaka, S.; Hiraoka, M. and Kimura, S. 2008. Near-infrared fluorescent labeled peptosome for application to cancer imaging. *Bioconjugate Chem.* 19: 109–17.

Thunig, D.; Semen, J. and Elias, H-G. 1977. Carbon dioxide influence on NCA polymerizations. *Makromol. Chem.* 178: 603–7.

Toennies, G. 1940. Sulfonium reactions of methionine and their possible metabolic significance. *J. Biol. Chem.* 132: 455–6.

Vayaboury, W.; Giani, O.; Cottet, H.; Deratani, A. and Schue, F. 2004. Living polymerization of alpha-amino acid N-carboxyanhydrides (NCA) upon decreasing the reaction temperature. *Macromol. Rapid Commun.* 25: 1221–4.

Uralil, F.; Hayashi, T.; Anderson, J. M. and Hiltner, A. 1977. New block copolymers of α-amino acids. *Polym. Eng. Sci.* 17: 515–22.

Voet, D. and Voet, J. G. 1995. *Biochemistry 2nd Ed*. New York: Wiley.

Waite, J. H. 1992. The DOPA ephemera: A recurrent motif in invertebrates. *Biol. Bull.* 183: 178–84.

Webster, O. W.; Hertler, W. R.; Sogah, D. Y.; Farnham, W. B. and Rajanbabu, T. V. 1983. Group-transfer polymerization.1. A new concept for addition polymerization with organosilicon initiators. *J. Amer. Chem. Soc.* 105: 5706–8.

Woodward, R. B. and Schramm, C. H. 1947. Synthesis of protein analogs. *J. Amer. Chem Soc.* 69: 1551–2.

Yakovlev, I. and Deming, T. J. 2014. Analogues of poly(L-phosphoserine) via living polymerization of phosphonate-containing N-carboxyanhydride monomers. *ACS Macro Lett.* 3: 378–81.

Yakovlev, I. and Deming, T. J. 2015. Controlled Synthesis of Phosphorylcholine Derivatives of Poly(serine) and Poly(homoserine). *J. Amer. Chem. Soc.* 137: 4078–81.

Yu, M.; Nowak, A. P.; Deming, T. J. and Pochan, D. J. 1999. Methylated mono- and diethyleneglycol functionalized polylysines: Nonionic, alpha-helical, water-soluble polypeptides. *J. Amer. Chem. Soc.* 121: 12210–1.

18 Synthetic Membranes from Block Copolymers, Recombinant Proteins, and Dendrimers

Daniel A. Hammer, Zhichun Wang, Ellen Reed, Chen Gao, and Kevin B. Vargo
University of Pennsylvania, Philadelphia, Pennsylvania

CONTENTS

18.1 Introduction 323
18.2 Membranes from Polymers 323
18.3 Membranes from Proteins 326
18.4 Membranes from Recombinant Proteins *In Vitro* 326
18.5 *In Situ* Assembly 332
18.6 Membranes from Dendrimers 332
18.7 Summary 333
Acknowledgement 333
Bibliography 333

18.1 INTRODUCTION

The plasma membranes of cells are typically assembled from bilayers of amphiphiles with large concentrations of embedded proteins. In nature, the bilayer itself is normally comprised of diacyl chain phospholipids, with various appendages to the head group. A family of solutes often found in mammalian bilayers are the steroidal amphiphiles, such as cholesterol. Vesicles can be assembled from pure components (such as steroyl oleoyl phosphatidylcholine, SOPC) or mixtures of lipids and steroids. Single acyl chain surfactants can also be assembled into bilayers, such as zwitterionic surfactants [34]. Normally, it is thought that amphiphiles must form certain characteristic shapes to pack appropriately to form membranes [32], though, as zwitterionic surfactants and numerous examples shown below will illustrate, there are numerous exceptions to the rule, owing to the diversity of chemical interactions that are found in membrane-forming amphiphiles.

The realization that membranes can be made from other molecules that, like lipids, form lamellar phases, has led to the development of synthetic membranes and vesicles from block copolymers and Janus dendrimers. Correspondingly, making membranes from synthetic amphiphiles allows the introduction of novel functionality and responsiveness. The assembly of membranes from block copolymers, for instance, allows the introduction of specific functional groups, such as unsaturated side chains that allow for crosslinking [14] or light-sensitive intrachain linkers that lead to chain scission upon illumination [37]. Carbohydrates that can specifically bind to lectins can be incorporated into dendrimer membranes to mimic glycan components of natural bilayers [66].

To develop bioresponsive materials, it is desirable to include peptide or protein sequences into these amphiphiles. There has been some success in using click chemistry and other chemical attachment strategies for incorporating peptides or proteins into polymersome membranes [45]. Another type of amphiphile that can be engineered to incorporate amino acid sequences is the peptide amphiphile, in which a peptide made by solid-phase synthesis is attached to a lipid head group [44]. Although it is laborious to make and purify these hybrid molecules, numerous peptide motifs have been incorporated into the molecules with great success. Their ability to assemble into spherical micelles has been useful for drug delivery and imaging [56], whereas their ability to assemble into worm-like micelles has been useful for tissue engineering [12]. In both cases, the peptides can include motifs that target receptors for adhesion or growth factor receptor activation.

18.2 MEMBRANES FROM POLYMERS

The idea that membranes could be made from block copolymers in aqueous solution seems almost quaint now, although in the late 1990s, it was not so obvious that it could be done. Pioneering work by Eisenberg had shown that various diblock copolymers could be assembled into lamellar phase vesicles in various organic solvents [64]. The synthetic acumen of Frank S. Bates' laboratory at the University of Minnesota, which first identified polymers that could form the lamellar phase in the melt and then in aqueous solution [23,24], paved the way for the facile assembly of vesicles from diblock copolymers of poly(ethylene oxide)-b-poly(ethyl ethylene). Although the first polymersomes were assembled using this chemistry [15] (see Figure 18.1), the bulk of the early work on the materials properties of polymersomes was done with poly(ethyleneoxide)-b-poly(butadiene) [6,14], which possesses the exquisitely wonderful property that it can be cross-linked through condensation of unsaturated side chains, leading to the formation of ultrastrong vesicles that could be dehydrated and rehydrated after cross-linking [14].

There have been extensive reviews of polymer vesicles (see, for example, Ref. [16]), and there is no space here for a comprehensive review. Our laboratory, in collaboration with Michael Therien (now at Duke University) and Jason Burdick (University of Pennsylvania), focussed on the construction of vesicles from poly(ethylene oxide)-b-poly(caprolactone). We chose this system because both poly(ethylene oxide) (PEO) and poly(caprolactone) (PCL) are biocompatible polymers, and it is unlikely that very many other polymer chemistries that yield polymersomes would meet with regulatory approval. There is an extremely small phase space for which PEO-b-PCL makes vesicles, perhaps owing to the high crystallinity of the PCL blocks [51]. One construct, PEO (2K)-b-PCL (12K), assembles into vesicles, and owing to an ester bond in the backbone, is subject to hydrolytic degradation at acidic pHs, as is seen within endosomes [20]. The release of entrapped doxorubicin, a neoplastic agent, at low pH was demonstrated [20].

Synthetic polymer chemistry can be used to make polymer chains with specific functional groups that might not be achievable with naturally occurring surfactants, such as lipids. One illustration was the insertion of a functional group between the hydrophilic PEO and hydrophobic PCL blocks. The scheme was based on inserting a molecule that had both terminal amine and a terminal carboxylic acid group [37], such as an amino acid or a modified amino acid. Using this scheme, we inserted either 2-nitrophenylalanine (NPA), a photolabile amino acid analogue with sensitivity to light at a wavelength of 345 nm, or a fluorescent fluorescein-conjugated lysine, at the junction between the blocks. The insertion of either moiety did not prevent the assembly of vesicles. Illumination of NPA-containing vesicles at 345 nm led to the disassembly of vesicles – opening up the possibility of light-activated release of luminal contents – and vesicles containing a polymer with an embedded fluorescein-linked amino acid analogue could be made fluorescent with an intensity that depended on the extent of incorporation of fluorescein polymers [37].

Another route to photodestruction of polymer vesicles is based on the incorporation of porphyrin dyes within the hydrophobic core of the membrane. The dyes we used have been synthesized by the Therien laboratory and have the useful feature that they can be excited to emit light in the near IR, which makes them suitable for *in vivo* imaging [55]. The porphyrins are constructed from monomers that are connected, and the precise spectra of excitation and emission depend on the number of groups connected, how they are connected chemically, and the identity of the side chains. The molecules range in size from just over 1 nm in length (for monomeric porphyrins) to 6 nm (for pentamers) [19]. However, it is difficult to package these molecules, since they are hydrophobic. In collaboration with the Therien lab, we demonstrated that the molecules can be assembled into the hyperthick membrane of polymersomes as a hydrophobic solute; in contrast, molecules greater in size than a monomer could not be solubilized in phospholipid membranes, owing to their small membrane thickness [19]. Very high concentrations of porphyrin, up to a 1:10 porphyrin to polymer ratio, could be included

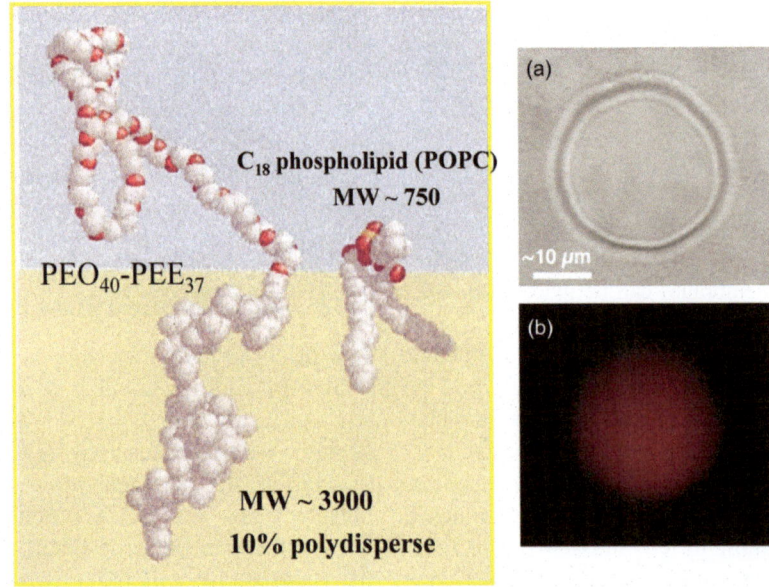

FIGURE 18.1 The original polymersome was made from poly(ethylene oxide)$_{40}$-b-poly(butadiene)$_{37}$ imaged immediately after electroformation (a) and was demonstrated to entrap a 10K-fluorecent Dextran (b) from Ref. [15].

while still maintaining the assembly of vesicles [21]. These vesicles ultimately are extremely bright optical probes that can be used for imaging science. For example, we used them to track dendritic cells during *in vivo* administration in a mouse model of cancer immune therapy [10].

However, by happenstance, we discovered an unusual and potentially useful optical switch when we tried to embed certain solutes within the lumen of porphyrin-containing vesicles. When embedding either ferritin [52] or Dextran [36] in the lumen of the vesicle, illumination of vesicles that contain porphyrin in the membrane shows materials defects, ranging from budding to tube formation and to complete failure [36]. The magnitude of the catastrophe depended on the molecular weight of the Dextran, with lower molecular-weight Dextrans leading to small surface defects (budding), where larger molecular-weight Dextrans led to complete failure [36]. Based on multiple lines of evidence, our interpretation of this effect is that because porphyrins have a low quantum yield, heat is generated upon their illumination in the membrane. When the inner leaflet is constrained during illumination, the outer leaflet expands and exposes the hydrophobic core, causing a defect. This interpretation is consistent with the increase in membrane stiffness in the presence of Dextran (observed with micropipette aspiration), and the finding that the fluorescently labelled Dextran associated with the membrane is immobile (as imaged with fluorescence recovery after photobleaching) [36]. The ability to illuminate vesicles and have them release their contents locally has potential for the local administration of drugs *in vivo*.

Another interesting property of porphyrin-containing polymersomes is that the optical emission of porphyrins in the membrane can be altered, through either the concentration of porphyrins or their orientation, which can be manipulated with stress. The origin of these effects is twofold. First, porphyrins have side chains that interact with polymer chains, which constrains the orientation of the porphyrins; this can prevent fully functional electrical continuity along the porphyrin backbone, which can lead to red- or blue-shifting of fluorescence [22]. Second, at higher concentrations, porphyrins can stack horizontally in the membrane, and the electron clouds of adjacent molecules can communicate – so-called z-stacking – which red-shifts the fluorescence [35]. This latter effect was used to make a soft-matter stress sensor, in which the tension in the membrane, regulated though micropipette aspiration, was directly correlated to the optical response of the porphyrins [35]. When the membrane is stressed, the conformational freedom of the porphyrins increases, owing to the significant areal strain achievable in polymersomes; the porphyrin is twisted, and the fluorescence blue-shifts. When the membrane is in a relaxed state, the porphyrin is planar and must z-stack due to limited space; in this case, the fluorescence is red-shifted. Therefore, the ratio of fluorescence between twisted and planar states gives a direct measure of the strain state of the membrane, and since PEO-b-PB is elastic in this regime, a direct measure of the stress in the membrane [35] (see Figure 18.2). This soft-matter stress sensor has numerous potential uses for the study of tribology, membrane biophysics, and the diagnosis of disease (such as monitoring the increase in shear stress during the occlusion of blood vessels).

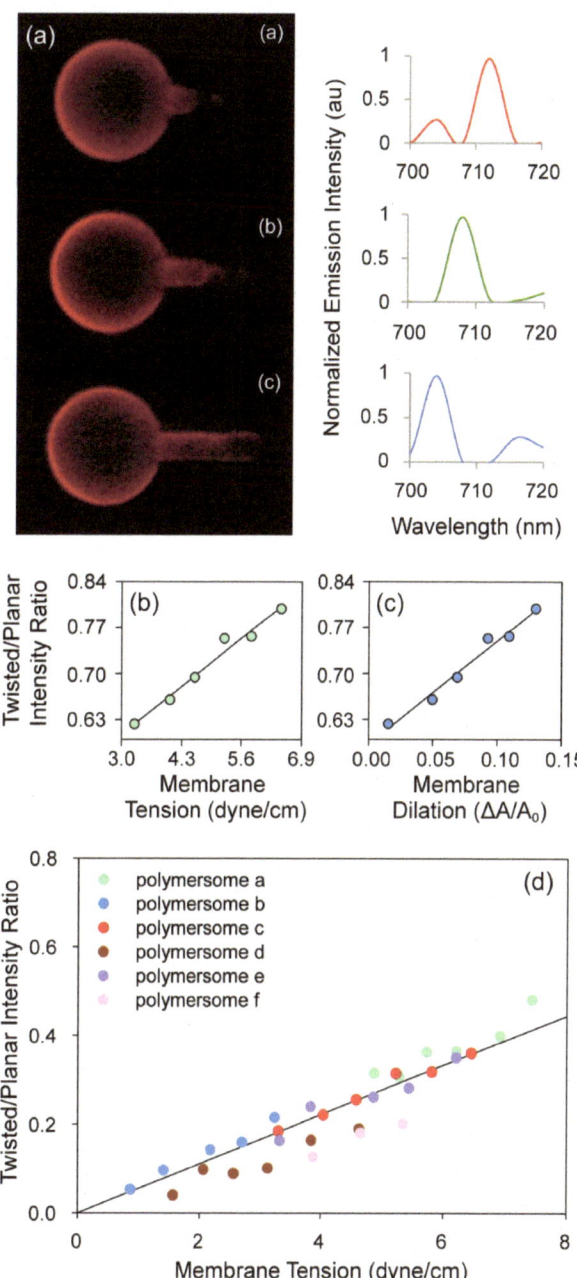

FIGURE 18.2 Polymersomes with embedded porphyrins were used to make a stress sensor that correlated the ratio of fluorescence between twisted and planar configurations to the level of stress, directly measured using micropipette aspiration. (a) Images of the aspiration of vesicles at different tensions, with tension increasing from top to bottom. (b) The ratio of fluorescence of the twisted to planar forms of the embedded porphyrins as a function of membrane tension. (c) The ratio of fluorescence of twisted and planar forms of the embedded porphyrins as a function of percent membrane dilation. (d) Plot of the fluorescence ratio of twisted to planar forms as a function of membrane tension. (Adapted from Kamat et al. (2011).)

18.3 MEMBRANES FROM PROTEINS

The suggestion that membranes could be made from purely proteinaceous materials first emerged from work on assembling membranes from block copolypeptides. Lecommandoux constructed vesicles from a diblock copeptide of poly(L-glutamic acid)-b-poly(L-lysine), $E_{15}K_{15}$ [53]. These vesicles possess interesting properties of pH switching, since between pH 5 and 9, both amino acid chains are charged, and the peptide chains are dispersed. However, at pHs below 5 or above 9, one of the two chains is neutralized, the other chain remains charged, and the peptide assembles into vesicles with an orientation dictated by pH [13]; these vesicles are called "schizophrenic." This work illustrates the strength of assembling structures from amino acid chains, in which switchable motifs can be readily incorporated into the shell owing to the ability to change the charge of the amino acid blocks.

Deming extended the repertoire of block copolymers of amino acids that make vesicles [13,28], making block copolymers of peptides using synthetic methods from polymer synthesis, yielding families of polymers with polydispersity indices between 1.1 and 1.3 [26,27]. A family of poly(L-lysine)-b-poly(L-leucine) diblock copeptide polymers (K_nL_m) were synthesized, and it was found that $K_{60}L_{20}$ assembled into membranes and formed vesicles. Other ratios of the two amino acids assembled into structures different from vesicles, such as fibrils and sheet-like membranes. Furthermore, it was shown that vesicles of $K_{60}L_{20}$ could be extruded through polycarbonate membranes into 100-nm vesicles, a size often useful for drug delivery [27]. Deming extended this work by making vesicles from poly(L-arginine)-b-poly(L-leucine) in the same molar ratio ($R_{60}L_{20}$) [28]. The advantage of this change is that it mimics the TAT peptide, which is rich in arginine, facilitating internalization [11]. Indeed, vesicles made from $R_{60}L_{20}$ were effectively internalized by HeLa cells in culture. It was also possible to randomly interdisperse 10% lysine residues within the arginine block, to facilitate attachment of fluorophores to track the progress of vesicles during internalization [28].

More recently, Deming was able to make vesicles that included the nonnatural amino acid dihydroxyphenylalanine, or DOPA; the multiple hydroxyl residues can achieve strong hydrogen binding to target surfaces and are used by mussels for strong adhesiveness in seawater [26]. Poly(L-lysine)-b-poly(L-dihydroxyphenylalanine) polymers of dispersity between 1.1 and 1.3 were synthesized, with and without leucine randomly interspersed through the hydrophobic block, and vesicles were assembled. Vesicles containing DOPA could also be cross-linked and lyophilized, generating vesicles of substantial stability [26]. To summarize, vesicles can be assembled from polypeptide chains, and the incorporation of amino acids can confer biological activity and function into the existing structures. A more comprehensive review of the work of the Deming lab appears in this compendium.

18.4 MEMBRANES FROM RECOMBINANT PROTEINS *IN VITRO*

As has been well documented by Tirrell, Chilkoti, and others, making materials from proteins recombinantly results in monodisperse proteins with amino acid composition dictated by the corresponding gene sequence [40, 43]. In biology, sequence controls function, and the introduction of specific protein sequences for targeting adhesion or growth factor receptors, assembly, protease cleavage, or nucleic acid binding is easily and straightforwardly achieved by genetic engineering. Thus, the assembly of vesicles (and other biologically relevant materials) using recombinant proteins is incredibly powerful.

Membranes and other suprastructures have been assembled primarily from the derivatives of two different recombinant proteins: oleosin and elastin. We will first describe our development of oleosin and then briefly summarize advances in making protein capsules and vesicles with elastin-like polypeptides.

Oleosin. Our rationale was to screen naturally occurring protein surfactants, to try to find a suitable amphiphile that could be made recombinantly and self-assembled into suprastructures, such as vesicles. There are very few naturally occurring proteins that act as surfactants. For stability, proteins generally embed their hydrophobic amino acids in their core and expose polar and charged amino acids at their surface. Of natural surfactant proteins, some display a hydrophobic patch such as hydrophobin [61] or rearrange at an interface to expose a hydrophobic domain such as latherin [5] and ranaspumin [25]. Although these proteins can act as surfactants, they are also globular, and therefore are not an ideal base to engineer as a free chain surfactant. Ideally, a chain-like protein would be desired to mimic the well-studied assemblies of lipids and polymers. A plausible candidate is the plant surfactant protein, oleosin [30].

Oleosins are a family of plant proteins whose biological role is to stabilize oil bodies [30]. They have an N-terminal hydrophilic segment, followed by a hydrophobic core (among the longest known natural hydrophobic stretches [4]) and another hydrophilic segment at the C-terminus [29, 38]. Although the crystal structure is unknown, the molecule is thought to resemble a hairpin with the hydrophobic domain bifurcated by a proline knot, a stretch containing three prolines that induce a 180° turn in the chain [1,2,4,30]. The two legs of the hairpin are helical, possibly forming a coiled coil [2,4]. Recently, recombinant oleosin and its derivatives have been shown to stabilize artificial oil bodies and emulsions [7–9,48,54]. We postulated that the surfactant-like block architecture of oleosin would make it a logical starting point for the creation of tunable self-assembled protein suprastructures.

Our first investigations into structures made from oleosin started with the wild-type sunflower protein. The numerical designations that follow are designations that refer to the number of residues in each of these blocks. Wild-type oleosin (Oleo-WT) is designated 43-87-63

(Figure 18.3); it has 43 hydrophilic amino acids in the N-terminus ("N" in Figure 18.3), 87 amino acids in the hydrophobic block ("H" in Figure 18.3, plus the conserved proline knot that separates them), and 63 hydrophilic amino acids in the C-terminus ("C" in Figure 18.3). Standard molecular biology techniques were used to create oleosin genes with truncations throughout the molecule. Proteins were recombinantly synthesized in *Escherichia coli* and purified by immobilized metal affinity chromatography using a terminal His tag. The resulting molecules were characterized by SDS-PAGE and MALDI-TOF-MS to verify their molecular weight.

With great fortune, we found oleosin can reliably be made recombinantly in bacteria and assembled into various structures, including vesicles, fibres, and sheets ([59]; see Figure 18.3 [57]; also Vargo and Hammer, unpublished). Generally, large, symmetric oleosins with substantial hydrophobic cores, either 87 or 65 residues, made vesicles. We believe this was one of the first demonstrations of the assembly of vesicles from a recombinant proteins; the work done by the Rodriguez-Cabello laboratory on elastins we describe later might have been the first, with caveats we note later [42]. Oleosins with substantial asymmetries, such as 42-87-7, led to assemblies with curvature, such as short worm-like micelles or longer nanofibres. Oleosins with substantial hydrophobic cores and small hydrophilic head groups, such as 7-87-7, packed tightly into two-dimensional sheets that extended for hundreds of microns in two dimensions yet were nanometric in the third dimension (Figure 18.3). These materials could be useful for making surface coatings and tissue engineering.

Because the head groups of oleosin are hydrophilic and potentially charged, we reasoned that the ability to shield electrical interactions among head groups would affect packing, curvature, and the resulting supramolecular structure. We made a subfamily in which we removed 22 amino acids in the hydrophobic block of Oleo-WT, 11 residues from each arm of the hydrophobic block. This family of mutants consisted of four proteins with various hydrophilic arm lengths and a constant hydrophobic block length (see Figure 18.4). This family of proteins ranged in molecular weight from 10.8 to 15.2 kDa and in hydrophobic fraction from 46% to 64%. All four mutants in the -65- family, 43-65-33, 33-65-23, 28-65-18, and 23-65-13, were found to assemble into vesicles under physiological conditions (100% PBS). Cryo-TEM was used to characterize the self-assembled structures as a function of hydrophilic block fraction and ionic strength of solution, ranging from 100% PBS to 0% PBS (distilled water). The resulting phase diagram (Figure 18.4) shows the relationship between protein architecture, buffer conditions, and the resulting supramolecular assemblies. Proteins with larger head groups form vesicles in PBS, while proteins with small head groups form sheets in PBS. For the vesicles, the membrane thicknesses match expected values based on the hairpin structure of oleosin and the secondary structure calculated in circular dichroism [59]. Interestingly, cryo-TEM revealed that the membrane thickness increased with the size of the head group within this family, even though the hydrophobic core remained constant [59]. We believe this is because increasing the size of the head group increases the lateral pressure through entropy, causing the embedded hydrophobic cores to partially separate, thickening the membrane. We also demonstrated assembly of giant vesicles of oleosin; these vesicles were visualized using fluorescent dyes that differentially labelled the specific phases [59].

Fibre formation is more favoured, and vesicle formation is less favoured when the charged head groups are repulsive at low ionic strength, because repulsion induces curvature. The formation of sheets is favoured when the

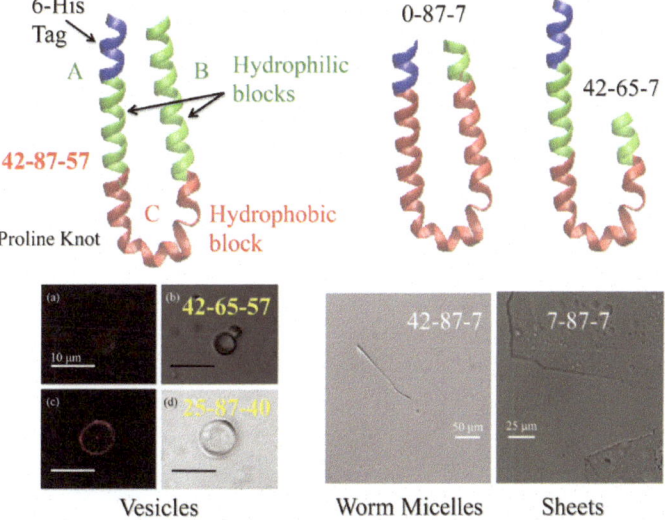

FIGURE 18.3 Wild-type oleosin is a triblock protein with 42-87-57 residues in the A, B, and C blocks, respectively. Panels (a) and (b) show vesicles made from 42-65-57, imaged with Nile Red in the membrane (a) and in phase contract (b). Panels (c) and (d) show vesicles made from 25-87-40, imaged with Nile Red in the membrane (c) and in brightfield (d). 42-87-7 makes worm micelles, imaged directly below, and 7-87-7 makes sheets. Different truncation mutants were made, which assembled into vesicles, worm-like micelles and sheets.

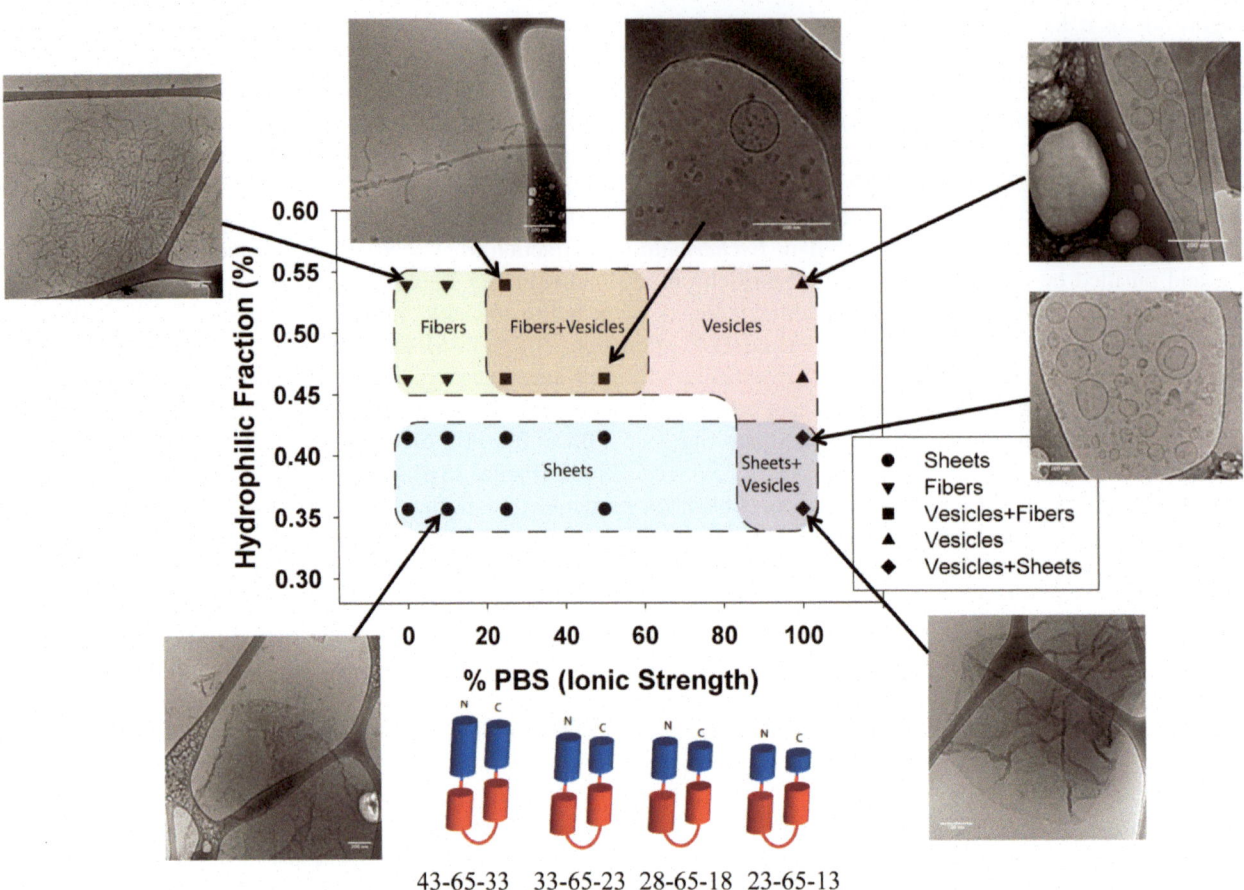

FIGURE 18.4 Phase behaviour of a small four-member family of -65- mutants of oleosin as a function of the ionic strength of the media. Structures were verified with cryo-TEM. (Adapted from Ref. [59].)

hydrophilic fractions are small and can tightly pack, reducing curvature [59].

The ability to change structure through ionic strength suggests changing pH might also alter the structure, since protonation of head group amino acids at pHs below their pKa should add charge and induce curvature. Indeed, with one -65- oleosin, 17-65-17, we observed the coexistence of worm-like micelles and vesicles a neutral pH, in the vicinity of the pKa of 7.4 we calculated for the head group amino acids. Cryo-TEM revealed that at pHs below the pKa, 17-65-17 formed worm-like micelles, consistent with the idea that the additional protons in the head group occupy space and induce interfacial curvature. At pHs above the pK, vesicles form, owing to the deprotonation of the head groups, and hence their tight packing (Vargo and Hammer, unpublished). These results suggest 17-65-17 possesses a natural pH switch, in which vesicles can be induced to disassemble and reassemble into worm-like micelles during acidification. Given that 17-65-17 has a pKa in the vicinity of 7, this means that the inter conversion from vesicles to worm-like micelles, and hence the release of any hydrophilic content, would occur within the acidic environment of endosomes.

A number of different avenues have since been pursued to broaden oleosin's use as a biological surfactant, particularly for drug delivery and imaging. First, oleosin is a molecule that is known to have substantial secondary structure, owing to the alpha-helicity of the hydrophobic core, which is stabilized by a series of six theonine residues that are mutually hydrogen bonded. We reasoned that removing the threonines, as well as replacing some of the bulky hydrophobic residues with glycines, might facilitate increased flexibility. Indeed, truncating the hydrophobic core down to 30 amino acids, which removed five of the six threonines, and replacing five bulky amino acids with glycines led to the formation of an oleosin variant we call "-30G-" that, by circular dichroism, was shown to have lost all secondary structure [60]. This protein acts as a true, unstructured surfactant in solution. The protein 43-30G-62 demonstrated self-assembly into spherical micelles with a CMC of 4.1 μM [60]. The resulting spherical micelles were 19 nm in diameter, as imaged by cryo-TEM and deduced by X-ray and dynamic light scattering [60].

Although 43-30G-62 does not assemble into vesicles, but rather spherical micelles, we have used this molecule to demonstrate some of the potential of recombinant proteins for delivery. Spherical micelles are used for drug delivery and imaging; spherical micelles can be targeted, enwrap encapsulated hydrophobic drugs, and with modification even carry hydrophilic drugs [56]. We have shown that -30G- can be modified in a variety of ways, to enhance targeting to cells. First, adding the integrin binding peptide RGD to -30G- facilitates receptor binding and uptake in a variety of cells, including an ovarian cancer cell line, OVCAR-5 [60]. Additionally, we have separate, unpublished results in which micelles comprised of 5% TAT peptide (GRKKRRQRRRPQ)-terminated-30G- and 5% RGD-terminated-30G- have greatly enhanced uptake over using micelles bearing either 5% TAT only or 5% RGD only (Vargo and Hammer, unpublished).

We have extended this concept further by making an oleosin micelle whose targeting can be activated by a protease. The logic was to hide a receptor-binding domain behind a protease cleavable domain, so that the receptor-binding domain would only be exposed upon the action of the specific protease. Such dual control would ensure that drugs would be delivered only at physiological sites at which the particular proteases were overexpressed, such as thrombin at sites of inflammation. To illustrate the potential of this approach, we placed the integrin-binding motif RGDS behind the thrombin cleavable domain (TCD) on a fluorescently labelled oleosin-30G that was assembled into spherical micelles [18]. Four constructs were made: RGDS–oleosin (a positive control that can always bind to the target regardless of protease availability); RADS–oleosin (a negative control with a scrambled binding motif); TCD–RGDS–oleosin, which should only be fully activated in the presence of thrombin; and TCD–RADS–oloesin, a negative control that tests the nonspecific effects of the thrombin on the adhesion of oleosin micelles. To facilitate the imaging of oleosin internalization, we labelled oleosin through NHS ester chemistry at lysines with DyLight 488. A schematic of the molecules we constructed is shown in Figure 18.5. After treatment with thrombin, the uptake of TCD–RGDS–oleosin is qualitatively similar to that of the positive control, RGDS–oleosin, at all times. Notably, without treatment with thrombin, TCD–RGDS–oleosin is endocytosed to less than a third of the extent of TCD–RGDS–oleosin after thrombin treatment. We believe this modest residual binding is due to the lateral access of the RGDS domain on the intact TCD–RGDS–oloesin chain. Nevertheless, the results conclusively show that TCD–RGDS–oleosin can be activated after thrombin treatment. Because the design of these constructs is modular – either the adhesive motif or the protease cleavable domain can be altered – the strategy embodied in this work can be used to make a wide range of "smart" nanocarriers [18].

Another use of oleosin is to stabilize interfaces, while simultaneously including targeting or other biofunctionality. One recent example is the stabilization of clusters of iron oxide nanoparticles with an affibody-containing oleosin targeted to the Her2neu receptor, which is overexpressed in a significant number of breast cancers [58]. Iron oxide nanoparticles have been proposed for magnetic resonance imaging, but they are hydrophobic and need to be stabilized by a surfactant. In collaboration with Andrew Tsourkas' lab at Penn, we stabilized iron oxide clusters of a few hundred nanometres in diameter with a functionalized oleosin that, through the terminal affibody, could bind to the Her2neu receptor with high affinity. The preferential binding to a cell line that overexpresses the Her2neu receptor was demonstrated, along with the superior T_2 signal upon binding. A second application was to stabilize monodisperse air bubbles that improve the signal-to-noise ratio of ultrasound imaging [3]. In collaboration with Daeyeon Lee at Penn, we made monodisperse air bubbles by microfluidics that were stabilized by mixtures of Pluronic and oleosin. The oleosin turned out to be absolutely essential for the stabilization of the bubbles, as bubbles would coalesce when stabilized by Pluronic alone. The presence of oleosin on the shell was verified by making a GFP–oleosin, which allowed fluorescent visualization of protein at the air–water interface [3]. In a later paper, the Hammer and Lee laboratories collaborated to measure the materials properties of the bubbles by micropipette aspiration [33], using a technique originally developed by Needham and coworkers [17]. The areal expansion modulus of gas bubbles, as determined by micropipette aspiration, depends on the structure as well as the concentration of Pluronic, and the relative amount of Pluronic in ratio to oleosin [33].

Elastin. Elastin-like polypeptides (ELPs) are protein polymers based on the repeat of pentameric amino acid blocks, VPGxG, where x is an adjustable residue, depending on the desired material properties and solubilities. (As is shown below, other residues in the pentamer may also be altered.) As a general class of molecules, ELPs generally exhibit a liquid–liquid phase transition at a lower critical solution temperature, or LCST, making them useful candidates for the engineering of assembly with temperature. The physics of this transition involves the liberation of water shells around critical amino acids by raising temperature, leading to the condensation of amino acids, driven primarily by hydrogen bonding around the L-proline residue [42].

Like oleosins, elastin derivatives can be made recombinantly, which affords the luxury of placing specific residues and biological motifs at any location with exquisite precision. Repeats of different pentamers, each with different guest residues, could be combined to create block copolymer proteins. These protein polymers have variable properties depending on the identity of the guest residue x in each block, the chain length of each block. With such a wide diversity of sizes and chemistries, it seems reasonable that some configuration of elastins could be engineered to self-assemble into membranes. While we focus here on the use of ELPs to make membranes and vesicles, a great deal of important work has been done with ELPs to deliver bioactive agents from other nanostructures (for a review, see Ref. [41]).

Rodriguez-Cabello and colleagues purported to make vesicles from ELPs, using temperature to trigger the assembly of a specific elastin into vesicle structures [42]. They made elastin-like block corecombiners (ELbcRs), based on the repeat of a block, called E, in which glutamic acid (E) was inserted in a central pentamer of the hydrophilic block: [(VPGVG)$_2$-(VPGEG)-(VPGVG)$_2$]. The hydrophobic

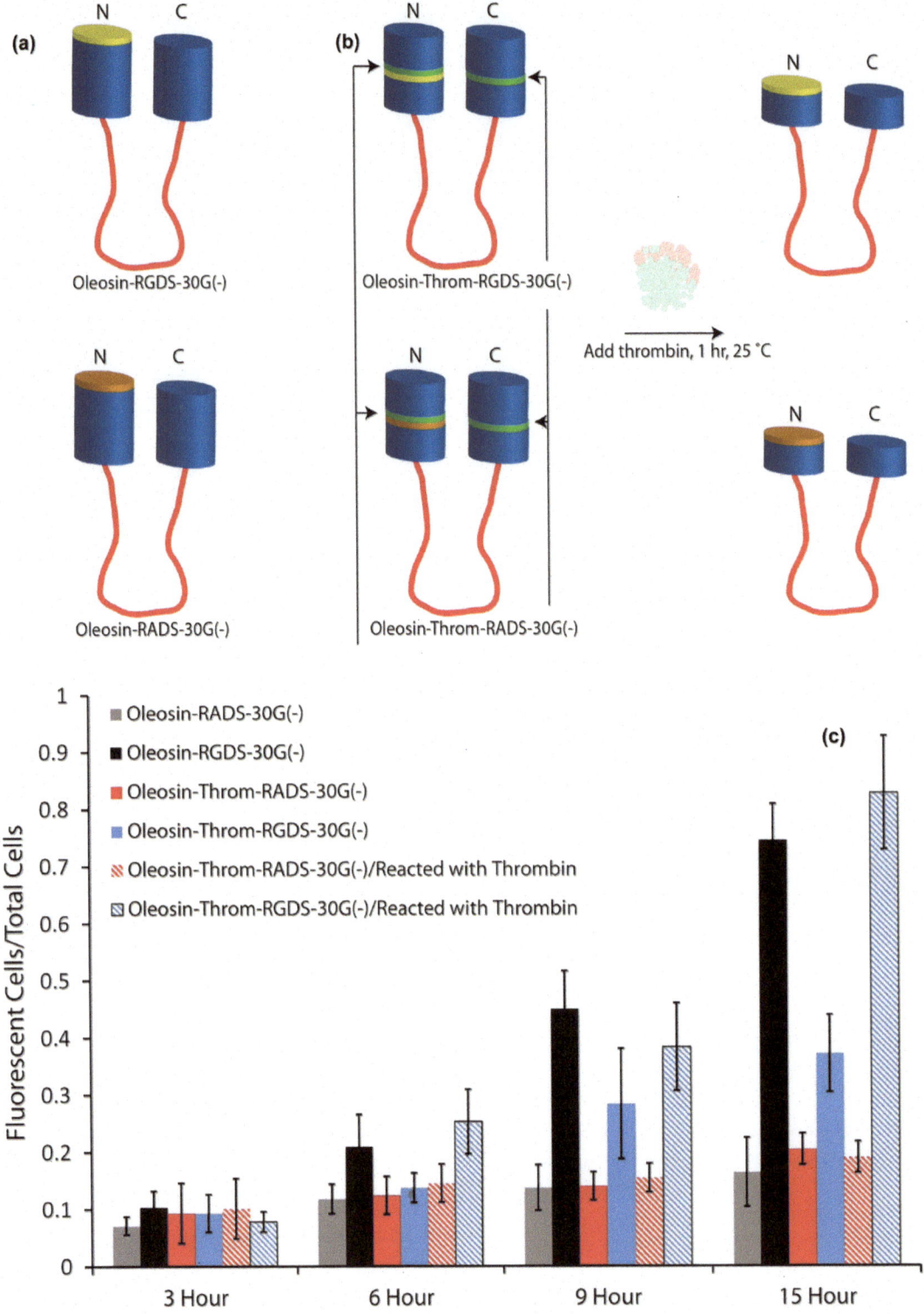

FIGURE 18.5 Molecular view of oleosin-30G(-) variants. (a) Illustrations depicting addition of motifs RGDS (yellow) and RADS (orange) to the variants oleosin–RGDS–30G(-) and oleosin–RADS–30G(-), respectively. (b) Oleosin–Throm–RGDS–30G(-) is mutated from oleosin–RGDS–30G(-) where the RGDS motif is moved to the centre of the N-terminus hydrophilic arm and protected by a thrombin cleavable domain (green). Another thrombin cleavable domain is added to the C-terminus hydrophilic arm so that the remaining protein molecules have two hydrophilic arms similar in length after the thrombin cleaving reaction. Oleosin–Throm–RADS–30G(-) is constructed in a similar way from oleosin–RADS–30G(-). (c) The addition of exposed RGDS led to a 4.6-fold increase in cellular uptake over 15 h, compared with the scrambled sequence of RADS. When RGDS motif was protected by 24 amino acids on the hydrophilic arm, increased uptake is observable starting from 9 h of incubation where it showed 2.0 and 1.8 times increase at 9 and 15 hours, respectively. Oleosin–Throm–RGDS–30G(-), when fully reacted with thrombin, resembled very closely to oleosin–RGDS–30G(-) smaller in micellar size, and showed a similar 4.4 times increase in cellular uptake compared with oleosin–Throm–RADS–30G(-) over 15 hours. (Taken from Ref. [18].)

block was based on repeats of the pentamer including alanine (A) at the third position and valine (V) at the fourth: VPAVG. Three ELbcRs were made, E50A40, E100A40, and a triblock, E50A40E50, where the number indicates the number of total residues in each block. Each of these proteins could be induced to assemble into suprastructures through a temperature-mediated phase transition; a temperature of 65°C was used to compare aggregate sizes, with DLS showing assembled structures with mean sizes greater than 200 nm seen for E100A40 and E50A40E50. The authors use TEM and AFM to claim that these two proteins assemble into vesicles; however, there was no conclusive evidence of a bilayer structure as imaging the nanoassemblies in their water-laden state using cryogenic-TEM – the gold standard for vesicle determination – was not performed. However, the results suggest and are not inconsistent with the possibility of the self-assembly of recombinant proteins into nanovesicles.

L4F is an alpha-helical peptide that binds lipids, inspired by the protein ApoA1, and thus has potential therapeutic effects in cardiovascular disease. The peptide is also amphipathic, owing to four hydrophobic phenylalanine residues that drive the formation of alpha-helices. The MacKay laboratory appended a peptide containing the L4F sequence to ELPs of different architectures – $(VPGAG)_{192}$ and $(VPGVG\ VPGAG\ VPGAG)_{64}$ – and demonstrated the resulting hybrid proteins assembled into small vesicles with a radius of roughly 50 nm [47] (see Figure 18.6). Heating causes the coacervation of proteins into aggregates, and cooling causes the assembly of the protein into vesicles. Vesicle formation was suggested by DLS and confirmed with cryo-TEM; vesicles possessed a membrane thickness of 8 nm. The condensation was driven by mutual L4F hydrophobicity on adjacent hybrids, forming the interior of the membrane, with the ELP chains extending into the hydrophobic space. The vesicles demonstrated the suppression of stellate cell activation *in vitro*, consistent with the activity of ApoA1, and 1 hour circulation times in a rat model [47].

Luo and Kiick made vesicles from a hybrid containing an elastin-like peptide and a triple helix-forming collagen-like peptide (CLP) [39]. The peptides were made by solid-phase synthesis in two sections, with a short, alkyne-terminated ELP $(VPGFG)_6$ and an azide-terminated CLP, and assembled using click chemistry. The addition of the CLP, $(GPO)_4GFOGER(GPO)_4GG$ which contained the collagen sequence GFOGER that binds several different integrins, led to the condensation of chains into a vesicular structure at temperatures lower than the melting temperature for CLP but greater than the LCST for the ELP chains. Heating the assemblies caused the disassembly of the vesicles, owing to the unfolding of the ELP domain. The assembly of the constructs into vesicles was verified with cryo-TEM at intermediate temperatures [39].

Park and Champion constructed single-layer vesicles, which can be thought of as shells, by combining an ELP segment linked with a coiled-coil domain, Z_R (ELP-Z_R) with a globular protein (either mCherry or GFP) linked to the complementary coiled-coil domain (Z_E) [46]. Various combinations of proteins led to the formation of a mCherry or GFP–coiled coil–ELP monolayer as temperature was raised from 4°C to 25°C, driven by the LCST-mediated phase transition of the ELP. The inner ELP monolayer is hydrophobic and could be stabilized by the protein coacervate phase, or fluorescein (which is hydrophobic) or alternatively polystyrene nanoparticles. It is hypothesized that the coiled-coil domains prevent curvature of the monolayer, stabilizing the layer and preventing its collapse. While not a bilayer in the strictest sense, these hollow structures provide the possibility of encapsulation. Also, because most

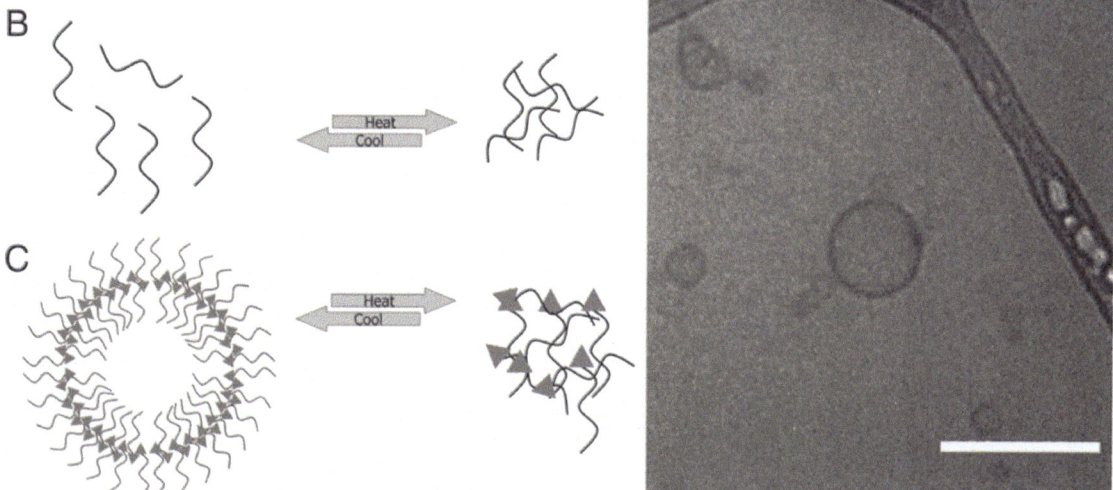

FIGURE 18.6 Appendage of L4F to the end of an elastin leads to the assembly of vesicles upon cooling, as imaged by cryo-TEM (length bar, 200 nm). (B) Indicates the behavior of elastins without L4F, which undergoes a condensation with heating, consistent with LCST behavior, (C) indicated the behavior with an attached L4F, which drives assembly in to vesicles, as imaged on the right. (Image modified with permission from Ref. [47].)

globular proteins can be equipped with coiled-coiled domains, this strategy allows the modular encapsulation of different globular proteins into the protein shell, which portends capsules with advanced enzymatic functionality [46].

18.5 IN SITU ASSEMBLY

ELPs have also been used to form vesicle-like organelles *in situ* within *E. coli* [31]. By introducing the gene for various elastin di- and triblocks into *E. coli* and inducing their expression, some of the polypeptides appear to assemble into vesicular structures. Two different pentameric elastin blocks, the hydrophilic VPGEG using glutamic acid (E) and the hydrophobic VPGFG using phenylalanine (F), were used. The monomeric enhanced green fluorescent protein (mEGFP) was appended to either the N- or C-terminus, to visualize the assembled structures. Expression of mEGFP–E20F20, in which the E and F pentamers were repeated 20 times, led to the assembly of organelle-like compartments *in situ*. Inverting the order of blocks, F20E20–mEGFP, led to the loss of assembly. Expression and purification of mEGFP–E20F20 *in vitro*, and imaging of the resulting self-assembled structures using cryo-TEM, showed that the molecules were capable, by themselves, of assembling into vesicles in solution [31], with membrane thicknesses that correspond to observed vesicles *in situ*. The ability to form assembled structures was sensitive to the block size ratio, as mEGFP–E40F20, in which the E block is repeated 40 times, lost all ability to form organelle structures. Furthermore, the ability of mEGFP–E20F20 to form organelles was not dependent on the GFP, because E20F20 formed membranes, even in the absence of mEGFP [31].

18.6 MEMBRANES FROM DENDRIMERS

Another recent advance in the development of membranes from synthetic amphiphiles is the construction of vesicles and other supramolecular assemblies from dendrimers – branched molecules of defined architecture. This effort, led by Virgil Percec's laboratory at Penn, has focussed on the construction of amphiphilic Janus dendrimers – molecules with one hydrophilic and one hydrophobic face. Because the molecules are amphiphilic like phospholipids, they would be expected to assemble in to suprastructures similar to those seen with lipids – membranes, micelles, etc. Our laboratory has collaborated on this effort, in some cases imaging the resultant assemblies at the nanoscale using cryo-TEM [63], and in other cases measuring the materials properties of the resulting membranes using micropipette aspiration [50].

The approach of the Percec laboratory is to make large libraries of molecules, through a modular approach in which hydrophilic and hydrophobic groups can be mixed and swapped. This approach allows one to ascertain the changes in the structure of the supramolecular assembly that sometimes result from small changes in dendrimer architecture. Like lipids and proteins, the molecules are unimolecular and not polydisperse; as a result, some of the supramolecular morphologies are very precise, owing to the homogeneity and architecture of the underlying amphiphile. In the initial paper on Janus dendrimers, over 140 molecules were made, and their assembly led to the formation of vesicles, disks, and twisted ribbons with specific chirality [50].

Recent efforts of the Percec laboratory have extended to Janus glycodendrimers, in which mono- or disaccharides are added to the head group of amphiphilic dendrimers. These amphiphiles are reminiscent of glycolipids found in biological membranes. Numerous glycodendrimers assemble into lamellar phases, including vesicles [49]. Glycodendrimersomes open the possibility of mimicking biological membranes with programmable glycan ligand presentations, to act as supramolecular lectin blockers, vaccines, and targeted delivery devices. The identity and multivalency of glycodendrimersomes with different sizes and their ligand bioactivity were demonstrated by selective agglutination with a diverse selection of sugar-binding protein receptors such as the plant lectins concanavalin A and the highly toxic mistletoe *Viscum album* L. agglutinin, the bacterial lectin PA-IL from *Pseudomonas aeruginosa*, and, of special biomedical relevance, human adhesion/growth-regulatory galectin-3 and galectin-4 [49]. In fact, a subset of Janus glycodendrimers (Janus-GDs) was used to make glycodendrimersomes (GDSs) that could discriminate between different forms of human adhesion/growth-regulatory lectin galectin-8 in an agglutination assay [66]. The naturally occurring forms of galectin-8, Gal-8S and Gal-8L, which differ by the length of linker connecting their two active domains and a single amino acid mutant (F19Y), were used as probes to study activity and sensor capacity. The highest level of activity occurred for a Janus-GD with six 3EO groups and one Lac. The bioactivity and sensor capacity was sensitive both to the density and identity of the carbohydrate. Both changes in topology of Lac presentation of the GDSs and seemingly subtle alterations in lectin structure resulted in different levels of bioactivity. Therefore, Janus-GDs can discriminate structure–activity relationships between programmable cell surface models and human lectins in a highly sensitive manner.

Interesting, glycodendrimers also demonstrate the propensity to make onion-like glycodendrimersomes, with repeated concentric lamellar phases [63,65] (see Figure 18.7), where the spacing between the layers is conserved. In the most recent work, the carbohydrate mannose was used to construct a family of Janus glycodendrimers. The spatial presentation of mannose is decisive for formation of either unilamellar or onion-like GDS vesicles. The mode of mannose presentation and the concentration of Janus GD determine the size of the dendrimersome and the number of bilayers. These multilamellar Janus dendrimersomes agglutinate in the presence of the lectin conconavalin A [63], a clear depiction of their bioactivity.

It has also been possible to embed functional receptors into glycodendrimer membranes [62]. Components of *E.*

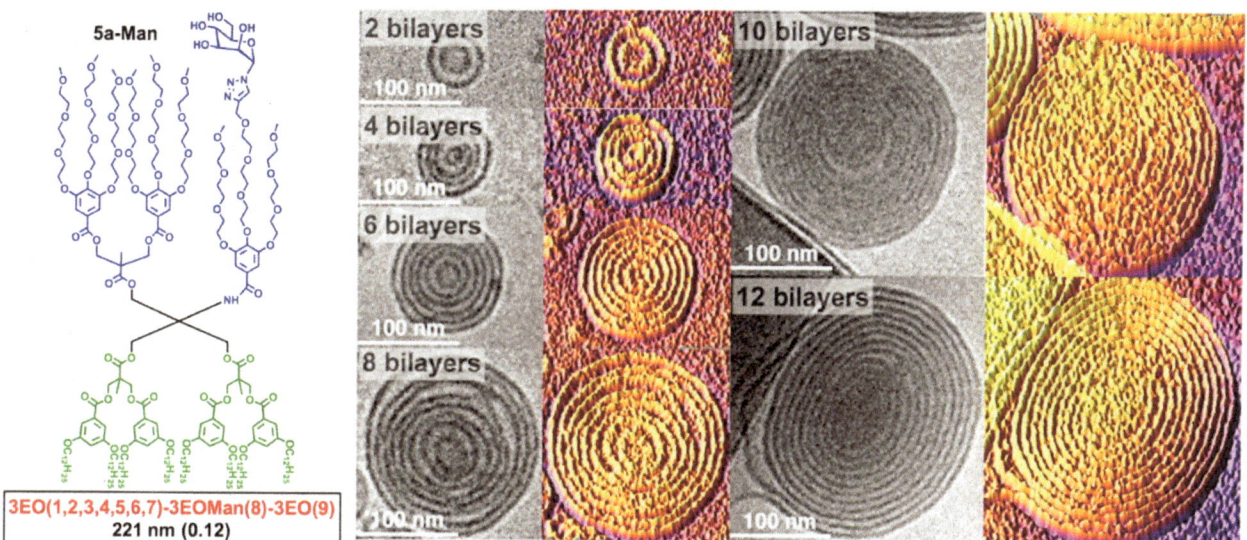

FIGURE 18.7 Left, one of a family of mannose-containing glycodendrimers, called 5a-Man [3EO (1,2,3,4,5,6,7)-3EOMan(8)-3EO(9)] in shorthand notation, synthesized by the Percec laboratory. Right, selected cryo-TEM images of onion-like GDs self-assembled from 0.1 mM Janus GD 5a-Man and their 3D intensity-plotting images with different numbers of bilayers and diameters. (Image modified and reproduced with permission from Xiao et al. 2016.)

coli inner membranes, which contained bacterial membranes receptors, were integrated into bacterial membranes by a dehydration followed by rehydration, leading to the formation of giant cell-like hybrid vesicles. These hybrid vesicles contain various bacterial transmembrane proteins as well as a small membrane protein, MgrB, tagged with a red fluorescent protein, as well as lipopolysaccharides and glycoproteins from the bacterium *E. coli*. Incorporation of different fluorescent probes in each of the assembled components allowed optical microscopy to identify components and demonstrate coassembly and the incorporation of functional membrane channels [62]. Owing to the carbohydrate composition of the glycodendrimer component in the membrane, the hybrid vesicles bind a human galectin, consistent with the display of sugar moieties from lipopolysaccharides or possibly glycosylated membrane proteins.

18.7 SUMMARY

We have summarized a variety of ways in which membranes can be made from novel surfactants, including block copolymers, block copolypeptides, recombinant proteins, and dendrimers. Each of these methods has its strengths and weaknesses. Many chemistries can be introduced into block copolymers, but the molecules are polydisperse, difficult to functionalize for specific application, and a limited number will meet standards for regulatory approval for use *in vivo*. Block copolypeptides illustrate that membranes can be made from amino acid chains, but these molecules are also polydisperse and incorporating functional motifs require additional ad hoc steps. A limited number of recombinant proteins can be assembled into vesicles, but the molecules are monodisperse, and the introduction of specific bioresponsive sequences is straightforward. Multiple sequences and therefore functionalities can be included, either coincident on the same protein, or in blends on different proteins assembled together. Recombinant proteins may generate immune responses for applications *in vivo*, although mutation could be used to mitigate those effects. Amphiphilic Janus dendrimers can be assembled into membranes, but incorporating chemical groups of biological relevance is thus far limited to small carbohydrates, although further developments are in process. The precision of the dendrimers leads to a precision in their supramolecular assemblies, which is promising for biomimicry, such as making filaments or artificial viruses.

ACKNOWLEDGEMENT

This work was supported by the U.S. Department of Energy, Office of Science, Basic Energy Sciences, under Award # DE-SC0007063.

BIBLIOGRAPHY

1. Abell, B.M., L.A. Holbrook, M. Abenes, D.J. Murphy, M.J. Hills, and M.M. Moloney, Role of the proline knot motif in oleosin endoplasmic reticulum topology and oil body targeting. *Plant Cell*, 1997. **9**(8): p. 1481–1493.
2. Alexander, L.G., R.B. Sessions, A.R. Clarke, A.S. Tatham, P.R. Shewry, and J.A. Napier, Characterization and modelling of the hydrophobic domain of a sunflower oleosin. *Planta*, 2002. **214**(4): p. 546–551.
3. Angile, F.E., K.B. Vargo, C.M. Sehgal, D.A. Hammer, and D. Lee, Recombinant protein-stabilized monodisperse microbubbles with tunable size using a valve-based microfluidic device. *Langmuir*, 2014. **30**(42): p. 12610–12618.
4. Beaudoin, F. and J.A. Napier, Targeting and membrane-insertion of a sunflower oleosin *in vitro* and in Saccharomyces cerevisiae: The central hydrophobic domain

contains more than one signal sequence, and directs oleosin insertion into the endoplasmic reticulum membrane using a signal anchor sequence mechanism. *Planta*, 2002. **215**(2): p. 293–303.
5. Beeley, J.G., R. Eason, and D.H. Snow, Isolation and characterization of latherin, a surface-active protein from horse sweat. *Biochemical Journal*, 1986. **235**(3): p. 645–650.
6. Bermudez, H., A.K. Brannon, D.A. Hammer, F.S. Bates, and D.E. Discher, Molecular weight dependence of polymersome membrane elasticity and stability. *Macromolecules*, 2002. **35**: p. 8203–8208.
7. Bhatla, S.C., V. Kaushik, and M.K. Yadav, Use of oil bodies and oleosins in recombinant protein production and other biotechnological applications. *Biotechnology Advances*, 2010. **28**(3): p. 293–300.
8. Chiang, C.J., C.C. Lin, T.L. Lu, and H.F. Wang, Functionalized nanoscale oil bodies for targeted delivery of a hydrophobic drug. *Nanotechnology*, 2011. **22**(41).
9. Chiang, C.J., L.J. Lin, C.C. Lin, C.H. Chang, and Y.P. Chao, Selective internalization of self-assembled artificial oil bodies by HER2/neu-positive cells. *Nanotechnology*, 2011. **22**(1).
10. Christian, N.A., F. Benencia, M.C. Milone, L. Guizhi, P.R. Frail, M.J. Therien, G. Coukos, and D.A. Hammer, In Vivo dendritic cell tracking using fluorescence lifetime imaging and near-infrared-emissive polymersomes. *Molecular Imaging and Biology*, 2009. **11**(3): p. 167–177.
11. Christian, N.A., M.C. Malone, S.S. Ranka, G. Li, P.R. Frail, K.P. Davis, F.S. Bates, M.J. Therien, P.P. Ghoroghchian, C.H. June, and D.A. Hammer, Tat-functionalized near-infrared emissive polymersomes for dendritic cell labeling. *Bioconjugate Chemistry*, 2007. **18**(1): p. 31–40.
12. Cui, H.G., M.J. Webber, and S.I. Stupp, Self-assembly of peptide amphiphiles: From molecules to nanostructures to biomaterials. *Biopolymers*, 2010. **94**(1): p. 1–18.
13. Deming, T.J., Synthetic polypeptides for biomedical applications. *Progress in Polymer Science*, 2007. **32**(8–9): p. 858–875.
14. Discher, B.M., H. Bermudez, D.A. Hammer, D.E. Discher, Y.Y. Won, and F.S. Bates, Cross-linked polymersome membranes: Vesicles with broadly adjustable properties. *Journal of Physical Chemistry B*, 2002. **106**(11): p. 2848–2854.
15. Discher, B.M., Y.-Y. Won, D.S. Ege, J.C.-M. Lee, F.S. Bates, D.E. Discher, and D.A. Hammer, Polymersomes: Tough, giant vesicles made from diblock copolymers. *Science*, 1999. **284**: p. 1143–1146.
16. Discher, D.E. and A. Eisenberg, Polymeric vesicles. *Science*, 2002. **297**: p. 967–973.
17. Duncan, P.B. and D. Needham, Test of the epstein-plesset model for gas microparticle dissolution in aqueous media: Effect of surface tension and gas undersaturation in solution. *Langmuir*, 2004. **20**(7): p. 2567–2578.
18. Gao, C., K.B. Vargo, and D.A. Hammer, Protease-triggered, integrin-targeted cellular uptake of recombinant protein micelles. *Biomolecular Materials*, 2016. **16**(9): p. 1398–1406.
19. Ghoroghchian, P.P., P. Frail, K. Susumu, D. Blessington, F.S. Bates, B. Chance, D.A. Hammer, and M.J. Therien, NIR-emissive polymersomes: Soft matter self-assembly meets in vivo optical imaging. *Proceedings of the National Academy of Sciences USA*, 2005. **102**: p. 2922–2927.
20. Ghoroghchian, P.P., P.R. Frail, K.P. Davis, F.S. Bates, M.J. Therien, and D.A. Hammer, Bioresorbable vesicles formed through spontaneous self-assembly of amphiphilic polyethyleneoxide-block-polycaprolactone. *Macromolecules*, 2006. **39**(5): p. 1673–1675.
21. Ghoroghchian, P.P., J.J. Lin, A.K. Brannan, P.R. Frail, F.S. Bates, M.J. Therien, and D.A. Hammer, Quantitative membrane loading of polymer vesicles. *Soft Matter*, 2006. **2**: p. 973–978.
22. Ghoroghchian, P.P., P.R. P. R. Frail, G. Li, J.A. Zupancich, F.S. Bates, D.A. Hammer, and M.J. Therien, Controlling bulk optical properties of emissive polymersomes through intramembranous polymer-fluorophore interactions. *Chemistry of Materials*, 2007. **19**(6): p. 1309–1318.
23. Hajduk, D.A., M.B. Kossuth, M.A. Hillmyer, and F.S. Bates, Complex phase behavior in aqueous solutions of poly(ethylene oxide)-poly(ethylethylene) block copolymers. *Journal of Physical Chemistry B*, 1998. **102**(22): p. 4269–4276.
24. Hillmyer, M.A., F.S. Bates, K. Almdal, K. Mortensen, A.J. Ryan, and J.P.A. Fairclough, Complex phase behavior in solvent-free nonionic surfactants. *Science*, 1996. **271**: p. 976–978.
25. Hissa, D.C., I.M. Vasconcelos, A.F.U. Carvalho, V.L.R. Nogueira, P. Cascon, A.S.L. Antunes, G.R. de Macedo, and V.M.M. Melo, Novel surfactant proteins are involved in the structure and stability of foam nests from the frog Leptodactylus vastus. *Journal of Experimental Biology*, 2008. **211**(16): p. 2707–2711.
26. Holowka, E.P. and T.J. Deming, Synthesis and cross linking of L-DOPA containing polypeptide vesicles. *Macromolecular Bioscience*, 2010. **10**(5): p. 496–502.
27. Holowka, E.P., D.J. Pochan, and T.J. Deming, Charged polypeptide vesicles with controllable diameter. *Journal of the American Chemical Society*, 2005. **127**(35): p. 12423–12428.
28. Holowka, E.P., V.Z. Sun, D.T. Kamei, and T.J. Deming, Polyarginine segments in block copolypeptides drive both vesicular assembly and intracellular delivery. *Nature Materials*, 2007. **6**(1): p. 52–57.
29. Hsieh, K. and A.H.C. Huang, Endoplasmic reticulum, oleosins, and oils in seeds and tapetum cells. *Plant Physiology*, 2004. **136**(3): p. 3427–3434.
30. Huang, A.H.C., Oil bodies and oleosins in seeds. *Annual Review of Plant Physiology and Plant Molecular Biology*, 1992. **43**: p. 177–200.
31. Huber, M.C., A. Schreiber, P. von Olshausen, B.R. Varga, O. Kretz, B. Joch, S. Barnert, R. Schubert, S. Eimer, P. Kele, and S.M. Schiller, Designer amphiphilic proteins as building blocks for the intracellular formation of organelle-like compartments. *Nature Materials*, 2015. **14**(1): p. 125–132.
32. Israelachvili, J.N., *Intermolecular and Surface Forces*. 2nd ed. 1991, London: Academic Press Limited. 450.
33. Jang, Y., W.S. Jang, C. Gao, T.S. Shim, J.C. Crocker, D.A. Hammer, and D. Lee, Tuning the mechanical properties of recombinant protein-stabilized gas bubbles using triblock copolymers. *ACS Macro Letters*, 2016. **5**(3): p. 371–376.
34. Kaler, E.W., A.K. Murthy, B.E. Rodriguez, and J.A.N. Zasadzinski, Spontaneous vesicle formation in aqueous mixtures of single-tailed surfactants. *Science*, 1989. **245**(4924): p. 1371–1374.
35. Kamat, N.P., Z.Z. Liao, L.E. Moses, J. Rawson, M.J. Therien, I.J. Dmochowski, and D.A. Hammer, Sensing membrane stress with near IR-emissive porphyrins. *Proceedings of the National Academy of Sciences of the United States of America*, 2011. **108**(34): p. 13984–13989.
36. Kamat, N.P., G.P. Robbins, M.J. Therien, I.J. Dmochowski, and D.A. Hammer, A generalized system for photoresponsive membrane rupture in polymersomes. *Advanced Functional Materials*, 2010. **20**: p. 2588–2596.

37. Katz, J.S., S. Zhong, B.G. Ricart, D.J. Pochan, D.A. Hammer, and J.A. Burdick, Modular synthesis of biodegradable diblock copolymers for designing functional polymersomes. *Journal of the American Chemical Society*, 2010. **132**(11): p. 3654–3655.
38. Lacey, D.J., N. Wellner, F. Beaudoin, J.A. Napier, and P.R. Shewry, Secondary structure of oleosins in oil bodies isolated from seeds of safflower (*Carthamus tinctorius* L.) and sunflower (*Helianthus annuus* L.). *Biochemical Journal*, 1998. **334**: p. 469–477.
39. Luo, T.Z. and K.L. Kiick, Noncovalent modulation of the inverse temperature transition and self-assembly of elastin-b-collagen-like peptide bioconjugates. *Journal of the American Chemical Society*, 2015. **137**(49): p. 15362–15365.
40. MacEwan, S.R. and A. Chilkoti, Elastin-like polypeptides: Biomedical applications of tunable biopolymers. *Biopolymers*, 2010. **94**(1): p. 60–77.
41. MacEwan, S.R. and A. Chilkoti, Applications of elastin-like polypeptides in drug delivery. *Journal of Controlled Release*, 2014. **190**: p. 314–330.
42. Martin, L., E. Castro, A. Ribeiro, M. Alonso, and J.C. Rodriguez-Cabello, Temperature-triggered self-assembly of elastin-like block Co-recombinamers: The controlled formation of micelles and vesicles in an aqueous medium. *Biomacromolecules*, 2012. **13**(2): p. 293–298.
43. Maskarinec, S.A. and D.A. Tirrell, Protein engineering approaches to biomaterials design. *Current Opinion in Biotechnology*, 2005. **16**(4): p. 422–426.
44. Matson, J.B., R.H. Zha, and S.I. Stupp, Peptide self-assembly for crafting functional biological materials. *Current Opinion in Solid State & Materials Science*, 2011. **15**(6): p. 225–235.
45. Pangburn, T.O., K. Georgiou, F.S. Bates, and E. Kokkoli, Targeted polymersome delivery of siRNA induces cell death of breast cancer cells dependent upon orai3 protein expression. *Langmuir*, 2012. **28**(35): p. 12816–12830.
46. Park, W.M. and J.A. Champion, Thermally triggered self-assembly of folded proteins into vesicles. *Journal of the American Chemical Society*, 2014. **136**(52): p. 17906–17909.
47. Pastuszk, M.K., X.D. Wang, L.L. Lock, S.M. Janib, H.G. Cui, L.D. DeLeve, and J.A. MacKay, An amphipathic alpha-helical peptide from apolipoprotein A1 stabilizes protein polymer vesicles. *Journal of Controlled Release*, 2014. **191**: p. 15–23.
48. Peng, C.C., J.C.F. Chen, D.J.H. Shyu, M.J. Chen, and J.I.C. Tzen, A system for purification of recombinant proteins in *Escherichia coli* via artificial oil bodies constituted with their oleosin-fused polypeptides. *Journal of Biotechnology*, 2004. **111**(1): p. 51–57.
49. Percec, V., P. Leowanawat, H.J. Sun, O. Kulikov, C.D. Nusbaum, T.M. Tran, A. Bertin, D.A. Wilson, M. Peterca, S.D. Zhang, N.P. Kamat, K. Vargo, D. Moock, E.D. Johnston, D.A. Hammer, D.J. Pochan, Y.C. Chen, Y.M. Chabre, T.C. Shiao, M. Bergeron-Brlek, S. Andre, R. Roy, H.J. Gabius, and P.A. Heiney, Modular synthesis of amphiphilic Janus glycodendrimers and their self-assembly into glycodendrimersomes and other complex architectures with bioactivity to biomedically relevant lectins. *Journal of the American Chemical Society*, 2013. **135**(24): p. 9055–9077.
50. Percec, V., D.A. Wilson, P. Leowanawat, C.J. Wilson, A.D. Hughes, M.S. Kaucher, D.A. Hammer, D.H. Levine, A.J. Kim, F.S. Bates, K.P. Davis, T.P. Lodge, M.L. Klein, R.H. DeVane, E. Aqad, B.M. Rosen, A.O. Argintaru, M.J. Sienkowska, K. Rissanen, S. Nummelin, and J. Ropponen, Self-assembly of Janus Dendrimers into uniform dendrimersomes and other complex architectures. *Science*, 2010. **328**(5981): p. 1009–1014.
51. Qi, W., P.P. Ghoroghchian, G.Z. Li, D.A. Hammer, and M.J. Therien, Aqueous self-assembly of poly(ethylene oxide)-block-poly(epsilon-caprolactone) (PEO-b-PCL) copolymers: Disparate diblock copolymer compositions give rise to nano- and meso-scale bilayered vesicles. *Nanoscale*, 2013. **5**(22): p. 10908–10915.
52. Robbins, G.P., M. Jimbo, J. Swift, M.J. Therien, D.A. Hammer, and I.J. Dmochowski, Photoinitiated destruction of composite porphyrin-protein polymersomes. *Journal of the American Chemical Society*, 2009. **131**(11): p. 3872–3874.
53. Rodriguez-Hernandez, J. and S. Lecommandoux, Reversible inside-out micellization of pH-responsive and water-soluble vesicles based on polypeptide diblock copolymers. *Journal of the American Chemical Society*, 2005. **127**(7): p. 2026–2027.
54. Scott, R.W., S. Winichayakul, M. Roldan, R. Cookson, M. Willingham, M. Castle, R. Pueschel, C.C. Peng, J.T.C. Tzen, and N.J. Roberts, Elevation of oil body integrity and emulsion stability by polyoleosins, multiple oleosin units joined in tandem head-to-tail fusions. *Plant Biotechnology Journal*, 2010. **8**(8): p. 912–927.
55. Susumu, K. and M.J. Therien, Decoupling optical and potentiometric band gaps in p-conjugated materials. *Journal of the American Chemical Society*, 2002. **124**: p. 8550–8552.
56. Trent, A., R. Marullo, B. Lin, M. Black, and M. Tirrell, Structural properties of soluble peptide amphiphile micelles. *Soft Matter*, 2011. **7**(20): p. 9572–9582.
57. Vargo, K.B., *Recombinant surfactants derived from the naturally occuring protein oleosin*, in *Chemical and Biomolecular Engineering*. 2014, University of Pennsylvania, Philadelphia, Pennsylvania, USA. p. 232.
58. Vargo, K.B., A. Al Zaki, R. Warden-Rothman, A. Tsourkas, and D.A. Hammer, Superparamagnetic iron oxide nanoparticle micelles stabilized by recombinant oleosin for targeted magnetic resonance imaging. *Small*, 2015. **11**(12): p. 1409–1413.
59. Vargo, K.B., R. Parthasarathy, and D.A. Hammer, Self-assembly of tunable protein suprastructures from recombinant oleosin. *Proceedings of the National Academy of Sciences, USA*, 2012. **109**(29): p. 11657–11662.
60. Vargo, K.B., N. Sood, T.D. Moeller, P.A. Heiney, and D.A. Hammer, Spherical micelles assembled from variants of recombinant oleosin. *Langmuir*, 2014. **30**(38): p. 11292–11300.
61. Wosten, H.A.B., O.M.H. Devries, and J.G.H. Wessels, Interfacial self-assembly of a fungal hydrophobin into a hydrophobic Rodlet layer. *Plant Cell*, 1993. **5**(11): p. 1567–1574.
62. Xiao, Q., S.S. Yadavalli, S.D. Zhang, S.E. Sherman, E. Fiorin, L. da Silva, D.A. Wilson, D.A. Hammer, S. Andre, H.J. Gabius, M.L. Klein, M. Goulian, and V. Percec, Bioactive cell-like hybrids coassembled from (glyco)dendrimersomes with bacterial membranes. *Proceedings of the National Academy of Sciences of the United States of America*, 2016. **113**(9): p. E1134–E1141.
63. Xiao, Q., S.D. Zhang, Z.C. Wang, S.E. Sherman, R.O. Moussodia, M. Peterca, A. Muncan, D.R. Williams, D.A. Hammer, S. Vertesy, S. Andre, H.J. Gabius, M.L. Klein, and V. Percec, Onion-like glycodendrimersomes from

sequence-defined Janus glycodendrimers and influence of architecture on reactivity to a lectin. *Proceedings of the National Academy of Sciences of the United States of America*, 2016. **113**(5): p. 1162–1167.
64. Zhang, L.F. and A. Eisenberg, Multiple morphologies of crew-cut aggregates of polystyrene-B-poly(acrylic acid) block-copolymers. *Science*, 1995. **268**(5218): p. 1728–1731.
65. Zhang, S.D., H.J. Sun, A.D. Hughes, R.O. Moussodia, A. Bertin, Y.C. Chen, D.J. Pochan, P.A. Heiney, M.L. Klein, and V. Percec, Self-assembly of amphiphilic Janus dendrimers into uniform onion-like dendrimersomes with predictable size and number of bilayers. *Proceedings of the National Academy of Sciences of the United States of America*, 2014. **111**(25): p. 9058–9063.
66. Zhang, S.D., Q. Xiao, S.E. Sherman, A. Muncan, A. Vicente, Z.C. Wang, D.A. Hammer, D. Williams, Y.C. Chen, D.J. Pochan, S. Vertesy, S. Andre, M.L. Klein, H.J. Gabius, and V. Percec, Glycodendrimersomes from sequence-defined Janus glycodendrimers reveal high activity and sensor capacity for the agglutination by natural variants of human lectins. *Journal of the American Chemical Society*, 2015. **137**(41): p. 13334–13344.

19 Amphiphilic Self-Assembly and the Origin of Life in Hydrothermal Conditions

Christos D. Georgiou
University of Patras, Patras, Greece

David W. Deamer
University of California Santa Cruz, Santa Cruz, California

CONTENTS

19.1 Introduction: Setting the Stage 337
19.2 Sources of Organic Carbon Compounds in the Prebiotic Environment................................ 338
19.3 Definition of Amphiphilic Molecules.. 338
19.4 Sources of Amphiphilic Compounds 339
19.5 Self-Assembly of Amphiphiles into Stable Membranes.. 339
19.6 Primitive Membrane Functions 341
    19.6.1 Permeability and Encapsulation of Polymers.. 341
    19.6.2 Comparison of Mineral Compartments and Amphiphilic Membranes 341
19.7 Energy Sources for Primitive Life.. 343
    19.7.1 Hydrothermal Vents as a Site for Chemiosmotic Energy Transduction.................... 343
    19.7.2 Hydrothermal Fields and Light-Driven Chemiosmotic Energy Transduction 344
19.8 Experimental Tests Simulating Hydrothermal Vent and Hydrothermal Field Conditions.....344
19.9 Summary 345
    19.9.1 Favourable Properties of Alkaline Hydrothermal Vents 345
    19.9.2 Limitations of Alkaline Hydrothermal Vents... 346
    19.9.3 Favourable Properties of Hydrothermal Fields.. 346
    19.9.4 Limitations of Hydrothermal Fields 346
References.. 346

19.1 INTRODUCTION: SETTING THE STAGE

There is a consensus that the Earth–Moon system was produced by a collision of a Mars-sized object with the Earth approximately 4.4 billion years ago. The energy of the collision left the Earth's surface in a state resembling molten lava. It is significant that much of the original atmosphere was eroded into space by the energy of this event, and no organic molecules could survive the elevated temperature. Over the next several hundred million years, the Earth cooled to the point at which water vapour could condense so that precipitation produced a global ocean. The source of the water was largely volcanic outgassing from the Earth's interior, with some contribution from the impacts of icy comets. Outgassing was also the source of a renewed atmosphere with a composition governed by the fugacity of the Earth's mineral crust, so that it contained a mixture of chemically inert nitrogen gas with a small percentage of carbon dioxide. There was little or no free molecular oxygen (for reviews, see Sleep, 2010 and Zahnle et al., 2010).

The specific geological sites we will consider are associated with volcanic activity generated by magma plumes rising to the Earth's crust. Contemporary examples include submarine hydrothermal vents such as black smokers that have been proposed as possible sites for the origin of life (Corliss et al., 1981). However, these are relatively short lived, so more recent attention has shifted to alkaline hydrothermal vents, one example being Lost City (Kelley et al., 2001). These exist for thousands of years and have substantial free energy available in sustained far-from-equilibrium conditions. Alkaline hydrothermal vents are characterized by a single interface between solid mineral surfaces and the aqueous phase of seawater.

The alternative sites we will consider are hydrothermal fields associated with volcanoes emerging through the global ocean. These more complex environments have three interfaces that undergo cyclic fluctuations:

FIGURE 19.1 Bumpass hell, a hydrothermal field on Mount Lassen, California.

chapter, we will use the organic material in carbonaceous meteorites as a useful guide to compounds available for the origin of life. Such meteorites contain a mixture of amino acids, nucleobases, sugars, and amphiphilic compounds that were clearly produced by nonbiological chemical processing (Sephton, 2005). A reasonable assumption is that similar mixtures would be available on the prebiotic Earth, either synthesized by geochemical reactions or delivered during late accretion. In a global ocean 5 km deep, organic solutes would be extremely dilute (Stribling and Miller, 1987), but organic compounds delivered to or synthesized in volcanic land masses would be dissolved by precipitation and accumulate in hydrothermal ponds. Such ponds represent containers in which a variety of physical and chemical processes can occur.

atmosphere:water, atmosphere:mineral, and mineral:water. In this chapter, we will argue that the increased complexity provided by three interfaces can drive self-assembly processes and chemical reactions that cannot occur in aqueous solutions or at mineral:water interfaces.

The existence of early terrestrial volcanoes is supported by the fact that volcanism was also occurring on the planet Mars over 3 billion years ago and it is reasonable to assume that the first landmasses on the Earth were volcanic in origin, resembling Hawaii and Iceland today. Precipitation on volcanic landmasses would produce hydrothermal fields characterized by freshwater geysers, hot springs, and clay-lined pools that undergo cycles of hydration and evaporation. One such site is illustrated in Figure 19.1.

19.2 SOURCES OF ORGANIC CARBON COMPOUNDS IN THE PREBIOTIC ENVIRONMENT

Given the two alternative hydrothermal sites described above, the next question to be addressed concerns the source and composition of organic compounds required for cellular life to emerge. All of the carbon now circulating in the biosphere was delivered during accretion of planetesimals as the Earth formed followed by late accretion responsible for the lunar cratering record. Because the original organic compounds delivered during primary accretion would have been destroyed by the Moon-forming event, the organic material necessary for the origin of life had to be replaced after the global ocean appeared. Two possible sources that are not mutually exclusive are delivery of organic compounds as interplanetary dust particles and comets over several hundred million years of late accretion (Chyba and Sagan, 1992), or geochemical reactions that could synthesize organic compounds (Bada and Lazcano, 2003).

It is still uncertain what the relative fractions of these two sources were, or their composition, but for the purposes of this

19.3 DEFINITION OF AMPHIPHILIC MOLECULES

The properties of amphiphilic molecules are a primary focus of this chapter. Amphiphiles are molecules containing a nonpolar hydrocarbon chain that is attached to a polar group such as a carboxyl or phosphate. The molecular structures of two common amphiphiles are illustrated in Figure 19.2.

An important property of certain amphiphilic compounds is that the combination of a nonpolar and polar group on the same molecule allows them to self-assemble into compartments bounded by a bimolecular membrane. All life depends on this property, and the goal of this chapter is to consider which amphiphilic compounds were likely to be available on the early Earth, what conditions would permit self-assembly into membranes, and how they could have participated in processes leading to the origin of cellular life in hydrothermal conditions.

Two lipid bilayer structures from molecular dynamics simulations are shown in Figure 19.3. The hydrocarbon chains form a nonpolar interior that has the dielectric constant of 2, essentially that of oil, while the polar head groups interact with the high dielectric aqueous phase on both sides of the membrane. The low dielectric interior is one of the most important properties of the bilayers of biological membranes in that it presents a very high Born energy barrier to the diffusion of ions and polar molecules (Parsegian, 1969).

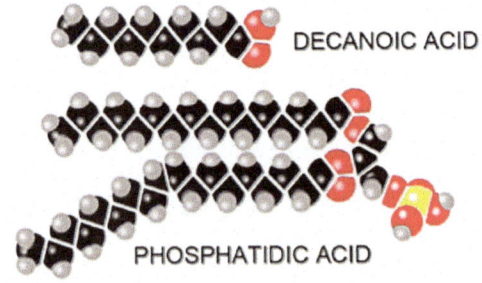

FIGURE 19.2 Structure of decanoic acid, one of the simplest amphiphilic fatty acids capable of forming bilayer membranes, and phosphatidic acid, the simplest species of a phospholipid having two fatty acid chains, one with a *cis* double bond.

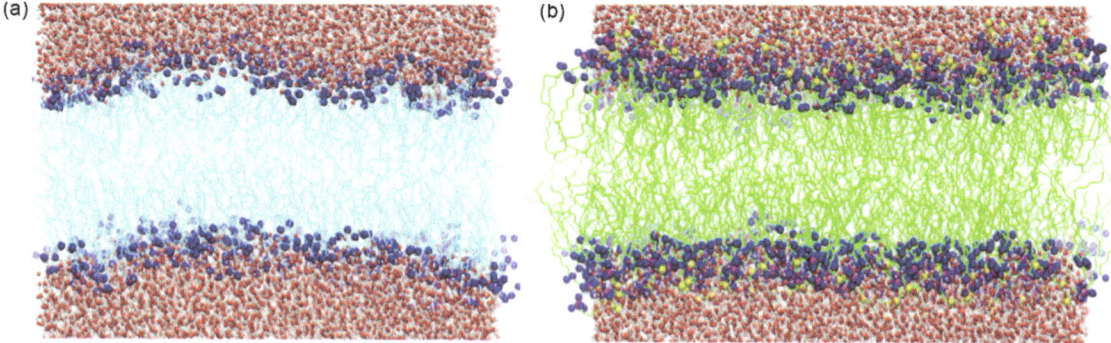

FIGURE 19.3 Structure of lipid bilayers from molecular dynamics simulations. Oleic acid (a) and phosphatidylcholine (b) both assemble into bilayer membranes. Hydrocarbon chains are present as a low-dielectric interior, and hydrophilic head groups interact with the aqueous phase at the membrane surface. Images kindly provided by Andrew Pohorille.

A second essential property of the lipid bilayers can be understood by considering a burning candle. The wax of the candle itself is a solid, but just below the flame the wax melts into a liquid. If it did not melt, the candle could not burn. The same is true of contemporary cell membranes, in that the hydrocarbon chains must be in a fluid state, rather than a gel state. All life today depends on membrane fluidity, so it is a reasonable assumption that a similar fluidity was required by the first cellular life. Which amphiphilic molecules may have been present that could assemble into fluid bilayer membranes which also represented a sufficient permeability barrier? That question is addressed in the following sections.

19.4 SOURCES OF AMPHIPHILIC COMPOUNDS

A well-known chemical reaction that produces hydrocarbons and their derivatives is referred to as the Fischer–Tropsch-type (FTT) synthesis. The industrial version passes carbon monoxide and hydrogen over a hot iron catalyst. The carbon monoxide transiently adsorbs to the iron surface where the carbons link and grow into chains while simultaneously being reduced by the hydrogen. The products are long-chain hydrocarbon derivatives of varying lengths.

Bernd Simoneit's research group has demonstrated hydrocarbon synthesis under much simpler conditions that do not require an iron catalyst (McCollom et al., 1999; Rushdi and Simoneit, 2001). In a typical reaction, a solution of oxalic acid was heated to 150°C in a stainless steel pressure vessel. Under these conditions, the oxalic acid decomposes into carbon monoxide and hydrogen. After a few hours, there is a significant yield of hydrocarbons derivatives ranging from C_{12} to $>C_{33}$ that included normal alcohols and monocarboxylic acids. A microscopic image of the oily products is shown in Figure 19.4. It is remarkable that the oil droplets were produced from nothing more than oxalic acid dissolved in water.

It seems feasible that FTT reactions may have synthesized long-chain hydrocarbons and their derivatives when hydrogen and carbon monoxide passed through high pressure and temperature conditions associated with

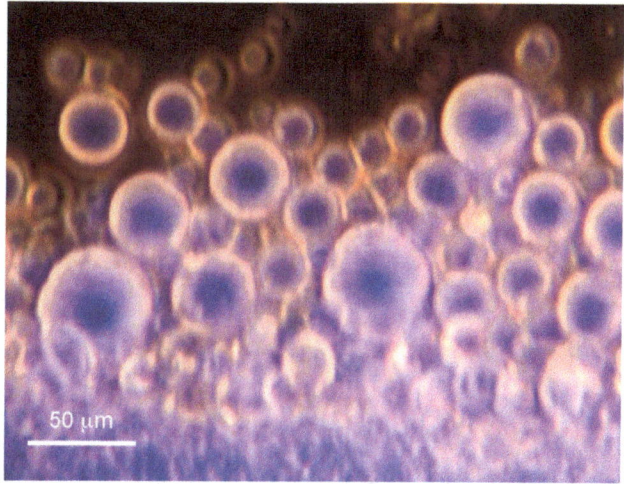

FIGURE 19.4 Oil droplets produced by FTT synthesis (Rushdi and Simoneit, 2001). The droplets display a blue fluorescence when illuminated with UV light, probably because small amounts of polycyclic aromatic compounds are also produced by the reaction. The largest droplets are ~50μm in diameter (unpublished micrograph by the authors).

volcanic activity on the early Earth. Fischer–Tropsch reactions (Lancet and Anders, 1970) have also been proposed to account for some of the alkanes and monocarboxylic acids present in carbonaceous meteorites, which range up to 12 carbons in length (Naraoka et al., 1999). As will be discussed in the next section, these can assemble into membranous vesicles, suggesting that similar mixtures of amphiphilic compounds were likely to be available to form boundary membranes of primitive life.

19.5 SELF-ASSEMBLY OF AMPHIPHILES INTO STABLE MEMBRANES

Amphiphiles such as phospholipids and cholesterol are essential components of cell membranes today, but these are relatively complex molecules that are synthesized by enzyme-catalysed metabolic pathways. Is it possible for membranes to assemble from simpler amphiphilic compounds that might be present in the prebiotic environment? In fact, a mixture of amphiphilic compounds extracted from the Murchison meteorite can assemble into membranous vesicles that have significant stability (Deamer, 1985; Deamer and Pashley, 1989). Figure 19.5 shows a series of micrographs illustrating vesicle formation from such an extract when a dried material interacts with an alkaline aqueous phase. The vesicles are fluorescent, as shown in Figure 19.5d, indicating that they contain polycyclic aromatic hydrocarbons (PAHs). The vesicles can encapsulate pyranine, a water-soluble fluorescent dye, so they do present a permeability barrier to the diffusion of ions.

Given that mixtures of amphiphilic compounds can be synthesized by abiotic reactions, the next aim was to establish the simplest pure compounds that can self-assemble into membranes. Hargreaves and Deamer (1978) made an extensive study of a variety of single-chain amphiphiles. The surprising result was that fatty acids could form stable vesicles at a pH where the head groups were half anionic, half protonated, that is, at the pKa of the carboxylic acid. This effect had been observed earlier with oleic acid by Gebicki and Hicks (1973). The other essential properties are related to stability, concentration, chain length, and fluidity. At ordinary temperature ranges of 20°C–40°C, fatty acids having 9–12 carbon chains readily form membranes but only if the concentration exceeds a certain value referred to as the critical vesicle concentration (CVC), which ranged from 88 mM for nonanoic acid (9 carbons) to 20 mM for lauric acid (12 carbons). If a double bond was present to maintain fluidity, chains as long as 18 carbons (oleic acid) formed membranes even at concentrations as low as a few millimolar. Fatty alcohols could not form membranes, but if mixed with fatty acids stable membranes readily formed at alkaline pH ranges, the first indication that mixed amphiphiles conferred increased stability.

Significantly, the presence of divalent cations such as magnesium and calcium strongly inhibited membrane assembly by fatty acids. This means that a pure fatty acid membrane would be unstable in marine environments due to the high concentrations of Mg^{2+} (54 mM) and Ca^{2+} (10 mM). This concern was addressed by Monnard and Deamer (2002) who reported that mixing a fatty acid with a monoglyceride ester of the fatty acid allowed membranes to assemble in the presence of divalent cations or seawater. Fatty acid membranes have been employed extensively by the Szostak research group by encapsulating nucleic acids as model systems of primitive life (Hanczyc et al., 2003, 2007). Magnesium ions

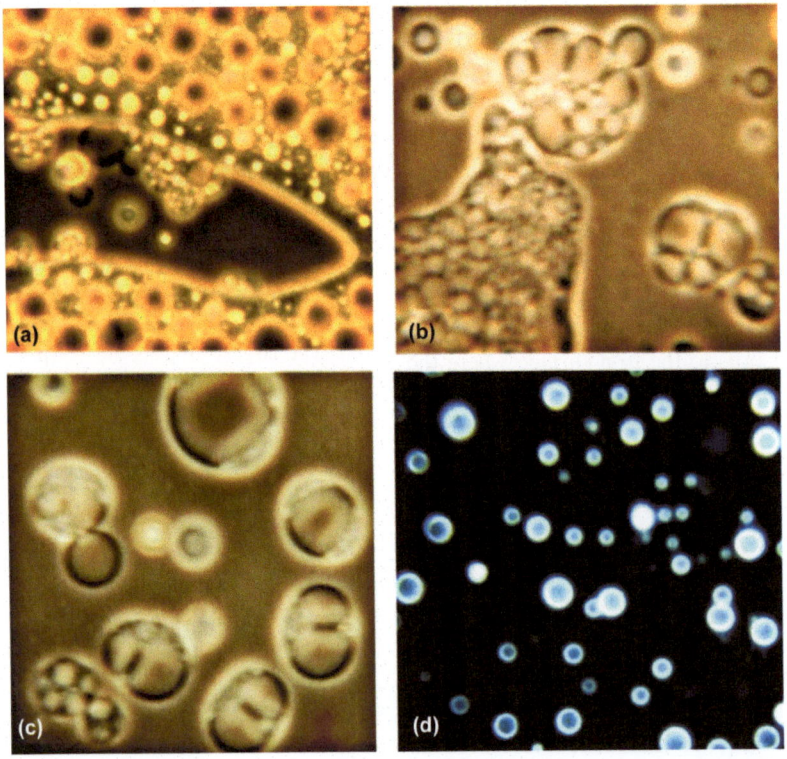

FIGURE 19.5 Self-assembly of amphiphilic compounds extracted from the Murchison carbonaceous meteorite. The lipid-like material was extracted in an organic solvent system, and the amphiphilic mixture was separated by thin-layer chromatography. A small amount of the extract was dried on a microscope slide, followed by addition of a dilute alkaline buffer. The image shown in panel (a) was taken a few minutes after buffer was added and membranous vesicles had begun to form within the extract. A few minutes later, the entire extract had vesiculated as water penetrated the dried material (b) and individual vesicles were released into the medium (c). The vesicles were highly fluorescent because polycyclic aromatic hydrocarbons (PAH) were also present in the membranes (d). Panels (a), (b), and (d) were 160× original magnification, and panel (c) was 400× original magnification. Typical vesicles ranged from 10 to 20 μm in diameter.

are required for some of the reactions but disrupted the membranes. However, Adamala and Szostak (2013) reported that addition of citrate to the mixture chelated the Mg^{2+} so that it would not damage the membranes but would still be available as an essential cofactor for RNA chemistry. Acidic pH ranges represent another limitation for fatty acid membranes, because the neutral protonated forms cannot assemble into stable membranes. Namani and Deamer (2008) reported that addition of a long-chain amine overcame this limitation by adding a positive charge at low pH ranges.

These observations put constraints on the composition of membranes that could assemble in the prebiotic environment. The monocarboxylic acid chain lengths detected in the Murchison meteorite range from 8 to 12 carbons in length, but the longer chain lengths are present only in very low amounts, a few ppm (Naraoka et al., 1999). Another concern is that membranes formed by pure single chain amphiphiles are relatively fragile compared with membranes formed by phospholipids having two chains. One way that additional stability could be imparted to prebiotic membranes is the observation of Budin et al. (2014) that membranes formed from fatty acid mixtures are more stable than those formed by a pure compound. Furthermore, Groen et al. (2012) found that small amounts of polycyclic aromatic hydrocarbons (PAHs) such as pyrene have a stabilizing effect resembling that of cholesterol in the membranes of cells today.

We conclude that the most plausible conditions under which membranes could be incorporated into the first life would be in moderately acidic (pH ~3–5) freshwater ponds so that high concentrations of divalent cations would not inhibit assembly. The membranes would not be a pure compound but instead would be a mixture of fatty acids with chains longer than 12 carbons together with other nonpolar compounds such as PAH that promote stability. It is important to note here that membrane compartments composed of mixed amphiphilic compounds have recently been demonstrated in laboratory simulations of hydrothermal vents (Jordan et al. 2019). These results show that the divalent cations of salty seawater are not an absolute limit to assembly of fatty acid membranes.

19.6 PRIMITIVE MEMBRANE FUNCTIONS

19.6.1 Permeability and Encapsulation of Polymers

Paula et al. (1996) measured the diffusion of potassium ions out of liposomes 0.2 μm in diameter and reported a permeability coefficient of $~10^{-11}$ cm s^{-1}. By way of contrast, the same value for water molecules is $~10^{-3}$ cm s^{-1} (Olbrich et al., 2000) giving half-times of exchange in liposomes measured in a few tens of milliseconds, nearly a billion times faster than potassium ions under the same conditions. The extreme permeability barrier to ions means that ionic concentration gradients produced by active transport can be maintained across the boundary membrane of a cell or subcellular compartments such as mitochondria and chloroplasts. This property, combined with ion-specific protein channels and active ion transport processes, allows living cells to maintain and regulate the composition of an internal volume that is very different from the external aqueous medium.

The earliest cells would not have had specialized protein-based transport systems, so how could they have access to nutrient solutes available in the environment? In fact, membranes composed of fatty acids and short-chain phospholipids are relatively permeable to small polar and charged molecules (Budin and Szostak, 2011) and to amino acids (Chakrabarti and Deamer, 1994). They even show selective permeability especially important for the function of early cells lacking specific transporters. For example, fatty acid (and phospholipid) membranes are approximately fivefold more permeable to ribose than its diastereomers arabinose, lyxose, and xylose (Sacerdote and Szostak, 2005). Moreover, such simple lipids have the properties necessary for spontaneous growth and division (Hanczyc et al., 2003; Zhu and Szostak, 2009). This is also consistent with a heterotrophic model for protocells, based on the passive diffusion across the cell membrane of chemical building blocks synthesized in the environment (Budin and Szostak, 2011; Monnard and Deamer, 2001).

Assuming that primitive membranes were sufficiently permeable so that small nutrient solutes such as amino acids and simple carbohydrates had access to the internal volume of vesicles, the next question concerns polymers such as RNA, which cannot permeate either phospholipid or fatty acid membranes. However, if polymers and amphiphilic compounds are both present in an aqueous medium, it is surprisingly simple for the polymers to be encapsulated when membranes assemble in solution (Hanczyc et al., 2003; Mansy et al., 2008) or undergo cycles of hydration and dehydration (Deamer and Barchfeld, 1982; Shew and Deamer, 1985). We define polymers encapsulated in membranous compartments as protocells, and each protocell can be considered to be a natural experiment. Most are inert, but a few by chance may have a system of polymers that happen to have catalytic properties, particularly the ability to catalyse polymerization of monomers into more polymers. One such example is a ribozyme with polymerase activity that can produce RNA polymers over 200 nucleotides in length (Attwater et al., 2013).

19.6.2 Comparison of Mineral Compartments and Amphiphilic Membranes

The origin of water-based cellular life requires some form of membrane encapsulation. This is considered to be an essential functional characteristic of the primitive cell-like structures, which could form spontaneously as new membranes self-assemble and encapsulate genetic molecules in solution (Szostak et al., 2001). Two different systems have been proposed to compose such compartments, and in the following discussion, we will compare their properties in terms of the ability to concentrate potential reactants, to maintain concentration gradients, to encapsulate polymers, and to generate energy.

The first sites to be considered are alkaline hydrothermal vents. These occur at relatively low temperature ranges, are stable for thousands of years, and produce porous mineral structures composed of mineral membranes through which warm seawater comes into contact with cold seawater. The alkaline fluids are saturated with calcium carbonate that precipitates as porous mineral structures upon cooling. Theoretical studies (Baaske et al., 2007) predicted that small organic molecules such as nucleotides and nucleic acids originating from very dilute external reservoirs could be concentrated by the strong thermal gradients present in hydrothermal vents developed within the thin channels produced by mineral precipitation. This has been confirmed by a laboratory simulation showing that subcritical concentrations of fatty acids can be concentrated (>1000-fold enrichment) and self-assemble into vesicles at the bottom of capillary channels, with a parallel concentration of DNA oligonucleotides and their encapsulation within the vesicles (Budin et al., 2009; Budin and Szostak, 2010).

The constant heating of the channel wall keeps the local concentration high in this steady-state system. Although this concentrating process works under laboratory conditions, marine environments are characterized by high salt concentration (0.5 M NaCl in today's ocean water) that may not be conducive to thermal diffusion. Furthermore, as noted earlier, divalent cations are also present in seawater and at even higher concentrations in alkaline hydrothermal vents composed of calcium carbonate. High salt concentrations and divalent cations tend to inhibit membrane assembly from amphiphilic compounds like monocarboxylic acids (Monnard and Deamer, 2002), a fact that weighs against the origin of cellular life under these conditions.

These constraints lead us to consider the second proposed site, which are the freshwater hydrothermal fields associated with volcanic activity in which cycles of dehydration and rehydration are readily apparent. It was shown more than 30 years ago that lipid vesicles mixed in aqueous solution with DNA and other macromolecules fuse upon drying to form a multilamellar film that effectively sandwiches and concentrates the macromolecules between lipid layers. A subsequent rehydration cycle produces large lipid vesicles that have encapsulated a significant fraction of the macromolecules (Shew and Deamer, 1985). We have extended this observation to a similar process that could occur in small pools associated with hydrothermal fields. In the first stage of the cycle, an aqueous phase consisting of a dilute solution of amphiphilic lipid-like compounds and potential monomers is dehydrated by evaporation and accumulates as a film on rough mineral surfaces. The dehydrated amphiphiles self-assemble into multilamellar layers that trap concentrated monomers between them, followed by condensation reactions linking the monomers into polymers (Deamer et al., 2006; DeGuzman et al., 2014; Rajamani et al., 2008). The elevated temperature of the hydrothermal site provides the required activation energy. In the hydration

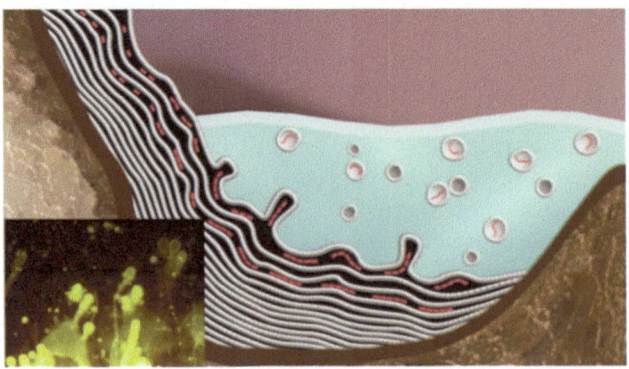

FIGURE 19.6 Protocell formation and genetic material encapsulation during hydration phase. Inset: Lipid vesicles with encapsulated DNA (160× original magnification). These were produced by a dehydration–rehydration cycle simulating a small pond in a volcanic hydrothermal site undergoing evaporation and refilling. Protocells containing the DNA are formed when the dried mixture of lipid and DNA is rehydrated. (Image adapted from Damer and Deamer 2015.)

stage of the cycle, water interacts with the dry multilamellar matrix and produces protocells having diameters similar to those of typical bacteria, some of which contain polymers of varying composition (Figure 19.6). The surviving protocells aggregate during the next dehydration cycle on the mineral surface and fuse again into a multilamellar matrix in which their contents can mix. The wet–dry cycling can continue indefinitely as long as the pool exists.

Although phospholipid was used in the studies of Rajamani et al. (2008) and DeGuzman et al. (2014), simpler amphiphiles such as fatty acids can also self-assemble into membranous vesicles when the amphiphiles are above a critical aggregate concentration (Budin et al., 2009). An example of encapsulation was provided by Hanczyc et al. (2003) who showed that RNA adsorbed on the surface of clay particles can be captured in fatty acid vesicles. This is significant because the clay mineral montmorillonite promotes RNA polymerization from activated nucleotides (Ferris et al., 1996; Huang and Ferris, 2006). Furthermore, such vesicles can grow by taking up fatty acids from nearby micelles and can be made to divide by extrusion through small pores without dilution of their content (Hanczyc et al., 2003). All of these reactions can potentially occur in freshwater pools associated with hydrothermal fields, particularly those undergoing cycles of hydration and dehydration (Damer and Deamer, 2015).

We finally note that protocells containing RNA can compete for "nutrients" in the form of fatty acids that compose their membranes. It was shown that upon encapsulation in fatty acid vesicles, RNA exerts an osmotic pressure on the vesicle membrane, which further grows by accumulating additional membrane components at the expense of relaxed vesicles lacking RNA, which shrink. Thus, encapsulated RNA undergoing efficient replication could cause faster cell growth and the emergence of Darwinian evolution at the cellular level (Chen and Szostak, 2004). It has also been demonstrated that protocells composed of phospholipid also compete for "nutrients" at the expense of protocells composed of fatty acids. This suggests that phospholipid membranes can emerge as a deterministic outcome of intrinsic lipid physical properties and related processes (Budin et al., 2012; Budin and Szostak, 2011).

19.7 ENERGY SOURCES FOR PRIMITIVE LIFE

A second comparison between hydrothermal vents and fields concerns the kinds of energy that each condition makes available. At the origin of life, the first protocells must have needed a continuous source of energy to drive their primitive metabolism and biosynthesis, because they would not yet have evolved specific enzymes to generate and store their energy as proton gradients and ATP. Most of the high-energy flux must have been wasted, with the only net gain being the maintenance of the required low entropy (ΔS) state and the negative free energy change (ΔG) state of protocells. This energy excess would allow the counterbalancing of the low ΔS state by an even larger decrease in protocell internal energy difference (ΔH), according to the Gibbs thermodynamic free energy formula $\Delta G = \Delta H - T \cdot \Delta S$. These thermodynamic considerations argue against other settings for life's origin such as a primordial soup formed by lightning strikes, UV radiation, or the delivery of organics from space, or for compartmentalized systems that cannot continually replenish chemically active precursors, such as ice or pumice (Lane and Martin, 2012; Sousa et al., 2013).

Lacking protein enzymes and metabolism to synthesize monomers, the first functional protocells must have assembled in naturally reactive chemical environments that thermodynamically favour the synthesis of life's building blocks, particularly amino acids and nucleobases. This has been shown with simple molecules such as H_2, CO_2, and NH_4^+ in submarine hydrothermal vents interfacing with ocean water (Amend and McCollom, 2009; Shock and Canovas, 2010; Shock et al., 1998). Moreover, it has been argued that the process of serpetinization releases H_2 by hydrothermal reaction of deep ocean minerals such as olivine with water together with formation of serpetinite and magnetite and may thereby generate natural proton gradients with magnitude and orientation found in modern cells (Russell et al., 1993; Russell and Hall, 1997). A reductive tricarboxylic acid cycle could fix CO_2 as energy-rich molecules has also been proposed and tested by demonstrating the production of pyruvate from lactate or α-ketoglutarate from pyruvate, both promoted by a ZnS mineral (Guzman and Martin, 2009).

In this regard, alkaline hydrothermal vents are of particular interest because of their sustained far-from-equilibrium conditions and their basic similarities with the carbon and energy metabolism of autotrophic cells. The best-known example is the Lost City Hydrothermal Field (Kelley et al., 2001). Black smoker hydrothermal vents (volcanic) are not ruled out thermodynamically. However, they last only for a few decades compared with more than 30,000 years for the alkaline vents, and their very high temperatures (>250°C) destabilize and degrade organic carbon compounds in contrast to the life-compatible temperature range (50°C–90°C) of alkaline vents. Furthermore, the acidic pH (~1) of black smokers would tend to hydrolyse polymers, while the alkaline vent pH (~9–10) would be less disruptive in this regard (Lane and Martin, 2012).

19.7.1 Hydrothermal Vents as a Site for Chemiosmotic Energy Transduction

Peter Mitchell first proposed that coupling membranes generated ATP by a chemiosmotic mechanism (Mitchell, 1961), and there is now abundant evidence supporting his original conjecture. A dramatic confirmation that chemiosmotic energy can drive ATP synthesis was presented by Racker and Stoeckenius (1974). They combined a light-driven proton pump (bacteriorhodopsin) with an ATP synthase in lipid vesicles and showed that the system could produce ATP from ADP and phosphate. A hypothetical mechanism has been proposed for protocells by which lipid compartments (protocells) can generate chemiosmotic potential using hydrogen as a source of electrons, which are donated to an internal acceptor to synthesize high-energy phosphate bonds (Koch, 1985; Koch and Schmidt, 1991).

Russell and Hall (1997) extended these concepts to a possible role of alkaline vents in generating chemiosmotic potentials in which proton gradients are sustained across thin mineral walls. Such gradients are similar in magnitude to the pH gradients and redox potential to modern autotrophic cells, leading to the suggestion that proton gradients in alkaline hydrothermal vents could have supported reduction of CO_2 by the pH-dependent reduction potential of transition metal sulphide (Fe-/Ni-S) minerals (Lane and Martin, 2012). An analogous process today involves the Fe^{2+}/Fe^{3+}-S protein ferredoxin in acetogens and methanogens that reduces CO_2 to methane or acetate. The electrons are derived from H_2 (a process known as flavin-based electron bifurcation) membrane potential used for carbon assimilation and ATP synthesis.

Lane and Martin (2012) further proposed that thin-walled inorganic compartments composed of semiconducting FeS in alkaline vents could transfer electrons from H_2 to CO_2. If so, the resulting hydrocarbon derivatives could then assemble into vesicular protocells within the microporous labyrinth of these vents. Inorganic protocells are defined as networks of inorganic compartments lined partially or completely with phase-separated alkanes, hydrophobic amino acids and peptides, fatty acids, and other amphiphiles. At evolutionary stages when membrane lipids and proteins were genetically encoded, cell-like structures may have begun to occupy vent pores. At all stages, protocells were immobilized as they were relying on both organic membranes and inorganic walls to harness the geochemical chemiosmotic potential. To escape from the vents as free-living cells they must have switched from relying on natural proton gradients to actively generating ion gradients on their own.

A possible scenario for the early stages of bioenergetic evolution has been described by Lane and Martin (2012) and Sousa et al. (2013). The main points are outlined below:

1. Reduced Fe^{2+}-/Ni^{2+}-S thin wall minerals (analogous to the catalytic FeS clusters in ferredoxin, thioesters, and acyl phosphates) catalyse CO_2 reduction by H_2, under a continuous flux of pH ~9 hydrothermal effluent, with subsequent organic synthesis.
2. These thermodynamic conditions (Amend and McCollom, 2009) could support the origin of a protomembrane lining the thin mineral wall, and also the development of an ion-gradient-harnessing ATP synthase together with thioester-dependent substrate-level phosphorylations.
3. Because early membranes are relatively permeable to protons compared to sodium ions (Nichols and Deamer, 1980), there would be selective pressure for protocells to evolve mechanisms by which the natural (free) proton gradients could be transduced into biochemical Na^+ gradients. These could be generated at no net energetic cost by developing a simple Na^+/H^+ transporter similar to the antibiotic nigericin.

19.7.2 Hydrothermal Fields and Light-Driven Chemiosmotic Energy Transduction

An alternative geochemical condition conducive to the origin of cellular life is based on the argument that intracellular ionic composition of modern cells weighs strongly in favour of a hydrothermal field (Mulkidjanian et al., 2012). In other words, the ion composition of modern cells might reflect the inorganic ion composition of the habitats of protocells. By combining geochemical data with phylogenetic analysis of the inorganic ion requirements of life today, it was proposed that protocells evolved in habitats with a high K^+/Na^+ ratio and relatively high concentrations of Zn, Mn, and phosphate. Geochemical reconstructions in the absence of oxygen and in a CO_2-rich atmosphere suggest that an ionic composition favouring the origin of cells could not have existed in marine settings. Instead, this is compatible with inland geothermal systems with ionic compositions resembling the internal environment of modern cells. In this scenario, precellular stages of life developed in shallow ponds lined with porous silicate minerals and enriched in K^+, Zn^{2+}, and phosphate.

Given that light is an abundant energy source used by cyanobacteria and other prokaryotes capable of photosynthesis, it has been suggested that early life quickly evolved mechanisms to capture and transduce light-driven chemiosmotic energy for reactions (Deamer, 1997; Deamer and Weber, 2010). In principle, such processes would involve a light-absorbing pigment such as ferrocyanide, porphyrins, proteinoids, or polycyclic aromatic hydrocarbons, either encapsulated as solutes or partitioned into boundary lipid membranes. If these could undergo a relatively simple photochemical reaction that releases or takes up protons, the result would be a proton gradient across the membrane that captures a portion of the original light energy. At some point in early evolution, such gradients could be coupled to biochemical processes to initiate a primitive version of chemiosmosis.

It has already been demonstrated in the laboratory that membranes composed of simple amphiphiles such as fatty acids can encapsulate mineral particles (Hanczyc et al., 2003, 2007). Such encapsulated minerals have the potential to capture light energy and thereby act as photocatalysts for electrochemical reactions, thus providing a simple energy transduction system for protocells. For instance, encapsulated TiO_2 particles have been shown to reduce the redox indicator, MV^{+2} or NAD^+ (to NADH) upon irradiation (Summers et al., 2009, 2010). Moreover, aqueous suspensions of the mineral sphalerite (ZnS) can photoreduce CO_2 to formate when illuminated with near-UV light (Zhou and Guzman, 2014), providing a mechanism of prebiotic carbon fixation through abiotic reactions. These examples illustrate the variety of possible models for primitive energy capture and transduction. However, their plausibility depends on whether they could be linked to a constant source of energy in an environment that provides initial sources of electrons and carbon and the appropriate inorganic catalysts for their conversion to the initial building blocks of life.

To summarize these points, there is no doubt that the earliest microbial life would be under strong selective pressure to evolve mechanisms for capturing light energy using pigment systems that were present in the local environment or produced as a by-product of primitive metabolic pathways. How this came about remains a major gap in our understanding of early life, but it is relevant to this chapter that a hydrothermal field would be bathed in abundant light energy, while life in deep alkaline vents would necessarily be entirely chemotropic with no potential to develop photosynthetic pigment systems.

19.8 EXPERIMENTAL TESTS SIMULATING HYDROTHERMAL VENT AND HYDROTHERMAL FIELD CONDITIONS

Two simulation experiments have been reported in the literature, one for alkaline hydrothermal vents and a second for hydrothermal fields. Herschy et al. (2014) constructed a reaction vessel in which an alkaline fluid can be slowly mixed with an acidic fluid. The acidic fluid (pH 5) was a dilute mixture of $FeCl_2$, $NaHCO_3$, and $NiCl_2$, and the alkaline fluid (pH 11) was composed of K_2HPO_4, $Na_2Si_3O_7$, and Na_2S. To simulate the mineral deposition associated with alkaline vents, the alkaline fluid was slowly injected (10–120 μL h^{-1}) into a larger volume of the acidic fluid in the reaction vessel.

Microscopic precipitates of ferrous silicates and phosphates formed, and these appeared to be hollow tubes when examined by scanning electron microscopy. The authors proposed that proton gradients across the mineral membranes of the tubes could drive the reduction of CO_2. Because CO_2 was present in the mixture from the equilibrium of bicarbonate and CO_2 at pH 5, measuring expected products of CO_2 reduction could test this proposal. After several hours, traces of formic acid were detected at 50 μM concentrations, and in some of the runs formaldehyde at 100 nM.

Although these were small yields, the authors were encouraged and decided to test an alkaline fluid with much higher concentrations of formaldehyde. This was not done in the simulation chamber, but instead 0.5 M formaldehyde was incubated at pH 12 and 60°C. After a 5-hour incubation time, small amounts of glyceraldehyde, erythrose and ribose were detected. Typical yields of sugars were low, in the range of 0.06% of the formaldehyde present.

The authors also tested whether thermal gradients in a porous material could concentrate otherwise dilute solutes. For this, a ceramic form was prepared and placed in the reactor vessel. A 75°C "hydrothermal" fluid containing 1 μM fluorescein was pumped slowly into the foam (15 μL h^{-1}) and cooling water at 20°C was pumped into the foam through 8 side ports to produce a thermal gradient of 50°C. After several hours, UV illumination showed that the fluorescein had become visibly concentrated in cooler regions of the foam. When samples were taken for fluorometric analysis, the fluorescein had been depleted in the warm region and concentrated 5000-fold in cooler regions. No concentrating effect was observed in the absence of a thermal gradient. The authors concluded these preliminary tests of the alkaline hydrothermal vent hypothesis are consistent with expectations.

Simulations of the alternative site – hydrothermal fields – were described by DeGuzman et al. (2014) and Da Silva et al. (2014) who tested whether polymerization of monomers could occur under hydrothermal field conditions. Reaction mixtures of 10 mM 5′-mononucleotides were prepared, and small volumes (0.2 mL) were exposed to multiple cycles of hydration and dehydration at elevated temperatures to simulate the wetting, evaporation, and drying that are characteristic of small pools in hydrothermal fields. DeGuzman et al. (2014) used a lipid as an organizing agent to promote polymerization (Deamer, 2012), but Da Silva et al. (2014) found that monovalent salts such as KCl, NaCl, and NH$_4$Cl were even more efficient in promoting polymerization. In both cases, the expectation is that during dehydration, the monomers become increasingly concentrated, and at some point, the water activity is sufficiently low so that condensation reactions form ester bonds between the phosphate and hydroxyl groups attached to the ribose moiety of the nucleotides.

Both studies reported that significant amounts of polymeric products were produced. Typical yields with lipid promoters were a few percent of the monomers, while with monovalent salts, yields as high as 50% were achieved after multiple cycles. Nanopore analysis of the products revealed that linear polyanions were present, as expected if RNA-like products had been synthesized. Because the monomers were mixtures of AMP and UMP in 1:1 mole ratios, it is possible that base pairing within or between strands should occur. Two experimental results support this proposition. First, in nondenaturing agarose gels, the products were stained with ethidium bromide, a dye that requires base stacking to exhibit maximum fluorescence. Second, the products were tested for temperature-dependent hyperchromicity, which occurs because single-stranded polymers have stronger absorbance at 260 nm than duplex polymers. When double-stranded nucleic acids are heated, the strands separate over a certain temperature range, and absorbance increases. The products of the hydrothermal field simulation all exhibited hyperchromicity.

The authors conclude that polymerization reactions relevant to the origin of nucleic acids can be driven by conditions simulating those that occur in hydrothermal fields. They speculate that such reactions could not occur if the monomers are simply heated and dried, because in the dry state the monomers are fixed in a solid glass and unable to diffuse. However, if the nucleotides are concentrated in a fluid state such as a multilamellar liquid crystal or the eutectic film between crystals of monovalent cations, diffusion of the highly concentrated monomers allows them to undergo elongation reactions. As a result, during multiple cycles, the oligomers continue to grow, ultimately reaching lengths well over 100 nt. It is possible that stacking of the concentrated nucleotide bases also helps to organize the monomers and thereby promotes the condensation reaction by bringing the reactive groups into close contact.

19.9 SUMMARY

Hydrothermal fields and alkaline hydrothermal vents are both attractive ideas for sites that would be conducive to the origin of life. We will conclude this chapter by weighing the relative merits and limitations of both sites, with the aim of stimulating further experimental tests of the two scenarios. In our judgement, the weight of evidence so far favours an origin of life in freshwater hydrothermal fields followed by a later migration and adaptation to a marine environment.

19.9.1 Favourable Properties of Alkaline Hydrothermal Vents

- A source of chemical energy is available from solutes and minerals at different redox states.
- The estimated life of alkaline vents can be thousands of years, providing a continuing supply of chemical energy at life-compatible temperature ranges (50°C–90°C).
- Vent minerals are a source of transition metals with potential catalytic activity when incorporated into peptides.

- Dissolved carbon dioxide has been demonstrated to be reduced to formic acid.
- A mechanism for concentrating dilute solutes has been demonstrated in a laboratory simulation of alkaline vents.

19.9.2 Limitations of Alkaline Hydrothermal Vents

- Self-assembly of amphiphiles into membranes is inhibited by high salt concentration and divalent cations present in alkaline vents.
- Condensation reactions leading to polymerization require activated substrates in an aqueous medium like seawater. Neither activated substrates nor polymerization processes have been demonstrated experimentally in hydrothermal vent conditions.
- The proposed chemiosmotic potential has not yet been demonstrated to provide an energy source in the vent environment.
- Cycles of wetting and drying cannot occur in hydrothermal vent conditions.
- Photosynthesis could not develop in vent conditions.

19.9.3 Favourable Properties of Hydrothermal Fields

- A source of chemical energy is available from reduced water activity during dehydration.
- Self-assembly of amphiphilic compounds readily occurs in low ionic strength freshwater.
- Concentration of dilute solutes on mineral surfaces occurs naturally upon drying.
- Condensation reactions and polymerization have been demonstrated in laboratory simulations.
- Cycles of hydration and dehydration drive increased complexity as products accumulate in closed systems of hydrothermal pools.
- Light energy is abundant, so photosynthesis can develop.

19.9.4 Limitations of Hydrothermal Fields

- Low pH ranges may inhibit certain reactions.
- Extensive clay deposits adsorb reactants and make them unavailable.
- Exposure to the ultraviolet component of sunlight can damage some molecules.

REFERENCES

Adamala, K., and Szostak, J.W. (2013) Nonenzymatic template-directed RNA synthesis inside model protocells. *Science* 342:1098–1100.

Amend, J.P., and McCollom, T.M. (2009) Energetics of biomolecule synthesis on early Earth. In *Chemical Evolution II: From the Origins of Life to Modern Society*. edited by L. Zaikowski, J.M. Friedrich and S.R. Seidels, American Chemical Society, Washington, DC, pp 63–94.

Attwater, J., Wochner, A., and Holliger, P. (2013) In-ice evolution of RNA polymerase ribozyme activity. *Nat. Chem.* 5:1011–1018.

Baaske, P., Weinert, F.M., Duhr, S., Lemke, K.H., Russel, M.J., and Braun, D. (2007) Extreme accumulation of nucleotides in simulated hydrothermal pore systems. *Proc. Natl. Acad. Sci. USA* 104:9346–9351.

Bada, J.L., and Lazcano, A. (2003) Prebiotic soup--Revisiting the Miller experiment. *Science* 300:745–746.

Budin, I., Bruckner, R.J., and Szostak, J.W. (2009) Formation of protocell-like vesicles in a thermal diffusion column. *J. Am. Chem. Soc.* 131:9628–9629.

Budin, I., Debnath, A., and Szostak, J.W. (2012) Concentration-driven growth of model protocell membranes. *J. Am. Chem. Soc.* 134:20812–20819.

Budin, I., Prywes, N., Zhang, N., and Szostak, J.W. (2014) Chain-length heterogeneity allows for the assembly of fatty acid vesicles in dilute solutions. *Biophys. J.* 107:1582–1590.

Budin, I., and Szostak, J.W. (2010) Expanding roles for diverse physical phenomena during the origin of life. *Annu. Rev. Biophys.* 39:245–263.

Budin, I., and Szostak, J.W. (2011) Physical effects underlying the transition from primitive to modern cell membranes. *Proc. Natl. Acad. Sci. USA* 108:5249–5254.

Chakrabarti, A., and Deamer, D.W. (1994) Permeation of membranes by the neutral form of amino acids and peptides: Relevance to the origin of peptide translocation. *J. Mol. Evol.* 39:1–5.

Chen, I.A., and Szostak, J.W. (2004) Membrane growth can generate a transmembrane pH gradient in fatty acid vesicles. *Proc. Natl. Acad. Sci. USA* 101:7965–7970.

Chyba, C., and Sagan, C. (1992) Endogenous production, exogenous delivery and impact-shock synthesis of organic molecules: An inventory for the origins of life. *Nature* 355:125–132.

Corliss, J.B., Baross, J.A., and Hoffman, S.E. (1981) An hypothesis concerning the relationship between submarine hot springs and the origin of life on earth. *Oceanol. Acta* 4:59–69.

Da Silva, L., Maurel, M.C., and Deamer, D. (2014) Salt-promoted synthesis of RNA-like molecules in simulated hydrothermal conditions. *J. Mol. Evol.* [Epub ahead of print].

Damer, B., and Deamer, D. (2015) *Life* (submitted, under revision).

Deamer, D.W. (1985) Boundary structures are formed by organic components of the Murchison carbonaceous chondrite. *Nature* 317:792–794.

Deamer, D.W. (1997) The first living systems: A bioenergetic perspective. *Microbiol. Mol. Biol. Rev.* 61:239–261.

Deamer, D.W. (2012) Liquid crystalline nanostructures: Organizing matrices for non-enzymatic nucleic acid polymerization. *Chem. Soc. Rev.* 41:5375–5379.

Deamer, D.W., and Barchfeld, G.L. (1982) Encapsulation of macromolecules by lipid vesicles under simulated prebiotic conditions. *J. Mol. Evol.* 18:203–206.

Deamer, D.W., and Pashley, R.M. (1989) Amphiphilic components of the Murchison carbonaceous chondrite: Surface properties and membrane formation. *Orig. Life Evol. Biosph.* 19:21–38.

Deamer, D.W., Singaram, S., Rajamani, S., Kompanichenko, V., and Guggenheim, S. (2006) Self-assembly processes in the prebiotic environment. *Phil. Trans. Roy. Soc. B-Biol. Sci.* 361:1809–1818.

Deamer, D.W., and Weber, A.L. (2010) Bioenergetics and life's origins. *Cold Spring Harb. Perspect. Biol.* 2:a004929.

DeGuzman, V., Vercoutere, W., Shenasa, H., and Deamer, D.W. (2014) Generation of oligonucleotides under hydrothermal conditions by non-enzymatic polymerization. *J. Mol. Evol.* 78:251–262.

Ferris, J.P., Hill, A.R.J., Liu, R., and Orgel, L.E. (1996) Synthesis of long prebiotic oligomers on mineral surfaces. *Nature* 381:59–61.

Gebicki, J.M., and Hicks, M. (1973) Ufasomes are stable particles surrounded by unsaturated fatty acid membranes. *Nature* 243:232–234.

Groen, J., Deamer, D.W., Kros, A., and Ehrenfreund, P. (2012) Polycyclic aromatic hydrocarbons as plausible prebiotic membrane components. *Orig. Life Evol. Biosph.* 42:295–306.

Guzman, M.I., and Martin, S.T. (2009) Prebiotic metabolism: Production by mineral photoelectrochemistry of α-ketocarboxylic acids in the reductive tricarboxylic acid cycle. *Astrobiology* 9:833–842.

Hanczyc, M.M., Fujikawa, S.M., and Szostak, J.W. (2003) Experimental models of primitive cellular compartmentalization: Encapsulation, growth, and division. *Science* 302:618–622.

Hanczyc, M.M., Mansy, S.S., and Szostak, J.W. (2007) Mineral surface directed membrane assembly. *Orig. Life Evol. Biosph.* 37:67–82.

Hargreaves, W.R., and Deamer, D.W. (1978) Liposomes from ionic, single-chain amphiphiles. *Biochemistry* 17:3759–3768.

Herschy, B., Whicher, A., Camprubi, E., Watson, C., Dartnell, L., Ward, J., Evans, J.R.G., and Lane, N. (2014) An origin-of-life reactor to simulate alkaline hydrothermal vents. *J. Mol. Evol.* 79:213–227.

Huang, W., and Ferris, J.P. (2006) One-step, regioselective synthesis of up to 50-mers of RNA oligomers by montmorillonite catalysis. *J. Am. Chem. Soc.* 128:8914–8919.

Jordan, S.F., Rammu, H., Zheludev, I.N., Hartley, A.M., Marechal, A., Lane, N. (2019) Promotion of protocell self-assembly from mixed amphiphiles at the origin of life. *Nat. Ecol. Evol.* 3:1705–1714.

Kelley, D.S., Karson, J.A., Blackman, D.K., Früh-Green, G.L., Butterfield, D.A., Lilley, M.D., Olson, E.J., Schrenk, M.O., Roe, K.K., Lebon, G.T. *et al.* (2001) An off-axis hydrothermal vent field near the Mid-Atlantic Ridge at 30° N. *Nature* 412:145–149.

Koch, A.L. (1985) Primeval cells: Possible energy-generating and cell-division mechanisms. *J. Mol. Evol.* 21:270–277.

Koch, A.L., and Schmidt, T.M. (1991) The first cellular bioenergetic process: Primitive generation of a proton motive force. *J. Mol. Evol.* 33:297–304.

Lancet, H.S., and Anders, E. (1970) Carbon isotope fractionation in the Fischer-Tropsch synthesis of methane. *Science* 170:980–982.

Lane, N., and Martin, W.F. (2012) The origin of membrane bioenergetics. *Cell* 151:1406–1416.

Mansy, S.S., Schrum, J.P., Krishnamurthy, M., S., T., Treco, D.A., and Szostak, J.W. (2008) Template-directed synthesis of a genetic polymer in a model protocell. *Nature* 454:122–125.

McCollom, T.M., Ritter, G., and Simoneit, B.R.T. (1999) Lipid synthesis under hydrothermal conditions by Fischer-Tropsch-Type reactions. *Orig. Life Evol. Biosph.* 29:153–166.

Mitchell, P. (1961) Coupling of phosphorylation to electron and hydrogen transfer by a chemi-osmotic type of mechanism. *Nature* 191:144–148.

Monnard, P.-A., and Deamer, D.W. (2001) Loading of DMPC based liposomes with nucleotide triphosphates by passive diffusion: A plausible model for nutrient uptake by the protocell. *Orig. Life Evol. Biosph.* 31:147–155.

Monnard, P.-A., and Deamer, D.W. (2002) Membrane self-assembly processes: Steps toward the first cellular life. *Anat. Rec.* 268:196–207.

Mulkidjanian, A.Y., Bychkov, A.Y., Dibrova, D.V., Galperin, M.Y., and Koonin, E.V. (2012) Origin of first cells at terrestrial, anoxic geothermal fields. *Proc. Natl. Acad. Sci. USA* 109(14):E821–E830.

Namani, T., and Deamer, D.W. (2008) Stability of model membranes in extreme environments. *Orig. Life Evol. Biosph.* 38:329–341.

Naraoka, H., Shimoyama, A., and Harada, K. (1999) Molecular distribution of monocarboxylic acids in Asuka carbonaceous chondrites from Antarctica. *Orig. Life Evol. Biosph.* 29:187–201.

Nichols, J.W., and Deamer, D.W. (1980) Net proton-hydroxyl permeability of large unilamellar liposomes measured by an acid-base titration technique. *Proc. Natl. Acad. Sci. USA* 77:2038–2042.

Olbrich, K.C., Rawicz, W., Needham, D., and Evans, E. (2000) Water permeability and mechanical strength of polyunsaturated phopshatidylcholine bilayers. *Biophys. J.* 79:321–327.

Parsegian, A. (1969) Energy of an ion crossing a low dielectric membrane: Solutions to four relevant electrostatic problems. *Nature* 221:844–846.

Paula, S., Volkov, A.G., Van Hoek, A.N., Haines, T.H., and Deamer, D.W. (1996) Permeation of proton, potassium ions, and small polar molecules through phospholipid bilayers as a function of membrane thickness. *Biophys. J.* 70:339–348.

Racker, E., and Stoeckenius, W. (1974) Reconstitution of purple membrane vesicles catalyzing light-driven proton uptake and adenosine triphosphate formation. *J. Biol. Chem.* 249:662–663.

Rajamani, S., Vlassov, A., Benner, S., Coombs, A., Olasagasti, F., and Deamer, D.W. (2008) Lipid-assisted synthesis of RNA-like polymers from mononucleotides. *Orig. Life Evol. Biosph.* 38:57–74.

Rushdi, A.I., and Simoneit, B.R. (2001) Lipid formation by aqueous Fischer-Tropsch-type synthesis over a temperature range of 100 to 400 degrees C. *Orig. Life Evol. Biosph.* 31:103–118.

Russell, M.J., Daniel, R.M., and Hall, A. (1993) On the emergence of life via catalytic iron-sulphide membranes. *Terra Nova* 5:343–347.

Russell, M.J., and Hall, A.J. (1997) The emergence of life from iron monosulphide bubbles at a submarine hydrothermal redox and pH front. *J. Geol. Soc. London* 154:377–402.

Sacerdote, M.G., and Szostak, J.W. (2005) Semipermeable lipid bilayers exhibit diastereoselectivity favoring ribose. *Proc. Natl. Acad. Sci. USA* 102:6004–6008.

Sephton, M.A. (2005) Organic matter in carbonaceous meteorites: Past, present and future research. *Phil. Trans. R. Soc. A* 363:2729–2742.

Shew, R., and Deamer, D.W. (1985) A novel method for encapsulating macromolecules in liposomes. *Biochim. Biophys. Acta.* 816:1–8.

Shock, E., and Canovas, P. (2010) The potential for abiotic organic synthesis and biosynthesis at seafloor hydrothermal systems. *Geofluids* 10:161–192.

Shock, E.L., McCollom, T., and Schulte, M.D. (1998) The emergence of metabolism from within hydrothermal systems. In *Thermophiles: The Keys to Molecular Evolution and the Origin of Life*. edited by J. Wiegel and M.W.W. Adamss, Taylor and Francis, Washington, DC, pp 59–76.

Sleep, N.H. (2010) The Hadean-Archaean environment. In *Origins of Life*. edited by D.W. Deamer and J.W. Szostak, Cold Spring Harbor Laboratory Press, NY, pp 35–48.

Sousa, F.L., Thiergart, T., Landan, G., Nelson-Sathi, S., Pereira, I.A., Allen, J.F., Lane, N., and Martin, W.F. (2013) Early bioenergetic evolution. *Philos. Trans. R. Soc. Lond. B Biol. Sci.* 368:1–30.

Stribling, R., and Miller, S.L. (1987) Energy yields for hydrogen cyanide and formaldehyde syntheses: The HCN and amino acid concentrations in the primitive ocean. *Orig. Life Evol. Biosph.* 17:261–273.

Summers, D.P., Noveron, J., and Basa, R.C. (2009) Energy transduction inside of amphiphilic vesicles: Encapsulation of photochemically active semiconducting particles. *Orig. Life Evol. Biosph.* 39:127–140.

Summers, D.P., Noveron, J., Basa, R.C., and Rodoni, D. (2010) Energy transduction inside vesicles by mineral particles: Formation of NADH. In: *Astrobiology Science Conference 2010: Evolution and Life: Surviving Catastrophes and Extremes on Earth and Beyond*, LPI Contribution No. 1538, p.5596, Lunar and Planetary Institute, League City, TX.

Szostak, J.W., Bartel, D.P., and Luisi, P.L. (2001) Synthesizing life. *Nature* 409:387–390.

Zahnle, K., Schaefer, L., and Fegley, B. (2010) Earth's earliest atmosphere. In *Origins of Life*. edited by D.W. Deamer and J.W. Szostak, Cold Spring Harbor Laboratory Press, NY, pp 49–65.

Zhou, R., and Guzman, M.I. (2014) CO_2 reduction under periodic illumination of ZnS. *J. Phys. Chem.* 118:11649–11656.

Zhu, T.F., and Szostak, J.W. (2009) Coupled growth and division of model protocell membranes. *J. Am. Chem. Soc.* 131:5705–5713.

Index

Abkarian, M. 186
acoustic biosensors 247
actin-binding domain of ezrin (EzrABD) 165
actin–membrane linkage 169
actin polymerization 165–166
Adamala, K. 341
adhesion protein, force measurement of
  cadherins
    binding mechanisms 92
    confinement reveals interactions 94
    crystal structures 92–93
    intercellular adhesion frequency measurements 93–94
    mechanical and kinetic properties 90–92
    molecular dynamics and Monte Carlo simulations 93
    superresolution imaging 93
  extracellular regions 83
  immunoglobulin family cell molecules
    neural cell adhesion molecule 88–90
    posttranslational modification 90
  intermembrane binding potentials 85–87
  intrinsically disordered proteins 87–88
  lectin
    conformational flexibility and protein recognition 85–86
    genetic length variants and receptor function 86–87
  structural diversity 83
  surface forces apparatus measurements 84
advanced X-ray scattering
  assemblies 65–68
  biological membranes 65
  coherent X-ray imaging 77–79
  3D reconstructions 65
  membrane fusion 70–73
  molecular resolution 65
  myelinated axons 75–77
  of synaptic vesicles 73–75
  time-resolved diffraction 68–71
Akinc, A. 217
Albanesi, J.P. 73
alkaline hydrothermal vents 337
Allovectin 7 219
Almeida, P.F.F. 16
ALN-TTR02 219
Als-Nielsen, J. 43
amphipathic peptide-membrane complexes 237–238
amphipatic helices 12, 137
  insertion of 137
amphiphilic dendrimers 332
amphiphilic fatty acids 338
amphiphilic Janus dendrimers 332
amphiphilic lipid-like compounds 342
amphiphilic molecules 50
amphiphilic polymers 332
amphiphilic self-assembly
  definition 338–339
  energy sources 343
    light-driven chemiosmotic energy transduction 344
  hydrothermal vent and field conditions 343–345

organic carbon compounds 338
primitive membrane functions
  mineral compartments and amphiphilic membranes 341–343
  permeability and encapsulation 341
sources 339
stable membranes 339–341
AMPs see antimicrobial peptides (AMPs)
Andelman, D. 12, 19
Angelova, M.I. 180
antimicrobial peptides (AMPs) 141, 143–144
  amino acid requirements 153–154
  background and history 148–150
  composite mechanisms 152–153
  induced curvature 150–151
  phase behaviour 150–151
  selective action 151–152
area difference elasticity (ADE) model 19, 136
artificial vesicles 324
asymmetric membranes 115–116; see also charged membranes
  attraction/repulsion crossover 118
  counterions 117–118
  Debye–Hückel regime 116–117
atomic force microscopy 92

Bangham, A.D. 129, 179, 195
BAR (Bin, Amphiphysin, Rvs)-domain 146
barrel-stave model 150
Bassereau, P. 21
bead-on-string model 198
Bear, R. 33
Beerlink, A. 77
Bell, G.I. 22
Bellocq, A.M. 43
Bensimon, D. 19
Bhattacharya, K. 188
bilayer coupling model 135–136
binding kinetics 93
bioanalytical sensors
  applications 242
  exploiting force-driven membrane component manipulation 260–261
  immobilization
    planar lipid membranes 243–245
    surface-tethered lipid vesicles 245–246
  lipid-membrane-mediated interactions 254
    glycolipid-mediated interactions 254–256
    peptide-mediated membrane disruption events 256
    single-vesicle assays 256
  liposomes 256–259
  membrane–proteins
    detect interactions 250–251
    functional integral membrane proteins 248–250
    peripheral membrane proteins 248
    transmembrane proteins 248
    transport 251–254
  nanoreactors 259–260
  surface-based biosensing techniques
    biomolecular recognition 246
    label-free biosensors 246–247
    probing transport 247–248

biological membranes 129, 136–138
  elastic shells
    bending excitations 20–21
    flickering 21–22
    molecular aspects 22
    morphology and shape transitions 18–19
    physiological significance 20–21
    red blood cell 19–22
  structure and function 2
bio-nanotechnology 197
biosensors 241
  to ion channels 252
Birgeneau, R.J. 43
Bjerrum length 100
black lipid membranes (BLMs) 3
block copolypeptides 305, 323–333
block liposomes (BLs) 130–131
Bloom, M. 10
Boltzmann distribution 103
Bragg reflection 72
Bragg rods 40
Bragg–Williams lattice models 8
brain adhesion proteins 88
Brezesinski, G. 234
Brochard, F. 20, 23
Bruinsma, R. 23
Burdick, J.A. 324

cadherin adhesion proteins
  binding mechanisms 92
  confinement reveals interactions 94
  crystal structures 92–93
  intercellular adhesion frequency measurements 93–94
  mechanical and kinetic properties 90–92
  molecular dynamics and Monte Carlo simulations 93
  structure and binding interactions 91
  superresolution imaging 93
Cahn–Hilliard theory 14
Caillé, A. 43, 44
Cameron, N.R. 310
cancer nucleotide therapy 218
carbohydrate recognition domains (CRDs) 85–86
cardiolipin (CL) 149
carpet model 150
carriers 271
catch bond 92
cationic lipid-DNA complexes 223, 231–233
cationic lipids 216
cationic liposomes (CLs)
  engineered viral vectors 197
  with long DNA
    $H_{II}C$ phase 198
    non-lamellar cationic liposome 200–201
    optical microscopy 197
    positive and negative ions 198
    self-assembled structures 197
    spontaneous curvature 199–200
    synchrotron studies 198
    transfection efficiency of lamellar and non-lamellar complexes 200–201

cationic liposomes (CLs) (cont.)
    in molecular delivery 197
    with sDNA 197
    with short nucleic acids
        gyroid cubic lipid phase 201
        three-dimensional columnar phase 202–204
        two-dimensional packing 201–202
    vectors 196
caveolin 147
cell adhesion 22–23
    biomimetic systems 23–24
    design model systems 24–25
    molecular aspects 23
cell binding kinetics, by micropipette techniques 94
cellular membranes 242
Cevc, G. 12
chain packing defects 18
chain packing frustration 51–52
chaotropes 224
Chapman, D.L. 105
charge density 39
charged macromolecules
    amphipathic peptide 237–238
    of anionic lipids 223
    continuum mean-field 224–227
    coupling parameter 125
    DNA complexes 231
        adsorption 234–235
        cationic lipid 231–233
        zwitterionic lipids and divalent cations 233–234
    interactions 223, 224
    microions 224
    protein adsorption 227
        charge regulation 230–231
        lipid mobility 227–229
        protein-induced membrane phase separation 229–230
charged membranes
    asymmetric membranes 115–116
        attraction/repulsion crossover 118
        counterions 117–118
        Debye–Hückel regime 116–117
    Bjerrum length 100
    charge regulation 118–121
        via free energy 121–122
    Debye length 100
    electrostatic interactions 100
    Gouy–Chapman length 100
    limitations and generalizations 125–126
    local electroneutrality 100
    one planar membrane
        added electrolyte 105–107
        boundary condition 104
        counterions 104–105
        Grahame equation 107
    Poisson–Boltzmann theory
        approaches 102
        aqueous solution 102
        Boltzmann distribution 103
        Debye–Hückel approximation 103–104
        electrostatic energy 103
        equation 103
        ionic association/dissociation 102
        modification 107–109
        phospholipids 101–102
        total charge density 102
    symmetric membranes 111
        added electrolyte 112–114

counterions 111–112
Debye–Hückel regime 114
intermediate regime 114–115
pressure regime 115
two-membrane system 109–111
van der Waals' interactions 122
    Derjaguin–Landau–Verwey–Overbeek theory 124–125
    Hamaker pairwise summation 122–123
    macroscopic theory 123–124
charge regulation 118–121
    via free energy 121–122
chemical properties of lipids 343
Chen, I.A. 189
Chong, J.Y.T.X. 55
Christensen, S.M. 259
clinical trials 219
coalescence 281
coat proteins 146
complex vesicle morphologies 130
composite nanocore-lipid vector 218
compression models 22
condensed ordered phase 8
confinement 92–93
Conn, C. 200
continuum models 134
    area difference elasticity model 136
    bilayer coupling model 135–136
    spontaneous curvature model 134–135
co-polypeptide 326
co-polypeptide synthesis
    amine initiators 307–309
    transition metal catalysis 306–307
co-polypeptide vesicles
    assemblies 310–314
    functional vesicles 314–317
    structural parameters 310
corneocytes 272
cosmetics and pharmaceutics
    lipid nanocarriers formulation 276
        emulsions 278–281
        liposomes and derivatives 276–278
        nanoparticles 281–283
    skin
        dermal applications 274–276
        sensitivity 273–274
        structure 272–273
counterion release mechanism 15
critical edge 37
critical micelle concentration (CMC) 50
crumpling transition 5
cryogenic electron microscopy 130–131
cubic complexes 198
cubic phases
    applications 59–60
    discontinuous inverse 53–54
    during lipid digestion 58–59
    for membrane proteins 60
    mesophase engineering 54–55
    types 53
    in vivo 57–58
    water content 53
cubosomes 55
curvature elastic energy 51–52
curvature elasticity 2
curvature sensing protein 147
cytoskeletal motor activity 137
cytoskeleton 137
    membrane curvature driven mechanism 162
    molecular dynamics simulations 169
    phosphoinositides 169–170

phospholipid–protein interactions 170–171
phosphoinositides 162
    curvature-dependent vesicle trafficking 166–169
    curvature-generating proteins 163
    nonrandom distribution 163–164
    structure, in eukaryotic cells 163
    regulation 164
        actin polymerization 165–166
        actomyosin dynamics 165
        cortical cytoskeleton perturbs 165
        submembrane actin assembly 164

Dalous, J. 186
Danielli, J.F. 2
Daniel, S. 299
Da Silva, L. 345
Davson, H. 2
Deamer, D.W. 340, 341
Debye–Hückel regime
    Poisson–Boltzmann theory 103–104
    two asymmetric membranes 116–117
Debye length 100
decanoic acid, structure of 338
DeGuzman, V. 345
Delbrück, M. 16
Demarche, S. 249
Dembo, M. 92
dendrimers 332–333
dendritic cell-specific intercellular adhesion molecule-3-grabbing nonintegrin (DC-SIGN) 85–86
density fluctuations 17
Derjaguin, B.V. 110
Derjaguin–Landau–Verwey–Overbeek (DLVO) theory 124–125; see also charged membranes
dermis 272
Deuling, H.J. 185
diacylglycerol (DAG) 14
dielectric contrast 124
differential adhesion hypothesis (DAH) 23
1,2-dilinoleyloxy-3-di-methylaminopropane (DLinDMA) 216
dilute lamellar phase 42–45
Dimitrov, D.S. 180
1,2-dimyristoyl-sn-glycero-3-phosphocholine (DMPC)
    chemical structure of 33, 34
    freely suspended film 39
    phase diagram of 34
    temperature-relative humidity phase diagram 41, 42
dioleoylphosphatidylcholine (DOPC) 73
dioleoylphos-phatidylethanolamine (DOPE) 73
discontinuous inverse cubic phases 53–54
dispersion interactions 123
divalent cation-zwitterion lipid interactions 233–234
DMPC see 1,2-dimyristoyl-sn-glycero-3-phosphocholine (DMPC)
DNA complexes 196
    spatial organization of 199–200
    transfection efficiency 200–201
Döbereiner, H.G. 185
DOTAP dioleoyl-trimethylammonium propane 214
Dowben, R.M. 180
drug delivery 59–60
    field of 211

Index

Ducic, T. 76
dynamic light scattering (DLS) 55

early membranes, hydrothermal vent 344
"effector" domain (ED) 12
Eibl, H. 11
Eisenberg, A. 323
elastic incoherent scattering factor (EISF) 16
elastic shells
    bending excitations 20–21
    flickering 21–22
    molecular aspects 22
    morphology and shape transitions 18–19
    physiological significance 20–21
    red blood cell 19–22
elastin-like polypeptides (ELPs) 329
electrochemical sensing 246, 247–248
electromagnetic fluctuation interactions 123
electrophoresis 299
emulsions 278
    destabilization 280–281
    lipid ingredients 280
    nanoemulsions and microemulsions 279–280
    surfactants 279
endosomal sorting complex required for transport (ESCRT) 147
engineering sensing platforms, based natural pores 252–254
enhanced permeability and retention (EPR) 218
epidermis 272
epithelial (E) cadherins 23
Evans, E. 19, 20, 23
ezrin, radixin, moesin (ERM) protein 165

Felgner, P. 196
Fermi–Dirac distribution 108
filopodia 137
Fischer–Tropsch-type (FTT) synthesis 339
Fluid Mosaic Model 2
fluorescence correlation spectroscopy (FCS) 14
fluorescence resonance energy transfer (FRET) 87, 88
Fong, C. 54
form factor 37, 40
freely suspended multilayer film 39–42, 45
Fricke, K. 2
Friedmann, T. 219
functional proteins, electrostatic switching of
    chain-melting transition 11–14
    counterion release 15
    macroion aggregation 14–15
    thermodynamics and kinetics 14

Gallop, J.L. 136
Gauss–Bonnet theorem 143
Gaussian curvatures 50, 51, 142
Gebicki, J.M. 340
gel phase 39–42
gel-phase membrane 228
gene delivery 196
generic lyotropic phase diagram 50–51
gene silencing 197, 201
geological site 337
geometrical Brownian ratchet 299
giant early endosomes (GEEs) 204
giant unilamellar vesicles (GUVs) 130, 182–183
    cellular osmoregulation 179
    electroformation 180
    formation 181

gentle hydration 180
    in hypertonic media 183–186
    in hypotonic media 186–189
    myriad challenges 182
Gingell, D. 116
Givli, S. 188
glycolipids (GLs) 254
Gorter, E. 2
Goulian, M. 10
Gouy–Chapman (GC) length 100, 105
Gouy–Chapman–Overbeek theory 12
Gouy, G. 105
Grahame equation 107, 108
grazing incidence diffraction (GID) 67
grazing incidence small-angle X-ray scattering (GISAXS) 67
Grendel, F. 2
Gromelski, S. 234
Groves, J.T. 302

Hall, A. 343
Hamaker pairwise summation 122–123
Hardy, G.J. 249
Hargreaves, W.R. 340
Heimburg, T. 18
Heise, A. 315
Helfrich curvature elastic energy density 143
Helfrich undulation forces 42–45
Helfrich, W. 2, 18, 20, 42, 134, 143, 185, 199
Helmholtz free energy 103, 110
Herschy, B. 344
Heterogeneous Shell Model 2
hexagonal complexes 198
hexamethyldisilazane (HMDS) 309
hexatic phase 5
hexosomes 55
Hicks, M. 340
Hirano-Iwata, A. 249
hisactophilin 231
Höök, F. 293
Horne, R.W. 195
Huxley–Hodgkin model 2
hydrophilic and lipophilic balance (HLB) 279
hydrostatic pressure, effect of 55–57
hydrothermal fields 337
hydrothermal vent 343–345
hypertonic media 183–186
hypodermis 272
hypotonic media 186–189

ideal-gas (IG) regime 115
immunogenicity 212
inorganic protocells 343
In situ assembly 332
integral membrane proteins 136–137
intrinsically disordered proteins (IDP) 87–88
inverse hexagonal phases 52–53
    applications 59–60
    during lipid digestion 58–59
    mesophase engineering 54–55
in vivo therapeutic applications
    clinical trials 219
    drug delivery 211
    lipid vectors 212–219
    applications 218–219
    for delivery of DNA, RNA, and peptide 214–218
    rational design of 212–214
ion channel 252
ion channel switch 253
ionizable lipid 214–215

ion specificity 224
Israelachvili, J.N. 142, 200

Jacobi elliptic functions 113
Jähnig, F. 12
Jahn, R. 73
Janmey, P. 18
Janus dendrimers 332

Kas, J. 185
Komura, S. 16
Korenstein, R. 21
kosmotropes 224
Kozlov, M.M. 73
Kozlovsky, Y. 73
Krafft boundary 50
Kramers–Kronig relation 123

label-free biosensors 246–247
lab-on-a-chip 244
lamellar complexes 200–201
lamellar fluid phase 38
Landau, L.D. 43
Landau–Peierls Effect 42–45
Lane, N. 343, 344
Langmuir–Blodgett technique 41
Langmuir–Davies isotherm 120, 121
Langmuir film balance 74–75
large pharma 211
Lau, A.W.C. 116, 117
Lecommandoux, S. 315, 326
lectin adhesion proteins
    conformational flexibility and protein recognition 85–86
    genetic length variants and receptor function 86–87
Lee, Y.K. 261
Leibler, S. 19, 20
Lennon, J.F. 20
lensless coherent imaging 65
Liepina, I. 169
life-threatening diseases 211
Lifshitz theory 123
light-driven chemiosmotic energy transduction 344
linear CR8C 218
lipid bilayer spreading, on glass surface 294
lipid calcium phosphate (LCP) nanoparticles 213
lipid diffusion 295
lipid lateral diffusion, and impact 15–17
lipid lateral organization 164–166
lipid nanoparticles (LNPs) 281–283
lipidome
    binary and ternary mixtures 8
    biological function 7
    Bragg–Williams lattice models 8
    cholesterol concentrations 7
    condensed ordered phase 8
    homeostatic control 7
    oligosugar-carrying gangliosides 6
    phospholipid–cholesterol mixtures 8
    pseudo-binary lipid mixtures 9
    small-angle neutron scattering 8
    sphingomyelins 6
    thermodynamic phase state 7
lipid packing shapes 133
lipid–protein
    interaction 9–10
    sorting and functional microdomains formation 11

lipids
- assemblies 242, 243
- bilayer 242
- bilayer spreading 294
- chain order 6
- interaction 9–10
- mobility 227–229
- monolayers 5–6

lipid shape evolution
- continuum models 134
  - area difference elasticity model 136
  - bilayer coupling model 135–136
  - spontaneous curvature model 134–135
- fluctuations and transitions 131–132
- physical concepts
  - lipid molecules 132–133
  - membrane composition 133–134
  - molecular interactions 134
- in scientific discovery 130–131
- from *in vitro* to *in vivo* 136–138

lipid vectors
- applications 218–219
- clinical trials 219
- for delivery of DNA, RNA, and peptide 214–218
- rational design of 212–214

lipid–water systems 3
- chain-melting transition 4–5
- chain order, in L_A phase 6, 7
- lipid monolayers 5–6
- lyotropic polymorphism 4
- thermotropic polymorphism 4

lipoplexes 216, 231
liposomal drug delivery 59–60
liposomes 129, 276
- formulation 277–278
- release mechanisms 276–277
- for signal amplification 256–259

Lipowsky, R. 18, 20, 23
liquid crystal phase behaviour 195–206
liquid-phase vesicles 130
Litster, J.D. 43
Liu, Q.T. 261
liver diseases 218
LNPs *see* lipid nanoparticles (LNPs)
local spontaneous curvature 19
Lohmüller, T. 302
Love, K.T. 217
Luzzati, V. 2, 33, 65
lyotropic polymorphism 4

macroscopic parameters 133
macroscopic theory 123–124
Markin, V.S. 73
Martin, S.T. 343, 344
Ma, Y. 256
McConnell, H. 8, 293
McLaughlin, S. 12
McMahon, H.T. 136
mean curvature 50, 142
mean-field approximation 10
mean-field electrostatics 224–227
mean-field (MF) equation 102
medium chain triglycerides (MCT) 58
membrane and intracellular trafficking 166
membrane-associated proteins 2
membrane barrier properties 338
membrane bending 235–237
membrane curvature elasticity 19, 23
membrane curvature generation mechanisms 136–138

antimicrobial peptides 141, 143–144
- amino acid requirements 153–154
- background and history 148–150
- composite mechanisms 152–153
- induced curvature 150–151
- phase behaviour 150–151
- selective action 151–152

composite mechanism 148
curvature sensing and curvature-mediated attraction 147
molecular crowding 147–148
partitioning and insertion 143–145
scaffolding 145–147
wrapping 148

membrane deformation 136–137
membrane fusion, intermediate structures of 71–73
membrane pores 237
membrane–proteins
- detect interactions 250–251
- functional integral membrane proteins 248–250
- peripheral membrane proteins 248
- transmembrane proteins 248
- transport 251–254

membrane wrapping 148
messenger RNA (mRNA) 213
metastable nonequilibrium shapes 131, 132
micelles 50
microemulsions 279–280
microscopic parameters 132
Milner, S.T. 21
Mitchell, P. 343
modified Poisson–Boltzmann (mPB) theory 107–109
Möhwald, H. 10
molecular crowding 147–148
molecular dynamics simulations 152, 169
- phosphoinositides 169–170
- phospholipid–protein interactions 170–171

molecular motor activity 137
Monnard, P.-A. 340
monolayer mean curvature 51
monomeric enhanced green fluorescent protein (mEGFP) 332
monovalent lipids 216
Montal, M. 3
Mouritsen, O.G. 10
mouse models 218
Mueller, P. 3
Mulet, L.J. 59
multicomponent lipid membranes 9
multivalent lipids (MVLs) 200
Munroe Bretcher analysis 11
MVLBG2 133–134
myelinated axons 75–77
myristoylated alanine-rich C-kinase substrate (MARCKS) protein 12–14

Namani, T. 341
nanocore-based lipid vectors 218
nanoemulsions 279–280
nano-focused beams 76
nanopatterned supported lipid bilayer 302
nanostructured lipid carriers (NLCs) 282–283
N-carboxyanhydrides (NCAs) 305–306
negatively mismatched 145
Netz, R.R. 125
neural (N) cadherins 23
neural cell adhesion molecule (NCAM) 88–90
neutral lipids 216

Nicolson, G.L. 2
Nielsen, C.H. 249
Ninham, B.W. 118
nonaqueous capillary electrophoresis (NACE) 308
non-lamellar phases 49, 50
nucleic acids 217

oil-in-water 49
oleosin 326
one planar membrane
- added electrolyte 105–107
- boundary condition 104
- counterions 104–105
- Grahame equation 107

Onsager, L. 201
optical biosensors 246–247
optical sensing 246
- of membrane transport 248
ordered membranes 35
origin of life 337–346
Orwar, O. 259
osmotically triggered water influx 187
osmotic pressure 109–111
osmotic stress
- cellular compartments 179
- compatible solutes 177
- consequences of 177
- giant unilamellar vesicles 182–183
  - in hypertonic media 183–186
  - in hypotonic media 186–189
- model membrane compartments 179–182
- osmoprotectants 178
- universal effects 177, 178

Ostwald ripening 281

packing geometry 132
packing parameter 50, 142
Palmer, K. 33
Parsegian, V.A. 22, 116, 118
Patching, S. 250
patisiran 219
Peierls, R. 43
permeability 142
Peterson, M. 19
Phan, S. 59
phase contrast tomography 75–78
phase diagram 34, 41, 42
phospatidylserine (PS) 223
phosphatidic acid (PA) 224
phosphatidylcholine (PC) 149
Phosphatidylethanolamine (PE) 149
phosphatidylglycerol (PG) 149
phosphoinositides (PPIs) 12, 162
- curvature-dependent vesicle trafficking 166–169
- curvature-generating proteins 163
- molecular dynamics simulations 169–170
- nonrandom distribution 163–164
- structure, in eukaryotic cells 163

phospholipid bilayer membranes 33
physical properties of lipids 343
Pick, H. 252
Pincus, P. 116, 117
Pink, D. 5
PIP2, submicron domains of 163, 164
planar lipid bilayers
- formation 243–244
- patterns 244–245
planar lipid membranes 243–245
plasma membrane 293

Index

pleckstrin homology (PH) 163
Poisson–Boltzmann (PB) equation 225, 226
Poisson–Boltzmann (PB) theory; see also charged membranes
    approaches 102
    aqueous solution 102
    Boltzmann distribution 103
    Debye–Hückel approximation 103–104
    electrostatic energy 103
    equation 103
    ionic association/dissociation 102
    modification 107–109
    phospholipids 101–102
    total charge density 102
polymeric drug delivery 59–60
polymers 323–325
polymersome 323–325
polymer-supported lipid bilayers 300–301
polypeptides 305
polysialic acid (PSA) chains 90
Poo, M.M. 260
pore formation 145, 150, 186, 189
porphyrin-containing polymersomes 325
porphyrins 325
positively mismatched 145
power laws 44
PPIs see phosphoinositides (PPIs)
prebiotic environment 338
pressure jump technique 57
principal curvatures 142
principal directions 142
proprotein convertase subtilisin/kexin type 9 (PCSK9) 219
proteins 326
proteins on membranes 228
PtdIns(4,5)P_2 170

quasi-elastic neutron scattering (QENS) 16

Racker, E. 343
Rafts 11
Rajamani, S. 342
Ramm, B. 299
Rapoport, T.A. 4
$R_\alpha^{sDNA,3D}$ phase 203, 204
reciprocal space mapping (RSM) 68
recombinant proteins 326–332
Reeves, J.P. 180
reflectance 37
Reiss-Husson, F. 65
reticuloendothelial system (RES) 213
Riegler, J. 10
rippled phase 35
ripple phase 5
Rodriguez-Cabello, J.C. 329
Roux, D. 43
Russell, M.J. 343

Sackmann, E. 15, 185
Saffmann–Delbrück model 16
Saffman, P.G. 16
Safran, S.A. 10, 21
Salditt, T. 76
Saxton, M.J. 16
"schizophrenic" vesicles 312
Schlaad, H. 308
Schmitt, F. 33
Schwille, P. 299
Schröder, H. 10
Scott, H.L. 5
Seddon, J. 200

Seifert, U. 18, 23
self-assembly 50, 310
Servuss, R.M. 42
Sherrer formula 38
short DNA (sDNA) 197
short RNA molecules 201
short interfering RNA (si-RNA) molecules 201, 202
Singer, S.J. 2
single DNA molecule 203, 297
single proton self-correlation function 16
single-vesicle assays 256
Skalak, R. 20
skin
    dermal applications 274–276
    sensitivity 273–274
    structure 272–273
sliding columnar phase 203
SLNs see solid lipid nanoparticles (SLNs)
small-angle neutron scattering (SANS) 8
small-angle X-ray scattering (SAXS) 38, 55, 58
    antimicrobial peptides 150
    CL–DNA complexes 198
    in lipid–protein assemblies 65
    myelinated axons 75–77
    of synaptic vesicles 73–74
small unilamellar vesicles (SUVs) 293–294
smectic-A phase 42
SNARE 87, 88
solid lipid nanoparticles (SLNs) 281–283
solid supported lipid bilayers 293
Sousa, F.L. 344
sphingomyelin (SM) 149
spontaneous curvature model 134–135
Stacked Lyotropic Liquid Crystalline Phases 42–45
Stamou, D. 256
starfishvehicles 130
Steck, T.L. 7
Steinberg, M.S. 23
Stelzle, M. 299
Stoeckenius, W. 343
Stokke, B.T. 19
Stora, T. 252
strand dimer 90
stratum corneum (SC) 272
strong coupling (SC) 125, 126
structure factor 38–40
supported lipid bilayers (SLBs) 293
    adsorption of macromolecules 296–298
    formation 293–295
    lipid and protein mobility in 295–296
    manipulation 298–300
    molecules to living cells 301–302
    physical properties of 299
    polymer 300–301
surface acoustic waves (SAWs) 69
surface-based biosensing techniques
    biomolecular recognition 246
    label-free biosensors 246–247
    probing transport 247–248
surface forces apparatus (SFA) 84
surface-tethered lipid vesicles 245–246
surfactants 279
Svetina, S. 19, 185
swelling process 180
symmetric membranes 111; see also charged membranes
    added electrolyte 112–114
    counterions 111–112
    Debye–Hückel regime 114

intermediate regime 114–115
pressure regime 115
synaptic vesicles (SVs) 73–75
synchrotron X-ray scattering
    dilute multilayer fluid membranes 42
    Landau–Peierls Effect 42–45
    stacked fluid membranes 42–45
    undulation forces, in multilayer L_α phase membranes 42
    geometry 36
    intensity 36
    length density 37
    phospholipid membranes
        freely suspended multilayer films 39–42
        lamellar L_α phase 38
    reflectivity 37–38
synthetic copolypeptides see copolypeptide synthesis
synthetic nonviral vectors 197
Szostak, J.W. 341

Takamori, S. 74
Takeichi, M. 23
tethered supported bilayers 300
Therien, M.J. 324
thermodynamic conditions, hydrothermal vent 344
thermotropic polymorphism 3
Tiefenauer, L. 249
time-resolved X-ray diffraction 66
    nonequilibrium dynamics 68–70
topological transformations, in lipid membranes 143
toroidal-pore model 150
traditional drug discovery 211
transappendageal delivery methods 274
transdermal delivery (TDD) methods 274
transepidermal delivery methods 274
transition metal catalysis 306–307
transmembrane ion channels 162
transmembrane (TM) proteins 145
Träuble, H. 11, 15
treatment of transthyretin (TTR) 219
two-dimensional bilayer sheet 34
2D polymer 298

undulation interaction 42

Van Deenen, L.L. 6
Van der Waals' interactions 122
    Derjaguin–Landau–Verwey–Overbeek theory 124–125
    Hamaker pairwise summation 122–123
    macroscopic theory 123–124
vascular endothelial growth factor (VEGF) 218
vascular permeability 218
Vaz, W.L.C. 16
Vega, B. 250
vehicles 59, 138, 231, 235, 271
vesicles
    co-polypeptide
        assemblies 310–314
        functional vesicles 314–317
        structural parameters 310
    phase behavior 135
    trafficking 166–169
vesicle SNARE (v-SNARE) 87
Viallat, A. 186
virus-derived vectors 212
virus-like particles (VLPs) 255

Viswanathan, S. 258
Vogel, H. 250, 252

Wang, Y.J. 87
water-in-oil 49
weak coupling (WC) 125, 126
wild-type oleosin (Oleo-WT) 326–327
Wortis, M. 19
Wu, S. 8

X-dimer 92
X-ray analysis 65–68
X-ray diffraction 33, 34, 39

X-ray reflectivity 37–38, 66, 67
X-ray scattering
  advanced
    assemblies 65–68
    biological membranes 65
    coherent X-ray imaging 77–79
    3D reconstructions 65
    membrane fusion 70–73
    molecular resolution 65
    myelinated axons 75–77
    of synaptic vesicles 73–75
    time-resolved diffraction 68–71
  synchrotron

    dilute multilayer fluid membranes 42–45
    geometry 36
    intensity 36
    length density 37
    phospholipid membranes 38–42
    reflectivity 37–38

Yanagisawa, M. 185

Zeks, B. 19, 185
Zhong, Z.Y. 258
zwitterionic lipid 8, 216